WITHDRAWN

BIG HISTORY

D1222620

DK

BIG HISTORY

MACQUARIE UNIVERSITY BIG HISTORY INSTITUTE | SYDNEY · AUSTRALIA

FOREWORD BY **DAVID CHRISTIAN**

PORTER MEMORIAL BRANCH LIBRARY
NEWTON COUNTY LIBRARY SYSTEM
6191 HIGHWAY 212
COVINGTON, GA 30016

DK | Penguin Random House

Lead Senior Editor Helen Fewster
Senior Editors Peter Frances, Dr. Rob Houston
Senior Art Editors Amy Child, Phil Gamble, Sharon Spencer
Project Art Editors Paul Drislane, Mik Gates, Duncan Turner, Francis Wong
Design Assistant Alex Lloyd
Project Editors Camilla Hallinan, Wendy Horobin, Andy Szudek
Editor Kaiya Shang
Editorial Assistant Francesco Piscitelli
US Editors Margaret Parrish, Christy Lusiak
Picture Researcher Liz Moore
Cartography Ron and Dee Blakey, Ed Merritt, Simon Mumford
Jacket Designers Mark Cavanagh, Suhita Dharamjit
Jacket Editor Claire Gell

Senior DTP Designer (Delhi) Harish Aggarwal
Jacket Design Development Manager Sophia MTT
Managing Jackets Editor (Delhi) Saloni Singh
Producer, pre-production Jacqueline Street-Elkayam
Producer Mary Slater

Managing Art Editor Michael Duffy
Managing Editor Angeles Gavira Guerrero
Art Director Karen Self
Design Director Phil Ormerod
Publisher Liz Wheeler
Publishing Director Jonathan Metcalf

DK DELHI
Senior Managing Art Editor Arunesh Talapatra
Art Editors Roshni Kapur, Meenal Goel
Managing Editor Chitra Subramanyam

Editor Ira Pundeer
Production Manager Pankaj Sharma
Pre-production Manager Balwant Singh
Senior DTP Designer Vishal Bhatia
DTP Designer Nand Kishore Acharya
Picture Research Manager Taiyaba Khatoon
Picture Researcher Sakshi Saluja

Illustrators
Rajeev Doshi (Medi-mation)
Peter Bull (Peter Bull Art Studio)
Arran Lewis
Dominic Clifford
Jason Pickersgill (Acute Graphics)
Mark Clifton

First American Edition, 2016
Published in the United States by DK Publishing
345 Hudson Street, New York, New York 10014

Copyright © 2016 Dorling Kindersley Limited
DK, a Division of Penguin Random House, LLC

16 17 18 19 10 9 8 7 6 5 4 3 2 1
001–287353–October/2016

All rights reserved. Without limiting the rights under the copyright reserved above, no part of this publication may be reproduced, stored in or introduced into a retrieval system, or transmitted, in any form, or by any means (electronic, mechanical, photocopying, recording, or otherwise), without the prior written permission of the copyright owner.

Published in Great Britain by Dorling Kindersley Limited.

A catalog record for this book is available from the Library of Congress.

ISBN: 978-1-4654-5443-0

Printed and bound in Hong Kong

DK books are available at special discounts when purchased in bulk for sales promotions, premiums, fund-raising, or educational use. For details contact: DK Publishing Special Markets, 345 Hudson Street, New York, New York 10014 or SpecialSales@dk.com

A WORLD OF IDEAS: SEE ALL THERE IS TO KNOW
www.dk.com

Big History Institute

MACQUARIE UNIVERSITY
BIG HISTORY
INSTITUTE | SYDNEY · AUSTRALIA

Macquarie University was founded with a unique purpose: to bring minds together unhindered by tradition. Created to challenge the education establishment—Macquarie has a rich track record of innovation—Big History is such an innovation. The Big History Institute builds upon the pioneering role that Macquarie University has played in the evolution of the new field of Big History. It brings together a community of scholars and students from both the sciences and the humanities that pursues research questions across disciplinary boundaries and discovers new ways of thinking. The Big History Institute is a global hub for educators, members of the public, and partners from the research, government, nonprofit, and business sectors.

Consultants

David Christian, Director, Big History Institute
Professor David Christian (DPhil Oxford) is the founder of Big History. David is a Distinguished Professor at Macquarie University, co-founder with Bill Gates of the Big History Project, host of one of the 11 classic TED Talks with over 6 million views, and instructor in the world's first Big History MOOC on COURSERA®.

David has given hundreds of presentations internationally, including Davos World Economic Forum in 2012, 2014, and 2015. He is a member of the Australian Academy of the Humanities and the Royal Holland Society of Sciences and Humanities, and a member of the editorial boards of the Journal of Global History and The Cambridge World History.

Andrew McKenna, Executive Director, Big History Institute
Andrew McKenna (BCom LLB UNSW, MIntRel Macquarie) coordinates Big History as an integrated initiative encompassing research, teaching, and outreach. Andrew leads the strategic growth and development of the Big History Institute globally.

Tracy Sullivan, Education Leader, Big History Institute
Tracy Sullivan (BA GradDipEd UWS, MA Macquarie) was on the curriculum development team for Big History Project, and oversees implementation in Australian schools. Tracy coordinates educational initiatives to support the growth and development of Big History globally.

Contributors

Introduction – Elise Bohan
Threshold 1 – Robert Dinwiddie
Threshold 2 – Jack Challoner
Thresholds 3 and 4 – Colin Stuart
Threshold 5 – Derek Harvey
Threshold 6 – Rebecca Wragg-Sykes
Threshold 7 – Peter Chrisp
Threshold 8 – Ben Hubbard
Timelines of world history – Philip Parker

CONTENTS

8 FOREWORD BY DAVID CHRISTIAN
10 INTRODUCTION

1 THE BIG **BANG**

16 GOLDILOCKS CONDITIONS
18 ORIGIN STORIES
20 THE NEBRA SKY DISK
22 ASTRONOMY BEGINS
24 EARTH ORBITS THE SUN
26 SEEING THE LIGHT
28 THE ATOM AND THE UNIVERSE
30 THE UNIVERSE GETS BIGGER
32 THE EXPANDING UNIVERSE
34 THE BIG BANG
36 RECREATING THE BIG BANG
38 BEYOND THE BIG BANG

2 STARS ARE **BORN**

42 GOLDILOCKS CONDITIONS
44 THE FIRST STARS
46 THE PUZZLE OF GRAVITY
48 THE FIRST GALAXIES
50 HUBBLE EXTREME DEEP FIELD

3 ELEMENTS ARE **FORGED**

54 GOLDILOCKS CONDITIONS
56 THE LIFE CYCLE OF A STAR
58 HOW NEW ELEMENTS FORM INSIDE STARS
60 WHEN GIANT STARS EXPLODE
62 MAKING SENSE OF THE ELEMENTS

4 PLANETS **FORM**

66 GOLDILOCKS CONDITIONS
68 OUR SUN IGNITES
70 THE PLANETS FORM
72 THE IMILAC METEORITE
74 THE SUN TAKES CONTROL
76 HOW WE FIND SOLAR SYSTEMS
78 EARTH COOLS
80 EARTH SETTLES INTO LAYERS
82 THE MOON'S ROLE
84 THE CONTINENTS ARE BORN
86 DATING EARTH
88 ZIRCON CRYSTAL
90 CONTINENTS DRIFT
92 HOW EARTH'S CRUST MOVES
94 OCEAN FLOOR

5 LIFE EMERGES

98 GOLDILOCKS CONDITIONS
100 STORY OF LIFE
102 LIFE'S INGREDIENTS FORM
104 THE GENETIC CODE
106 LIFE BEGINS
108 HOW LIFE EVOLVES
110 HISTORY OF EVOLUTION
112 MICROBES APPEAR
114 LIFE DISCOVERS SUNLIGHT
116 OXYGEN FILLS THE AIR
118 COMPLEX CELLS EVOLVE
120 SEX MIXES GENES
122 CELLS BEGIN TO BUILD BODIES
124 MALES AND FEMALES DIVERGE
126 ANIMALS GET A BRAIN
128 ANIMAL LIFE EXPLODES
130 ANIMALS GAIN A BACKBONE
132 RISE OF THE VERTEBRATES
134 JAWS CREATE TOP PREDATORS
136 PLANTS MOVE ONTO LAND
138 WENLOCK LIMESTONE
140 ANIMALS INVADE LAND
142 REINVENTING THE WING
144 THE FIRST SEEDS

6 HUMANS EVOLVE

180 GOLDILOCKS CONDITIONS
182 THE PRIMATE FAMILY
184 HOMININS EVOLVE
186 APES BEGIN TO WALK UPRIGHT
188 GROWING A LARGER BRAIN
190 THE NEANDERTHALS
192 KEBRA NEANDERTHAL
194 EARLY HUMANS DISPERSE
196 ANCIENT DNA
198 THE FIRST HOMO SAPIENS
200 BRINGING UP BABIES
202 HOW LANGUAGE EVOLVED
204 COLLECTIVE LEARNING
206 THE BIRTH OF CREATIVITY
210 HUNTER-GATHERERS EMERGE
212 PALEOLITHIC ART
214 THE INVENTION OF CLOTHING
216 HUMANS HARNESS FIRE
218 BURIAL PRACTICES
220 HUMANS BECOME DOMINANT

7 CIVILIZATIONS DEVELOP

224 GOLDILOCKS CONDITIONS
226 CLIMATE CHANGES THE LANDSCAPE
228 FORAGERS BECOME FARMERS
230 AFFLUENT FORAGERS
232 HUNTERS BEGIN TO GROW FOOD
234 FARMING BEGINS
236 WILD PLANTS BECOME CROPS
238 POLLEN GRAINS
240 FARMERS DOMESTICATE ANIMALS
242 FARMING SPREADS
244 MEASURING TIME
246 NEW USES FOR ANIMALS
248 INNOVATIONS INCREASE YIELDS
250 SURPLUS BECOMES POWER
252 POPULATION STARTS TO RISE
254 THE FENTON VASE
256 EARLY SETTLEMENTS
258 SOCIETY GETS ORGANIZED
260 RULERS EMERGE
262 LAW, ORDER, AND JUSTICE
264 THE WRITTEN WORD
266 WRITING DEVELOPS

8 INDUSTRY RISES

302 GOLDILOCKS CONDITIONS
304 THE INDUSTRIAL REVOLUTION
306 COAL FUELS INDUSTRY
308 STEAM POWER DRIVES CHANGE
310 THE PROCESS OF INDUSTRIALIZATION
312 INDUSTRY GOES GLOBAL
314 GOVERNMENTS EVOLVE
316 CONSUMERISM TAKES OFF
318 EQUALITY AND FREEDOM
320 NATIONALISM EMERGES
322 THE INDUSTRIAL ECONOMY
324 THE WORLD OPENS TO TRADE
326 WAR DRIVES INNOVATION
328 COLONIAL EMPIRES GROW
330 SOCIETY TRANSFORMS
332 EDUCATION EXPANDS
334 MEDICAL ADVANCES
336 ROAD TO GLOBALIZATION
338 ENGINES SHRINK THE WORLD
340 NEWS TRAVELS FASTER
342 SOCIAL NETWORKS EXPAND
344 GROWTH AND CONSUMPTION
346 FINDING THE ENERGY
348 NUCLEAR OPTIONS
350 ENTERING THE ANTHROPOCENE
352 CLIMATE CHANGE

146 SHELLED EGGS ARE BORN
148 HOW COAL FORMED
150 LIZARD IN AMBER
152 THE LAND DRIES OUT
154 REPTILES DIVERSIFY
156 BIRDS TAKE TO THE AIR
158 CONTINENTS SHIFT AND LIFE DIVIDES
160 THE PLANET BLOSSOMS
162 MASS EXTINCTIONS
164 PLANTS RECRUIT INSECTS
166 THE RISE OF MAMMALS
168 GRASSLANDS ADVANCE
170 EVOLUTION TRANSFORMS LIFE
172 HOW WE CLASSIFY LIFE
174 ICE CORES
176 EARTH FREEZES

268 WATERING THE DESERT
270 CITY-STATES EMERGE
272 FARMING IMPACTS THE ENVIRONMENT
274 BELIEF SYSTEMS
276 GRAVE GOODS
278 CLOTHING SHOWS STATUS
280 USING METALS
282 ÖTZI THE ICEMAN
284 CONFLICT LEADS TO WAR
286 AGE OF EMPIRES
288 HOW EMPIRES RISE AND FALL
290 MAKING MONEY
292 UNHEALTHY DEVELOPMENTS
294 TRADE NETWORKS DEVELOP
296 EAST MEETS WEST
298 TRADE GOES GLOBAL

354 ELEMENTS UNDER THREAT
356 THE QUEST FOR SUSTAINABILITY
358 WHERE NEXT?
360 TIMELINES OF WORLD HISTORY
362 PREHISTORIC WORLD, 8 MYA–3000 BCE
367 CALENDAR SYSTEMS
368 ANCIENT WORLD, 3000–700 BCE
374 CULTURE AND CREATIVITY
376 CLASSICAL WORLD, 700 BCE–600 CE
383 GREAT BUILDINGS
384 MEDIEVAL WORLD, 600–1450
392 PHILOSOPHY AND FAITH
394 EARLY MODERN WORLD, 1450–1750
402 INVENTION AND DISCOVERY
404 WORLD OF EMPIRES, 1750–1914
411 ASTRONOMY AND SPACE
412 MODERN WORLD, 1914 ONWARD
422 MILESTONES IN MEDICINE
424 INDEX AND ACKNOWLEDGMENTS

FOREWORD

I vividly remember a globe map of the world sitting in a classroom when I was a child. I also remember a geography class, taught in a school in Somerset in England, where we learned how to draw sections through the earth, showing the various layers of soil beneath our feet, and how they connected to other parts of England. For me, the most exciting thing in school was always the sudden connections, realizing that layers of chalk beneath our feet were made from the remains of billions of tiny organisms—called coccolithophores—that had lived millions of years ago, and that the same remains could also be found in layers of chalk in other parts of England and other countries much farther away. What was Somerset like when the coccolithophores were alive? For that matter, where was Somerset back then? That's a question I couldn't even ask when I was at school because at that time scientists didn't know for sure that the continents moved around the surface of the earth.

For me, the globe in the corner of my classroom was a key to all this knowledge. It helped me see the place of Somerset in Britain, of Britain in Europe—so *that's* where the Vikings came from!—and of Europe in the world. Big History is like the globe, but it's much bigger: it includes all the observable universe and all observable time, so it reaches back in time for 13.8 billion years to the astonishing moment of the Big Bang, when an entire universe was smaller than an atom. Big History includes the story of stars and galaxies, of new elements from carbon—the magical molecule that made life possible—to uranium, whose radioactivity enabled us not just to make bombs, but also to figure out when our earth was formed. It is like a map of all of space and time. And once you start exploring that map, you will be able, eventually, to say: "So that's what I'm a tiny part of! That's my place in the grand scheme of things! So what's next?"

Today, more and more schools and universities are teaching Big History, and it's a story we all need to know. In the book you are holding in your hands, you will find a beautifully illustrated account of this story, a sort of globe in words and pictures that links knowledge from many different disciplines. *Big History* shows how our world developed, threshold by threshold, from a very simple early universe, to the emergence of stars and chemistry, and on to a cosmos that contained places like our earth on which life itself could emerge.

And you'll also see the strange role played by our own species, humans, in this huge story. We appear at the very end of the story, but our impact has been so colossal that we are beginning to change the planet. We have done something else that is perhaps even more astonishing: from our tiny vantage point in the vast universe, we have figured out how that universe was created, how it evolved, and how it became as it is today. That is an amazing achievement, and in this book you will explore the discoveries that allowed us to piece together this story. This is the world globe that we need today, early in the 21st century, as we try to manage the huge challenges of maintaining our beautiful planet and keeping it in good condition for those who will come after us.

DAVID CHRISTIAN
FOUNDER OF BIG HISTORY
DIRECTOR, BIG HISTORY INSTITUTE
CO-FOUNDER OF THE BIG HISTORY PROJECT

> “Big History provides a framework for understanding literally all of history, ever, from the Big Bang to the present day. So often subjects in science and history are taught one at a time—physics in one class, the rise of civilization in another—but Big History breaks down those barriers. Today, whenever I learn something new about biology or history or just about any other subject, I try to fit it into the framework I got from Big History. No other course has had as big an impact on how I think about the world.”

BILL GATES, WWW.GATESNOTES.COM
CO-FOUNDER OF THE BIG HISTORY PROJECT

WHAT IS **BIG HISTORY?**

BIG HISTORY IS THE STORY OF HOW YOU AND I CAME TO BE.

It is a modern origin story for a modern age. This grand evolutionary epic rouses our curiosity, confronts our ingrained intuitions, and marries science, reason, and empiricism with vivid and dynamic storytelling. Best of all, Big History provides the scope and scientific foundations to help us ponder some of the most exciting and enduring questions about life, the universe, and everything.

These universally compelling questions include: How did life on Earth evolve? What makes humans unique? Are we alone in the universe? Why do we look and think and behave the way we do? And what does the future hold for our species, our planet, and the cosmos? Throw a dart at any point in the history of the universe and it will land on a page of the Big History story. No matter how obscure this page, or how far removed it may seem from the world we know, it will invariably describe a fragment of this grand scientific narrative, in which all events and all chapters are connected.

In this volume we traverse the stars, the galaxies, the cells inside your body, and the complex interactions between all living and nonliving things. We stretch our minds to the limits of human understanding in order to see reality from many angles, and on many scales. What is truly remarkable about looking at the world from such an expansive perspective is that we begin to engage with many facets of the natural world that we often miss, or take for granted.

How often do we think about the fact that every atom inside each of our

HOW OFTEN DO WE THINK ABOUT THE FACT THAT EVERY ATOM INSIDE EACH OF OUR BODIES WAS MADE INSIDE A DYING STAR?

bodies was made inside a dying star? Or that ancient celestial implosions gave rise to the kinds of chemistry that makes life possible? How frequently do we zoom out far enough in our historical musings to see connections that transcend the actions of kings, armies, politicians, and peasants?

Our minds do not instinctively follow the threads of our evolutionary history to the point where all national, tribal, and species boundaries fall away. But when we allow ourselves to explore beyond these domains we come face to face with a single family tree, which shows that every one of us shares a common ancestor with every living organism on the planet: from worms, to fish, to reptiles, to chimpanzees, to a bird singing on the other side of the world, and the strangers who sleep through its refrain.

naked eye. It is also important to remember that Big History is not a static tale that proclaims how things are and will be for all time. It is a provisional narrative that is constantly being updated as our knowledge about the natural world grows, and as our needs as a species evolve.

From a cosmic perspective, we see that humans are a novel species that appear on the scene very late in this evolutionary history. We were not there at the beginning, and we are almost certainly not the species with whom the evolutionary buck stops. Yet Big History is still very much a human story, written by humans, for humans. At a certain point in this tale we choose to focus on our species and our corner of the galaxy, because from our point of view, this is where the action and the meaning is.

In the grand scheme of space and time, humanity may seem like little more than a cosmic footnote. But when we look closely at our blue planet we see that our species is responsible for some very remarkable things that no other species has achieved in the 3–4 billion years that life has existed on Earth. As far as we know, *Homo sapiens* is the first and only species to represent the universe becoming self-aware. Humans are now the dominant force altering the planetary biosphere, and we have kicked the pace of terrestrial evolution into a dramatic new gear.

BIG HISTORY HELPS US TO QUESTION EVERYTHING WE SEE, AND EVERYTHING WE THINK WE KNOW.

Big History helps us question everything we see, and everything we think we know. In the process, we discover that the universe is far stranger than we often imagine, and that the shape of history is molded by forces that are often surprising, and hard to see with the

As you explore this remarkable narrative, you will discover that our species has been so successful in expanding and colonizing the globe, in large part because of our capacity for what big historians call collective learning. Although we cannot impart our accumulated knowledge and experiences to new generations via DNA, we have developed the means to transmit this information culturally. Such a radical innovation in information sharing was made possible by the human invention of symbolic language.

At first this meant sharing ideas through the oral tradition. But eventually we developed writing, which reduced the error rate in the transmission of information and left humans in possession of a tool resembling a crude external hard drive. For the first time we could store large bodies of information without having to use the limited memory capacity of our brains.

With the ability to build upon existing information over many generations, humans learned ever faster, and knowledge and innovation proliferated. While many civilizations collapsed and some discoveries were lost for centuries, the overall trend was a feedback loop of accelerating cultural change: the invention of ever faster and more accurate methods of information sharing generated rapid bursts of innovation, and vice versa.

While the oral tradition persisted for tens of thousands of years, it only took a few hundred years for humans to transition from the age of the printing press to the digital world of today. If the pace of cultural evolution continues at such a rate we may see the emergence of a new evolutionary paradigm in mere decades.

Because of our astounding capacity for collective learning and cultural development, humans have made a giant evolutionary leap in a relatively short period of time. We have transitioned from our initial role as one of evolution's many simple players, to a fledgling director engaged in the task of consciously shaping the trajectory of evolution on Earth. While this is a very exciting role, it also presents immense challenges.

WITH THE ABILITY TO BUILD UPON EXISTING INFORMATION OVER MANY GENERATIONS, HUMANS LEARNED EVER FASTER, AND KNOWLEDGE AND INNOVATION PROLIFERATED.

It is sobering to look back at our extensive family tree and recall that 99 percent of species that have ever lived are now extinct. In light of this, it is natural and beneficial to consider whether our species will be able to live sustainably and prosperously for many years into the future. And if we can achieve this, how might it be possible?

Will we reduce our consumption of energy and live more simply? Or will we harness our immense collective brainpower to engineer more sophisticated ways of producing clean energy and sustainable products and services? Will our modern technological arms race leave us liberated or enslaved? And how long will most of us continue to exist as fully biological beings, unenhanced by technological modifications?

These are the kinds of questions that the Big History story prompts us to consider. There is no doubt that in terms of its scope, content, and method, Big History is a truly modern story, fit for the needs of a modern age.

Like all origin stories of previous ages, this narrative is designed to help orient us with where we come from, what we are, and where we might be going. But unlike ancient origin stories that were built upon myth and intuition, this evolutionary epic relies on the theories of modern science to help us get to grips with the world around us.

For most of us, thinking about things that are very big, very small, and very old does not come naturally. But pursuing big ideas and chasing the answers to profound universal questions

BIG HISTORY IS A TRULY MODERN ORIGIN STORY, FIT FOR THE NEEDS OF A MODERN AGE.

does! We cannot help wanting to know what else is out there: whether it be among the stars, inside black holes, or in the mysterious workings of our brains, our DNA, or the remarkable bacterial ecosystems that live on, around, and inside us.

The Big History story helps to facilitate our exploration of these and other exciting domains. It allows us to focus on an array of subjects and historical moments and encourages us to ponder the nature of reality on many different scales. We learn to relate the details to the big picture, and observe how broad trends can contextualize local phenomena and events. By exploring the viewpoints of both the generalist and the specialist, we are able to think more carefully and creatively about cause and effect, and devise more innovative responses and solutions to the many challenges we face in the world today.

Big History's unified perspective also helps us to see the present in dynamic terms, and shows us that we are not only the successors of previous evolutionary thresholds, but also the possible progenitors of those to come.

Our story is divided into eight thresholds of increasing complexity, each of which highlight some of the key transitionary phases in this cosmic evolutionary history. As we move from threshold to threshold, you will see how profoundly each stage is connected, and how matter and information in the universe grow denser and more complex in various pockets of cosmic order. This story helps us to see that our planet and our species emerged among a rare set of goldilocks conditions, where the balance and stability of elements was "just right" to sustain life.

Once you explore this book and get a feel for the big picture it presents, we hope you will be left pondering many new and rousing questions. As you sit, poised to embark on this journey of discovery, there is one question in particular that we hope you will consider.

What role will you play in determining how events unfold in the next threshold of this great cosmic drama?

"BENEATH THE AWESOME DIVERSITY AND COMPLEXITY OF MODERN KNOWLEDGE, THERE IS AN UNDERLYING UNITY AND COHERENCE, ENSURING THAT DIFFERENT TIMESCALES REALLY DO HAVE SOMETHING TO SAY TO EACH OTHER."

DAVID CHRISTIAN, BIG HISTORIAN

THRESHOLD

THE BIG **BANG**

What are the origins of our universe? It is a question that has captivated our species, probably since we emerged. Centuries of observation, investigation, and scientific endeavor have led us to the Big Bang theory—but that too leaves questions unanswered, and our quest for further explanation continues.

GOLDILOCKS CONDITIONS

The universe formed in the Big Bang. We do not know if anything existed before it, and we only have a glimpse of what happened in the fraction of a second immediately afterward. But over the next 380,000 years, the universe expanded and cooled, and the fundamental forces and forms of matter that we know today emerged.

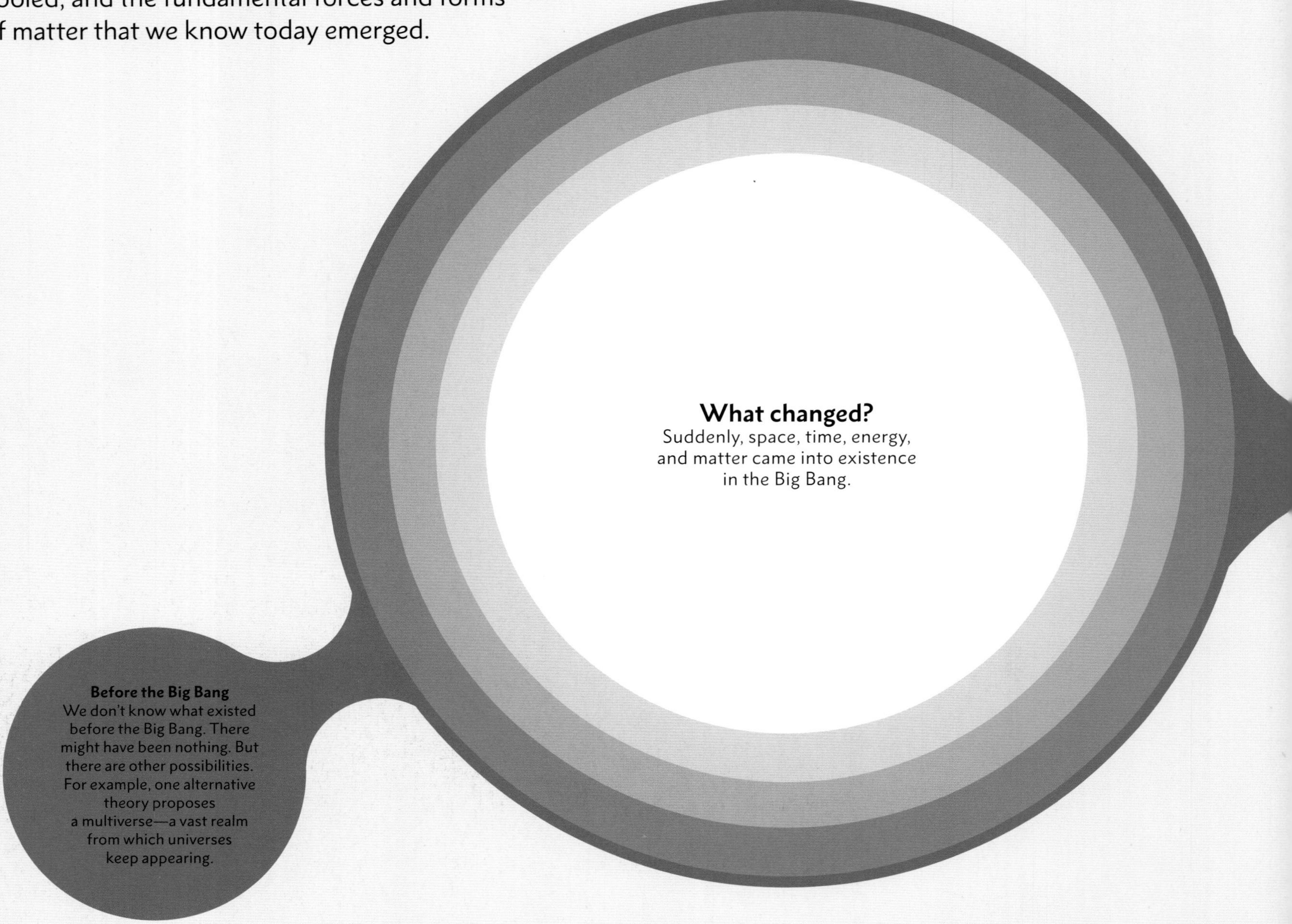
What changed?
Suddenly, space, time, energy, and matter came into existence in the Big Bang.
Before the Big Bang
We don't know what existed before the Big Bang. There might have been nothing. But there are other possibilities. For example, one alternative theory proposes a multiverse—a vast realm from which universes keep appearing.

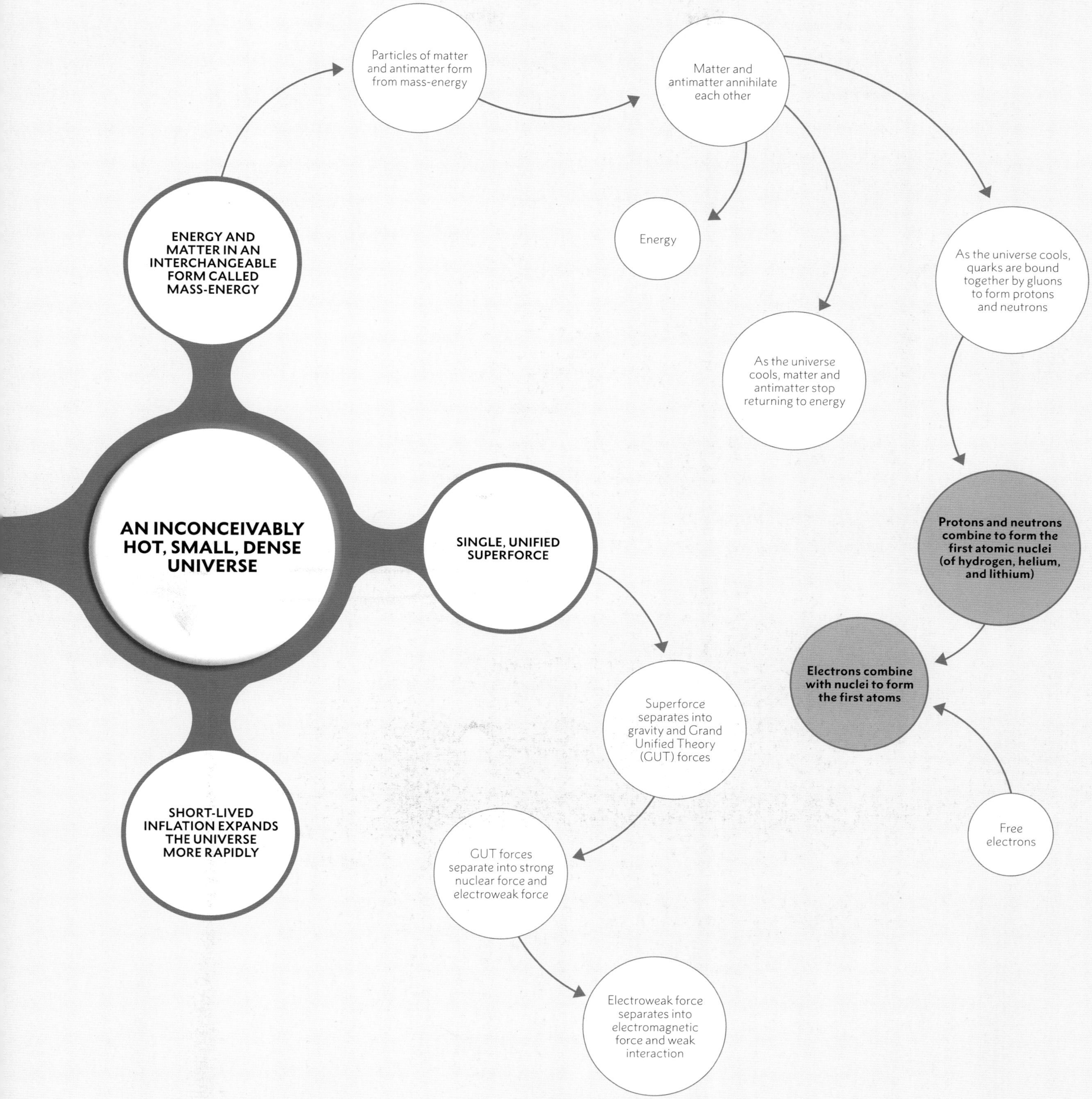
AN INCONCEIVABLY HOT, SMALL, DENSE UNIVERSE
ENERGY AND MATTER IN AN INTERCHANGEABLE FORM CALLED MASS-ENERGY
Particles of matter and antimatter form from mass-energy
Matter and antimatter annihilate each other
Energy
As the universe cools, matter and antimatter stop returning to energy
As the universe cools, quarks are bound together by gluons to form protons and neutrons
Protons and neutrons combine to form the first atomic nuclei (of hydrogen, helium, and lithium)
Electrons combine with nuclei to form the first atoms
Free electrons
SINGLE, UNIFIED SUPERFORCE
Superforce separates into gravity and Grand Unified Theory (GUT) forces
GUT forces separate into strong nuclear force and electroweak force
Electroweak force separates into electromagnetic force and weak interaction
SHORT-LIVED INFLATION EXPANDS THE UNIVERSE MORE RAPIDLY

ORIGIN **STORIES**

Nearly all human cultures and religious traditions have nurtured origin stories—symbolic accounts that describe how the world came about. These stories or narratives were most often passed from one generation to the next in the form of folk tales or ballads, and sometimes through writing or pictures.

Origin stories are extremely varied in detail, but they tend to include some common themes. Often they tell how the universe acquired order from an original state of either darkness or deep chaos. In several versions, including the Old Testament's *Book of Genesis*, this order is imposed by a supreme being or deity. In some stories, creation is a cyclical process. For example, in Hindu thought, order is generated only to be destroyed and then regenerated. Many stories begin with Earth. In some, people and gods emerge from the Earth. In others, an animal dives into a boundless primeval ocean and retrieves a portion of Earth from which the cosmos is created.

ORIGINS OF THE SKY, SUN, AND MOON

Many origin stories describe how the sky was created along with Earth, often by splitting off from another primeval object. In a common form of the Maori creation myth, the universe is created from nothing by a supreme being, Io. He also creates Ranginui (Rangi) and Papatuanuku (Papa), the Sky Father and Earth Mother. Rangi and Papa remain physically cleaved together until pushed apart by their six offspring to create the separate realms of Earth and sky. Many stories also account for the creation of celestial bodies such as the sun and moon. For example, in a story from China, the first living being, Pangu, hatches from a cosmic egg. Half the shell lies under him as the Earth; the rest arcs above him as the sky. Each day for thousands of years he grows, gradually pushing Earth and sky apart until they reach their correct places. But then Pangu disintegrates. His arms and legs become mountains, his breath the wind, his eyes turn into the sun and moon. Often celestial objects originate as physical representations of gods. For example, an origin story from ancient Egypt begins with Nun, the primeval ocean, from which the god Amen rises. He takes the alternative name Re and breeds more gods. While his tears become mankind, Amen-Re retires to the heavens, to reign eternally as the sun.

Origin stories such as these developed because early humans needed to find an explanation for their own existence and for everything that they saw around them. The cultures that fostered these stories regarded them as true, and for their adherents they usually carried great importance and emotional power. But such perceptions were based on faith and not on accurate observations or scientific reasoning.

> WE HAVE INHERITED FROM OUR FOREFATHERS THE KEEN LONGING FOR **UNIFIED, ALL-EMBRACING KNOWLEDGE**.
>
> **Erwin Schrödinger**, Austrian theoretical physicist, 1887–1961

MORE THAN **100 DISTINCT ORIGIN STORIES** HAVE BEEN IDENTIFIED FROM VARIOUS PEOPLES AND CULTURES ACROSS THE WORLD

THE EARLIEST ASTRONOMERS

At points in history that vary according to the culture, but typically from about 4000 BCE in Europe and the Middle East, it seems that humans began to tire of merely gazing at, and devising stories about, objects such as the stars, sun, and moon. Instead some individuals began making detailed recordings of celestial phenomena. These investigations were carried out for a variety of mostly practical reasons. An ability to identify a few stars, and to understand sky movements, proved useful for navigation. It was also realized that the sky is a sort of clock that could be used, for example, to tell farmers when to sow crops or to give warning of important natural events. In ancient Egypt, for example, the rising of the bright star Sirius around the same time as the sun heralded the annual flooding of the Nile. A final reason for studying the heavens was to predict solar eclipses. Chinese astronomers are thought to have attempted this as long ago as 2500 BCE, but it was not until the 1st century BCE that the ancient Greeks reached the level of astronomical sophistication needed to do it accurately. Successful eclipse prediction had little specific practical use but it did confer on the predictor very significant mystical powers and, as a result, considerable peer respect.

In some early cultures, accurate observation not only had practical uses but was also intertwined with religion. Some of the most sophisticated observations before the invention of the telescope were made by the Maya, who colonized parts of Central America between 250 and 900 CE. They made accurate calculations of the length of the solar year, compiled precise tables of the positions of Venus and the moon, and were able to predict eclipses. They used their calendar to time the sowing and harvesting of crops. But they also saw a link between the cycles they observed and the place of their gods in the natural order. Specific events in the night sky were seen as representing particular deities. The Maya also practiced a form of astrology, drawing a connection between cycles in the sky and the everyday life and concerns of the individual.

ASTRONOMERS IN CHINA RECORDED OBSERVATIONS OF **MORE THAN 1,600 SOLAR ECLIPSES** FROM 750 BCE ONWARD

A MODERN NARRATIVE

Big History is a modern-day origin story. Part of this story is an account of how the universe formed provided by the Big Bang theory of cosmology. The theory describes the formation of a universe with a beginning and a structure. Modern cosmology as a whole also contains an account of a universe that changes over time, as matter and energy take on different forms, new particles come into existence, space itself expands, and structures such as stars and galaxies emerge. The Big Bang theory, as part of the Big History narrative, shares some other features with traditional origin stories. For example, in common with several of the stories, it proposes that everything—all matter, energy, space, and time—originated from nothing. Big Bang theory and the traditional stories also set out to answer many of the same questions—including how did the universe begin? The theory does not give a complete account of how the universe came to be the way it is now. For example, it does not explain the origin of life or the evolution of humans. But it does form part of the larger framework of Big History that attempts to answer these and other questions.

However, in one crucial respect, Big Bang theory, like Big History in general, differs from traditional origin stories in that it seeks to provide a literal and accurate account of the universe's origins. It represents the current state of scientific thinking, arrived at after many centuries of both gradual change and sudden leaps forward. Like other scientific theories in Big History, the theory also makes predictions that can be tested against evidence, allowing it to be refined or even disproved and overturned. Some questions remain unanswered by Big Bang theory. But, at least for now, it offers the most convincing account of when and how the universe began.

▶ **Brahma the creator** According to some older forms of Hinduism, the god Brahma, who is usually depicted with four heads, was born from a golden egg and created Earth and everything in it.

> “THERE WAS NEITHER **NONEXISTENCE NOR EXISTENCE** THEN; THERE WAS NEITHER THE REALM OF **SPACE NOR THE SKY** WHICH IS BEYOND.”
>
> **The Rig Veda**, a collection of Sanskrit hymns, 2nd millennium BCE

THE NEBRA SKY DISK

During the European Bronze Age, people developed their knowledge of astronomy and put it to practical uses. The Nebra Sky Disk is a key piece of evidence for observation of the sky at this time. Analysis of the disk's materials also reveals information about metalworking and trade.

The Bronze Age in Europe began around 3200 BCE. Dug up near Nebra in central Germany in 1999, the 3,600-year-old Nebra Sky Disk depicts the sun, moon, and 32 stars, including possibly the Pleiades star cluster. It is the oldest known portrayal of such a variety of sky objects. The disk also reveals that its owners had measured the angle between the rising and setting points of the sun at the summer and winter solstices—the days of greatest and least daylight each year.

There are two schools of thought as to what the disk was used for or represents. Some archaeologists think that it was an astronomical clock, which could have been used to indicate times for sowing and harvesting crops and to coordinate the solar and lunar calendars. Alternatively, the objects on the disk may illustrate a significant astronomical event—a solar eclipse on April 16, 1699 BCE. On that date, the sun, as it was eclipsed by the moon, was close in the sky both to the Pleiades and to a tight grouping of three planets—Mercury, Venus, and Mars.

Whatever its exact use, the Nebra Sky Disk provides clear evidence that some Bronze Age people had made detailed sky observations and also developed tools to help them mark the passage of time and the seasons.

Small disks may denote stars, but most appear to be decorative, as they do not match known star patterns
Large gold disk probably represents the sun
Holes were punched into the rim after other additions for an unknown purpose

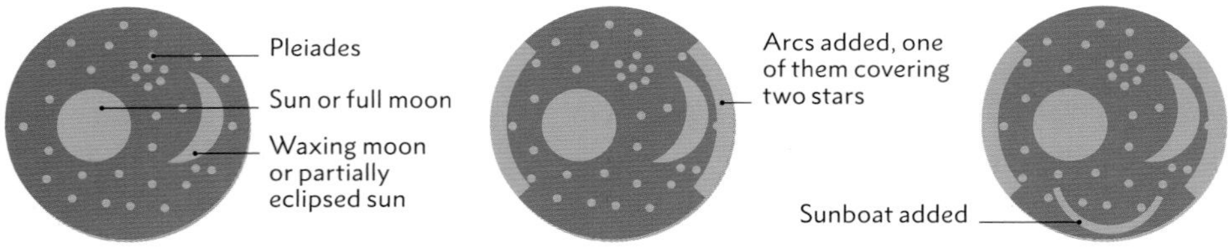
Pleiades
Sun or full moon
Waxing moon or partially eclipsed sun
PHASE 1
Arcs added, one of them covering two stars
Sunboat added
PHASE 2
PHASE 3

▲ Phases in construction
The disk was made in three phases, significantly separated in time, suggesting it underwent some repurposing. The addition of the sunboat indicates that it may have taken on religious significance.

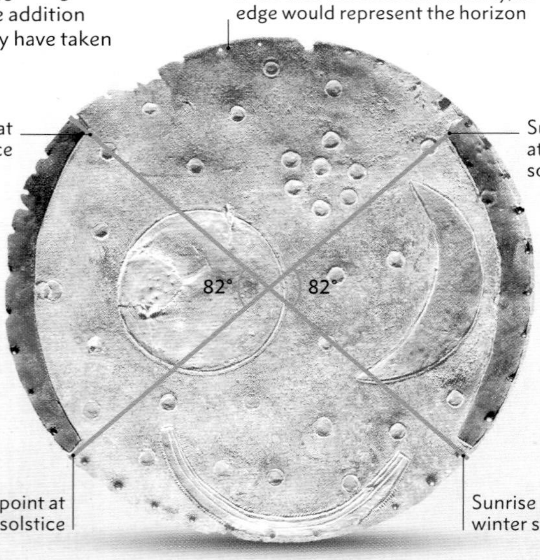
If the disk was held horizontally, its edge would represent the horizon
Sunset point at summer solstice
Sunrise point at summer solstice
82°
82°
Sunset point at winter solstice
Sunrise point at winter solstice

► The golden arcs
The two arcs on the disk span 82°, the angle between the points on the horizon where the sun sets (or rises) at the summer and winter solstices for the location where the disk was found.

Metal sources

The disk's copper came from the Austrian Alps. Its tin—used with copper to make bronze—and its original gold were from Cornwall, England. The gold in the arcs and sunboat came from the Carpathian Mountains in eastern Europe. Evidently there were well-established trade routes across Europe at the time.

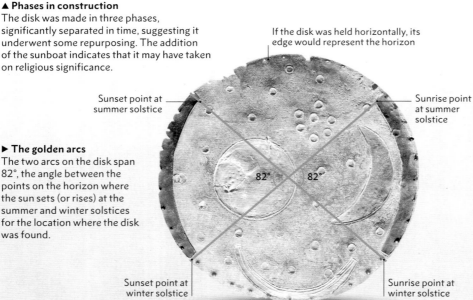

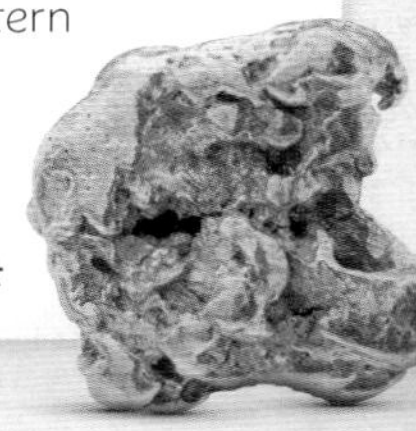

Gold nugget

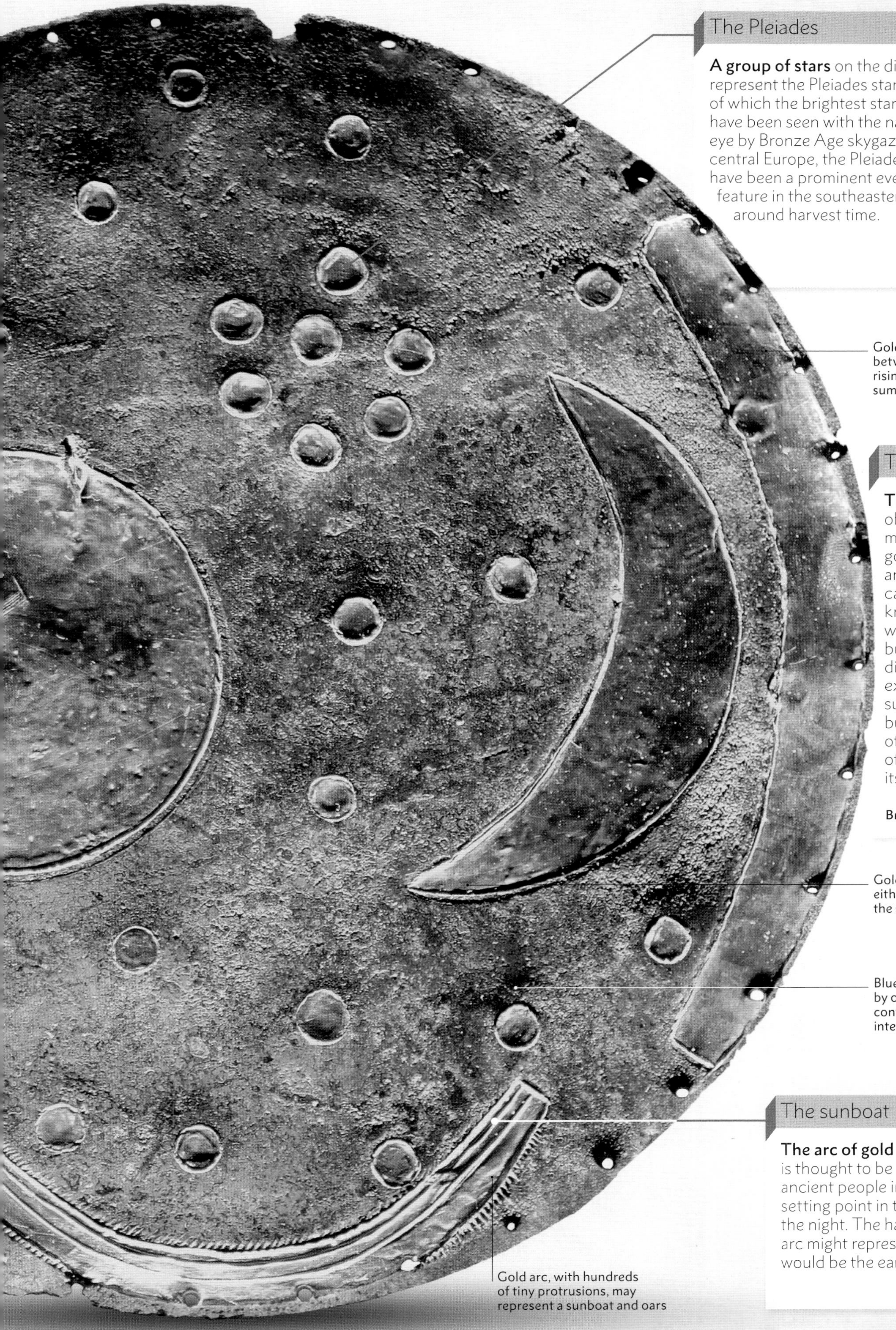
Golden arcs span the angle between the setting (or rising) points of the sun at summer and winter solstices
Gold crescent may signify either a crescent moon or the sun during a solar eclipse
Blue-green patina, caused by oxidation of disk's copper content, was probably an intentional decorative feature
Gold arc, with hundreds of tiny protrusions, may represent a sunboat and oars

The Pleiades

A group of stars on the disk may represent the Pleiades star cluster, of which the brightest stars could have been seen with the naked eye by Bronze Age skygazers. In central Europe, the Pleiades would have been a prominent evening feature in the southeastern sky around harvest time.

Stars and dust in the Pleiades

The Nebra hoard

The disk was buried with other objects, including two swords made of bronze with copper and gold inlays, a chisel, two axeheads, and two armbands, collectively called the Nebra hoard. It is not known why the disk was placed with these objects. The hoard was buried in around 1600 BCE, but the disk could be older. When first examined by archaeologists, it was suspected to be an elaborate fake, but corrosion tests, excavation of the burial site, and examination of the other artifacts pointed to its authenticity.

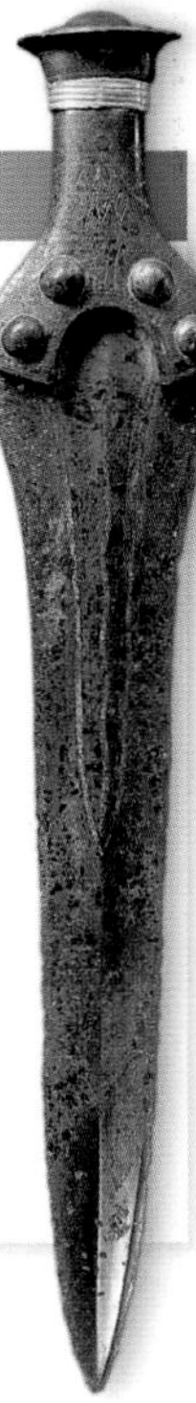

Bronze Age sword from the Nebra hoard

The sunboat

The arc of gold at the bottom of the Nebra sky disk is thought to be a sunboat—the means by which some ancient people imagined the sun was conveyed from its setting point in the west to its rising point in the east during the night. The hairlike protrusions around the edge of the arc might represent oars. If the arc is indeed a sunboat, it would be the earliest known representation of one.

ASTRONOMY BEGINS

For most of human history, people were too busy surviving to spend much time thinking about the world's underlying nature and origins. But from around 1000 BCE, a few began to try answering key questions about the universe without recourse to supernatural explanations.

These thinkers—initially concentrated in Mediterranean lands, especially Greece—realized that to understand the world it is necessary to know its nature, and that natural phenomena should have logical explanations. Although they did not always find the correct answers, this leap marked the start of a 3,000-year journey that has led in the modern world to such key theories key as the Big Bang model of the universe.

THE NATURE OF MATTER

The fundamental questions of what the world is made of, and where matter came from, are some of the oldest. In the 6th century BCE, Greek philosophers such as Thales and Anaximenes proposed that all substances were modifications of more intrinsic substances, the main candidates being water, air, earth, and fire. In the 5th century BCE, Empedocles claimed that everything was a mixture of all four of these substances, or elements. His near-contemporary Democritus developed the idea that the universe is made of an infinite number of indivisible particles called atoms. Finally, in the 4th century BCE the influential scholar Aristotle added a fifth element, ether, to Empedocles' four. Although Aristotle was skeptical of the idea of atoms, it is remarkable that the concepts of both atoms and elements had been proposed more than 2,000 years before either was proved to exist.

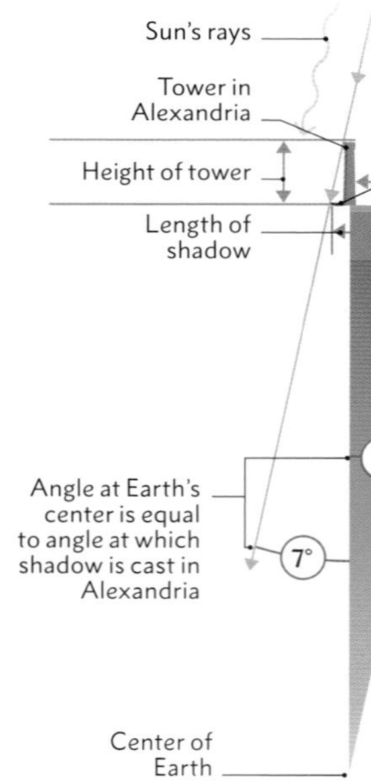

◄ **Estimating Earth's circumference**
When the sun was overhead above a well at Syene, it cast a shadow at an angle of about 7° in Alexandria. Dividing this angle into 360°, then multiplying the result by the distance between the two places gave an estimate of Earth's circumference of about 25,000 miles (40,000km).

EARTH'S SHAPE AND SIZE

Among the many other ideas that Aristotle gave his views on was the concept that Earth is a sphere. Earlier Greek scholars, such as Pythagoras, had already argued this, but Aristotle was the first to summarize the main points of evidence. Chief among them was that travelers to southern lands could see stars that could not be seen by those living farther north—explainable only if Earth's surface is curved. In 240 BCE, by comparing how the sun's rays reach Earth at Syene and Alexandria, the mathematician Eratosthenes was able to estimate Earth's circumference. He came up with a figure of about 25,000 miles (40,000km)—close to the true value known today.

THE IDEA THAT **EARTH IS FLAT** WAS STILL THE PREVAILING VIEW IN **CHINA UP TO THE EARLY 17TH CENTURY**

EARTH AND THE SUN

Aristotle thought that Earth was at the center of the universe and that the sun, planets, and stars move around it. This seemed like common sense given that every night various celestial objects (and during the day, the sun) could be seen moving across the sky from east to west, whereas Earth itself did not seem to move. An alternative view, put forward by the astronomer Aristarchus, was that the sun is at the center and that Earth orbits it, but this idea did not gain much credence. In 150 CE, Claudius Ptolemy—an eminent Greek scholar living in Alexandria—published a book called

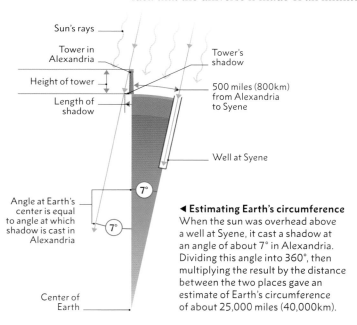

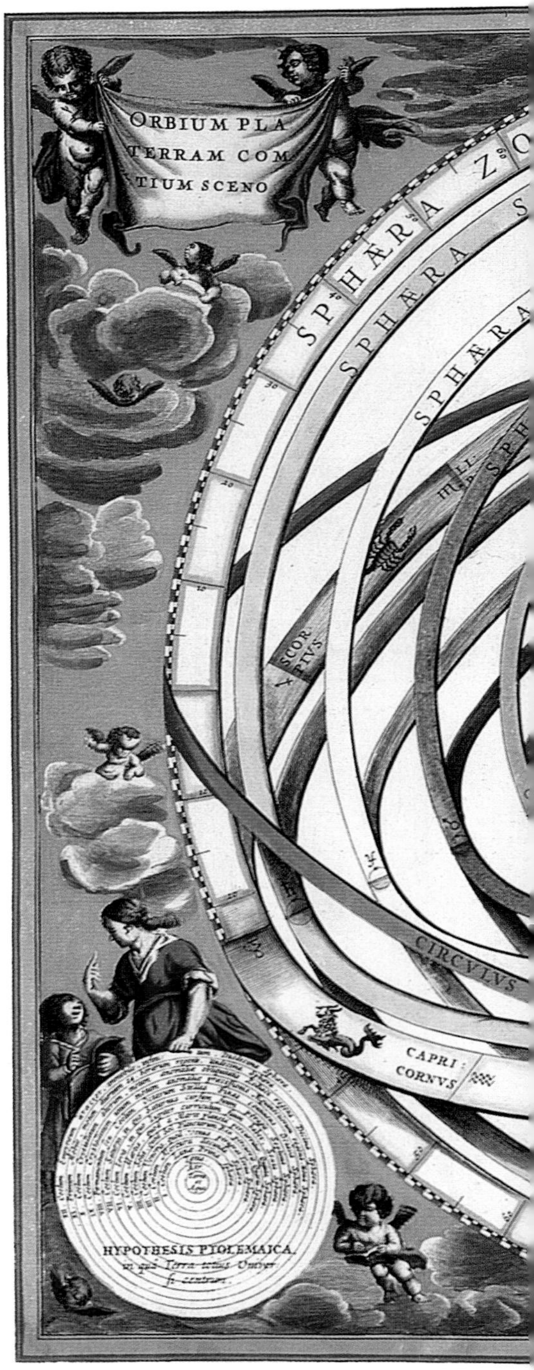

▲ **Earth-centered universe**
This 17th-century illustration by Andreas Cellarius depicts Aristotle and Ptolemy's model. Working out from the center, the Moon, Mercury, Venus, the Sun, Mars, Jupiter, Saturn, and the stars move in circular orbits around Earth.

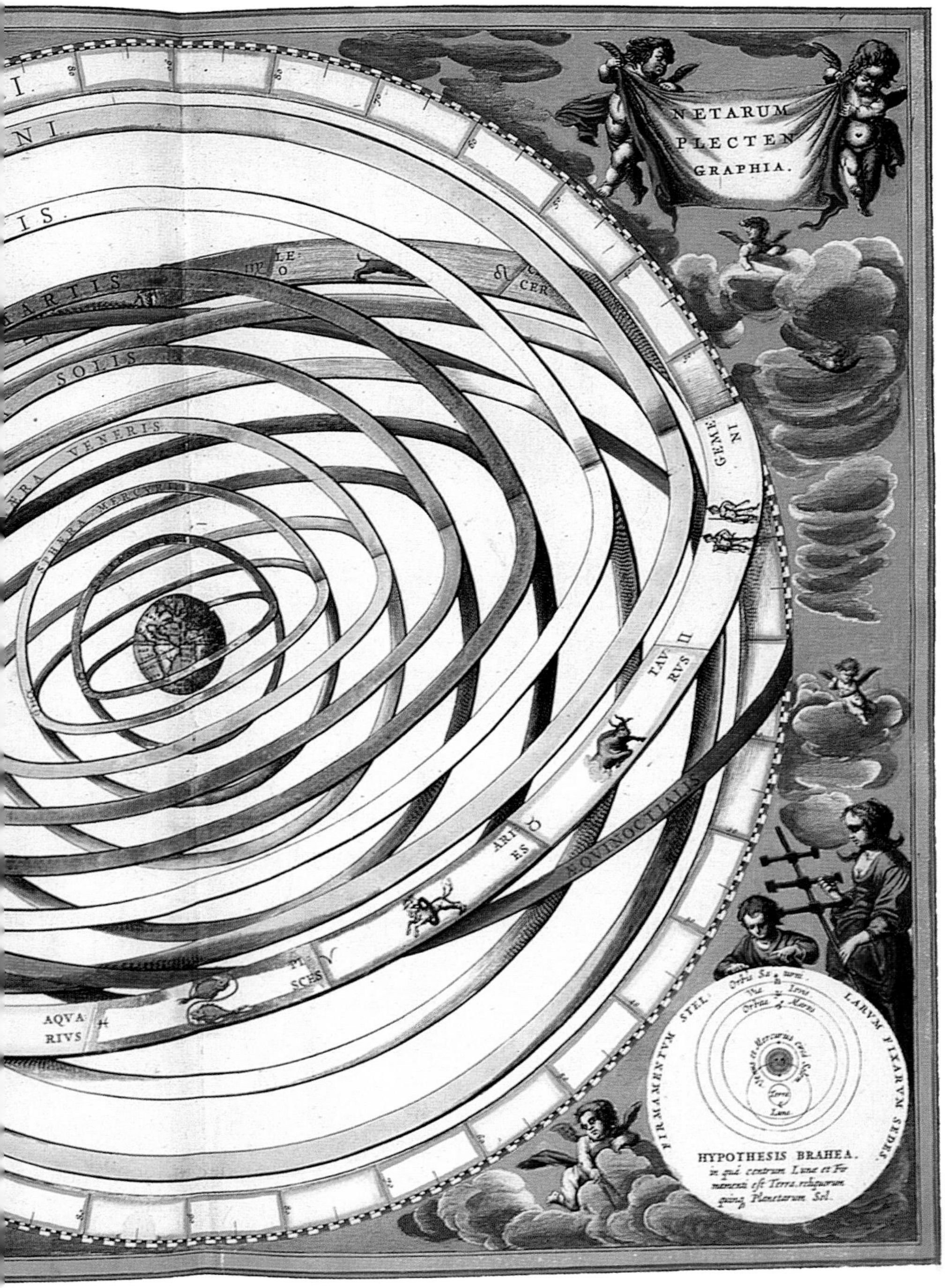

the *Almagest*, which affirmed the prevailing view that Earth is at the center. Ptolemy's detailed model fit with all known observations but in order to do so contained complex modifications to Aristotle's original ideas. For about the next 14 centuries, the Earth-centered view of Aristotle and Ptolemy totally dominated astronomical theory, and it was adopted throughout Europe by medieval Christianity. During this time, Islamic astronomers such as Ulugh Beg (who worked from a great observatory in Samarkand, in what is now Uzbekistan, during the 15th century) made major contributions to knowledge of the solar system and in particular to cataloging star positions.

A STATIONARY OR A SPINNING EARTH?

Linked to the issue of what is at the center of the universe, the question of whether or not Earth rotates was debated for around 2,000 years up to the 17th century CE. The prevailing view was that Earth does not spin, as this fit best with the idea of an Earth-centered universe. However, there were opponents to this view, including Greek philosopher and astronomer Heraclides Ponticus in the 4th century BCE, as well as an Indian astronomer, Aryabhata, and Persian astronomers (Al-Sijzi and Al-Biruni) between the 5th and 15th centuries CE. Each proposed that Earth rotates and that the stars' apparent movement is just a relative motion caused by Earth's spin. But it was not until the Copernican Revolution (see pp.24–25) that Earth's rotation became accepted as fact, and it was not until the 19th century that it was categorically proved.

▲ Ulugh Beg
Working at his observatory at Samarkand, Ulugh Beg and other astronomers determined matters such as the tilt of Earth's spin axis and an accurate value for the length of the year.

THE SIZE AND AGE OF THE UNIVERSE

A final popular subject for speculation among early philosophers was the question of whether the universe is finite (limited) or infinite, both in extent and in time. Aristotle proposed that the universe is infinite in time (so it has always existed) but finite in extent—he believed that all the stars were at a fixed distance, embedded in a crystal sphere, beyond which was nothing. The mathematician Archimedes made a reasoned estimate of the distance to the fixed stars and realized it was vast—at least what we would now call 2 light years—but stopped short of claiming it to be infinite. In the 6th century CE, Egyptian philosopher John Philoponus opposed the prevailing Aristotelian view by arguing that the universe is finite in time. It was not until the 20th century that scientists began to find answers to these questions.

> “IN POSITION **EARTH LIES IN THE MIDDLE OF THE HEAVENS** VERY MUCH LIKE ITS CENTER.”
>
> **Claudius Ptolemy**, astronomer and geographer, 90–168 CE

BIG IDEAS

EARTH **ORBITS THE SUN**

In the 16th and early 17th centuries, the prevailing view of an Earth-centered, or geocentric, universe, as first put forward by the Greek scholars Aristotle and Ptolemy, was challenged by a simpler sun-centered, heliocentric, model. This single idea eventually led to the scientific revolution, a whole new way of thinking about the universe.

To the people of medieval Europe up to the mid-16th century, the question of how the universe is organized had been answered centuries before by Ptolemy, in his modifications to ideas first asserted by Aristotle (see pp.22–23). According to Ptolemy, Earth stood still at the center of the universe. Stars were "fixed" or embedded in an invisible, distant sphere that rotated rapidly, approximately daily, around Earth. The sun, moon, and planets also revolved around Earth, attached to other invisible spheres. For most people, this explanation seemed reasonable—after all, looking up at the sky at night, it did seem that Earth was quite still, while all other objects in the sky, including the sun and stars, rose up in the east, moved across the sky, and then set below the western horizon.

DOUBTS ABOUT GEOCENTRISM

The geocentric model of the universe did not satisfy everyone, however. A serious doubt focused on what it predicted about the planets. According to the original Aristotelian version of geocentrism, the planets rotated around Earth in perfect circles, each at its own steady speed. But if this was true, the planets should move across the sky with unvarying speed and brightness because they were always the same distance from Earth—and this wasn't what was observed. Some planets, such as Mars, varied hugely in brightness over time, and when their movements were compared with those of the outer sphere of fixed stars, the planets sometimes reversed direction—a behavior called retrograde motion.

To deal with these problems, Ptolemy had modified the Aristotelian model. For example, he had planets attached not to the spheres themselves, but to circles called epicycles attached to the spheres. To some astronomers, these modifications looked like "fixes" to fit the model to observational data. From time to time, they suggested alternative ideas, such as that Earth orbits the sun. But supporters of geocentrism had what seemed like an excellent reason for ruling this out. They argued that if Earth moves, the stars should be seen shifting a little relative to each other over the course of a year—but no such shifts could be detected and so, they answered, Earth cannot move.

COPERNICUS WAS A DOCTOR, CLERIC, DIPLOMAT, AND ECONOMIST AS WELL AS AN **ASTRONOMER**

THE COPERNICAN MODEL

In the face of these arguments—and the power of the Catholic Church, which supported the established view—for centuries there was little opposition to the idea of a geocentric universe. However, around 1545, rumors began circulating in Europe that a new and convincing challenge—in the form of a sun-centered theory of the universe—had appeared in a book,

▼ The solar system in miniature
This model of the solar system, called an armillary sphere, is a Copernican version, showing the sun at the center and the planets revolving around it.

> I THINK THAT IN THE DISCUSSION OF **NATURAL PROBLEMS** WE OUGHT TO BEGIN NOT WITH THE **SCRIPTURES**, BUT WITH **EXPERIMENTS AND DEMONSTRATIONS**.
>
> **Galileo Galilei**, astronomer, 1564–1642

De Revolutionibus Orbium Coelestium ("On the Revolutions of the Celestial Spheres"), by a Polish scholar, Nicolaus Copernicus.

Copernicus based his theory on several assumptions. The first was that Earth spins on its axis, and it is this rotation that accounts for most of the daily motion of the stars, planets, moon, and sun across the sky. Copernicus considered that it was inconceivable that thousands of stars were rotating rapidly around Earth. Instead, he proposed that their apparent motion was an illusion caused by Earth's spin. He discounted an objection that this would create catastrophic winds, pointing out that Earth's atmosphere was part of the planet and so part of the motion.

Copernicus's core assumption was that the sun, not Earth, is at or near the center of the universe, and that the planets—including Earth, just another planet—circle the sun at differing speeds. This system could explain, in a simpler way, the motions and variation in brightness of the planets without recourse to any of Ptolemy's "fixes." A third important assumption was that the stars are much farther from Earth and the sun than had previously been accepted. This explained why the relative positions of the stars as seen from Earth appeared to remain fixed over the course of a year.

THE THEORY DEVELOPS

De Revolutionibus was published when Copernicus was dying, and it was a century or more before his theory became widely accepted. One problem was that his model contained misconceptions that had to be corrected by later astronomers. Copernicus clung to the idea that all movements of celestial bodies occur with the objects embedded in invisible spheres. In 1576, the English astronomer Thomas Digges suggested modifying the Copernican system by removing the outermost sphere, in which stars are embedded, and replacing it with a star-filled unbound space. In the 1580s, the Danish astronomer Tycho Brahe banished the rest of the spheres in favor of planets moving freely in orbits. Brahe had observed comets apparently passing through the spheres, which convinced him that they did not actually exist. He also observed a supernova, contradicting a long-held idea that no change takes place in the heavens.

Another shortcoming in Copernicus's theory was his belief that all celestial objects must move in circles, which forced him to retain some of Ptolemy's "fixes." But in the 1620s, the work of the German astronomer Johannes Kepler showed that orbits were elliptical, not circular. By removing most of the remaining "fixes," this simplified and improved the heliocentric model. In the late 17th century, Isaac Newton expanded on Kepler's work, and with his laws of motion and newly introduced force of gravity (see pp.46–47), Newton was able to explain exactly why celestial objects move in the way they do. His work *Principia* effectively removed the last doubts about heliocentrism.

These improvements in the Copernican theory took place against the backdrop of other important advances in cosmology. In the early 17th century, the development of telescopes helped establish that stars are far more distant than planets and exist in vast numbers. It was even suggested that the universe could be infinite. Kepler, however, pointed out that it cannot be infinite, static, and eternal, otherwise the night sky would look uniformly bright due to there being a star emitting light from every direction.

CHURCH REACTION

In 1616, the Roman Catholic Church banned *De Revolutionibus*—a ban that was enforced for more than 200 years. This probably came about as a result of a dispute the Church was having with the astronomer Galileo Galilei, a champion of the Copernican theory who had made discoveries that supported heliocentrism. In particular, in about 1610, Galileo had discovered moons circling Jupiter, proving that some celestial objects do not orbit Earth. The dispute with Galileo caused *De Revolutionibus* to undergo intense scrutiny by the Church, and because some of its ideas seemed to go against biblical statements, it was banned. In 1633, Galileo himself was eventually put on trial and forced to recant his views.

> GALILEO NAMED JUPITER'S MOONS **THE MEDICEAN STARS** AFTER THE **MEDICI FAMILY**

THE SCIENTIFIC REVOLUTION

Banned by the Catholic Church and viewed ambivalently at first by astronomers, the Copernican theory took time to catch on. More than 150 years passed before some of its basic assumptions were shown to be true beyond dispute. But what was important about the theory was that it established cosmology as a science and represented a serious blow to some old, traditional ideas about how the universe works, many dating from Aristotle. As such, it is often viewed as ushering in the scientific revolution—a series of advances between the 16th and 18th centuries that transformed views of nature and society in the early modern world.

> **AT REST**, HOWEVER, IN THE **MIDDLE OF EVERYTHING** IS **THE SUN**
>
> 
>
> **Nicolaus Copernicus**, astronomer and mathematician, 1473–1543

TIMELINES

SEEING THE **LIGHT**

The telescope, and a lesser-known instrument, the spectroscope, are the main tools with which astronomers have expanded knowledge of the universe and its beginnings.

The first telescopes were designed to gather visible light only, and within 100 years had split into two main types—refractors and reflectors. In the 19th century, the spectroscope—which can be used to study the composition and motions of celestial objects—was invented. During the 20th century, bigger optical telescopes came along, followed by radio telescopes. Innovations since the 1970s have included launching telescopes into space and, in the case of radio telescopes, arranging them on the ground in arrays.

VISIBLE LIGHT TELESCOPE TECHNOLOGY

VISIBLE LIGHT TELESCOPE ASTRONOMY

1600

Patent filed for a refracting telescope, 1608, by Hans Lippershey.

Galileo builds a 20-power telescope, 1609. Some of his observations made by telescope later throw him into conflict with the Catholic Church.

Telescopic sight and micrometer invented by William Gascoigne, 1638, facilitating more accurate plotting and measurement of celestial objects.

Refractor 11½ft (3.5m) long built by Johannes Hevelius, 1647. Hevelius subsequently builds even longer telescopes and uses them to make the first accurate map of the moon.

New record for the largest telescope set in 1686 by Christiaan Huygens, who builds a 310ft (95m) long aerial, tubeless refractor.

First working reflecting telescope produced by Isaac Newton, 1668. The design avoids a drawback of refractors called chromatic aberration, but another problem, spherical aberration, renders it impractical.

Newton's reflector

1700

First practical reflecting telescope built by John Hadley, 1721. A parabolic mirror avoids spherical aberration, improving image quality.

Reflecting telescopes improved by James Short, 1730s–1760s. Several are used to observe transits of Venus.

Various nebulous objects, now known to be mostly star clusters and galaxies, identified by Charles Messier, 1760s–1780s.

William Herschel completes a 40ft (12m) long reflecting telescope, 1789.

1800

Fraunhofer's spectroscope

Spectroscope invented, 1814, by Joseph von Fraunhofer, who notices dark (absorption) lines in the sun's spectrum. Later known as Fraunhofer lines, they help astronomers to determine the chemical makeup of stars.

First measurement of a star's parallax, made by Friedrich Bessel, 1838. This becomes the standard method for measuring distances to nearby stars.

First use of photography in astronomy, 1839, allowing fainter objects to be seen and efficient, permanent recording.

Massive reflector, with a 5ft (1.5m) wide mirror, built by William Parsons, 1845.

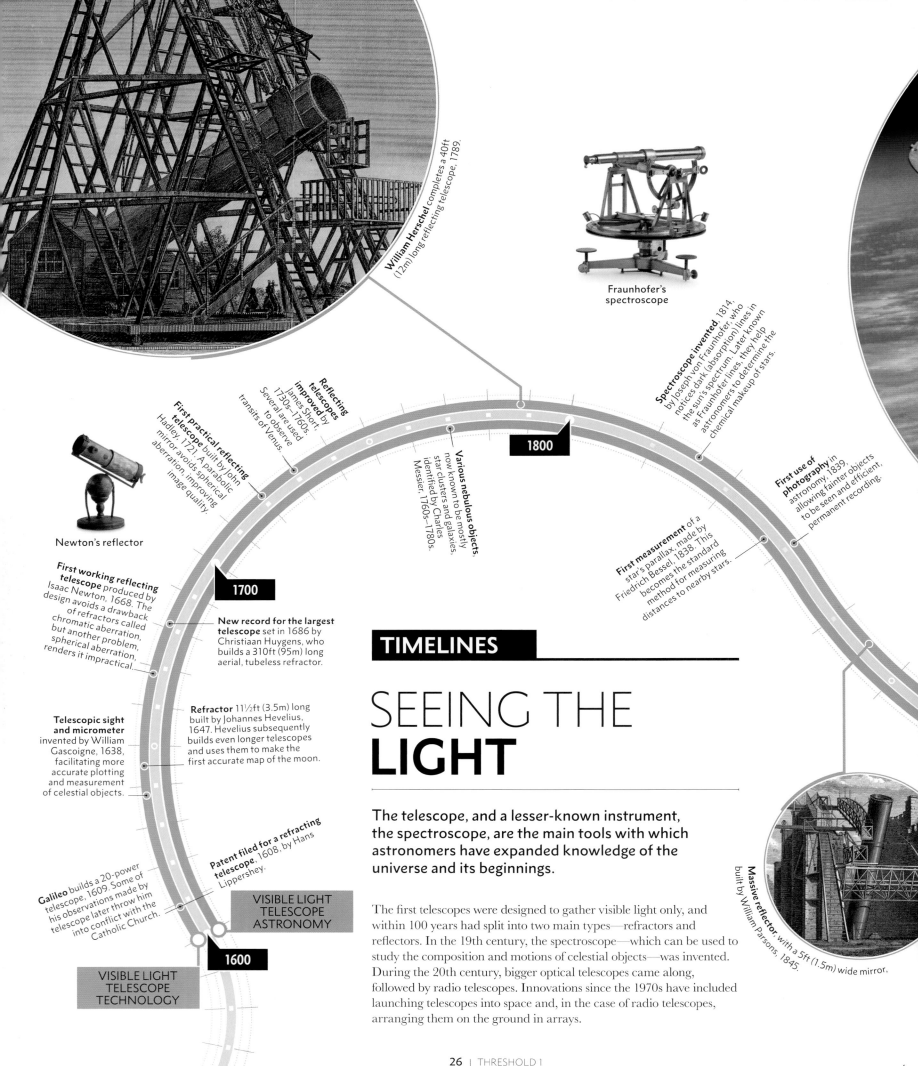

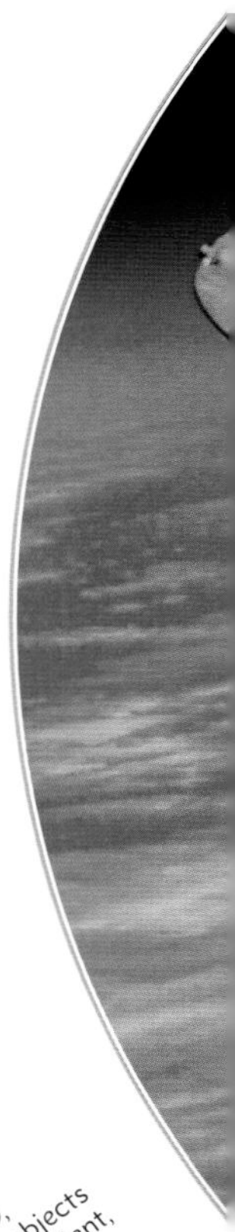

Spectra of nebulae, stars, and galaxies are studied by William and Margaret Huggins, 1860s. They measure the redshifts of stars, showing how fast they are moving.

Elements in the sun's atmosphere identified by Custav Kirchoff, 1861. He notices dark lines in its spectrum match wavelengths of light emitted by elements burned in a flame.

First photograph of a star's spectrum taken by Henry Draper, 1872. The star is Vega, and the photograph shows its absorption lines.

Yerkes Refractor is completed in Wisconsin, 1895. The largest refractor ever used for research, it is involved in discoveries such as the spiral nature of the Milky Way.

1900

Edwin Hubble uses the Hooker Telescope in the 1920s to show that there are galaxies outside our own and to relate the distances and recessional velocities of galaxies—leading to the discovery that the universe is expanding.

Pluto discovered, 1930, at the Lowell Observatory, Arizona, by Clyde Tombaugh, who makes his discovery after comparing photographic plates, taken about a week apart.

BEYOND VISIBLE LIGHT

First parabolic dish radio telescope, built by Grote Reber in Wheaton, Illinois, 1937. Reber then makes an all-sky map of radio emissions from space.

Cosmic microwave background radiation detected, 1964, by Arno Penzias and Robert Wilson using a radio telescope at Bell Telephone Laboratories, New Jersey. It helps to confirm the Big Bang theory.

First pulsar discovered, 1967, by Jocelyn Bell and Antony Hewish using a radio telescope at the University of Cambridge.

The Very Large Array (VLA), a group of 27 radio telescopes that work together to form images, begins operations in New Mexico, 1980.

SPACE TELESCOPES

Hipparcos Satellite makes its first observations, 1989. Hipparcos makes highly accurate measurements of the positions of stars.

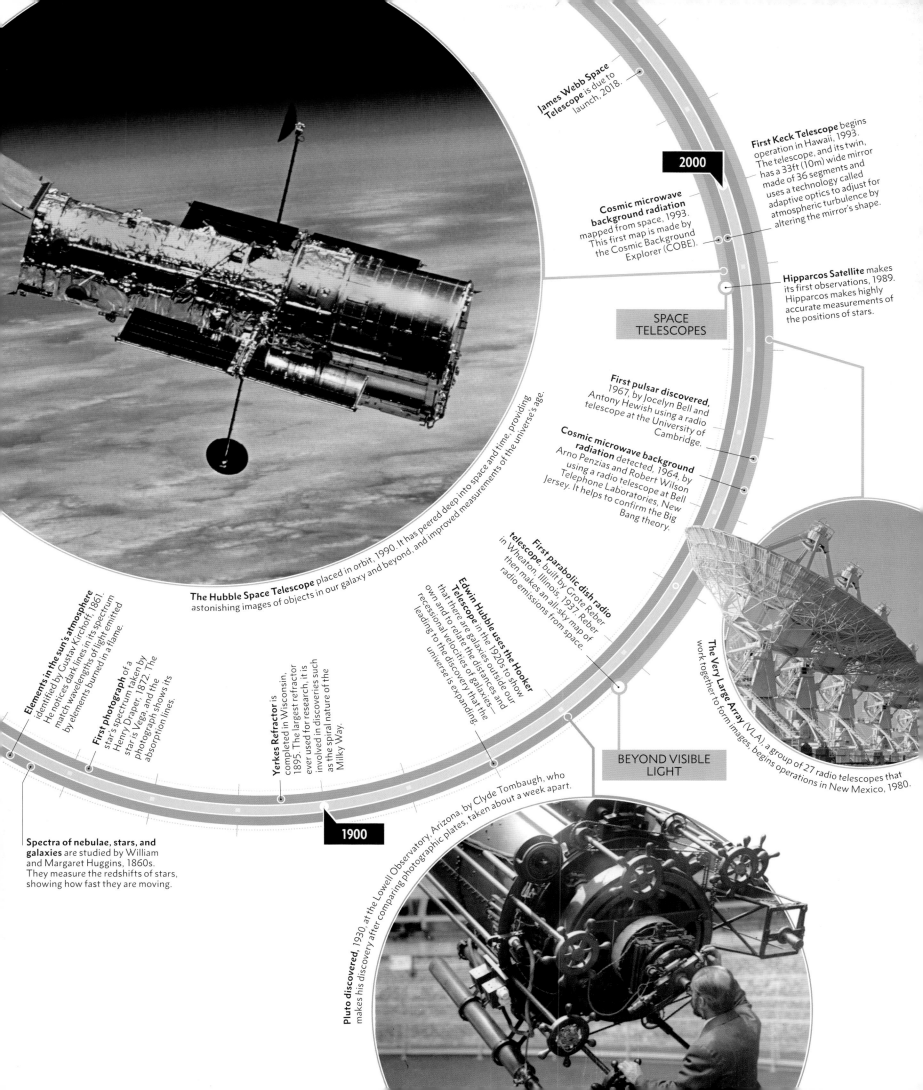

The Hubble Space Telescope placed in orbit, 1990. It has peered deep into space and time, providing astonishing images of objects in our galaxy and beyond, and improved measurements of the universe's age.

Cosmic microwave background radiation mapped from space, 1993. This first map is made by the Cosmic Background Explorer (COBE).

First Keck Telescope begins operation in Hawaii, 1993. The telescope, and its twin, has a 33ft (10m) wide mirror made of 36 segments and uses a technology called adaptive optics to adjust for atmospheric turbulence by altering the mirror's shape.

2000

James Webb Space Telescope is due to launch, 2018.

THE **ATOM** AND THE **UNIVERSE**

From the early 19th century to the late 1920s, a series of breakthroughs occurred in the physical sciences. They transformed our understanding of the workings and structure of the world at both infinitesimally small scales and at the very largest, raising the possibility of an infinite cosmos.

◀ Henrietta Leavitt Over 20 years, Leavitt studied 1,777 variable stars at the Harvard College Observatory before stumbling upon her key discovery.

These discoveries paved the way for the advances of the 1930s to the 1950s, from the realization that the universe is expanding to the development of ideas on how energy and matter interact at the subatomic level. Through the coming together of ideas in cosmology and particle physics, these breakthroughs eventually led to the development of the Big Bang theory.

PROBING MATTER AND ENERGY

The idea that matter consists of atoms was first suggested by the ancient Greek, Democritus (see p.22). In the early 1800s, an Englishman, John Dalton, revived the idea. Dalton regarded atoms as indivisible, but around the turn of the 20th century, experiments by scientists such as the New Zealander Ernest Rutherford proved that they have a substructure. Around the same time, the German theoretical physicist Albert Einstein showed that matter and energy have an equivalence. Simultaneously, a new field of physics, quantum theory, was proposing (among other things) that light can behave either as a wave or as a stream of particles. By the late 1920s, it was known

> WHAT WE OBSERVE AS **MATERIAL BODIES AND FORCES** ARE NOTHING BUT **SHAPES AND VARIATIONS IN THE STRUCTURE OF SPACE**.
>
> **Erwin Schrödinger**, Austrian theoretical physicist, 1887–1961

▶ Understanding the atom From around 1800 until the mid-1920s, a step-by-step evolution occurred in the understanding of atomic structure. Later, from the late 1920s, physicists found that atomic nuclei have a substructure.

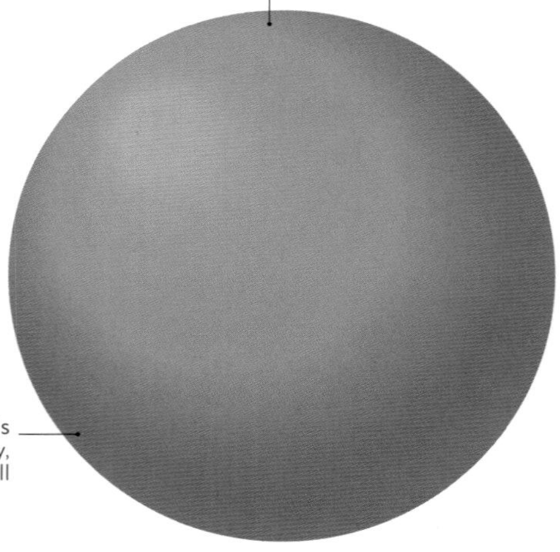

Dalton's atom (1803) English chemist John Dalton pictures atoms as extremely small spheres, like tiny billiard balls, that have no internal structure and cannot be subdivided, created, or destroyed.

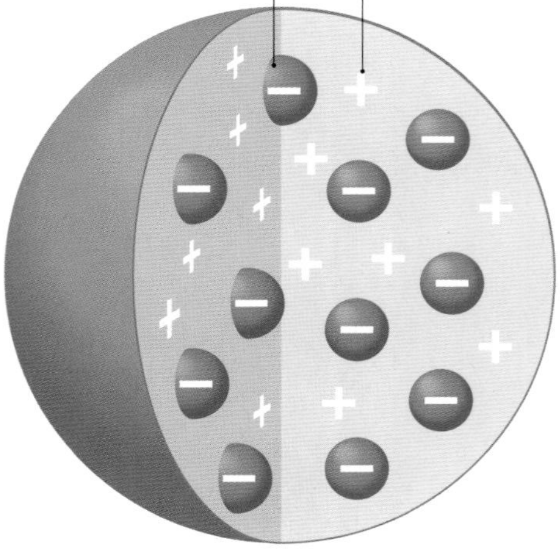

Thomson's plum pudding (1904) The discoverer of the electron, British physicist J.J. Thomson, suggests a "plum-pudding" model, with negatively charged electrons embedded in a positively charged sphere.

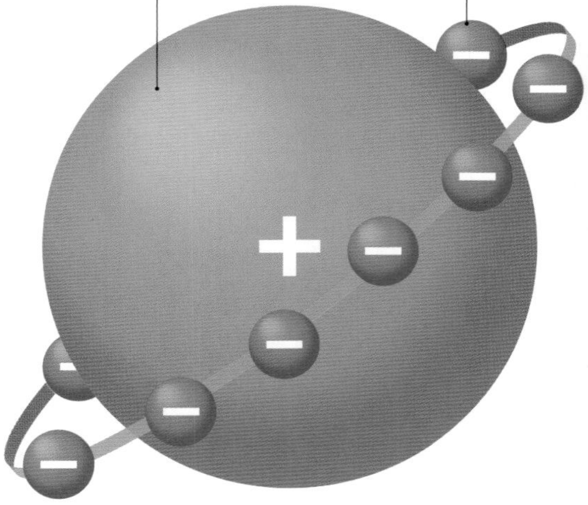

Nagaoka's Saturnian model (1904) Japanese physicist Hantaro Nagaoka proposes an atom has a central nucleus, around which the electrons orbit in one or more rings, like the rings of Saturn.

that atomic nuclei consist of protons and neutrons and are held together by a newly detected force, the strong force. Also discovered at this time was antimatter—subatomic particles that are identical to their matter equivalents except for opposite electrical change—and that the coming together of matter and antimatter can annihilate both, producing pure energy.

THE DISTANCES TO STARS

During roughly the same period, great advances were made in understanding the true scale of the cosmos. In 1838, the German astronomer Friedrich Bessell made the first reliable measurement of the distance to a star other than the sun, using a method called stellar parallax. The star, although one of the closest to the sun, seemed at the time almost unimaginably far off—what would now be called 10.3 light-years away. It was 1912 before a system was discovered for estimating the distance to many more remote stars. The discoverer was an American named Henrietta Leavitt. Her breakthrough concerned a class of star called Cepheid variables, which cyclically vary in brightness. Leavitt found a link between the cycle period and brightness of these stars, meaning that if both could be measured, a good estimate could be made of their distance from Earth. Within a few years, it became apparent that some stars are tens of thousands of light-years away, while some vaguely spiral-shaped nebulous patches in the sky, known at the time as "spiral nebulae," seemed to be millions of light-years away.

A LIGHT-YEAR—THE DISTANCE LIGHT TRAVELS THROUGH SPACE IN A YEAR—IS ABOUT **6 TRILLION MILES (9.5 TRILLION KILOMETERS)**

SHIFTING NEBULAE

Between 1912 and 1917, the American astronomer Vesto Slipher studied several "spiral nebulae" and realized that many were moving away from Earth at high speed, while a few were approaching Earth. He found this out by measuring a property of the light from the nebulae called redshift or blueshift. It seemed odd that the nebulae were moving at such speed relative to the rest of the galaxy. Partly prompted by Slipher's findings, in 1920 a formal debate was held in Washington, DC on whether these nebulae might be separate galaxies outside our own. The debate was inconclusive. But within a few years, the answer had been found—by another American astronomer named Edwin Hubble (see pp.30–31).

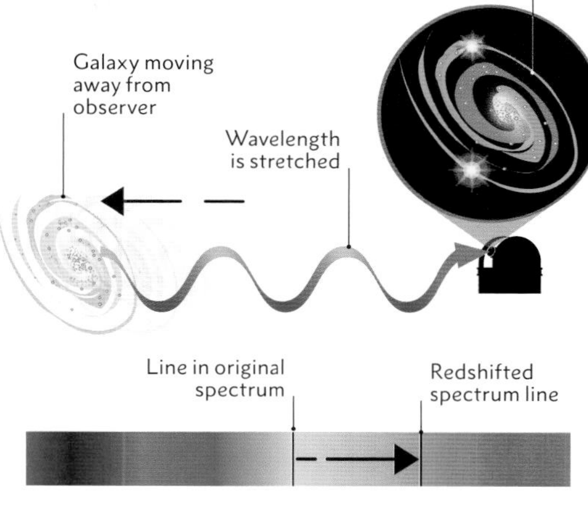

◀ Redshift When an object such as a galaxy is receding at high speed, light waves from it appear stretched. This causes features in the galaxy's spectrum, such as prominent lines, to shift toward the red (long wavelength) end. This is a redshift.

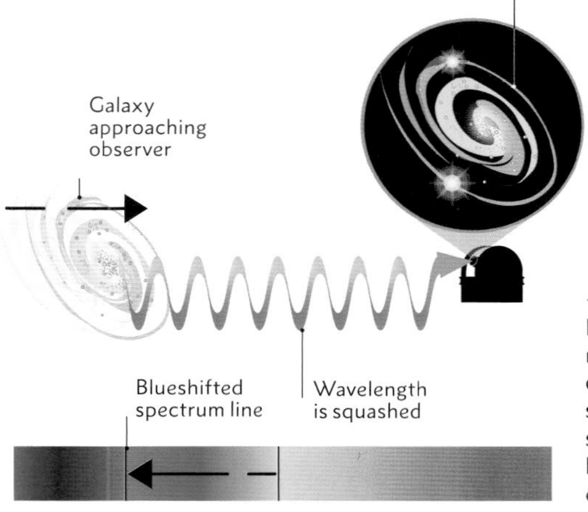

◀ Blueshift Light waves from a rapidly approaching object appear squashed, shifting features in the spectrum toward the blue (short wavelength) end. This is a blueshift.

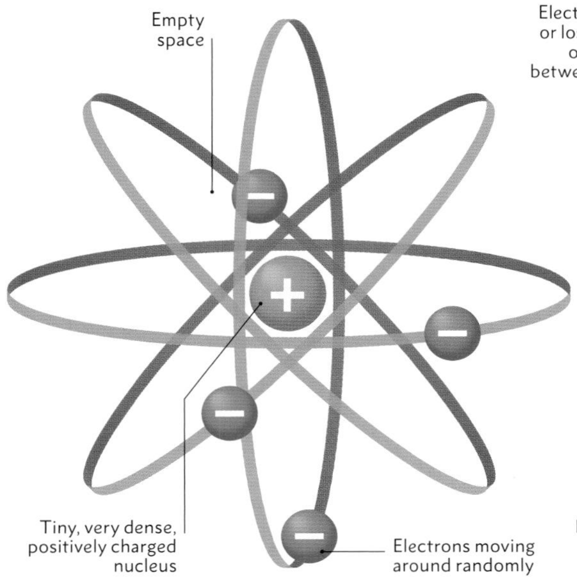

Rutherford and the nucleus (1911) Rutherford proves experimentally that an atom's nucleus is much smaller and denser than previously thought—and that much of an atom is empty space.

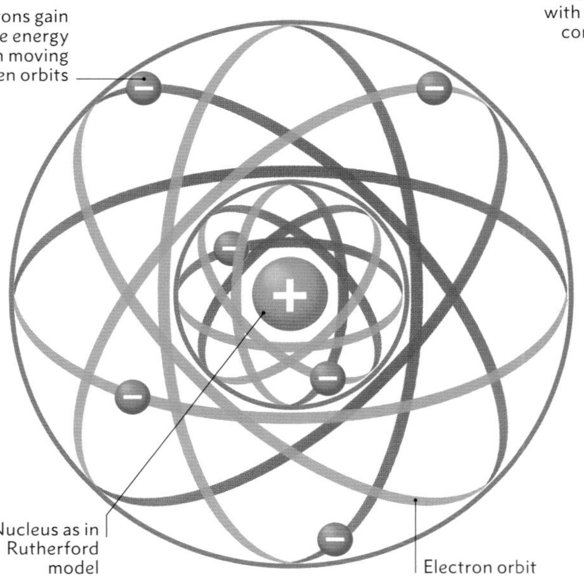

Bohr's electron orbits (1913) Danish physicist Niels Bohr proposes that electrons can move in spherical orbits, at fixed distances from the nucleus, and can "jump" between orbits.

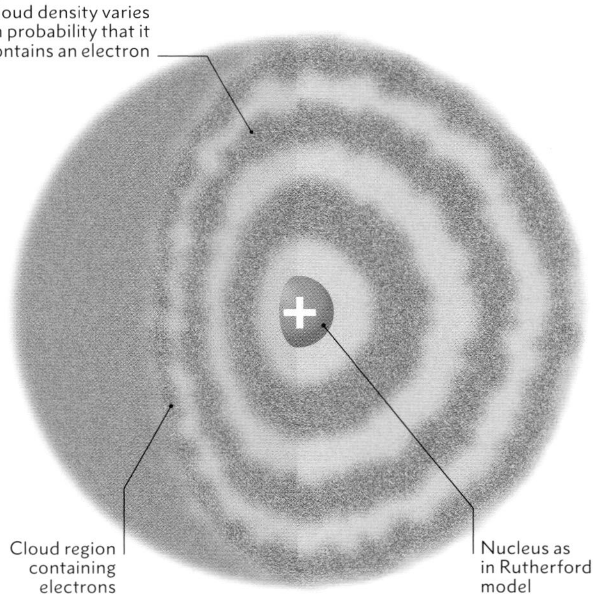

Schrödinger's electron cloud model (1926) According to Austrian physicist Erwin Schrödinger's model, the locations of electrons in an atom are never certain and can be stated only in terms of probabilities.

THE UNIVERSE **GETS BIGGER**

During the 1920s, two key breakthroughs led to a revolution in understanding the size and nature of the universe. Both were the result of discoveries made by the astronomer Edwin Hubble.

In 1919, Hubble arrived at Mount Wilson Observatory in California at age 30. His arrival coincided with the completion of what was then the largest telescope in the world, a reflector with a 100in (2.5m) wide mirror, called the Hooker Telescope.

ENDING THE GALAXY DEBATE

At that time, the prevailing view was that the universe consisted of just the Milky Way Galaxy, although in 1920 a famous debate (see p.29) had considered whether or not some vaguely spiral-shaped nebulae—fuzzy, star-containing objects—in the night sky might be collections of stars outside our own galaxy. Hubble, who had been studying these nebulae, already strongly suspected that they were outside our galaxy. In 1922–23, he used the Hooker Telescope to observe a class of stars called Cepheid variables in some of the nebulae, including what today is called the Andromeda Galaxy. Cepheid variables are stars whose distances can be estimated by measuring their average brightness and the lengths of their cycles of brightness variation. As a result of his observations, in 1924 Hubble was able to announce that the Andromeda nebula and other spiral nebulae were far too distant to be part of the Milky Way and so must be galaxies outside our own. Almost overnight, the universe had become a much bigger place than anyone had previously imagined.

RECEDING GALAXIES

Hubble next studied a phenomenon that had already been noted by an astronomer called Vesto Slipher: many of the spiral galaxies had large "redshifts" in their spectra, meaning that they were moving away from Earth at high speed (see p.29). Again by observing Cepheid variables, Hubble began measuring the distances to these galaxies and compared the distances to their redshifts. He noticed something remarkable: the more distant a galaxy was, the greater was its recessional velocity—a relationship that became known as Hubble's Law. Hubble published his results in 1929. Although he himself was initially skeptical, to other astronomers it was clear that only one conclusion could be drawn—the whole universe must be expanding!

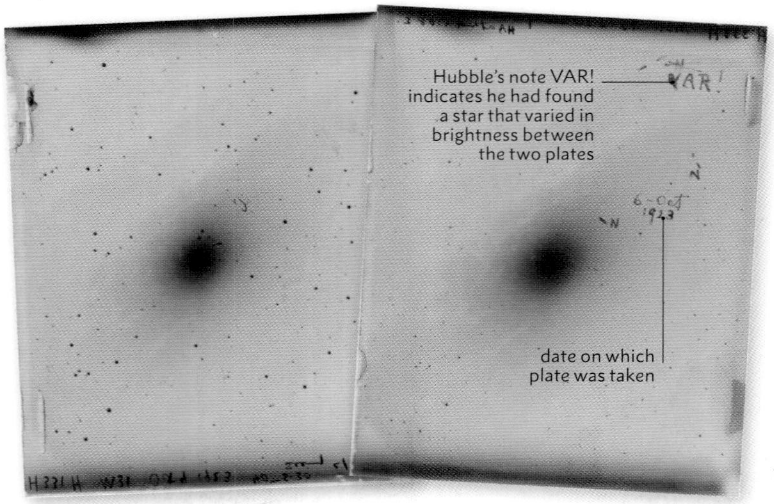
Hubble's note VAR! indicates he had found a star that varied in brightness between the two plates
VAR!
date on which plate was taken

▼ Photographic evidence
These two (negative) photographic plates were used by Hubble to identify a specific Cepheid variable star in the Andromeda Galaxy. Studies on this star were crucial in confirming that the Andromeda Galaxy is outside the Milky Way.

> THE **HISTORY OF ASTRONOMY** IS A **HISTORY OF RECEDING HORIZONS**.
>
> **Edwin Hubble**, American astronomer, 1889–1953

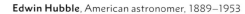

The world's largest telescope
Completed in 1917, the Hooker Telescope was the world's largest telescope for about 30 years. Its glass mirrors, which had to be cast to an accuracy of a few millionths of an inch, had to be kept cool to prevent them from warming up and becoming distorted.

THE **EXPANDING UNIVERSE**

Edwin Hubble's work showed that many galaxies are receding from us at a rate proportional to their distance. It was soon deduced that the universe must be expanding, but astronomers still had to understand the nature of this expansion and what the universe is expanding from.

By the beginning of the 1930s, scientists were also starting to address a question that philosophers had been pondering for several millennia—has our universe always existed or did it have a beginning? Physicists, mathematicians, and astronomers were now in a position to try answering this question.

> "THE **RADIUS OF SPACE BEGAN AT ZERO**; THE FIRST STAGES OF THE EXPANSION CONSISTED OF A **RAPID EXPANSION** DETERMINED BY THE MASS OF THE INITIAL ATOM."
>
> **Georges Lemaître**, astronomer, 1894–1966

EINSTEIN'S POSSIBLE UNIVERSES

The story of how scientists came to realize that the universe is expanding began in 1915 with the publication of Albert Einstein's general theory of relativity. This theory is a description of how gravity works at the largest scales, and it defines what possible universes can exist. Part of Einstein's theory consists of a set of equations that have to be solved to give a description of the long-term, large-scale behavior of the universe.

Einstein's initial solution to his equations suggested that the universe is contracting, but he could not believe this, so he introduced a "fix"—an expansion-inducing factor called the cosmological constant—into his theory to allow for a static universe. In 1927, the Belgian astronomer Georges Lemaître, who had studied Einstein's equations and heard of Hubble's measurements of galaxy distances, proposed that the whole of space is expanding—but his hypothesis failed to attract widespread attention. After Hubble released his findings about receding galaxies in 1929, it became clear to many astronomers that the universe must indeed be expanding, although neither Hubble nor Einstein was convinced at first. Despite this, for many years credit for the discovery was given to Hubble, but today most experts agree it should be equally shared with Lemaître.

▼ Georges Lemaître
Arguably the first person to propose that the universe is expanding, Lemaître was a priest and physicist as well as an astronomer.

DISCOVERING THE BIG BANG

If the universe is expanding, and the clock is run backward, then the farther back in time you look, the denser the universe becomes. But, as Lemaître reasoned, one can only go so far before the universe is crushed into an infinitely dense point. So in 1931, he suggested that the universe was initially a single, extremely dense particle—a "primeval atom" as he called it—that disintegrated in an explosion, giving rise to space and time and the expansion of the universe. By 1933, Einstein (who had by now abandoned his cosmological constant) was in full agreement with Lemaître's theory, calling it "the most beautiful and satisfactory explanation of creation to which I have ever listened."

Simple physics dictates that the universe compressed into a tiny point would be extremely hot. During the 1940s, Russian-American physicist George Gamow, and colleagues, worked out details of what might have happened during the exceedingly hot first few moments of a Lemaître-style universe. This included working out how the nuclei of light elements, such as helium, might have been forged, starting with just protons and neutrons. The work showed that a "hot" early universe, evolving into what is observed today, is at least theoretically feasible. In a 1949 radio interview, the British astronomer Fred Hoyle coined the term "Big Bang" for the model of the universe Lemaître and Gamow had been developing. At last, Lemaître's startling hypothesis had a name, which has stuck ever since.

► Expanding space
The universe's expansion is most accurately thought of in terms of space itself expanding and carrying objects with it—called cosmological expansion—rather than galaxies and galaxy clusters moving away from each other "through" space.

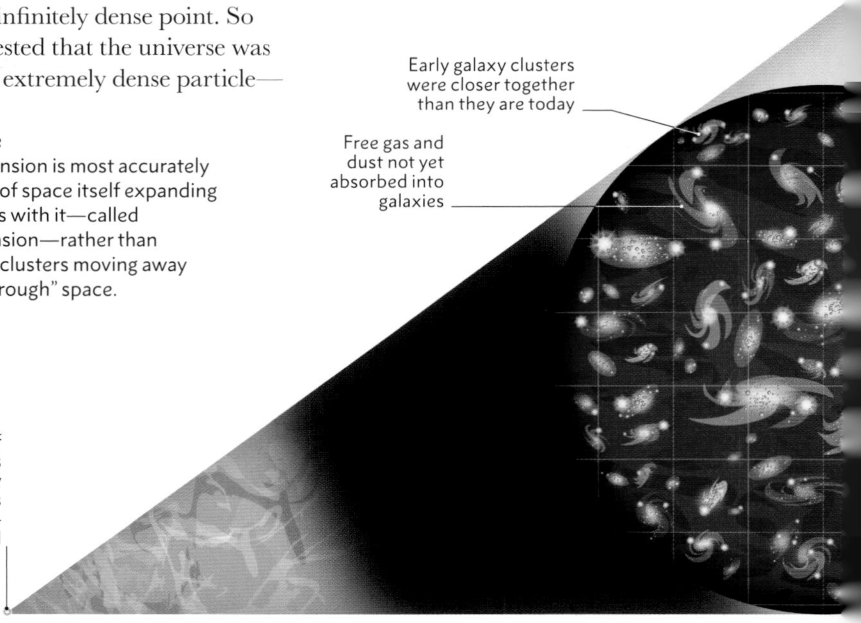
Early galaxy clusters were closer together than they are today
Free gas and dust not yet absorbed into galaxies
At the beginning of time, all matter was concentrated in a tiny particle –Lemaître's "primeval atom"—that exploded

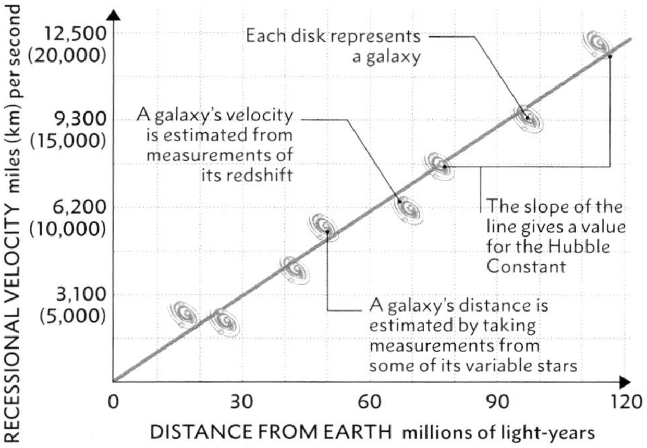
RECESSIONAL VELOCITY miles (km) per second
12,500 (20,000)
9,300 (15,000)
6,200 (10,000)
3,100 (5,000)
0
30
60
90
120
DISTANCE FROM EARTH millions of light-years
Each disk represents a galaxy
A galaxy's velocity is estimated from measurements of its redshift
The slope of the line gives a value for the Hubble Constant
A galaxy's distance is estimated by taking measurements from some of its variable stars

▲ The Hubble Constant
If the velocities of a number of galaxies are plotted against their distances, a "best fit" line can be drawn close to all the plotted points. The slope of the line is an estimate of the Hubble Constant, itself a measure of the rate of the universe's expansion.

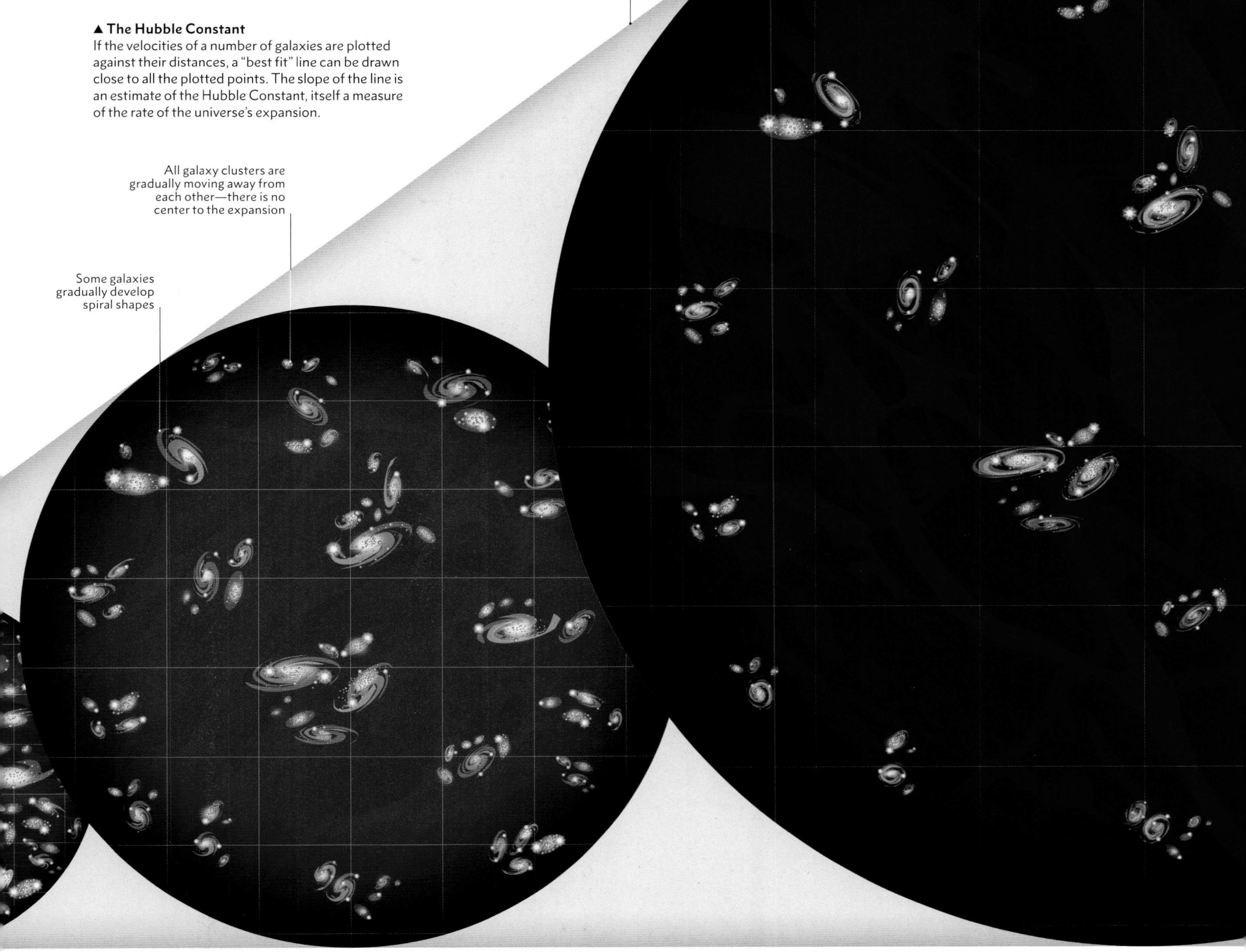
On a local scale, gravity dominates over expansion, holding galaxy clusters together
Over time, the space between galaxies and galaxy clusters empties out as free gas and dust are pulled into galaxies
In the 1930s, it was assumed that the rate of expansion was at or near a uniform rate—with just slight slowing due to gravity
All galaxy clusters are gradually moving away from each other—there is no center to the expansion
Some galaxies gradually develop spiral shapes

THE **BIG BANG**

Since the 1930s, when the Big Bang theory was first proposed, physicists and cosmologists have been testing and developing the theory and filling in the details of the first moments of the universe.

Part of the work to improve the Big Bang theory has been carried out by experiments in which high-energy particles are collided to recreate Big Bang-like conditions (see pp.36–37), and part has been purely theoretical, involving the formulation of equations and models. During the experimental side of this journey, many new subatomic particles have been discovered. Another focus of research has been the fundamental forces that govern particle interactions. It has been known since the 1930s that there are four of these forces: gravity, the electromagnetic force, the strong force, and the weak interaction. During the Big Bang, it is theorized that these forces were initially unified. Then, as the universe cooled, they split off, possibly triggering new phases of the Big Bang. Gradually, physicists have fitted all the known particles and the forces into a scheme called the Standard Model of particle physics.

One important change to the original theory was made in the 1980s by the American physicist Alan Guth. He proposed that at a very early stage a part of the universe underwent an extremely fast expansion called cosmic inflation. Guth's idea helped explain some aspects of the universe today, including why at the largest scales matter and energy seem to be distributed very smoothly. The reality of cosmic inflation is now widely accepted.

Up quark, **Down quark** — There are six types of quark. Up and down quarks are the most stable and common.

Electron — This tiny subatomic particle has a negative electrical charge.

Gluon — By carrying the strong nuclear force, gluons hold quarks together.

Photon — A photon is a tiny packet of light or other electromagnetic radiation.

Higgs boson — This particle is associated with a field that gives mass to other particles.

▲ Fundamental particles
These particles are not, so far as is known, made of smaller particles. Some, such as quarks, are building blocks of matter. Others, like gluons and photons, are force-carrier particles.

Proton — A proton is made of two up quarks and one down quark plus gluons.

Neutron — Two down quarks and one up quark, plus gluons, make up a neutron.

▲ Composite particles
These particles are composed of other smaller particles. Scores of different composite particles have been identified, but protons and neutrons are the only stable types.

Up antiquark, **Down antiquark** — For each of the six types of quark there is a corresponding type called an antiquark.

Positron — A positron is the positively charged equivalent of the electron.

Anti-proton — Two up antiquarks and one down antiquark, plus gluons, form an antiproton.

Anti-neutron — This consists of two down antiquarks and one up antiquark, plus gluons.

▲ Antiparticles
These are particles with the same mass but an opposite electric charge to their equivalent particles

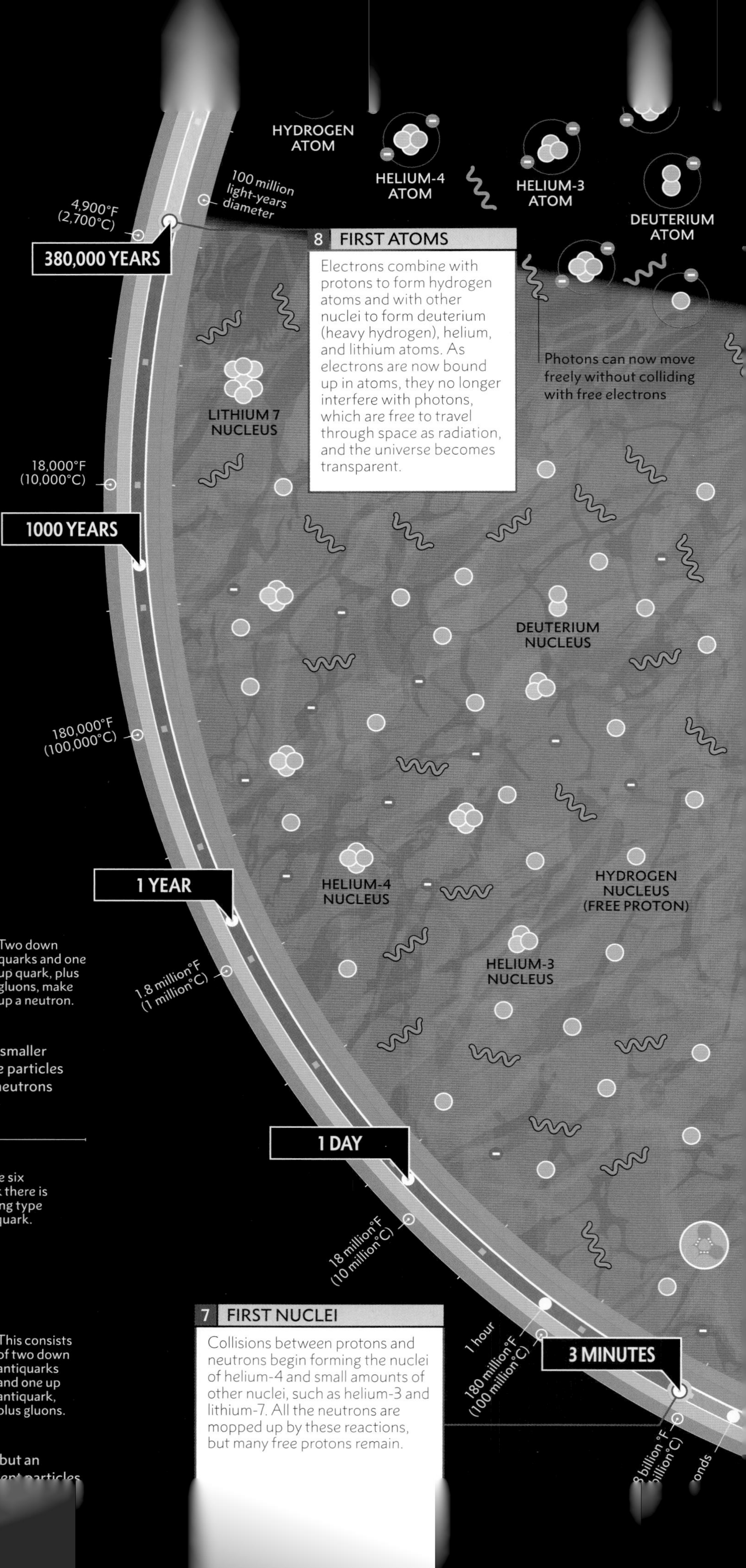

HYDROGEN ATOM
HELIUM-4 ATOM
HELIUM-3 ATOM
DEUTERIUM ATOM
100 million light-years diameter
4,900°F (2,700°C)
380,000 YEARS
8 FIRST ATOMS
Electrons combine with protons to form hydrogen atoms and with other nuclei to form deuterium (heavy hydrogen), helium, and lithium atoms. As electrons are now bound up in atoms, they no longer interfere with photons, which are free to travel through space as radiation, and the universe becomes transparent.
Photons can now move freely without colliding with free electrons
LITHIUM 7 NUCLEUS
18,000°F (10,000°C)
1000 YEARS
DEUTERIUM NUCLEUS
180,000°F (100,000°C)
1 YEAR
HELIUM-4 NUCLEUS
HYDROGEN NUCLEUS (FREE PROTON)
1.8 million°F (1 million°C)
HELIUM-3 NUCLEUS
1 DAY
18 million°F (10 million°C)
7 FIRST NUCLEI
Collisions between protons and neutrons begin forming the nuclei of helium-4 and small amounts of other nuclei, such as helium-3 and lithium-7. All the neutrons are mopped up by these reactions, but many free protons remain.
1 hour
180 million°F (100 million°C)
3 MINUTES

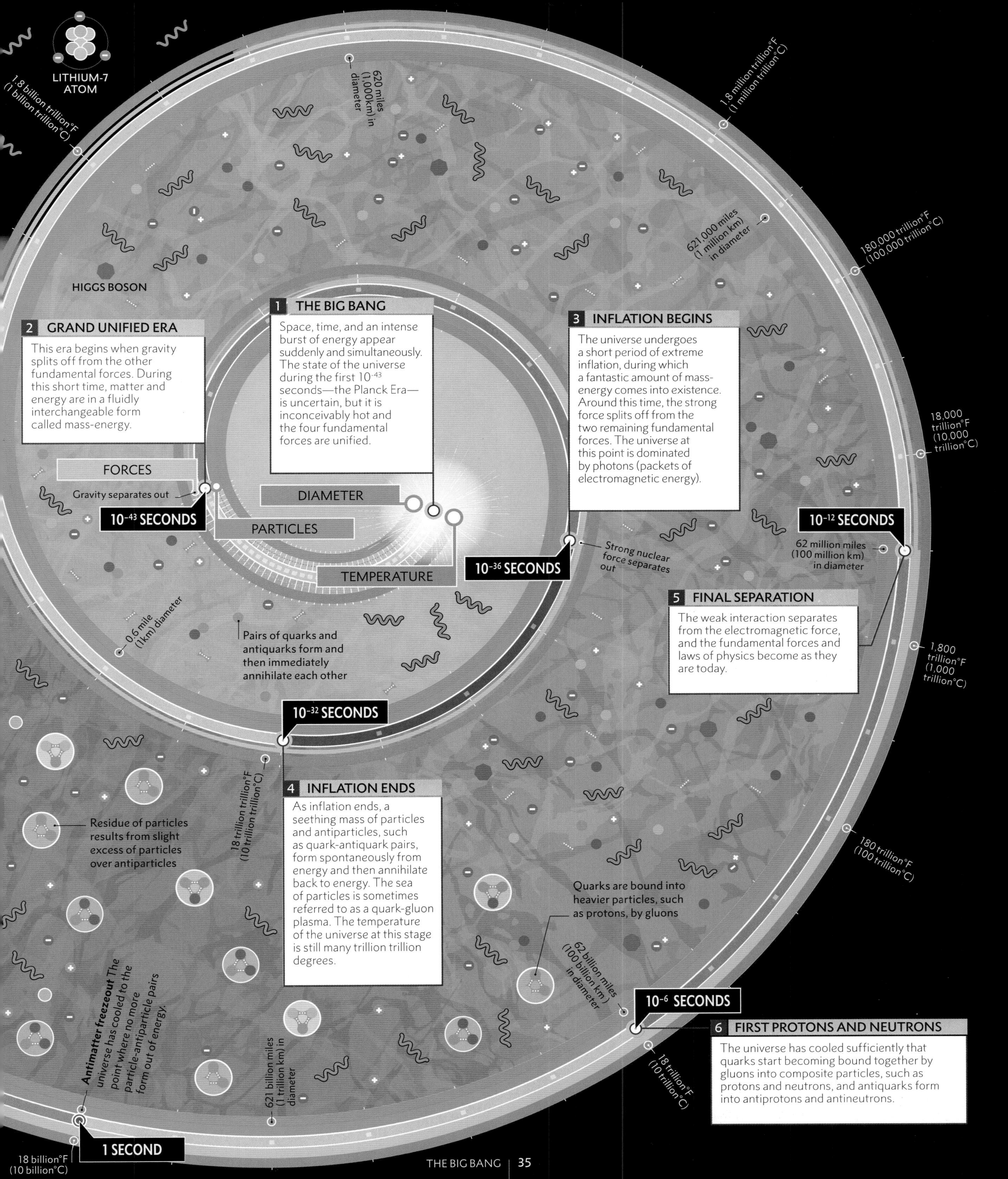

1 THE BIG BANG

Space, time, and an intense burst of energy appear suddenly and simultaneously. The state of the universe during the first 10^{-43} seconds—the Planck Era—is uncertain, but it is inconceivably hot and the four fundamental forces are unified.

2 GRAND UNIFIED ERA

This era begins when gravity splits off from the other fundamental forces. During this short time, matter and energy are in a fluidly interchangeable form called mass-energy.

3 INFLATION BEGINS

The universe undergoes a short period of extreme inflation, during which a fantastic amount of mass-energy comes into existence. Around this time, the strong force splits off from the two remaining fundamental forces. The universe at this point is dominated by photons (packets of electromagnetic energy).

4 INFLATION ENDS

As inflation ends, a seething mass of particles and antiparticles, such as quark-antiquark pairs, form spontaneously from energy and then annihilate back to energy. The sea of particles is sometimes referred to as a quark-gluon plasma. The temperature of the universe at this stage is still many trillion trillion degrees.

5 FINAL SEPARATION

The weak interaction separates from the electromagnetic force, and the fundamental forces and laws of physics become as they are today.

6 FIRST PROTONS AND NEUTRONS

The universe has cooled sufficiently that quarks start becoming bound together by gluons into composite particles, such as protons and neutrons, and antiquarks form into antiprotons and antineutrons.

Physics on a grand scale
The large, barrel-shaped machine undergoing a refit here is a part of the LHC called the electromagnetic calorimeter. It measures the energies of electrons and photons to a high degree of accuracy.

RECREATING THE **BIG BANG**

For years, researchers at the European Organization for Nuclear Research (CERN) have used the world's largest particle accelerator—the Large Hadron Collider (LHC)—to smash particles together at extreme speeds to recreate conditions that existed shortly after the Big Bang.

The LHC is the largest, most sophisticated scientific instrument ever built. Located underground on the French-Swiss border, it accelerates two beams of high-energy particles, moving in opposite directions, through pipes connected in a ring with a circumference of almost 17 miles (27km). From time to time, the beams are made to collide, and the results—which typically include the appearance of short-lived, exotic particles—are recorded by detectors around the ring. The purpose of the LHC is to study the range of subatomic particles that can exist and the laws governing their interactions.

Physicists hope these experiments will refine their ideas about what happened in the Big Bang and help them to investigate some poorly understood cosmic phenomena. The Big Bang-type conditions are recreated only in miniature—so there is no chance the experiments could trigger a new Big Bang and the appearance of a new universe.

NEW DISCOVERIES

One success of the LHC has been to create a quark-gluon plasma, a maelstrom of free quarks and gluons (see p.34) that is thought to have existed for up to a microsecond (a millionth of a second) after the start of the Big Bang. This was achieved in 2015 by colliding protons with lead nuclei, creating minuscule fireballs in which everything broke down momentarily into quarks and gluons.

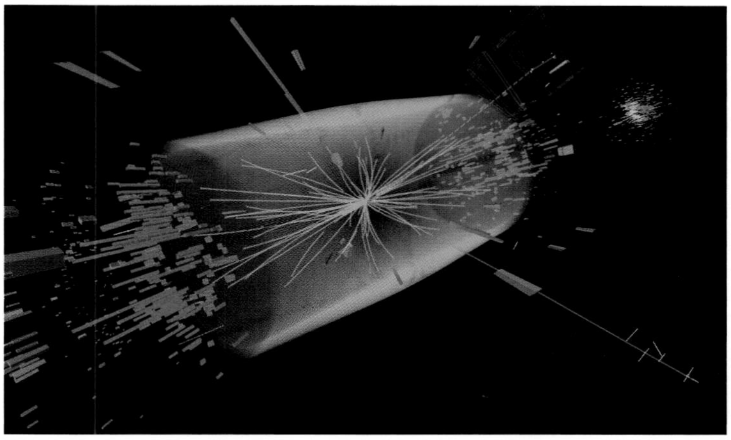

▲ Seeking the Higgs boson
This computer graphic shows a particle collision recorded during the search for the Higgs boson. It displays features that could be expected from the decay of a Higgs boson into two other bosons. One of these decays to a pair of electrons (green lines) and the other to a pair of particles called muons (red lines).

In 2012, a long sought-after, high-mass, extremely short-lived particle called the Higgs boson was detected. Its existence confirmed the presence of an energy field, the Higgs field, that imparts mass to particles passing through it. The significance of this for the Big Bang is that it explains how in the first moments of the universe particles such as quarks acquired mass, causing them to slow down and combine to form composite particles, such as protons and neutrons.

Other notable successes include the detection in 2014 of a pentaquark (consisting of four quarks and an antiquark). This discovery may allow scientists to study in more detail the strong force that holds quarks together.

“WE HAVE MADE THE DISCOVERY OF **A NEW PARTICLE**—A **COMPLETELY NEW PARTICLE**—WHICH IS MOST PROBABLY VERY DIFFERENT FROM ALL THE OTHER PARTICLES.”

Rolf-Dieter Heuer, Director of CERN, 1948–, on the discovery of the Higgs boson

BEYOND THE **BIG BANG**

Although the Big Bang model is now accepted by the vast majority of astronomers, additional evidence is continually being sought to support it. There are also some problems with the theory that need to be addressed and some aspects that have yet to be understood.

A general point in favor of the Big Bang model is that an important assumption on which it is based, the cosmological principle (see opposite page), has so far held true. The model also works within the framework of general relativity (see p.32), which is today considered a pillar of cosmology. However, these facts do not necessarily mean the Big Bang theory is correct. To be sure of its validity, specific positive evidence is needed—but there is no shortage of this.

SPECIFIC EVIDENCE

The most important positive evidence for the Big Bang is an extremely faint but uniform thermal radiation coming from the sky called the cosmic microwave background (CMB). Early supporters of the Big Bang theory predicted that this radiation should exist, and in 1964 it was detected by two American radio astronomers. The CMB arose soon after the Big Bang, when photons (small packets of radiant energy) were freed from interacting with matter and began to travel unhindered through space.

Further strong evidence comes from observations of deep space, looking back billions of years in time. Such observations have revealed objects called quasars (the highly energetic centers of galaxies) that no longer seem to exist today. Furthermore, the most distant galaxies—that is, galaxies as they existed 10–13 billion years ago—look different from closer, modern galaxies. These observations suggest the universe is of a finite age and has evolved over time rather than been static and unchanging.

One other important piece of evidence comes from the predominance and proportions of the chemical elements hydrogen and helium in the universe. The ratios of these two elements in their different forms (called isotopes) agree very closely with what is predicted by the Big Bang theory.

▼ Dark matter
In this image of a galaxy cluster over 7 billion light-years from Earth, called El Gordo ("The Fat One"), the blue haze indicates the distribution of dark matter—hard-to-detect matter that appears to bind galaxy clusters together gravitationally. The pink haze indicates X-ray emissions.

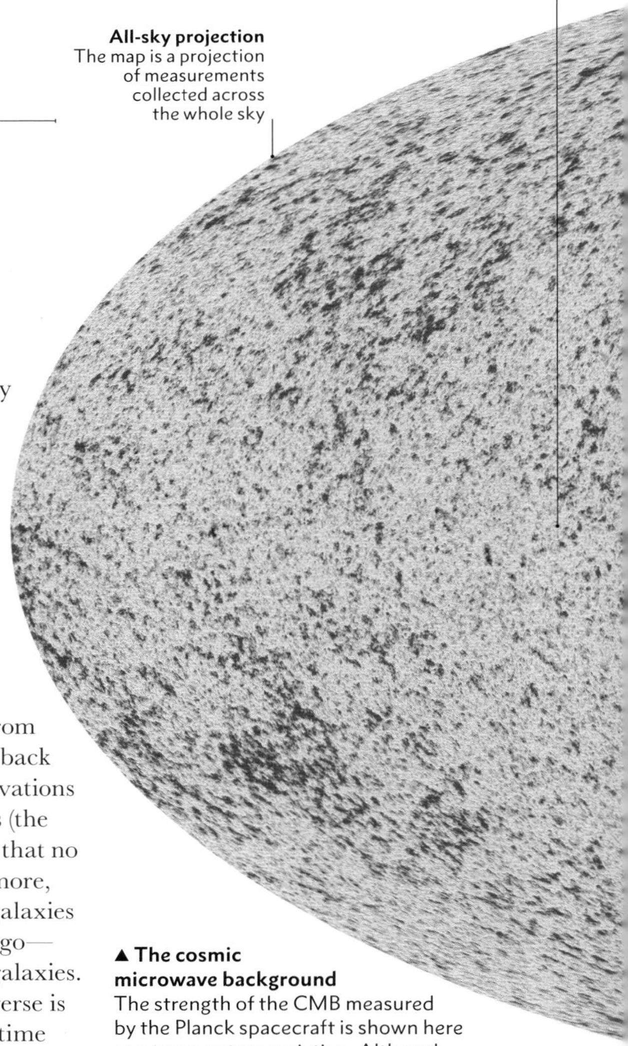
Red-orange spots
These have a temperature just 32°F (0.0002°C) higher than the average CMB temperature
All-sky projection
The map is a projection of measurements collected across the whole sky

▲ The cosmic microwave background
The strength of the CMB measured by the Planck spacecraft is shown here as a temperature variation. Although the CMB is uniform across the sky, a finely graded scale has been used to show tiny variations as colored spots.

> WE CAN TRACE THINGS BACK TO **THE EARLIER STAGES OF THE BIG BANG**, BUT WE STILL DON'T KNOW **WHAT BANGED** AND **WHY IT BANGED**. THAT'S A CHALLENGE FOR **21ST-CENTURY SCIENCE**.
>
> **Martin Rees**, British cosmologist, 1942–

UNANSWERED QUESTIONS

One major problem in cosmology in general is to shed light on the nature of "dark matter" and how it may have arisen in the Big Bang. Dark matter is an unknown substance that emits no light, heat, radio waves, nor any other kind of radiation—making it extremely hard to detect—but it does interact with other matter. Another challenge is to understand "dark energy." In 1998, it was discovered that the expansion of the universe has been accelerating over the past 6 billion years. The reason for the acceleration is not known, but the mysterious phenomenon of dark energy has been proposed as the cause. Very little is known about it at present, but if dark energy exists, it must permeate the whole universe.

10^{-6} SECONDS AFTER BIG BANG
THE FIRST PROTONS AND NEUTRONS FORM
3 MINUTES AFTER BIG BANG
THE FIRST ATOMIC NUCLEI FORM
380,000 YEARS AFTER BIG BANG
THE UNIVERSE BECOMES TRANSPARENT
13.6 BYA
THE FIRST STARS FORM

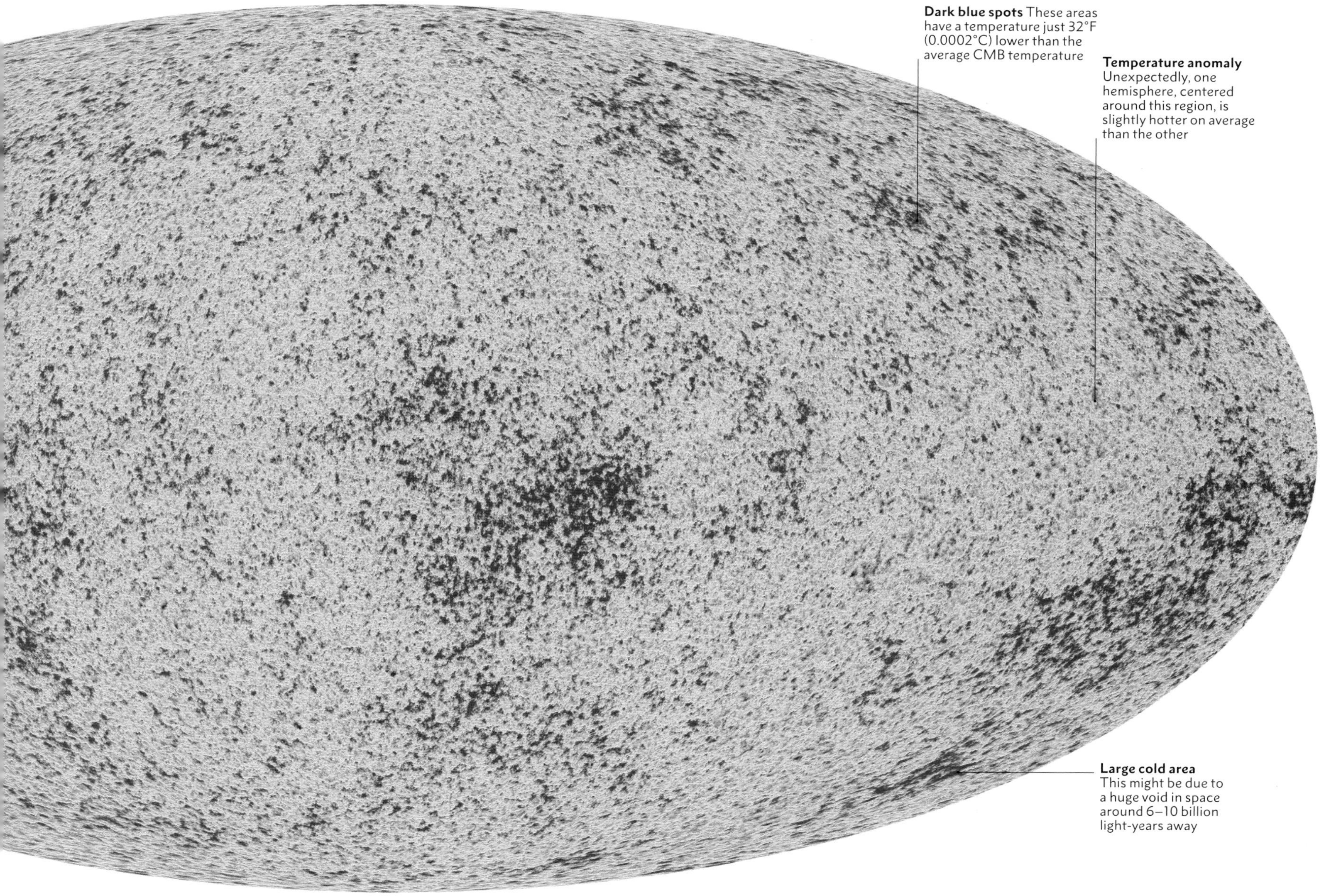

Dark blue spots These areas have a temperature just 32°F (0.0002°C) lower than the average CMB temperature
Temperature anomaly Unexpectedly, one hemisphere, centered around this region, is slightly hotter on average than the other
Large cold area This might be due to a huge void in space around 6–10 billion light-years away

Other unanswered questions include why an excess of matter over antimatter appeared during the universe's first few moments—without it, no atoms could ever have formed—and what caused the cosmic inflation that produced the smooth distribution of matter that we see in the universe today. The final question is "what triggered the Big Bang?" and this, of course, may never be answered.

▶ The cosmological principle
This principle states that when viewed on a sufficiently large scale, the universe is uniform, although on small scales there are clear variations in the distribution of objects such as galaxies. It follows from the cosmological principle that the universe has no center and no edges.

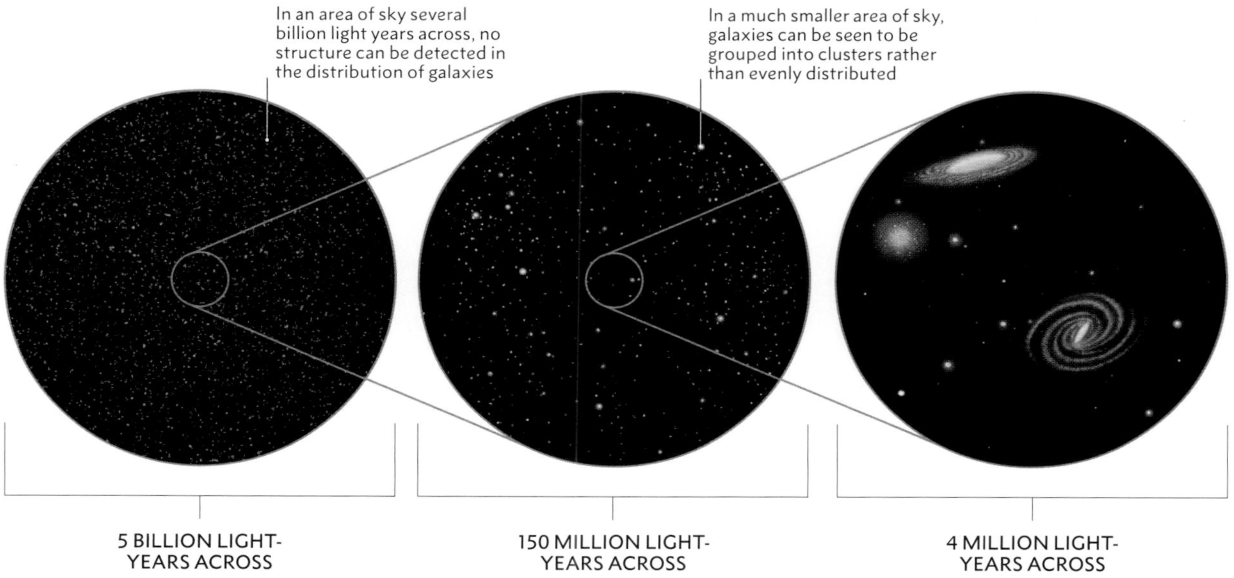

In an area of sky several billion light years across, no structure can be detected in the distribution of galaxies
In a much smaller area of sky, galaxies can be seen to be grouped into clusters rather than evenly distributed
5 BILLION LIGHT-YEARS ACROSS
150 MILLION LIGHT-YEARS ACROSS
4 MILLION LIGHT-YEARS ACROSS

THRESHOLD

2

STARS ARE **BORN**

With space, time, matter, and energy in place after the Big Bang, new powerhouses start to appear—stars. These form as matter is packed tighter and tighter together under the influence of gravity. The extremely high temperatures that result cause atoms to fuse together, releasing a huge amount of energy and opening the door to a new level of complexity in the universe.

GOLDILOCKS CONDITIONS

The early universe was shaped by two ingredients, both of which emerged while it was less than a second old. Gravity acted on tiny variations in the density of matter, setting into motion processes that led to the formation of the first stars and galaxies and, ultimately, a far more complex universe.

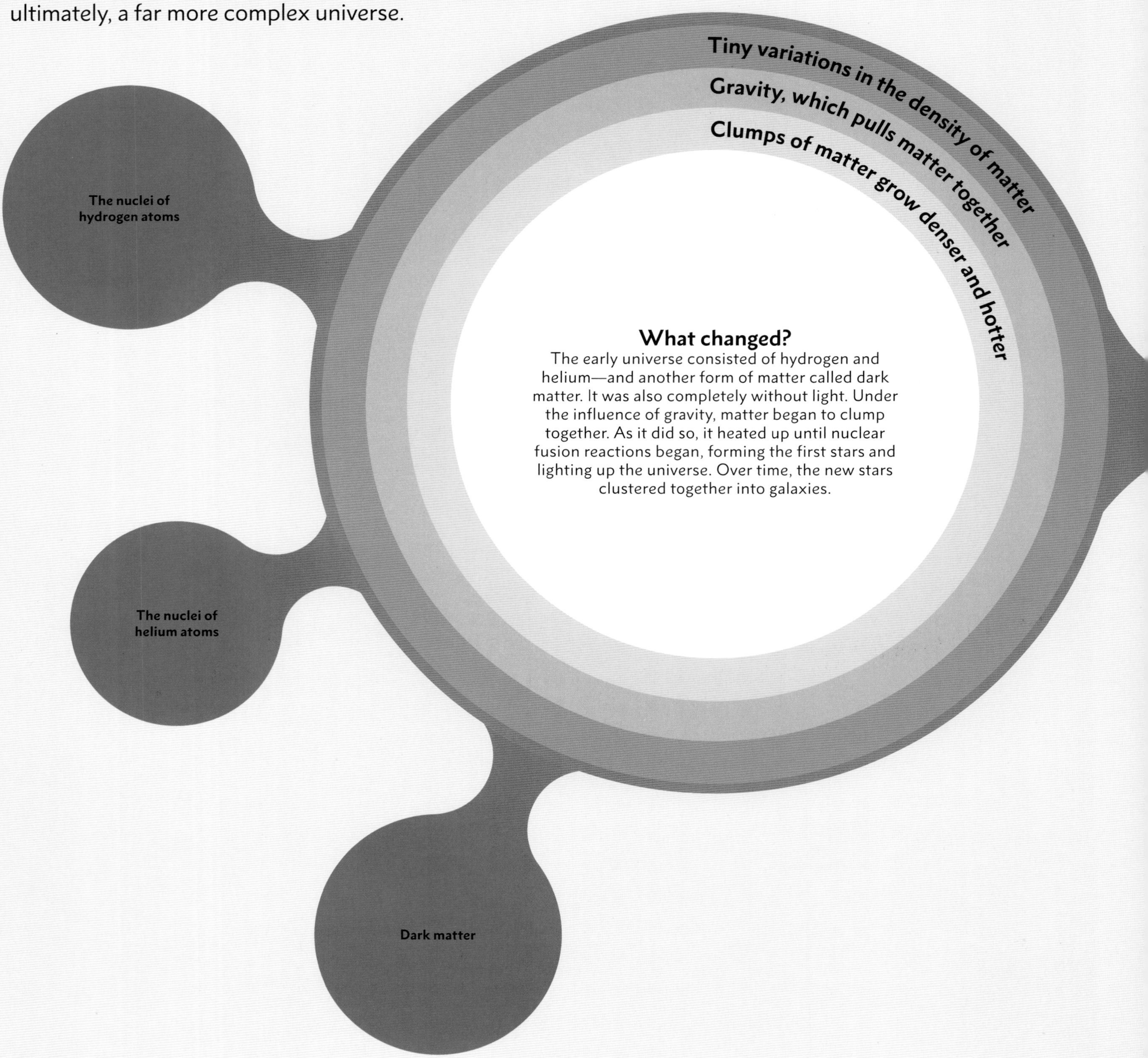

What changed?

The early universe consisted of hydrogen and helium—and another form of matter called dark matter. It was also completely without light. Under the influence of gravity, matter began to clump together. As it did so, it heated up until nuclear fusion reactions began, forming the first stars and lighting up the universe. Over time, the new stars clustered together into galaxies.

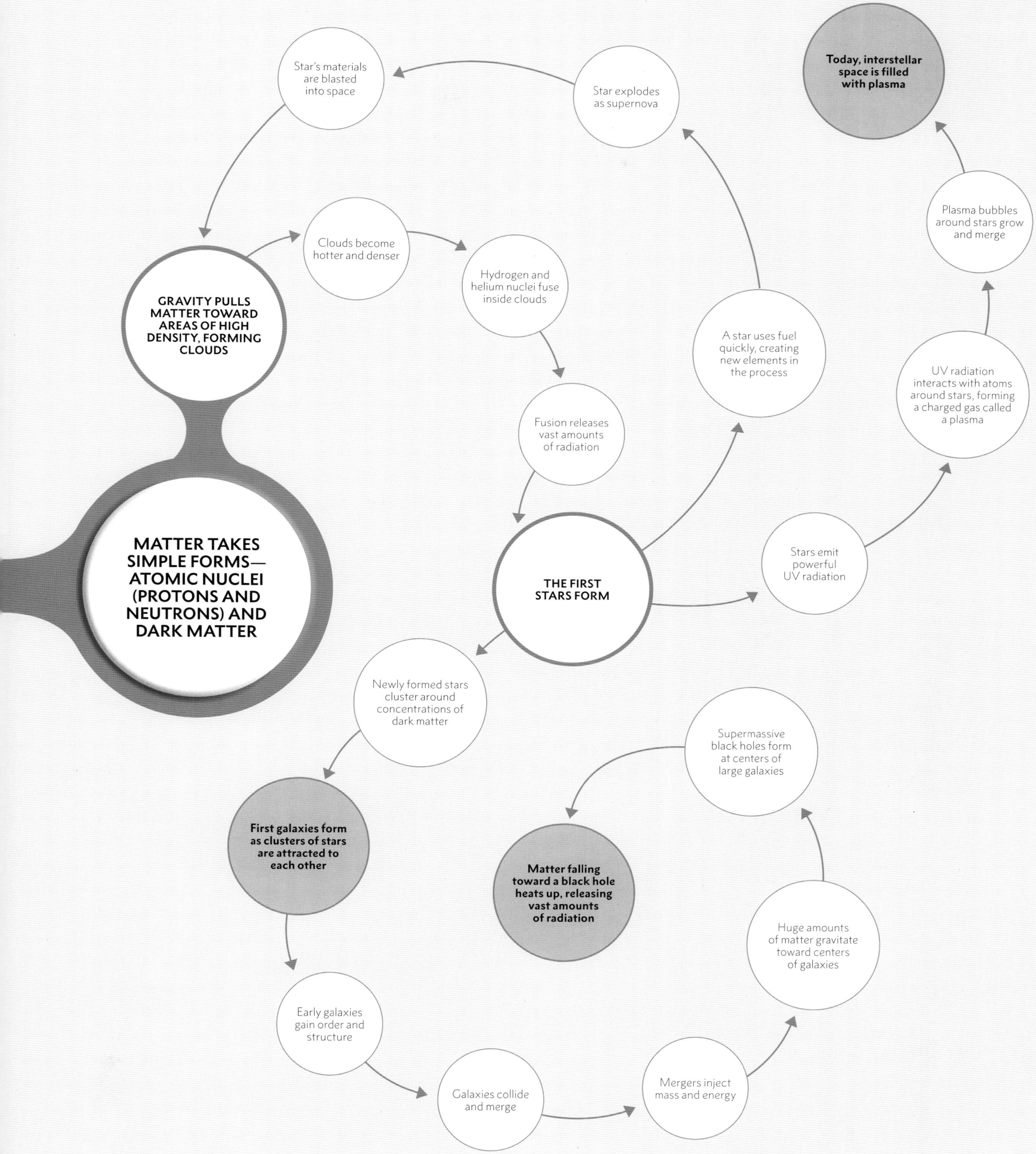

MATTER TAKES SIMPLE FORMS—ATOMIC NUCLEI (PROTONS AND NEUTRONS) AND DARK MATTER
GRAVITY PULLS MATTER TOWARD AREAS OF HIGH DENSITY, FORMING CLOUDS
Clouds become hotter and denser
Hydrogen and helium nuclei fuse inside clouds
Fusion releases vast amounts of radiation
THE FIRST STARS FORM
A star uses fuel quickly, creating new elements in the process
Star explodes as supernova
Star's materials are blasted into space
Stars emit powerful UV radiation
UV radiation interacts with atoms around stars, forming a charged gas called a plasma
Plasma bubbles around stars grow and merge
Today, interstellar space is filled with plasma
Newly formed stars cluster around concentrations of dark matter
First galaxies form as clusters of stars are attracted to each other
Early galaxies gain order and structure
Galaxies collide and merge
Mergers inject mass and energy
Huge amounts of matter gravitate toward centers of galaxies
Supermassive black holes form at centers of large galaxies
Matter falling toward a black hole heats up, releasing vast amounts of radiation

THE **FIRST STARS**

For its first 200 million years, the universe was a dark place. But things changed dramatically when clouds of gas collapsed to form the first stars. Inside, new chemical elements formed, and at the ends of their short lives the stars exploded, dispersing the elements into space.

During the Epoch of Recombination, 380,000 years after the Big Bang (see p.34), positively charged hydrogen and helium nuclei combined with negatively charged electrons to form neutral (uncharged) atoms. Until this point, collisions with free electrons had prevented photons of light from moving any distance in a straight line. Now the universe became transparent to light, although it was also dark, for there were no sources of light. It was a time cosmologists refer to as the Cosmic Dark Ages. Amid the dark soup of neutral gas was even darker stuff: dark matter. Scientists have little idea about the nature of dark matter, although they do know there is lots of it and that it is affected by gravity but doesn't interact with light or any other form of radiation.

HOW STARS FORM

Tiny variations in the density of the dark matter and the hydrogen and helium gases caused vast clouds of gas to collapse under the influence of gravity to form huge spherical clumps of matter. This would have happened without dark matter but much more slowly—so slowly that no stars would have formed to this day.

The enormous energy liberated in the collapse heated the balls of gas. At the increasing densities deep inside the balls of gas and as a result of the high temperatures

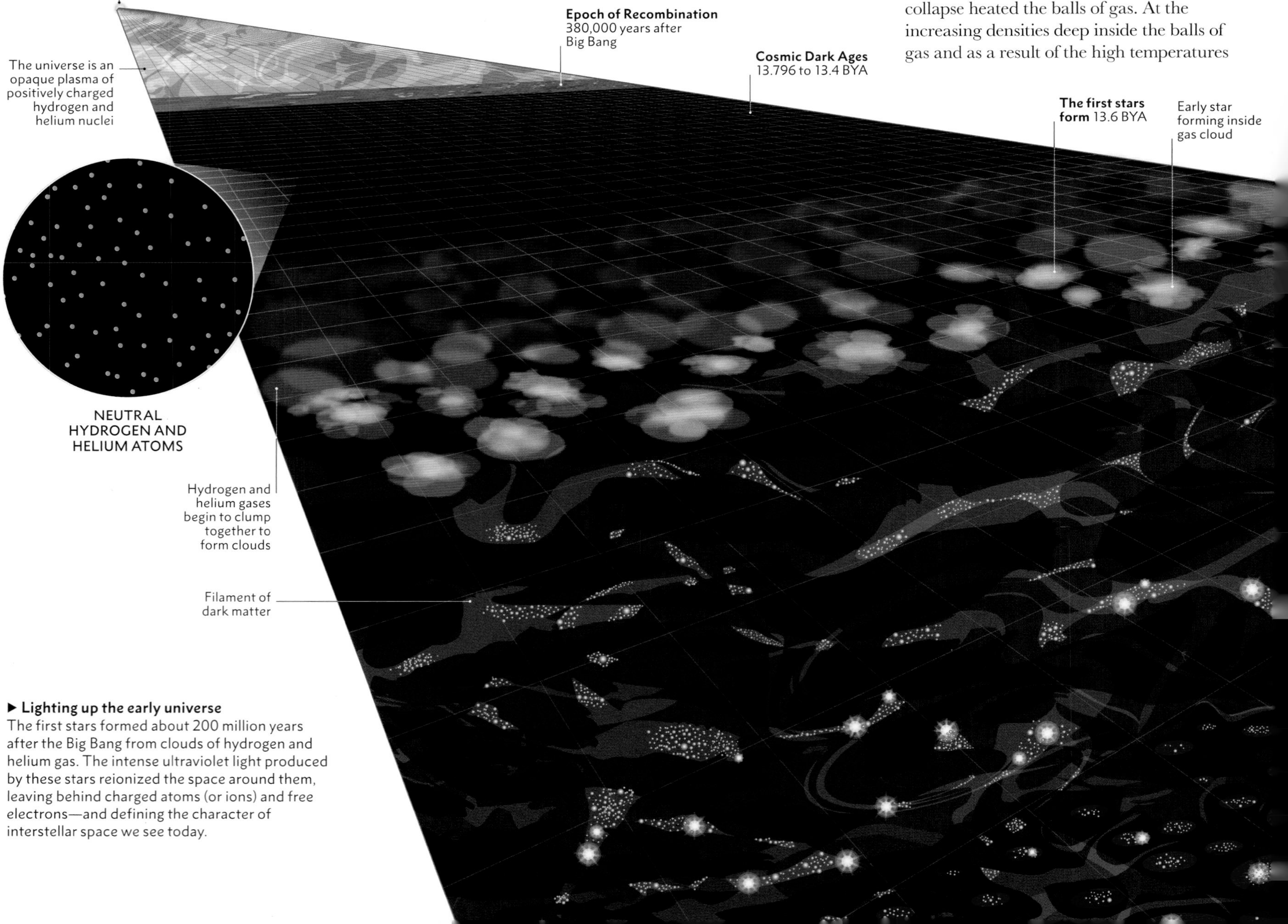

▶ **Lighting up the early universe**
The first stars formed about 200 million years after the Big Bang from clouds of hydrogen and helium gas. The intense ultraviolet light produced by these stars reionized the space around them, leaving behind charged atoms (or ions) and free electrons—and defining the character of interstellar space we see today.

▲ The size of early stars
According to astrophysicists' best models, most early stars were much larger than the sun and hundreds of times as massive.

at their cores, hydrogen and helium nuclei collided, and some of them joined together, or fused. This nuclear fusion resulted in the production of more helium nuclei from the hydrogen nuclei, and new, heavier elements—including boron, carbon, and oxygen—from the helium nuclei (see pp.58–59).

The nuclear fusion inside the collapsing balls of gas released a huge amount of energy, enough to heat the gas to incredibly high temperatures. That made the gas expand, buoying it up against further collapse. The high temperature also made the balls of gas glow brightly—and become the first stars.

The extremely hot first stars emitted large amounts of powerful ultraviolet radiation that had far-reaching effects. When the intense radiation hit neutral hydrogen and helium atoms still in space, its energy separated the electrons from their nuclei—just as they had been before the Epoch of Recombination. This "reionization" created a plasma bubble of hydrogen ions, helium ions, and free electrons in the space around each star. Interstellar space today is an extremely tenuous plasma that was created by this reionization, and nearly all radiation can pass through it.

SHORT LIVES

The first stars were large and massive: probably dozens of times the diameter of the sun and with hundreds of times as much mass. Such stars burn out quickly. The first generation of stars probably only lived for a few million years, compared to several billion years for an average star in later generations. As the hydrogen and helium "fuel" began to dwindle at the cores of the stars, they cooled, enabling the collapse to begin again, eventually causing the stars to explode as supernovas (see pp.60–61). The explosions threw a cocktail of new elements and the remaining unfused hydrogen and helium out into space. This cocktail formed the ingredients of a second generation of stars.

FIRST-GENERATION STARS LIVED ONLY **A FEW MILLION YEARS** BEFORE EXPLODING AS **VIOLENT SUPERNOVAS**

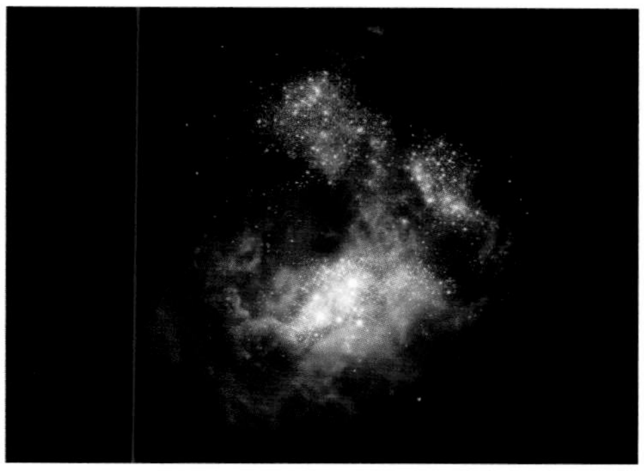

◄ Early light
This is an artist's impression of CR7, a small, bright galaxy. At 12.7 billion light years away, CR7 appears as it was about a billion years after the Big Bang. It represents the best evidence so far of first-generation stars.

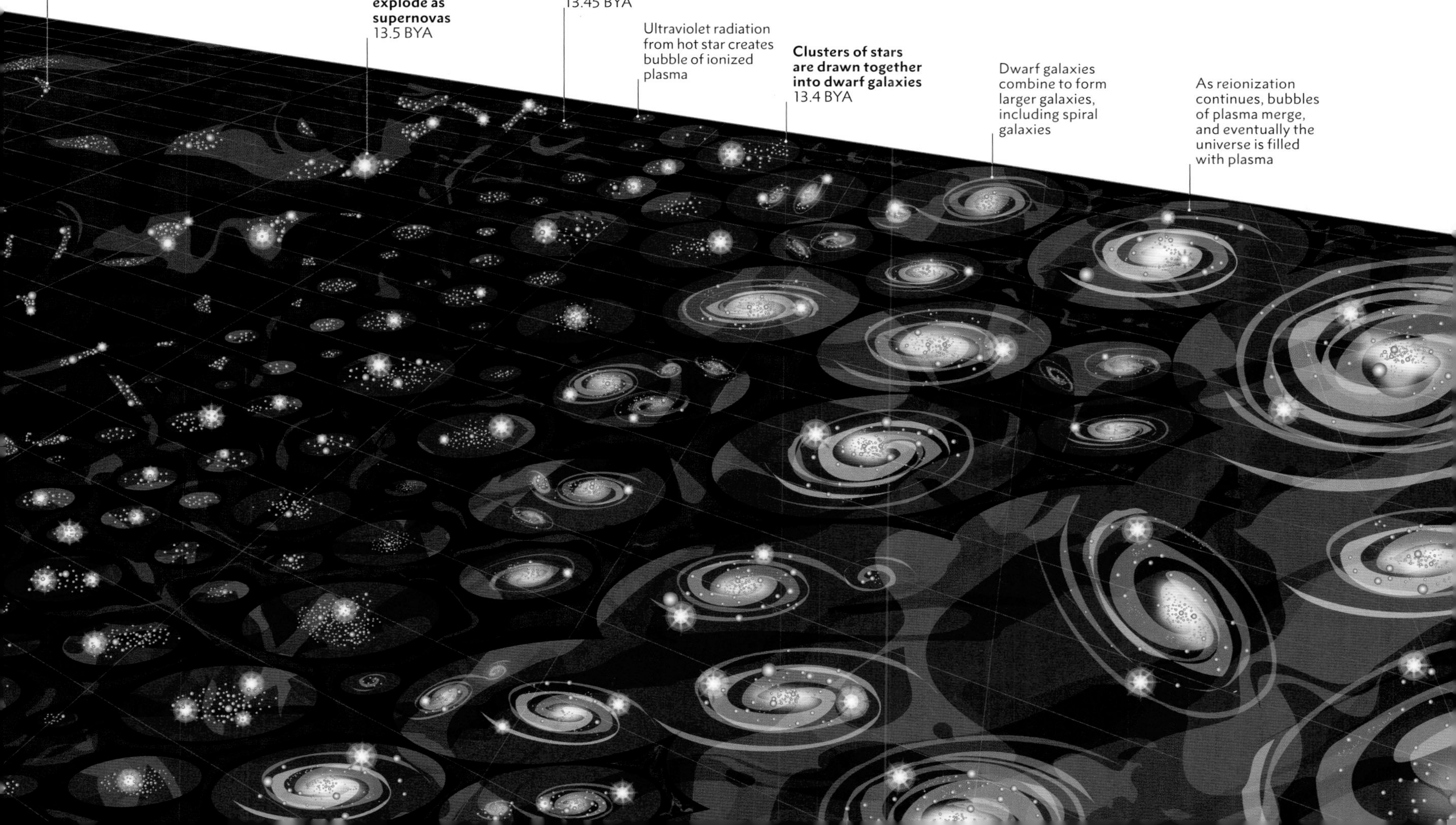
Stars form in clusters that coincide with concentrations of dark matter
The first stars explode as supernovas 13.5 BYA
Reionization starts 13.45 BYA
Ultraviolet radiation from hot star creates bubble of ionized plasma
Clusters of stars are drawn together into dwarf galaxies 13.4 BYA
Dwarf galaxies combine to form larger galaxies, including spiral galaxies
As reionization continues, bubbles of plasma merge, and eventually the universe is filled with plasma

THE **PUZZLE OF GRAVITY**

Gravity, or gravitation, plays a crucial role in the formation of stars and planets because it causes matter to clump together. The modern theory of gravity, Einstein's general theory of relativity, accurately explains its effects. Nevertheless, the true nature of gravity remains a mystery.

▼ Isaac Newton
In the late 1680s, Newton published both his Universal Law of Gravitation—the first scientific theory of gravity—and his three laws of motion.

The ancient Greek philosopher Aristotle supposed that Earth is at the center of the and that everything has a natural tendency to move toward it. According to Aristotle, heavier things have more of this tendency and so fall faster. Although Aristotle's simple notion was superficially supported by observations, experiments by Italian scientist Galileo Galilei in the 17th century showed that he was wrong. Galileo's experiments led him to predict, correctly, that in the absence of air resistance, all falling objects would accelerate downward at the same rate. English scientist Isaac Newton made sense of Galileo's prediction with his Universal Law of Gravitation.

NEWTON'S GRAVITY

Newton realized that what makes things fall to the ground here on Earth also keeps the moon in orbit. He proposed that gravity is a force and derived an equation that could predict the strength of the force between any two objects. According to Newton's law, the force depends on the masses of the objects and the distance between their centers.

By combining his law of gravitation with his laws of motion, Newton was able to account for the motions of any object under the influence of gravity—from projectiles on Earth to planets in space. His theory was accepted for over 200 years—and scientists still use his equation in most situations where they need to calculate the effects of gravity. However, in the 19th century,

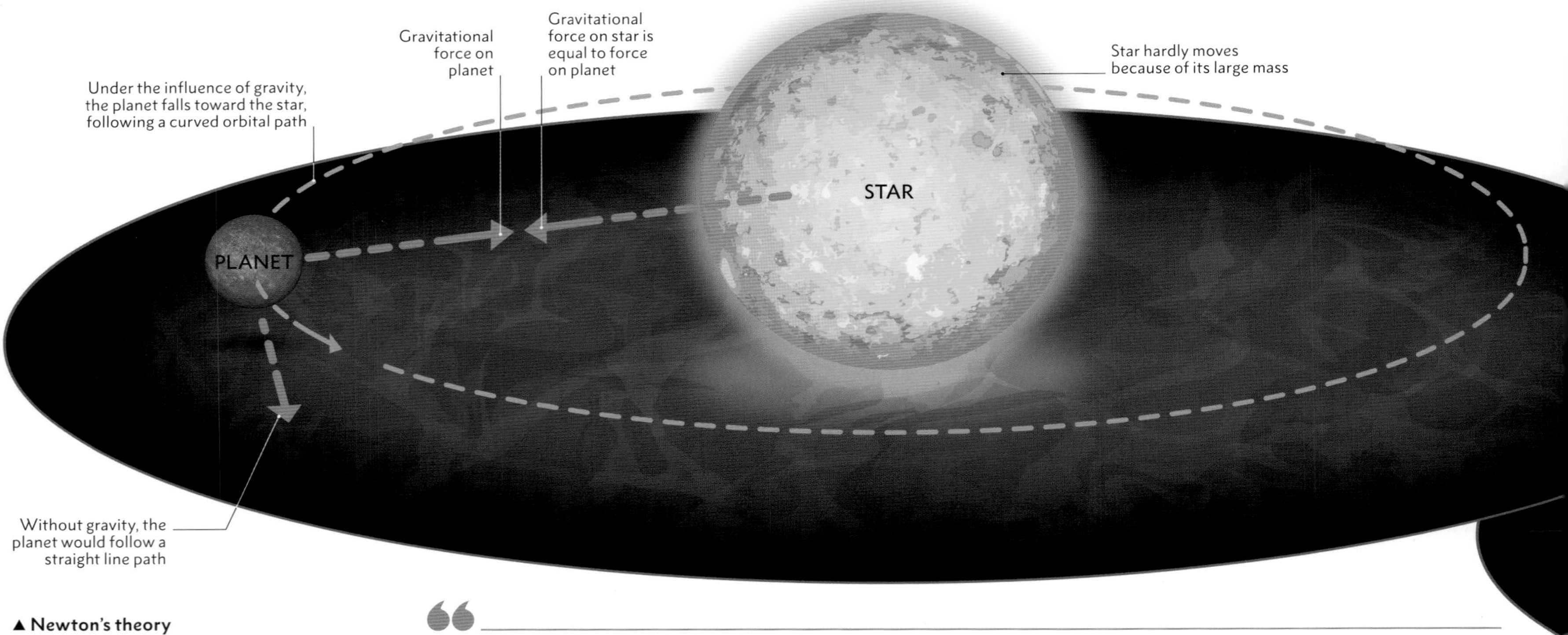
Gravitational force on planet
Gravitational force on star is equal to force on planet
Star hardly moves because of its large mass
Under the influence of gravity, the planet falls toward the star, following a curved orbital path
STAR
PLANET
Without gravity, the planet would follow a straight line path

▲ Newton's theory
In Newton's theory, a star and planet exert an attractive force on each other. Both are subject to an equal force, but the effect on the planet is more obvious because it has a lower mass.

> **NEWTON** HIMSELF WAS BETTER AWARE OF THE **WEAKNESSES IN HIS INTELLECTUAL EDIFICE** THAN THE GENERATIONS OF LEARNED SCIENTISTS WHICH FOLLOWED HIM.
>
> **Albert Einstein**, German physicist, 1879–1955

calculations of the orbit of planet Mercury, at odds with observations, showed Newton's theory to be flawed. In 1915, German physicist Albert Einstein proposed a radical new theory of gravitation—the general theory of relativity—that could accurately predict the orbit of Mercury. And according to Einstein's theory, gravity is not a force at all.

EINSTEIN'S GRAVITY

General relativity is an extension of special relativity, a theory Einstein published in 1905. Special relativity was an attempt to reconcile Newton's laws of motion with the theory of electromagnetism, developed in the 1860s. To do that, Einstein had to abandon the idea that space and time are absolute: people in motion relative to each other measure distances and intervals of time differently—the differences only become significant at extremely high relative speeds. One of the direct consequences of special relativity was the realization that time is a dimension, just like the three dimensions of space, and that all four exist in a four-dimensional grid called spacetime; objects therefore move through spacetime, not space.

In order to generalize special relativity to include gravity, Einstein realized that objects with mass distort spacetime. The more massive an object, the greater the distortion. Objects traveling freely through distorted spacetime follow curved paths. So projectiles and planets are simply following the equivalent of straight line paths, but in distorted spacetime. A force is needed to change an object's path. For example, the ground pushes upward on a person's feet, which stops the person from following a path that would take him or her "freefalling" toward the center of Earth. For a star, the expansion of the hot gas of which it is made provides the force necessary to keep it from collapsing—expansion that lasts as long as the star produces heat (see pp.56–57).

EINSTEIN'S PREDICTIONS

The general theory of relativity has been tested many times, to extremely high precision. It has also made several important predictions, such as the idea that light must also follow the curved paths of distorted spacetime. The result is a phenomenon called gravitational lensing, which is evident in the distorted views of distant galaxies whose light has been bent as it passed close to nearby galaxies. Another key prediction is the existence of gravitational waves: ripples in spacetime emanating at the speed of light from any very energetic event. In 2015, scientists detected the first hard evidence of the existence of gravitational waves, produced by the merging of two black holes.

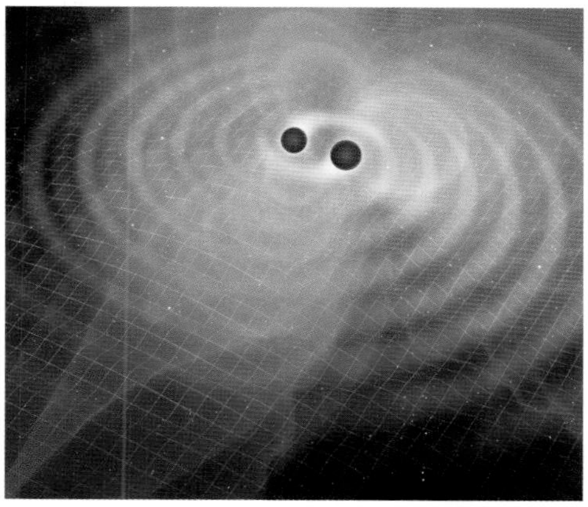

◀ **Gravitational waves**
The first gravitational waves ever detected resulted from the merger of two black holes. Here, the waves are represented as ripples in a two-dimensional sheet of spacetime. These ripples were detected by sensitive equipment on Earth.

Despite the success of general relativity, the theory is at odds with quantum mechanics, an equally well-tested cornerstone of modern science. Quantum mechanics accurately describes the behavior of matter at the atomic and subatomic scales, while gravity accurately describes the behavior of matter at much larger scales—but the two theories are incompatible. The search for a quantum theory of gravity is a major concern of modern physics, and it is likely that Einstein's theory of gravity will be reinterpreted or superseded as part of a grand theory that can describe the behavior of matter at all scales. One thing is certain: the puzzle of gravity is not yet solved.

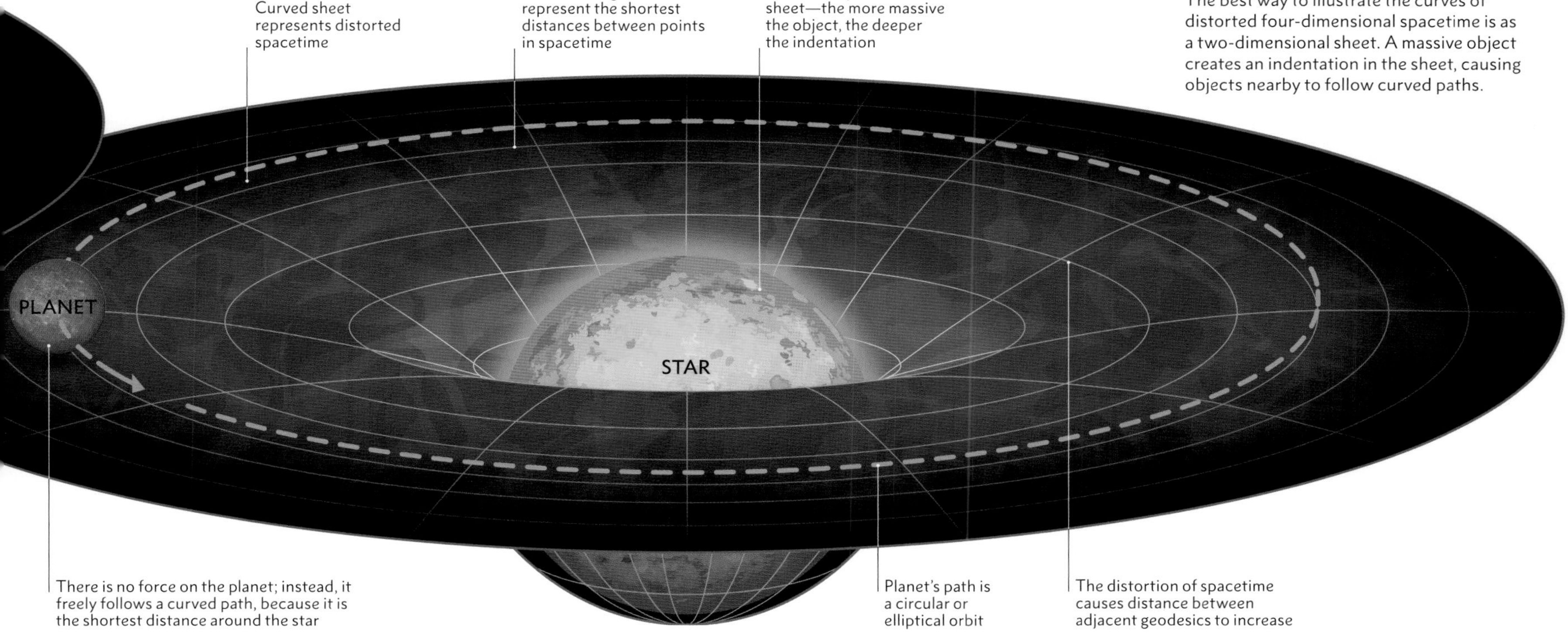
Curved sheet represents distorted spacetime
Lines called geodesics represent the shortest distances between points in spacetime
Massive objects indent the sheet—the more massive the object, the deeper the indentation
PLANET
STAR
There is no force on the planet; instead, it freely follows a curved path, because it is the shortest distance around the star
Planet's path is a circular or elliptical orbit
The distortion of spacetime causes distance between adjacent geodesics to increase

▼ **Einstein's theory**
The best way to illustrate the curves of distorted four-dimensional spacetime is as a two-dimensional sheet. A massive object creates an indentation in the sheet, causing objects nearby to follow curved paths.

THE **FIRST GALAXIES**

A galaxy is a vast congregation of stars orbiting a common center. The first galaxies began to form soon after the first stars, around clumps of dark matter. Mutual gravitational attraction caused these small galaxies to merge, each merger sparking new flurries of star birth.

▼ Galaxy evolution In the absence of direct observations, astrophysicists construct simulations to test their theories of how the first galaxies formed. The images below are snapshots from one of those simulations.

Dark matter was crucial in the creation of the first galaxies, just as it was for the formation of the first stars (see pp.44–45). Slight variations in the density of dark matter in the early universe caused the dark matter and ordinary matter—in the form of hydrogen and helium gas—to clump together. The dark matter formed a network of sinuous filaments and nodes, or haloes, at various scales. The clumping process drove the formation of individual stars as the concentrations of matter began to rotate and heat up, eventually resulting in nuclear fusion (see pp.56–57). At a larger scale, the same process also produced clusters of stars. Each star cluster, plus its surrounding gas, was attracted to neighboring clusters, and the universe's first galaxies were born.

GROWING GALAXIES

As matter fell toward matter, the dark matter haloes grew in size, and so did the galaxies. Like water draining down a plug hole, much of the matter began to spin as it fell, so that it went into orbit around the most dense, central part of the halo. As a result, galaxies that began as irregularly shaped masses began to gain order and

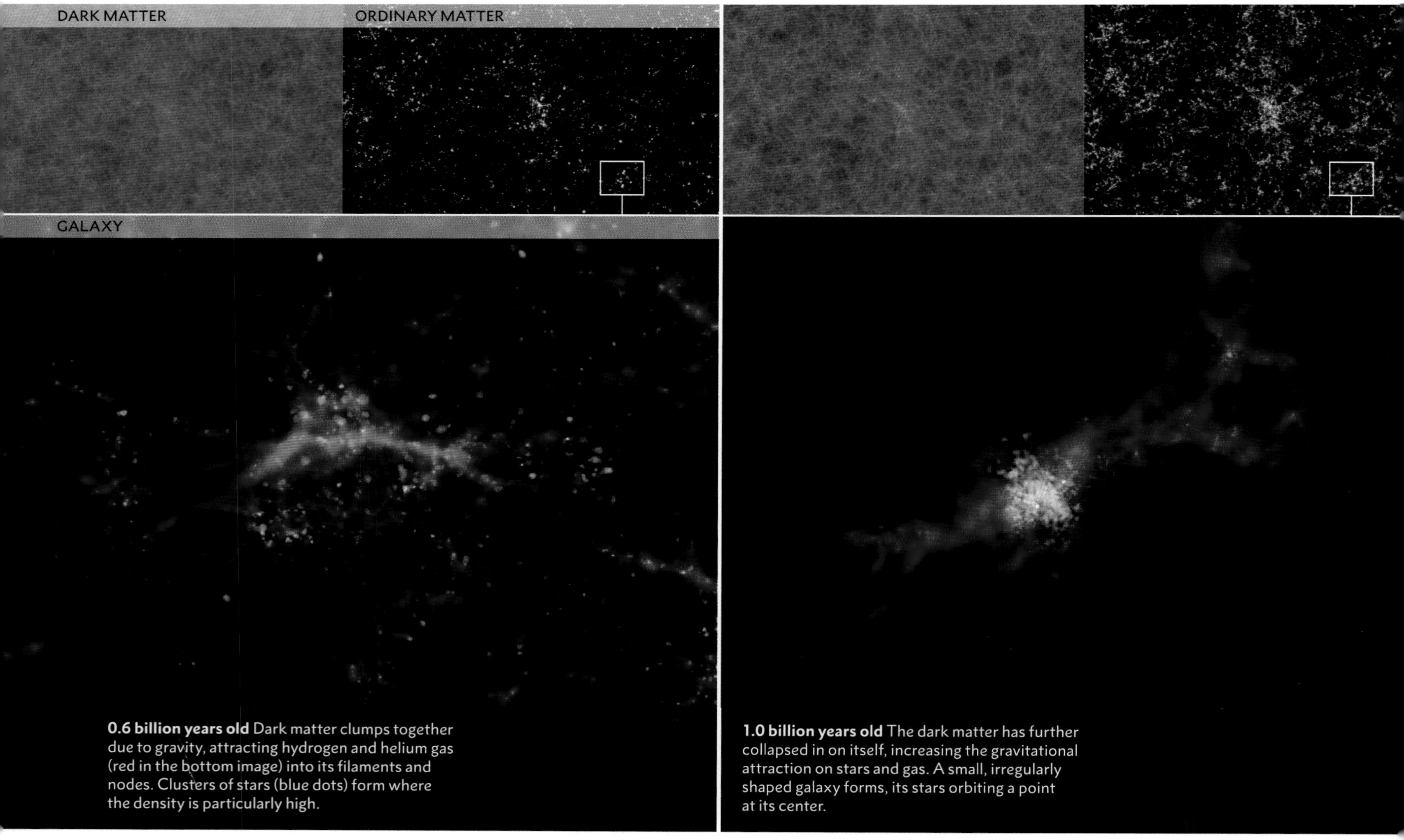

0.6 billion years old Dark matter clumps together due to gravity, attracting hydrogen and helium gas (red in the bottom image) into its filaments and nodes. Clusters of stars (blue dots) form where the density is particularly high.

1.0 billion years old The dark matter has further collapsed in on itself, increasing the gravitational attraction on stars and gas. A small, irregularly shaped galaxy forms, its stars orbiting a point at its center.

structure. Many formed spinning disks, with spiral arms; others were egg-shaped elliptical galaxies. But with each merger, the structure was disrupted, only to be regained or developed millions or billions of years later. The mergers injected energy and mass, too, and the rate of star formation and star death increased. Each star inside a young galaxy inevitably ended its life in a powerful supernova explosion that filled the galaxy with the elements that would seed the next generation of stars and even planets.

SUPERMASSIVE BLACK HOLES

Although much of the gas and many of the stars stayed in orbit around the center of each galaxy, huge amounts of the matter fell toward the center. In large galaxies, the density at the center increased so much that a supermassive black hole (see p.47) formed there. As matter jostled its way in toward the growing black hole, friction heated it to extremely high temperatures, releasing vast amounts of energy as high-energy (short wavelength) X-rays, ultraviolet radiation, and bright visible light. Astronomers first detected these energetic galaxies in the 1950s; they made the discoveries with early radio telescopes, since the short-wavelength radiation has been stretched to such an extent by the expansion of space that it arrives as long-wavelength infrared and radio waves. Most large galaxies in the universe today, including our own, still have supermassive black holes at their centers.

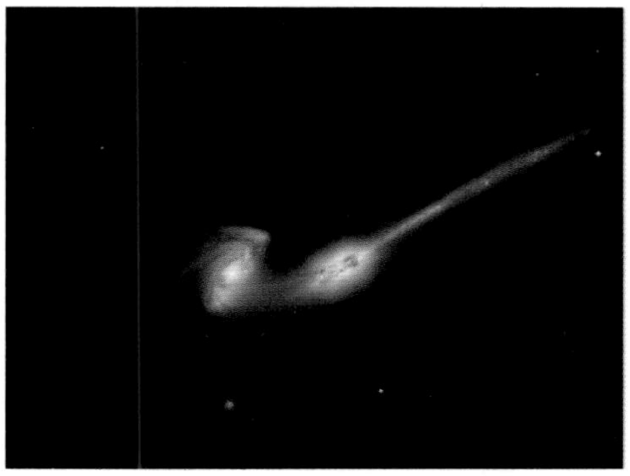

◀ Merging galaxies Astronomers observe many merging galaxies. Shown here is NGC 4676—also known as the Mice Galaxies—a pair of colliding galaxies around 290 million light-years away.

> [IN SIMULATIONS] YOU CAN MAKE **STARS AND GALAXIES** THAT LOOK LIKE THE REAL THING. BUT IT IS THE **DARK MATTER** THAT IS **CALLING THE SHOTS**.
>
> **Professor Carlos Frenk**, cosmologist, 1951–

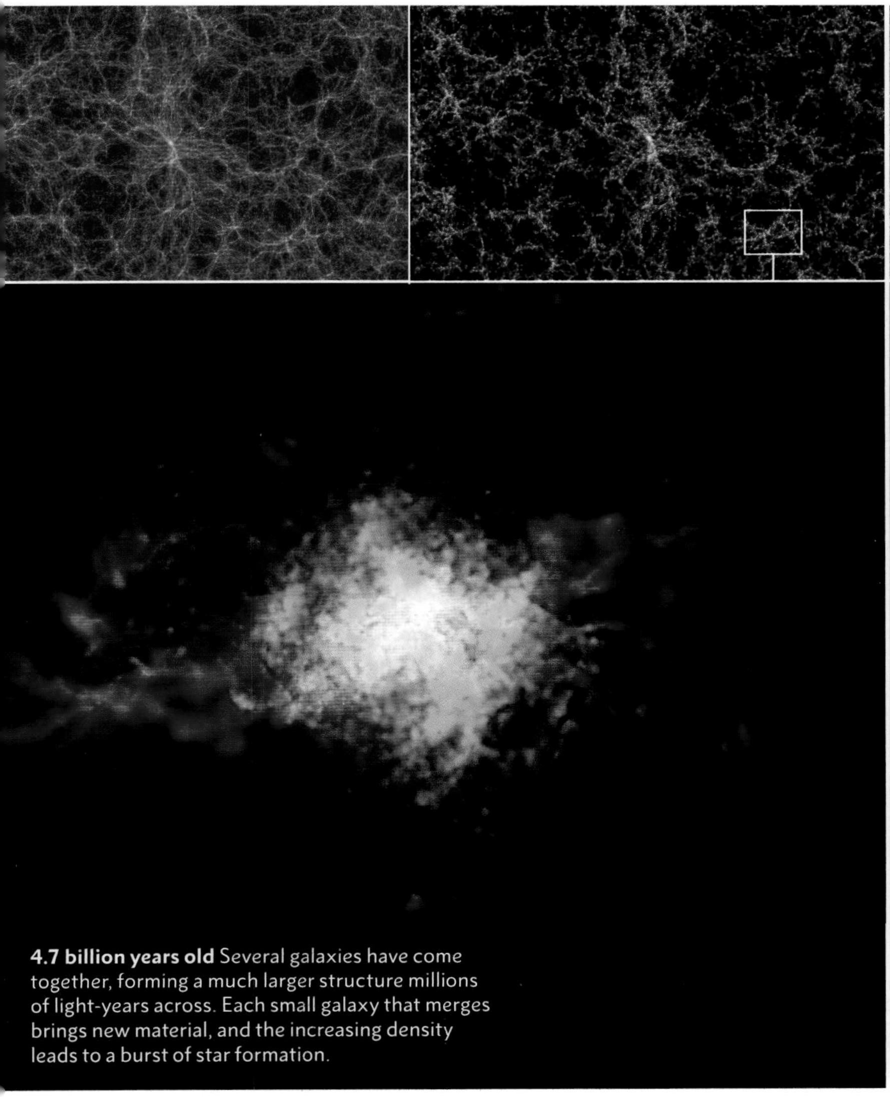

4.7 billion years old Several galaxies have come together, forming a much larger structure millions of light-years across. Each small galaxy that merges brings new material, and the increasing density leads to a burst of star formation.

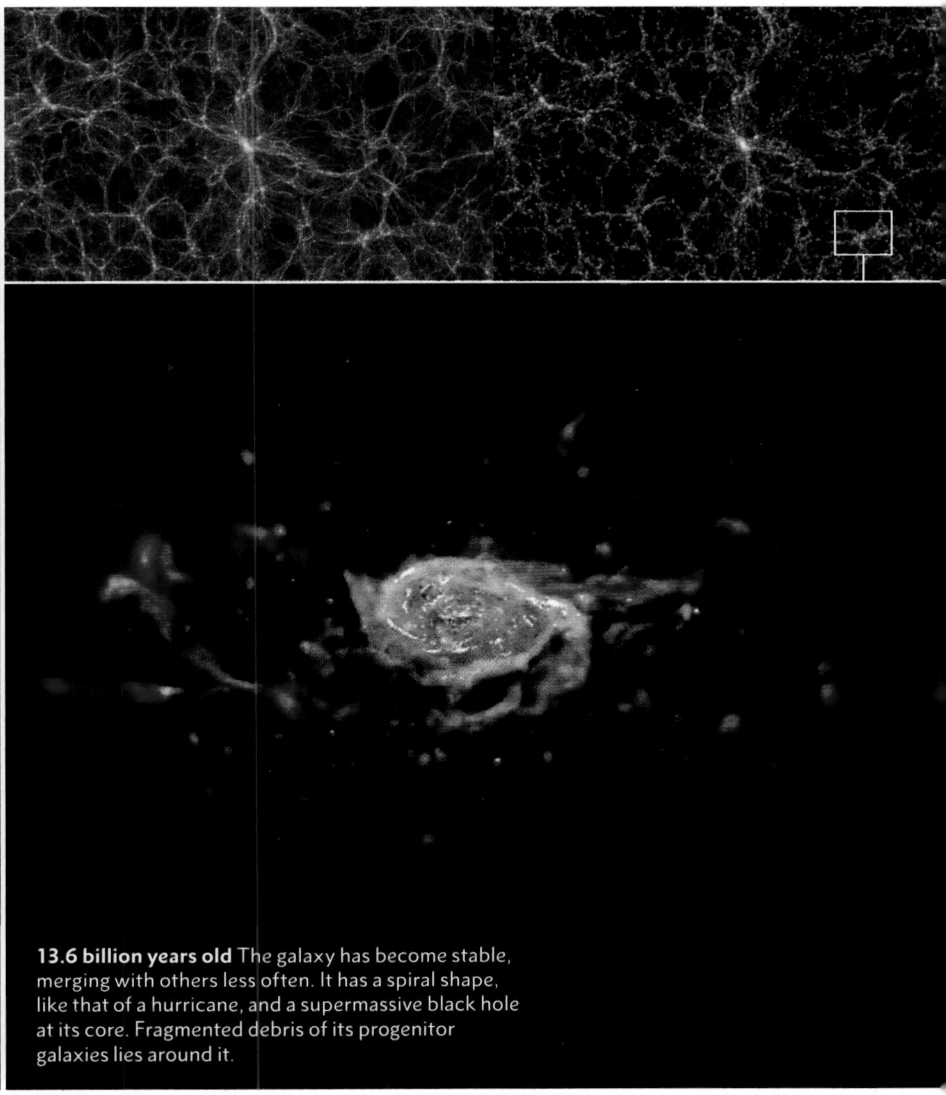

13.6 billion years old The galaxy has become stable, merging with others less often. It has a spiral shape, like that of a hurricane, and a supermassive black hole at its core. Fragmented debris of its progenitor galaxies lies around it.

HUBBLE EXTREME DEEP FIELD

Taken by the Hubble Space Telescope, the eXtreme Deep Field records faint light from thousands of galaxies in a small area of sky. The deepest view of space ever captured, it provides the best evidence we have about the early universe's stars and galaxies.

When we look out into space, we are looking back in time, because the light from distant objects left a long time ago. Light that left a galaxy 5 billion years ago will appear extremely faint, however bright the galaxy was at the time. Imaging such a dim object requires a long exposure time—not a fraction of a second, like a typical photograph, but millions of seconds.

In 1995, astronomers pointed NASA's Hubble Space Telescope at a tiny patch of sky for over 140 hours and combined a total of 342 images into a single, remarkable image called the Hubble Deep Field. In 2004, NASA scientists produced the even more remarkable Hubble Ultra Deep Field—an image with an even longer exposure, on a different patch of sky. Observations on that area continued over the next eight years, and the addition of an infrared camera to the telescope in 2009 meant that objects whose light has been redshifted (see p.29) beyond the visible spectrum and into the infrared could also be seen. The new observations were combined with the Ultra Deep Field, and the result was published in 2012 as the Hubble eXtreme Deep Field (XDF). Light from the most distant galaxies in the XDF took more than 13 billion years to reach us, and they appear one ten-billionth as bright as the dimmest thing visible to the naked eye.

Containing evidence of galaxy mergers (see p.49), extreme redshifting, and gravitational lensing (see p.47), the Hubble XDF is a significant piece of evidence in support of the most convincing theories we have about the evolution of the universe.

Relatively nearby galaxy looks red as its stars are running low on hydrogen fuel
This foreground star is in our own galaxy
Light from this very faint galaxy, called UDFj-39546284, took 13.4 billion light-years to reach Earth
This relatively nearby object is a spiral galaxy, like the Milky Way, seen front-on

► **Looking back** The largest, brightest objects in the XDF include mature galaxies that appear as they were about 5–9 billion years ago—when they had grown by merging and were populated by second- or third-generation stars. Galaxies in the background are smaller: young, irregular galaxies seen as they were over 9 billion years ago. The foreground is relatively empty because the XDF team chose an area almost devoid of nearby galaxies and stars in our own galaxy.

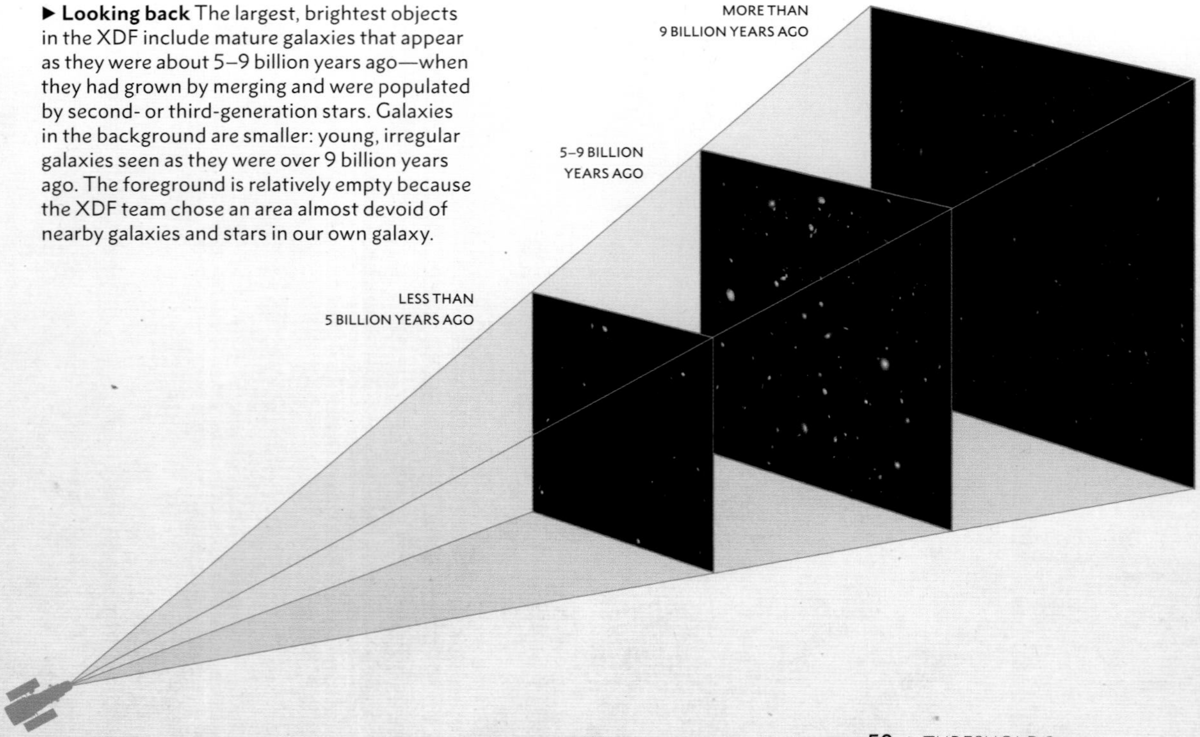
MORE THAN 9 BILLION YEARS AGO
5–9 BILLION YEARS AGO
LESS THAN 5 BILLION YEARS AGO

More recent galaxies are the result of mergers of smaller, older galaxies

Distant galaxy appears red due to redshifting of its light

Field of view

Next to the full moon, the Hubble eXtreme Deep Field covers a tiny area: less than one twenty-millionth of the area of the whole sky. To see the image at its true size, you would need to hold this page about 1,000ft (300m) away. It is remarkable that more than 7,000 galaxies can be seen in such a small field of view—and to think that each tiny dot in the image is a collection of millions or billions of stars frozen in time.

XDF's field of view, with the moon for comparison

Early galaxies

The XDF gives astronomers a unique view of galaxies as they were during the universe's first few hundred million years, when they were relatively small, irregularly shaped groups of stars. As they collided and merged, most became spiral shaped because the collisions resulted in rotation. The universe was smaller when the light captured in the XDF left the young galaxies. As space has expanded, the light has been "stretched," shifting its frequencies toward or even beyond the red end of the spectrum, which is why so many of the XDF galaxies appear reddish.

Close-up of heavily redshifted galaxy merger

THRESHOLD

ELEMENTS ARE **FORGED**

We all come from dying stars. All the elements that make up our world originated there. Stars are hungry, and as they use up their fuel, age, and finally die, some of them collapse and go out with a tremendous explosion of energy. But from star death comes new building blocks—the elements—pushed out into the universe to start something new.

GOLDILOCKS CONDITIONS

The formation of the first stars had profound consequences. As well as lighting up the universe, stars act as chemical factories, producing new chemical elements that provide the raw materials for everything else in the universe, including living things.

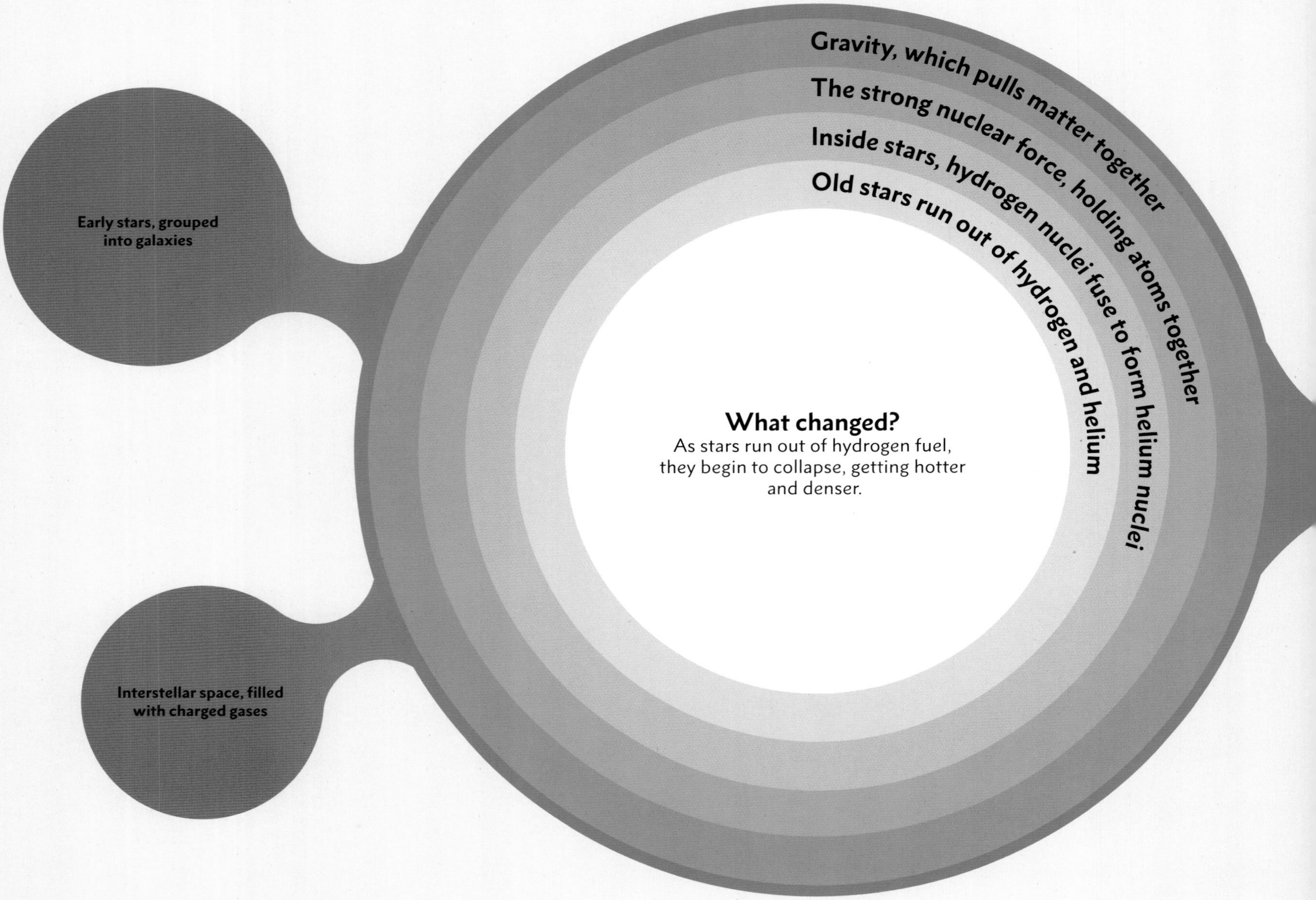
Gravity, which pulls matter together
The strong nuclear force, holding atoms together
Inside stars, hydrogen nuclei fuse to form helium nuclei
Old stars run out of hydrogen and helium
Early stars, grouped into galaxies
Interstellar space, filled with charged gases
What changed?
As stars run out of hydrogen fuel, they begin to collapse, getting hotter and denser.

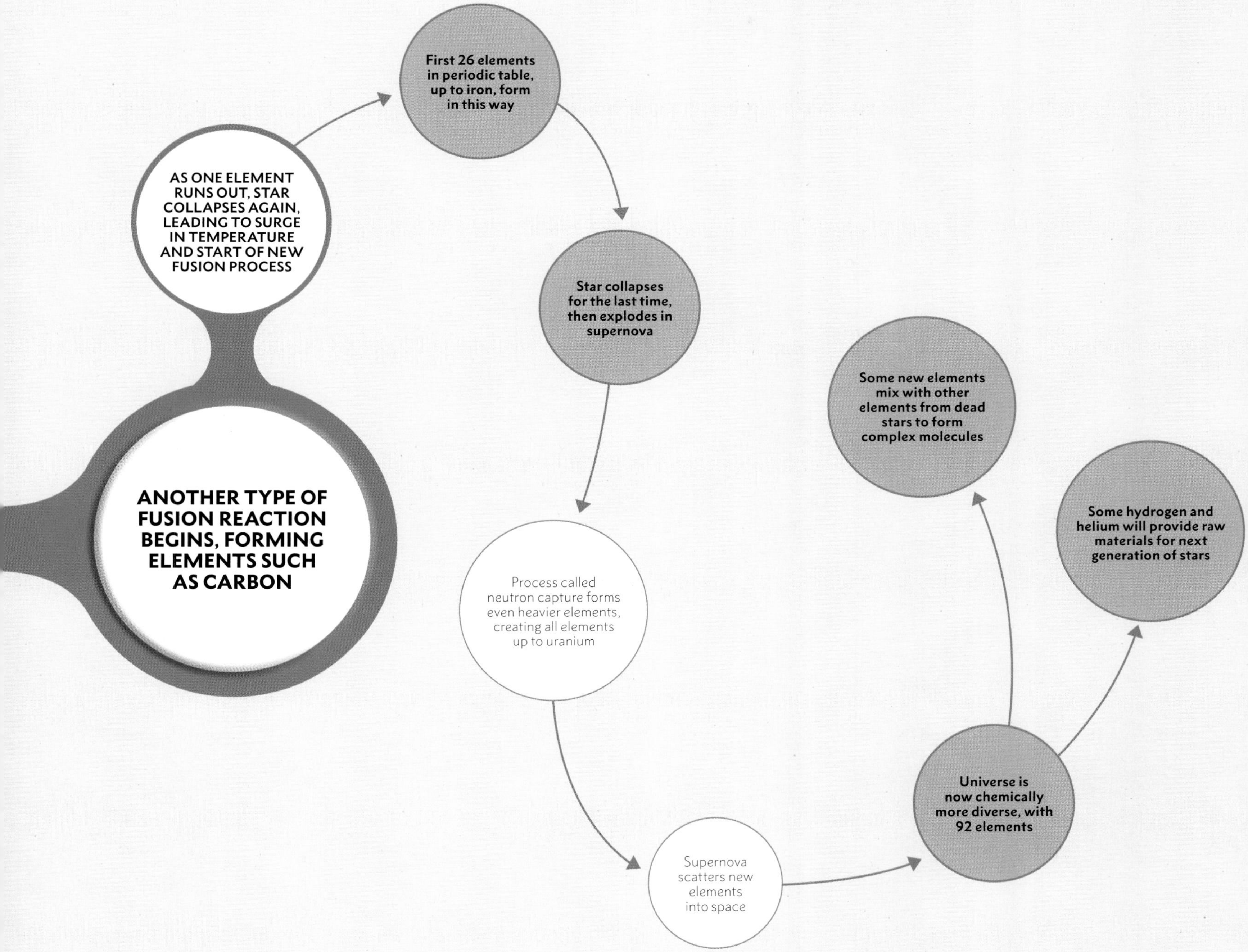

AS ONE ELEMENT RUNS OUT, STAR COLLAPSES AGAIN, LEADING TO SURGE IN TEMPERATURE AND START OF NEW FUSION PROCESS
ANOTHER TYPE OF FUSION REACTION BEGINS, FORMING ELEMENTS SUCH AS CARBON
First 26 elements in periodic table, up to iron, form in this way
Star collapses for the last time, then explodes in supernova
Process called neutron capture forms even heavier elements, creating all elements up to uranium
Supernova scatters new elements into space
Universe is now chemically more diverse, with 92 elements
Some new elements mix with other elements from dead stars to form complex molecules
Some hydrogen and helium will provide raw materials for next generation of stars

THE **LIFE CYCLE** OF A **STAR**

Just like humans, stars are born, grow old, and die. The way a star ends its days depends on its mass, with the largest stars exploding as supernovas. These detonations furnished, and continue to furnish, the universe with heavier elements, recycling material ready for it to be turned into new stars.

Consequently, the life cycle of stars also played a crucial role in the emergence of life on Earth. Essential ingredients—including the calcium in your bones and the iron in your blood—were forged inside stars, only for supernovas to spread them far and wide.

Stars come in a vast array of sizes. Astronomers classify them into seven main groups from largest to smallest denoted by the letters O, B, A, F, G, K, and M. Our sun is a G star, meaning there are bigger and smaller stars out there than our own. The smallest stars, known as dwarfs, are the most common. M stars, for example, make up more than 75 percent of all stars. By contrast, O stars account for just 0.00003 percent.

The size of a star also governs how long it will live. The larger the star, the quicker it will consume its nuclear material. O stars live fast and die young, often dying out within just a few million years, whereas the smallest stars can eke out their existence for trillions of years.

A **SUPERGIANT STAR** CAN HAVE A VOLUME **8 BILLION** TIMES THAT OF **THE SUN**

LIFE STAGES

Stars begin their lives as protostars, formed from clouds of interstellar dust (see pp.44–45). Nuclear processes in a star's core then shore it up against gravitational collapse. For most of a star's life, this balance is maintained, but things change when fusion eventually stops. Astronomers refer to a star still fusing hydrogen into helium as a main-sequence star. Once this fusion ceases, the star evolves off the main sequence.

For all but the smallest stars, the core contracts and the temperature rises to around 180 million°F (100 million°C). This is hot enough for helium to fuse into carbon, which creates enough energy to upset the balance the other way and the star bloats outward. Then, depending on size, it will either turn into a planetary nebula with a white dwarf at the center, or detonate as a supernova, leaving behind a neutron star or black hole.

▼ Sunlike star
Stars like the sun typically live for around 10 billion years. After entering a red giant phase, they form a planetary nebula—and usually do not explode as supernovas.

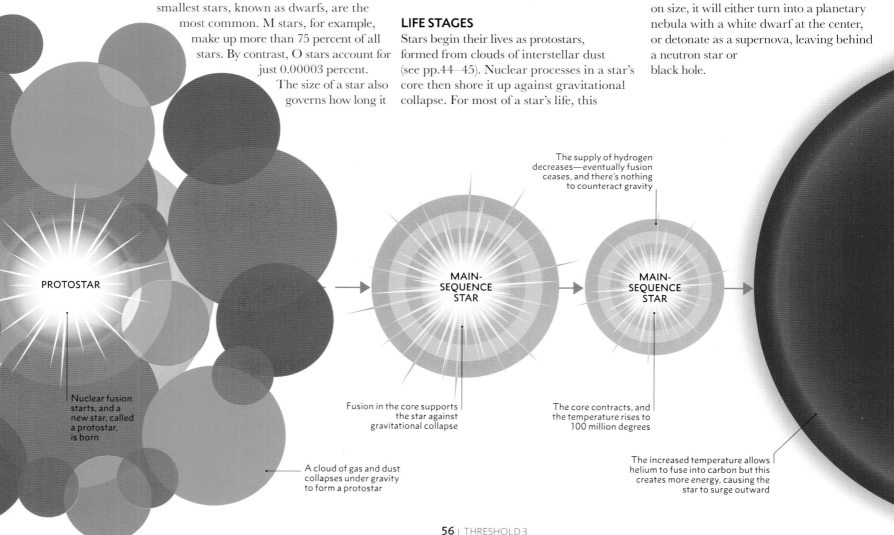

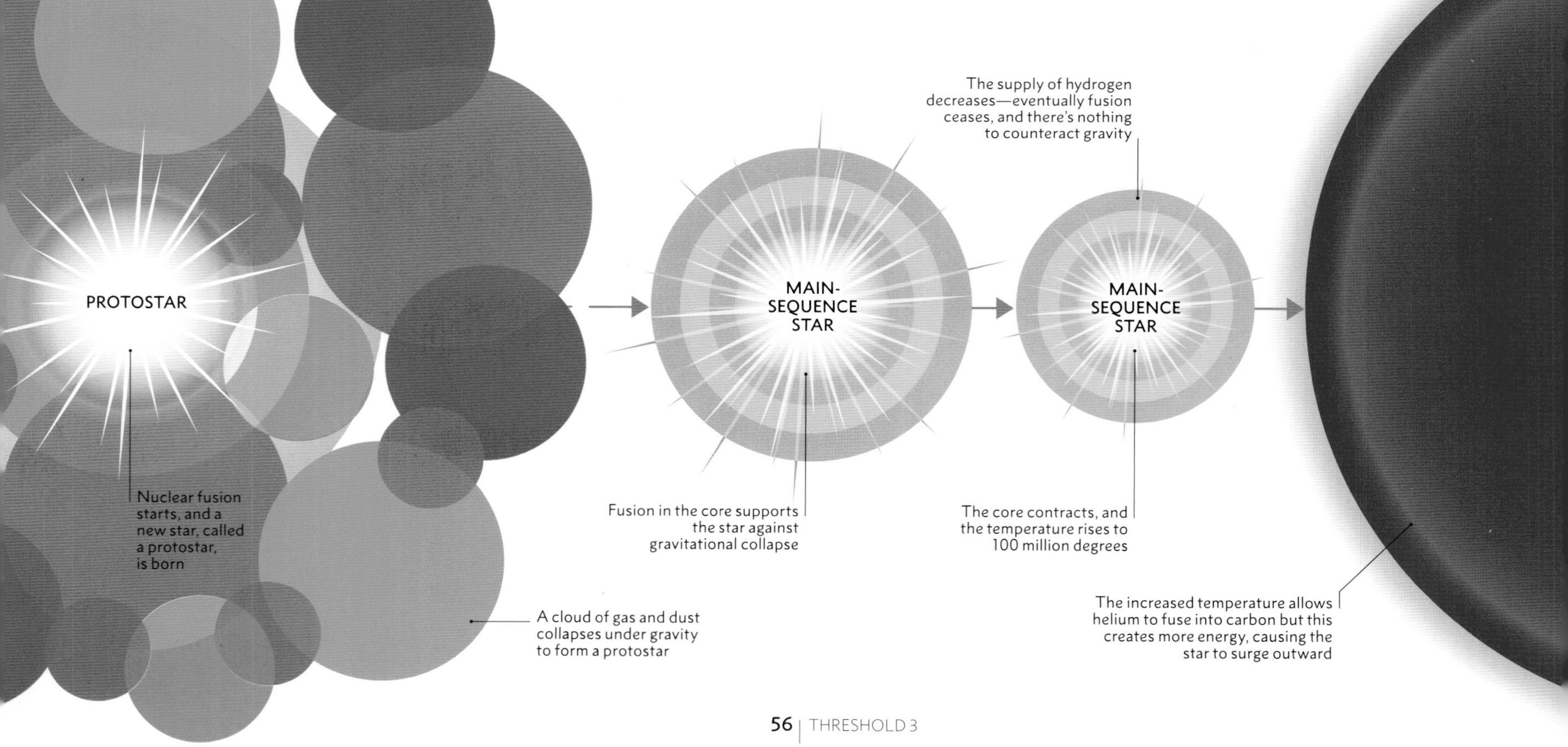
PROTOSTAR
Nuclear fusion starts, and a new star, called a protostar, is born
A cloud of gas and dust collapses under gravity to form a protostar
MAIN-SEQUENCE STAR
Fusion in the core supports the star against gravitational collapse
The supply of hydrogen decreases—eventually fusion ceases, and there's nothing to counteract gravity
MAIN-SEQUENCE STAR
The core contracts, and the temperature rises to 100 million degrees
The increased temperature allows helium to fuse into carbon but this creates more energy, causing the star to surge outward

► **Low-mass star**
These smaller stars are able to mix their interiors, meaning that the core's supply of hydrogen gets replenished by the outer layers falling toward the center—so the core doesn't contract to start helium fusion.

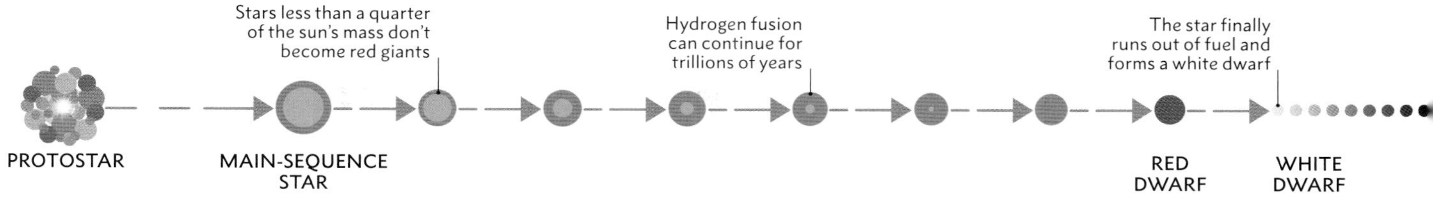
Stars less than a quarter of the sun's mass don't become red giants
Hydrogen fusion can continue for trillions of years
The star finally runs out of fuel and forms a white dwarf
PROTOSTAR
MAIN-SEQUENCE STAR
RED DWARF
WHITE DWARF

► **High-mass star**
The evolution of more massive stars is initially similar to that of sunlike stars. But they form red supergiants, instead of red giants, and eventually supernovas. The star's ultimate fate depends on its mass.

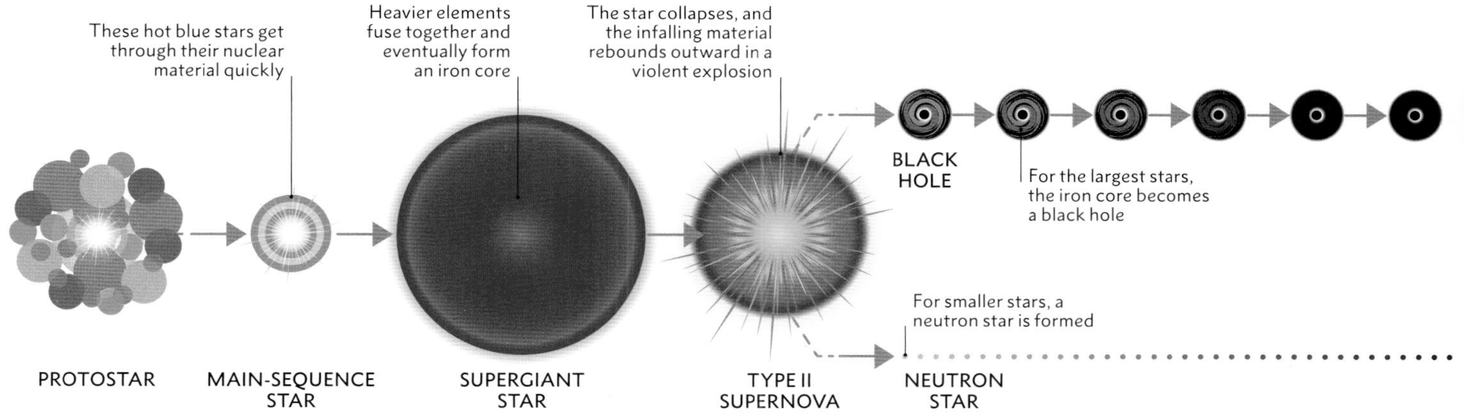
These hot blue stars get through their nuclear material quickly
Heavier elements fuse together and eventually form an iron core
The star collapses, and the infalling material rebounds outward in a violent explosion
BLACK HOLE
For the largest stars, the iron core becomes a black hole
For smaller stars, a neutron star is formed
PROTOSTAR
MAIN-SEQUENCE STAR
SUPERGIANT STAR
TYPE II SUPERNOVA
NEUTRON STAR

> STARS ARE **BORN, LIVE**—OFTEN FOR BILLIONS OF YEARS—AND **DIE**... SOMETIMES IN A **SPECTACULAR MANNER**.
>
> **Carl Sagan**, American astronomer, 1934–1996

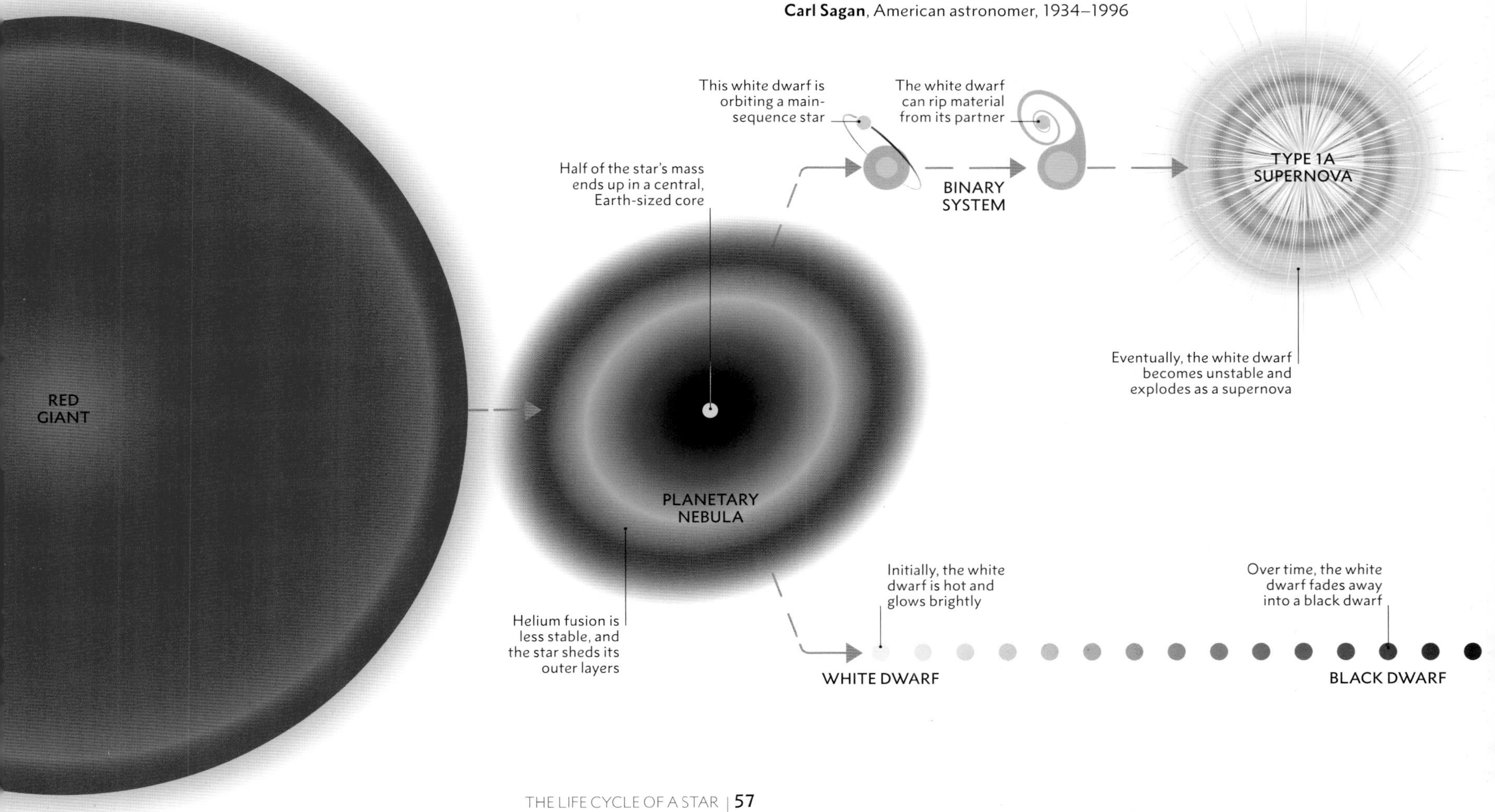
This white dwarf is orbiting a main-sequence star
The white dwarf can rip material from its partner
Half of the star's mass ends up in a central, Earth-sized core
BINARY SYSTEM
TYPE 1A SUPERNOVA
Eventually, the white dwarf becomes unstable and explodes as a supernova
RED GIANT
PLANETARY NEBULA
Initially, the white dwarf is hot and glows brightly
Over time, the white dwarf fades away into a black dwarf
Helium fusion is less stable, and the star sheds its outer layers
WHITE DWARF
BLACK DWARF

HOW **NEW ELEMENTS** FORM INSIDE **STARS**

Before the first stars shone, the universe was just a sea of hydrogen, helium, and residual energy from the Big Bang. The chemical diversity in the universe today is due to stars—effectively, vast atom factories—churning primitive materials into more complex elements and then flinging them outward when they die.

Inside stars, the temperature is high enough to rip electrons away from the nuclei of atoms. In the case of hydrogen, this leaves solitary protons (and electrons) wandering around the star's interior. Matter in this state is known as plasma. Due to their like electric charges, protons repel each other, rather like similar poles of a magnet.

NEW ELEMENTS IN STARS

However, deep in the core of the star, the temperature and pressure are high enough to squash protons together. Known as nuclear fusion, this process releases energy and is the star's power source. It also exerts an outward pressure that counters the inward pull of gravity.

The simplest fusion mechanism is called the proton-proton (or pp) chain. In the first step, one of the fused protons turns into a neutron, creating a new proton-neutron pair called a deuteron. This is bombarded by another proton to create the nucleus of a helium-3 atom. When two of these helium-3 atoms collide, they create a helium-4 nucleus, along with two protons, which can start the whole process again. The German–American physicist Hans Bethe was a key player in uncovering this process and was awarded the 1967 Nobel Prize in Physics for his work. Crucially, the total mass of the products of the pp-chain is less than the mass of the ingredients entering into it. In the sun, for example, 620 million tons of hydrogen (protons) is turned into 616 million tons of helium every second. The missing four million tons of mass is converted into energy according to Einstein's famous equation $E = mc^2$.

Eventually, the hydrogen in the star's core runs out and gravity contracts the core. The resulting temperature surge allows a new fusion mechanism to take over—the triple alpha process—one which uses helium-4 nuclei (alpha particles) as its main ingredient. This enables two helium nuclei to fuse into beryllium and then, with the addition of a third helium nucleus, into carbon. In smaller stars, such as the sun, the atom construction process ends here.

However, larger stars can go on increasing the diversity of chemical elements; once one fusion path runs out, the core contracts and the temperature spikes to kick-start another. Next, carbon fuses with helium to form oxygen, which is bombarded by another helium nucleus to forge neon, which itself is fashioned into magnesium by a similar process. The sheer range of possible reactions is vast. Eventually, carbon and oxygen fuse together to form silicon. At this point, the temperature in the core has soared to 5.4 billion°F (3 billion°C), which is enough to force two silicon nuclei together to form iron. In this way, a wealth of elements builds up in shells within the star, resembling the layers of an onion, with

> FINALLY, I GOT TO **CARBON**, AND AS YOU ALL KNOW, IN THE CASE OF CARBON THE REACTION **WORKS OUT BEAUTIFULLY**.
>
> **Hans Bethe**, German–American physicist, 1906–2005

▼ The triple alpha process
In this process, two helium-4 nuclei fuse into beryllium-8, which becomes carbon-12 when struck by a third helium-4 nucleus. Helium-4 nuclei are also called alpha particles, and so this mechanism is known as the triple alpha process.

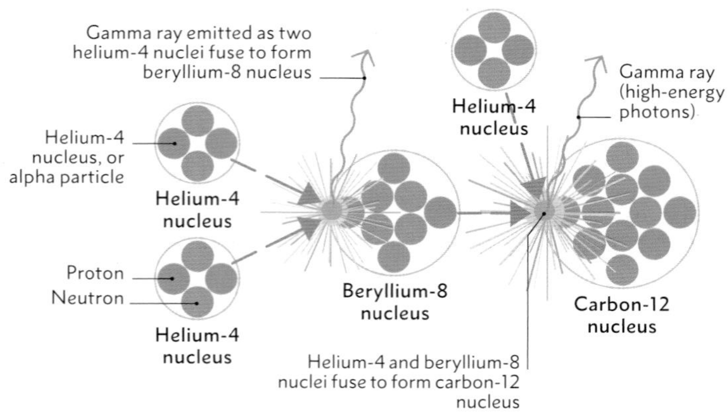

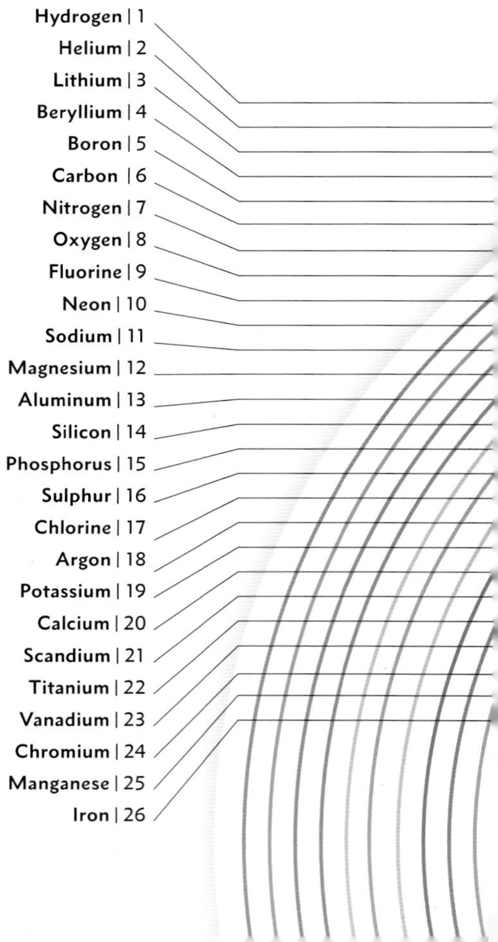

iron at its heart. However, because iron is the most stable of all the elements, it cannot be fused into anything else and fusion ceases. As heavier elements form, the process gathers pace—it can take millions of years for a star to exhaust its hydrogen, but the fusion of silicon nuclei to form iron takes just a single day.

NEW ELEMENTS IN SUPERNOVAS

Elements heavier than iron can only be created when a massive star explodes in a supernova. The next heaviest elements are formed by the s-neutron-capture process—"s" stands for slow, as it typically takes hundreds of years. This process actually begins inside stars, but in stars the interactions are extremely slow—they only speed up once a supernova gets going. The earlier transformation of carbon into oxygen, and neon into magnesium, created a wealth of additional neutrons. The gradual combination of these excess particles with existing nuclei allows elements as heavy as bismuth to form. However, this process cannot produce any elements heavier than bismuth, because bismuth decays away into polonium before it can combine with a neutron. A much faster neutron capture mechanism is required—the r-process ("r" stands for rapid). The r-process can only happen in the extreme conditions of a supernova. The density of neutrons increases greatly during the explosion, and new elements can be formed in a fraction of a second. Some of these r-process nuclei later decay away, creating new elements not fashioned directly by either neutron capture process.

COMPLEX CHEMISTRY

This profusion of material is dispersed into the wider universe by the force of the supernova. It then mixes with interstellar material and debris from other dead stars to form giant molecular clouds that will eventually collapse to form new stars. Individual atoms can combine with others in the clouds to form complex molecules, some of which are crucial for life. Astronomers and astrochemists have already found evidence of these molecules. The simplest amino acid—glycine—has been detected in a cloud of gas toward the center of our Milky Way galaxy, as well as in the nearby Orion Nebula. Amino acids are regarded as life's building blocks, so it is possible that the basic ingredients for life were fashioned long before the sun lit up.

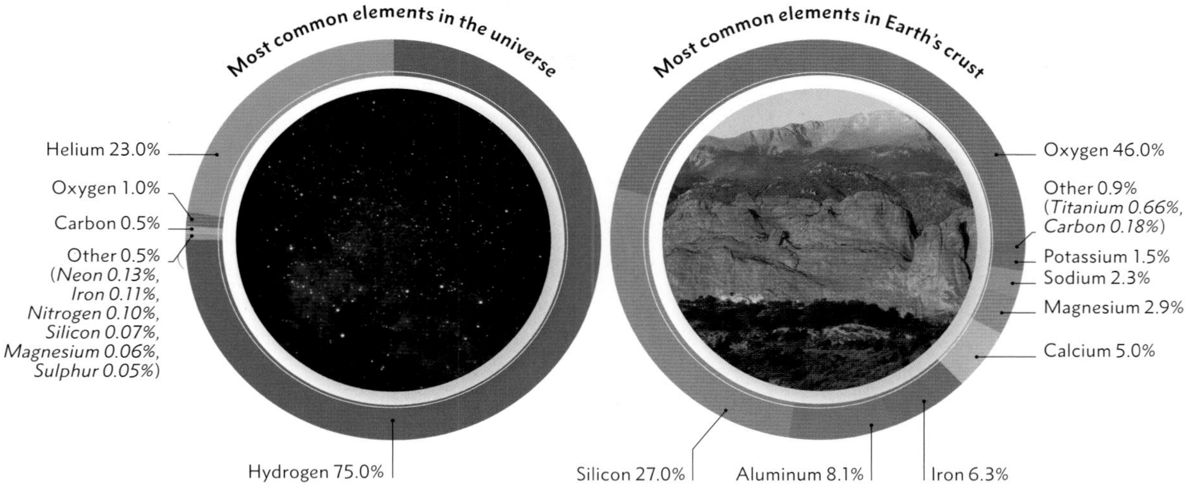

▲ The distribution of the elements
The combination of elements found on Earth differs greatly from the universe at large. The lightest elements, hydrogen and helium, were expelled from Earth's orbit by the young sun. Oxygen, the crust's most abundant element, was created as life turned carbon dioxide into sugar via photosynthesis.

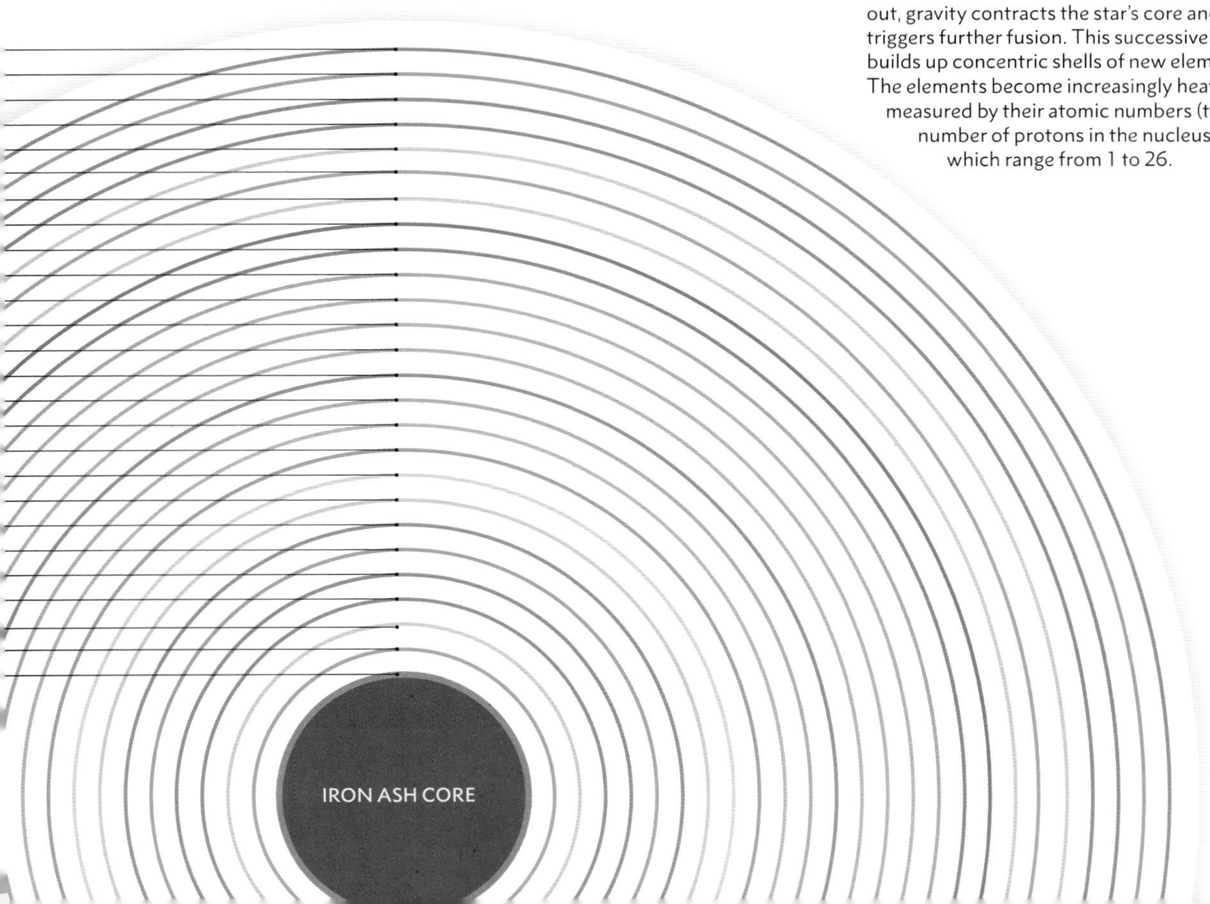

◀ New elements in dying stars
As one source of fusion material runs out, gravity contracts the star's core and triggers further fusion. This successively builds up concentric shells of new elements. The elements become increasingly heavy, as measured by their atomic numbers (the number of protons in the nucleus), which range from 1 to 26.

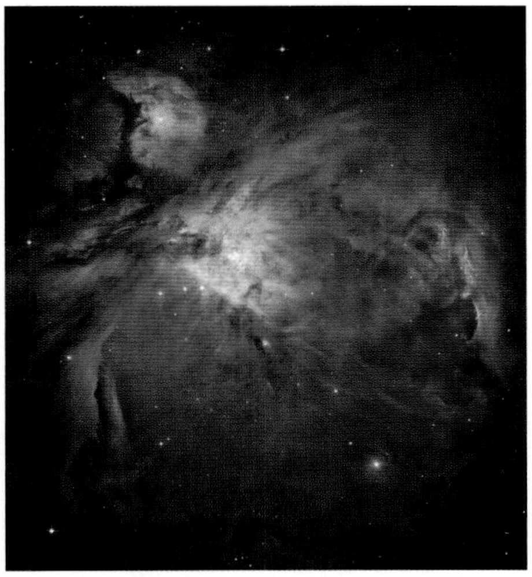

▲ Life's cosmic origins
The building blocks of life have been found in the nearest star-forming region to our solar system, the Orion Nebula. Amino acids combine to create proteins and are a key component of DNA.

WHEN GIANT STARS **EXPLODE**

Today we know that supernovas pepper the universe with elements heavier than iron. But our quest to understand these searing explosions dates back to a time long before the advent of our astronomical understanding. We've been documenting them for almost 2,000 years.

The earliest recorded evidence of an observed supernova dates back to Chinese astronomers in 185 CE. They documented the appearance of a sudden bright light in the sky that took eight months to fade from view. A similar event occurred in 393 CE, and up to 20 other potential events appear in Chinese records, although modern astronomers haven't been able to confirm they were all supernovas.

JUST BEFORE A **SUPERGIANT STAR EXPLODES AS A SUPERNOVA**, ITS TEMPERATURE REACHES ABOUT **180 BILLION°F (100 BILLION°C)**

One definitive explosion—perhaps the most famous of the pre-telescope age—was seen to detonate in 1054. It was observed in Japan and the Middle East, as well as in China. Luminous enough to be seen during daylight hours for nearly a month, it was a guest in the night sky for almost two years. The remnant of this colossal explosion is the spectacular Crab Nebula in the constellation Taurus.

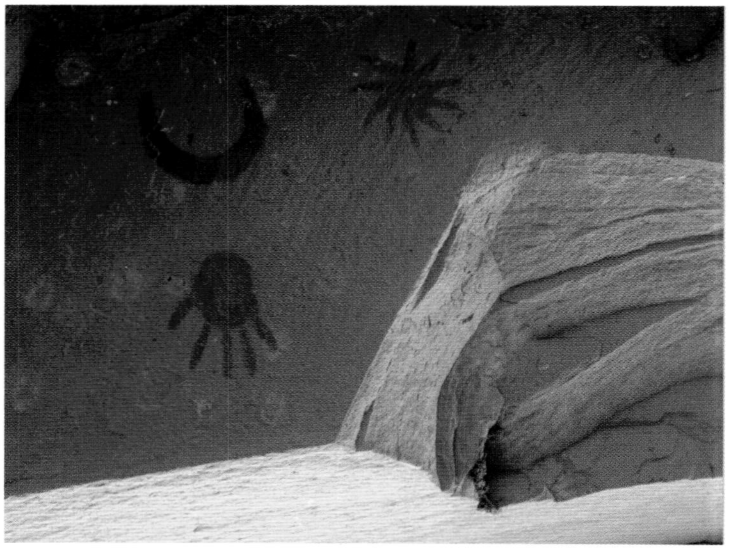

▼ **Chaco Canyon** These wall markings in a New Mexico cave show a large star, a crescent moon, and a handprint. It has been suggested that the local Anasazi people drew it as a record of the 1054 supernova.

ENTER THE TELESCOPE

The 1054 event was followed nearly six centuries later by the supernovas of 1572 and 1604, the last in the pre-telescope age. The latter, known as Tycho's supernova, was the last observed to explode in our Milky Way galaxy.

However, in more recent times, light reached us in 1987 from an explosion in one of our galaxy's satellites—the Large Magellanic Cloud. By then, astronomers were able to observe it with telescopes within days of detonation. The Voyager probe, then on its way to the farthest planets, was also pointed toward the explosion for a closer look. Designated SN 1987A, it surprised astronomers because the best theories of the day said the star that exploded shouldn't have done so. Consequently, it has become a valuable source of evidence against which astronomers can test their theories. Some of their ideas were backed up by SN 1987A, particularly that the radioactive decay of cobalt atoms keeps the supernova remnant bright long after the initial explosion. But some mysteries remain. For example, astronomers have yet to find the neutron star that should have formed at the heart of the dying star.

The 1054 supernova and SN 1987A were both Type II supernovas, formed by the core collapse of massive stars. In recent years, astronomers have also been able to pick out some relatively close Type 1a supernovas, which are formed by stars of lower mass. These include SN 2011fe in the Pinwheel Galaxy and SN 2014J in the nearby Cigar Galaxy.

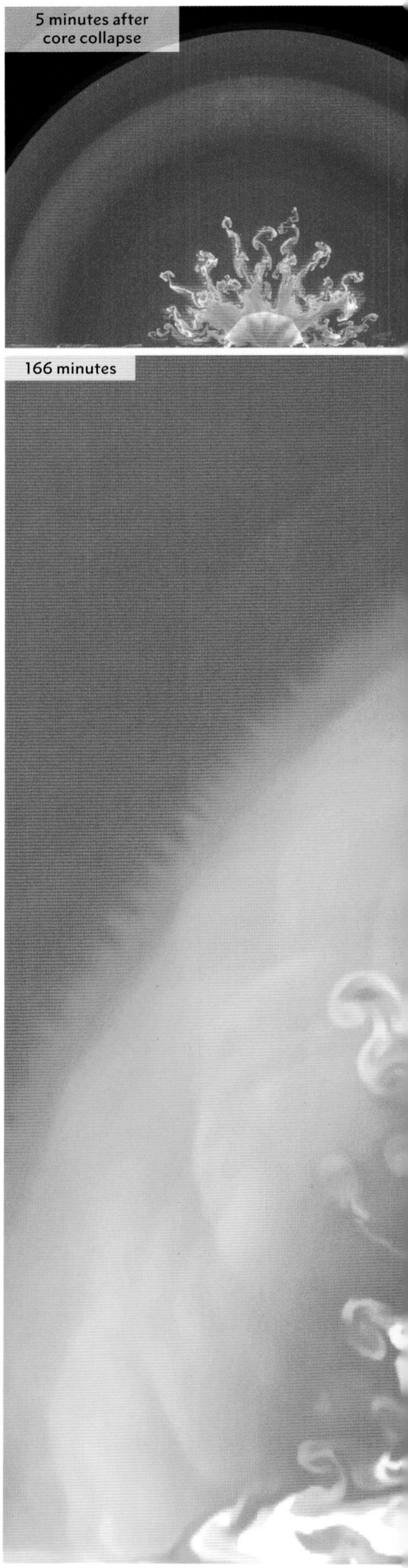

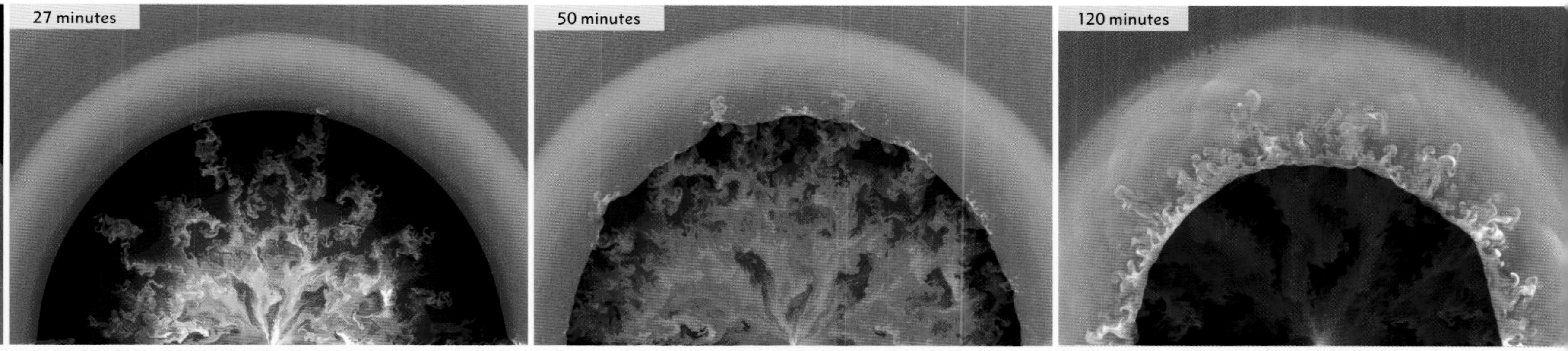

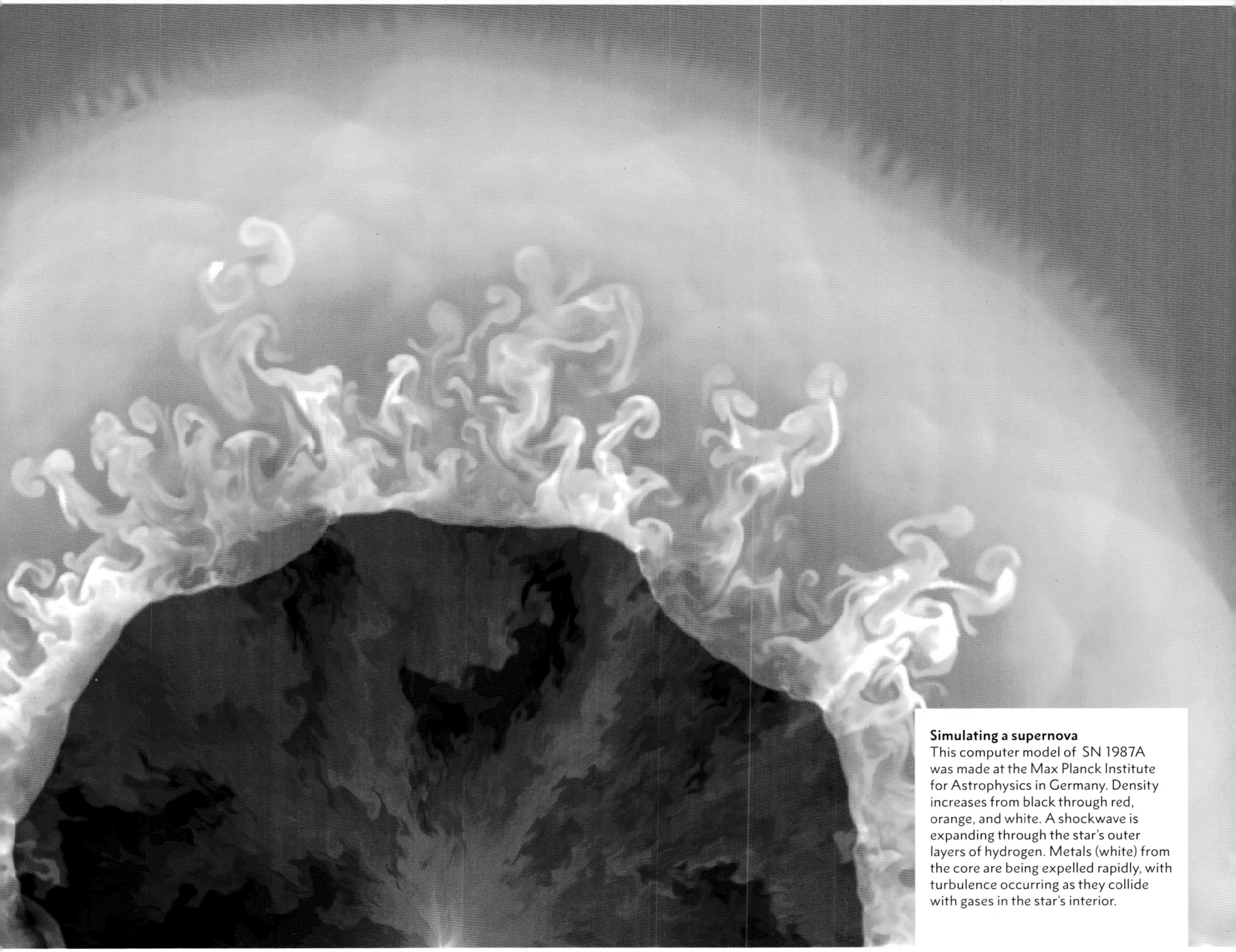

Simulating a supernova
This computer model of SN 1987A was made at the Max Planck Institute for Astrophysics in Germany. Density increases from black through red, orange, and white. A shockwave is expanding through the star's outer layers of hydrogen. Metals (white) from the core are being expelled rapidly, with turbulence occurring as they collide with gases in the star's interior.

▼ The Periodic Table
First presented to the Russian Chemical Society on March 6, 1869 as "the period system," this famous depiction of the primary components of matter organizes the elements in an incredibly useful way.

Missing elements By arranging the table in terms of the behavior and structure of elements, Mendeleev was able to spot gaps that suggested as-yet-unseen elements, including germanium

Atomic number This is the number of protons in the nucleus—just one in the case of hydrogen

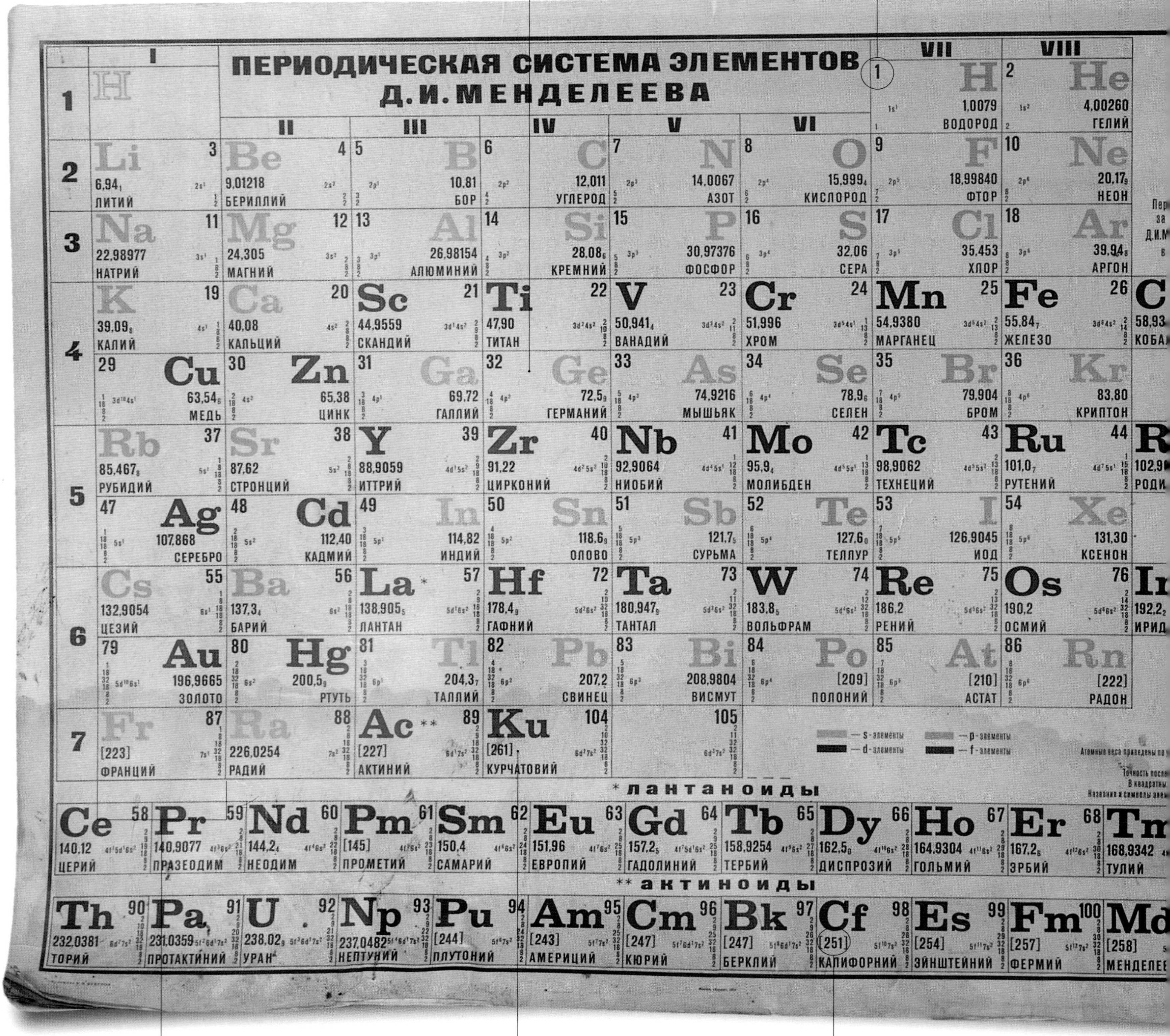

Group Vertical columns are called groups. Group members have similar electron configurations and so exhibit similar chemical properties. Today, 18 groups are officially recognized

Unstable elements Some elements are not stable and decay over time. Even the most stable form of kurchatovium (now called rutherfordium) will decay to half the original amount in just 1 hour 20 minutes

Relative atomic mass This is measured in atomic mass units (amu), where 1 amu is equal to 1/12 of the mass of a carbon atom. This is why it is called relative—it helps compare the masses of different elements

MAKING SENSE OF THE ELEMENTS

The periodic table of the elements is one of the most recognizable icons in science. By organizing the elements according to their atomic structure, it provides a standard way to order and classify them. Of the 118 elements in the table, 92 form inside stars and supernovas.

Dmitri Mendeleev Mendeleev is the name most associated with the Periodic Table. He didn't win the Nobel Prize, but he does have an element named after him (Mendelevium), as well as a crater on the moon

Period Rows are known as periods. Their main function is to make sure that elements with similar chemical properties appear in the correct group. There are currently seven periods

Tile Each tile displays a chemical symbol for the element (either one or two letters), along with information including atomic number and relative atomic mass number

As the scientific revolution gathered pace, so did the rate at which new elements were discovered. Over time, a pattern in their chemical behavior was found. The first attempt to organize the elements into groups came in the late 18th century, when French chemist Antoine Lavoisier sorted them into four categories: gases, non-metals, metals, and earths. In 1829, the German Johann Döbereiner noted that trios of elements had similar chemical properties. Crucially, he realized that the attributes of one could be predicted from those of the other two. By the 1860s, the British chemist John Newlands had devised his Law of Octaves, which said that every eighth element exhibited similar chemical behavior. However, on occasion he had to squeeze two elements into the same box, and he did not leave gaps for as-yet-undiscovered elements. This problem explains why the Russian Dmitri Mendeleev is often regarded as the father of the periodic table. In 1869, Mendeleev published a primitive version of the famous table, leaving gaps based on the "periodicity" of the known elements.

HOW THE TABLE WORKS

The elements are organized in order of increasing atomic mass. The horizontal rows are known as periods—a new period begins when the behavior of an element repeats. For example, a new period starts after neon to ensure that sodium is in the same column as lithium (both are highly reactive). These columns, or groups, are the real key to the table. Mendeleev's table only had seven groups, but the power of his system was confirmed in the 1890s when the noble gases were discovered and fitted in perfectly as an eighth group.

WHERE THE ELEMENTS ARE FORGED

The searing heat in the first minutes after the Big Bang turned some of the cosmos's nascent hydrogen into helium via nuclear fusion (see p.58). After just 20 minutes, fusion stopped and the basic composition of the universe was set down as about 75 percent hydrogen and 25 percent helium. It took millions of years for more elements to appear. The elements up to and including iron form by fusion in stars, whereas many beyond iron can only be made in the cataclysm of a supernova.

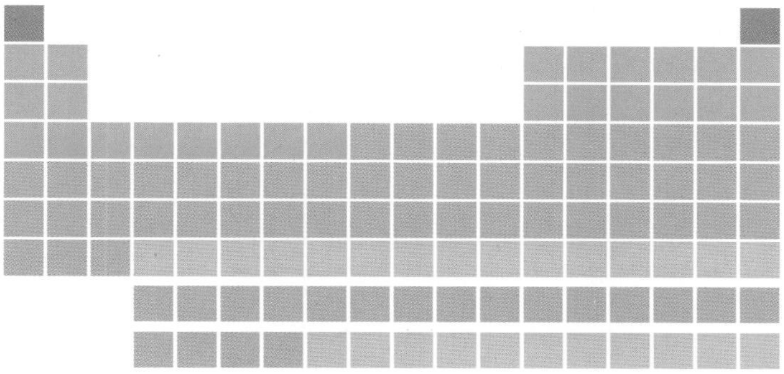

▲ **Organizing the elements** The elements can be grouped according to how they formed. Most of the elements up to uranium formed as a result of nuclear reactions in stars or supernovas. Elements heavier than uranium are unstable and rarely encountered.

KEY
- Formed in Big Bang (hydrogen and helium)
- Formed in stars by fusion (lithium to iron)
- Formed in stars by neutron capture (cobalt to uranium)
- Unstable elements

> IT IS THE **FUNCTION OF SCIENCE** TO DISCOVER THE EXISTENCE OF A GENERAL REIGN OF ORDER IN NATURE AND **TO FIND THE CAUSES GOVERNING THIS ORDER**.

Dmitri Mendeleev, Russian chemist, 1834–1907

THRESHOLD

4

PLANETS **FORM**

As our own star—the sun—ignites, its gravitational pull sweeps up the elements into orbit around it. As they crash together, planets begin to form. While the lighter elements are blown to the outer regions, forming gas giants, close to the sun the heavier elements remain and form rocky planets, including Earth: our home is born.

GOLDILOCKS CONDITIONS

When stars were born from the debris of former stars, some chemically rich material was left in orbit. This debris clumped into balls of matter stuck together by gravity and chemical bonds. These structures were planets, and they were far more complex than anything seen before. We now know that this first happened long ago in solar systems far older than our own.

A newly formed sunlike star

New chemical elements and clouds of chemically rich matter orbiting the new star

Gravity, accretion, and random collisions

Matter from dying stars
The death of stars builds up an ever-increasing supply of heavier elements, supplementing the hydrogen and helium of the early Universe. This results in a more chemically complex world with 92 elements that can combine to form compounds.

What changed?
After the formation of a star, material was left orbiting in a disk. The star's fierce radiation blasted light, volatile material, particularly hydrogen and helium, far from the star. These gases would go on to form distant gas-giant planets. Nearer to the star, the heavier, chemically rich materials from the death of previous generations of stars remained solid or liquid and clumped into rocky planets. In our own solar system, one of these planets was Earth.

Star nurseries
Clouds of dead star material, rich in heavy elements, such as carbon, oxygen, nitrogen, aluminum, nickel, and iron, gather under weak forces of gravity and electromagnetism. They become sites of new star formation.

Shockwave from a supernova
A disturbance, such as a shockwave from a neighboring exploding star, may trigger a cloud to begin contracting to form a star. As it slowly collapses, the cloud begins spinning faster and faster, and takes on a disk shape.

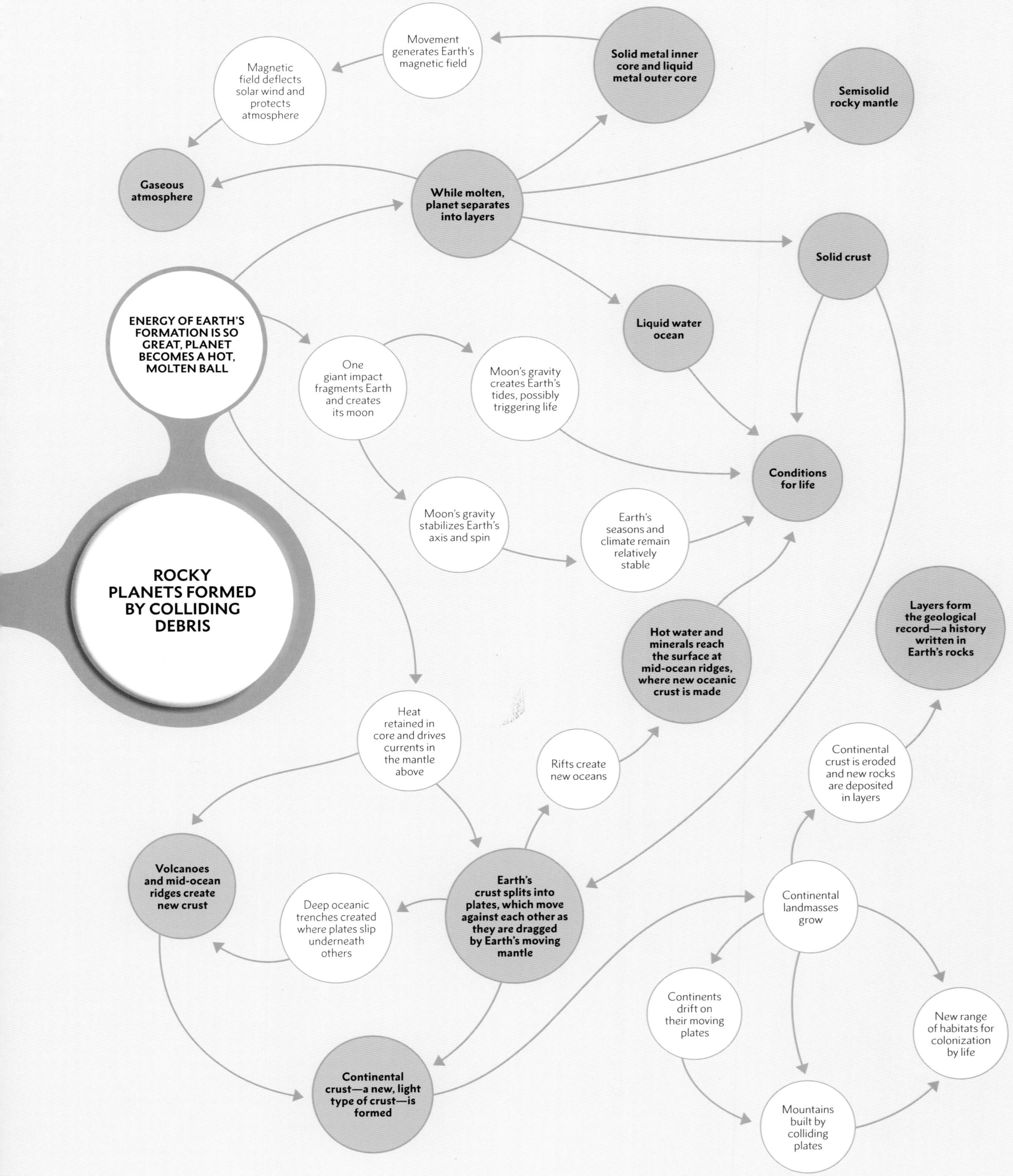
ROCKY PLANETS FORMED BY COLLIDING DEBRIS
ENERGY OF EARTH'S FORMATION IS SO GREAT, PLANET BECOMES A HOT, MOLTEN BALL
While molten, planet separates into layers
Solid metal inner core and liquid metal outer core
Movement generates Earth's magnetic field
Magnetic field deflects solar wind and protects atmosphere
Gaseous atmosphere
Semisolid rocky mantle
Solid crust
Liquid water ocean
One giant impact fragments Earth and creates its moon
Moon's gravity creates Earth's tides, possibly triggering life
Conditions for life
Moon's gravity stabilizes Earth's axis and spin
Earth's seasons and climate remain relatively stable
Hot water and minerals reach the surface at mid-ocean ridges, where new oceanic crust is made
Layers form the geological record—a history written in Earth's rocks
Heat retained in core and drives currents in the mantle above
Rifts create new oceans
Continental crust is eroded and new rocks are deposited in layers
Volcanoes and mid-ocean ridges create new crust
Deep oceanic trenches created where plates slip underneath others
Earth's crust splits into plates, which move against each other as they are dragged by Earth's moving mantle
Continental landmasses grow
Continents drift on their moving plates
New range of habitats for colonization by life
Continental crust—a new, light type of crust—is formed
Mountains built by colliding plates

OUR **SUN IGNITES**

In an otherwise inconspicuous region of our Milky Way galaxy, a giant cloud of matter began to coalesce. Our sun had a tempestuous birth, heating up and spinning until it exploded into life.

An unassuming mass of gas and dust, measuring only a few gas molecules per cubic centimeter, floated aimlessly in space. Eventually, it started to collapse under the weight of its own gravity.

It is likely that this collapse was kick-started by a shockwave from a nearby supernova. A rare type of aluminum can be found across the solar system, which may be a potential trace of this supernova.

UNSTOPPABLE FORCE

Whatever the cause, what we do know is that over tens of millions of years the cloud progressively became more dense. In the center, the cloud was at its densest and hottest—this was the protosun, and it was composed of about 75 percent hydrogen and 25 percent helium. Extreme temperatures and pressures counteracted its own gravitational force, blasting ice, rock, and gas away from the center. These materials flattened in a spinning disk that began to orbit the protosun.

Entering a new phase of intense activity, the protosun began to eject jets of radiation from its poles. Fierce winds blasted lighter elements such as hydrogen and helium to the edge of the protosun's orbit. Soon, the protosun's temperature, pressure, and size rose even higher, until it had absorbed 99.9 percent of material from the original solar nebula.

Despite these events occurring almost 5 billion years ago, we can gather clues as to how our sun was born because we can watch new stars being created elsewhere in the galaxy.

> THE **SUN**, WITH ALL OF THOSE PLANETS REVOLVING AROUND IT... CAN STILL **RIPEN A BUNCH OF GRAPES** AS IF IT HAD **NOTHING ELSE** IN THE **UNIVERSE TO DO**.
>
> **Galileo Galilei**, astronomer, 1564–1642

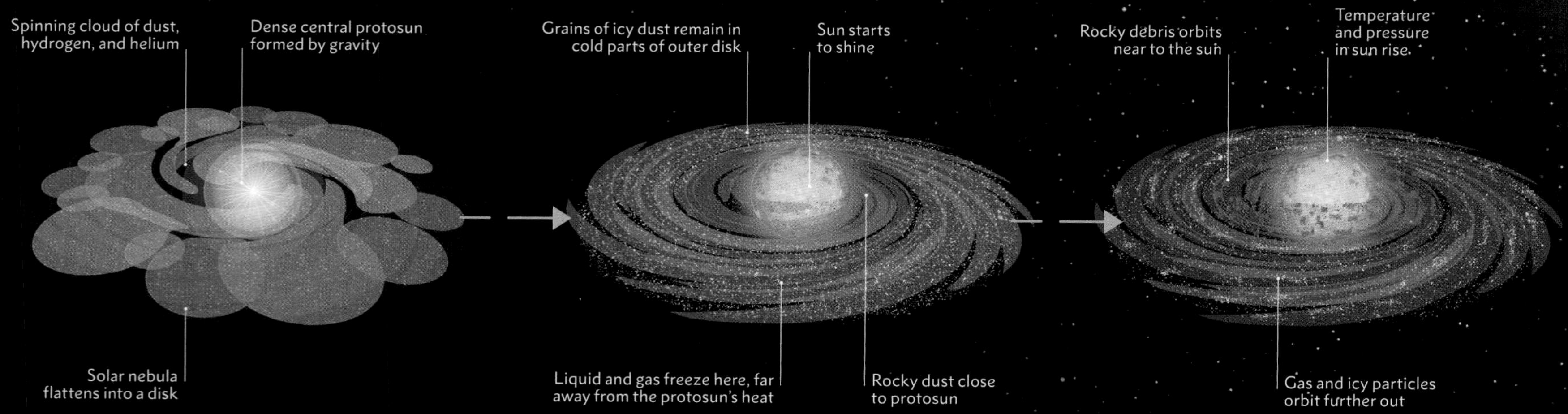
Spinning cloud of dust, hydrogen, and helium
Dense central protosun formed by gravity
Solar nebula flattens into a disk
Grains of icy dust remain in cold parts of outer disk
Sun starts to shine
Liquid and gas freeze here, far away from the protosun's heat
Rocky dust close to protosun
Rocky debris orbits near to the sun
Temperature and pressure in sun rise.
Gas and icy particles orbit further out

An interstellar cloud of gas and dust begins to collapse under gravity, spinning and heating up as it does so. In the hot, dense center, a protosun forms.

Extreme temperatures inside the protosun generate energy that counteracts its own gravity. Ice and gas near the protosun burn away, leaving rocky dust particles.

The protosun's temperature and internal pressure rise, and it becomes an early sun. Lumps of rock and ice orbiting the sun start to collide.

Bursting into life
Intense jets of radiation erupt from the protosun's poles as it swallows up dust and gas. Fierce winds collide with the surrounding rock and ice that will later form planets.

Rocky worlds are born
As planetesimals within the frost line orbited the young sun, incoming materials approached with ever greater speed. The constant impacts accelerated their growth, pulling even more material toward them.

THE PLANETS **FORM**

The planets in our solar system started their lives as gas and tiny grains of dust. Formed into a whirling disk by the young sun's gravitational pull, millions of years of violent collisions would eventually mold the gas and dust into impressive planets, one of which would become our home.

Before the modern planets came the planetesimals—the building blocks from which planets are made. The gathering together of smaller chunks to form larger ones is a process known as accretion.

ASSEMBLING A PLANET

The irregular orbits of the mostly solid materials around the young sun led to frequent impacts, causing accretion. Initially, grains that were a fraction of an inch grew to foot-sized lumps. It took tens to hundreds of millions of years for their collective gravity to accumulate materials that resulted in planetesimals that stretched miles across.

The largest planetesimals had enough gravitational power to attract additional material relentlessly. The planetesimals formed by this process of runaway accretion created the embryos of planets.

DIFFERENT TYPES OF WORLDS

The distance at which these planetary seeds formed from the sun determined whether the eventual planet was made primarily of rock or gas.

In the hot ring of the inner solar system, only materials with very high melting points, such as iron, nickel, and silicon, could survive to be incorporated into the rocky planets, Mercury, Venus, our home planet Earth, and Mars.

In the outer solar system, beyond what astronomers refer to as the frost line, materials such as water and methane froze in the frigid temperatures. With more solid material available, the gravitational pulls of these larger planetesimals were stronger. Consequently, lighter elements such as hydrogen and helium were more easily captured, resulting in the vast gaseous atmospheres typical of Jupiter, Saturn, Uranus, and Neptune.

> “THE **FORMATION OF THE PLANETS** IS LIKE A **GIGANTIC SNOWBALL FIGHT**... A PLANET-BALL THAT HAS **GATHERED ALL THE SNOWFLAKES** IN THE SURROUNDING AREA.”

Claude Allègre, scientist and politician, 1937–

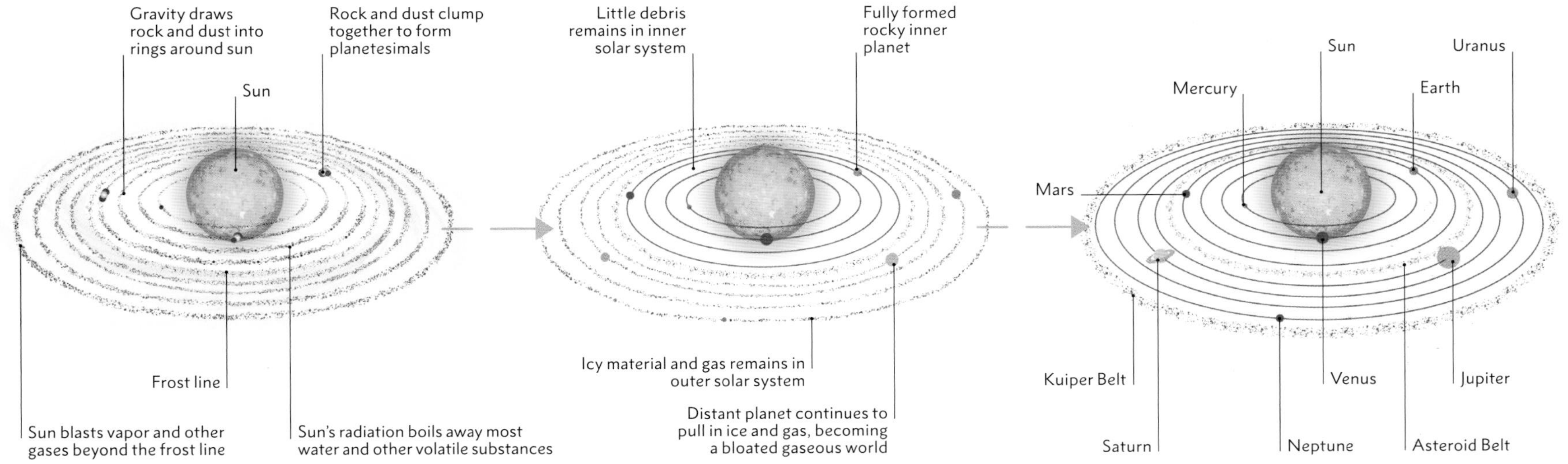
Gravity draws rock and dust into rings around sun
Rock and dust clump together to form planetesimals
Sun
Frost line
Sun blasts vapor and other gases beyond the frost line
Sun's radiation boils away most water and other volatile substances
Little debris remains in inner solar system
Fully formed rocky inner planet
Icy material and gas remains in outer solar system
Distant planet continues to pull in ice and gas, becoming a bloated gaseous world
Sun
Uranus
Mercury
Earth
Mars
Kuiper Belt
Venus
Jupiter
Saturn
Neptune
Asteroid Belt

Materials and debris left over from the sun's formation orbited the young sun in rings. The inner rings were composed of metals and rock; outer rings beyond the frost line held rock, frozen water, and gases.

Large planetesimals attracted smaller particles. Their gravitational fields grew stronger as they continued to grow larger. Most of the orbiting material was eventually swept up.

Stabilization of the solar system took hundreds of millions of years (see pp.74–75). The gravitational interactions of the infant planets settled, eventually forming the stable orbits we see today.

HARD EVIDENCE

THE IMILAC **METEORITE**

Meteorites—pieces of material that have flown through space and landed on Earth—deliver small time capsules of ancient data. They have drifted since the birth of the solar system, so the information they contain is often older than Earth.

Artifacts that were around after the solar system formed are still orbiting our sun today as comets and asteroids. They are relics of the early solar system that have remained relatively unchanged due to the absence of geological activity. When they land on Earth as meteorites, studying them allows us to journey into the past and test out our theories of how our solar system, and our planet, came to be. Tens of thousands of meteorites weighing more than 1/4oz (10g) land on Earth every year, each parachuting down precious information on what the solar system was like billions of years ago.

This sample is a slice of a meteorite named "Imilac", which was itself a small fragment of almost a ton of material that fell into the Atacama Desert, Chile, as part of a single impact event. Imilac is classified as a pallasite meteorite due to its matrix of metal encapsulating its crystals. Like all pallasites, it originated from the boundary between the metallic core and the rocky mantle of a planetesimal or asteroid, which broke apart during the formation of our solar system, possibly due to the early sun's gravitational pull. Some small pieces of the mantle fell into the molten core during this process. It then took at least a million years for these chunks to cool into the crystals scattered throughout the metal you can see here.

Not only can pallasite meteorites help determine the age of the solar system, they can also provide clues as to its early chemical composition. Pallasites such as this one are incredibly rare in our Earthly collection—they make up just 0.4 percent of the meteorites scientists have gathered up.

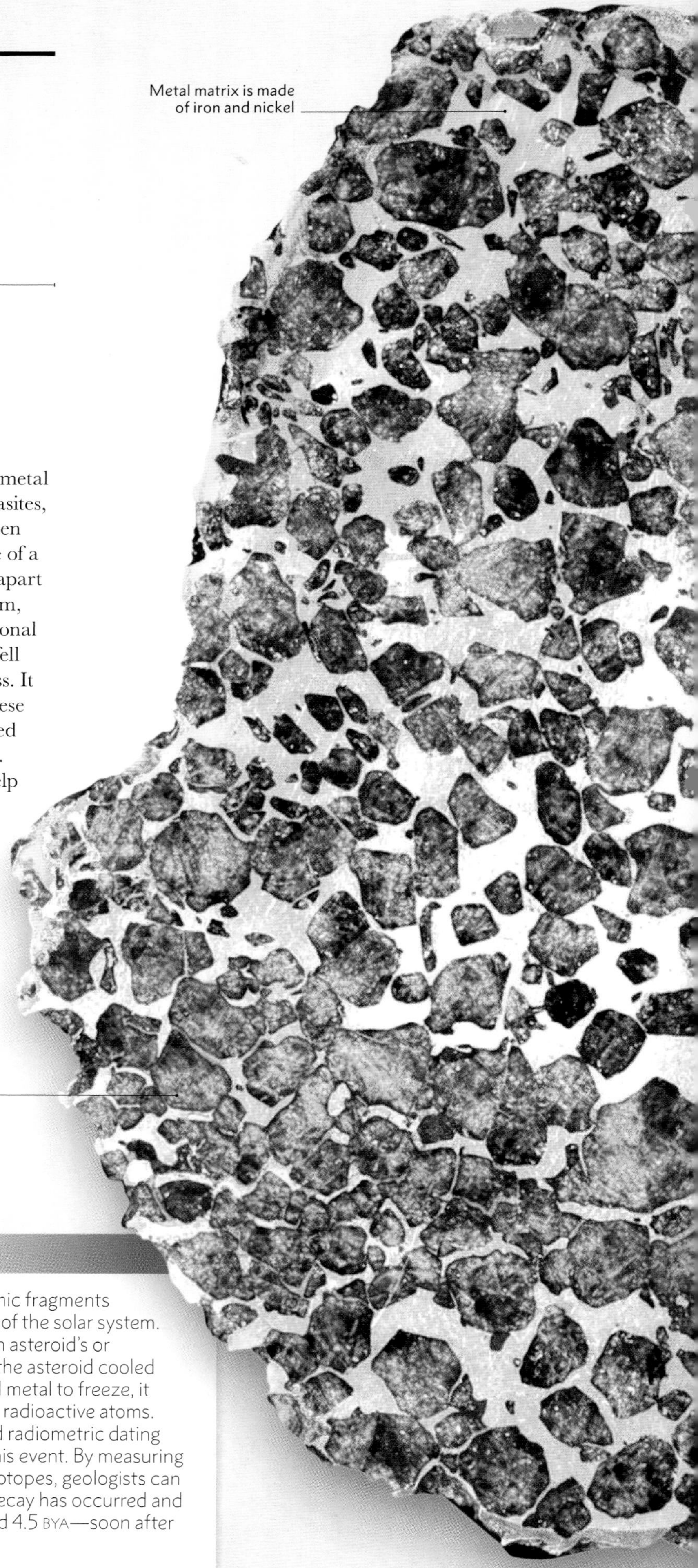
Metal matrix is made of iron and nickel
See-through parts are olivine crystals

▲ Orbiting evidence
These ice mountains on Comet 67p, studied by probes in 2014–15, are as old as our solar system. The presence of ice in the comet's interior demonstrates that water or ice was present during the solar system's formation.

How do we know its age?

Calculating the age of these cosmic fragments allows geologists to date the birth of the solar system. This meteorite was once part of an asteroid's or planetesimal's hot interior. When the asteroid cooled sufficiently for its molten rock and metal to freeze, it also sealed in isotopes—unstable, radioactive atoms. Scientists can use a process called radiometric dating (see pp.88–89) to put a date on this event. By measuring the present-day densities of the isotopes, geologists can calculate how much radioactive decay has occurred and estimate that the asteroid solidified 4.5 BYA—soon after the birth of the sun.

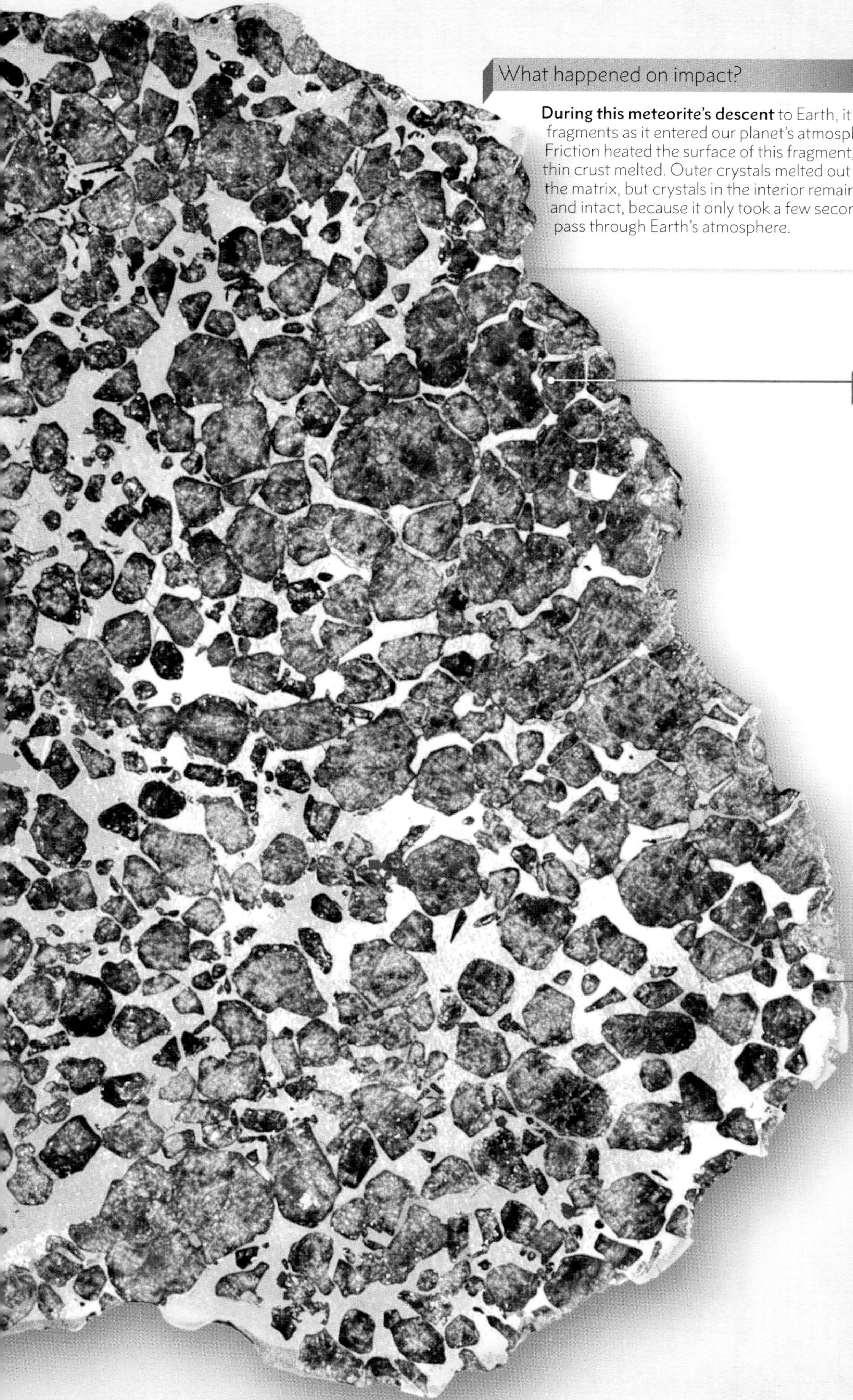

What happened on impact?

During this meteorite's descent to Earth, it split into fragments as it entered our planet's atmosphere. Friction heated the surface of this fragment, and a thin crust melted. Outer crystals melted out of the matrix, but crystals in the interior remained cool and intact, because it only took a few seconds to pass through Earth's atmosphere.

Earth's building block?

By comparing the composition of these meteorites to the composition of Earth, geologists can identify the type of planetesimals that came together to form our planet. Like Earth, this meteorite contains iron and nickel—both of which are thought to constitute Earth's core. Asteroids, dwarf planets, and this pallasite meteorite have remained unchanged since the early solar system and therefore can be key pieces of evidence in determining its history.

Planetesimal forming from smaller bodies

Crystals from the rocky mantle

The crystals are made of olivine and peridot—materials found in tetrataenite, a mineral that can record magnetic fields. Microscopic analysis of these particles demonstrates that when the meteorite was part of an asteroid, the asteroid had a magnetic field—until the core solidified.

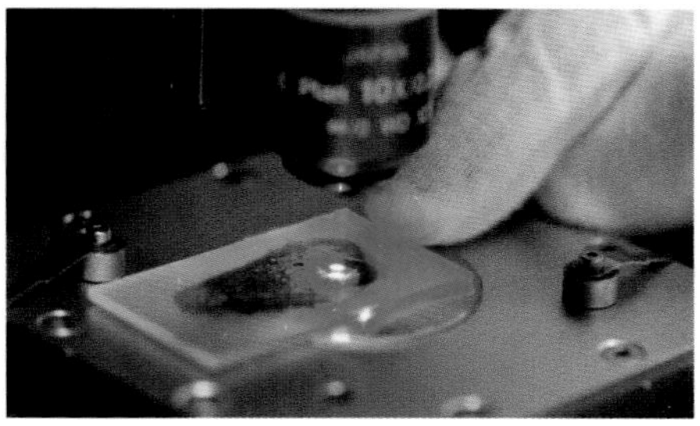

A thin slice of meteorite under a microscope

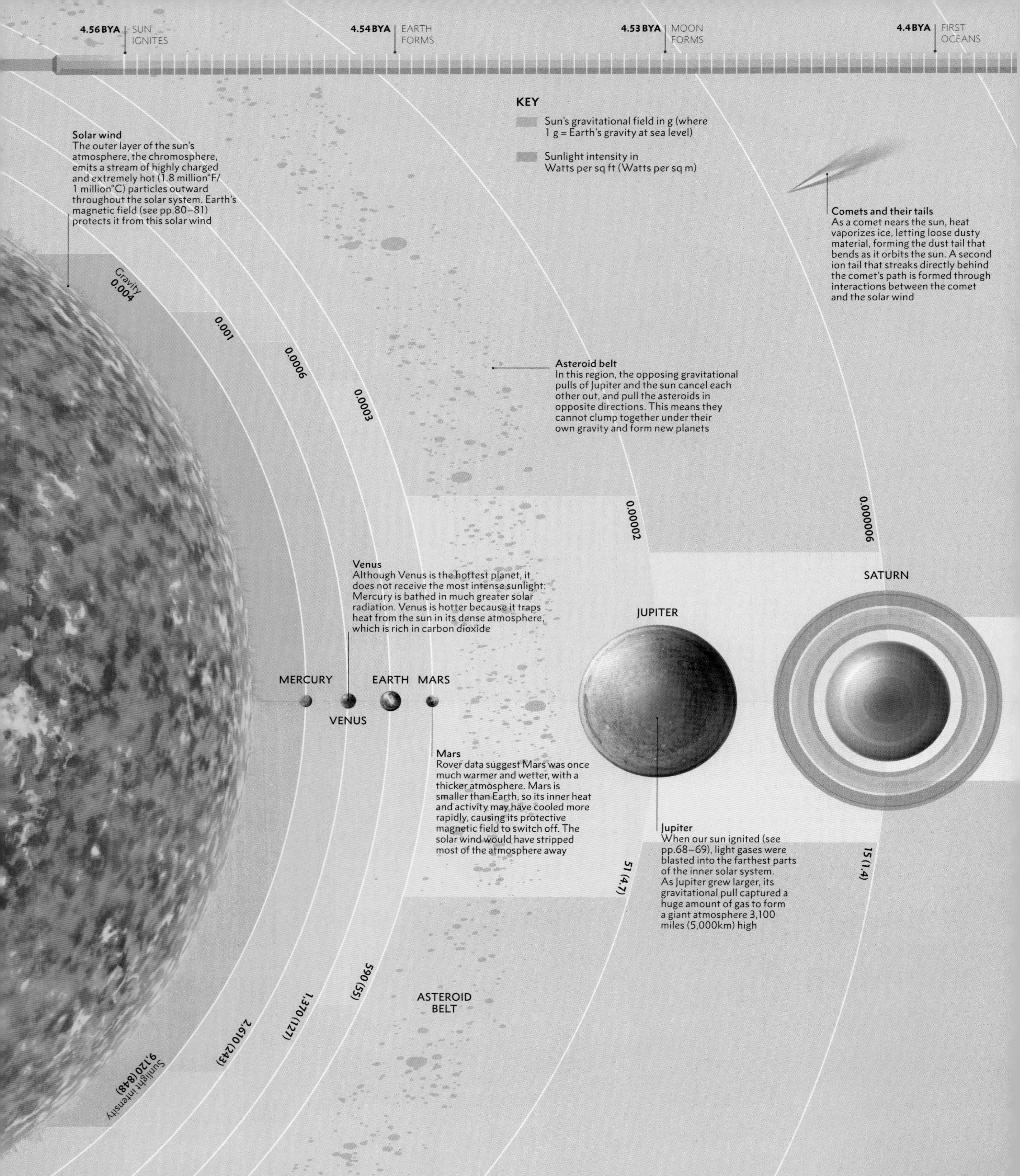

Solar wind
The outer layer of the sun's atmosphere, the chromosphere, emits a stream of highly charged and extremely hot (1.8 million°F/ 1 million°C) particles outward throughout the solar system. Earth's magnetic field (see pp.80–81) protects it from this solar wind

Asteroid belt
In this region, the opposing gravitational pulls of Jupiter and the sun cancel each other out, and pull the asteroids in opposite directions. This means they cannot clump together under their own gravity and form new planets

Comets and their tails
As a comet nears the sun, heat vaporizes ice, letting loose dusty material, forming the dust tail that bends as it orbits the sun. A second ion tail that streaks directly behind the comet's path is formed through interactions between the comet and the solar wind

Venus
Although Venus is the hottest planet, it does not receive the most intense sunlight: Mercury is bathed in much greater solar radiation. Venus is hotter because it traps heat from the sun in its dense atmosphere, which is rich in carbon dioxide

Mars
Rover data suggest Mars was once much warmer and wetter, with a thicker atmosphere. Mars is smaller than Earth, so its inner heat and activity may have cooled more rapidly, causing its protective magnetic field to switch off. The solar wind would have stripped most of the atmosphere away

Jupiter
When our sun ignited (see pp.68–69), light gases were blasted into the farthest parts of the inner solar system. As Jupiter grew larger, its gravitational pull captured a huge amount of gas to form a giant atmosphere 3,100 miles (5,000km) high

THE SUN **TAKES CONTROL**

Between 4.1 and 3.8 BYA, planets shifted their orbits in a cascade of gravitational disruption. The process left eight major planets in orbits that remain stable to this day. However, the sun controls much more in its neighboring space than just these planets.

Scientists have long grappled with the problem of how the modern solar system came to be. When modeling the evolution of the sun's environment, it was hard to explain its present form if the planets had always been where they are now.

NICE MODEL

The present arrangement of the solar system fits with the explanation that the four gas giants started out much closer together: Jupiter moved inward while the other three backed away from the sun. It is even possible that Uranus and Neptune may have swapped order. The outward migration of Neptune would have scattered many of the solar system's smaller objects into a region known as the Kuiper Belt.

This simulation is known as the Nice Model, after the city in France where it was devised. If the migration of the gas giants took place about 600 million years after the formation of the solar system, then it might also account for the event known as the Late Heavy Bombardment. This occurred when a sudden shift in the movements of the gas giants and their gravitational fields caused a catastrophic torrent of asteroids to fall on the inner solar system, including Earth. Lunar rock samples returned to Earth by the Apollo astronauts point to a clustering of meteor impacts around 3.9 BYA. According to the Nice Model, the giant planet migration was to blame.

A MISSING PLANET

Simulations of the solar system's infancy also suggest that our sun once had more planets. By adding a fifth gas planet to the model, researchers found they could get a much better match for the modern arrangement of planets. We do not have five gas planets today, however, so the fifth must have been ejected from the solar system. Given that astronomers have recently found rogue planets that wander through empty space with no host star, the idea is not as bizarre as it may at first appear.

◀ Inner solar system
The realm of the eight planets is referred to as the inner solar system. However, that is by no means the end of the sun's family of orbiting objects. There are many objects beyond Neptune, including dwarf planets and comets. Light and gravity spread out from the sun in all directions—each rapidly losing intensity with distance

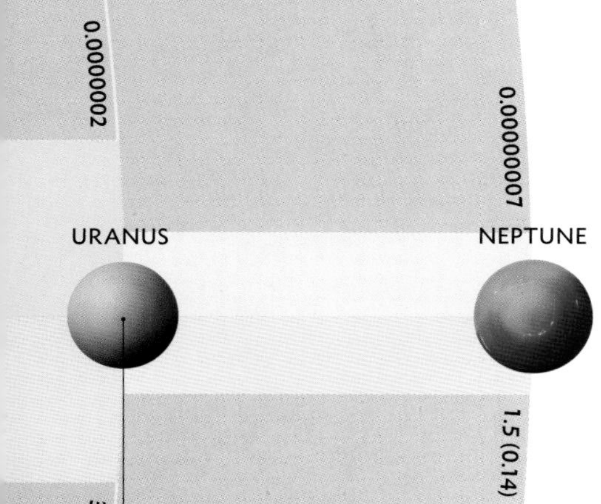

Uranus
The intensity of light fades over distance: at twice the distance, sunlight is four times weaker. Uranus's orbit is 20 times farther from the sun than Earth's, so the intensity of sunlight is just 1/400th of that on Earth

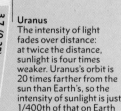

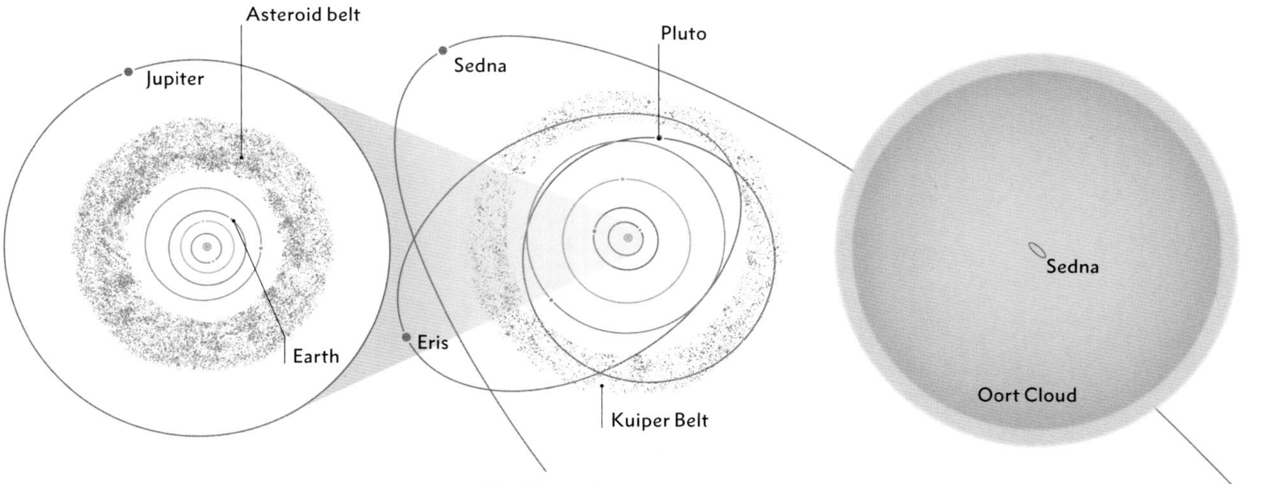

▲ Central solar system
The sun's gravity holds four rocky planets—Mercury, Venus, Earth, and Mars—and an asteroid belt. Beyond that, the gas giants Jupiter, Saturn, Uranus, and Neptune also orbit the sun.

▲ The Kuiper Belt
The band of icy objects – including Pluto – that sits 30–50 times further from the Sun than Earth is known as the Kuiper Belt. Objects including Eris and Sedna orbit even further out.

▲ Outer solar system
The Oort Cloud is a large, spherical region sparsely populated by comets. The sun's gravity controls their orbits up to one light year away, which is the extent of our solar system.

HOW WE FIND SOLAR SYSTEMS

For centuries, astronomers have recognized the stars as distant versions of our own sun. The stars are so far away that it took until the late 20th century to tease out the presence of planets orbiting them and to discover new solar systems.

Stars are often millions of times bigger than planets, and their considerable brightness easily overwhelms any light their suites of planets happen to reflect. The stars themselves appear only as tiny flecks of light from Earth due to their vast distances—the closest one is over 25 trillion miles (40 trillion km) away. It is only in the last few decades that scientists have developed the technology to spot the alien worlds orbiting them.

BLOCKING THE LIGHT

While too small and dark to be observed directly, a planet blocks some of its host star's light when passing, or "transiting," in front of it. Astronomers can glean a wealth of information from this simple event. The planet's size, for example, is betrayed by the amount of light that is blocked out. A transiting Earth would cause a 0.01 percent change in the brightness of the sun.

The time between successive transits reveals the duration of the planet's orbit, which in turn discloses its orbital distance: shorter orbits mean closer planets. Consequently, astronomers use this distance to estimate the planet's temperature and whether it might be habitable.

GRAVITATIONAL WOBBLE

The other main way of finding other solar systems is to exploit the two-way nature of gravity. While stars famously pull on planets, planets also pull back on their suns. This slight tugging causes the star to wobble slightly on the spot. These small changes in the star's motion have an effect on the way we see the light it emits. If wobbling toward us, the star's light is shifted toward the blue end of the color spectrum. Conversely, if it is moving away from us, the shift is toward the red end (see pp.28–29). As more massive planets pull on their stars with a greater gravitational force, these color shifts are more pronounced for heavier planets, allowing astronomers to estimate the planet's mass.

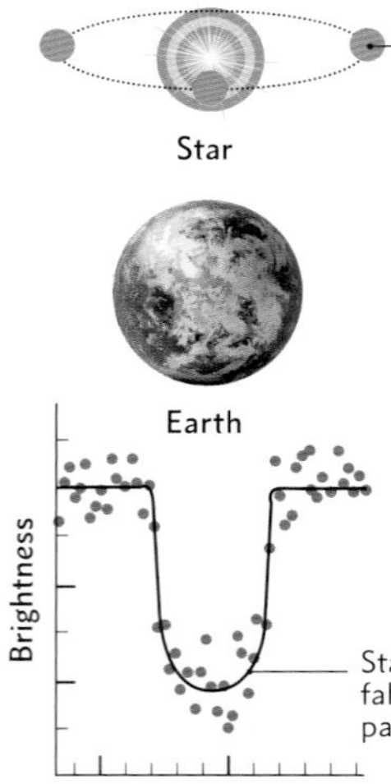
Planet blocks some of its star's light reaching Earth as it orbits
Star
Earth
Brightness
Star brightness falls as planet passes in front
Time

▲ Finding distant planets
Star brightness (red dots) is sampled many times. The line shows the average as it dips due to the passing planet.

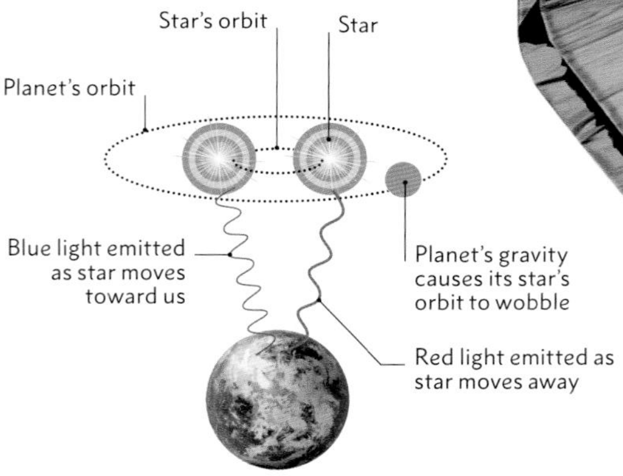
Star's orbit
Star
Planet's orbit
Blue light emitted as star moves toward us
Planet's gravity causes its star's orbit to wobble
Red light emitted as star moves away
Earth

▲ Tracking distant stars
As the star wobbles, color shifts in its light tell us its speed of travel toward or away from us.

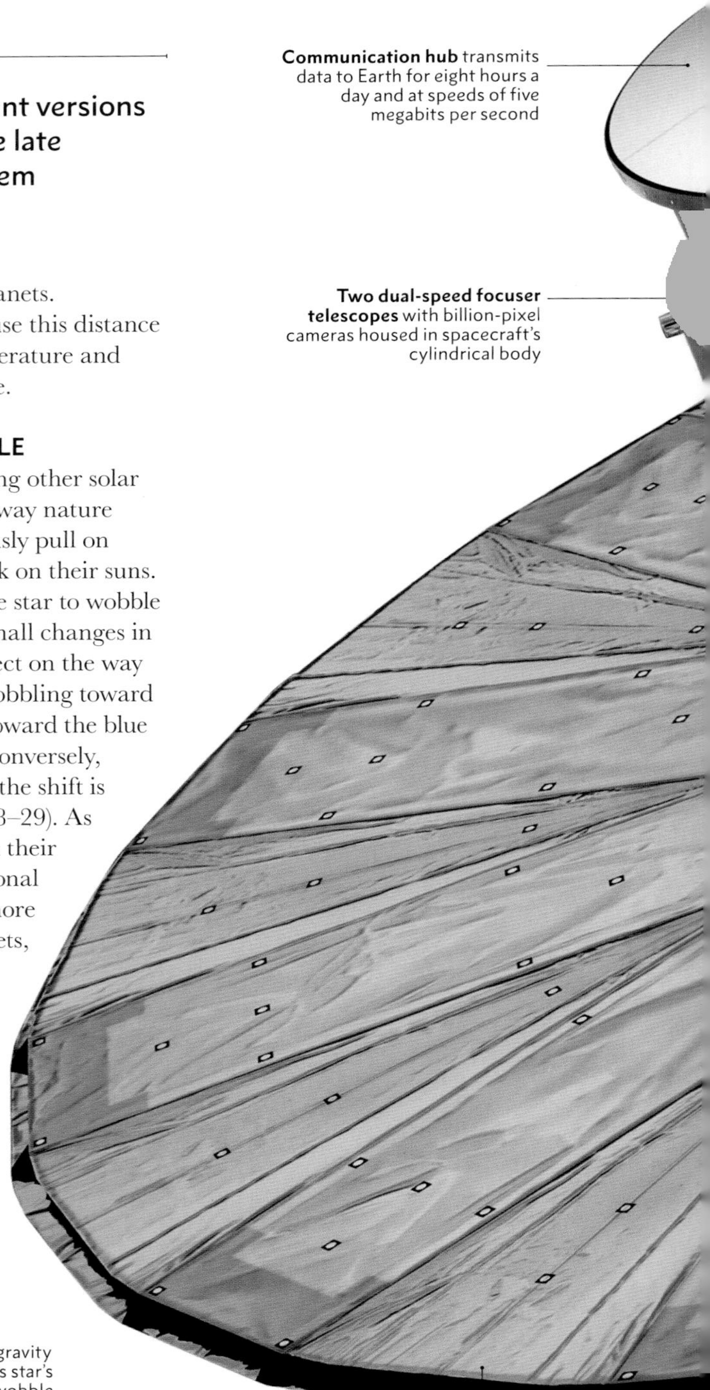
Communication hub transmits data to Earth for eight hours a day and at speeds of five megabits per second
Two dual-speed focuser telescopes with billion-pixel cameras housed in spacecraft's cylindrical body
Temperature-resistant materials cope with conditions between -270°F and 160°F (-170°C and 70°C)

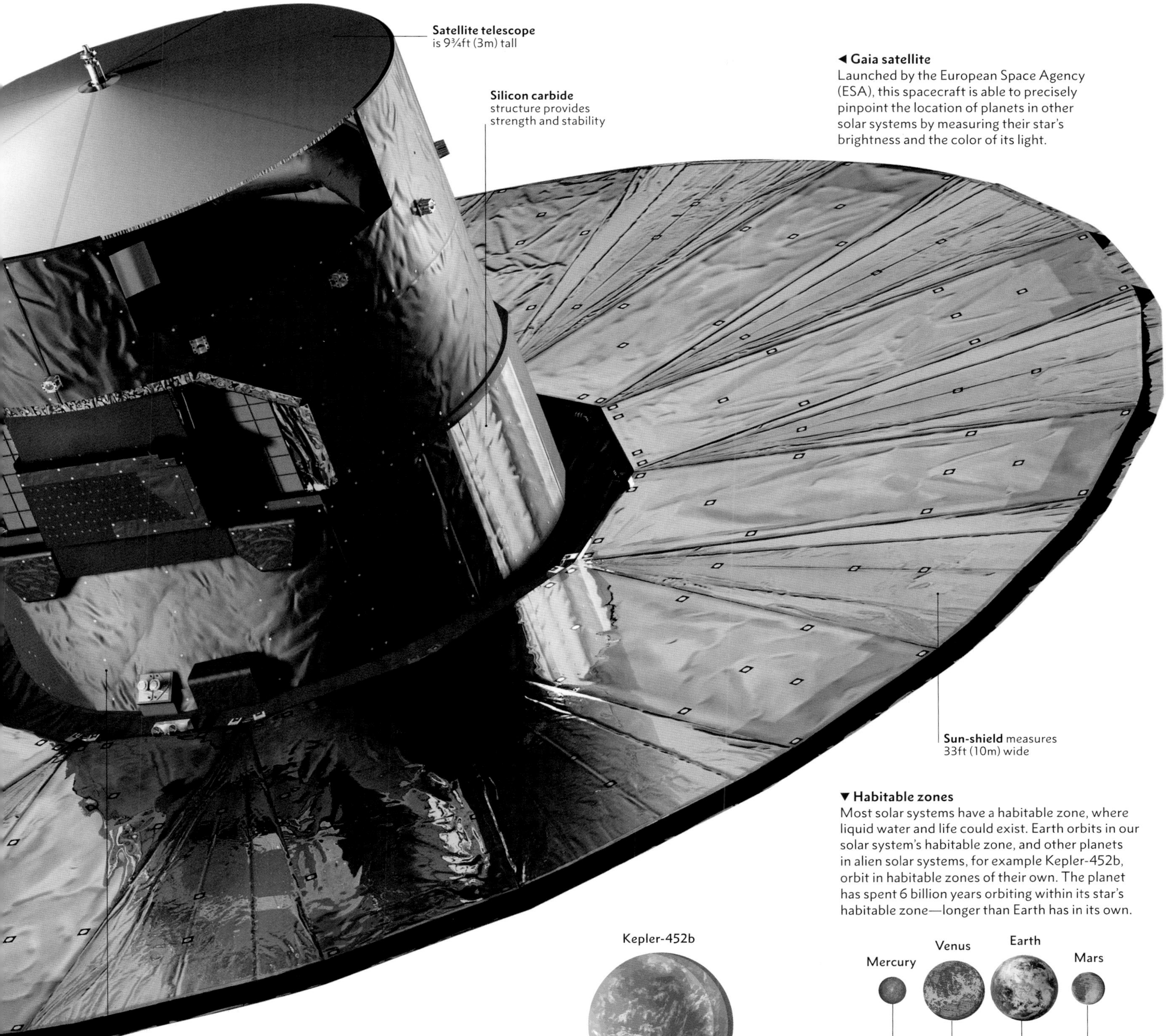
Satellite telescope is 9¾ft (3m) tall
Silicon carbide structure provides strength and stability
Sun-shield measures 33ft (10m) wide
Sensor within cylinder can detect stars 400,000 times dimmer than human eyes can see

◄ Gaia satellite
Launched by the European Space Agency (ESA), this spacecraft is able to precisely pinpoint the location of planets in other solar systems by measuring their star's brightness and the color of its light.

▼ Habitable zones
Most solar systems have a habitable zone, where liquid water and life could exist. Earth orbits in our solar system's habitable zone, and other planets in alien solar systems, for example Kepler-452b, orbit in habitable zones of their own. The planet has spent 6 billion years orbiting within its star's habitable zone—longer than Earth has in its own.

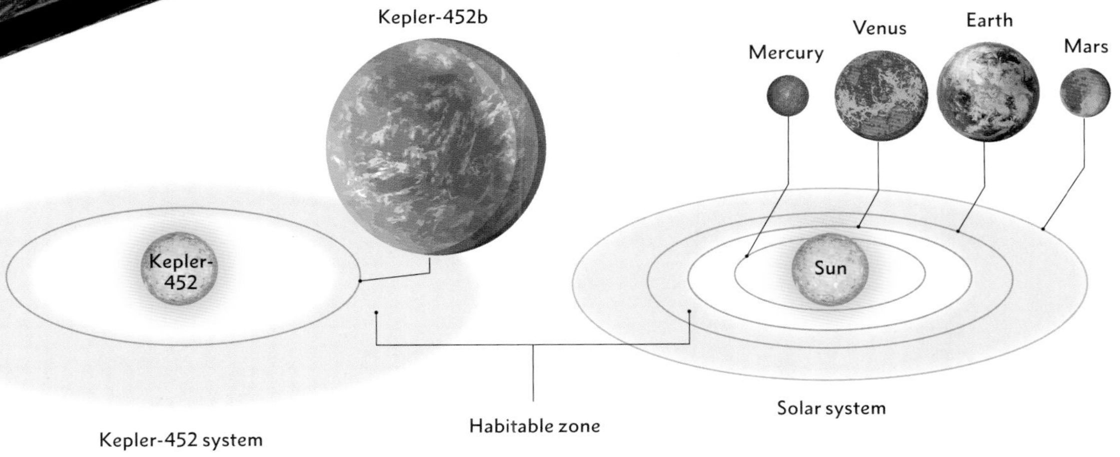
Kepler-452b
Kepler-452
Mercury
Venus
Earth
Mars
Sun
Habitable zone
Solar system
Kepler-452 system

EARTH **COOLS**

Early Earth was very different from the warm, blue planet we know today. Its tumultuous first years were dominated by almost constant collisions from elsewhere in the solar system. Initially a giant molten ball of magma, it gradually became a world fit for life.

Around 4,560 million years ago, rock and ice orbiting the early sun collided into a small, rocky planet under the force of gravity. Earth would have looked very different, with no atmosphere and no oceans. The collisions were far from over—our infant planet was still being battered by many objects, some the size of planets. One collision, with an impactor about the size of Mars, is thought to have formed our moon 100 million years later (see pp.82–83).

BOMBARDMENT OF EARTH

The energy of these collisions, along with that emitted by the radioactive decay of heavier elements, kept early Earth incredibly hot. Much of its material remained molten. This allowed heavier materials, such as iron and nickel, to sink deep toward the planet's core. Less dense, rocky materials, such as molten magnesium and silicon oxides, floated to the surface. Geologists call this process "differentiation" and it would stabilize Earth's structure (see pp.80–81).

HELLISH PLANET

Earth's earliest period was once believed to be so hellish that it is named the Hadean Era—after Hades, the god of the underworld. It was thought that much of Earth's surface remained molten for hundreds of millions of years, but recent findings are overturning this notion and suggest our planet began to cool more rapidly. It may have had oceans less than 200 million years after it formed, as vapor released by volcanic activity condensed into water.

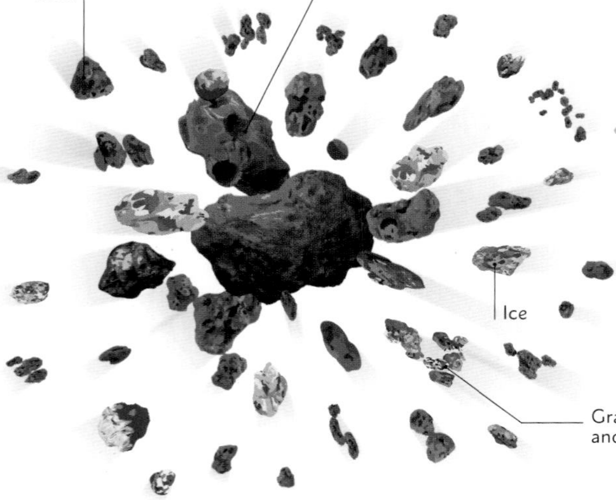

▲ **Accretion** over many millions of years pulled increasingly large clumps of rock and ice (planetesimals) together. They formed a planetary embryo, which then attracted more material. Lumps of ice that remained intact despite the sun's heat would later become the initial source of water on Earth.

▲ **A tiny Earth began to form,** bearing the scars of continual impacts. Its bumpy surface was a result of recent additional material. Gravity molded it into a roughly spherical shape.

▲ **The gravitational potency** of early Earth increased and it attracted impactors, such as asteroids, that were hurtling around the solar system. Each impactor that joined Earth added to the planet's mass and gravitational force. This increased the acceleration and energy of the next impactor.

THE **HADEAN ERA**, IN WHICH **EARTH FORMED,** AND IN WHICH ITS **LAYERS STARTED TO STABILIZE**, OCCURRED **4.6–4 BYA**

4.1 BYA | FIRST TRACE OF POSSIBLE LIFE

4 BYA | EARTH SETTLES INTO LAYERS

3.8 BYA | EARTH'S CONTINENTS START TO FORM

Force of impacts grew as gravitational force increased
Primordial planet
Rock pulled toward planet by its gravity

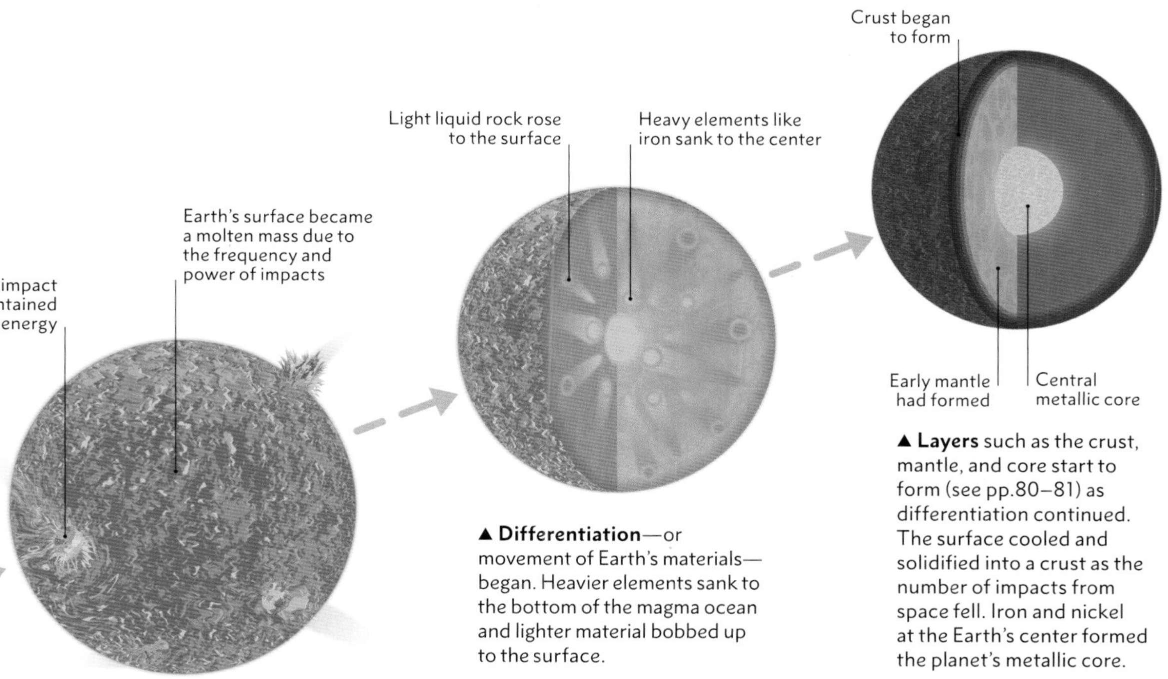

Each impact now contained immense energy
Earth's surface became a molten mass due to the frequency and power of impacts
Light liquid rock rose to the surface
Heavy elements like iron sank to the center
Crust began to form
Early mantle had formed
Central metallic core

▲ **Molten surfaces** caused by relentless asteroid strikes created a vast ocean of magma. As it was then primarily made of liquid, Earth's materials had the potential to move.

▲ **Differentiation**—or movement of Earth's materials—began. Heavier elements sank to the bottom of the magma ocean and lighter material bobbed up to the surface.

▲ **Layers** such as the crust, mantle, and core start to form (see pp.80–81) as differentiation continued. The surface cooled and solidified into a crust as the number of impacts from space fell. Iron and nickel at the Earth's center formed the planet's metallic core.

> “EVEN UNDER THE **MOST EXTREME CONDITIONS**... **EARTH** WOULD NOT HAVE BEEN **COMPLETELY STERILIZED** BY THE **BOMBARDMENT**.”

Oleg Abramov, scientist and astronomer, 1978–

◄ **Hadean Earth**
During the Hadean Era, molten lava dominated the surface, and Earth's atmosphere was devoid of oxygen. The moon, far nearer than it is today, caused huge tides, as a deluge of impactors rained from above.

EARTH SETTLES INTO **LAYERS**

The Earth is formed of distinct layers, and each is made of different materials. The processes responsible for this structure began billions of years ago and continue to shape and influence our planet today.

For hundreds of millions of years after the planet formed, Earth was a molten mass. It was still contracting under its own gravity and material left over from the solar system's formation was still bombarding it. Both processes generated heat. Earth's crust solidified, but the planet continued to differentiate, settling into its present layers.

FROM CORE TO ATMOSPHERE

Material in the center hardened to form a solid inner core, surrounded by a largely liquid outer core. The fluid in the outer core flowed easily, and turbulence within it is thought to contribute to Earth's magnetic field to this day. Above the outer core sits the thickest of the layers—the mantle. The next layer, formed by molten rock erupting from the mantle, is the crust, which accounts for only 0.5 percent of the planet's thickness.

TEMPERATURES IN **EARTH'S CORE** ARE ESTIMATED TO BE HIGHER THAN **12,000°F (6,700°C)**

Differentiation continued as water vapor released by early volcanic activity condensed into water and became the first oceans. The Late Heavy Bombardment about 4.1–3.9 BYA (see pp.74–75) saw a significant, secondary spike in the number of impacts thumping into Earth. These asteroids and comets are thought to have added much of the water that contributed to the primordial oceans.

The lightest materials—gases—escaped from the mantle via volcanoes and became part of our planet's carbon dioxide-rich atmosphere. Hydrogen and helium were blasted away by the solar wind, but Earth's gravity was strong enough to hold onto carbon dioxide, nitrogen, water vapor, and argon. Gaseous oxygen was absent from the atmosphere—all of Earth's oxygen was bound into its rocks and water.

EXPLORING INSIDE EARTH

Our planet's depths are so hot and under such extreme pressure that we have never even penetrated the crust. Instead, scientists have used other methods to deduce what is inside Earth. They knew that there must be significantly heavier material at the center, because the average density of Earth is greater than the density at its surface. Studies of the way earthquakes travel and how our magnetic field emerges provide additional clues about the inner structure of Earth.

▼ Earth's layers
Layers began to form 4.4–3.8 BYA. Our planet is divided here into six layers: the solid inner core, liquid outer core, semi-solid mantle, solid crust, liquid ocean, and gaseous atmosphere.

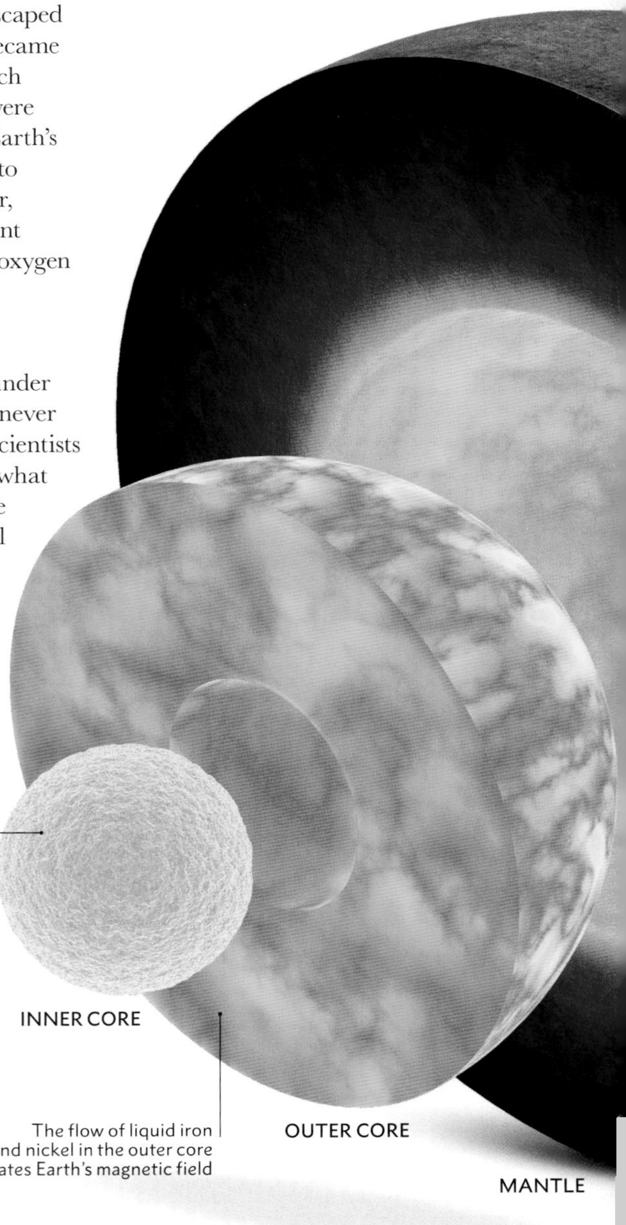
Solid core made of iron and nickel sank to the center soon after Earth formed
INNER CORE
The flow of liquid iron and nickel in the outer core creates Earth's magnetic field
OUTER CORE
MANTLE

▶ Seismic waves
Vibrations from earthquakes are either primary (P) waves or secondary (S) waves. The speed at which they travel through the planet during seismic events can help to determine Earth's structure.

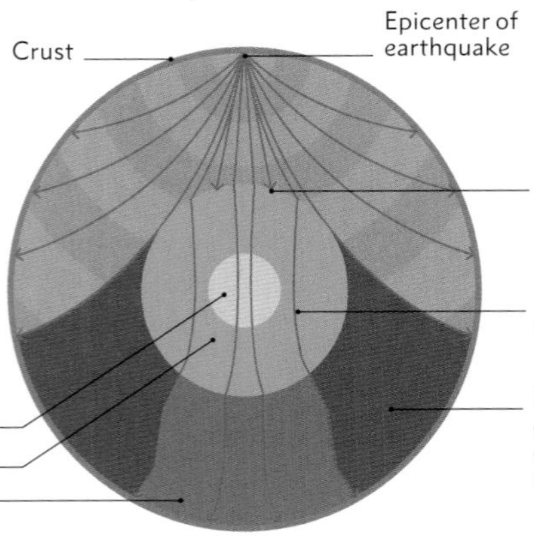
Crust
Epicenter of earthquake
S-waves cannot travel through liquid outer core
Paths of P-waves refract and wobble as they travel through each layer
Shadow zone where no waves can travel due to a change in the direction of travel at the boundary between mantle and core
Inner core
Outer core
Mantle

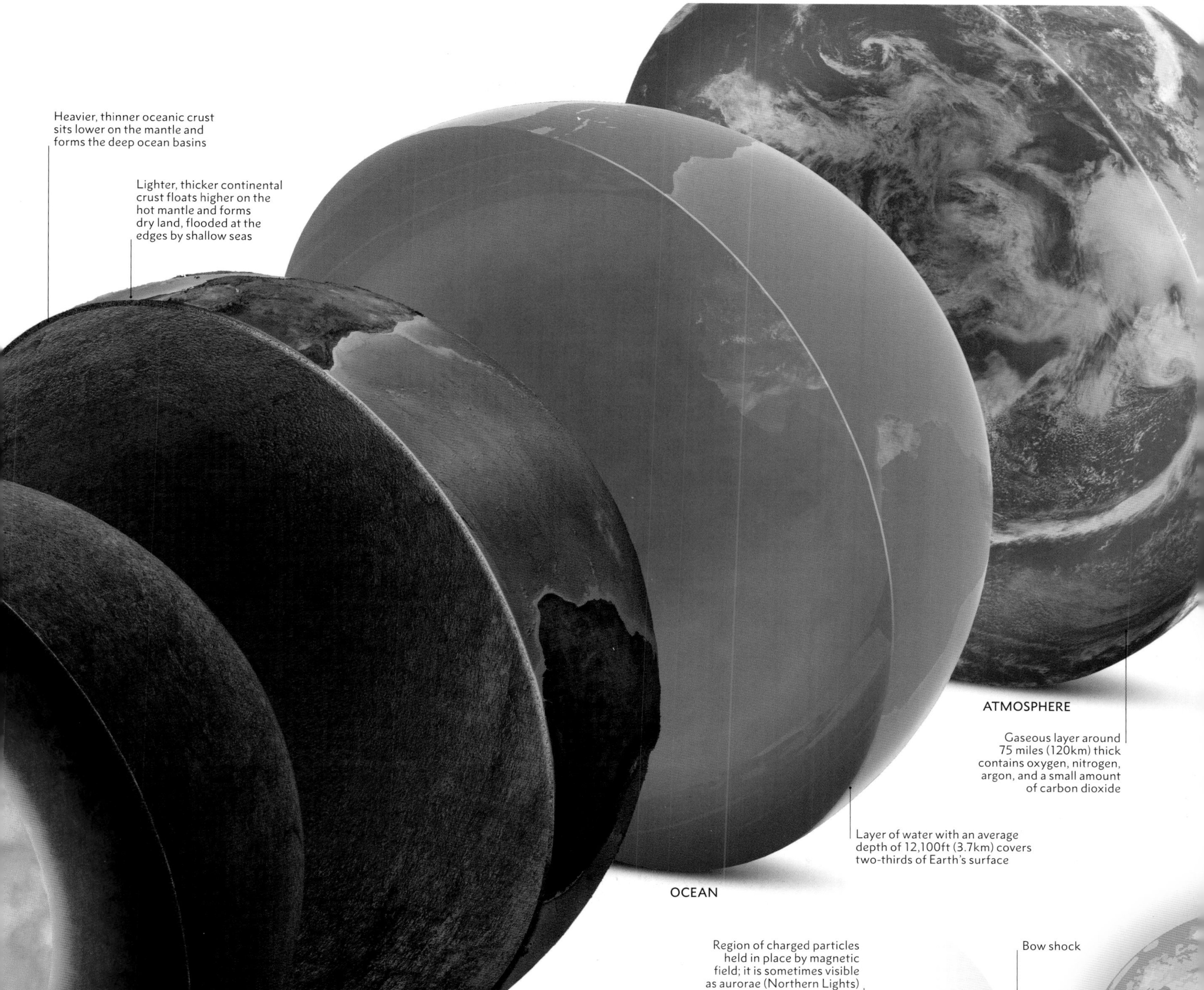
Heavier, thinner oceanic crust sits lower on the mantle and forms the deep ocean basins
Lighter, thicker continental crust floats higher on the hot mantle and forms dry land, flooded at the edges by shallow seas
ATMOSPHERE
Gaseous layer around 75 miles (120km) thick contains oxygen, nitrogen, argon, and a small amount of carbon dioxide
Layer of water with an average depth of 12,100ft (3.7km) covers two-thirds of Earth's surface
OCEAN
CRUST
Semisolid rock in the mantle flows very slowly in convection currents, which cause plate movements in the crust (see pp.92–93)

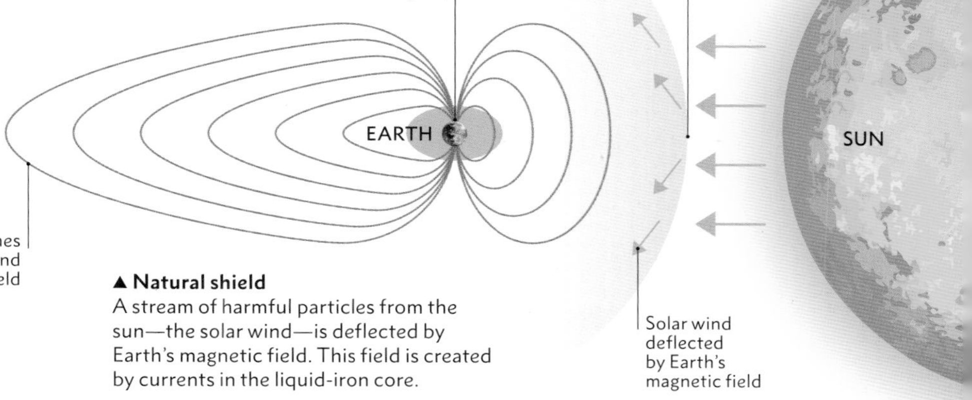
Region of charged particles held in place by magnetic field; it is sometimes visible as aurorae (Northern Lights)
Bow shock
EARTH
SUN
Magnetic field lines show shape and strength of field
Solar wind deflected by Earth's magnetic field

▲ Natural shield
A stream of harmful particles from the sun—the solar wind—is deflected by Earth's magnetic field. This field is created by currents in the liquid-iron core.

THE **MOON'S ROLE**

Despite being a relatively small planet, Earth is blessed with a particularly large moon—the fifth largest in the solar system. The moon is our only natural satellite and has had such a significant influence on our planet that it may even have played a role in kick-starting life on Earth.

If the length of Earth's existence was condensed into a single day, the moon would have formed when the Earth was 10 minutes old. The moon is our planet's steadfast partner and it is likely that we would not be here without it.

It is thought that a giant piece of rock smashed into our infant planet during its early days. Rock from the impact, while in Earth's orbit, gathered together to form the moon. As it formed, it was 10 times closer to Earth than it is currently.

THE MOON AND LIFE

During Earth's childhood, the moon's close proximity would have created a considerably mightier gravitational pull than we feel now. Tides were extreme, and biologists have speculated that the intense churning during these super tides was a key factor in the mixing of ingredients that led to life in the first oceans. Over millions of years, the moon retreated from Earth due to the moon's gradually increasing orbital velocity. Today, the moon is the main driver of the roughly daily cycle of high and low tide, and continues to drift away from Earth at a rate of 1.5in (3.8cm) per year. As it edges further away, tidal strength falls.

The tides swirled the oceans, and this helped to spread heat from polar to equatorial regions, regulating the young Earth's temperature. The moon's gravity also keeps the tilt of Earth's axis constant, which means our seasons are steady and repeat predictably. The moon stabilized Earth over time and this has given life a chance to thrive.

PULLING ON THE PLATES

Geologists have speculated that Earth is the only planet with plate tectonics (see pp.92–93) because of the early moon's strong gravitational pull. During Earth's hellish Hadean Era, our moon would have pulled on the primordial oceans of magma. Theories suggest that the wrench of the moon on the cooling liquid rock helped separate it into the distinct pieces of crust our planet possesses today.

▼ Extreme tides
The Bay of Fundy on Canada's Atlantic coast boasts the widest tidal ranges on Earth. The water rises and falls twice each day by up to 52ft (16m), regularly submerging the Hopewell Rocks.

▼ Pull of the moon
The moon's gravitational force creates tidal bulges on both sides of Earth. On the side facing the moon, the moon's gravity pulls the oceans toward it, resulting in high tides. As well as attraction, however, gravity exerts a stretching force on Earth. Counterintuitively, this results in a second high tide facing away from the moon.

> "THE POSSIBILITY DESERVES CONSIDERATION THAT **THE FORMATION OF THE MOON**... **PROVOKED THE ORIGIN OF LIFE** ON EARTH."
>
> **Richard Lathe**, molecular biologist, c.1950–

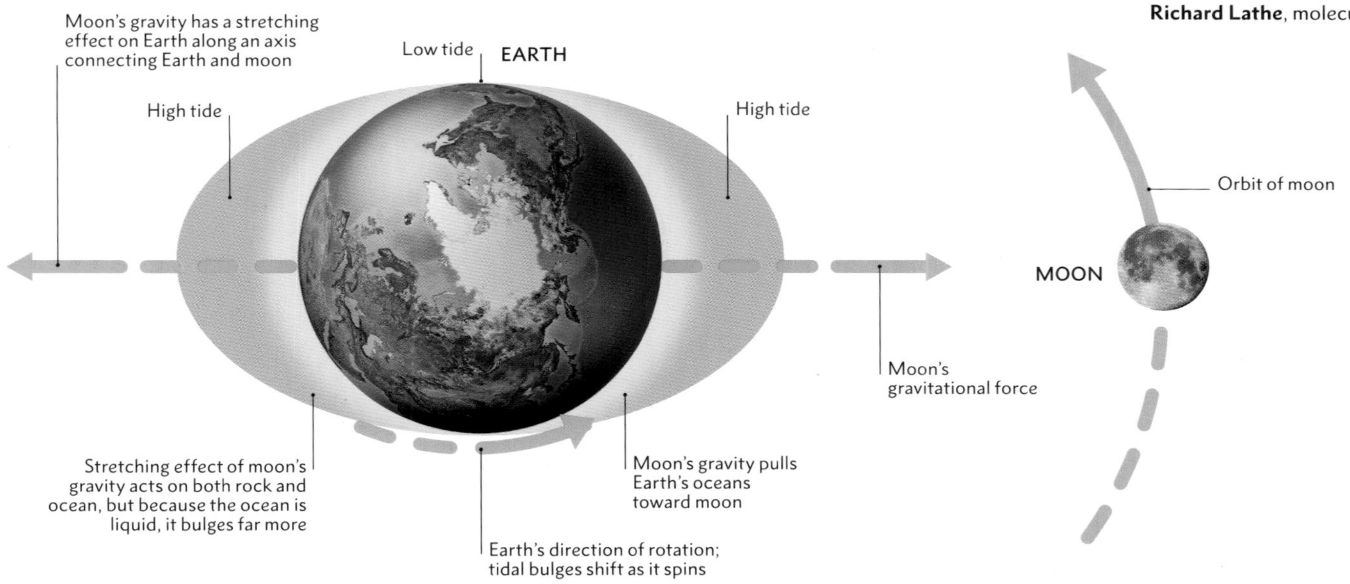
Moon's gravity has a stretching effect on Earth along an axis connecting Earth and moon
Low tide
EARTH
High tide
High tide
Orbit of moon
MOON
Moon's gravitational force
Stretching effect of moon's gravity acts on both rock and ocean, but because the ocean is liquid, it bulges far more
Moon's gravity pulls Earth's oceans toward moon
Earth's direction of rotation; tidal bulges shift as it spins

THE **CONTINENTS** ARE **BORN**

At some time around 4 BYA, Earth's crust began moving, forcing some crust down into the mantle. Magma erupted and cooled into a new, lighter kind of crust—continental crust. It bobbed up higher than the surrounding rock, creating the first land masses. The process continues today, with 30 percent of our planet's surface now made of continents.

Before continents came cratons—the seedlings from which greater swathes of land would grow. Cratons in turn were made from strings of islands formed from the first continental crust. The process began in the Archean era (4–2.5 BYA). Although Earth had cooled since the Hadean era, the planet was still much hotter than it is today. Earth's layers had settled, however, and oceans had formed on a solid crust.

Today, Earth's crust is made of both heavy oceanic crust and continental crust, which is lighter and thicker. The primordial crust was uniform, but when currents in Earth's mantle began dragging on its underside (see pp.92–93), it began moving, splitting into plates. When these plates collided, one plate was forced under the other. This triggered a further stage of

Primordial crust initially covered Earth. When two plates of the moving crust met head-on, one was forced underneath. In the mantle, its lighter materials were melted first, and these bubbled to the surface.

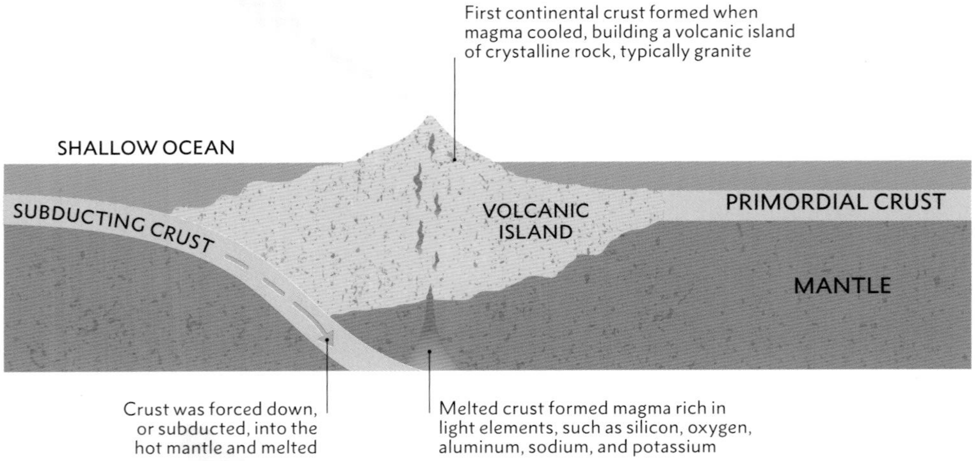
First continental crust formed when magma cooled, building a volcanic island of crystalline rock, typically granite
SHALLOW OCEAN
SUBDUCTING CRUST
VOLCANIC ISLAND
PRIMORDIAL CRUST
MANTLE
Crust was forced down, or subducted, into the hot mantle and melted
Melted crust formed magma rich in light elements, such as silicon, oxygen, aluminum, sodium, and potassium

Movements of Earth's crust pushed adjacent islands together and formed progressively larger masses of light rock called cratons. But two more processes were at work: heavy material rose to the surface where cratons split, and new. heavy oceanic crust was also created where plates separated in oceans.

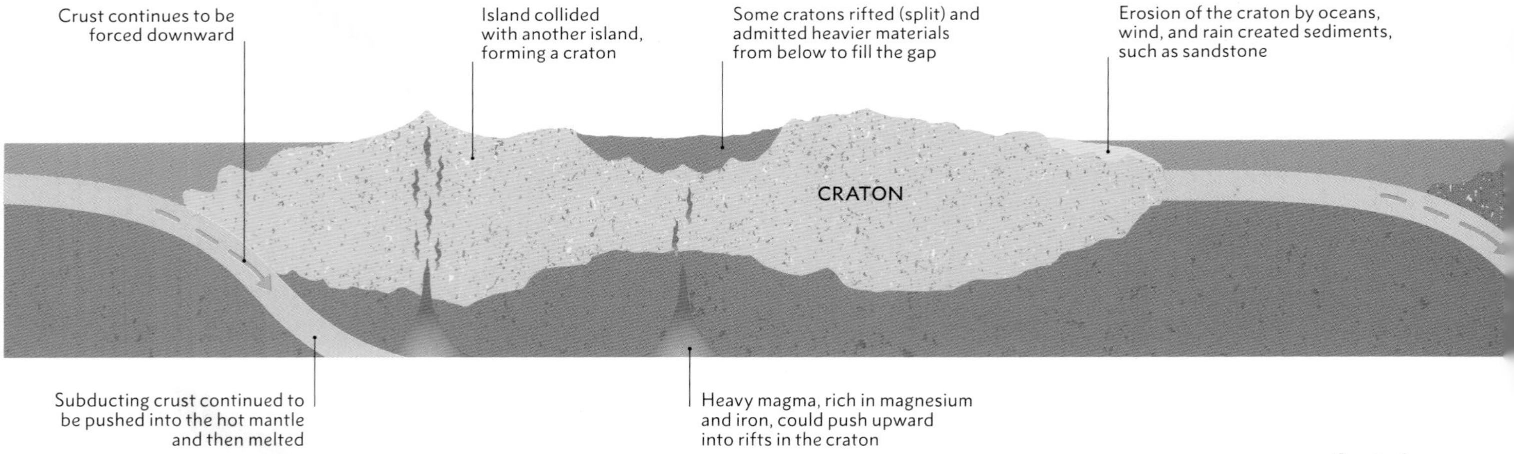
Crust continues to be forced downward
Island collided with another island, forming a craton
Some cratons rifted (split) and admitted heavier materials from below to fill the gap
Erosion of the craton by oceans, wind, and rain created sediments, such as sandstone
CRATON
Subducting crust continued to be pushed into the hot mantle and then melted
Heavy magma, rich in magnesium and iron, could push upward into rifts in the craton

The first continents eventually formed from colliding cratons and islands. Because they were light, they stayed on the surface, but became composed of a growing variety of rock. Oceanic crust, being heavier, is continuously subducted, and never gets old and complex. It is replaced by new crust at spreading ridges.

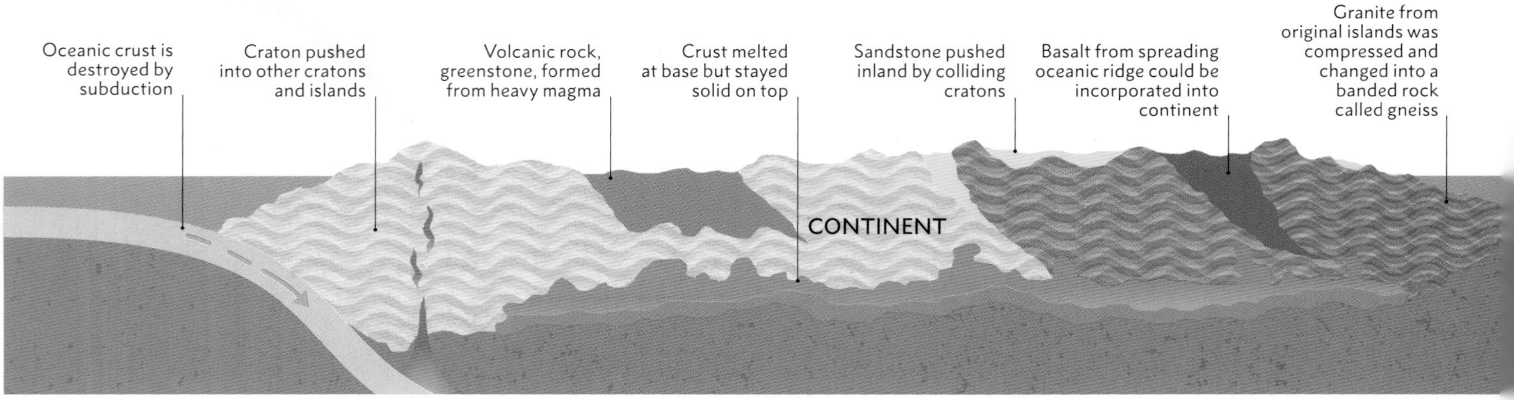
Oceanic crust is destroyed by subduction
Craton pushed into other cratons and islands
Volcanic rock, greenstone, formed from heavy magma
Crust melted at base but stayed solid on top
Sandstone pushed inland by colliding cratons
Basalt from spreading oceanic ridge could be incorporated into continent
Granite from original islands was compressed and changed into a banded rock called gneiss
CONTINENT

differentiation, in which some primordial crust melted and created lighter material that bobbed to the surface and solidified, forming islands. Over millions of years, the movement of Earth's crust pushed the islands together to form cratons—small protocontinents. Eventually, these cratons collided and coalesced to create successively larger land masses—the first continents.

THE FIRST SUPERCONTINENT

By the end of the Archean Era, 2.5 BYA, the Earth's surface had 80 percent of the land mass it does today, largely gathered together into a supercontinent called Vaalbara. Vaalbara was formed by colliding cratons called Kaapvaal and Pilbara. These survive today, but Kaapvaal is now in South Africa and Pilbara is in Australia, and each has rocks dated to 3.6–2.7 BYA. In fact we now know these land masses have split and rejoined more than once (see pp.158–59), and that the cratons that formed the first continents are now scattered across the modern continents. Even though continents change, cratons remain as their stable cores.

Continent formation is still occurring. Oceanic crust continues to subduct under other oceanic crust, causing magma to push to the surface and cool into arcs of volcanic islands—such as those in the Caribbean.

THE OLDEST CONTINENT WHOSE ROCKS STILL EXIST TODAY **IS CALLED "UR"** AFTER THE ANCIENT SUMERIAN CITY

◀ Nishinoshima
In 2013, a new island was discovered off the coast of Japan. It appeared when lava broke through Earth's crust in a burst of volcanic activity and then cooled, following the same process that created continents 4 BYA.

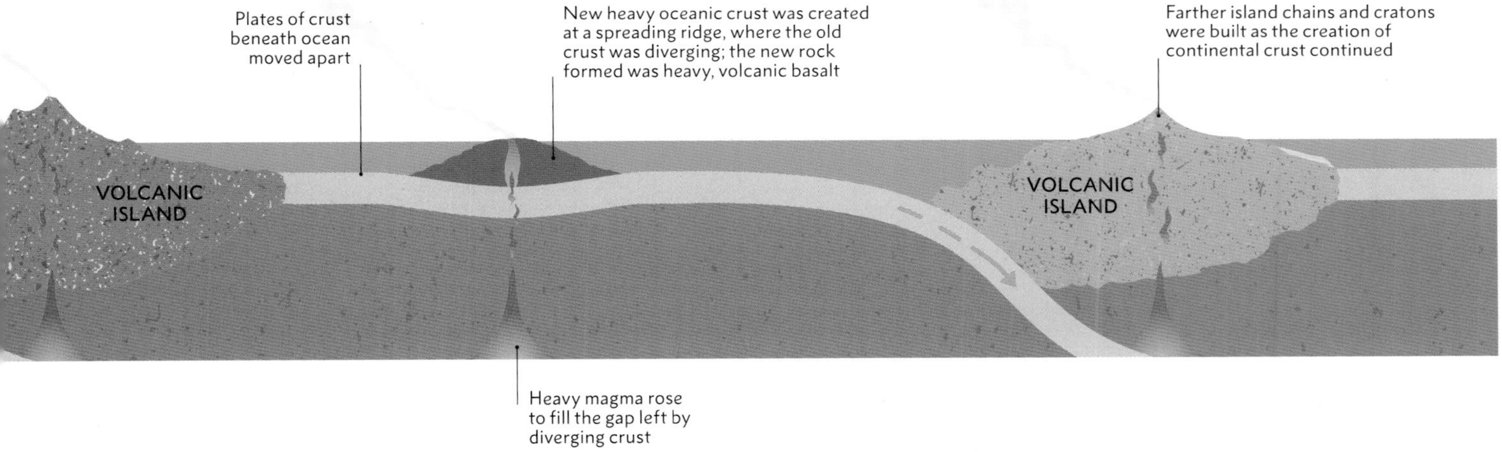
Plates of crust beneath ocean moved apart
New heavy oceanic crust was created at a spreading ridge, where the old crust was diverging; the new rock formed was heavy, volcanic basalt
Farther island chains and cratons were built as the creation of continental crust continued
VOLCANIC ISLAND
VOLCANIC ISLAND
Heavy magma rose to fill the gap left by diverging crust

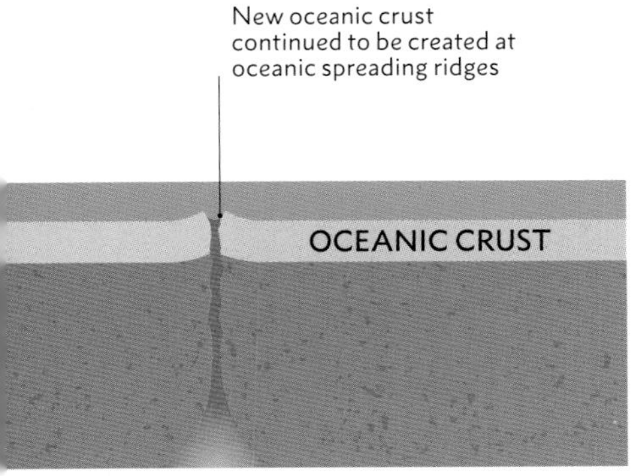
New oceanic crust continued to be created at oceanic spreading ridges
OCEANIC CRUST

> THE **CORES OF CONTINENTS**... MAKE UP THE STABLE LITHOSPHERE. **THEIR FORMATION... OCCURRED BILLIONS OF YEARS AGO**.

Nicholas Wigginton, *Science* editor, c.1970–

DATING EARTH

The question of Earth's age has only been resolved in the last few decades. As knowledge increased and scientific techniques were honed, estimates of the age of our planet increased from thousands of years to billions. We now know Earth to be around 4.54 billion years old.

It was not always clear if Earth had an origin at all. Ancient Greek philosophers including Aristotle believed that our planet was eternal—it has always been here and always will be. Most civilizations had their own origin stories (see pp.18–19), and before the onset of modern science, religious texts were the main sources of ideas about Earth's origins. In 1645, Irish Bishop James Usher famously used the genealogy in the Bible to calculate the date of Earth's creation as October 23, 4004 BCE.

EARLY SCIENTIFIC IDEAS

Not everyone believed the idea of a young Earth. Back in the 16th century, French thinker Bernard Palissy argued that if the erosion of rocks was caused by the gradual battering of wind and rain, then Earth must be much older than a few thousand years. French natural historian Benoît de Maillet tried to explain why marine fossils were found at high elevations by wrongly concluding that Earth's sea level must have been much higher in the past. This was long before the discovery of plate tectonics (see pp.90–91). This idea of rates of erosion was revisited by Scottish geologist James Hutton in the late 18th century as the tide of opinion began to turn toward a greater age for the planet. Hutton argued that Hadrian's Wall, despite being built by Romans in England more than 1,000 years previously, had barely eroded. Therefore, other rocks that had been significantly eroded must have been around much longer. Hutton also noted that layers of rock had not been laid down continuously, but in separate episodes of deposition, leading to "unconforming" layers that would have taken millions, not thousands of years, to form. Victorian geologist Charles Lyell agreed with Hutton, but emphasized the idea of Earth in a state of slow, perpetual change. Rates of change observed in modern times could then be used to estimate rates of change in the past.

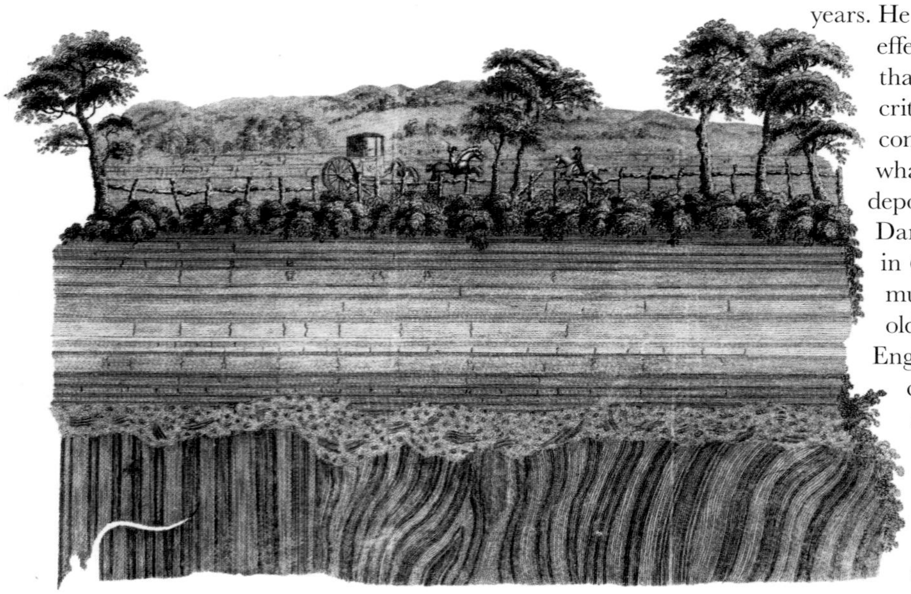

▼ Clues in the rocks
A sketch from 1787 of rock layers in Jedburgh, Scotland, shows horizontal layers of rock that sit on top of vertical layers, each from different periods. This unconformity served as geologist James Hutton's evidence that Earth was very ancient.

THE DEBATE INTENSIFIES

By the middle of the 19th century, attempts to determine Earth's age had picked up steam, and scientists from many different disciplines made estimates. In 1862, physicist William Thompson (later Lord Kelvin), imagined our infant planet as a ball of molten rock and calculated how long it would have taken to cool to its present temperature, concluding 20–400 million years. He did not take into account the effect of radioactivity, a phenomenon that had yet to be discovered. Lyell criticized his ideas for being too conservative and inconsistent with what he had learned about the deposition of rock layers. Charles Darwin joined the debate, stating in *On the Origin of Species* that Earth must be at least 300 million years old in order for chalk deposits in England to have eroded to their current state. Charles's son, astronomer George Darwin, believed that the moon was formed from Earth. If so, he reasoned it would have taken at least 56 million years for the moon to reach its current distance. By the 20th century, the general consensus for the age of Earth had leapt from thousands of years to tens, if not hundreds, of millions of years.

▲ Dangerous beliefs
Bernard Palissy (1509–1589) worked as a potter for most of his life, but he was also a scientist. He put forward his then-radical belief that fossils were prehistoric animals, and not from the biblical flood. The Catholic Church ultimately imprisoned him.

AGE OF RADIOACTIVITY

The discovery of radioactivity by Marie Curie in 1903 would enable scientists to find concrete evidence for Earth's age. The decay of radioactive atoms in rocks occurs over millions of years, and the proportion of unstable atoms remaining can be measured to provide an age for the rock (see pp.88–89). Over the next 30 years, many scientists used radiometric dating methods to analyze rocks from all over the world—arriving at ages between 92 million years and 3 billion years.

By the 1960s, the number of ways to use radioactivity to date rock samples started to rise. The precision of these techniques and the accuracy of the calculated ages steadily increased. We know now that Earth has been around for close to 4.54 billion years, give or take one percent. Such figures are supported by the age of meteorites that we think are slightly older than Earth.

FOSSILIZED TREES ON TOP OF A **PREHISTORIC SEA BED** 6,000FT (1,800M) **HIGH IN THE ANDES** CONVINCED CHARLES DARWIN THAT **EARTH WAS VERY OLD**

▼ History in the rocks
Rock such as this limestone on a Greek coast, with its evident long history of deposition, followed by crumpling, followed by erosion, is the sort of evidence that, in the 18–19th centuries, set the minds of pioneering geologists thinking—how much time is needed for all of this geological change?

> WITH RESPECT TO HUMAN OBSERVATION, **THIS WORLD** HAS **NEITHER A BEGINNING NOR AN END**.

James Hutton, geologist, 1726–1797

ZIRCON **CRYSTAL**

Some ancient crystals have survived 4.4 billion years on Earth. Their persistence provides an excellent opportunity to probe into our planet's history, and learn more about the origins of life and the first oceans.

The Jack Hills of Western Australia are home to the oldest material ever found on Earth. These tiny zircon crystals are each only the size of a dust mite, yet hold within them the secrets of our planet's turbulent infancy. The oldest crystals date from 4.4 BYA—100 million years after a giant impact struck Earth and created the moon—which means that Earth's solid crust, in which they formed, must be at least the same age. Zircon is a mineral that contains the element zirconium. It has a similar hardness to diamond, its more illustrious cousin—which means zircon crystals can survive erosion and other geological processes, making them an excellent record keeper of Earth's history.

Normally zircon crystals are red, but when scientists bombard them with electrons in order to study them, they take on a blue hue. Analysis of these crystals is subverting previous ideas of the conditions on early Earth. It was long thought that our planet's infancy was a hellscape, one much too fierce to support liquid water and life, but opinions are beginning to shift to an Earth that cooled relatively quickly, because the crystals needed those cool conditions to form.

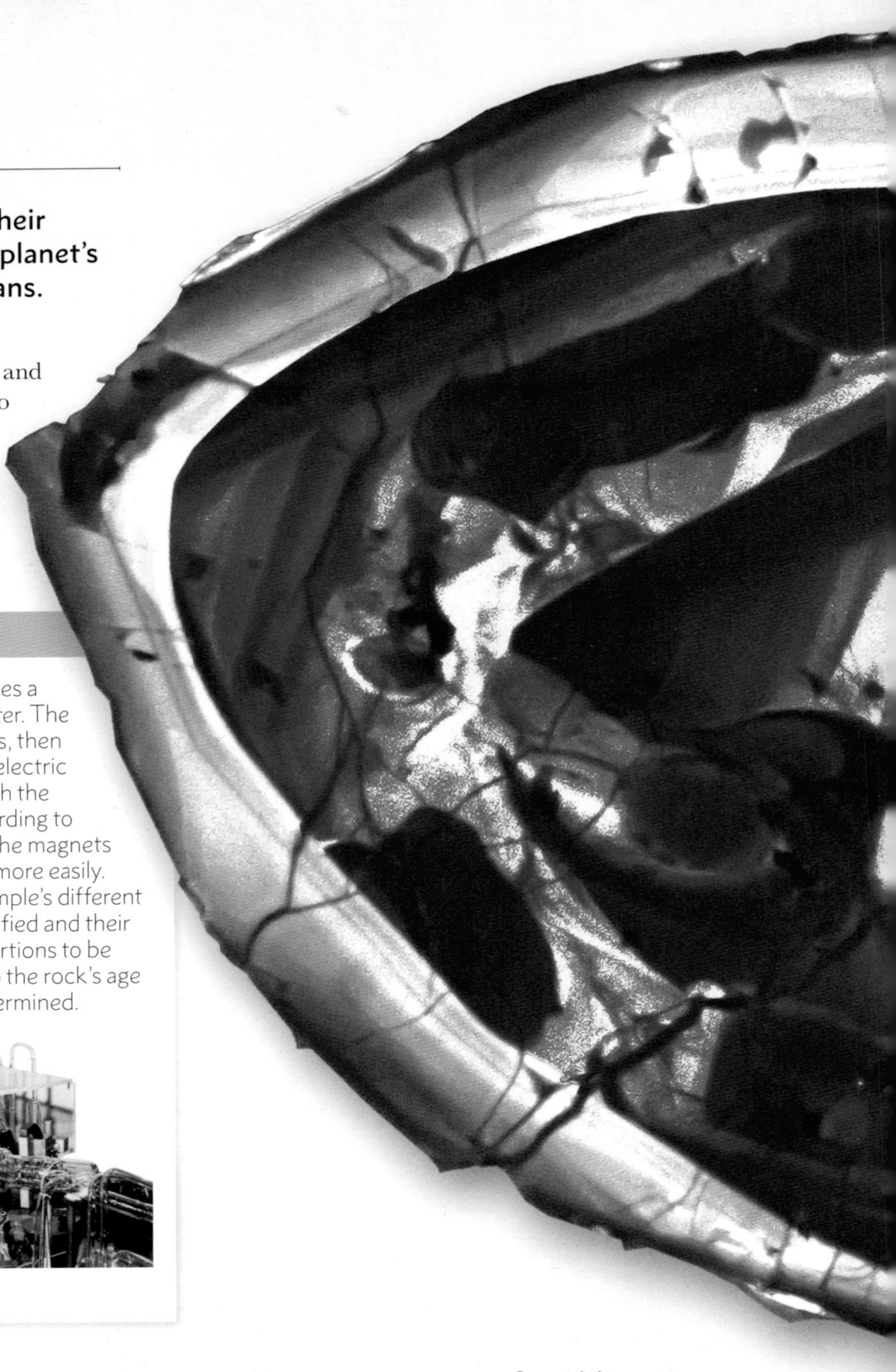

Crystal composition

Radiometric dating analysis uses a device called a mass spectrometer. The rock sample is broken into atoms, then the atoms are ionized (given an electric charge). As the ions pass through the device, magnets sort them according to their mass, because the magnets deflect lighter ions more easily. This allows the sample's different ions to be identified and their precise proportions to be measured so the rock's age can be determined.

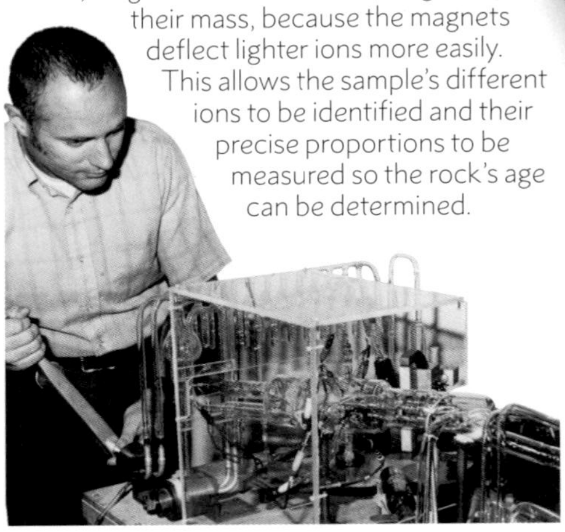

Mass spectrometer

▼ How radiometric dating works
Uranium atoms are so large and unstable that they decay—they give off radiation and change into more stable atoms—and they do this at a known rate. Measuring the ratio of uranium in rock to its final decay product (lead) tells us how much radioactive decay has occurred since the rock formed, and therefore how much time has passed.

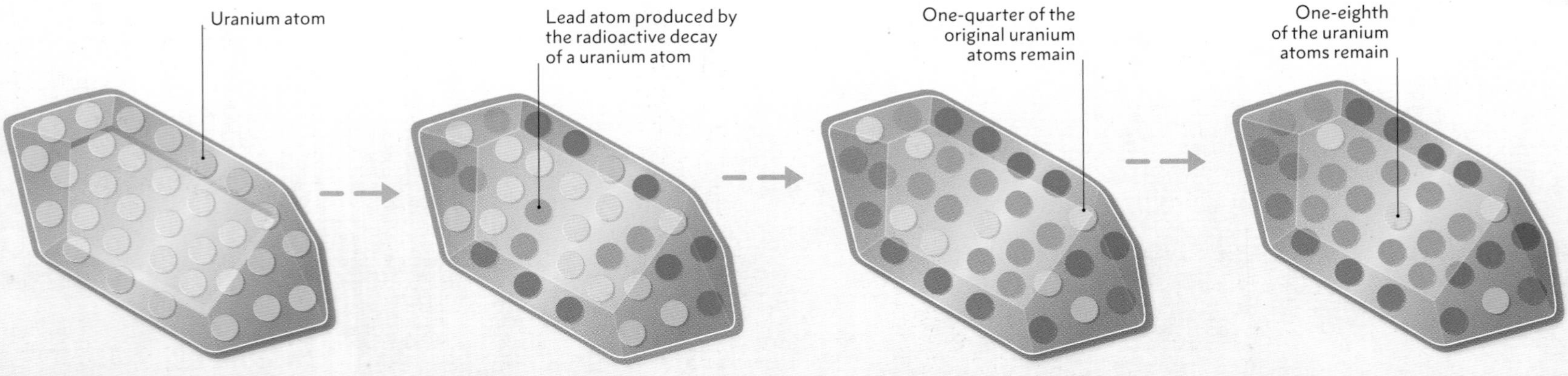

When the rock formed, the sample contained only uranium as it solidified from molten rock and crystallized.

704 million years later, the uranium atoms have decayed, giving off radiation and changing into lead atoms.

After 1.406 billion years, more uranium atoms have decayed. The more lead found in the rock, the older the sample.

Today, a geologist measures the ratio of uranium to lead remaining in the rock and dates this rock to 2.112 billion years old.

This particular zircon crystal is 4.4 billion years old
This zircon crystal is incredibly small—measuring just ⅛in (0.4mm)—and is barely visible to the naked eye

Evidence of early oceans

By comparing the ratio of oxygen isotopes found within the Jack Hills zircon crystals, scientists have concluded that oceans of liquid water may have been present on Earth as early as 4.4 billion years ago. Isotopes are versions of an atom with differing atomic weight. The ratio of oxygen-18 to oxygen-16 isotopes found in the crystals indicates the presence of liquid water.

Earth in the Archean Era, 3.5 BYA

Signs of life

Earth was previously thought to be inhospitable until 3.8 billion years ago, but isotope analysis of graphite flecks found inside zircon crystals dating back to 4.1 billion years ago suggests that life was present at this earlier time. Graphite is made of carbon, and the ratio of carbon-12 to carbon-13 isotopes in the graphite is characteristic of the ratio produced by living organisms.

Protecting the crystals

Around 200,000 zircons have been unearthed in the Jack Hills since the 1980s, and 10 percent of them are more than 3.9 billion years old. The geology of the area is so important that the Australian government has declared the region a geoheritage site, to protect it from future mining activity and preserve its scientific treasures.

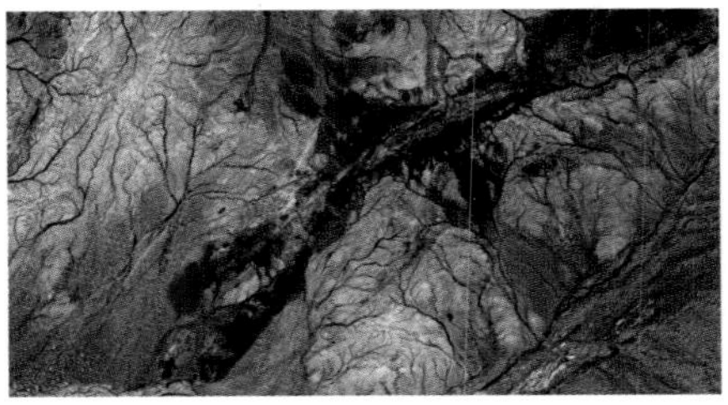

Jack Hills, Australia

CONTINENTS **DRIFT**

The map of our modern world is a familiar image, but this arrangement of continents is a relatively recent development in our planet's history. Entire continents have split and moved apart over hundreds of millions of years. This idea wasn't accepted until the late 20th century.

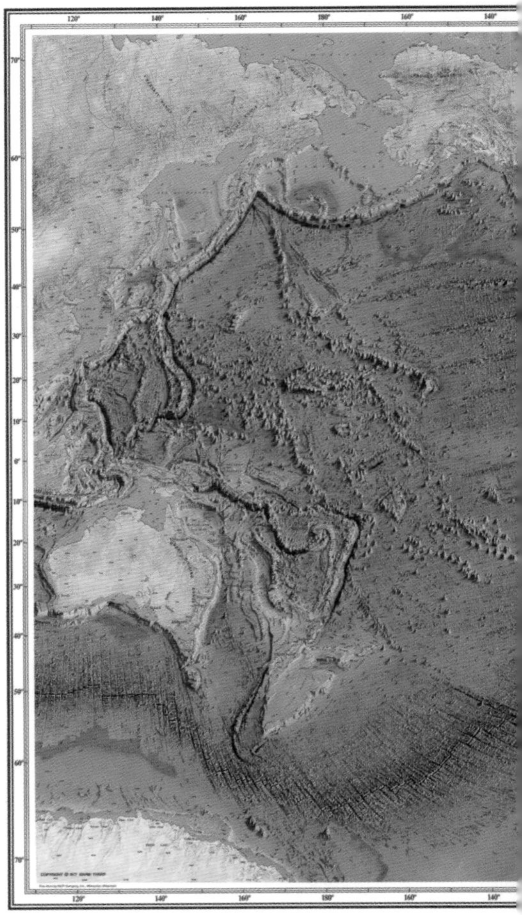

The fact that Earth's land masses have shifted over time makes sense when looking at a map of the world. Some continents appear to fit together, like puzzle pieces. However, the notion that these vast land masses could move was long considered outrageous to the scientific community. Despite scientists' reservations, the idea has been around for centuries, with the Flemish cartographer Abraham Ortelius widely credited as being the first to express such thoughts at the end of the 16th century.

BRIDGING THE GAP

In the 19th century, Antonio Snider-Pellegrini created two maps showing the ease with which the meandering coastlines of the various continents appear to slot into place to form one giant supercontinent. Further evidence that the far-flung continents had once been conjoined came from the fossil record (see pp.158–59). Scientists were beginning to discover that the fossilized remains of similar animals, and in particular plants, were cropping up in places now separated by vast oceans. This was explained away by the idea that continents were once connected via vast land bridges, which have since been eroded away or submerged deep beneath the sea.

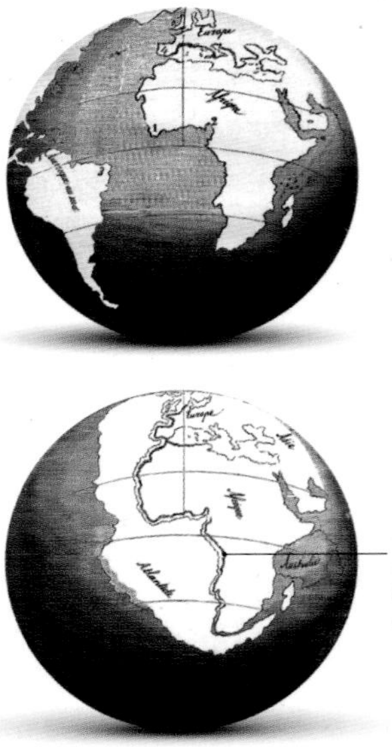
Corner of Africa appears to fit snugly with South America's coastline

▲ **First clues**
Explorers noticed that the east coast of South America and the west coast of Africa appeared to fit together. These maps were drawn by geographer Antonio Snider-Pellegrini in 1858.

▶ **Bold ideas**
German scientist Alfred Wegener (1880–1930) hoped to collect solid evidence for his continental drift theory on his fourth expedition to Greenland, but he died while collecting supplies for his camp.

Another thorny issue perplexing geologists was the origin of mountain ranges, such as the Himalayas. The leading idea in the 19th century was that the peaks were formed as wrinkles, as Earth cooled and shrank. If that were true, mountain chains should be spread evenly across the planet's surface—and that is not the case.

Ideas continued to develop at the turn of the 20th century. George Darwin, Charles's son, proposed that the moon had once formed part of Earth and its absence accounted for the vast, landless Pacific Ocean. His theory suggested that the continents separated as the moon broke away, explaining their present positions. Another theory was that Earth was expanding. As the planet got bigger, its land masses were forced to spread out. Both of these ideas gradually lost support as the precise physical mechanisms behind them could not be found.

A NEW IDEA

In 1912, German scientist Alfred Wegener argued in favor of continental drift. He not only showed matching fossil evidence on disparate continents, but also concluded that the types of rock and other geological structures were similar too. He decided that this idea could not coexist with the theory of now-submerged land bridges, so he suggested that the continents themselves had moved apart. This offered a potential solution to the mountain conundrum. If continents were free to roam, then over time some could collide. If India had smashed into mainland Asia, the Himalayas would be the result of continental crumpling.

Wegener published his findings the same year, suggesting that Earth's land masses plowed through the sea over time. His work met with a lukewarm reception from the scientific community, in part because

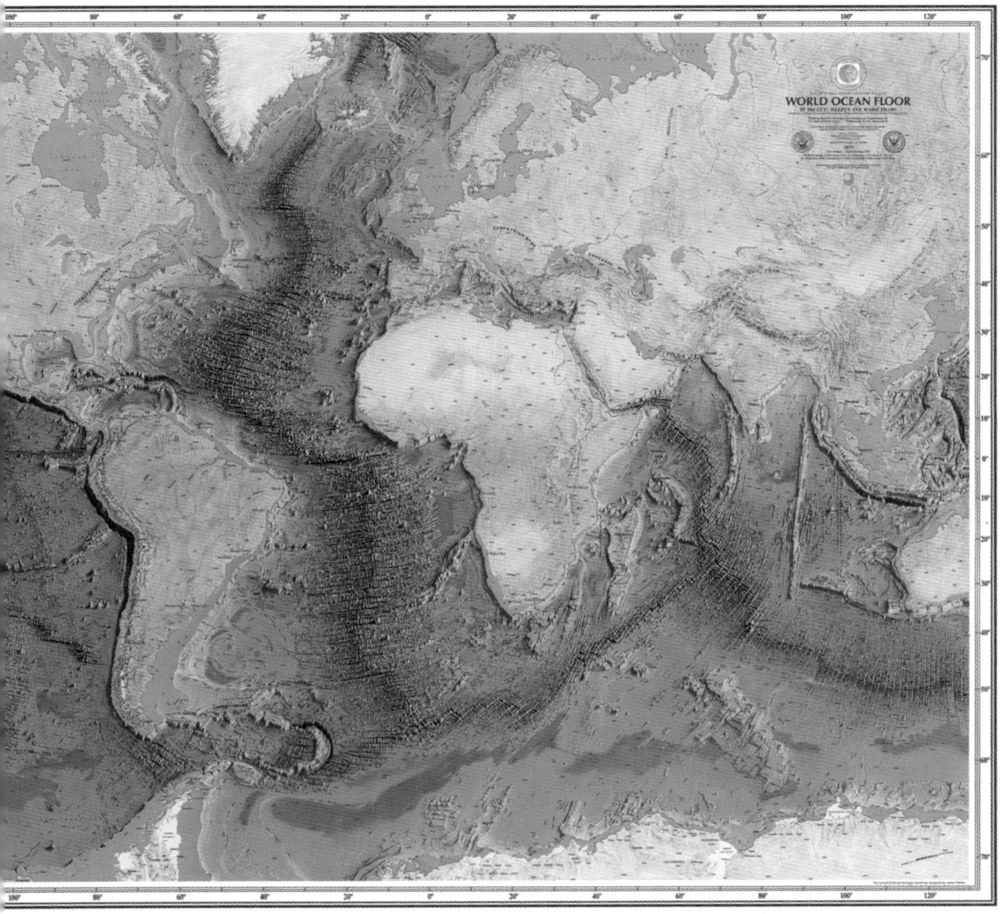

◀ Continent scars
In 1977, this map, the result of a lifetime's work by oceanographers and cartographers Marie Tharp and Bruce Heezen, revealed the ocean floor in new detail, providing conclusive evidence for plate tectonics.

he could not provide a plausible reason as to why the continents would drift. He incorrectly calculated the rate of their movement and overestimated by a factor of 100 compared to today's accepted value, which did not help his cause.

Wegener's academic background was also a hindrance. Given his training as an astronomer and meteorologist, many in the geological community suggested he did not have the expertise required to be taken seriously. He was not without some support, however—British geologist Arthur Holmes backed his ideas, arguing as early as 1931 that Earth's mantle contained currents that helped move parts of the crust.

CLUES FROM THE SEAFLOOR

It was not until the 1950s that evidence emerged to turn the tide of opinion in Wegener's favor. In 1953, analysis of rocks in India suggested that it was once in the Southern Hemisphere, bolstering Wegener's mountain formation argument. Around the same time, a huge underwater mountain range—the Mid-Ocean Ridge—was discovered. It is Earth's longest mountain range and extends through all of its oceans. The geologists of the day now had to explain the presence of this ridge, too. It would fall to former US Navy officer turned geologist Harry Hess to tie all these ideas together. Having used sonar to map the ocean during World War II, by the early 1960s, Hess's research led him to propose that the continents did indeed drift apart thanks to a process called "seafloor spreading." In 1958, Australian geologist Samuel Carey had suggested that Earth's surface, its crust, was constructed from plates. Hess ran with the idea, arguing that the planet's crust ruptures at plate boundaries, allowing magma to well up from the mantle. As this material solidifies, it forms a ridge, pushing the existing seafloor apart. So it is not that the continents plow through the ocean crust as Wegener had suggested, but rather that the seafloor itself is growing, carrying away the continents, which are part of moving tectonic plates (see pp.92–93).

IT TOOK **OVER 300 YEARS** FOR THE IDEA OF **CONTINENTAL DRIFT** FINALLY TO **BE ACCEPTED AS FACT**

Today, these ideas are brought together as the theory of plate tectonics. It is supported by observations of Earth from space using geodesy, which maps small changes in Earth's gravity to locate concentrations of mass. Studies of the polarity of Earth's magnetic field, which is known to have flipped frequently over time (north becoming south, and vice versa), also lends weight. The reversals leave stripes of magnetic rock on the ocean floor (see pp.94–95) that allow us to date the bands and show how fast the seafloor is spreading.

Plate tectonics was not widely accepted until the 1970s, when maps of the ocean floor, such as that made by Marie Tharp and Bruce Heezen, left no doubt that the seafloor was spreading, accounting for continental drift.

> “I ONCE ASKED ONE OF MY LECTURERS… I WAS TOLD, SNEERINGLY, THAT **IF I COULD PROVE** THERE WAS A **FORCE THAT COULD MOVE CONTINENTS**, THEN HE MIGHT THINK ABOUT IT. **THE IDEA WAS MOONSHINE**.”
>
> **David Attenborough**, natural history broadcaster, 1926–

HOW EARTH'S **CRUST MOVES**

The surface of our planet is sculpted by extremely slow convection currents in the mantle layer below. Earth's system of plate tectonics sets it apart from the other rocky planets in the solar system, since its surface is constantly changing and is alive with geological activity.

Earth's surface layer, the crust, is formed of seven major tectonic plates—African, Antarctic, Eurasian, North American, South American, Pacific, and Indo-Australian—along with several smaller ones. These solid plates float on a semisolid layer called the mantle. Plates move incredibly slowly, typically at about the rate that fingernails or human hair grow. Since Earth's layers stabilized 4 BYA, these plates have been constantly moving.

▼ Volcanic eruption
The Eyjafjallajökull volcano in Iceland erupts molten magma, along with black clouds of ash that fall on the ground as added layers atop Earth's crust.

TECTONIC PHENOMENA

Where plates meet, a range of tectonic activity may occur, but exactly what depends on the crust material and the direction of movement. There are three main types of plate boundary: transform boundaries, where plates slide or grind past one another; divergent boundaries, where they slide apart, allowing magma to cool into new crust; and convergent boundaries, where two plates collide head on. Parts of the crust sink and melt at subduction zones, but new crust is made elsewhere by volcanoes and at mid-ocean ridges, where oceanic crust diverges.

Earthquakes, sudden movements of Earth's crust, occur at plate boundaries. At divergent and transform boundaries, they tend to be shallow, whereas collisions at convergent boundaries cause the deepest earthquakes.

Where two plates collide, they can push up continental crust to form a mountain range, such as the Himalayas. Those particular mountains were created when the Indian plate slammed into the Eurasian plate around 50 million years ago.

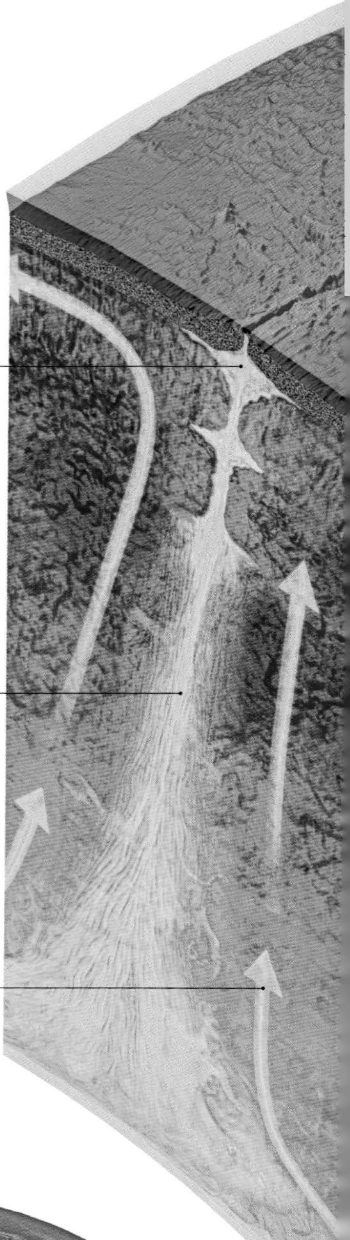
Underwater volcanoes spew molten lava, which cools into new oceanic crust
Convection current causes an upwelling of molten magma
Heat in the core causes convection currents in the mantle that drive the movement of tectonic plates

EARTH'S SURFACE MOVES

Convection currents in the mantle are generated by heat in the core that filters into the mantle. Although the mantle is almost solid, it flows slowly, tugging at the base of the crust and moving the plates. The crust is of two kinds: oceanic crust, which is made of dense rock rich in magnesium and iron, and continental crust, made of rock with lighter elements including aluminum. Where the edge of a plate is made of oceanic crust, its greater density makes it subduct, or slip underneath, the lighter crust. It then sinks deep into the hot mantle, causing an upwelling of molten magma that breaks the surface of the crust as a volcano.

► Dynamic surface
Earth's crust is constantly changing as the plates move, buoyed by currents in the mantle below. Depending on how the plates interact, earthquakes can occur and volcanoes and mountain chains can form.

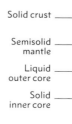

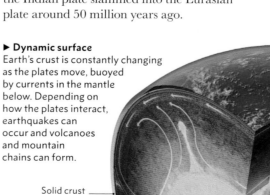

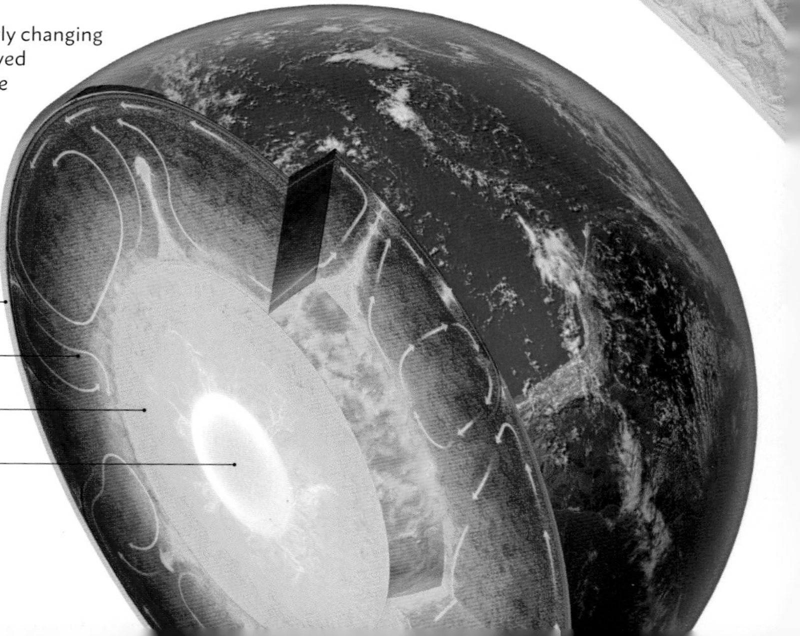
Solid crust
Semisolid mantle
Liquid outer core
Solid inner core

"CONTINENTAL DRIFT… EARTHQUAKES, VOLCANICITY… ARE **UNDOUBTEDLY CONNECTED** ON A GRAND SCALE."

Alfred Wegener, geologist and meteorologist, 1880–1930

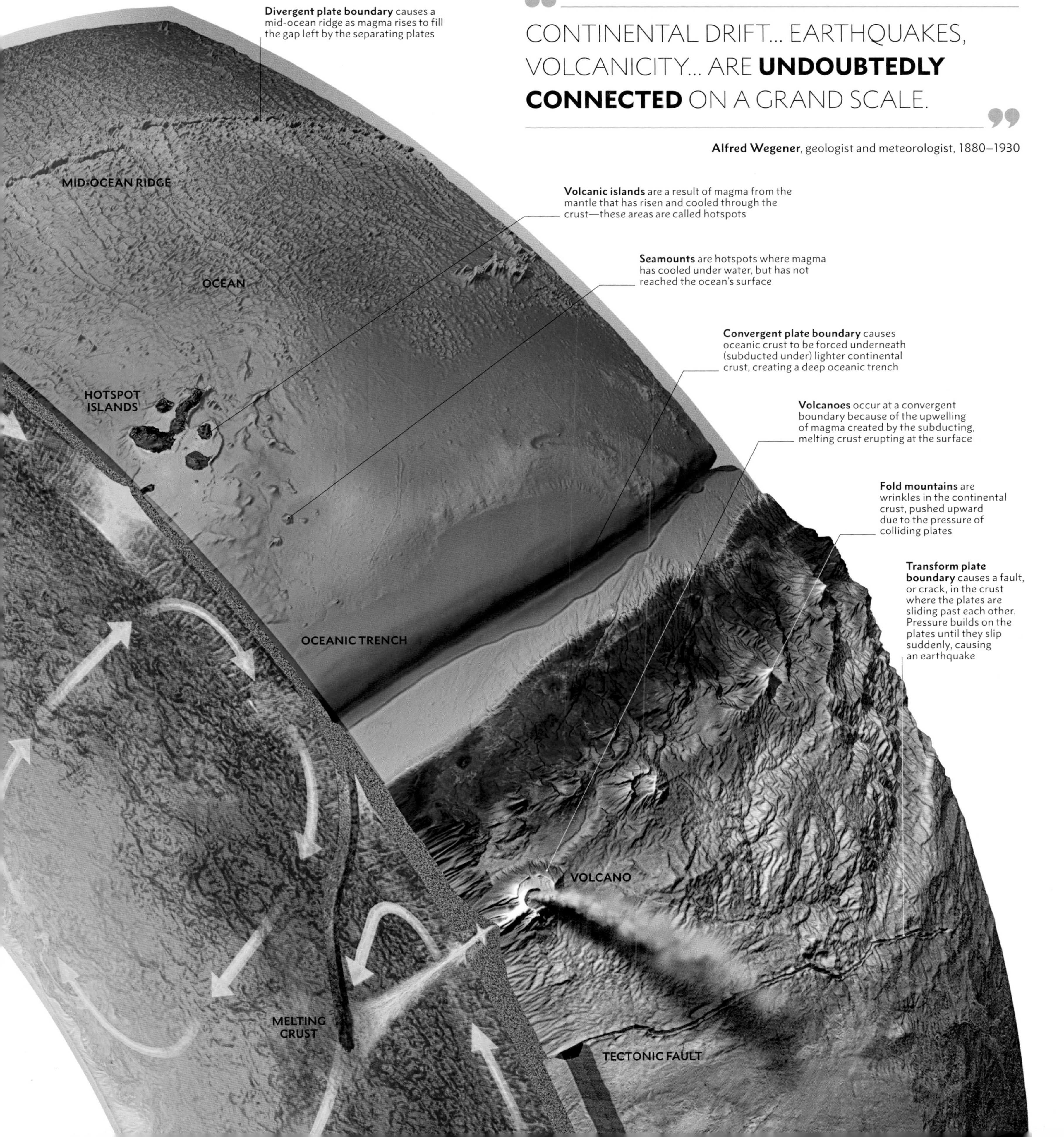
Divergent plate boundary causes a mid-ocean ridge as magma rises to fill the gap left by the separating plates
MID-OCEAN RIDGE
OCEAN
HOTSPOT ISLANDS
OCEANIC TRENCH
MELTING CRUST
VOLCANO
TECTONIC FAULT
Volcanic islands are a result of magma from the mantle that has risen and cooled through the crust—these areas are called hotspots
Seamounts are hotspots where magma has cooled under water, but has not reached the ocean's surface
Convergent plate boundary causes oceanic crust to be forced underneath (subducted under) lighter continental crust, creating a deep oceanic trench
Volcanoes occur at a convergent boundary because of the upwelling of magma created by the subducting, melting crust erupting at the surface
Fold mountains are wrinkles in the continental crust, pushed upward due to the pressure of colliding plates
Transform plate boundary causes a fault, or crack, in the crust where the plates are sliding past each other. Pressure builds on the plates until they slip suddenly, causing an earthquake

OCEAN **FLOOR**

In many ways, the ocean floor is a guide to Earth's history—studying it helps us decipher the mysteries of our planet's past. Exploring it has even given scientists clues about how life originated. Mapping the ocean floor reveals a diverse, active landscape full of tectonic phenomena.

The depths of the ocean are cold, dark, and incredibly hostile. At its deepest point, there are 8.4 tons of water pressing down on every square inch (1.2 metric tons per square centimeter). Such extremes mean oceanographers resort to imaging the seabed using sonar from the surface. It is easier for us to get images from Mars than map parts of our own seabed.

Despite its inaccessibility, the ocean floor holds clues that are vital in understanding the development of Earth's crust, and also life. Deep ocean exploration is sharpening our ideas on plate tectonics (see pp.90–91). The chemically-rich material and heat generated by underwater volcanoes found on the ocean floor have led biologists to believe that these areas are where the first life-forms appeared (see pp.106–07).

The deepest places of the ocean floor are where two oceanic plates meet and form an underwater valley—one plate slips underneath (subducts beneath) the other, creating a V-shaped trench. The deepest ocean trench is the Mariana Trench in the Pacific Ocean: its deepest point is at 36,070ft (10,994m) below sea level. It could accommodate Mount Everest with about 6,560ft (2,000m) of water to spare.

The Puerto Rico Trench in the Atlantic Ocean has depths greater than 27,560ft (8,400m). The underwater boundary between the Caribbean and North American plates, where the Puerto Rico Trench is found, is a particularly active area of the ocean floor. Its unique plate boundary and unusual phenomena provide a rich resource for scientific research: oceanographers, biologists, seismologists (who study earthquakes), and bathymetrists (who study the underwater terrain of lakes and oceans) all work here, hoping to unlock the secrets of the ocean floor.

Caribbean plate is sliding toward the east
Muertos Trough
▲ WEST
CARIBBEAN PLATE
◀ SOUTH
ANTILLES ARC
The Antilles islands have been formed due to both folding and volcanism at this plate boundary

How sonar surveys work

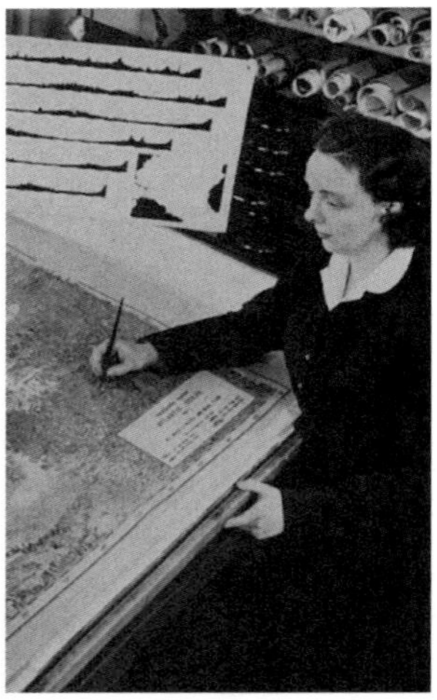

Marie Tharp, oceanographer

Multibeam sonar records the time taken for sound to bounce back from the seafloor in order to measure ocean depth. Oceanographers can use this data to create a colored map of the seafloor, showing its terrain. Side-scan sonar is more accurate in that the intensity of its echoes can reveal whether the ocean floor is rocky (strong) or sandy (weak). Marie Tharp and Bruce Heezen mapped Earth's ocean floor in the 1950s (see pp.90–91).

▶ Clues on the ocean floor
Magma from the mantle breaks through the crust and forces tectonic plates apart (see pp.92–93). As the magma cools to form new crust, minerals in the magma orient themselves in line with Earth's magnetic field. For reasons unknown, Earth's north-south polarity reverses from time to time, and over millions of years these reversals are etched into the ocean floor as a series of stripes.

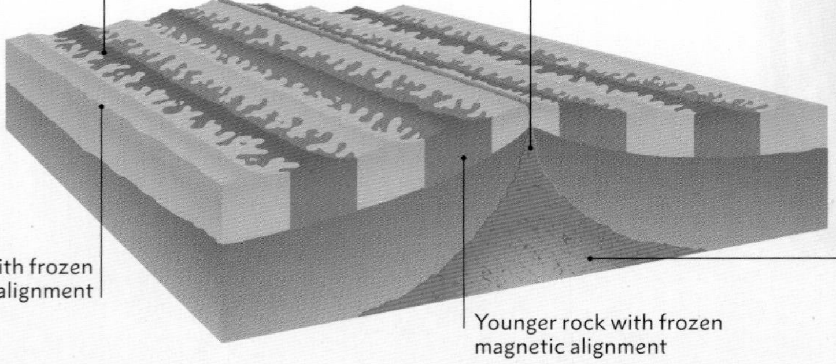
Magnetic field reversals create stripes
Cooling material forces plates apart
Older rock with frozen magnetic alignment
Younger rock with frozen magnetic alignment
Molten material from mantle breaks through crust

▼ Exploring the ocean floor
A bathymetry map generated by sonar reveals the underwater terrain of the northeastern corner of the Caribbean plate. Differences in relief are represented by color, highlighting deep sea trenches in purple.

KEY

Above sea level	
Sea level - 0m	0 ft
2,000m	6,562ft
3,000m	9,843ft
4,000m	13,123ft
5,000m	16,404ft
6,000m	19,685ft
7,000m	22,966ft

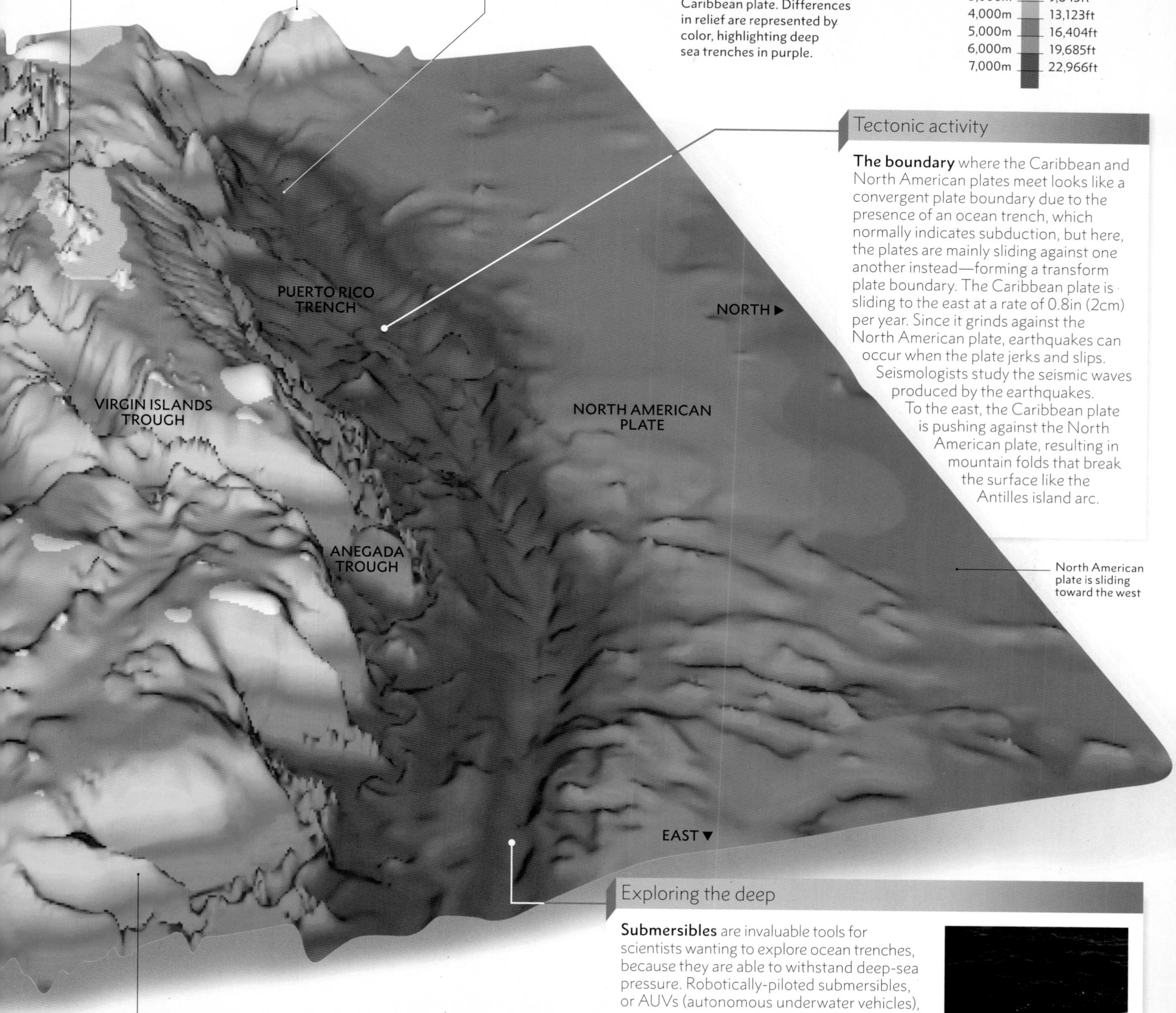

The island of Puerto Rico rises above the ocean surface
Seamount emerging from ocean surface becomes one of the southernmost islands of the Bahamas
The most negative gravity anomaly on Earth is found in the Puerto Rico Trench; this indicates an active downward force
PUERTO RICO TRENCH
NORTH ▶
VIRGIN ISLANDS TROUGH
NORTH AMERICAN PLATE
ANEGADA TROUGH
North American plate is sliding toward the west
EAST ▼
Compression of the Caribbean plate's crust from the slightly subducting North American plate creates folds

Tectonic activity

The boundary where the Caribbean and North American plates meet looks like a convergent plate boundary due to the presence of an ocean trench, which normally indicates subduction, but here, the plates are mainly sliding against one another instead—forming a transform plate boundary. The Caribbean plate is sliding to the east at a rate of 0.8in (2cm) per year. Since it grinds against the North American plate, earthquakes can occur when the plate jerks and slips. Seismologists study the seismic waves produced by the earthquakes.

To the east, the Caribbean plate is pushing against the North American plate, resulting in mountain folds that break the surface like the Antilles island arc.

Exploring the deep

Submersibles are invaluable tools for scientists wanting to explore ocean trenches, because they are able to withstand deep-sea pressure. Robotically-piloted submersibles, or AUVs (autonomous underwater vehicles), are pre-programmed with instructions on where to explore and what to measure. Some submersibles also allow scientists to visit the ocean floor in order to examine and collect samples of both rocks and life-forms for analysis at the surface.

Example of an AUV

5

LIFE **EMERGES**

Earth has a privileged position in the solar system—in a band that's not too cold and not too hot to support liquid water. It is in this vital ingredient that life first emerges. And through a process of natural selection, life evolves from simple bacteria to complex vertebrates, shaping our planet and filling it with astounding diversity.

GOLDILOCKS CONDITIONS

Living organisms emerged from nonliving complex chemicals. Life-forms could metabolize, meaning they were able to extract energy from their surroundings. They could also copy themselves and adapt to their environment—through the process of natural selection.

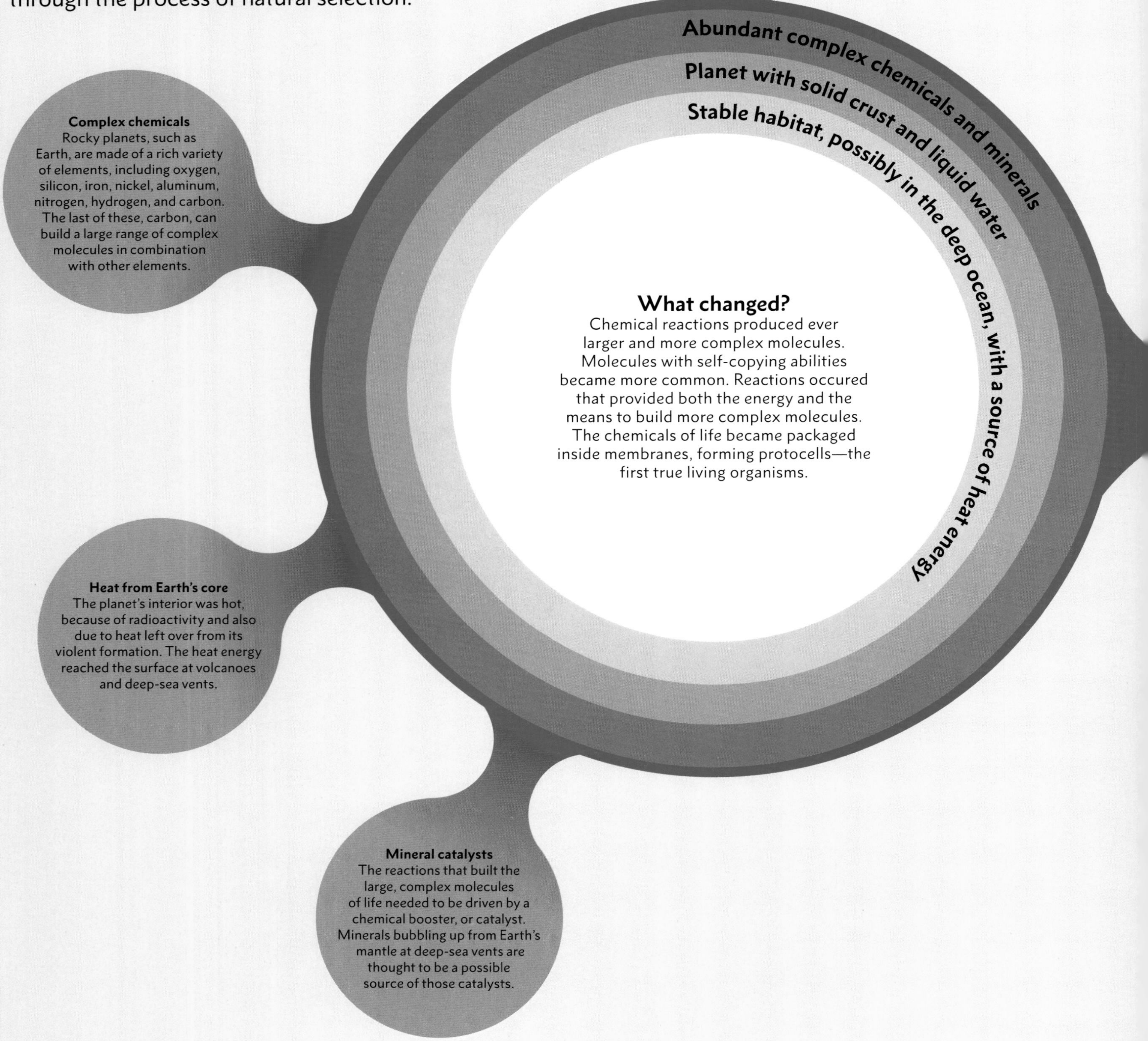
Abundant complex chemicals and minerals
Planet with solid crust and liquid water
Stable habitat, possibly in the deep ocean, with a source of heat energy
What changed?
Chemical reactions produced ever larger and more complex molecules. Molecules with self-copying abilities became more common. Reactions occured that provided both the energy and the means to build more complex molecules. The chemicals of life became packaged inside membranes, forming protocells—the first true living organisms.
Complex chemicals
Rocky planets, such as Earth, are made of a rich variety of elements, including oxygen, silicon, iron, nickel, aluminum, nitrogen, hydrogen, and carbon. The last of these, carbon, can build a large range of complex molecules in combination with other elements.
Heat from Earth's core
The planet's interior was hot, because of radioactivity and also due to heat left over from its violent formation. The heat energy reached the surface at volcanoes and deep-sea vents.
Mineral catalysts
The reactions that built the large, complex molecules of life needed to be driven by a chemical booster, or catalyst. Minerals bubbling up from Earth's mantle at deep-sea vents are thought to be a possible source of those catalysts.

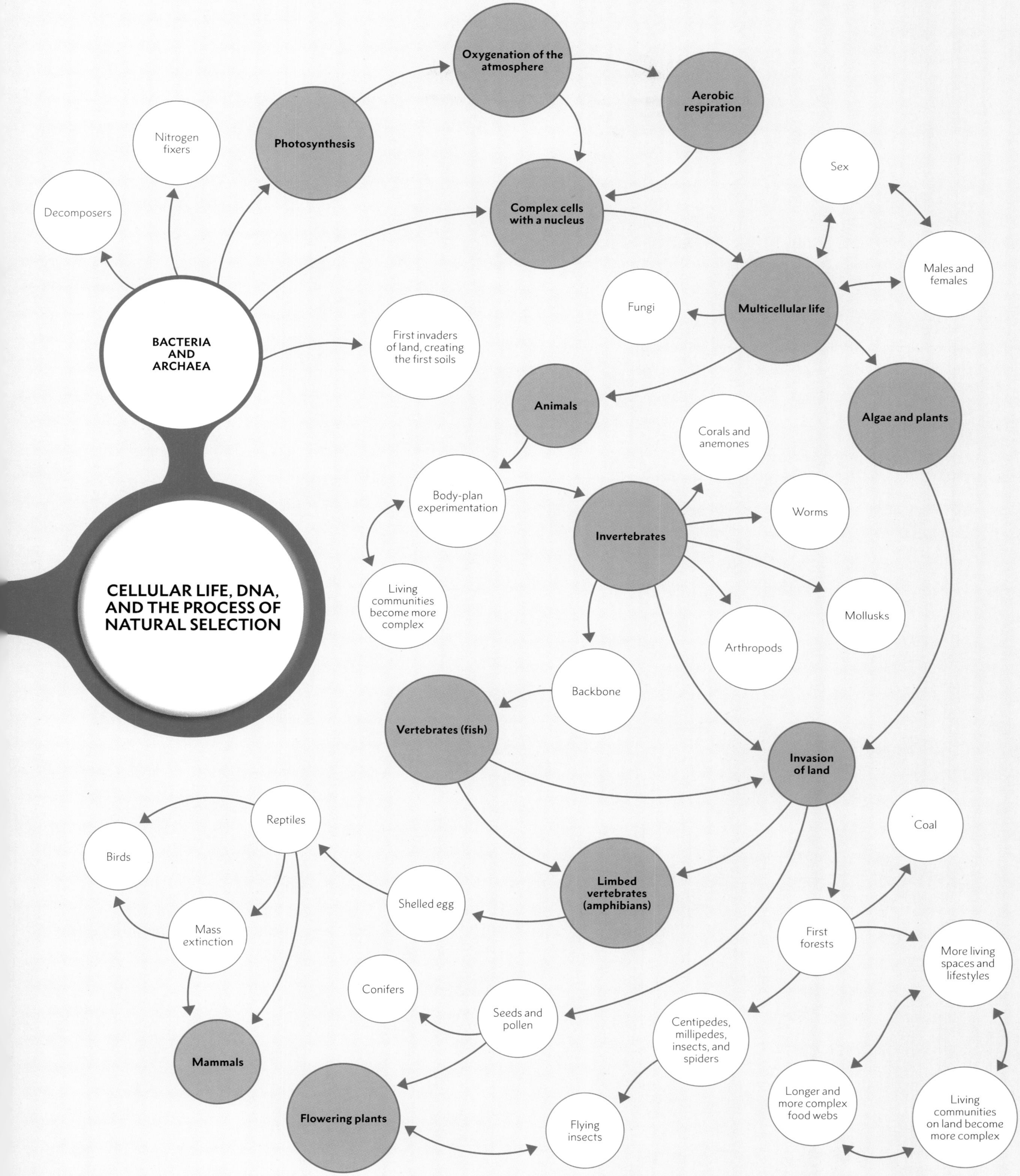

CELLULAR LIFE, DNA, AND THE PROCESS OF NATURAL SELECTION
BACTERIA AND ARCHAEA
Decomposers
Nitrogen fixers
Photosynthesis
Oxygenation of the atmosphere
Aerobic respiration
Complex cells with a nucleus
First invaders of land, creating the first soils
Sex
Males and females
Fungi
Multicellular life
Algae and plants
Animals
Corals and anemones
Body-plan experimentation
Living communities become more complex
Invertebrates
Worms
Mollusks
Arthropods
Backbone
Vertebrates (fish)
Invasion of land
Coal
Reptiles
Birds
Shelled egg
Limbed vertebrates (amphibians)
Mass extinction
First forests
More living spaces and lifestyles
Conifers
Seeds and pollen
Centipedes, millipedes, insects, and spiders
Mammals
Flowering plants
Flying insects
Longer and more complex food webs
Living communities on land become more complex

STORY OF **LIFE**

Life is over 4 billion years old, first emerging when Earth was only one-tenth of its present age. But even at the very beginning, life, although microscopic, was already the most complex thing in the known universe.

The Earth's violent birth left a planet where life was not only possible, but perhaps inevitable. As the land cooled and permanent oceans formed, the first protocells emerged—probably deep under water around chemically-rich fissures in the young ocean floor. Within a few million years, these protocells had become microbes—and for billions of years after that, the world belonged only to them. They evolved ways of getting energy either from sunlight or by eating other microbes, laying the foundations for the rest of life's diversity.

The biggest, most complex of life-forms—multicellular life—evolved only in the last billion years of Earth's history. These are the organisms that evolved into the familiar plants and animals of recent times. It was then that life could emerge from the microscopic and fill the oceans and land with greenery and fast-moving creatures.

1 ORIGIN OF LIFE

A carbon trace in Australian rocks 4.1 billion years old could be life's oldest "signature." DNA evidence (from living organisms) leads to a slightly earlier estimate of life's beginnings and predicts that all organisms alive today can trace their ancestry back to a hypothetical microbe called LUCA—our Last Universal Common Ancestor.

EARTH

MICROBES

First permanent oceans possibly form, 4.4 BYA, providing the first habitat for life.

Bacteria and archaea split from their common ancestor, LUCA, 4.2 BYA, according to DNA evidence from organisms alive today.

4 BYA

The Late Heavy Bombardment—a peak in space impacts 4.1–3.9 BYA—possibly strips away the atmosphere and kills all early life.

A type of archaea emerges, 3.8 BYA, according to modern DNA evidence, that goes on to form the nucleus of complex cells.

Stromatolites, mounds formed 3.55 BYA by bacterial colonies, provide the earliest fossil evidence of life.

Fossil stromatolite

Bacteria invade land 3.18 BYA, according to DNA estimates of the date of origin of landliving bacteria.

3 BYA

Soils rich in organic matter, 2.9 BYA, provide evidence of early life on land.

Complex cells with a nucleus evolve 2.73 BYA according to DNA evidence.

Traces of an oxygenated habitat, 2.7 BYA, indicate that photosynthetic organisms have started to fill the air with oxygen.

Fossil soil-surface bacteria are the first trace of life on land, 2.6 BYA.

The Great Oxygenation completes the flooding of Earth's atmosphere with oxygen, 2.4 BYA.

Stromatolites—colonies of bacteria—are still forming on today's Australian coastline.

2 COMPLEX CELLS

Eukaryotes have complex cells with a nucleus and include plants, animals, fungi, and many microbes. Traces of steroid-like substances – unique to eukaryotes—have been found in rocks 2.4 billion years old, but direct evidence comes from the fossil *Diskagma*—a possible fungus—from 2.2 BYA.

2 BYA

Mitochondria—the energy-producing factories of complex cells—evolve, 2 BYA.

Eukaryotes (organisms with complex cells) split into plantlike and animal-like groups, 1.6 BYA.

Chloroplasts appear 1.5 BYA, allowing complex cells to collect energy from sunlight.

1 BYA

3 MULTICELLULAR LIFE

The oldest fossil of a multicellular organism is 1.2 billion years old and belongs to a seaweed, *Bangiomorpha*. The fossil is complete with a possible stalklike "hold-fast" and reproductive organs. This is also the earliest complex organism (eukaryote) that can be attributed to a group still alive today—the red algae.

PLANTS AND GREEN ALGAE

Plants appear 934 MYA, according to modern DNA evidence.

ANIMALS

Sponges, the first animals, evolve 750 MYA, DNA evidence suggests.

Rifting of supercontinents, 650 MYA, creates Iapetus Ocean and may have triggered Ediacaran and Cambrian explosions of evolution.

The earliest known animal embryos and cnidarians (relatives of jellyfish and anemones) are fossilized, 635 MYA.

"Experimental" Ediacaran animals, including *Charnia* (above), appear, 550 MYA

“

THERE ARE **4 MILLION** DIFFERENT KINDS OF **ANIMALS AND PLANTS** IN THE WORLD—4 MILLION DIFFERENT **SOLUTIONS** TO THE **PROBLEMS** OF **STAYING ALIVE**.

”

David Attenborough, natural history broadcaster, 1926–
Life on Earth, 1979

600 MYA

Ediacaran fossil, possibly a jellyfish

Animal trackways show evidence from 530 MYA of unidentified arthopods on land; oldest known fish are fossilized: *Myllokunmingia* and *Haikouichthys*.

Bonelike tissue is fossilized 510 MYA: dentine in a fish.

Experimental animal body plans fossilized 505 MYA in Canada's Burgess Shale.

500 MYA

DNA evidence suggests insects evolve 483 MYA.

Spores of land plants, fossilized 470 MYA, prove mosslike plants have invaded land.

Jaws first preserved as a fossil 436 MYA, in a placoderm fish.

Air breathing and internal fertilization—advances vital for animals living on land—are shown in a fossil millipede, 428 MYA.

Ground cover on land was provided by the first plants with strong transport tissues 420 MYA, such as *Cooksonia*. Towering above them was the enigmatic, funguslike *Prototaxites*.

400 MYA

Fossil trees are preserved 380 MYA and provide evidence of forests.

Fossil tetrapod (limbed vertebrate) shows that vertebrates are invading land 375 MYA.

Fossil seeds, 372 MYA, prove plants have become free of watery habitats and can complete their life-cycle on dry land.

Fossil reptile, 318 MYA, shows some vertebrates were breaking free of watery habitats.

Mammal-like reptile, or synapsid, is fossilized 312 MYA.

Fossil cones show that plants 305 MYA have evolved structures that protect their seeds.

300 MYA

Fossil cynodont, 260 MYA, shows diverse teeth, a muscular jaw structure, and a large braincase—all features that will characterize its descendants: mammals.

The Great Dying kills all trilobites and more than 70 percent of known animal and plant species at the end of the Permian period, 250 MYA.

Earliest fossil dinosaur, 231 MYA, heralds the age of giant reptiles.

Mammals, such as *Adelobasileus*, evolve from their reptilian cynodont ancestors 225 MYA.

Milk production evolves in mammals 210 MYA.

200 MYA

Seed-bearing plants diversify 190 MYA.

Trace of fossil fur proves mammals have evolved fur 165 MYA.

Fossil *Archaeopteryx* shows that birds have developed flight 150 MYA.

Flowering plants, including *Archaefructus* and *Montsechia*, appear 120 MYA.

100 MYA

Asteroid impact, 65 MYA, causes extinction of dinosaurs, pterosaurs, and large marine reptiles, allowing the explosive evolution of large mammals and birds.

Primates first appear in the fossil record 56 MYA.

Messel lake pit in Germany preserves a time capsule of mammal evolution, including early primate *Darwinius*, 47 MYA.

Apes evolve 25 MYA.

0 MYA

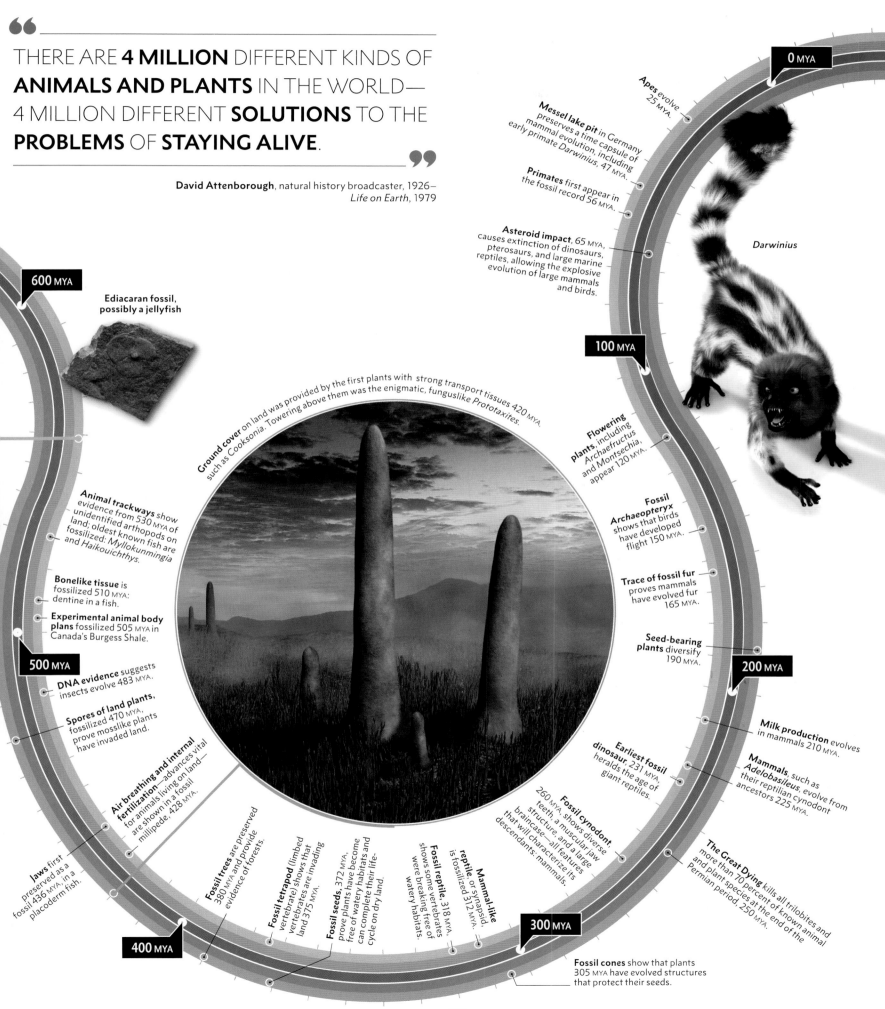

Darwinius

LIFE'S **INGREDIENTS** FORM

Earth's crust is made of dozens of chemical elements, but only some—including carbon, hydrogen, oxygen, and nitrogen—are the stuff of living things. Their atoms lock together into complex molecules and it was this kind of chemical assembly that precipitated the origin of life.

Earth has an iron core surrounded by mostly silicon-based rocks. Carbon is comparatively scarce, but all known life is carbon-based. Both silicon and carbon atoms bond prolifically with others, but while silicon's affinity is mainly with oxygen (making up the silicon dioxide that dominates Earth's rocks), carbon is versatile. It bonds with other elements, such as hydrogen, nitrogen, and phosphorus.

Complex life needs complex molecules. Earth—with its rocks still cooling in the wake of its violent birth, and liquid water condensing into the first oceans—provided just the right conditions for them to form.

Earth's first atmosphere was thick with unbreathable gases, such as carbon dioxide, hydrogen, nitrogen, and water vapor—but these were sources of life's elements. In a world without oxygen gas to react with it, hydrogen joined to other elements, making methane (CH_4) and ammonia (NH_3). In 1953, American chemists Stanley Miller and Harold Urey simulated early Earth in the lab with electrical sparks to imitate lightning. They showed that with enough heat and energy, the chemicals in Earth's atmosphere could make simple organic molecules—life-giving, carbon-based chemicals.

EVEN BIGGER MOLECULES

But life needed more—proteins, which are long chains of amino acids, and DNA. Today, pools rich with protein would be cleared by hungry organisms. But early Earth was energized by warmth and full of minerals that acted as catalysts, boosting specific chemical reactions. Giant molecules could persist long enough to get trapped in membranes—precursors of the first cells.

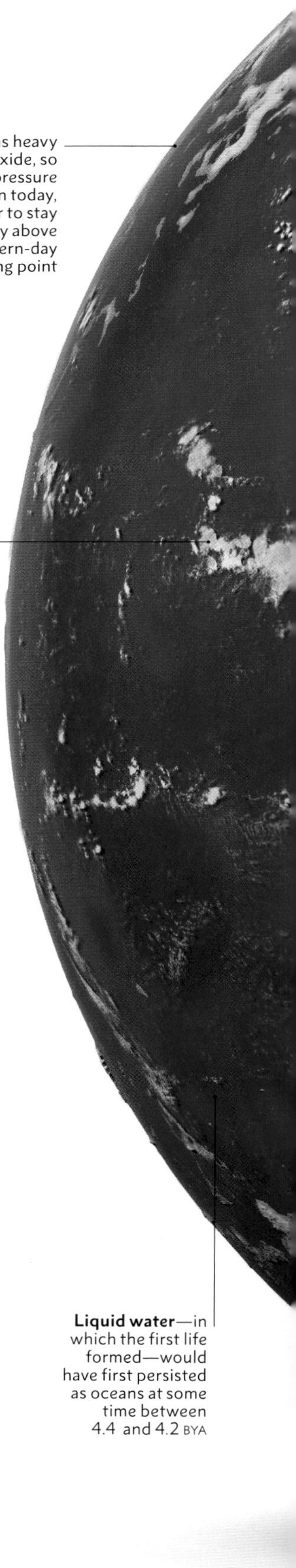
Atmosphere was heavy with carbon dioxide, so atmospheric pressure was higher than today, allowing water to stay liquid way above its modern-day boiling point
Clouds of water droplets would have filled the sky, as they do today
Liquid water—in which the first life formed—would have first persisted as oceans at some time between 4.4 and 4.2 BYA

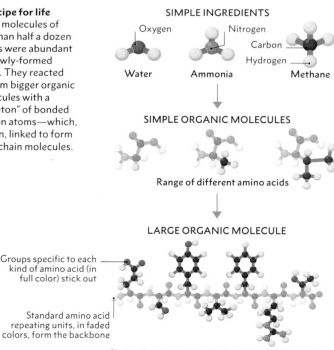

► Recipe for life
Small molecules of less than half a dozen atoms were abundant on newly-formed Earth. They reacted to form bigger organic molecules with a "skeleton" of bonded carbon atoms—which, in turn, linked to form long-chain molecules.

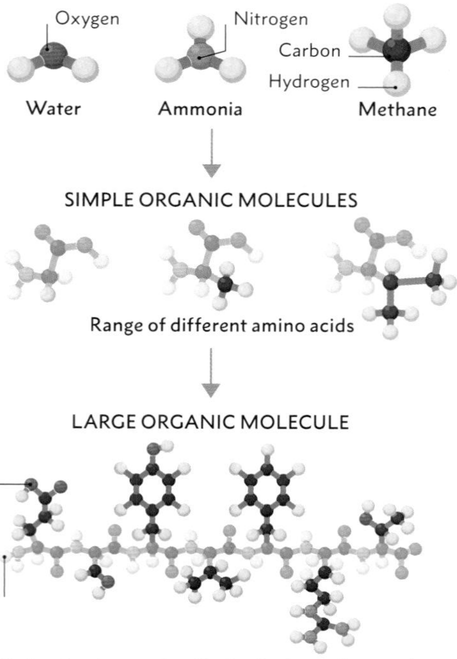
SIMPLE INGREDIENTS
Oxygen
Nitrogen
Carbon
Hydrogen
Water
Ammonia
Methane
SIMPLE ORGANIC MOLECULES
Range of different amino acids
LARGE ORGANIC MOLECULE
Groups specific to each kind of amino acid (in full color) stick out
Standard amino acid repeating units, in faded colors, form the backbone

Chain of amino acids—the beginnings of a protein

▼ Life wrapped up
The chemicals that created life needed compartments, where they could become concentrated. These may have been provided by oily molecules called phospholipids (present in all cell membranes today), which naturally aggregate into membranes in water. When they form spheres, they can trap the life-giving chemicals inside.

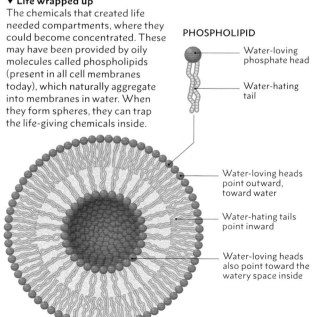

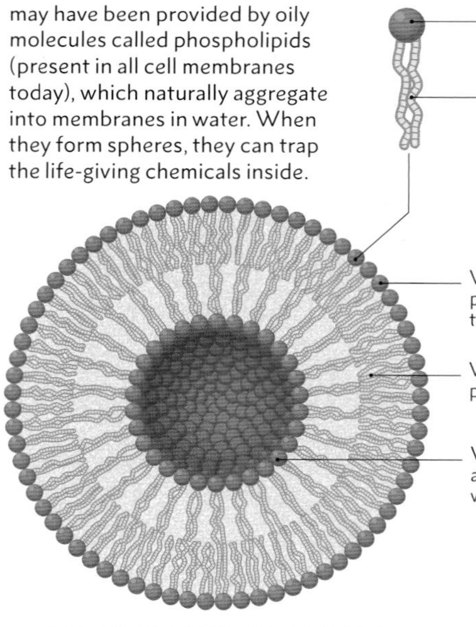
PHOSPHOLIPID
Water-loving phosphate head
Water-hating tail
Water-loving heads point outward, toward water
Water-hating tails point inward
Water-loving heads also point toward the watery space inside

MEMBRANE FORMING A SPHERE

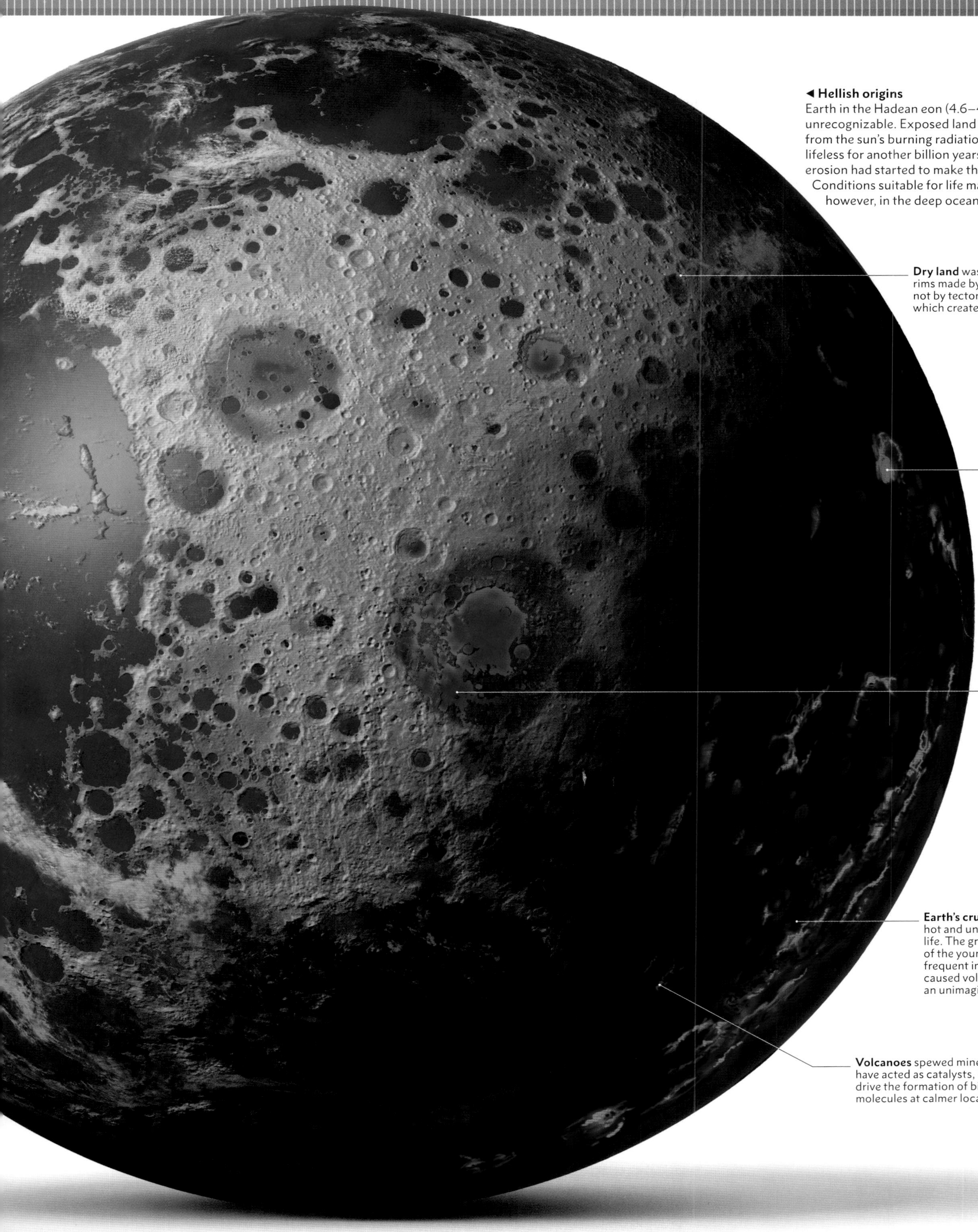
Dry land was formed by crater rims made by asteroid impacts, not by tectonic movements, which create mountains today
Lava could fill craters as asteroid impacts triggered volcanic activity
Cooling seas, cut off from the violence elsewhere, may have provided the conditions needed for newly-formed complex organic molecules to persist
Earth's crust was mostly too hot and unstable to nurture life. The greater internal heat of the young Earth, and the frequent impacts from space, caused volcanic activity on an unimaginable scale
Volcanoes spewed minerals that may have acted as catalysts, helping to drive the formation of bigger organic molecules at calmer locations

◀ **Hellish origins**
Earth in the Hadean eon (4.6–4.0 BYA) was unrecognizable. Exposed land was unprotected from the sun's burning radiation and would remain lifeless for another billion years—by which time erosion had started to make the first soils. Conditions suitable for life may have existed, however, in the deep ocean or shallow seas.

THE **GENETIC** CODE

A living organism is the most precisely ordered thing in the known universe. The assembly and upkeep of a living body need direction and control. The entire operation is guided by self-replicating molecules of nucleic acid (DNA and its ancestors) that were present at the dawn of life.

Until the discovery of DNA's precise shape in 1953, it was a mystery how life-forms passed on genetic information to the next generation. Once revealed, the double-stranded structure of DNA hinted at how information was inherited whenever one cell splits into two. In the next years, experiments confirmed not only that DNA carried units of heredity (called genes), but also that it exerted its influence in an astonishingly intricate way.

> **DNA** IS LIKE A **COMPUTER PROGRAM** BUT FAR, FAR **MORE ADVANCED** THAN ANY SOFTWARE **EVER CREATED.**
>
> **Bill Gates**, technology pioneer and philanthropist, 1955–

INFORMATION CARRIERS

DNA is a giant long-chain molecule—just like protein, cellulose, and many other biological molecules. But whereas cellulose is a monotonous fiber of identical subunits, those of DNA—and protein—come in different kinds. Different subunits follow in an information-carrying sequence—just as letters form a word. DNA belongs to a class of long-chain information-carriers called nucleic acids. The sugars and other components of their structure would have been among life's primordial ingredients. The first nucleic acids, possibly a type called RNA, were probably capable of boosting their own replication reactions. Their chains could have acted as templates guiding the assembly of new parallel chains. Copying from a template is also used by DNA in living organisms today, but it happens only when the two chains of the double helix separate in preparation for cell splitting. Otherwise, one chain is fixed to another like the sides of a ladder. The copying results in two double helices, each with identical information destined for a new daughter cell. In this way, genetic information is copied and inherited.

USING THE INFORMATION

DNA cannot carry out any tasks alone. It instructs other molecules—the proteins—to do the work of maintaining and developing a living organism. A single DNA molecule carries hundreds of sections—genes—each carrying instructions to make a specific protein. In a living cell, sections of DNA are continually being unwound and wound—as genes are exposed for protein manufacture.

▼ They were simpler times...
Today, DNA needs protein to replicate and RNA to make protein to carry out all its other functions. At the origin of life, there was no such complexity. The earliest replicating molecules, possibly RNA, had the ability both to carry data and multiply unaided.

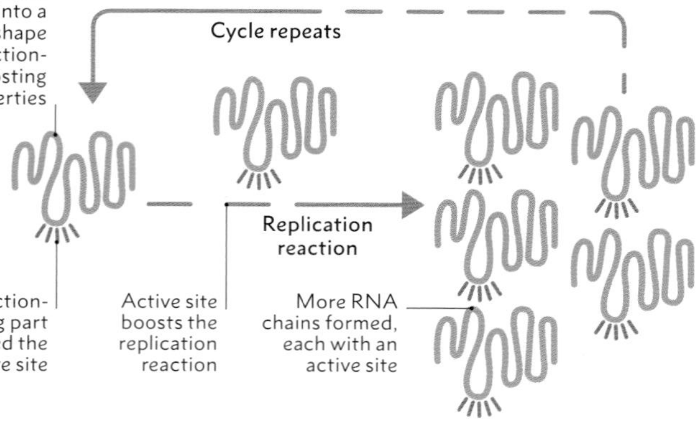
RNA chain folds into a specific shape with reaction-boosting properties
Cycle repeats
Replication reaction
Reaction-boosting part is called the active site
Active site boosts the replication reaction
More RNA chains formed, each with an active site

► Reading the code
In a living cell's nucleus, DNA's double helix is unzipped so that genes can be used to make RNA, and then protein. Here, a strand of RNA is being built by matching bases (chemical components), copying the sequence. This RNA strand will go on to make a specific protein useful to the life-form. The sequence of RNA bases is the code for a specific sequence of chemical components that makes just the right protein.

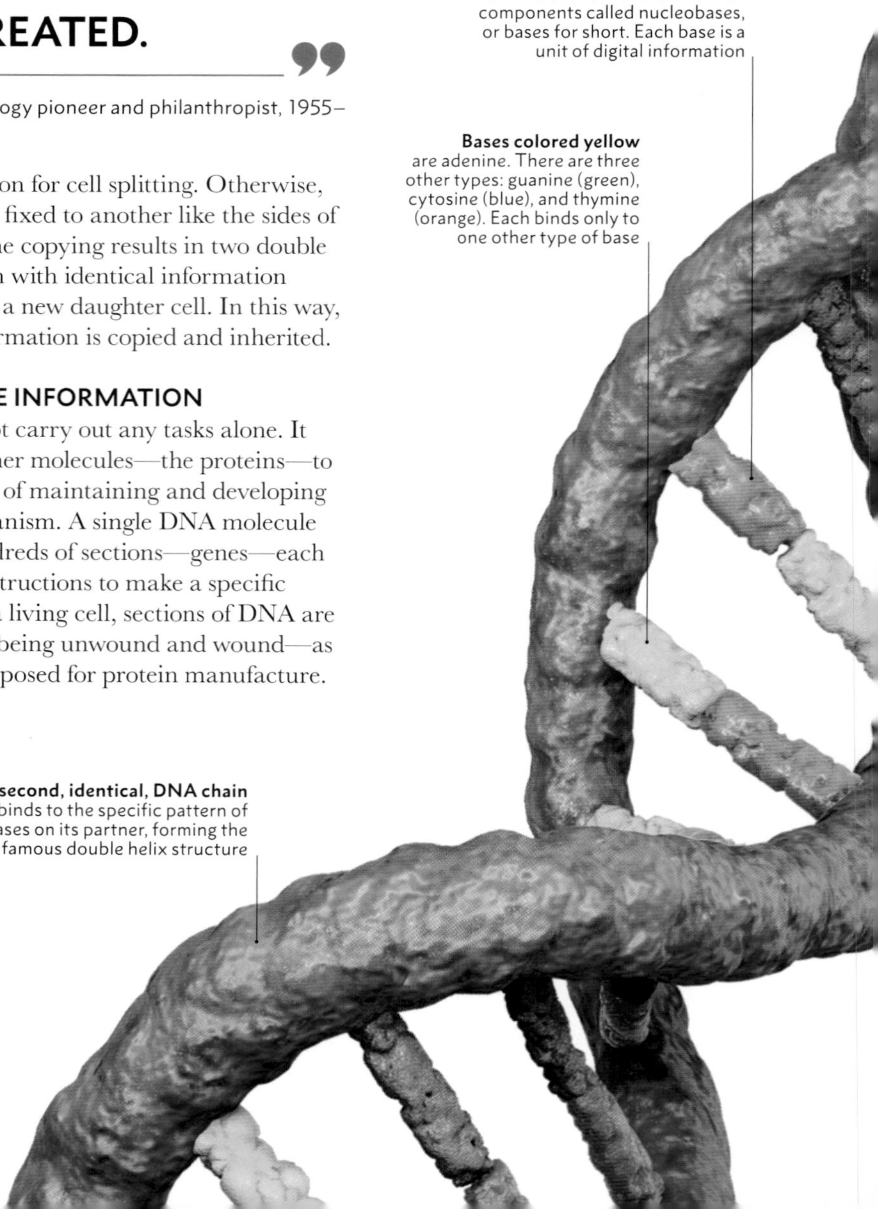
The rungs that connect DNA's backbones are chemical components called nucleobases, or bases for short. Each base is a unit of digital information
Bases colored yellow are adenine. There are three other types: guanine (green), cytosine (blue), and thymine (orange). Each binds only to one other type of base
A second, identical, DNA chain binds to the specific pattern of bases on its partner, forming the famous double helix structure

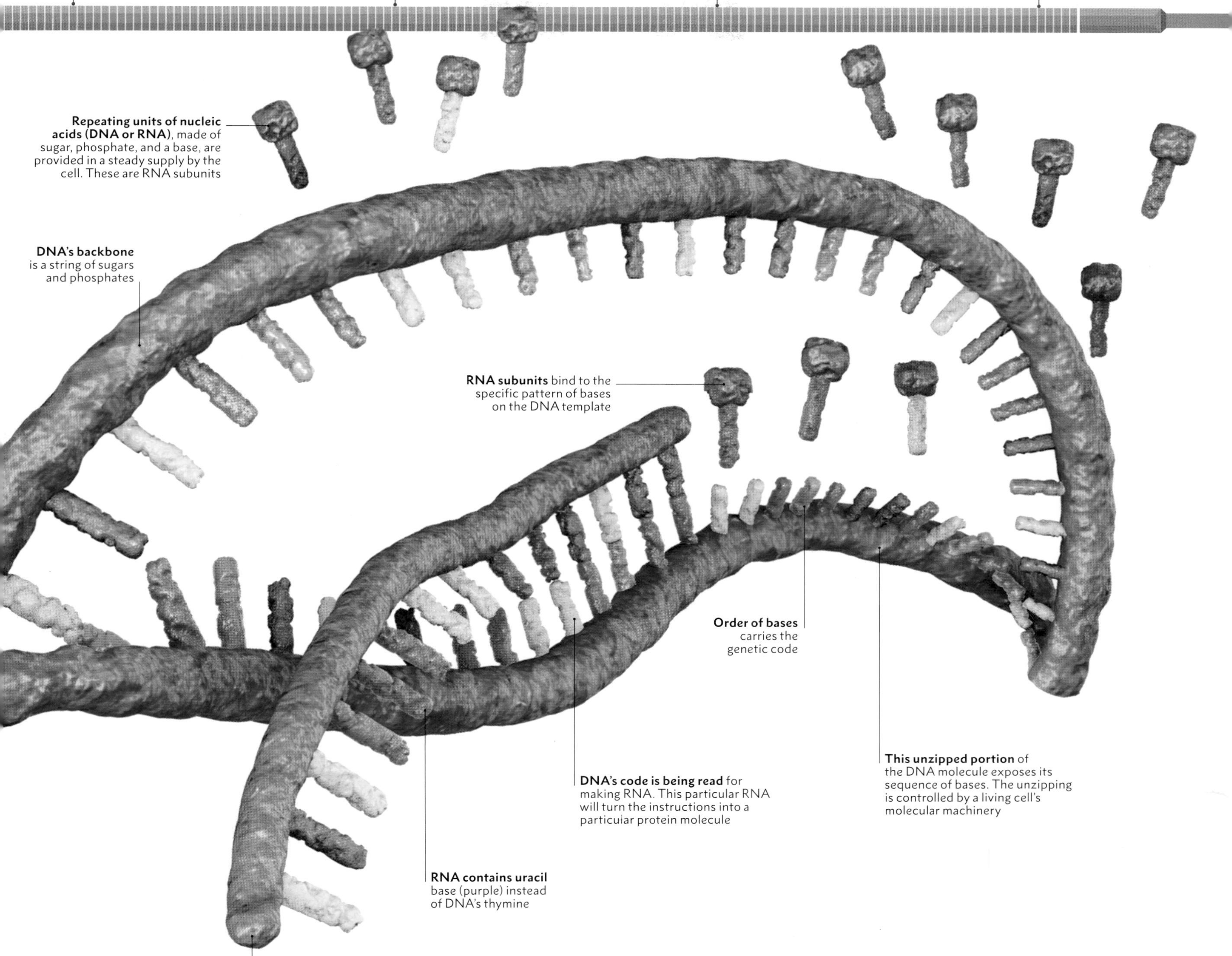
Repeating units of nucleic acids (DNA or RNA), made of sugar, phosphate, and a base, are provided in a steady supply by the cell. These are RNA subunits
DNA's backbone is a string of sugars and phosphates
RNA subunits bind to the specific pattern of bases on the DNA template
Order of bases carries the genetic code
This unzipped portion of the DNA molecule exposes its sequence of bases. The unzipping is controlled by a living cell's molecular machinery
DNA's code is being read for making RNA. This particular RNA will turn the instructions into a particular protein molecule
RNA contains uracil base (purple) instead of DNA's thymine
RNA backbone has a different sugar than DNA

DNA IS AMONG THE **LONGEST OF MOLECULES**—CHAINS IN HUMANS ARE **UP TO** 3.3IN (**8.4CM**) **LONG** AND CONTAIN **249 MILLION** BASE PAIRS

◀ **Discovering DNA**
In 1953 in Cambridge University, scientists made a breakthrough. American biologist James Watson (far left) and British biologist Francis Crick deduced that DNA (deoxyribonucleic acid) had a regular double-helical shape and properties that would allow it to pass on genetic information.

LIFE **BEGINS**

Life arose from nonliving matter by processes of gathering complexity. As self-replicating molecules mixed with catalysts—substances that drive chemical reactions—self-assembly snowballed into the first cells: organisms with familiar characteristics of life.

All life consists of cells with the chemicals of life contained inside a membrane. A living organism is continually dynamic, resisting collapse into disorder and death. How such a system emerged from the nonliving Earth is a mystery, but scientists apply what they know about biochemistry and conditions on early Earth to deduce what might have happened. The transition demands a special setting, and conditions may have been just right around 4 BYA.

PAID TO EAT A FREE MEAL

Deep-sea volcanic vents were rich in chemicals and were warm, but not so hot as to break apart big molecules. Billions of years ago, they were also a safe haven from bombarding asteroids and fierce solar rays. Vents today get encrusted with metal sulfides as the water cools. These minerals boost, or catalyze, reactions—some of which convert carbon dioxide into acetate. Acetate has a pivotal position in the metabolism of all life today. What is more, one sort of acetate-forming reaction can even generate energy. This combination of food manufacture and payment in energy—all trapped within the catalytic encrustation—could have been a "hatchery" for life.

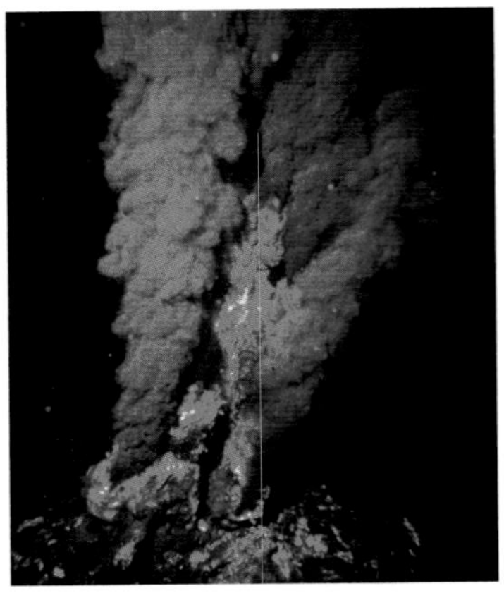

▲ Hot habitat
As water emerges from a deep-sea vent, encrusting minerals build up "chimneys," some of which appear to smoke with dark iron sulphide. These habitats support bizarre life-forms today—entirely dependent on the mineral effluent.

> **TODAY'S DNA**, STRUNG THROUGH ALL THE **CELLS OF THE EARTH**, IS ... AN EXTENSION AND **ELABORATION** OF THE **FIRST MOLECULE**.
>
> **Lewis Thomas**, physician, writer, and educator, 1913–1993

ESCAPING THE CHIMNEYS

The first "protocells" formed when oily membranes encapsulated chemicals that were generated in the chimneys. Sea water helped protocells disperse from the chimneys, and the catalytic minerals in it helped maintain their primitive metabolism.

The versatility of the element carbon—which forms the skeleton of acetate—means that its atoms can assemble into a wide range of molecules. Some of the molecules generated by mineral-catalysis may have developed catalytic abilities of their own—and could even drive their own assembly. It is possible that these molecules may have been related to RNA—a material found in all cells today. RNA—or molecules like it—marked the emergence of biological information, too. Such molecules could control how cells maintained the emerging qualities of life.

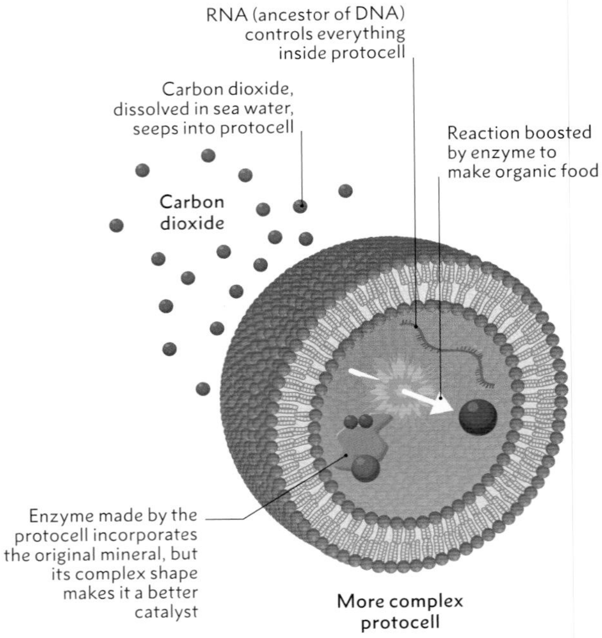
RNA (ancestor of DNA) controls everything inside protocell
Carbon dioxide, dissolved in sea water, seeps into protocell
Reaction boosted by enzyme to make organic food
Carbon dioxide
Enzyme made by the protocell incorporates the original mineral, but its complex shape makes it a better catalyst
More complex protocell

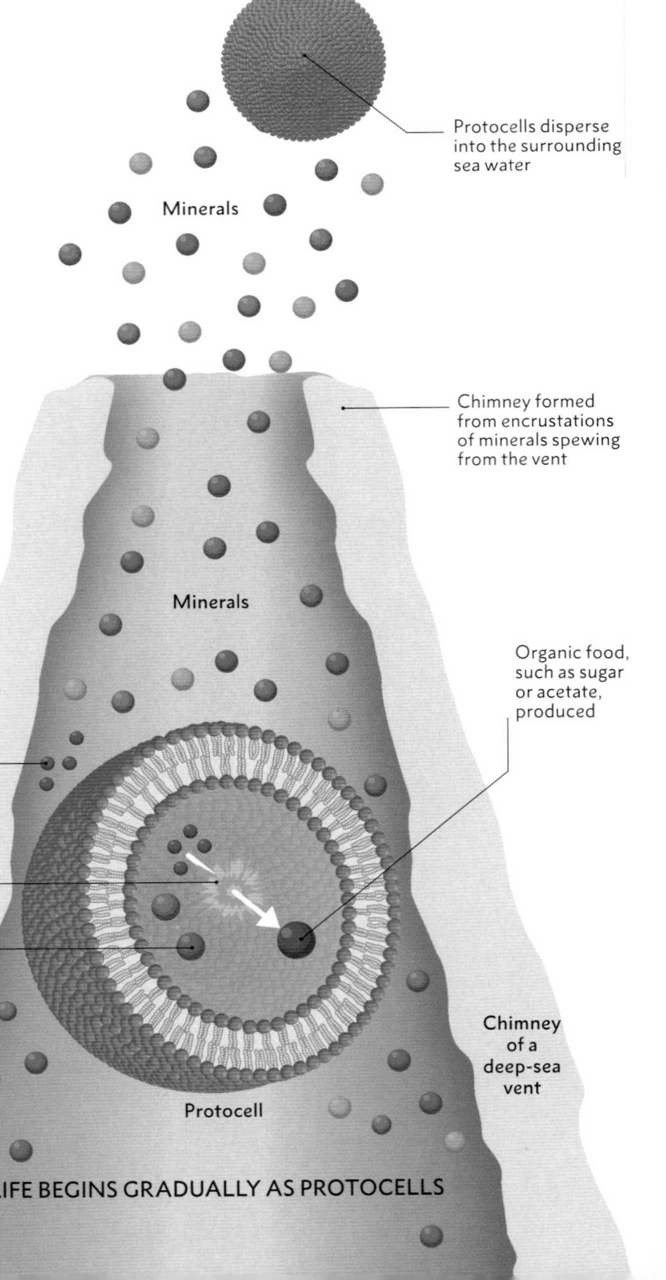
Protocells disperse into the surrounding sea water
Minerals
Chimney formed from encrustations of minerals spewing from the vent
Minerals
Organic food, such as sugar or acetate, produced
Carbon dioxide seeps into protocell
Energy released when carbon dioxide forms organic food
Minerals catalyze (boost) the reaction
Chimney of a deep-sea vent
Protocell
LIFE BEGINS GRADUALLY AS PROTOCELLS

► The origin of life
A chemical reaction boosted by minerals inside a deep-sea chimney, and contained inside a membrane, may have been the basis for the first life—a "protocell." More complex protocells later started to make their own catalysts, which drove their reactions. To begin with, these catalysts may have been RNA. But eventually, protocells developed protein catalysts called enzymes. RNA (and eventually DNA) took over the job of controlling the entire assembly.

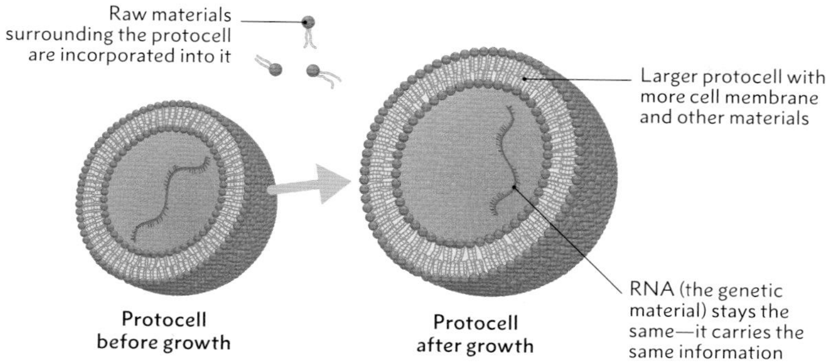
Raw materials surrounding the protocell are incorporated into it
Larger protocell with more cell membrane and other materials
RNA (the genetic material) stays the same—it carries the same information
Protocell before growth
Protocell after growth

▲ Growth
As protocells acquired and made more organic molecules, these became incorporated into their structure—allowing them to grow. Membranes became more expansive, but kept the same two-molecule-thick structure that is common to all cell membranes to this day.

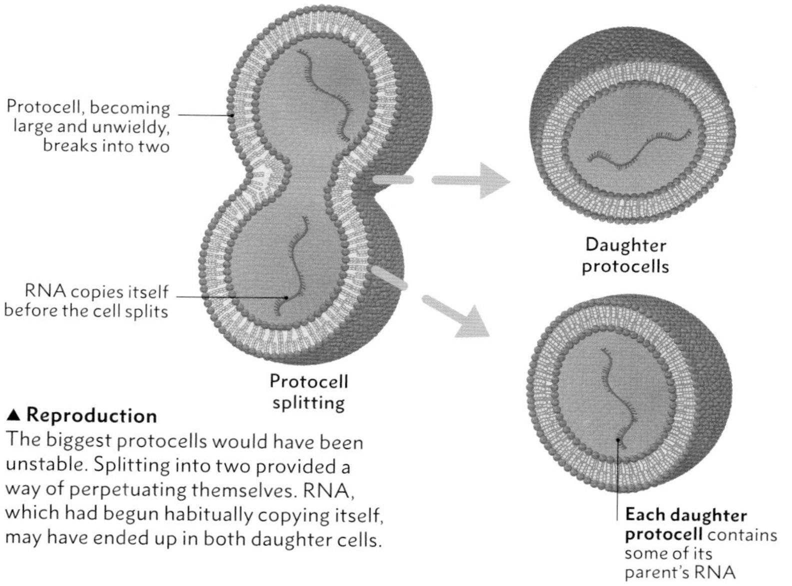
Protocell, becoming large and unwieldy, breaks into two
RNA copies itself before the cell splits
Protocell splitting
Daughter protocells
Each daughter protocell contains some of its parent's RNA

▲ Reproduction
The biggest protocells would have been unstable. Splitting into two provided a way of perpetuating themselves. RNA, which had begun habitually copying itself, may have ended up in both daughter cells.

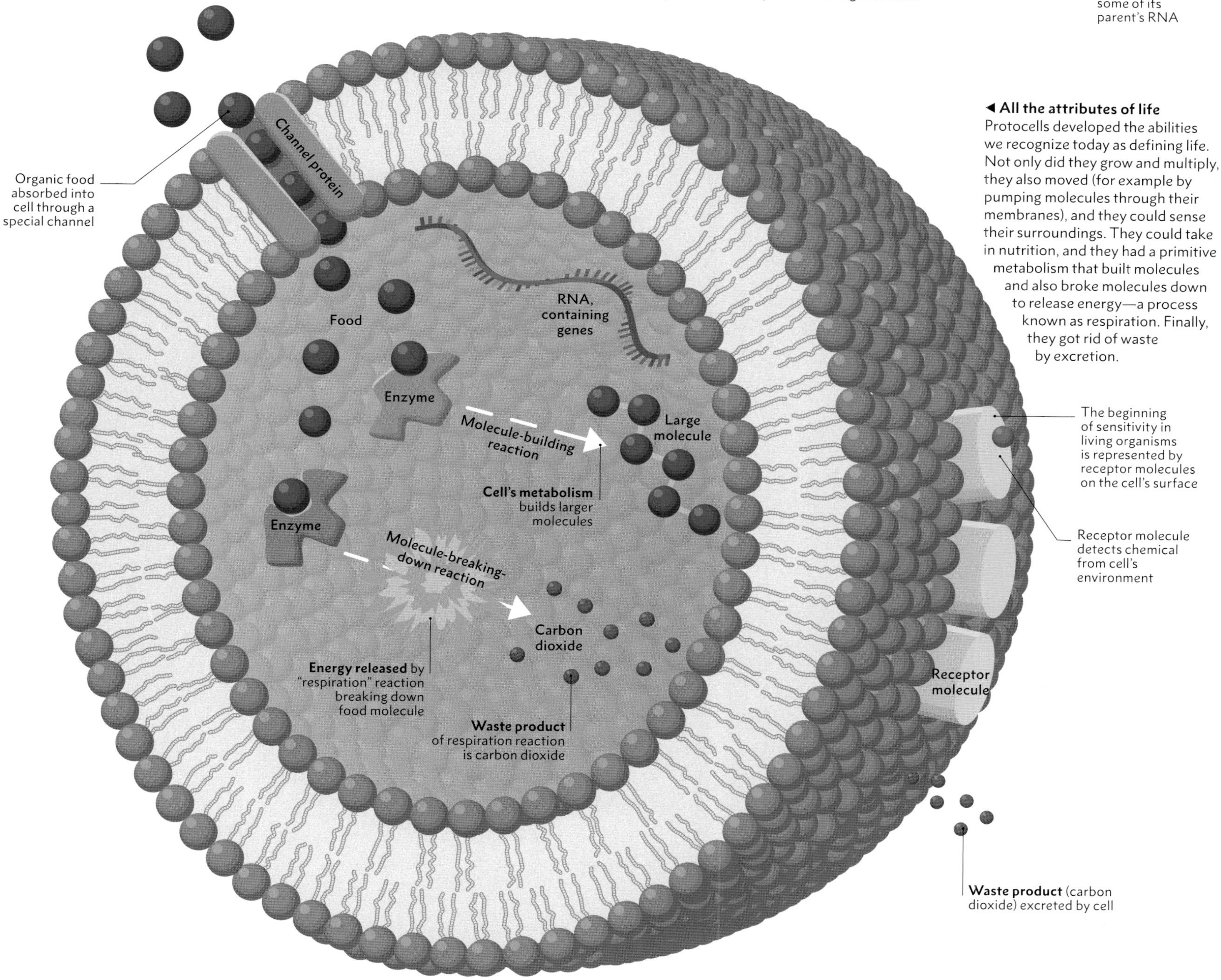
Organic food absorbed into cell through a special channel
Channel protein
Food
RNA, containing genes
Enzyme
Molecule-building reaction
Large molecule
Cell's metabolism builds larger molecules
Enzyme
Molecule-breaking-down reaction
Carbon dioxide
Energy released by "respiration" reaction breaking down food molecule
Waste product of respiration reaction is carbon dioxide
The beginning of sensitivity in living organisms is represented by receptor molecules on the cell's surface
Receptor molecule detects chemical from cell's environment
Receptor molecule
Waste product (carbon dioxide) excreted by cell

◀ All the attributes of life
Protocells developed the abilities we recognize today as defining life. Not only did they grow and multiply, they also moved (for example by pumping molecules through their membranes), and they could sense their surroundings. They could take in nutrition, and they had a primitive metabolism that built molecules and also broke molecules down to release energy—a process known as respiration. Finally, they got rid of waste by excretion.

PROTOCELLS ACQUIRE FULL CHARACTERISTICS OF LIFE

HOW **LIFE EVOLVES**

Even at the dawn of life, the process of evolution was under way. Life was changing, and at the root of every novelty was mutation—imperfections in DNA's copy-making process. The mistakes produced variety, and on a changeable planet some variations succeeded, while others failed.

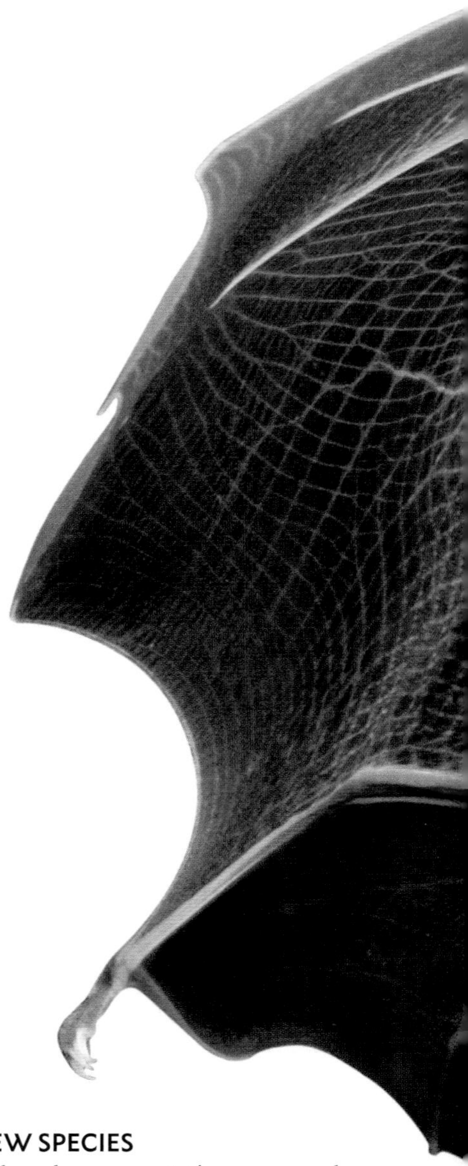

All organisms change during their lifetime. But a grander scale of change, at the level of populations, happens through generations. When an organism reproduces, it copies its entire DNA, which ranges from under a million to many billions of digital "bits" of information. The enterprise represents a monumental turnover of molecular data. Even with natural system-checks in place, copying errors, called mutations, happen.

> **EVOLUTION** HAS **NO LONG-TERM GOAL**. THERE IS NO LONG-DISTANCE TARGET, **NO FINAL PERFECTION** TO SERVE AS A CRITERION FOR SELECTION.
>
> **Richard Dawkins**, evolutionary biologist, 1941–

Mutation produces the raw material of variation. Some mutations have scarcely any effect, but others can abort development, while a few are beneficial.

SELECTION BY THE ENVIRONMENT

While mutation is haphazard, evolution is far from random. The mutations are subject to a selection process. Life-forms with beneficial mutations are selected—they proliferate and pass on their "good genes" to at least some of their offspring. Those with mutations that harm their survival or ability to reproduce will diminish and may die out.

The changing environment, and a life-form's habitat and survival strategy within it, determine whether its mutations are helpful or harmful. Deep-sea fish have big eyes and glowing devices that allow them to hunt in the dark, while desert cacti have water stores defended with spines. Cactus spines and luminous lures need genetic diversity to appear, but it is the environment that selects them for the right places. Chance can play a role in spreading mutations, especially in small populations, but only natural selection can explain adaptation—the fitting of an organism to its environment.

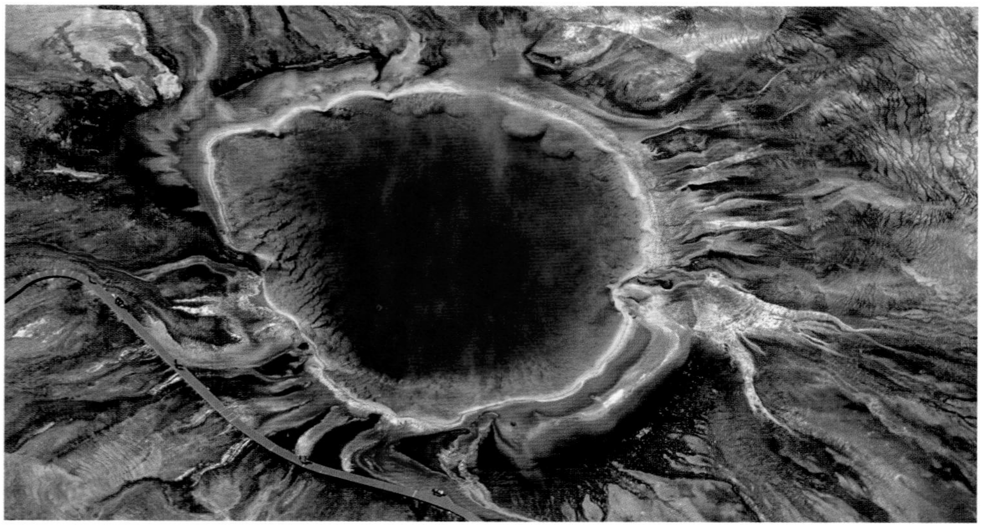

▶ Reaching the limit
A few microbes, that today stand out as bright colors at the edges of hot acidic pools, are a testament of the extent to which genetic variation and adaptation allows life to live in extremes.

NEW SPECIES

Although some mutations can produce sudden, distinct novelties, evolutionary change is generally slow and gradual. Selection typically works on sets of genes that work together to control broad features such as size or shape. But living diversity is not continuous—it occurs in discrete units called species. New species arise when two populations can no longer interbreed. They cannot exchange genes, and their evolutionary paths drift apart. This divergence might happen across an emerging barrier—such as a river or mountain range. But mutations themselves, such as those involving whole chromosomes, especially among plants, can prevent interbreeding and isolate populations.

There are millions of species living today, but all—including countless more that lived in the past—are products of evolutionary change shaped by the environment.

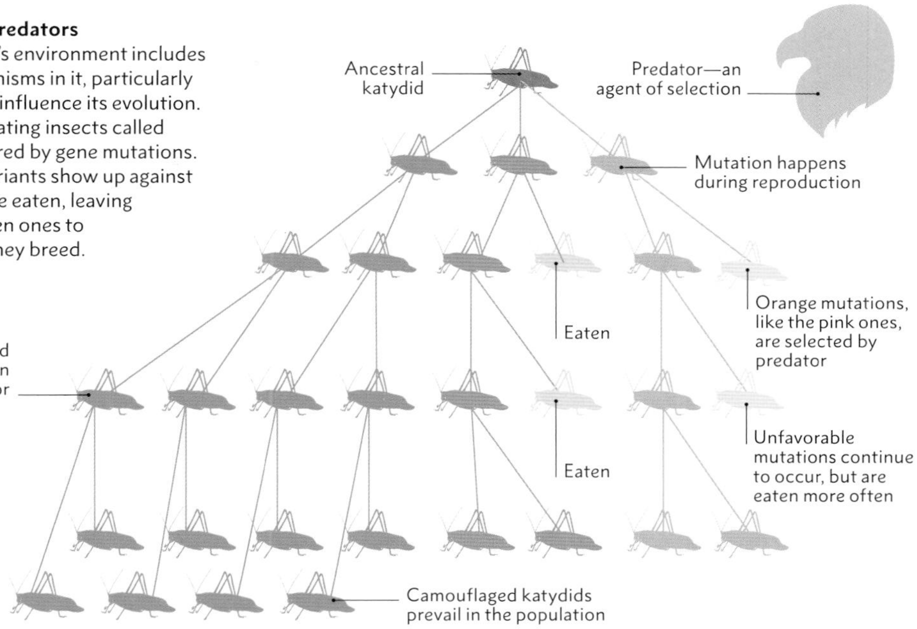
Ancestral katydid
Predator—an agent of selection
Mutation happens during reproduction
Eaten
Orange mutations, like the pink ones, are selected by predator
Camouflaged katydid hidden from predator
Eaten
Unfavorable mutations continue to occur, but are eaten more often
Camouflaged katydids prevail in the population

▶ Selection by predators
A living organism's environment includes all the other organisms in it, particularly predators, which influence its evolution. Cricketlike leaf-eating insects called katydids are colored by gene mutations. Pink or orange variants show up against vegetation and are eaten, leaving camouflaged green ones to predominate as they breed.

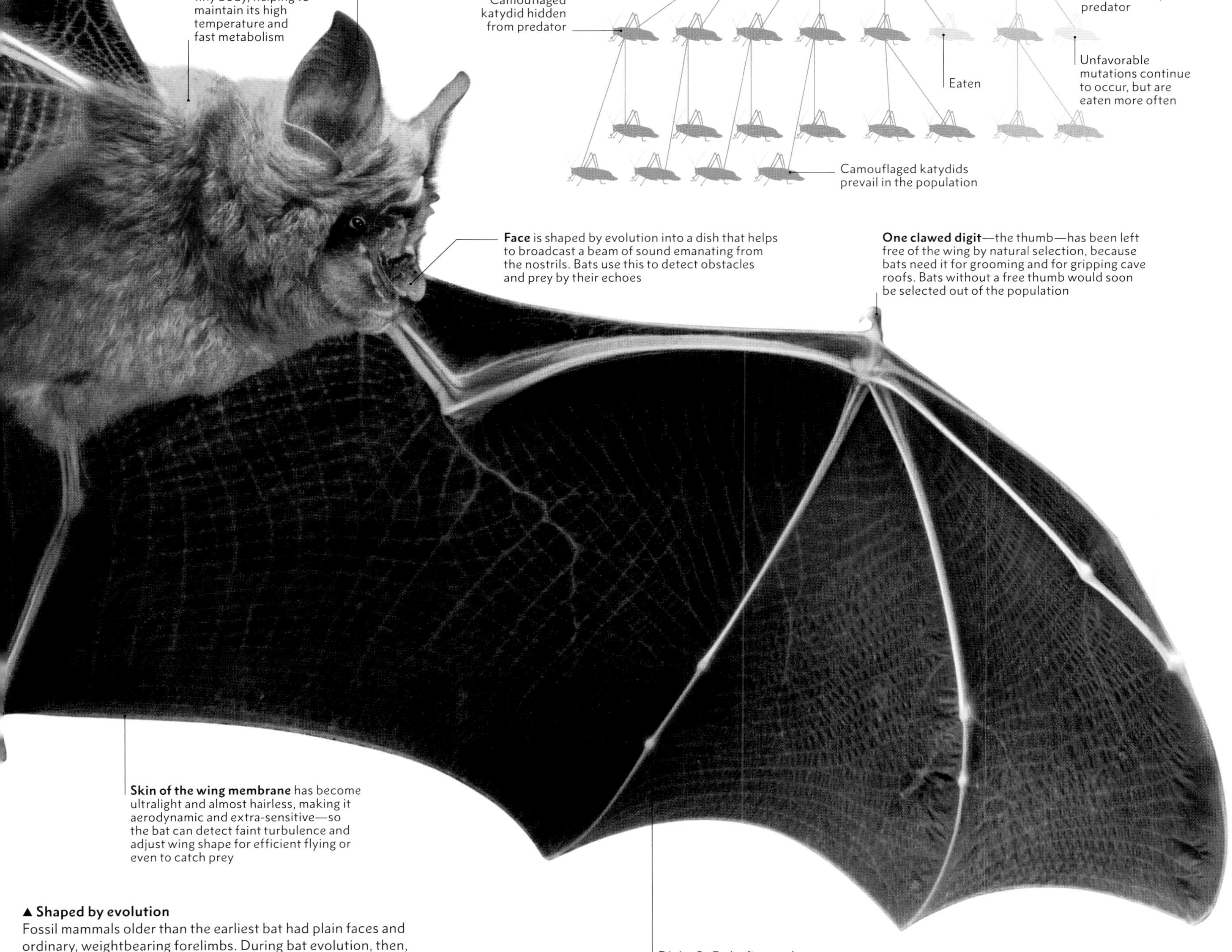
Ears are elaborately sculpted by the need to receive faint echoes at specific frequencies coming from prey or obstacles
Fur insulates the tiny body, helping to maintain its high temperature and fast metabolism
Face is shaped by evolution into a dish that helps to broadcast a beam of sound emanating from the nostrils. Bats use this to detect obstacles and prey by their echoes
One clawed digit—the thumb—has been left free of the wing by natural selection, because bats need it for grooming and for gripping cave roofs. Bats without a free thumb would soon be selected out of the population
Skin of the wing membrane has become ultralight and almost hairless, making it aerodynamic and extra-sensitive—so the bat can detect faint turbulence and adjust wing shape for efficient flying or even to catch prey
Digits 2–5, the fingers, have been vastly lengthened by natural selection

▲ Shaped by evolution
Fossil mammals older than the earliest bat had plain faces and ordinary, weightbearing forelimbs. During bat evolution, then, selection must have favored the drastic elongation of their fingers, allowing them to support wings. In horseshoe bats, it also favored the extraordinary modification of their faces, which improved their echolocation (their natural sonar).

BIG IDEAS

HISTORY OF **EVOLUTION**

Some have called it the biggest idea of all time: that everything that has ever lived on Earth—dodos and diatoms, cabbages and kings—has descended from a single common ancestor. The possibility of life's evolution occupied some of the greatest minds, but it took one gentleman's lifetime pursuit of "the species problem" to explain how it could happen.

▶ Galápagos finches These finches were collected from the Galápagos islands and Darwin noticed that slight differences in the size and shape of their beaks might have arisen from a common ancestor.

Life changes over the course of thousands, and millions, of years. From one form of life another will arise, modified in some way by the environment in which it lives. The second form of life is more adapted to survive in its environment, and it retains some aspects of its previous form. This is evolution by natural selection, and we can track its progress through the fossil record.

EARLY CLUES

Philosophers of antiquity had anticipated evolutionary thought: some considered the possibility that all life could be ranked in a hierarchy—with humans at the top.

In the 17th and 18th centuries, western naturalists explored the world and filled museums with fossils. Those that named these extinct animals did so from a religious point of view. Animals were assumed to have been created in their current form by God. Every species on Earth had always been there, and they could not be changed. Fossils could be explained away as animals that had died during the Great Flood. Scientists who compared the anatomy of various animals saw plenty of parallels between species. These similarities supported the idea of an affinity between certain groups of animals. For instance, African baboons were undoubtedly closer to Asian macaques than they were to diminutive South American marmosets. Likewise, chimpanzees seemed close to humans. What did this closeness mean, if anything?

ALTERNATIVE WORLDVIEW

For Charles Darwin—born into a reverent society—these anatomical affinities caught his attention. He was recommended for a five-year voyage aboard the HMS *Beagle*. During his journey, he collected specimens from across the globe.

Darwin pondered on the unexpected regional similarities in his specimens. Similarities between species that lived thousands of miles away from each other seemed to go against the idea of a single, spontaneous Creation event. Animals on

> HISTORY WARNS US... THAT IT IS THE **CUSTOMARY FATE** OF **NEW TRUTHS** TO **BEGIN AS HERESIES**...

Thomas Henry Huxley, biologist, 1825–1895

the Galápagos Islands resembled those in nearby South America, and the unusual wildlife in Australia seemed to belong to a different Creation altogether. Upon Darwin's return to England, ornithologist John Gould examined his collection of Galápagos birds. Darwin assumed they belonged to multiple families, but Gould showed how they were in fact species of closely related finches within one family. Darwin's experiences were persuading him that not only were these new species modified from a former generalized species, but perhaps that was the case with all forms of life—that there is one common ancestor for all. Darwin ruminated on his theory that evolution happened by infinitesimally small changes over many, many years and animals with traits that aided survival were more likely to breed and pass these "favorable" characteristics on to the next generation.

In 1858, English naturalist Alfred Russell Wallace wrote to Darwin with the same idea. A year later, Darwin published his ideas in a book, his famous *On the Origin of Species* in 1859, which caused a stir in the scientific community. He faced outrage, since it essentially challenged Biblical Creation as fact. Nevertheless, Darwin's theories gained respectable supporters, including the English naturalist Thomas Henry Huxley, a friend of Darwin's who championed his cause in the scientific community. Within a few years, evolution by natural selection was being lauded in textbooks. In his *Principles of Biology*, the philosopher Herbert Spencer coined an expression that became synonymous with Darwin's ideas: "survival of the fittest."

A UNIFIED THEORY

Darwin's *On the Origin of Species* was exhaustive in its catalog of evidence, but the mystery of inheritance remained. Darwin understood that life changed over time, but how exactly did these changes occur? The popular view was that hereditary qualities blended from two parents—akin to mixing paints of different colors. No one knew if these qualities physically existed. In reality, this blending led to a dilution of varieties, not the emergence of new ones, and so was not a sufficient explanation.

> **CHARLES DARWIN** WAITED **23 YEARS** BEFORE PRESENTING HIS IDEAS TO THE PUBLIC, **DUE TO THEIR CONTROVERSY**

The breakthrough came from an unlikely source: an Augustinian friar in Austria. In the 1860s, Gregor Mendel's experiments in breeding different varieties of pea plants allowed him to deduce that inheritance was due to particles, later called genes. Sexual reproduction remixed genes to produce unique combinations, some of which may express themselves in later generations. This explained two mysteries: the appearance of characteristics that skip generations, and the perpetuation of characteristics that aided survival (natural selection). When he bred yellow and green peas together, Mendel saw that the next generation of peas were uniformly yellow. Therefore, some traits were more likely to express themselves than others. When this generation was interbred, the result were a group of peas with mixed colors, indicating that traits could also skip generations.

Mendel's discoveries not only augmented Darwin's, despite each having no knowledge of the other's work, but also debunked popular rival theories—such as "Lamarckism." The French naturalist Jean-Baptiste Lamarck had proposed that features acquired through life, such as larger and stronger muscles, could be transmitted to offspring. Mendelism was finally rediscovered in 1900 and more scientists began thinking about evolution with genetic inheritance in mind. With genetics as the exciting new discipline of natural science, it became clear that new varieties of genes arise by a process of spontaneous mutation. Natural selection acts upon these varieties by choosing, and keeping, the most useful. By the 1940s German-American biologist Ernst Mayr showed that if populations fragmented, evolution could take different courses away from a single ancestor—and create new species.

Fossils record evolution in progress: fish fins morphing into amphibious limbs, limbs into wings, mammalian limbs back into finlike flippers, and so on. Today, DNA analysis proves beyond doubt that even the lowliest and loftiest life-forms share the same origins.

> **EVOLUTION** COULD... **BE DISPROVED** IF... A **SINGLE FOSSIL** TURNED UP IN THE **WRONG DATE ORDER**. EVOLUTION HAS PASSED THIS TEST WITH FLYING COLORS.

Richard Dawkins, biologist, 1941–

MICROBES APPEAR

Bacteria have been around far longer than any other kind of organism. They were the first to photosynthesize, the first to consume food—and are still the only living things capable of making their food in the absence of light. Billions of years ago, they were pioneers of both oceans and land.

Bacteria are the simplest cellular organisms, but also by far the most abundant and widespread. They are far smaller than the cells of plants and animals—most are about one-tenth the size of a human skin cell. They are called prokaryotic ("pro" meaning before, and "karyon" meaning kernel), because their cells lack the dense nucleus that contains DNA in more complex cells.

Bacteria seem uniform in structure, but this belies remarkable chemical diversity. In 1977, biologists recognized some kinds of prokaryotes as an entirely new life-form, called archaea. These archaea—mostly living in hostile environments, such as salt lakes or hot acidic pools—had unique, ether-based membranes unlike any other living thing. Some performed bizarre chemical processes, spewing out methane.

BACTERIA ARE SO **WIDESPREAD**, SOME LIVE 2 MILES (3KM) **DEEP IN EARTH'S CRUST**, LIVING ON ENERGY FROM **RADIOACTIVE URANIUM**

BANKS OF DEFENSES

Early bacterial evolution happened in a world teeming with other microbes—and many of these early life-forms produced repelling substances, so-called antibiotics, as they competed for food and space. Bacteria, therefore, have layers of defenses. Outside their thin cell membrane, which is common to all life, they have a tough cell wall, and most types also have a second membrane that helps stop antibiotics from penetrating—and still today, bacteria with a wall sandwiched between inner and outer membranes are most resistant to antibiotics.

CHEMICAL DIVERSITY

Bacterial nutrition spans the full range of types seen in plants and animals—and more besides. Many have retained the food-making capability of life's earliest ancestors, deriving energy from minerals. Some of these bacteria invaded soils and became critical for other life by recycling elements such as nitrogen. Others—the cyanobacteria—evolved photosynthesis, making food from sunlight, and were the first organisms to pour oxygen into the atmosphere. But as microbial communities evolved to be more complex, many became food-eaters—absorbing nourishment from their surroundings. It was bacteria like these that—billions of years later—would invade the dead and living bodies of plants and animals, becoming decomposers or disease-causing parasites.

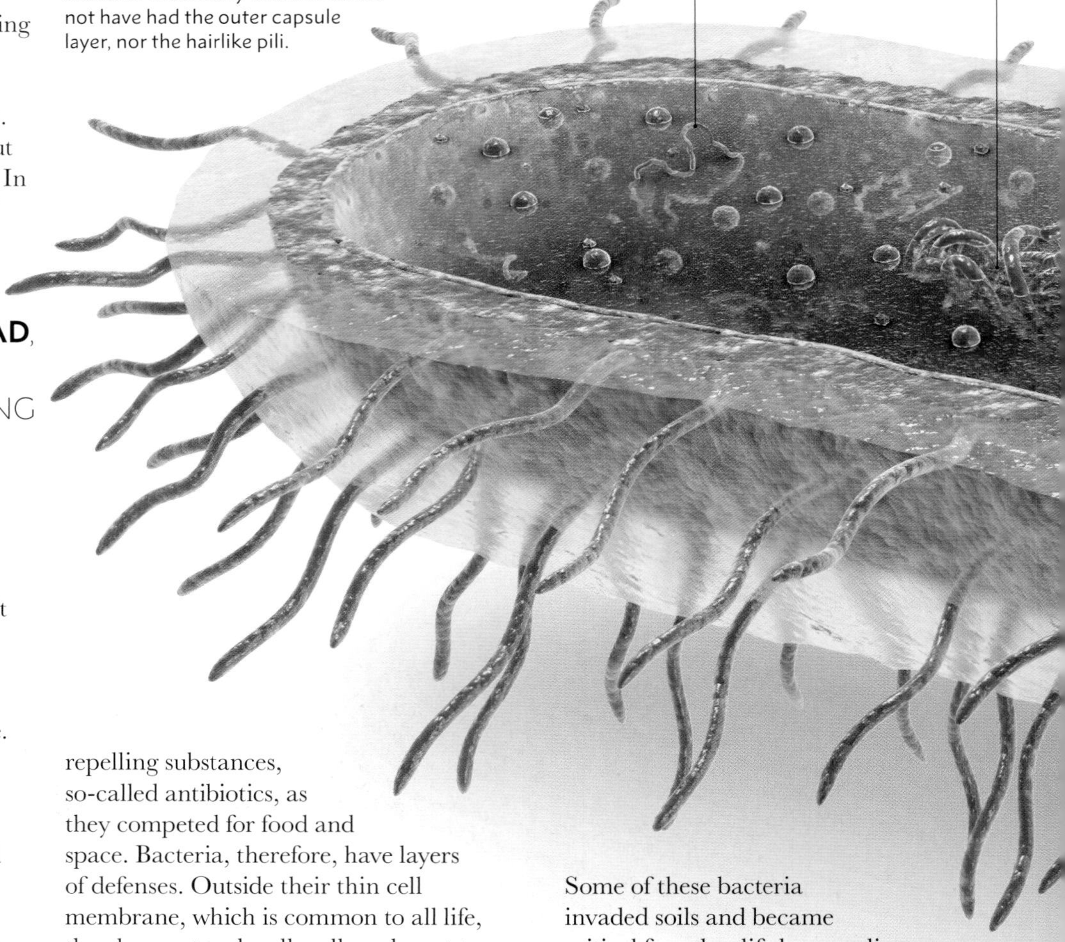
Plasmid—one of many short loops of DNA
Main genome—a long, twisted, closed loop of DNA, containing a few thousand genes, loosely bound to the center of the cell

▼ **Bacillus**
The shapes of bacteria vary from spherical to spiral-shaped, but this rod-shaped type, called a bacillus, is very common. It shows a range of features present in some modern bacteria. Most early bacteria would not have had the outer capsule layer, nor the hairlike pili.

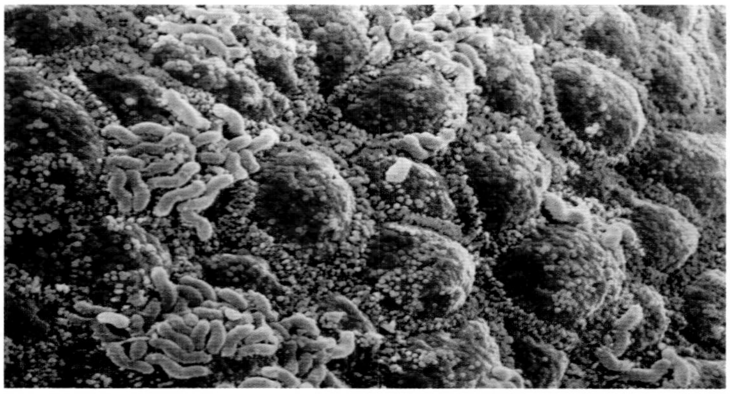

▼ **Bacteria inside animals**
Many food-eating bacteria live inside the guts of animals—such as these on the lining of a human colon. Most maintain a cooperative relationship with their host by exchanging nutrients –in humans, they are essential to digestion. But a few cause disease.

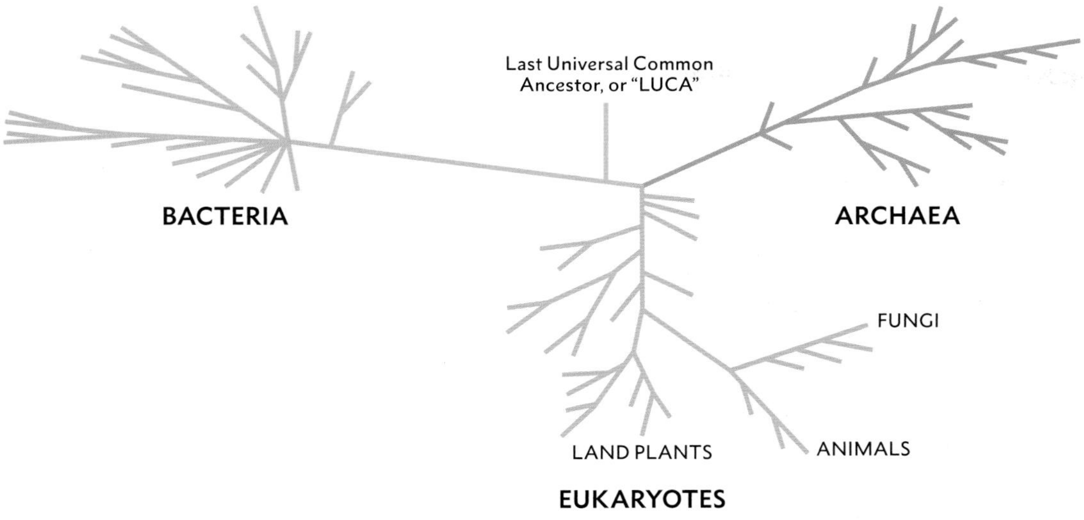
Last Universal Common Ancestor, or "LUCA"
BACTERIA
ARCHAEA
FUNGI
LAND PLANTS
ANIMALS
EUKARYOTES

◀ Tree of life
This tree shows the branching relationships among all forms of life, according to DNA analysis. The analysis suggests that all cellular life alive today has a common origin—it evolved just once, from an unknown ancestor dubbed "LUCA," and that it has three main branches, or domains: bacteria, archaea, and eukaryotes.

KEY

- Bacteria are prokaryotes—all simple, single-celled microbes.
- Archaea are prokaryotes, like bacteria. They resemble bacteria, but at a chemical level they are utterly different, and only distantly related.
- Eukaryotes are much more complex (see pp.118–19), but most branches are also microbes. The plants, animals, and fungi are just small twigs within the eukaryote limb of the tree of life.

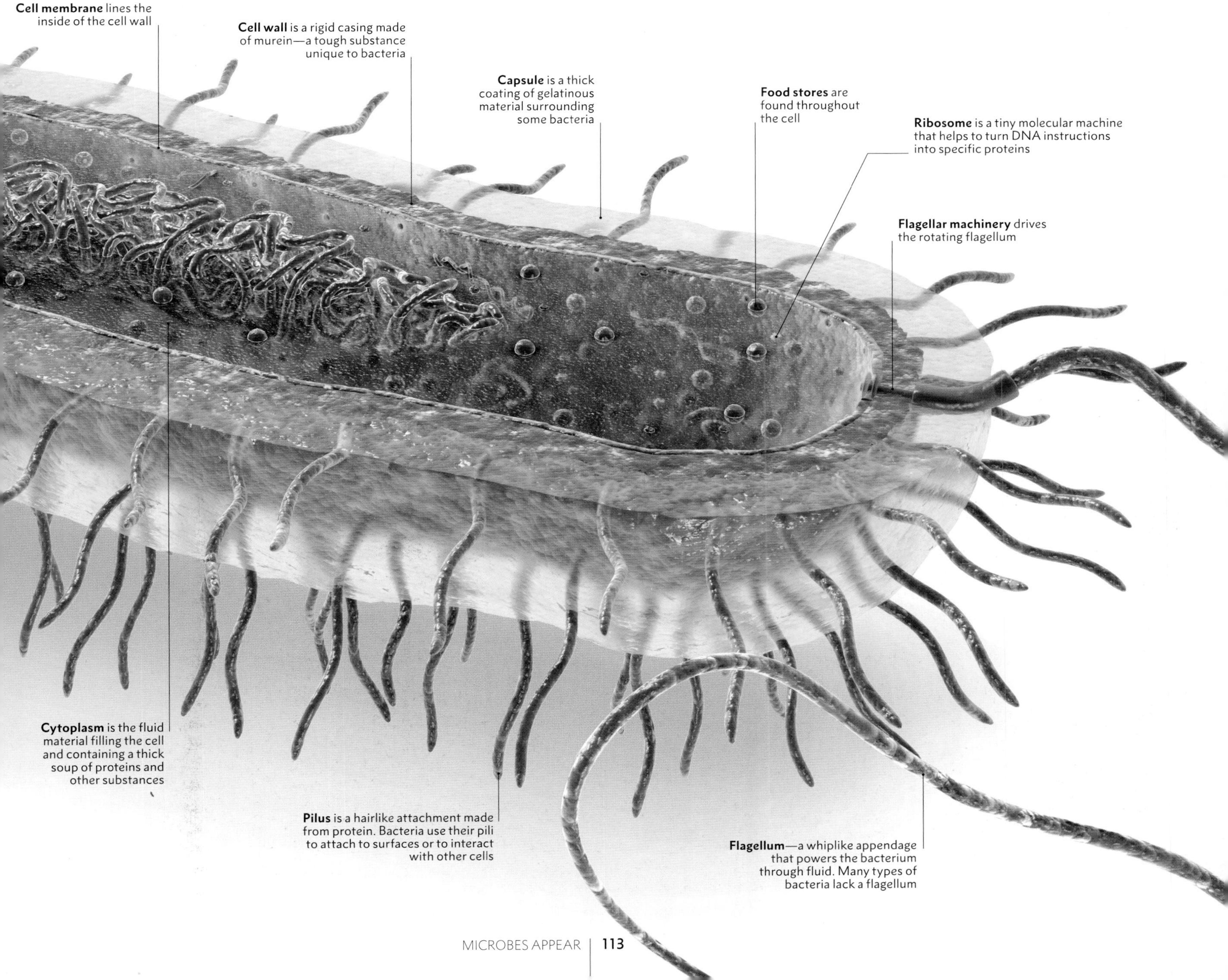
Cell membrane lines the inside of the cell wall
Cell wall is a rigid casing made of murein—a tough substance unique to bacteria
Capsule is a thick coating of gelatinous material surrounding some bacteria
Food stores are found throughout the cell
Ribosome is a tiny molecular machine that helps to turn DNA instructions into specific proteins
Flagellar machinery drives the rotating flagellum
Cytoplasm is the fluid material filling the cell and containing a thick soup of proteins and other substances
Pilus is a hairlike attachment made from protein. Bacteria use their pili to attach to surfaces or to interact with other cells
Flagellum—a whiplike appendage that powers the bacterium through fluid. Many types of bacteria lack a flagellum

LIFE DISCOVERS SUNLIGHT

Life needs energy, and the first living things drew it from minerals and made their food in the darkness of the deep ocean. Those that followed found energy in other places—and, as ancestors of plants and animals, they captured sunlight in the shallows or ate food made by other cells.

Every living thing—from a microbe to the tallest tree—consumes energy that changes small molecules into big ones, pumps life-giving matter into cells, and resists decay. The immediate energy source for this is food. Energy-rich substances, such as sugars and fats, go through a kind of controlled combustion inside cells—in the same way that chemical fuel can be burned to power any machine. But instead of ignition, cells use molecular catalysts (called enzymes) to tease the energy from their nutritive fuel in a safe and manageable way. The process is called respiration.

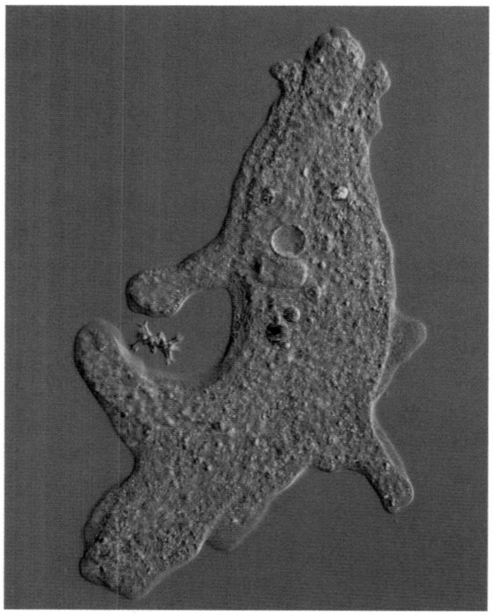

▶ **Predator in miniature**
Amoebas get food by engulfing smaller organisms, such as algae, and breaking them down using digestive enzymes. It means amoebas can live in darkness but need prey to stay alive.

The most self-sufficient strategy for nutrition is to make food, such as sugar, fat, and protein, from nonfood materials. Carbon dioxide in air or dissolved in water provides the carbon and some of the oxygen. Water can provide the hydrogen—and minerals such as nitrates, phosphates, and sulfates deliver nitrogen, phosphorus, and sulfur. Today the world is covered in plants that use the sun's energy to do just that—but the full scope of food-making life is far greater.

MAKING FOOD

Plants are not the only food producers. The most self-sufficient organisms of all can live without light and survive on nothing but water dosed with minerals. These life-forms—all of them bacteria or archaea—can extract energy from chemical processes involving these minerals—and use it to manufacture their food. Organisms that perform this chemical nutrition were among the first life-forms to thrive in the deep, mineral-rich oceans. Some are now the unseen recyclers of nature, their mineral-changing abilities helping to return the nitrogen in dead plants and animals to other living things.

A significant shift in the abilities of prehistoric microbes came when they invaded sunlit shallow waters. These new bacteria used sunlight to make food—in the process of photosynthesis. They could only get nourishment in daylight—but the reward for doing so far outweighed that of making food in darkness: sunlight contains much more energy than minerals. These microbes therefore thrived as they basked in coastal seas. They reorganized and reinvented chemical processes, changing energy-giving reactions into new ones that used solar radiation. They did it with pigments, such as chlorophyll, that absorbed and trapped the light energy. The first photosynthesizers converted carbon dioxide to sugar by adding the hydrogen from hydrogen sulfides. Scientists know this due to the yellow deposits of sulfur this process left behind in rock. But a later refinement to photosynthesis helped life-forms get hydrogen from water instead. The substance left over this time—oxygen—eventually filled the atmosphere (see pp.116–17), and later helped cells burn

▲ **Energy from sunlight**
A thin mat of cyanobacteria on a living stromatolite uses green chlorophyll to trap sunlight. The energy is used to make organic food from carbon dioxide and water, and oxygen bubbles off as a byproduct.

> “BY BLENDING **WATER AND MINERALS** FROM **BELOW** WITH **SUNLIGHT** AND CARBON DIOXIDE FROM **ABOVE**, GREEN PLANTS **LINK THE EARTH TO THE SKY**.”
>
> **Fritjof Capra**, physicist (1939–)

their food in respiration more efficiently. These pioneers were probably like today's cyanobacteria. They grew into sticky films of cells that trapped sediment. Over thousands of years, these colonies formed rocky mounds called stromatolites ("stroma," bed; "lithos," rock). Stromatolites still live in a few warm coastal seas, where extra-salty conditions suppress grazing animals—but they are abundant in the fossil record.

CONSUMING FOOD

As soon as some life-forms started producing food, the opportunity for a shortcut existed. Instead of being producers, organisms could evolve a new strategy—they could eat food produced by others. These organisms abandoned food-making and became consumers—collecting their nourishment in ready-made form. Those that consume organic food in this way are represented by animals, fungi, and a whole range of microbes. The earliest food-eaters probably acquired dissolved food—such as sugars—simply by absorbing it from the vicinity. Decomposers, such as fungi, still get nourishment this way—producing digestive juices to break down any organic materials that are close by so they become more absorbable. Active hunting, in which one organism eats and digests another, became an obvious next step, and complex cells, such as amoebas, evolved the means to engulf tinier organisms. It was the appearance of this predatory behavior that marked the start of microscopic food chains.

Today, producers and consumers are linked by the transfer of energy along bigger food chains. Ocean and land life starts with the solar-powered algae and plants that now provide almost all of the world's food. Herbivores and predators are voracious in the scale of their consumption, while all these living things are, in turn, dependent on the fungi and bacteria that—in their various ways—recycle dead matter.

▼ Where photosynthesis is happening
Photosynthesis is the principal food-making process for modern life. Plants and algae are the producers of food chains that support animals on land and in oceans.

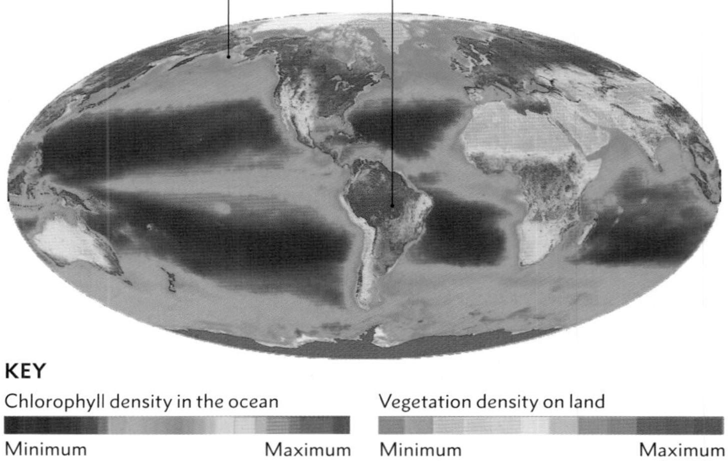
Marine algae are concentrated in seasonally recycled, nutrient-rich waters far from the equator or near to coasts
Tropical rain forests have especially high productivity on land
KEY
Chlorophyll density in the ocean
Minimum
Maximum
Vegetation density on land
Minimum
Maximum

OXYGEN FILLS THE AIR

Nearly two and a half billion years ago, Earth's air underwent a dramatic change: it became oxygenated. This momentous event was caused by new kinds of microbes, and it was incredibly important for the future of all life.

These microbes, bubbling away in the ocean's sunlit shallows, produced oxygen, and ensured that the organisms that followed would never be the same again.

Oxygen is a remarkable element. It causes fire, which turns organic material to cinder—but it is also a component of complex molecules, such as DNA. Most living things need it to breathe and stay alive. Today, oxygen gas makes up about one-fifth of the atmosphere's chemical composition, but for the first half of Earth's history, there was practically no gaseous oxygen at all. Instead, all oxygen lay chemically bound in water and rocks. Photosynthesizing microbes were the first organisms to release oxygen by splitting it away from water as they made their food (see pp.114–15).

POISON TURNED PROFIT

Early life was so unaccustomed to growing levels of oxygen that the response was cataclysmic. The same oxygen that can corrode metal to rust wreaked havoc on the delicate machinery of cells ill-protected from it. Much of early life, having evolved in habitats devoid of oxygen, died in the new poisonous oxygen onslaught. A few microbes had the means to survive—they had enzymes that locked the oxygen away inside their molecules where it could do no damage. But one kind of life-form went a stage further by exploiting the fact that oxidation can be productive as well as destructive.

The eagerness with which oxygen reacts means that oxidation releases energy. So much energy is released during combustion that the reaction grows hot. For billions of years, cells had been honing ways of capturing energy to drive the processes of life. The presence of oxygen opened up a new avenue of metabolism—aerobic respiration—by reacting oxygen with organic molecules (see pp.102–03) and harnessing the energy that was released. It was such an efficient mechanism for creating energy that within another billion years, virtually all life on Earth was breathing oxygen.

Layer of chert
Rich iron layer

◀ Bands of evidence Excavation of rocks dating back to before the Great Oxygenation Event reveals bands of red iron ore. They formed in the seas in which oxygen was being released.

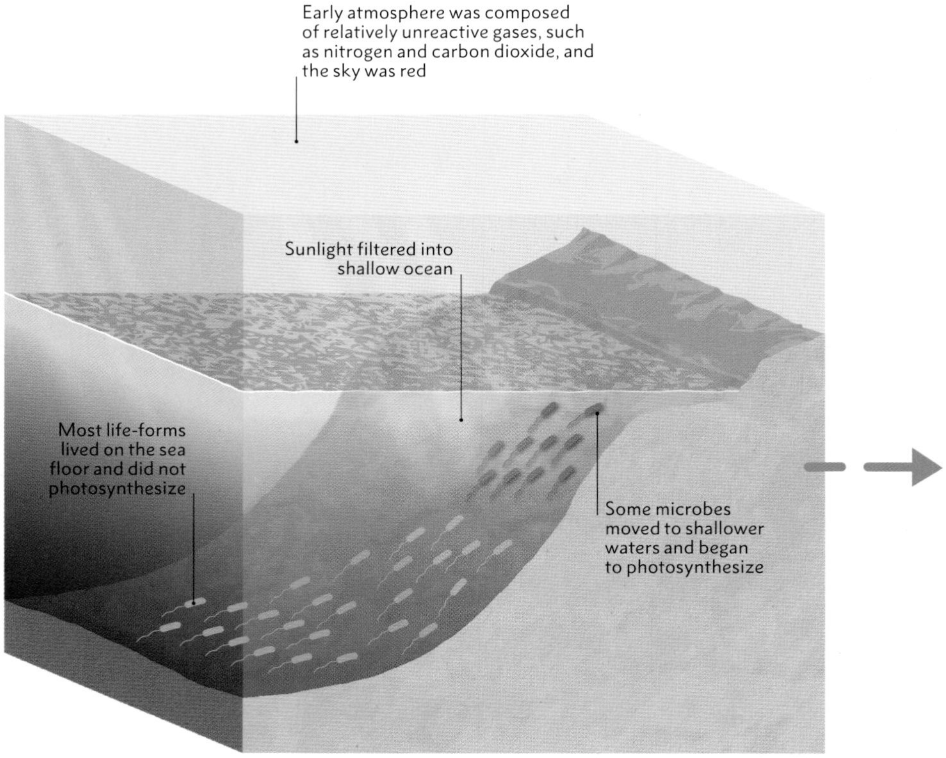
Early atmosphere was composed of relatively unreactive gases, such as nitrogen and carbon dioxide, and the sky was red
Sunlight filtered into shallow ocean
Most life-forms lived on the sea floor and did not photosynthesize
Some microbes moved to shallower waters and began to photosynthesize

Life probably originated in the deep ocean, beyond the reach of sunlight. As early life-forms dispersed into new habitats, those in the sunlit shallows found a new source of energy for making food: light energy from the sun.

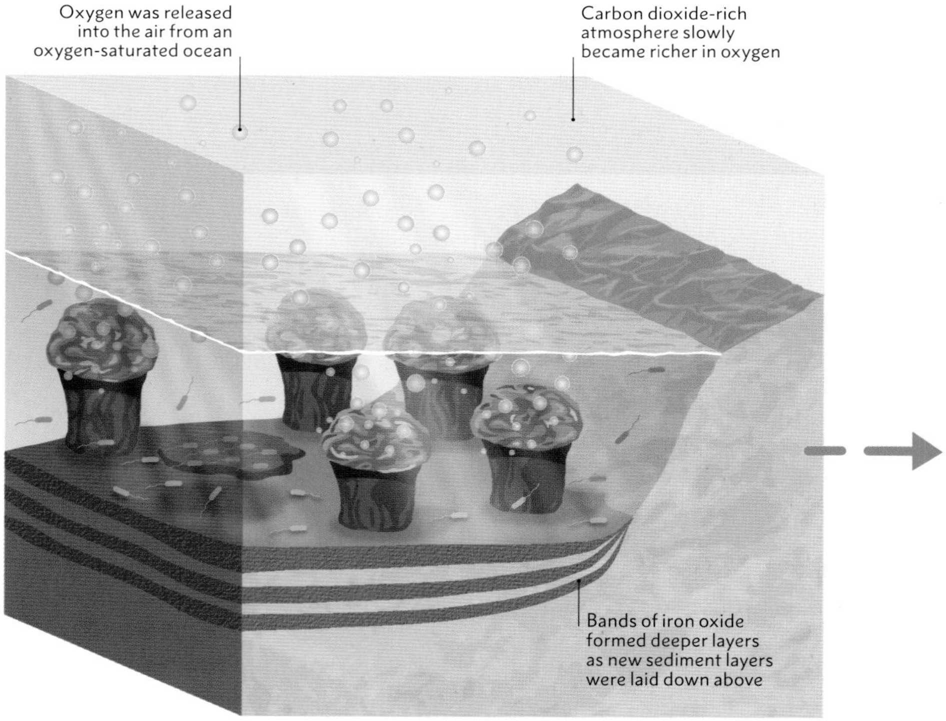
Oxygen was released into the air from an oxygen-saturated ocean
Carbon dioxide-rich atmosphere slowly became richer in oxygen
Bands of iron oxide formed deeper layers as new sediment layers were laid down above

For hundreds of millions of years, oxygen produced by photosynthesis was soaked up by the ocean's iron and laid down in rusty bands that today comprise an important part of the world's reserves of iron ore. When the ocean's dissolved iron ran out around 2.4 BYA, oxygen saturated the water, then started to fill the atmosphere.

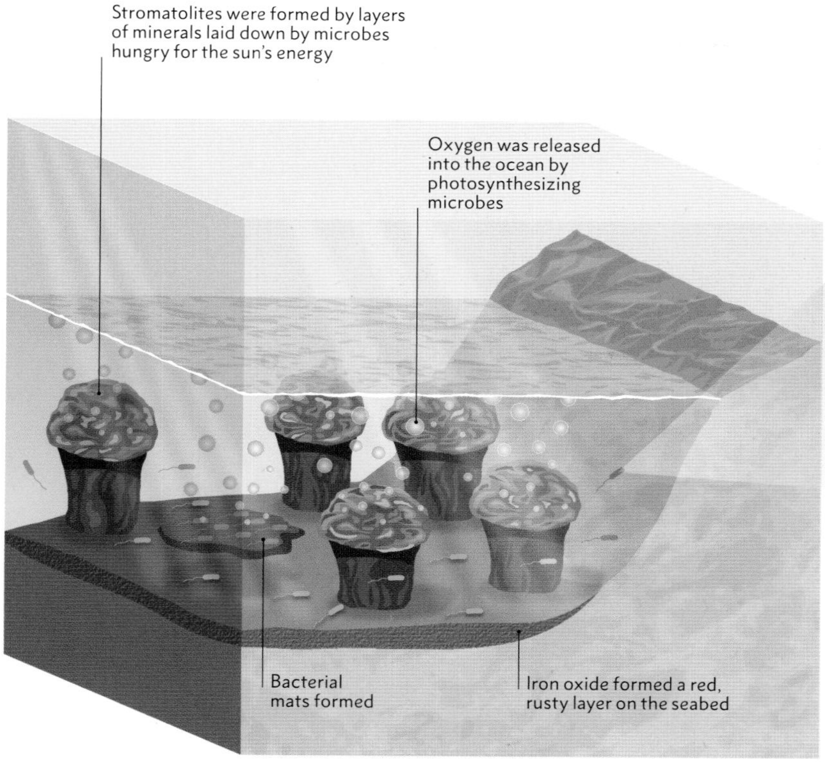
Stromatolites were formed by layers of minerals laid down by microbes hungry for the sun's energy
Oxygen was released into the ocean by photosynthesizing microbes
Bacterial mats formed
Iron oxide formed a red, rusty layer on the seabed

Microbes in shallow seas began photosynthesizing between 3.8 and 3.2 BYA. They formed colonies, collecting as bacterial mats and building stromatolites. By extracting hydrogen from water, they released oxygen, but it did not escape into the atmosphere. It reacted with the ocean's dissolved iron, turning it to iron oxide on the seabed.

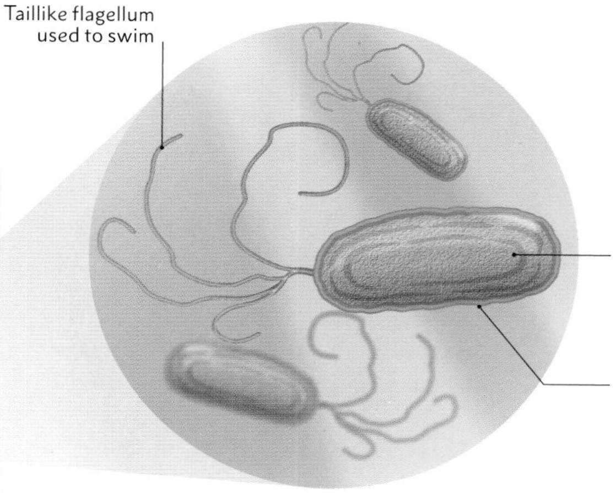
Taillike flagellum used to swim
Green pigment packed onto membranes inside microbe
Cell wall surrounds simple cell body

◀ **Solar-powered microbe**
The photosynthesizing microbes evolved pigments, colored green, that absorbed the sun's energy. They harness this energy to make organic food, such as sugars, in the process of photosynthesis (see pp.114–15).

“ IT IS THIS CONDITION THAT MAY HAVE **SET THE ENVIRONMENTAL STAGE** AND ULTIMATELY THE CLOCK **FOR THE ADVANCE OF... ANIMALS**. ”

Timothy Lyons, biogeochemistry professor, c.1960–

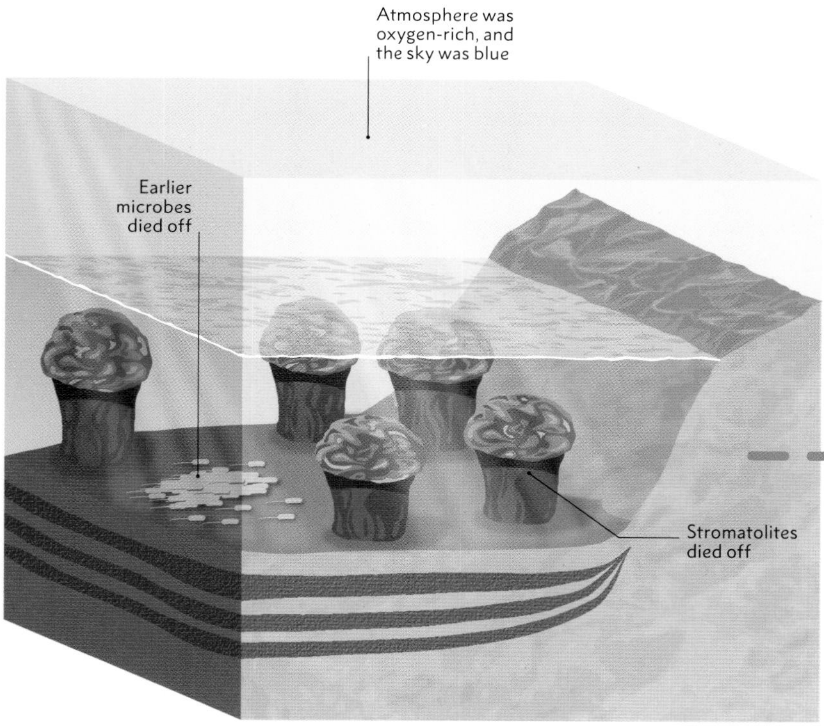
Atmosphere was oxygen-rich, and the sky was blue
Earlier microbes died off
Stromatolites died off

After 2.4 BYA, the ocean's water was full of oxygen and the atmosphere was oxygen-rich. Since organisms had evolved in habitats low in oxygen, these new conditions poisoned most of them. Only a few had the means to detoxify the oxygen, and so could survive.

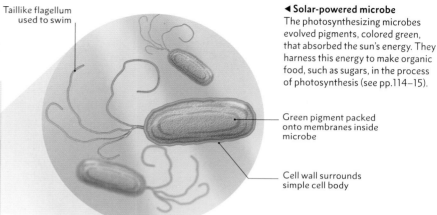

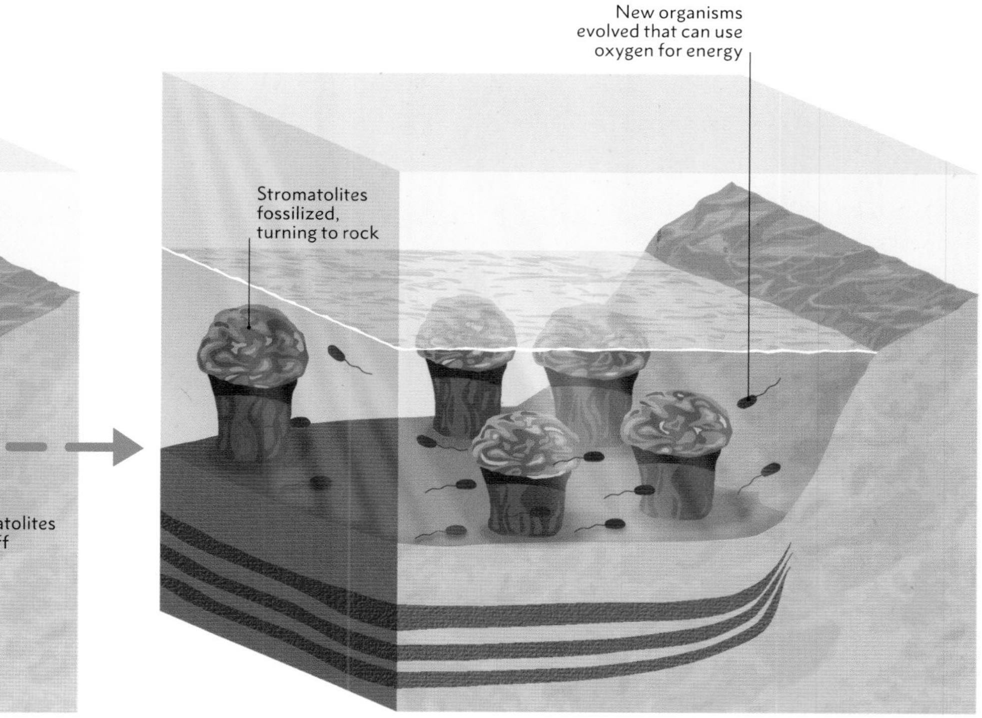
New organisms evolved that can use oxygen for energy
Stromatolites fossilized, turning to rock

New microbes evolved and could now use oxygen to extract more energy from food and went on to be the dominant life-forms in the new oxygen-rich habitat. A few oxygen-hating microbes persisted where oxygen could not reach them—such as in thick muds.

COMPLEX **CELLS EVOLVE**

In a world 2.7 BYA, teeming with microbes, life found a way to move forward. Simple bacteria were joined by bigger cells to form microscopic cooperatives, merging and collaborating to form complex new cells. Such cells would become the living units of plants and animals.

The abilities of bacteria are limited by their simple structure. Although they can perform chemical tricks impossible in more complex life, they are restricted in how they move and socialize. Greater possibilities opened up when bigger microbes swallowed smaller ones—and kept them alive inside them.

CELLULAR COMPARTMENTS

Plant and animal cells are eukaryotic ("eu," true; "karyon," kernel), meaning they have a central compartment called the nucleus. This, together with many other membrane-bound chambers, distinguishes these complex cells from bacteria. The chambers are called organelles, because their uses in a cell are comparable to the functions of organs in a larger body. Some, notably chloroplasts and mitochondria, are reminiscent of some free-living bacteria. It suggests they came to be when microbes in prehistoric communities engulfed smaller cells for food, but instead of eating them, held them captive, preserving their life processes. In this way, some photosynthetic bacteria of yesterday became the chloroplasts of today. And mitochondria, which respire using oxygen, came from oxygen-breathing bacteria. Even the nucleus may have begun like this, although little remains to hint at its probable archaea ancestors. In each case, the prisoners were "cultivated" and passed down whenever their hosts reproduced. Over millions of years, host and organelle became entirely codependent.

Eukaryotes expanded more than bacteria ever could. Some used their photosynthesizing chloroplasts to become algae and plants. The food-eaters became amoebas, fungi, and animals. A few, such as *Euglena*, could even switch between photosynthesis in sunshine and absorbing food in darkness. But it was cell-to-cell interaction that continued to be the driving force in escalating complexity—so, in time, eukaryotes evolved into the largest and most elaborate organisms on the planet.

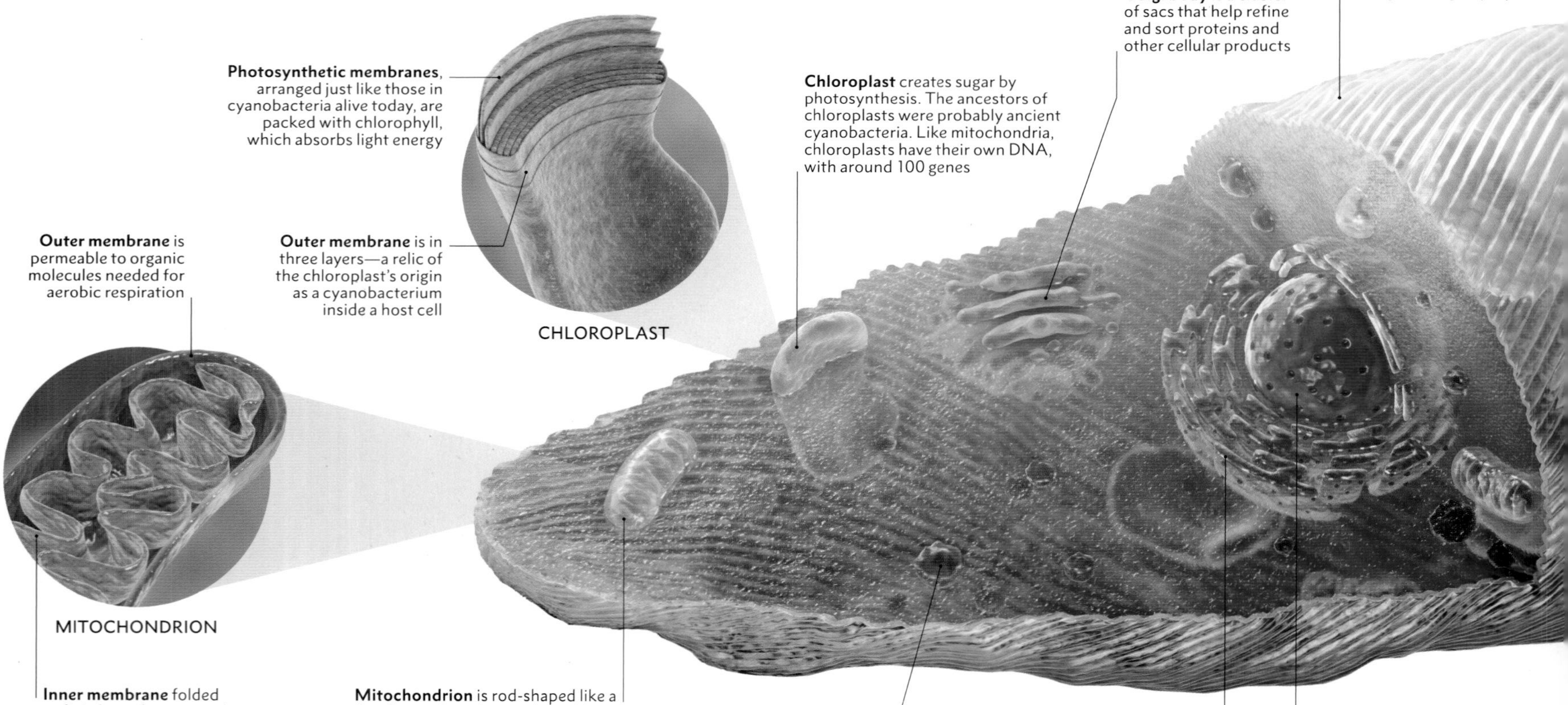
Ridged surface, or pellicle, is tough enough to protect the organism but flexible enough to let *Euglena* engulf prey
Golgi body is a cluster of sacs that help refine and sort proteins and other cellular products
Photosynthetic membranes, arranged just like those in cyanobacteria alive today, are packed with chlorophyll, which absorbs light energy
Chloroplast creates sugar by photosynthesis. The ancestors of chloroplasts were probably ancient cyanobacteria. Like mitochondria, chloroplasts have their own DNA, with around 100 genes
Outer membrane is permeable to organic molecules needed for aerobic respiration
Outer membrane is in three layers—a relic of the chloroplast's origin as a cyanobacterium inside a host cell
CHLOROPLAST
MITOCHONDRION
Inner membrane folded to fit in lots of enzymes that perform aerobic respiration—the process that uses oxygen to break down food for energy
Mitochondrion is rod-shaped like a bacillus and is a descendant of free-living bacteria. It even retains some DNA, containing more than 30 genes, from its time as a free-living organism
Food particles engulfed by the cell are digested inside by enzymes
Endoplasmic reticulum is a stack of membranous envelopes. Most are studded with protein-making granules called ribosomes; others help make oily substances
Nucleus contains DNA inside its double membrane. Nuclei may have originated as free-living archaea—microbes that survived in hot acidic pools

Neither animal nor plant
A marvel of microscopic intricacy, *Euglena* is a single-celled pond dweller that can photosynthesize like a plant, but can also eat food like an animal. Its whiplike flagellum helps it to move into sunlight or toward nourishment in darkness.

Contractile vacuole squeezes out excess water from *Euglena* to keep its body fluids in balance. This allowed its ancestors to move from the salty ocean to fresh water

Light-sensitive bulge at base of flagellum tells *Euglena* the direction of light

Main flagellum beats, triggered by the bulged light receptor at its base, so that the cell moves toward light

Stigma, or eye-spot—a patch of orange pigment that casts a shadow on *Euglena*'s light-sensitive bulge

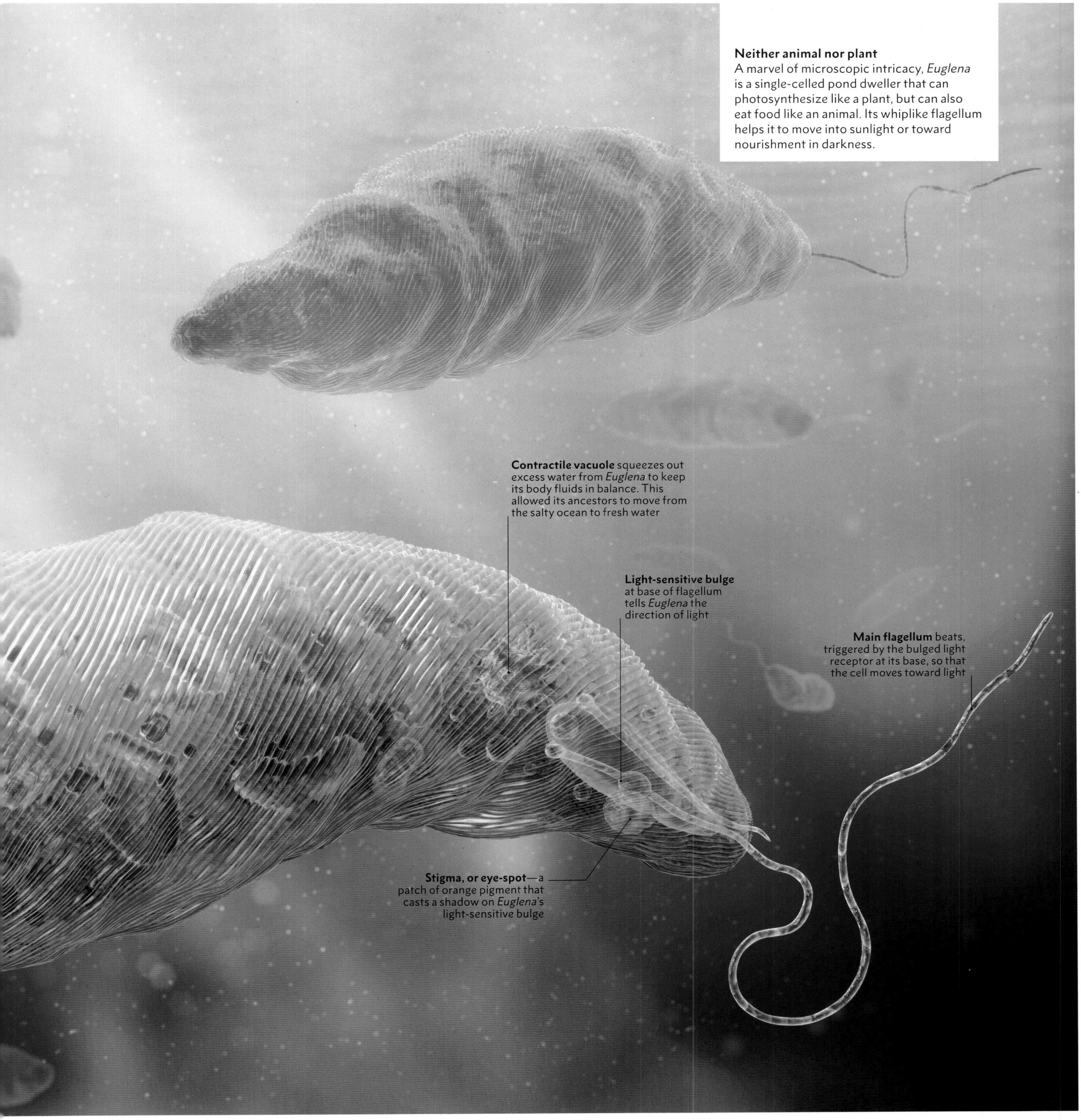

SEX MIXES GENES

Mistakes in the copying of genetic material, known as mutations, create new genes and characteristics—but it is the sexual behavior of life that mixes them up, creating unique individuals. Sexuality is a basic property of all known life and it is likely that it emerged very early on in evolution.

Some organisms reproduce without sex, so offspring carry exact copies of their parent's genes. The only way they can change over generations is when mutation produces variety. But most organisms, because of their sexuality, can vary much more. Sex mixes up DNA, enriching a population with new combinations. A plant species might have genetically-determined white or purple flowers, as well as tall or dwarf statures. These variants are produced by mutations (see pp.108–09), but sex mixes them up, so both tall or dwarf plants can produce flowers of either color.

The simplest kind of sex happens when bacteria exchange bits of DNA. When they separate, each partner is genetically changed, but no new cells are made. So bacteria have evolved sex, but not sexual reproduction.

HOW TO SHUFFLE A HUGE GENOME

Complex, or eukaryotic, cells (see pp.118–19), including those of all plants, fungi, and animals, cannot exchange their genes as bacteria do: their long, unwieldy chains of DNA prevent it. Instead, they rely on first making special sex cells containing only half their DNA, and then fuse, or fertilize, them with half the DNA of another individual.

To achieve the halving and fertilization, they need two "doses" of each kind of gene. The halving process, called meiosis, separates the doses into sex cells (usually sperm and eggs), and fertilization restores the double dose. This ensures that each gene gets inherited and no information is lost.

THERE ARE **8 MILLION** GENETIC COMBINATIONS POSSIBLE IN THE **SPERM OR EGGS** PRODUCED BY **EVERY HUMAN**

VARIETY IN SPERM AND EGGS

Fertilization mixes genes from different individuals, but meiosis ensures that all the sex cells coming from a single parent are different, too. As a prelude to meiosis, DNA is shuffled around in the cells of the sex organs, so that all the sperm or egg cells made by one parent are genetically different.

Plants, animals, and other complex organisms evolved sexual lifecycles that were molded by their capabilities. Fungi—which grow as microscopic threads—adopted a method reminiscent of bacteria: their threads fuse in places without producing true sperm or eggs. Plants—rooted to the ground—evolved cycles that used dispersive spores or pollen. But in all these organisms, sex served to multiply the raw material for natural selection—variation.

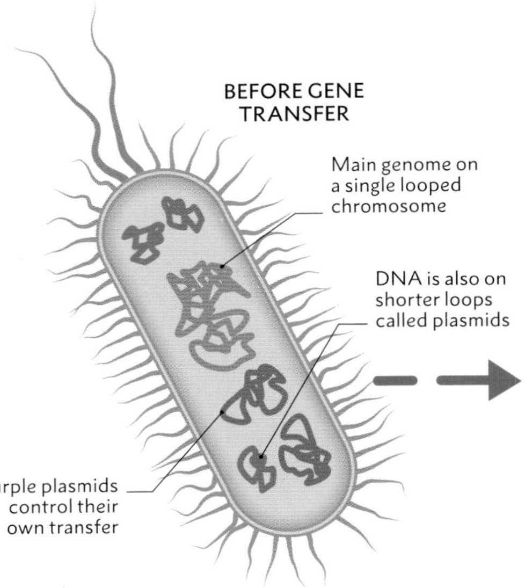
BEFORE GENE TRANSFER
Main genome on a single looped chromosome
DNA is also on shorter loops called plasmids
Purple plasmids control their own transfer

▲ Bacterium
Bacteria have sex by transferring DNA to another individual. Some of the genes that control the exchange are actually on the DNA that is moved, so the DNA strand controls its own transfer, a little like an independent life-form.

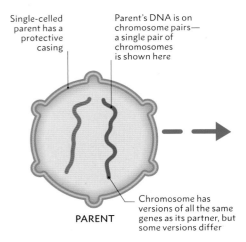

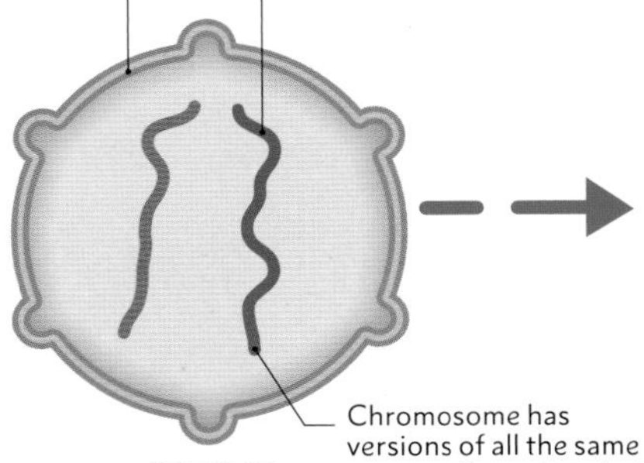
Single-celled parent has a protective casing
Parent's DNA is on chromosome pairs—a single pair of chromosomes is shown here
PARENT
Chromosome has versions of all the same genes as its partner, but some versions differ

▲ Complex microbe
Chlamydomonas is a single-celled microbe, but as a complex cell (eukaryote), it has a double-dose of DNA divided into pairs of equivalent chromosomes. Each member of a chromosome pair has equivalent genes to its partner, but these genes may differ, due to mutations accumulated over millions of years.

▼ Spawning
Production of sex cells can be prolific. Corals release millions of sperm and eggs simultaneously—increasing the chance of fertilization in the open ocean water.

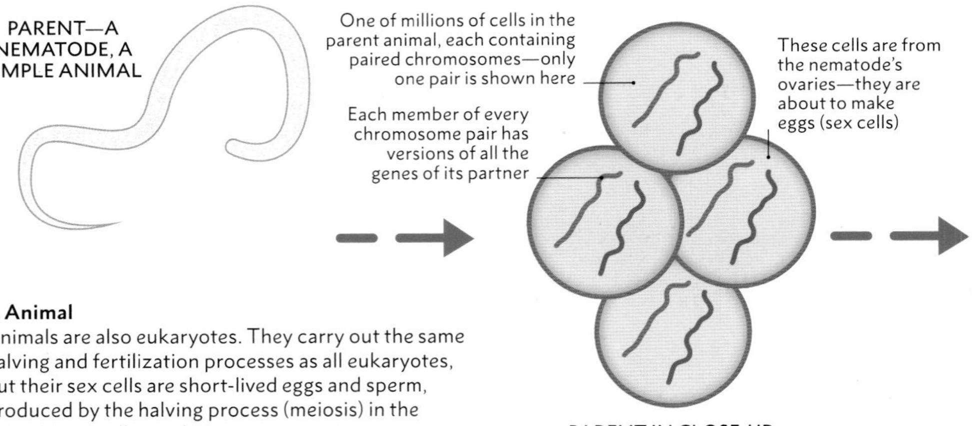
PARENT—A NEMATODE, A SIMPLE ANIMAL
One of millions of cells in the parent animal, each containing paired chromosomes—only one pair is shown here
Each member of every chromosome pair has versions of all the genes of its partner
These cells are from the nematode's ovaries—they are about to make eggs (sex cells)
PARENT IN CLOSE-UP

▲ Animal
Animals are also eukaryotes. They carry out the same halving and fertilization processes as all eukaryotes, but their sex cells are short-lived eggs and sperm, produced by the halving process (meiosis) in the animals' sex cells—either ovaries or testes.

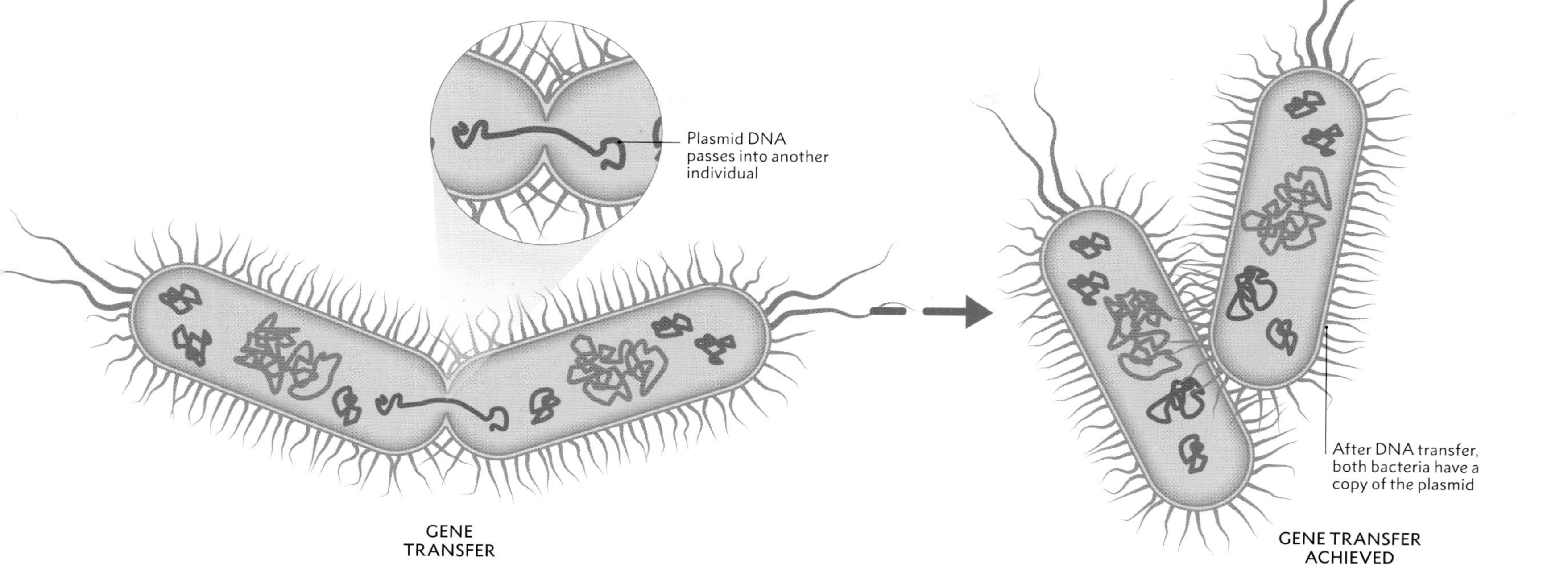
Plasmid DNA passes into another individual
GENE TRANSFER
After DNA transfer, both bacteria have a copy of the plasmid
GENE TRANSFER ACHIEVED

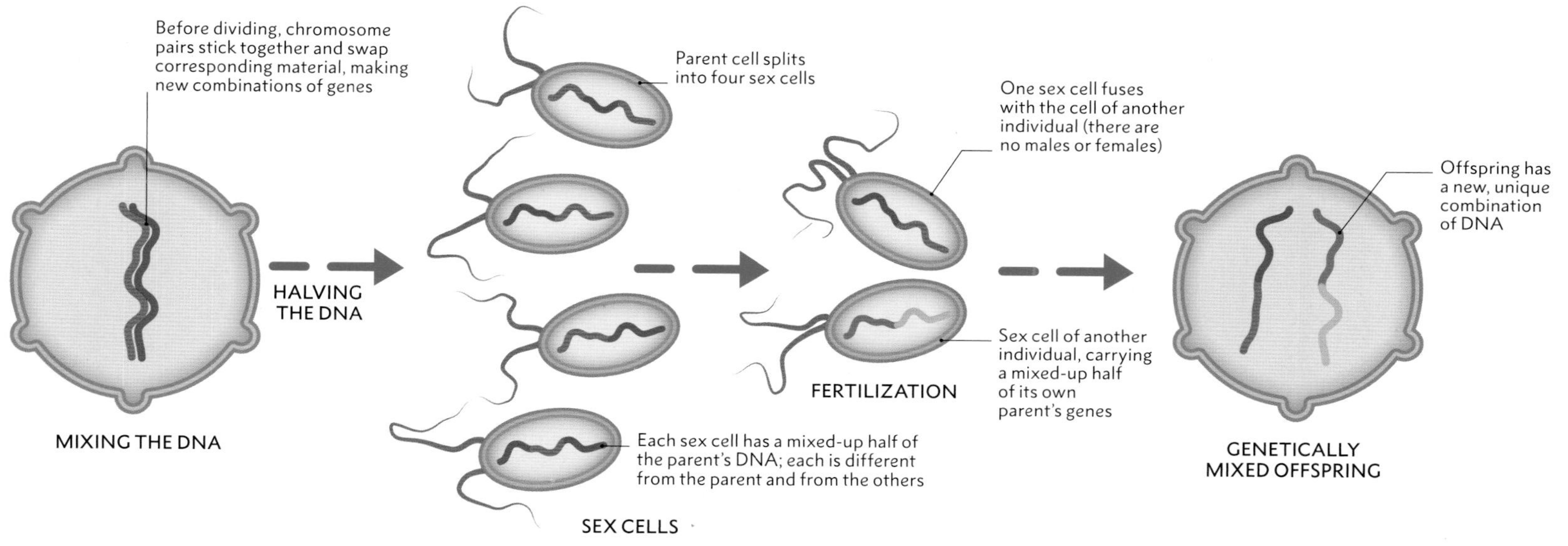
Before dividing, chromosome pairs stick together and swap corresponding material, making new combinations of genes
MIXING THE DNA
HALVING THE DNA
Parent cell splits into four sex cells
Each sex cell has a mixed-up half of the parent's DNA; each is different from the parent and from the others
SEX CELLS
One sex cell fuses with the cell of another individual (there are no males or females)
Sex cell of another individual, carrying a mixed-up half of its own parent's genes
FERTILIZATION
Offspring has a new, unique combination of DNA
GENETICALLY MIXED OFFSPRING

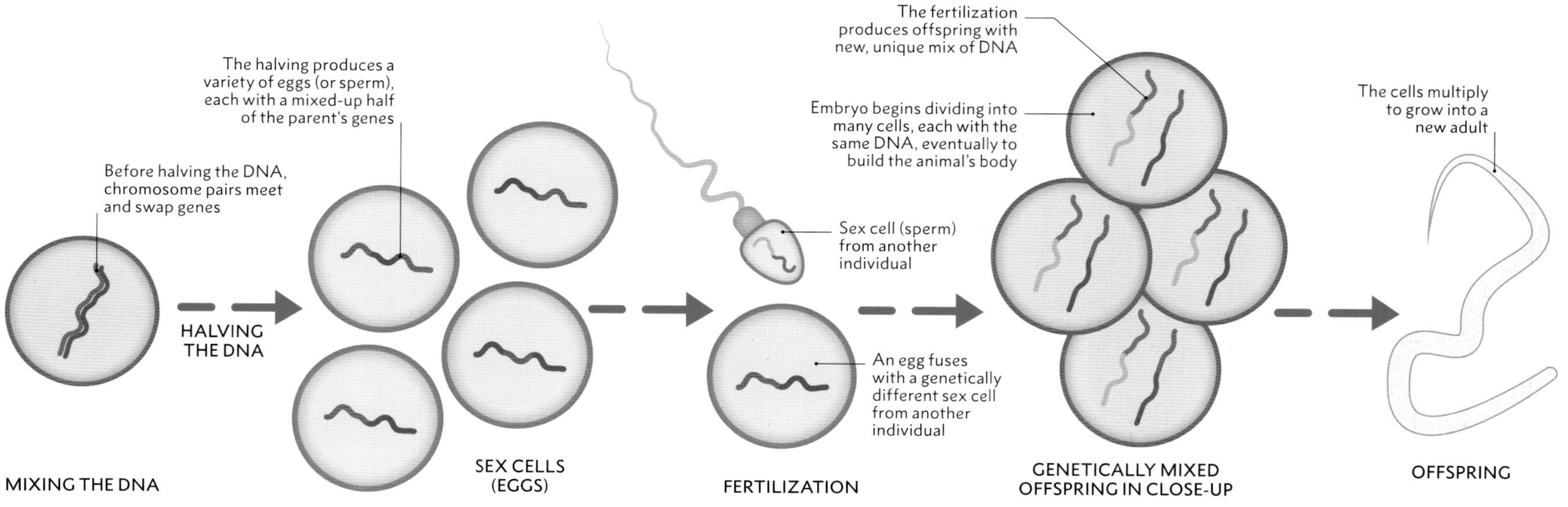
Before halving the DNA, chromosome pairs meet and swap genes
MIXING THE DNA
HALVING THE DNA
The halving produces a variety of eggs (or sperm), each with a mixed-up half of the parent's genes
SEX CELLS (EGGS)
Sex cell (sperm) from another individual
An egg fuses with a genetically different sex cell from another individual
FERTILIZATION
The fertilization produces offspring with new, unique mix of DNA
Embryo begins dividing into many cells, each with the same DNA, eventually to build the animal's body
GENETICALLY MIXED OFFSPRING IN CLOSE-UP
The cells multiply to grow into a new adult
OFFSPRING

CELLS BEGIN TO **BUILD BODIES**

The step from microscopic, single-celled microbes to organisms such as plants and animals, with up to trillions of cells, was another quantum leap in the complexity of life. Maintaining order in a multicellular organism demands that cells not only stick together in the right way, but also communicate so that the entire body develops properly.

There are limits to the capabilities of a single-celled microbe. Cells cannot grow beyond a certain size without becoming unmanageable—using diffusion, materials for life pass in and out of their bodies only over microscopic distances, and the oily cell membrane breaks up if a cell gets too big. Cells divide when they reach a certain stage, so microbes stay microscopic.

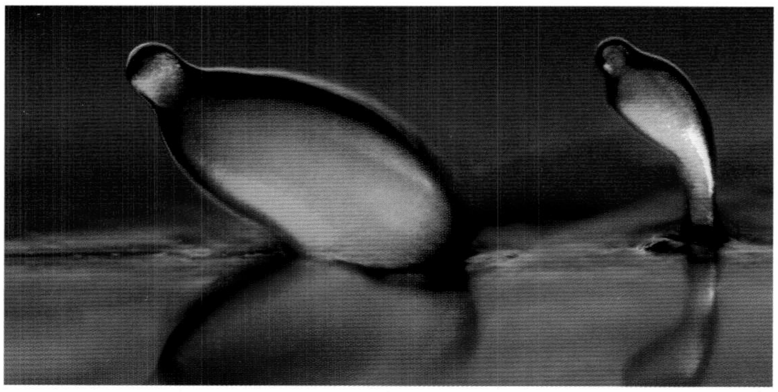

▲ Temporary body
Slime molds are on the cusp of multicellularity. They are usually solitary, amoebalike single cells, but in times of stress, they band together and form multicellular fruiting bodies such as these.

Bigger organisms with cooperating working parts can evolve new ways to live, but they must become multicellular to do so. Some microbes refuse to separate after division, so their cells remain attached in a colony. The simplicity of this arrangement—division without segregation—suggests that multicellularity in itself is not such a monumental achievement—but getting body parts (and therefore cells) focused on different tasks is another matter.

DIVISION OF CELLULAR LABOR

True multicellularity happens when a colony's cells work together and specialize, relying on chemical cues from their neighbors to do so. All cells in a colony carry copies of DNA made by replication during each cellular division. Although they keep identical genetic blueprints, cells switch off selected genes as they forego certain functions to concentrate on specific jobs—and increasingly rely on other cells around them to supply their deficiencies.

In the Precambrian oceans, filter-feeding sponges were among the first multicellular animals, although they are just a step away from being a loose colony. A sponge passed through a sieve can sprout new individuals from each separated cell, and the same is true of some simple algae. Later, more complex, animals and plants had cells more committed to their specific roles. Their fate—to become skin, muscle, or another tissue—is set by their location in the early embryo. Cooperating tissues then become organs, such as solar-powered leaves or beating hearts, and their cells no longer survive alone.

Multicellularity might make cells forever dependent, but it reaps enormous benefits for the bigger body. It allowed life to evolve working parts, such as stinging tentacles and sex organs. The variety of body sizes now possible multiplied the complexity of natural communities, leading to elaborate food webs and habitats built from the bodies of larger organisms, from corals to trees.

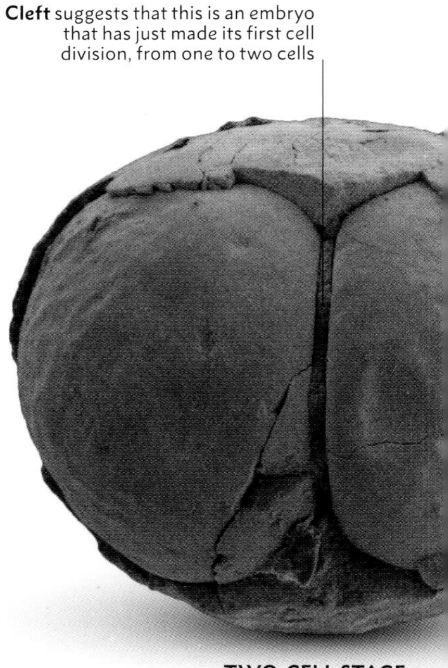
Cleft suggests that this is an embryo that has just made its first cell division, from one to two cells

TWO-CELL STAGE

16-CELL STAGE

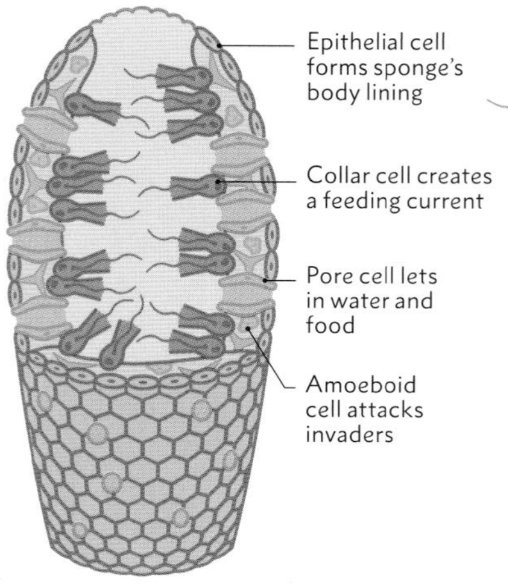
Epithelial cell forms sponge's body lining
Collar cell creates a feeding current
Pore cell lets in water and food
Amoeboid cell attacks invaders

Sponge

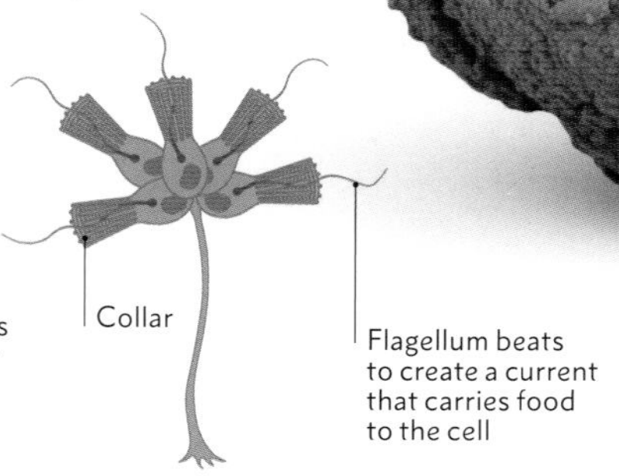
Collar
Flagellum beats to create a current that carries food to the cell

Choanoflagellate colony

◄ Creature or colony?
The distinction between colonies of cells and true multicellular life is not always clear. Single-celled microbes called choanoflagellates form stalked colonies. Many cells in a sponge look and behave in much the same way. What makes the animal more than a colony are its different, specialized cell types, which must cooperate in an integrated way to survive.

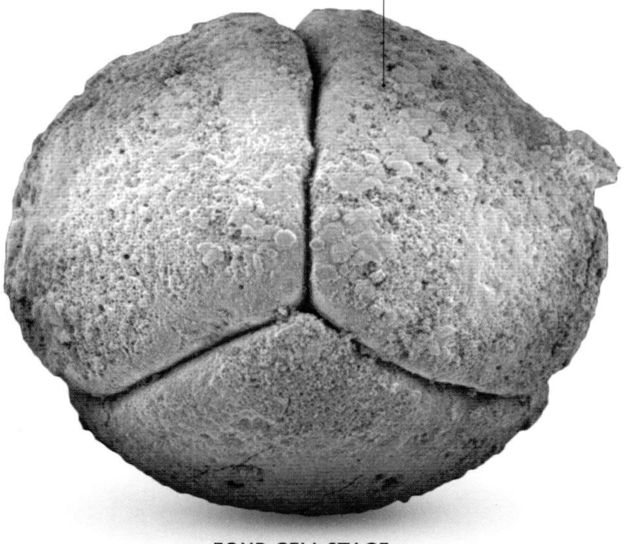
One of four cells in this fossil, suggesting it is an embryo that has divided twice

FOUR-CELL STAGE

Granular texture is due to the mineralization process of fossilization
A membrane encloses the cells, just as it would in an animal embryo

EIGHT-CELL STAGE

Cells in this "embryo" are more rounded—perhaps due to loss of its enveloping membrane

32-CELL STAGE

This membrane encloses what looks like a ball of cells—called a blastula in animal embryos

BLASTULA STAGE

> "THE **ANCESTORS** OF THE **HIGHER ANIMALS** MUST BE ... **ONE-CELLED BEINGS**, SIMILAR TO THE **AMOEBAE** WHICH ... OCCUR IN OUR RIVERS, POOLS, AND LAKES."
>
> **Ernst Haeckel**, evolutionary biologist, 1834–1919
> *The History of Creation*

▲ Arrested development
Astonishing fossils from the Doushantuo Formation of China appear to show embryos frozen in time at their very earliest stages of cell division, as they change from a single egg cell to form first two, then four, and eight cells, and so on. This act of cell division without separating is at the root of multicellularity; it may be that these fossils represent very early multicellular animals beginning life around 635 million years ago.

Showing off
Many males use color to impress females in species that have good daytime vision, such as big-eyed jumping spiders. This male peacock jumping spider combines color with choreography in his courtship display.

MALES AND **FEMALES DIVERGE**

As well as evolving complex, multicellular bodies, plants and animals also diverged into two sexes. In each species of animal, half became females and—through yolky eggs or pregnancy—focused on nourishing their offspring. The other half—the males—became fighters and show-offs.

Contrast between the sexes can be very pronounced indeed. A female elephant seal can be five times smaller than her mate—and an anglerfish female 40 times bigger. All sexual organisms have a shared genetic investment in producing offspring, but males and females have dissimilar—although complementary—interests in the way they help create the next generation.

> WE CAN HARDLY BELIEVE THAT ... THE **FEMALE** ... IS NOT **INFLUENCED** BY THE **GORGEOUS COLORS** OR OTHER ORNAMENTS WITH WHICH THE **MALE ... IS DECORATED**.
>
> **Charles Darwin**, biologist, 1809–82, *The Descent of Man and Selection in Relation to Sex*

MATING TYPES AND SEXES

The lowliest of organisms manages to be sexual without having males and females at all. Many microbes and fungi have multiple, but identical-looking, "mating types." Subtle chemical differences dictate whether they can fuse to mix their genes.

Mating types have equal reproductive responsibilities. But the evolution of different sexes changes this. Although each sex contributes the same amount of genetic information, the female sex supplies hers as an egg provided with nourishing yolk, while males make lightweight sperm devoted to racing to fuse with that egg. The battle of the sexes began when sperm started swimming toward food-packed eggs.

CHOOSY FEMALES, SHOWY MALES

Some females—such as many insects and fish—deposit tiny amounts of yolk in each egg so can still afford to produce hundreds. Others make fewer, yolkier eggs or give birth to young after a costly pregnancy. Either way, high bodily investment in the next generation makes a female choosy when it comes to selecting mates and passing on her genes.

The cost of sperm production is far lower. In the drive to pass on their genes, males invest more in beating other males to fertilize eggs, either in competition, such as a race or fight, or by wowing females with advertisement displays. This has resulted in extravagant male features, from the giant jaws of stag beetles to a bird-of-paradise's plumes. Fossil evidence—such as the crests of male pterosaurs—suggests that this is nothing new. But male displays relying on color, voice, or behavior leave no trace; today these attributes provide some of the most dazzling natural spectacles—as males fight, dance, or sing their way to mating success.

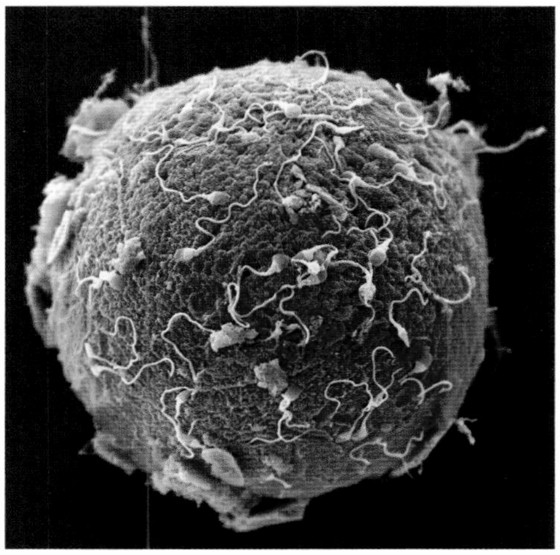

◀ **Size contrast**
An egg's package of cytoplasm and yolk makes it one of the biggest kinds of cells. A sperm—one of the smallest—has a whiplike flagellum that helps it swim, powered by a single mitochondrion.

ANIMALS GET A BRAIN

All animals have a nervous system that detects and responds to change. But only some evolved more complex behavior. The animals that did are those that started swimming or crawling forward. They developed a battery of sense organs and a decision-making brain to lead the way.

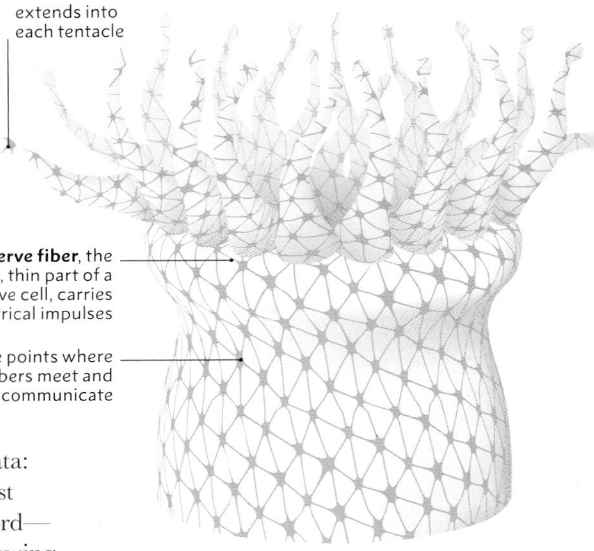
Nerve net extends into each tentacle
Nerve fiber, the long, thin part of a nerve cell, carries electrical impulses
Nodes are points where nerve fibers meet and communicate

▲ Nerve net
An anemone does not have any nerve cells concentrated in a brain. Instead, they are arranged into a net, with sensory ones collecting information and deeper ones communicating with muscles. Behavior is in its simplest stimulus-response form.

Some of the first animals, such as jellyfish, moved with tentacles radiating from the body in all directions. Their body had a top and bottom, but no front or back—so no head and tail. It was enough to respond to food and danger, and they had a nervous system for that, made up of long, interconnecting nerve cells. A stimulus, which can be any prompt from the environment, triggered their system to fire electrical impulses along the nerve cells' fibers—and when the signal reached a muscle, the muscle contracted to pull on a part of the body. But complex behavior was impossible: they had no brain to analyze sensory input and make decisions.

A HEAD FOR THINKING

More than 600 MYA, forward-moving animals introduced a key innovation. If they moved in one direction consistently, one part of the body—the front end—always encountered new territory first. Animals concentrated sense organs at this end and developed a corresponding mass of nerve cells that processed all the incoming data: they evolved the first heads with the first brains. A central conduit—a nerve chord—carried impulses through the body, allowing communication between brain, muscles, and sense organs. It meant a fundamental rearrangement. Two sides of the newly elongated body developed as mirror images of each other, giving the new kind of animal a single line of symmetry down the midline of its body. This body plan came to dominate animals from the simplest flatworms to the most complex vertebrates.

Brain power allows complex behavior, so spiders, for instance, can spin webs to catch prey. But behavior can still be "hardwired" and fixed by genes. Genuine versatility would come where traces of the brain's electrical activity left memories that affected behavior. Big-brained animals, such as mammals and birds, can learn from experience. And among them, a few gained foresight—the ultimate expression of brain power that foreshadowed human creativity.

▼ Fossil brain
Soft tissues, such as the brain, rarely fossilize, but the fossil head of a Cambrian shrimplike animal called *Fuxianhuia* shows a detailed brain impression. The large optic lobes suggest the animal relied on vision.

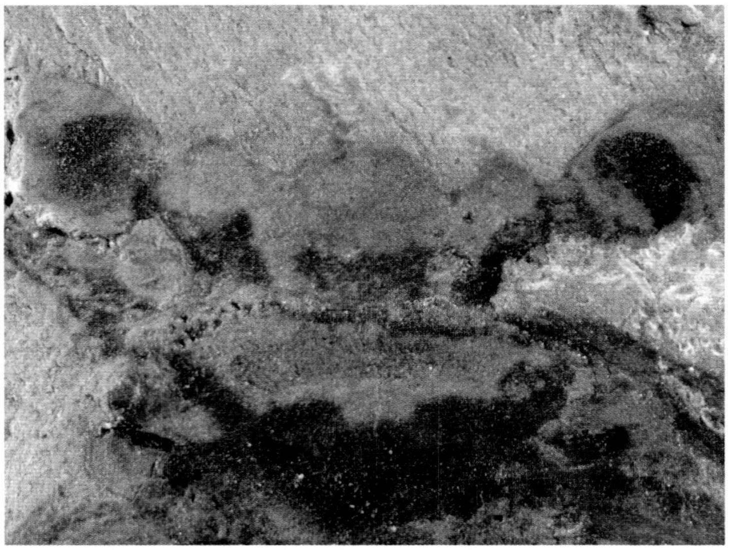

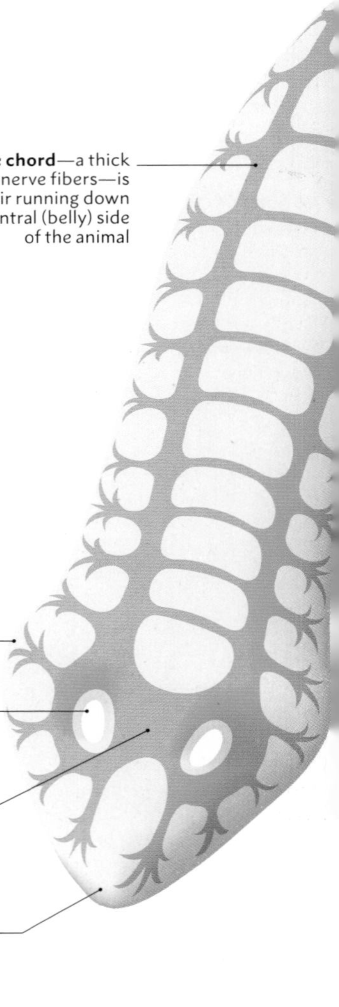
Nerve chord—a thick bundle of nerve fibers—is one of a pair running down the ventral (belly) side of the animal
Auricle is a projection on the side of the head that is sensitive to chemicals and is used to find food
Eyespot, or ocellus, responds to light but doesn't produce detailed images
Brain is simply a concentration of the biggest ganglia at the head end of the body
Snout is the first part of the body to encounter new things and is touch-sensitive

> PARTS OF THE BRAINS OF DIFFERENT ANIMALS HAVE **EXPANDED AND PROSPERED IN IMPORTANCE**... ALL IN ACCORDANCE WITH THE DEMANDS OF THE **LIFESTYLE OF THE SPECIES**.
>
> **Susan Greenfield**, neuroscientist, 1950–

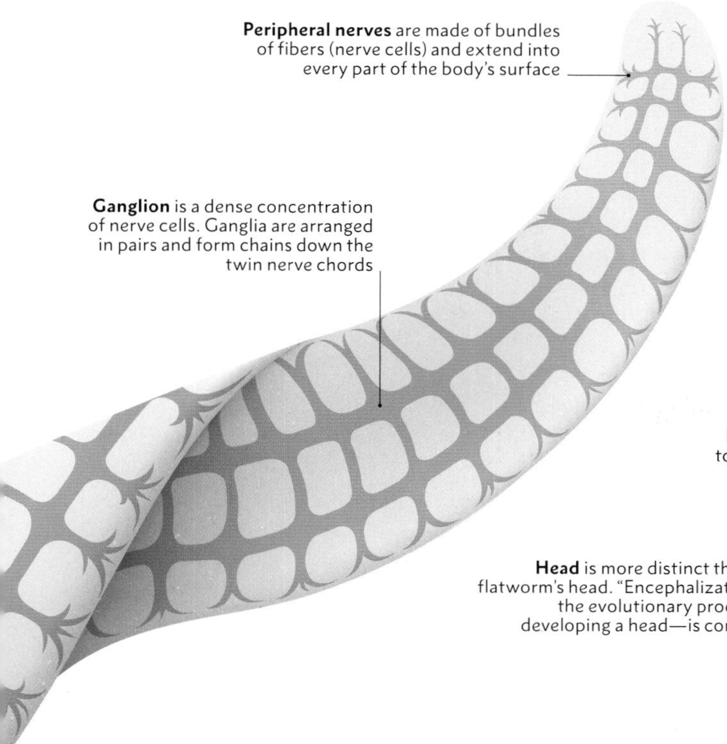
Peripheral nerves are made of bundles of fibers (nerve cells) and extend into every part of the body's surface
Ganglion is a dense concentration of nerve cells. Ganglia are arranged in pairs and form chains down the twin nerve chords

▲ Simplest brain
Flatworms are among the simplest of living animals with a head and a brain. The central nervous system consists of a cluster of ganglia (bulges of nerve cells) making up a primitive brain, while nerve chords communicate with the rest of the body. Further neurons carry impulses to and from sense organs and muscles.

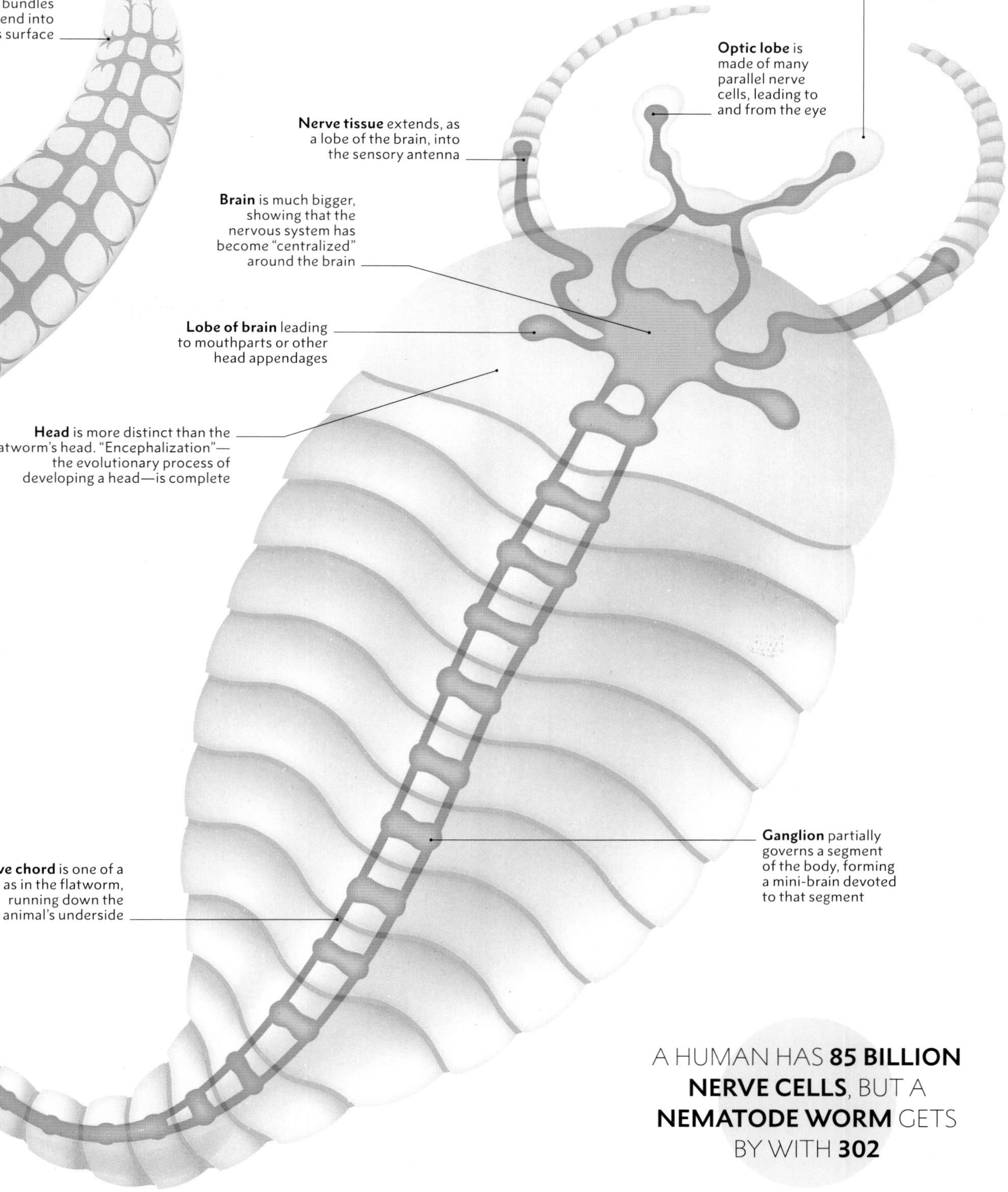
Compound eye is much more complex than an eyespot, and delivers detailed images to the brain
Optic lobe is made of many parallel nerve cells, leading to and from the eye
Nerve tissue extends, as a lobe of the brain, into the sensory antenna
Brain is much bigger, showing that the nervous system has become "centralized" around the brain
Lobe of brain leading to mouthparts or other head appendages
Head is more distinct than the flatworm's head. "Encephalization"—the evolutionary process of developing a head—is complete
Ganglion partially governs a segment of the body, forming a mini-brain devoted to that segment
Nerve chord is one of a pair, as in the flatworm, running down the animal's underside

► Bigger brain
This arthropod, based on fossil *Fuxianhuia* (see opposite), shows how complex nervous systems had become by the early Cambrian period (515 MYA). *Fuxianhuia* had a segmented body, with a pair of ganglia in each segment. The brain was much larger and features fat superhighways of nerve cells extending into the head's appendages and sense organs.

A HUMAN HAS **85 BILLION NERVE CELLS**, BUT A **NEMATODE WORM** GETS BY WITH **302**

ANIMAL LIFE **EXPLODES**

The first big explosion of animal life occurred just over 600 MYA—in oceans already alive with algae and microbes. From modest beginnings as creepers and grazers on the seabed, animals quickly evolved into all the main groups alive today.

▼ Colonizing the ocean floor
The earliest animals hugged the ocean floor, but their diversity and ecology escalated as some of them dug deeper into the mud and others grew upward into the water, discovering new survival strategies and building complex communities.

The oldest full-body fossils seem to appear so suddenly in the geological record that the first chapter in the evolution of animals has been called an "explosion." A fuller picture actually reveals what might be a series of explosions. An early wave of evolution left behind fossils worldwide, but notably in Newfoundland, Canada, and in Australia's Ediacara Hills, which gave their name to this period, the Ediacaran (635–541 MYA). The animals preserved are unrecognizable—some are disk-shaped, others frondlike—and scientists cannot place them in any modern groups. These were not the first animals. DNA evidence points to an even earlier pre-Cambrian origin, but the earliest forms left little more than tracks and traces. Those fossil traces can be a rich source of data themselves, however, telling us about animal lifestyles and communities.

EARLY RECYCLERS

Animals evolved from single-celled organisms. The pre-Cambrian track marks show that the lives of these first animals were tied to sediments on the ocean bed. Some crawled over the surface or grew into spongelike mats. Animals had evolved muscle systems, which distinguish them from other multicellular life. Their muscles helped them play an active role in shaping their environment. In their search for dissolved food, some of these pioneers of the sediment evolved into burrowers and began churning the sediment in ways that had never happened before. This swirled materials between the ocean water and the bottom muds—adding oxygen to the sediment and exchanging organic matter and minerals between the two habitats.

FROM THE **BEGINNING** TO THE **END** OF THE CAMBRIAN PERIOD, ANIMAL **BURROWING DEPTH INCREASED** FROM ½IN (1CM) TO 39IN (**1M**)

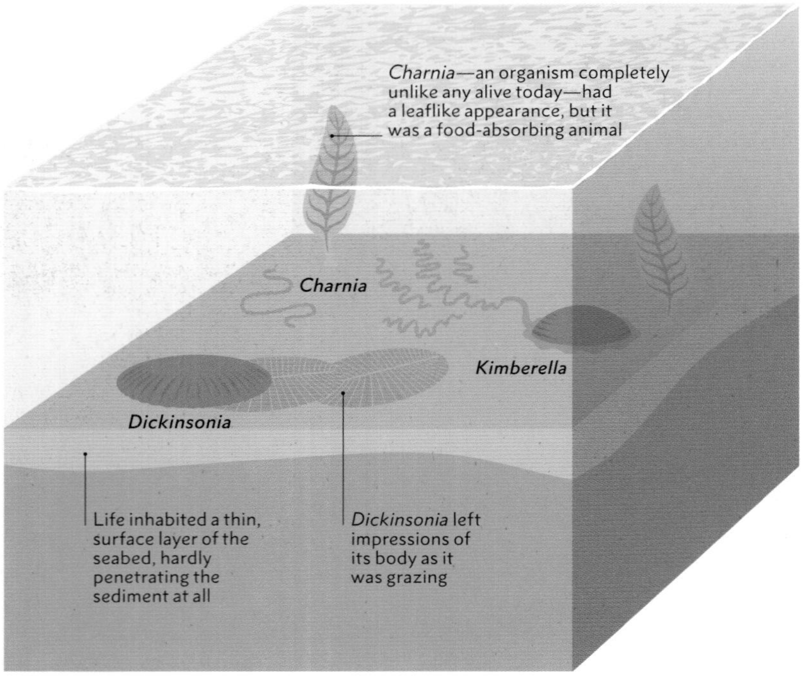
Charnia—an organism completely unlike any alive today—had a leaflike appearance, but it was a food-absorbing animal
Charnia
Kimberella
Dickinsonia
Life inhabited a thin, surface layer of the seabed, hardly penetrating the sediment at all
Dickinsonia left impressions of its body as it was grazing

In the Ediacaran period (about 560 MYA), the seabed was colonized by surface mats of algae, microbes, and possibly sponges. Scratch marks were made by early animals, possibly including *Kimberella*, as they grazed the algae.

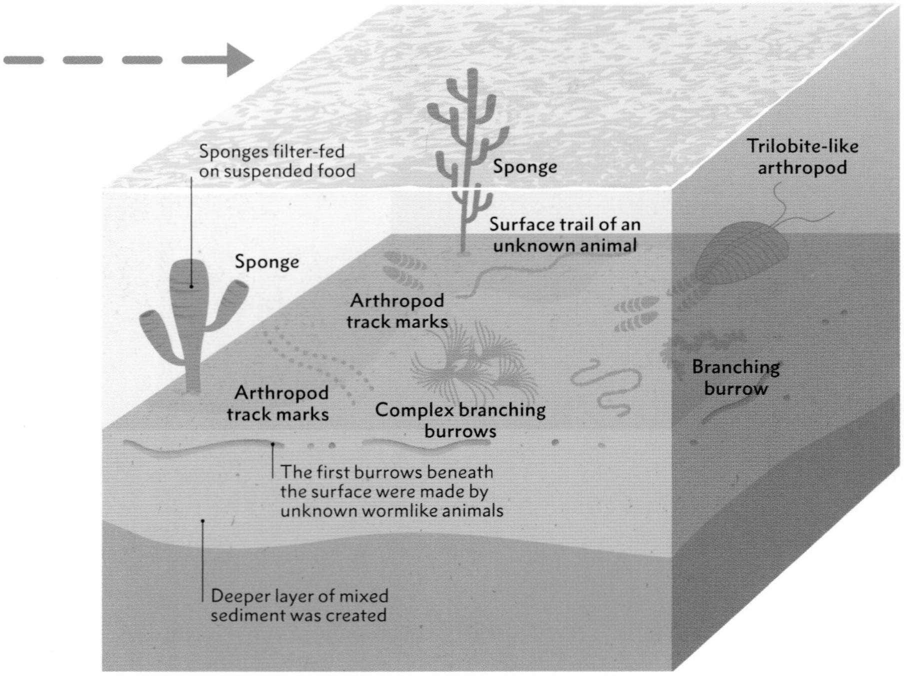
Sponges filter-fed on suspended food
Sponge
Trilobite-like arthropod
Surface trail of an unknown animal
Sponge
Arthropod track marks
Branching burrow
Arthropod track marks
Complex branching burrows
The first burrows beneath the surface were made by unknown wormlike animals
Deeper layer of mixed sediment was created

Early in the Cambrian period (about 540 MYA), a deeper layer of mixed, recycled sediment was created by animals burrowing and digging. The earliest known arthropods, probably resembling trilobites, left tracks—long before the first trilobite body was fossilized.

SEABED COMMUNITIES

By the early Cambrian, animal communities were flourishing on and around the seabed. The fossil record of this time is less incomplete, as many animals had chalky exoskeletons—protection from others but also able to support taller bodies and colonies. As plankton became richer with bigger organisms, their dead bodies and waste were more likely to sink. For the first time, life-forms in the water column were strongly linked to those on the ocean floor by a primitive food chain. Deposit-feeders came to depend on this rain of food.

Now was the time of the full Cambrian Explosion, documented most famously by Canada's Burgess Shale fossil assemblage (505 MYA). All the major kinds of living animals—flatworms, mollusks, and arthropods included—had evolved. But other, less familiar, types evolved alongside them. Some fossils suggest the existence of animals very unlike anything alive today, and many scientists have described this period as a time of experimentation in body shaping. Many of these ancient types disappeared without leaving lasting descendants, but others went on to fill the planet with animal life.

◀ **Experimental body** *Opabinia* is an example of an experimental body plan from the Burgess Shale. This creature is not related to any animal alive today, and some experts regard it as a failed body-plan experiment that soon died out.

> “SOME **15–20 BURGESS SPECIES** CANNOT BE ALLIED WITH ANY **KNOWN GROUP**. MAGNIFY SOME OF THEM... AND YOU ARE ON THE SET OF A **SCIENCE-FICTION FILM**.”
>
> **Stephen Jay Gould**, paleontologist and evolutionary biologist, 1941–2002
> *Wonderful Life: The Burgess Shale and the Nature of History*

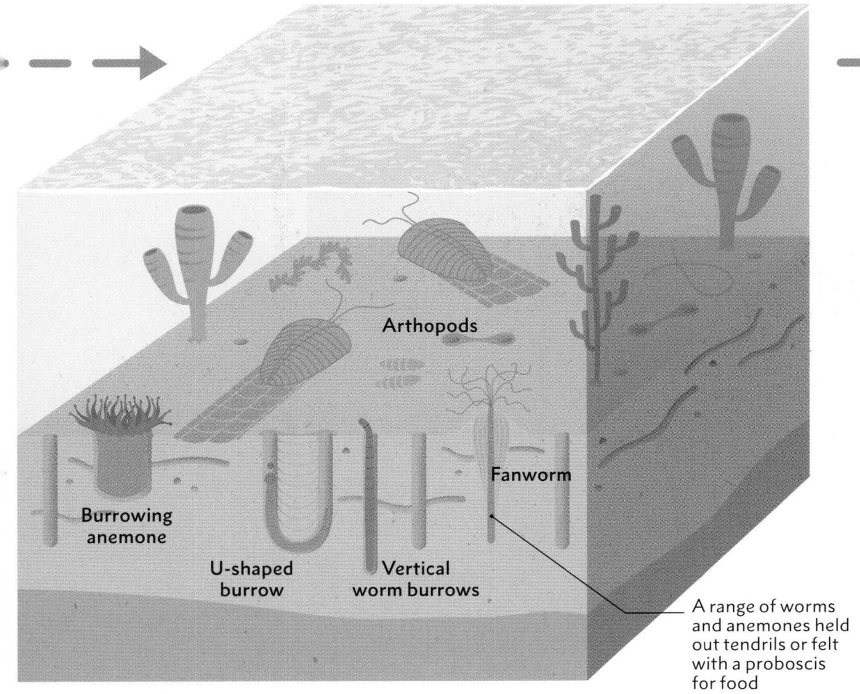

Later in the Cambrian (529 MYA), deposit-feeders subsisted on the “rain” of detritus from plankton above. They included animals with food-grabbing tentacles, including burrowers similar to the fanworms of today, and a diversity of trilobite-like arthropods, which left different types of tracks as they patrolled the seabed.

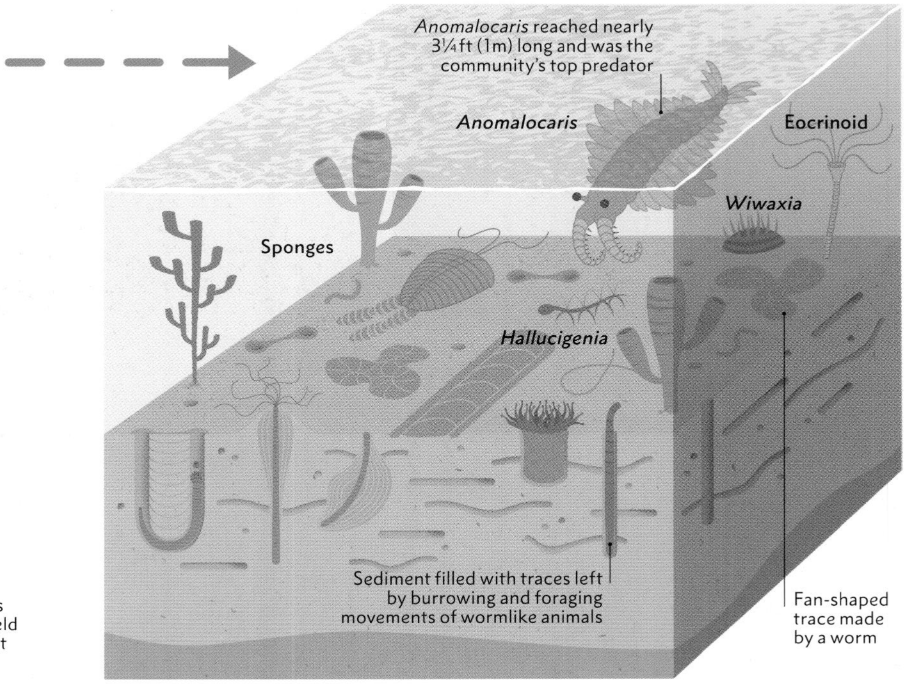

With the Cambrian Explosion in full flow (520–505 MYA), new lifestyles and experimental body plans really took off. Unique animals such as *Anomalocaris, Wiwaxia*, and *Hallucigenia* evolved, but left no successful descendants.

ANIMALS GAIN A **BACKBONE**

Backboned animals—from fish to mammals—have a history stretching back to small larvalike filter-feeders emerging in the evolutionary explosion of the Cambrian period. The internal skeleton they evolved went on to support animal bodies far larger than was possible before.

Vertebrates (animals with a spine, or vertebral column) emerged from small muscular swimmers in the Cambrian seas, before 500 MYA. They had a rubbery rod—a notochord—running through the back of a tapering body and blocks of flexing muscle that curved the rod from side to side. Fish use the same technique to swim today—but in most, the rod grows only in the embryo and is replaced with a harder backbone by adulthood. The Cambrian rod-backs were modest filter-feeders, but a backbone gave their descendants dramatically new ways to live their lives.

CARTILAGINOUS BEGINNINGS

The earliest elements of a skeleton were made from cartilage: tough, but flexible, tissue packed with collagen. Cartilage grew in the head of the first fish, such as *Haikouichthys*, and protected the brain and supported arches between their gill slits. In later animals, cartilage grew over the notochord and protected the spinal cord too, becoming the first true vertebral column. The column allowed stronger swimming, while fins—with cartilaginous supports of their own—improved control and stability.

Bodies with supporting cartilage could get bigger and more agile—but demanded more food and oxygen, too. The earliest fish got both by straining water through their gills—but feeding functions were later taken over by the mouth and throat, leaving the gills free to become better at extracting oxygen. This happened in bottom-living, armored, jawless fish called ostracoderms, which used throat muscles to suck food in from mud. But ostracoderms were also pioneers for another reason: they had the first bone.

BONY BODIES

Bone has its collagen hardened with at least 70 percent mineral. It may have evolved as a reservoir for the extra calcium and phosphate needed to trigger fast-acting muscles and nerves. But it had obvious mechanical benefits, too. Ostracoderms ("ostrakon," shell; "derma," skin) used bone as outer armor, packed with so much mineral it excluded living cells. Later fish permeated their bone with life-supporting microscopic channels, meaning it could grow from within to make an internal skeleton. Most vertebrates alive today have a bony skeleton, with cartilage largely around joints. A few—such as sharks and rays—reverted to a more lightweight cartilaginous skeleton, but bony fish diversified more, counteracting their heavier bone with a buoyant gas-filled swim bladder. And a bony skeleton was critical for the evolution of the land vertebrates that followed. Only giant bones could bear the weight of the biggest dinosaurs.

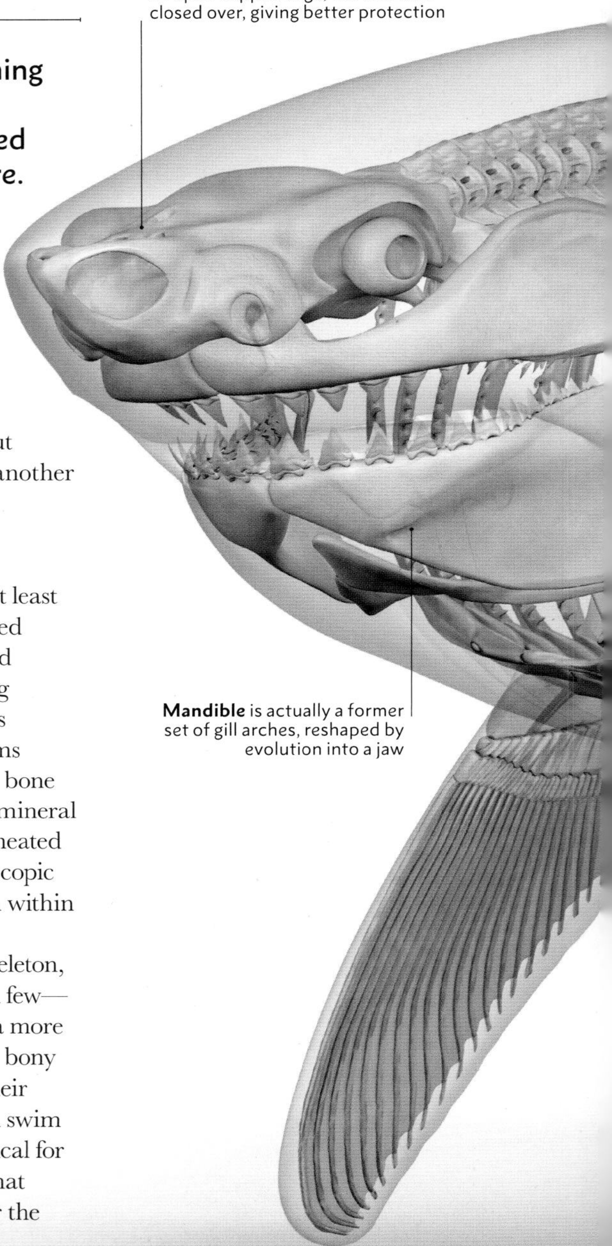
Cranium, or braincase, surrounds the brain. In early vertebrates, this formed an open-topped cage, but this later closed over, giving better protection
Mandible is actually a former set of gill arches, reshaped by evolution into a jaw

▼ Step by step
Fossils show that the evolution of a spine happened during the Cambrian period, 541–485 MYA. The story began with a back stiffened by a notochord (a rubbery rod) and passed through stages where vertebrae were made first from cartilage then mineralized as a true backbone.

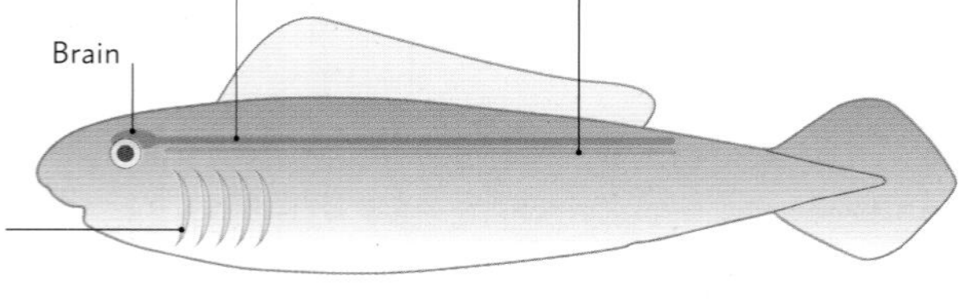
Nerve chord
Notochord—a stiff supportive rod
Brain
Gill slit—used for filter feeding

Chordate—a protofish with only a notochord

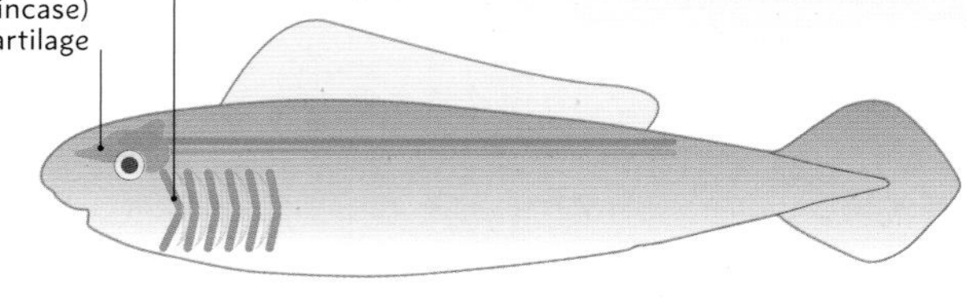
Cartilage gill arch stiffens gill, helping to hold it open for feeding
Cranium (braincase) of cartilage

Craniate—protofish with a braincase

◄ Modern vertebrate
A body reinforced by an internal jointed skeleton has enormous evolutionary promise. The great white shark is a highpoint of this potential: one of the most formidable top predators in the world today. Its skeletal elements are made from cartilage reinforced with mineral deposits. Sharks share the main skeletal elements of most other living vertebrates, except that they lack ribs.

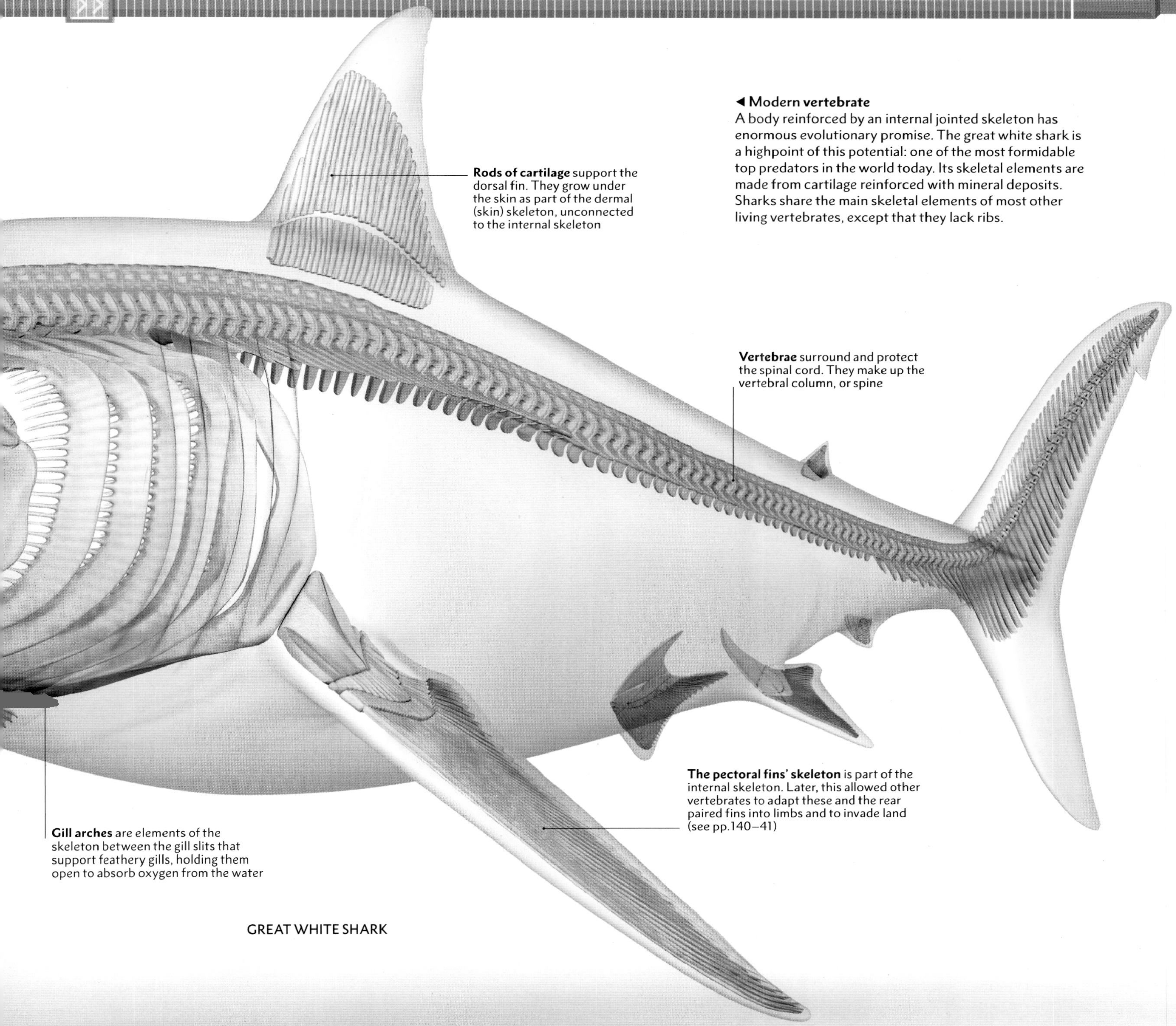
Rods of cartilage support the dorsal fin. They grow under the skin as part of the dermal (skin) skeleton, unconnected to the internal skeleton
Vertebrae surround and protect the spinal cord. They make up the vertebral column, or spine
The pectoral fins' skeleton is part of the internal skeleton. Later, this allowed other vertebrates to adapt these and the rear paired fins into limbs and to invade land (see pp.140–41)
Gill arches are elements of the skeleton between the gill slits that support feathery gills, holding them open to absorb oxygen from the water

GREAT WHITE SHARK

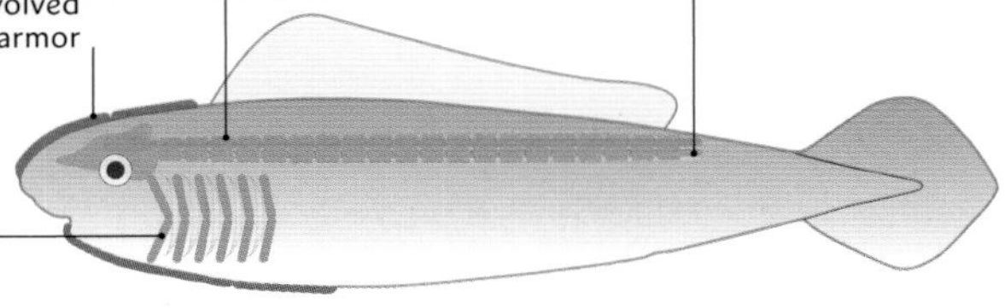
Vertebrae made of cartilage, making a chain called a vertebal column, or spine
Notochord still present in some early fish
Bony plates covering head—fish first evolved bone as external armor
Cartilage gill arch

Vertebrate—a jawless fish with cartilage internal skeleton and bony armor

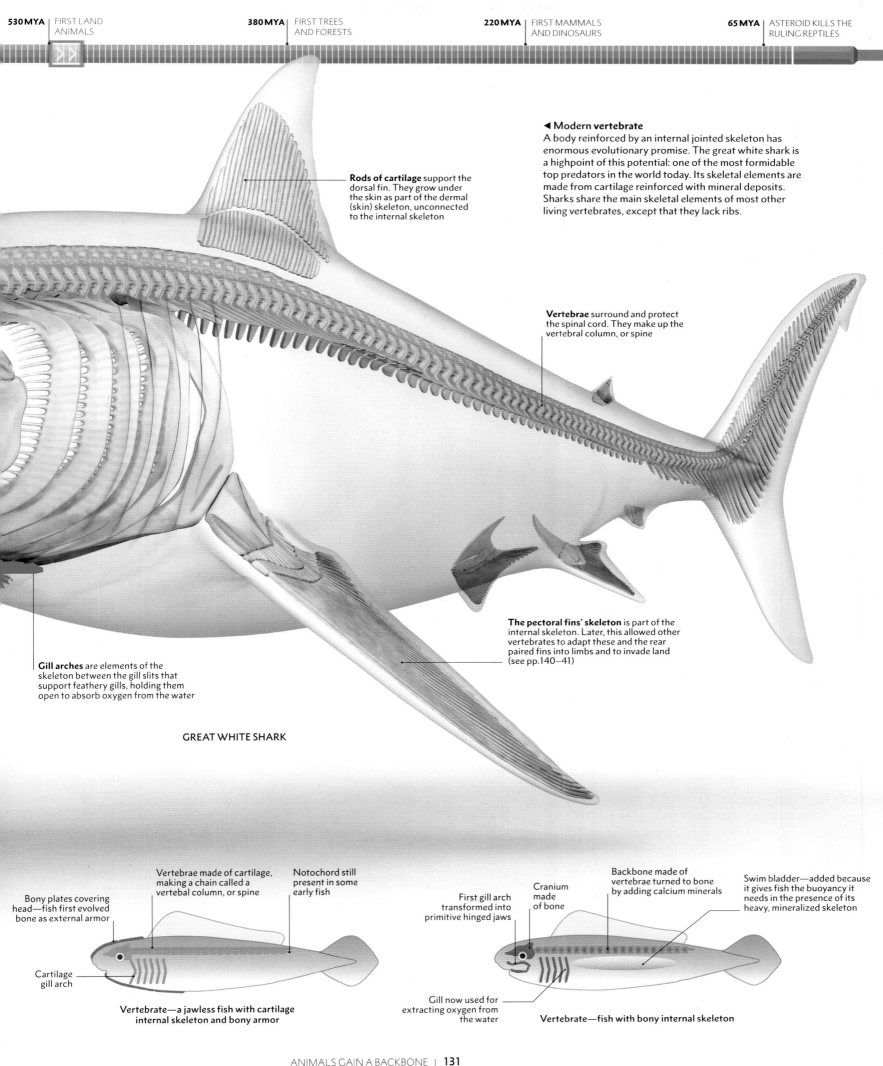

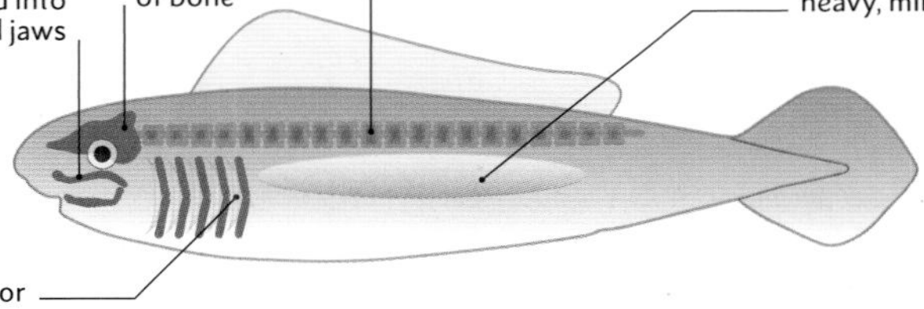
Backbone made of vertebrae turned to bone by adding calcium minerals
Swim bladder—added because it gives fish the buoyancy it needs in the presence of its heavy, mineralized skeleton
Cranium made of bone
First gill arch transformed into primitive hinged jaws
Gill now used for extracting oxygen from the water

Vertebrate—fish with bony internal skeleton

TIMELINES

RISE OF THE **VERTEBRATES**

Backboned animals span only one-eighth of the entire evolutionary history of life, but include some of the most remarkable of its products—including, of course, humans.

The first 100 million years of vertebrate evolution is justifiably described as the "Age of Fish." During this time, these animals evolved the key innovations that made vertebrates so successful—an internal skeleton, biting jaws, and complex organs of sight, hearing, and taste. Some vertebrates that followed used these adaptations to be successful on land. Their strong internal skeleton supported their weight; strongly jointed, bony fins became walking limbs; buoyant gas bladders became lungs; and an outer skin, in some, evolved feathers and fur.

FISH

Myllokunmingia* and *Haikouichthys, larvalike, jawless animals with notocord and cranium but no spine, are preserved in what is now China as the earliest known fossil fish, 530 MYA.

Conodonts—mysterious wormlike creatures thought to be primitive vertebrates—leave abundant toothlike fossils, 530 MYA.

Bonelike tissue is fossilized for the first time, in the form of dentine in fish scales, 510 MYA.

500 MYA

Metaspriggina, a larvalike jawless fish with possible rudiments of vertebral column, is preserved in Canada's Burgess Shale, 505 MYA.

Lampreys and hagfishes, the only jawless fish living today, evolve 482 MYA, according to modern DNA evidence.

Whole body of an ostracoderm, a bony-armored, jawless fish that uses gills for breathing, is fossilized for the first time, 465 MYA. Until now, gills were used for filter feeding.

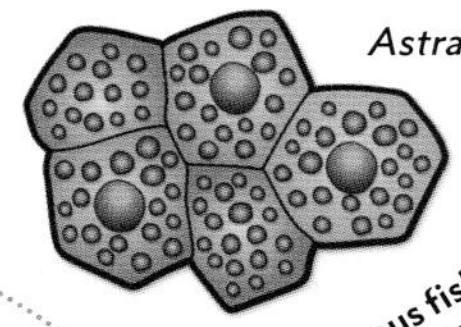

Astraspis scales

450 MYA

Cartilaginous fish (sharks and rays) evolve 450 MYA, as estimated by modern DNA; bony scales of the ostracoderm *Astraspis* are fossilized.

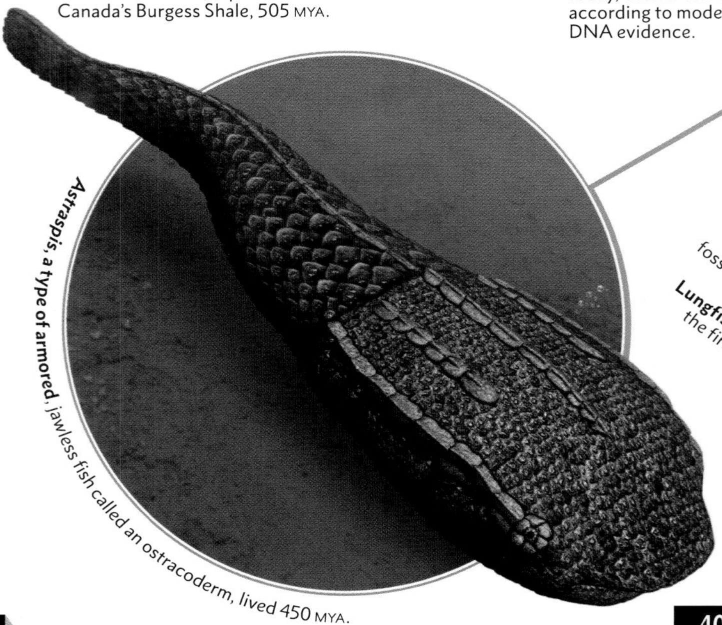

***Astraspis*, a type of armored**, jawless fish called an ostracoderm, lived 450 MYA.

First possible vertebrae fossilized 438 MYA—cartilaginous ones in the fish *Jamoytius*.

Jaws and modern-style, living bone are preserved in the earliest known placoderm ("plate-skinned") fish, 436 MYA.

Fully bony internal skeleton is fossilized for the first time, 420 MYA, in the fish *Guiyu*.

Teeth are preserved in a fossil placoderm 416 MYA.

Lungfish is fossilized for the first time 415 MYA.

400 MYA

TETRAPODS

First trace on land is left by a vertebrate 395 MYA: track marks of an unknown tetrapod (limbed vertebrate).

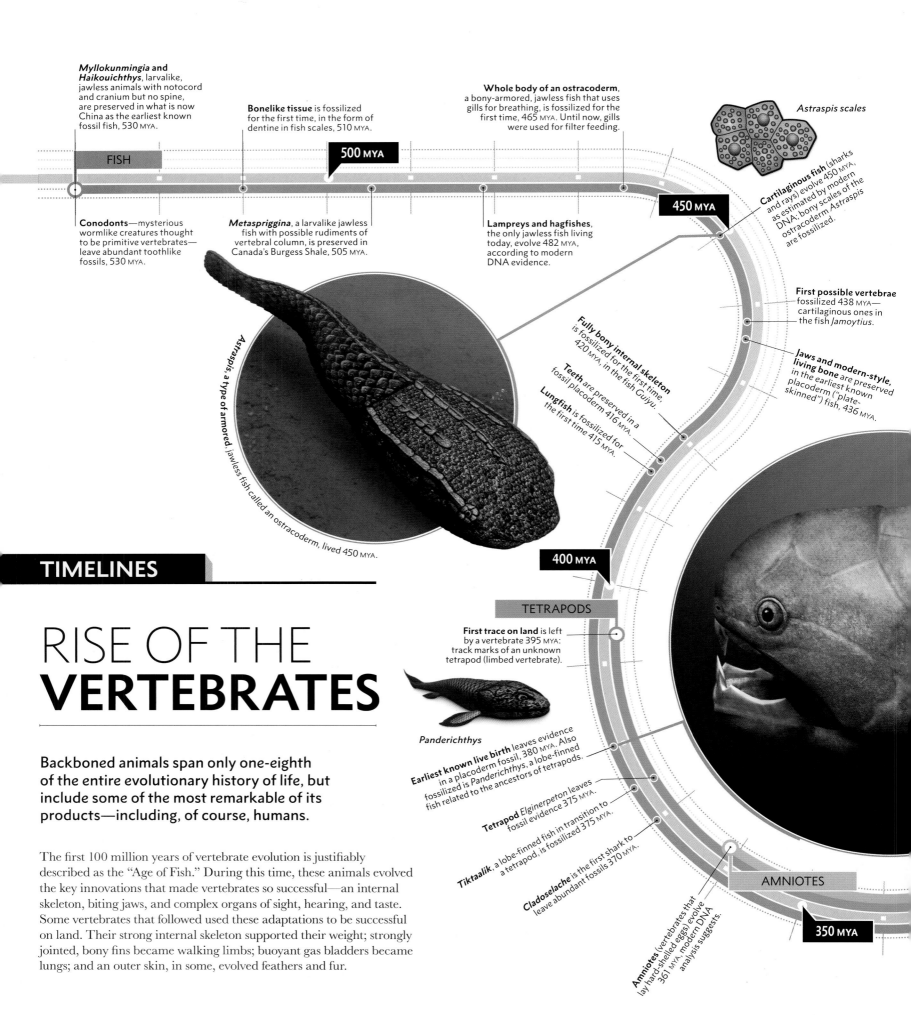

Panderichthys

Earliest known live birth leaves evidence in a placoderm fossil, 380 MYA. Also fossilized is *Panderichthys*, a lobe-finned fish related to the ancestors of tetrapods.

Tetrapod *Elginerpeton* leaves fossil evidence 375 MYA.

Tiktaalik, a lobe-finned fish in transition to a tetrapod, is fossilized 375 MYA.

Cladoselache is the first shark to leave abundant fossils 370 MYA.

AMNIOTES

Amniotes (vertebrates that lay hard-shelled eggs) evolve 361 MYA, modern DNA analysis suggests.

350 MYA

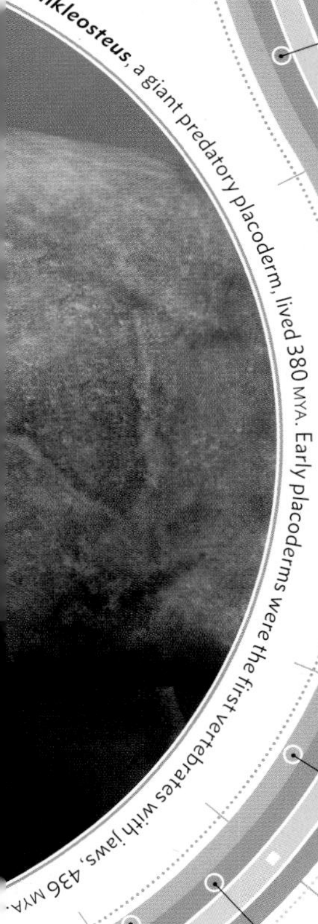

Dunkleosteus, a giant predatory placoderm, lived 380 MYA. Early placoderms were the first vertebrates with jaws, 436 MYA.

Reptile *Hylonomus* is fossilized 318 MYA.

Synapsid—the type of reptile leading to the evolution of mammals—is preserved 312 MYA.

Age of amphibians, during which tetrapods such as *Amphibamus* thrived in the Carboniferous coal forests, draws to a close with a mass extinction, 305 MYA.

Amphibamus

300 MYA

Varanops was a synapsid reptile living 275 MYA. Other synapsids would eventually give rise to mammals.

Modern amphibians (frogs and salamanders) evolve 264 MYA, according to DNA evidence.

Cynodont, a type of reptilian "proto-mammal," leaves fossils 259 MYA.

250 MYA

Ichthyosaur is fossilized 248 MYA.

DINOSAURS AND BIRDS

Earliest dinosaur leaves fossils 231 MYA.

First turtle is fossilized 230 MYA.

MAMMALS

First mammal is preserved 225 MYA: *Adelobasileus*.

Pterosaur leaves fossils for the first time 220 MYA.

Milk production and deciduous teeth (like those of a modern mammal) leave fossil evidence 210 MYA.

Mammals exist as a variety of small shrewlike animals, including *Megazostrodon*, 200 MYA.

Megazostrodon

200 MYA

Marsupials and modern mammals with a placenta evolve 176 MYA, DNA suggests.

Fur leaves a trace in the fossil record of mammals for the first time, 165 MYA.

Placental mammal is fossilized 156 MYA.

Archaeopteryx, the oldest known bird, is preserved 150 MYA.

Archaeopteryx

150 MYA

Marsupial mammal is fossilized 125 MYA.

First monotreme (modern egg-laying mammal, like today's platypus or echidna) leaves fossils 120 MYA.

100 MYA

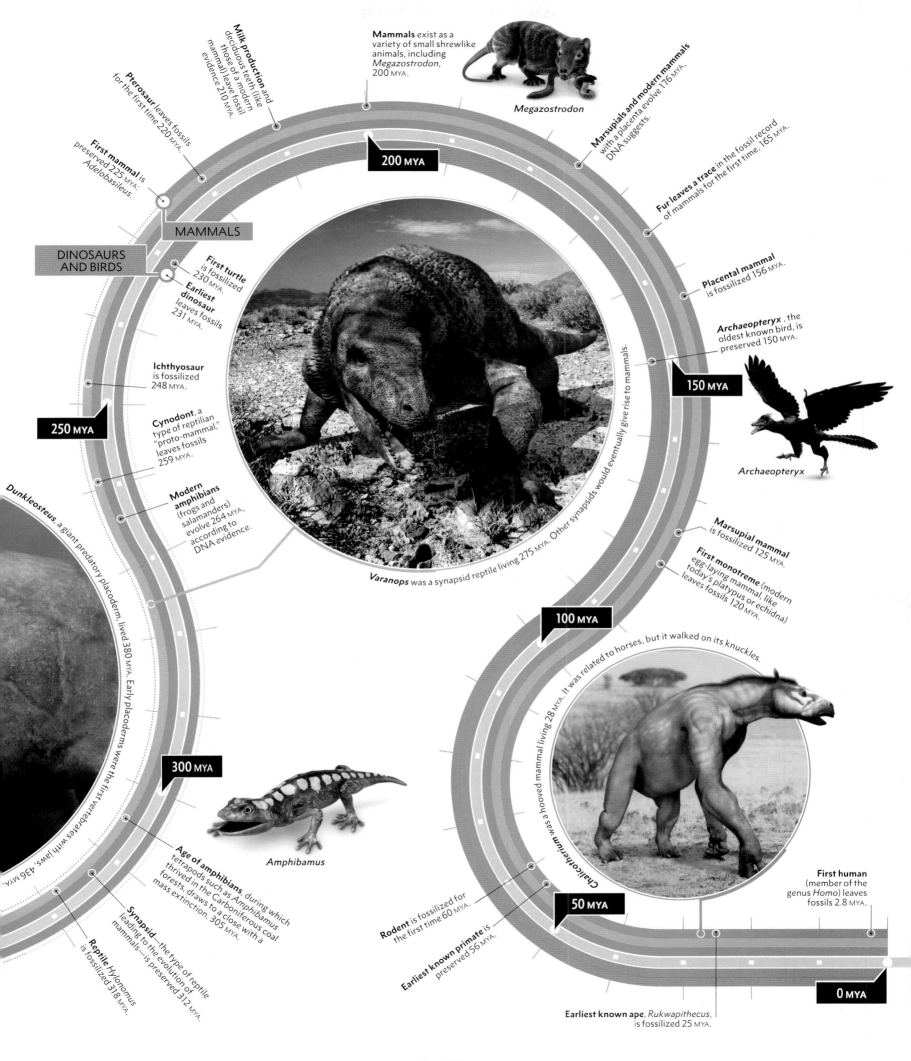

Chalicotherium was a hooved mammal living 28 MYA. It was related to horses, but it walked on its knuckles.

Rodent is fossilized for the first time 60 MYA.

Earliest known primate is preserved 56 MYA.

50 MYA

Earliest known ape, *Rukwapithecus*, is fossilized 25 MYA.

First human (member of the genus *Homo*) leaves fossils 2.8 MYA.

0 MYA

Neck joint between the shieldlike bones of the thorax and skull was not a true neck, but, unusually in fishes, it was flexible

Jaw muscles pulled the head back here, aided by the flexible joint at the back of the head, to help open up the jaws for a wider bite.

Jagged edge of jaw bone could slice through prey—100 million years before sharks evolved their bladed teeth

Joint connecting upper and lower jaws was powered by strong, fast-acting muscles to pull the jaws closed and give a formidable bite

Thoracic shield was a plate of bone that anchored muscles that pulled on the lower jaw to rapidly open the mouth

▼ Terror of the Devonian seas
Dunkleosteus was among the first predators to catch fast-moving prey with snapping jaws. Studies of its fossilized skull suggest it could have had one of the strongest bites in the history of vertebrate life.

JAWS CREATE TOP **PREDATORS**

Predators have been a part of the natural world ever since organisms evolved the ability to eat one another. However, backboned animals started as filter-feeders that sucked mud from the ocean floor. It was not until they evolved jaws that they could sit at the top of long food chains.

Many invertebrates—such as predatory worms, sea scorpions, and centipedes—have evolved sharp-edged jaws that can grab prey. But vertebrates, using cartilage and bone, made their jaws bigger and more muscular. The first jawed vertebrates did so through an evolutionary rearrangement of the arches that support the gills. Over generations, the front arches were shifted forward into the roof and floor of the mouth and met toward the back of the skull, forming a hinged joint.

SUPER-PREDATORS

Reshaping the gill arches into moveable jaws may have helped to fill the gills with more oxygen, but the development of stronger muscles allowed the jaws to bite, too. This helped fish both to catch prey and also to kill and dismember it. Natural selection would have favored the evolution of bigger fish with more powerful jaws—opening up more ambitious avenues of predation.

The earliest-known jawed vertebrates were placoderms: mostly armor-plated fish that flourished during the Devonian period (419–359 MYA). One of the largest known was *Dunkleosteus*, whose fossil remains are found throughout the world—evidence of its success. Growing twice the length of a car, *Dunkleosteus* was the biggest predator of its time—and its jaws could easily puncture the armor of its contemporaries. Its size and strength meant that it could prey on bigger animals, including other predators. Devonian oceans had an extra link to their food chains: a top predator.

DIETARY DIVERSITY

Despite their apparent supremacy, the placoderms did not last. They disappeared in the Late Devonian mass extinction—an event that was probably triggered by a drop in oxygen levels. But other jawed vertebrates—notably the sharks—had evolved in the meantime, and they survived. Although their jaws were built from flexible cartilage, they had blade-edged teeth that could be serially replaced—something that placoderms probably could never do. But it was bony vertebrates that took jaws and their especially hard, enamel-coated teeth to a new level. Crocodiles, dinosaurs, and mammals developed deeply-rooted teeth that could better resist struggling prey. Dentition was also modified in animals lower in the food chain. Grazing mammals developed grinding teeth, and their biting jaws became chewing jaws—extending the ecological range of vertebrates more than ever.

> “THE **VERTEBRATES** THAT CAME **STORMING THROUGH**… SWEEPING MOST OF THE [JAWLESS FISH] ASIDE **DURING THE DEVONIAN**, WERE **THE ONES WITH JAWS**.”
>
> **Colin Tudge**, biologist and writer, 1943–

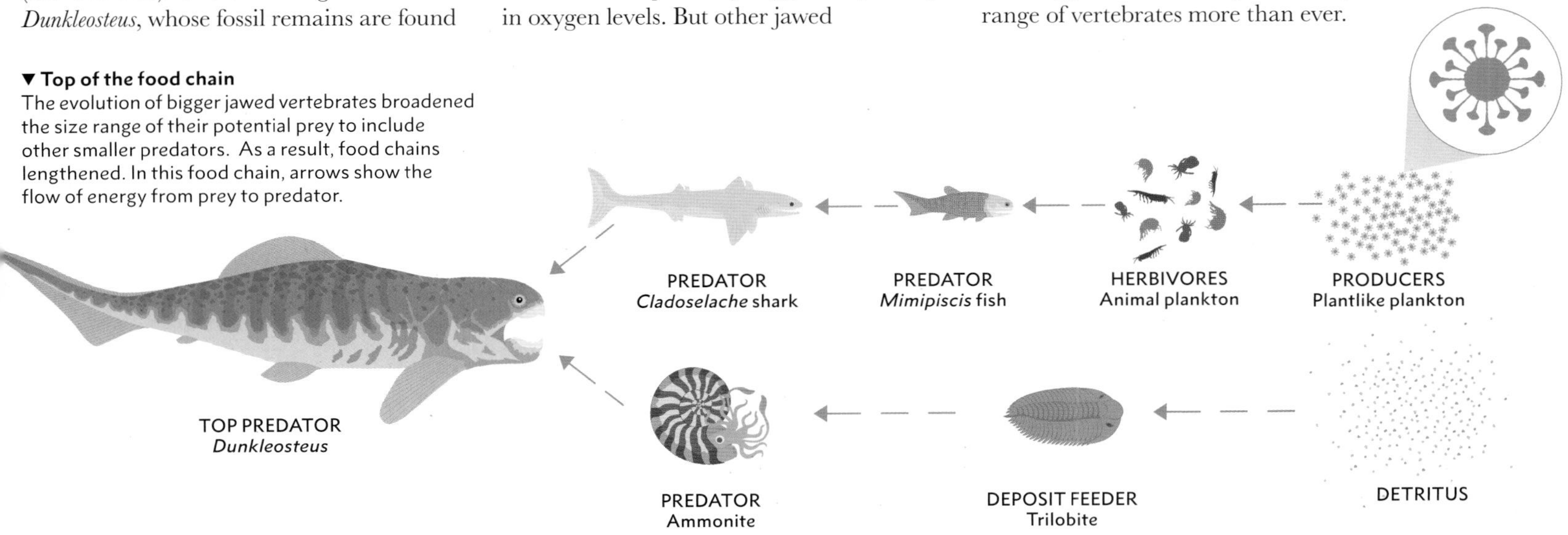

▼ Top of the food chain
The evolution of bigger jawed vertebrates broadened the size range of their potential prey to include other smaller predators. As a result, food chains lengthened. In this food chain, arrows show the flow of energy from prey to predator.

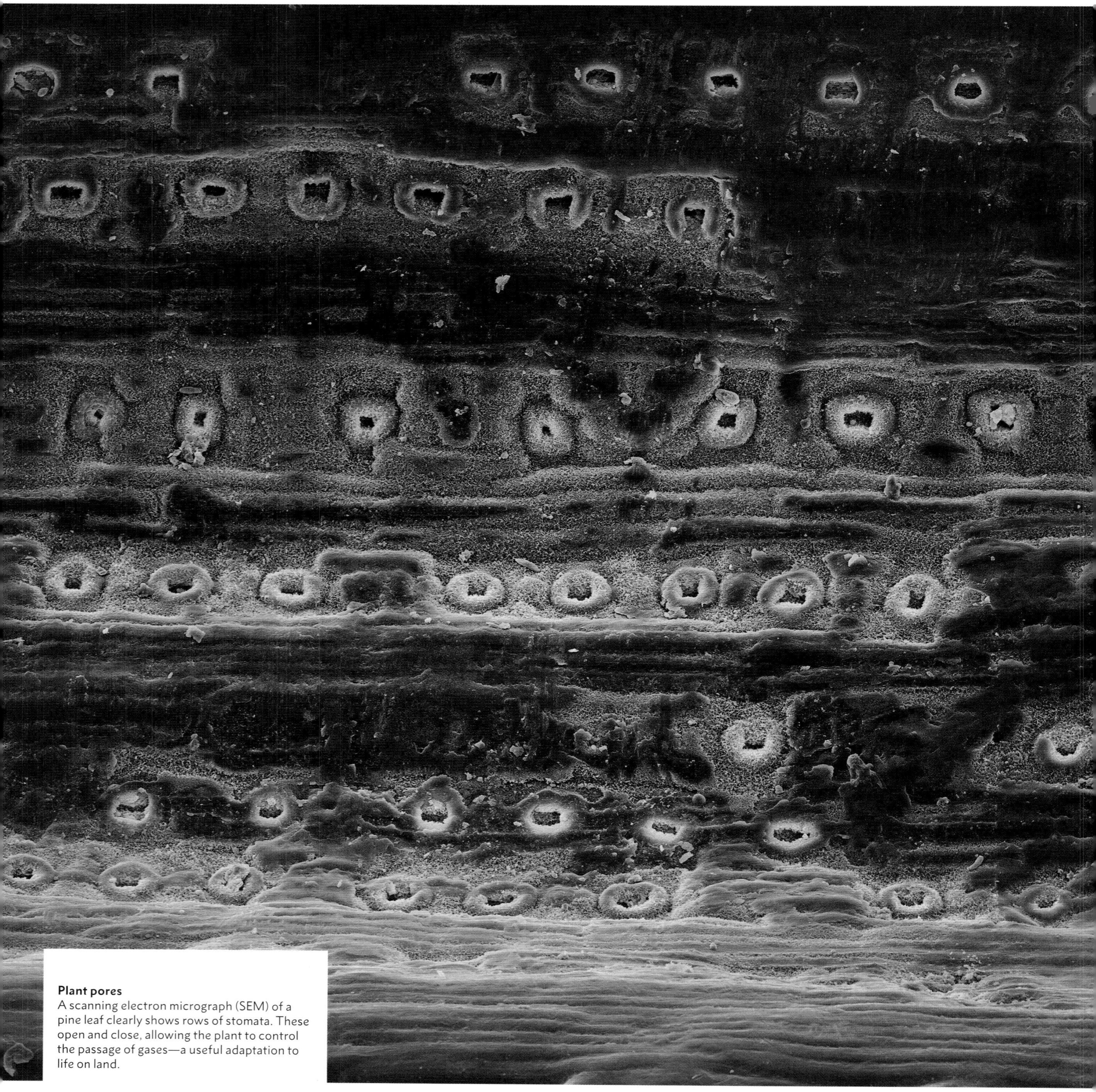

Plant pores
A scanning electron micrograph (SEM) of a pine leaf clearly shows rows of stomata. These open and close, allowing the plant to control the passage of gases—a useful adaptation to life on land.

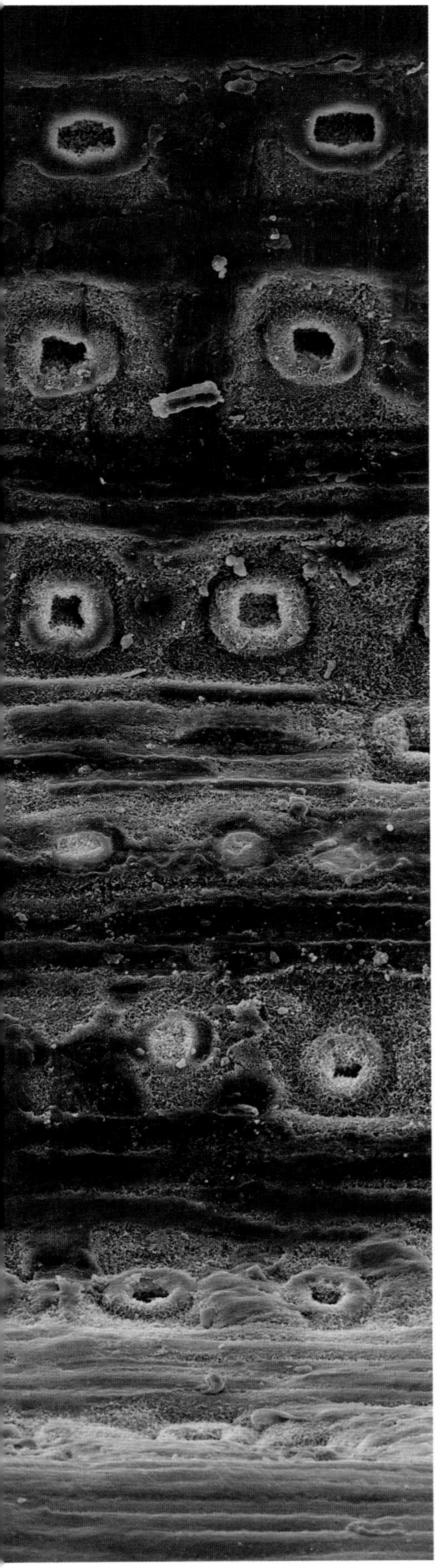

PLANTS MOVE ONTO LAND

The first sign that the land was turning green probably came when algae crept above the tidal zone along ocean shores. However, the move to permanently drier environments farther inland required plants with roots anchored in soil, and shoots that could grow upright in dry air.

Vegetation grew in water long before it invaded the land. Algae had evolved broad fronds that intercepted the sun's light energy and a "holdfast" that stuck the body to rock. These seaweeds still live in the ocean today. Many survive periodic exposure at low tide, but they are too flimsy to last long on dry land.

WATERPROOFING THE LEAVES

Water screens out some of the sun's energy. On land, although plants bask in stronger radiation, they risk drying out. Land plants evolved a waxy waterproof coating on their epidermis—the surface "skin" of cells. Pores in the epidermis called stomata helped to keep gases moving for processes such as photosynthesis (see pp.114–15) and respiration. The earliest land plants, like the mosses and liverworts of today, could only hug the land with creeping stems. They clung there with rhizoids—microscopic hairs that scarcely penetrated the ground to function as primitive roots.

STANDING TALL

Strength is required to stand upright. Plant cells are surrounded by a scaffolding of tough fibrous cellulose, and the thickening of this wall in places helps stems bear some weight. Although mosses can do this, they can rise no more than a few inches. Other plants managed to grow taller because of the evolution of a complex substance called lignin. By coating their microscopic transportation vessels, lignin helped to form watertight tubes that could deliver water and minerals up the stem. Lignified vessels were also physically strong, so these new plants grew and branched vertically. Tough vessels also grew downward as stronger, branching roots penetrated the soil to anchor the weight and absorb dissolved minerals. Many of these taller plants were already better suited to life on land by producing seeds. But thickened lignified tissue, called wood, helped trunks get thicker and trees became taller still.

> "...THE **FIRST ZOOLOGICAL LANDFALL** WAS CONTINGENT ON THE **GREENING OF THE TERRESTRIAL LANDSCAPE** BY **PLANT LIFE**...WHICH WAS... MORE AN INVASION OF AIR."
>
> **Karl Niklas**, professor of plant science, 1945–

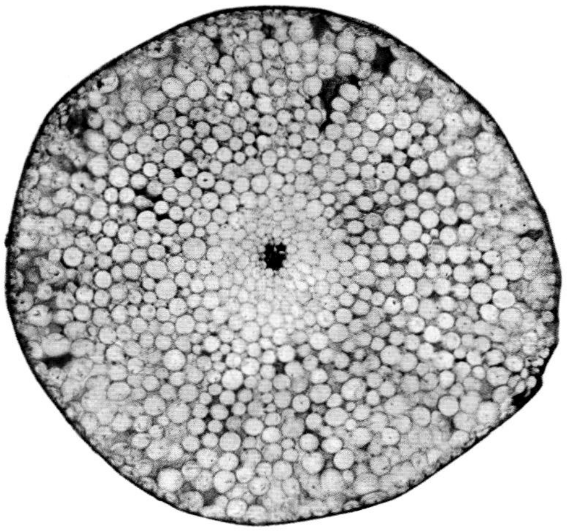

◀ Rigid stem
A cross-section of a fossilized plant (*Rhynia gwynne-vaughanii*) from the Devonian period about 410 MYA reveals watertight tubes that conducted water and nutrients.

WENLOCK LIMESTONE

Few organisms leave any kind of fossilized trace, but in some locations conditions have preserved extraordinary snapshots of entire communities. Their wonderful fossils—rich in species and finely detailed—offer rare insights into the ways groups of animals and plants lived and died.

Wenlock Edge—an outcrop of limestone on the Welsh-English border—holds an example of such a fossil assemblage, or Lagerstätte. It is packed with animals of a tropical reef from more than 420 MYA. At this time, the site straddled the coastline of the ancient Iapetus Ocean, where many of Earth's animals had evolved. The fossils show that corals, sponges, trilobites, and brachiopods flourished in the warm shallows.

Lagerstätten form under certain conditions that favor preservation. The Wenlock assemblage includes hard-shelled animals that have been broken or uprooted—suggesting that crashing waves left debris in mud at the bottom of a slope. It means a single Wenlock slab could contain animals from scattered localities. Other Lagerstätten may preserve communities intact. The Burgess Shale in the Canadian Rockies has soft-body impressions of animals that were smothered in mudslides 508 MYA. Although their orientation is chaotic, the postures suggest they were killed instantly. But not all Lagerstätten result from violent slaughter. North America's Green River formation comprises 50 million-year-old sediments left in lake basins that contain fish, leaves, insects, and even small birds complete with feathers. The oxygen-poor conditions in the lakebed muds slowed bacterial decomposition and allowed fragile parts to fossilize intact. The same process happened in Germany's Messel lake at a similar time.

Rugose coral was a solitary horn-shaped extinct relative of the corals that live today
Clam was a free-swimming filter-feeder like today's scallops
A piece of colonial coral has been ripped from its position on the reef, like many other attached Wenlock animals

Identifying extinct animals

An abundance of fossils from the same age not only helps to reconstruct the interacting lives of prehistoric animals, but helps to resolve their diversity too. Species are described on the basis of specimens—but fossil specimens are frequently incomplete. When so many individuals are preserved together, biologists have a better, more representative, view of anatomy—helping them to divide one species from another.

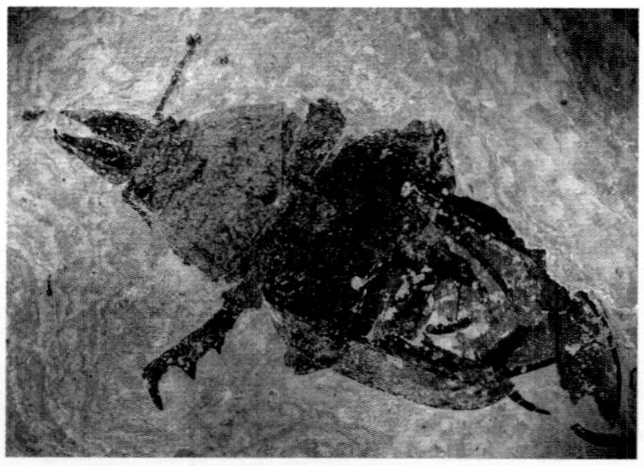

Giant flying ant with iridescent colors intact

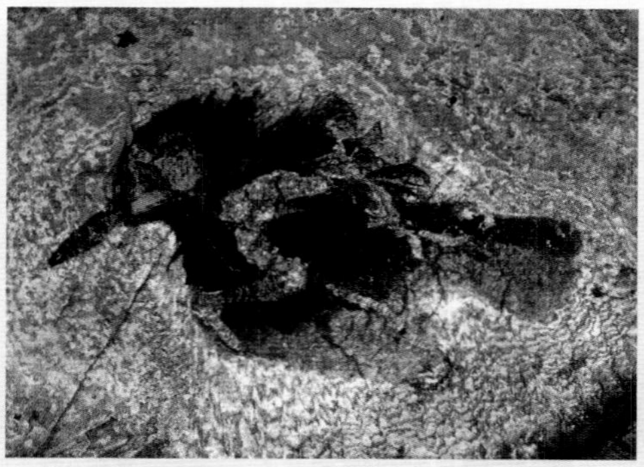

Bird complete with fossilized feathers

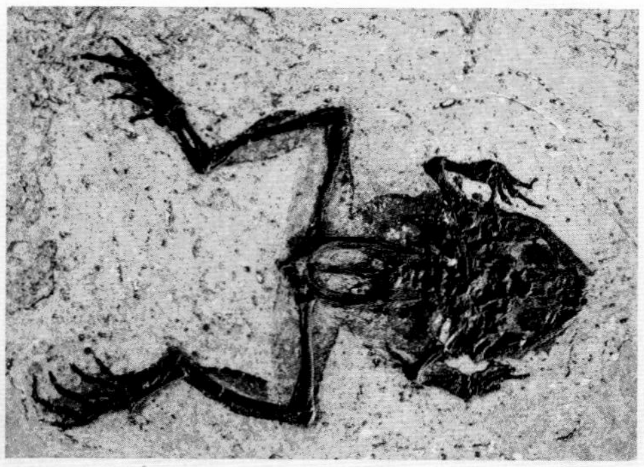

Frog, including an outline of soft body parts

◀ Messel lake pit
The site of Messel, Germany, shows very fine preservation of a community living 47 MYA. Its special conditions included poisonous gas emanating from the lake, which not only killed animals instantly, but also ensured no living scavengers ate the fallen remains before they were mineralized.

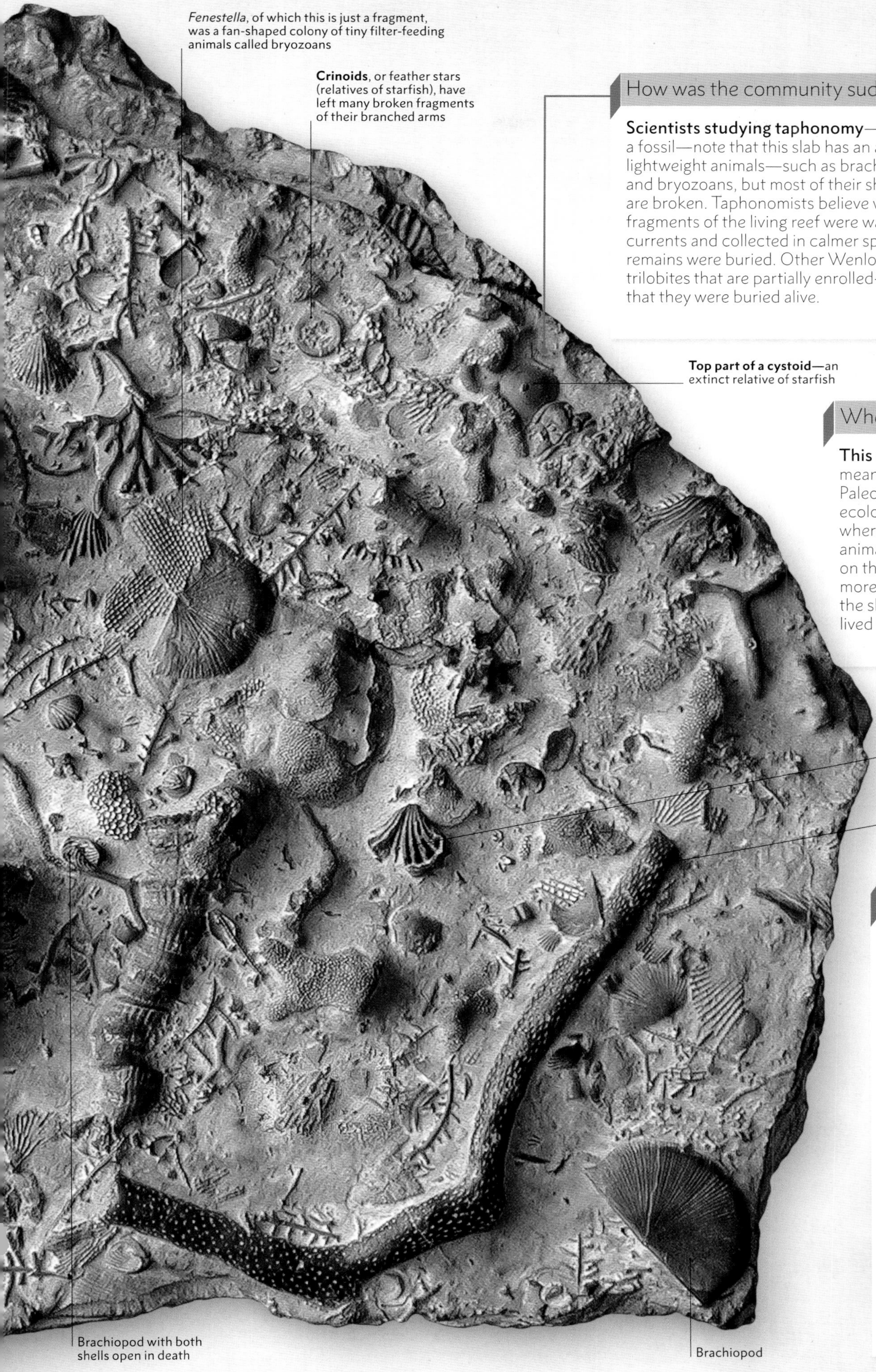

Fenestella, of which this is just a fragment, was a fan-shaped colony of tiny filter-feeding animals called bryozoans

Crinoids, or feather stars (relatives of starfish), have left many broken fragments of their branched arms

Top part of a cystoid—an extinct relative of starfish

Brachiopods had two shells connected by a hinge, just like a clam, but they are not related to clams

Supporting stems of crinoids were easily broken by strong currents and are abundant in some limestone rocks

Brachiopod with both shells open in death

Brachiopod

How was the community suddenly buried?

Scientists studying taphonomy—the history of a fossil—note that this slab has an abundance of lightweight animals—such as brachiopods, crinoids, and bryozoans, but most of their shells and cases are broken. Taphonomists believe wave-smashed fragments of the living reef were washed away on currents and collected in calmer spots, where their remains were buried. Other Wenlock fossils show trilobites that are partially enrolled—suggesting that they were buried alive.

Trilobite in defensive rolled posture

Where did the animals live?

This slab shows a death assemblage, meaning it includes animals that died together. Paleoecologists (scientists who study ancient ecology) need further fossils to build a picture of where the organisms lived. From remains of the animals fossilized as in life, they have found that on the ancient reef, harder-shelled brachiopods—more resistant to wave action—lived higher on the shoreline, whereas free-swimming animals lived in deeper waters.

Restoring the past

To reconstruct prehistoric life-forms, paleontologists use all the fossil evidence to put forward a hypothesis about how they looked and behaved, although they can never be certain about their conclusions. The Wenlock fauna consisted of attached animals—such as crinoids and honeycomb corals—that formed a reef habitat, which also contained bottom-feeding trilobites and predatory *Orthoceras*—a shelled relative of squid.

Restoration of *Orthoceras* in Ordovician seas

ANIMALS INVADE LAND

For billions of years, much of life was confined to oceans, lakes, and rivers. Such an ancient aquatic heritage meant that the first complex organisms also lived only in water. Dry land offered so many new opportunities that terrestrial colonization happened not just once, but many times.

◀ **First air-breather**
This modern millipede has armoured segments similar to those of *Pneumodesmus*, a millipede that lived 428 MYA. *Pneumodesmus* is the earliest body fossil of an animal known to walk on land and breathe air. Fragments of its exoskeleton show that it had spiracles, or breathing holes.

It is likely that the first microbes were invading land within a billion years of life's origin. For these bacteria, the wet coastal rocks and moist sediments where oceans lapped the shore were a natural extension of their range. As erosion and detritus formed the first soil over 3 BYA, bacteria began to live between its particles. The earliest burrowing organisms would have churned coastal sediment and added more organic material that served as food for fungi and other decomposers. Soils were becoming so enriched, that by 470 MYA, land was becoming an inviting place for plants, too.

LIVING ON LAND

Above ground, colonization was less straightforward. All living cells, whether of single- or multicellular organisms, must be surrounded by moisture. Land plants survived by evolving a thick, waxy outer layer (cuticle) that both retained water and let gases in and out (see pp.136–37).

The first land animals had a cuticle, too, that served the same water-retaining function, but there were other challenges to overcome in just getting around. In the Cambrian period (541–485 MYA), marine animals had evolved into some gigantic forms, but size was a liability on land. A body is buoyed in water, effectively weighing less, but on land, the same animal may be too heavy to move. Early land animals needed stronger muscles and supporting skeletons, and compensated for this extra baggage by getting smaller. At first, wormlike land animals probably survived underground or in rocky crevices, where, in moist microhabitats, these small animals might have used their skin to breathe air.

The early terrestrial colonists also included jointed-limbed arthropods. Prehistoric arthropods, relatives of today's crabs and spiders, were already thriving in the oceans. Their jointed limbs and armor gave them the potential to succeed on land. Fossil and DNA evidence indicates that millipedes and centipedes were part of the first big wave of land colonists, possibly more than 500 MYA. Their articulated, armored bodies helped them crawl over land without dehydrating and they evolved breathing holes in this armor, called spiracles, and got oxygen straight from the air. Millipedes would have been among the first grazers of land plants and centipedes the first predators of the terrestrial ecosystem.

FILLING THE FORESTS

Fossil evidence shows that by 380 MYA, the land was already supporting its first trees. By the beginning of the Carboniferous period (359 MYA), Earth was home to rich, swampy forests teeming with life. Plants could grow taller because of the evolution of tougher

> “THE **TETRAPODS**, WITH THEIR LIMBS AND FINGERS AND TOES, **INCLUDE OURSELVES AS HUMANS**, SO THAT THIS **DISTANT DEVONIAN EVENT** IS PROFOUNDLY **SIGNIFICANT FOR HUMANS** AS WELL AS **FOR THE PLANET**.”
>
> **Jennifer Clack**, paleontologist, 1947–
> *Gaining Ground: the Origin and Evolution of Tetrapods*

Track between the footprints suggests the creature dragged its abdomen
Small, thin prints suggest at least eight pairs of legs

◀ **Life's first steps**
These fossilized marks made in sand dunes in the early Cambrian period, 530 MYA, represent the oldest trace of animals on land that we have discovered. They were made by an arthropod that divided its time between land and sea.

supporting materials, such as woody lignin. Forests gave height to the land ecosystems—providing new niches for tree-climbing and flying animals. In particular, they encouraged the biggest radiation of land animals of them all: the insects. The evolution of life on land was producing entirely new kinds of animals with new ecological interactions: web-spinning predatory spiders, browsing insects, and grazing snails. In terms of diversity and abundance, the organisms roaming the land were rivaling anything swimming in the water of the oceans.

OUR ANCESTORS REACH LAND

When invertebrate life conquered land, vertebrates were still confined to water habitats. As with invertebrates, when vertebrates started to move onto land 395–375 MYA during the Devonian period, their bodies needed to change.

Fishes use their paired fins for stabilizing their swimming and although a few use them secondarily to "walk" on the sea bed, for most of them, fins are not strong enough to fashion into legs. One group, however—the "lobe-fins"—had an advantage. A few,

TIKTAALIK ROSAE LIVED **375 MYA; THE FIRST FOSSILS** WERE DISCOVERED IN **2004** IN THE **CANADIAN ARCTIC**

such as lungfishes and coelacanths, are still alive today, but in the Devonian, there were many different forms. They differed from all other fishes in having a stronger bony support for each of their paired fins. Their flexible joints allowed these fins to be used to walk under water and later helped them emerge from the water and crawl across land. Lobe-finned fishes may have done this in times of drought, just as lungfishes do today. As lobe-fins wandered farther ashore, their fins evolved into limbs with fingers and toes.

Fishes had other features that prepared them for a terrestrial life. Most species have a gas-filled bag—the swim bladder—used for controlling buoyancy. Modifications of this swim bladder in some modern fishes mean that the sac can communicate directly with air, helping the fish to breathe and supplement the supply of oxygen it extracts from the water with its gills. In early lobe-fins, this new kind of breathing mechanism, later powered by chest muscles such as the diaphragm, evolved to become the first air-breathing lungs.

The first vertebrates with lungs are often called amphibians, but these long-extinct creatures are only distantly related to today's frogs and newts. They were the first four-legged vertebrates, or "tetrapods," and ancestors of all reptiles, birds, and mammals, as well as modern amphibians. An astonishingly complete fossil record documents the transition from fish to tetrapod, via intermediate forms sometimes called "fishapods."

The four-legged, air-breathing plan was a major evolutionary step. Although some legs have since been lost or turned to arms or wings, it is the basis for most land-based vertebrates today.

▲ Transitional fossil
Tiktaalik rosae is an evolutionary wonder. Although it resembled a fish, its neck was more flexible than that of true fishes and its fins, although small, had strong joints that may have supported its weight on land.

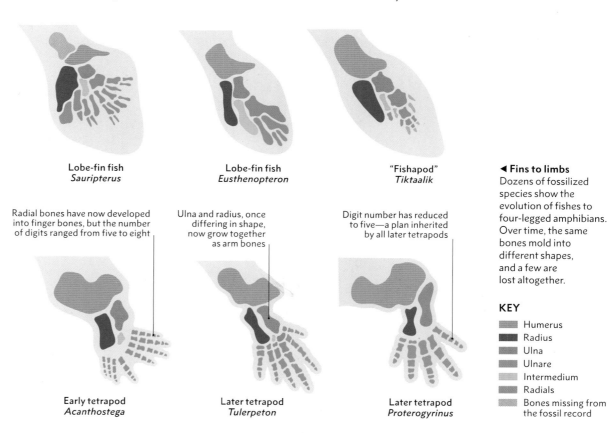

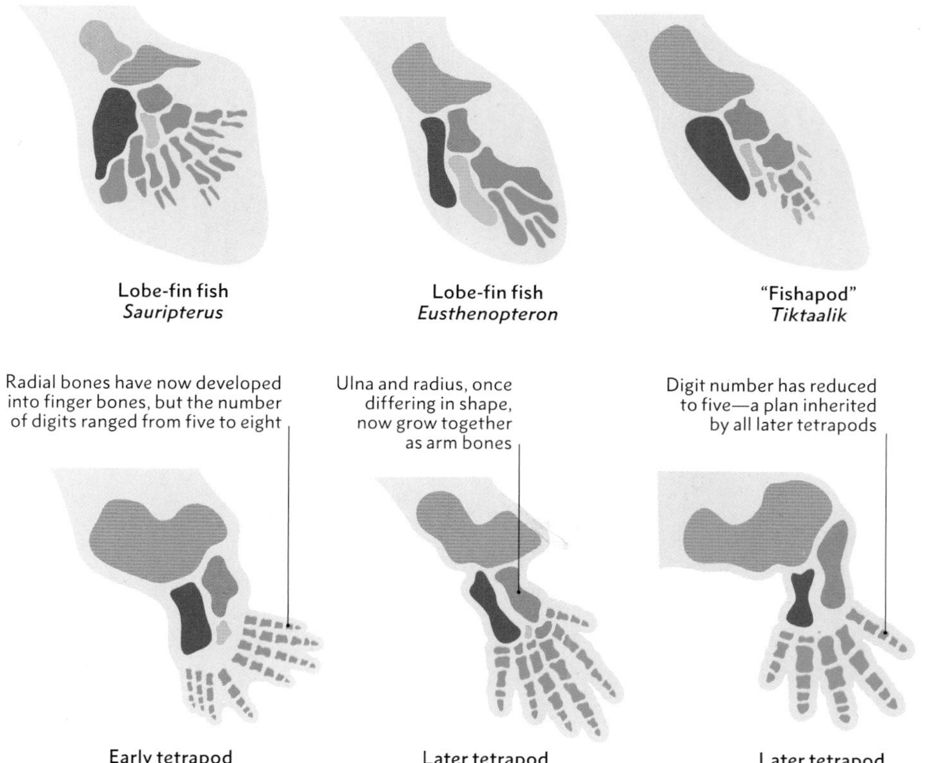
Lobe-fin fish
Sauripterus
Lobe-fin fish
Eusthenopteron
"Fishapod"
Tiktaalik
Radial bones have now developed into finger bones, but the number of digits ranged from five to eight
Ulna and radius, once differing in shape, now grow together as arm bones
Digit number has reduced to five—a plan inherited by all later tetrapods
Early tetrapod
Acanthostega
Later tetrapod
Tulerpeton
Later tetrapod
Proterogyrinus

◀ Fins to limbs
Dozens of fossilized species show the evolution of fishes to four-legged amphibians. Over time, the same bones mold into different shapes, and a few are lost altogether.

KEY
- Humerus
- Radius
- Ulna
- Ulnare
- Intermedium
- Radials
- Bones missing from the fossil record

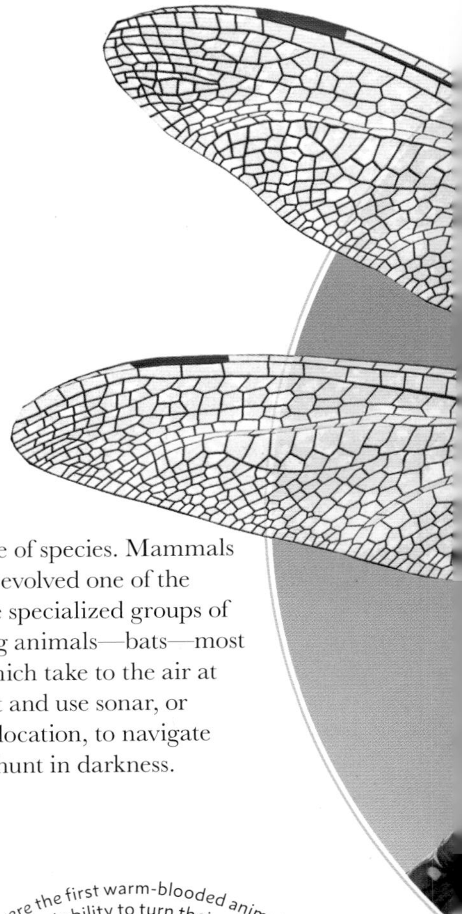

REINVENTING **THE WING**

Often, the similarities seen in life are the result of a common ancestor, but not always. For instance, flapping wings required for flight evolved independently in at least four groups of animals at separate points in time, allowing them to take to the air.

Organisms evolve adaptations that make them better suited to their lifestyles. Sometimes, natural selection can produce the same innovation in separate, unrelated groups. This is convergent evolution.

SHARING CHARACTERISTICS

All plants that produce seeds share a common ancestor—in the same way that the stingers of jellyfish and coral are related too. But sometimes, natural selection can produce a similar adaptation in unrelated groups—such as the flippers of swimming ichthyosaurs (reptiles) and dolphins (mammals).

When different forms of life, living in separate environments or even time periods, share an anatomical or behavioral similarity, it is often because they live in similar environments that demand certain adaptations. Despite living millions of years apart, both ichthyosaurs and dolphins needed to be fast swimmers in order to escape predators and catch fast prey, and therefore evolved flippers.

EVOLVING FLIGHT

Insects were the first animals to fly, and they are the only fliers to evolve wings that were not commandeered from existing limbs. Vertebrates became fliers by refashioning their existing limbs. Their forelimbs and hands, over time, evolved into different types of wings. Pterosaurs probably achieved this first and became the most well-known of the reptilian fliers, before becoming extinct along with the dinosaurs. Birds evolved from bipedal dinosaurs and fared better. They survived the same extinction event, possibly due to their warm-bloodedness, to thrive alongside mammals and diversify into a wide range of species. Mammals later evolved one of the more specialized groups of flying animals—bats—most of which take to the air at night and use sonar, or echolocation, to navigate and hunt in darkness.

Birds were the first warm-blooded animals to fly. This lanner falcon shows birds' ability to turn their wings into airbrakes by spreading their feathers.

Pterosaurs, cousins of the dinosaurs, were the first vertebrates to evolve wing-flapping flight. *Rhamphorhynchus* is an early example from the Jurassic (166–145 MYA).

Bats are the only mammals to have mastered true flight. "Flying" squirrels merely glide.

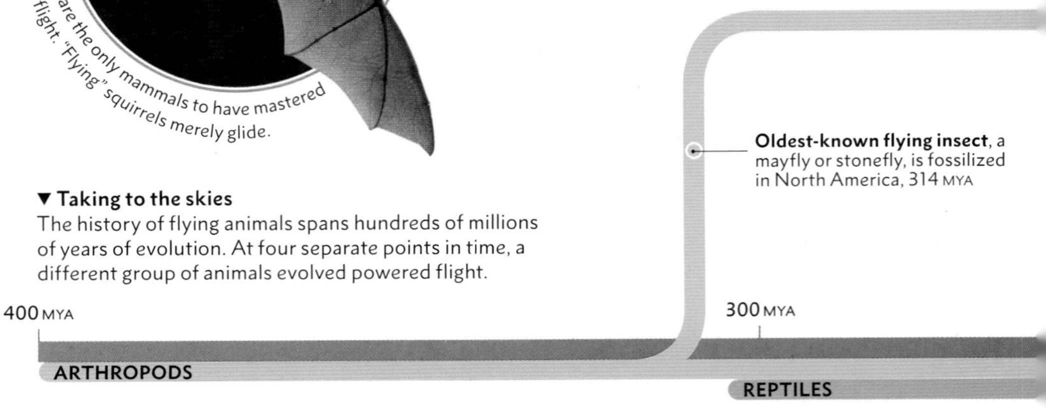
400 MYA
300 MYA
ARTHROPODS
REPTILES
Oldest-known flying insect, a mayfly or stonefly, is fossilized in North America, 314 MYA

▼ Taking to the skies
The history of flying animals spans hundreds of millions of years of evolution. At four separate points in time, a different group of animals evolved powered flight.

Insect wings, like those of a modern dragonfly, may have evolved when a peak in atmospheric oxygen triggered more active lifestyles.

▲ How did wings evolve?
The earliest proto-wings were flaps on insects' bodies that previously served as gills, or as oars allowing them to skim over water. Climbing insects may have used them to parachute through air before evolving flapping wings, which gave them better aerial control. Pterosaurs, birds, and finally bats became airborne not by growing new appendages, but by adapting existing ones.

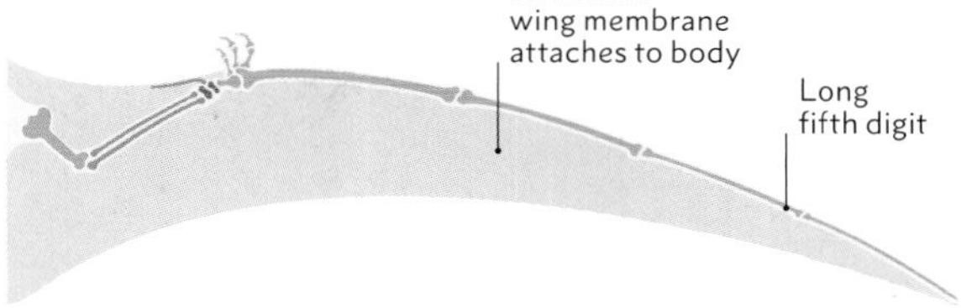
Stretched wing membrane attaches to body
Long fifth digit

Pterosaur wings were relatively rudimentary compared to a bird's or a bat's. Wing membranes were supported mainly by a single finger. Pterosaurs did not have the musculature for coordinated flight, and probably relied on soaring on air drafts to get around.

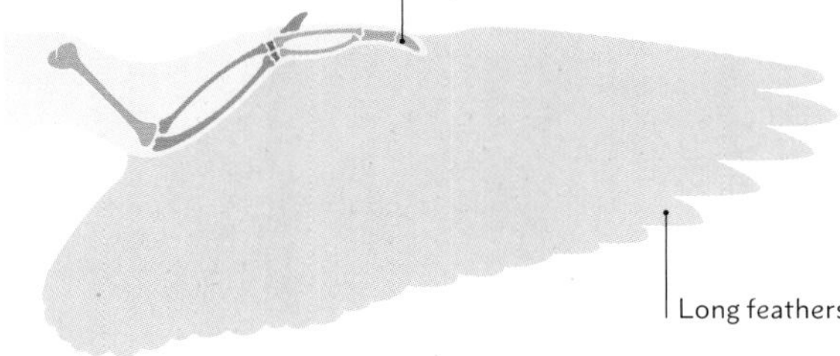
Shortened, fused finger bones
Long feathers

Birds have shorter arm and finger bones that bring the wing under better control. More powerful muscles also mean that the wing, equipped with long flight feathers, can be flapped at wider angles, providing birds with stronger flight than pterosaurs.

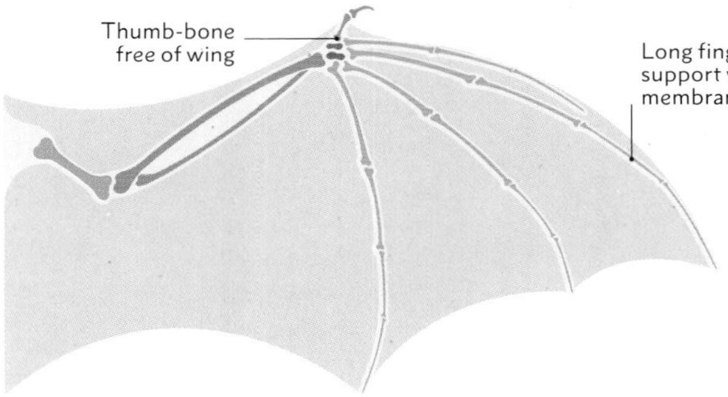
Thumb-bone free of wing
Long fingers support wing membrane

Bats have more finger bones than other flying vertebrates. Their wings are supported by four fingers. This makes their shape more flexible than the wings of birds, helping them move in ways that improve maneuverability and save energy.

◀ Anatomy of a wing
The wings of a pterosaur, bird, and bat each use bones of the arm and hand, but evolution has molded the bones differently in each case. The shape of pterosaur and bat wings depends on how the bones hold out the flight membrane, made of skin. A bird's flight surface is made of feathers and its shape depends on the form of its feathers.

KEY
- Humerus
- Radius and ulna
- Wrist bones
- Finger bones
- Thumb bones

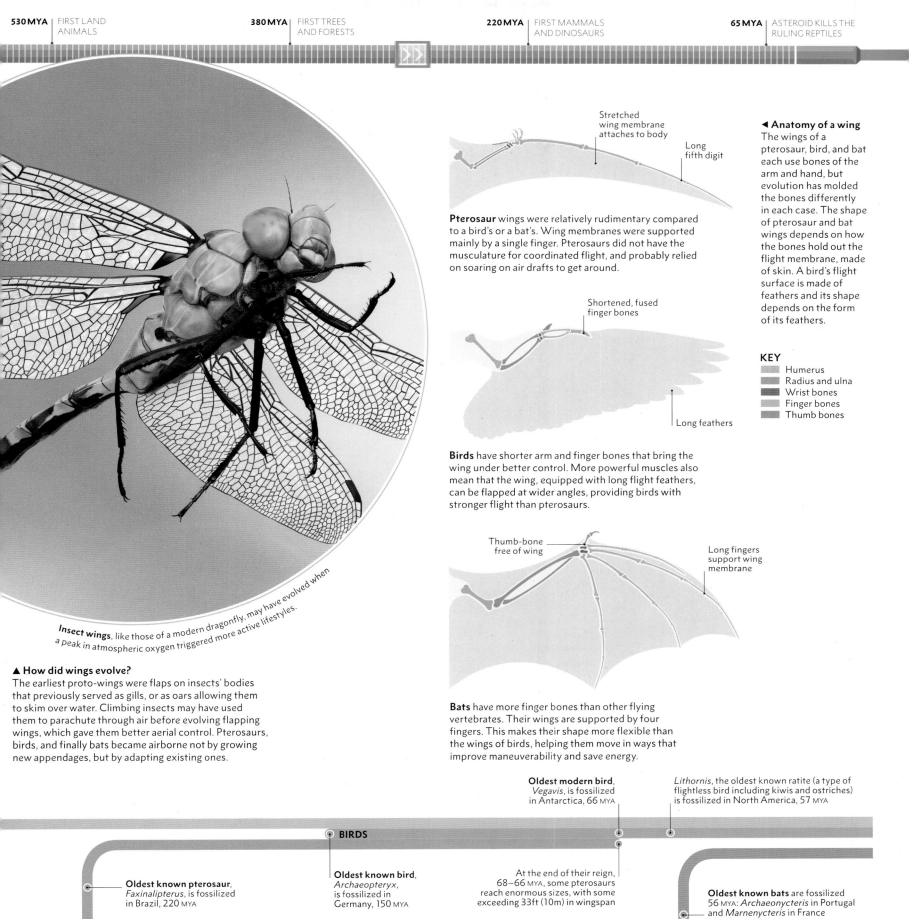

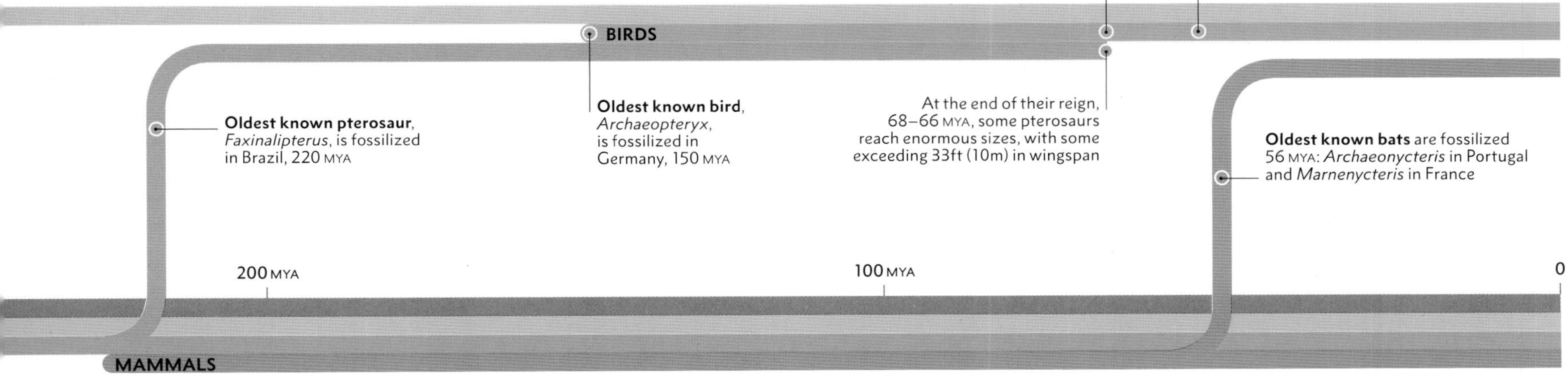
Oldest modern bird, *Vegavis*, is fossilized in Antarctica, 66 MYA
Lithornis, the oldest known ratite (a type of flightless bird including kiwis and ostriches) is fossilized in North America, 57 MYA
BIRDS
Oldest known pterosaur, *Faxinalipterus*, is fossilized in Brazil, 220 MYA
Oldest known bird, *Archaeopteryx*, is fossilized in Germany, 150 MYA
At the end of their reign, 68–66 MYA, some pterosaurs reach enormous sizes, with some exceeding 33ft (10m) in wingspan
Oldest known bats are fossilized 56 MYA: *Archaeonycteris* in Portugal and *Marnenycteris* in France
200 MYA
100 MYA
0
MAMMALS

THE **COMPLETELY PROTECTED EMBRYO** WITHIN **THE SEED** GIVES… A GREAT **ADVANTAGE**.

Douglas Houghton Campbell, American botanist, 1859–1953

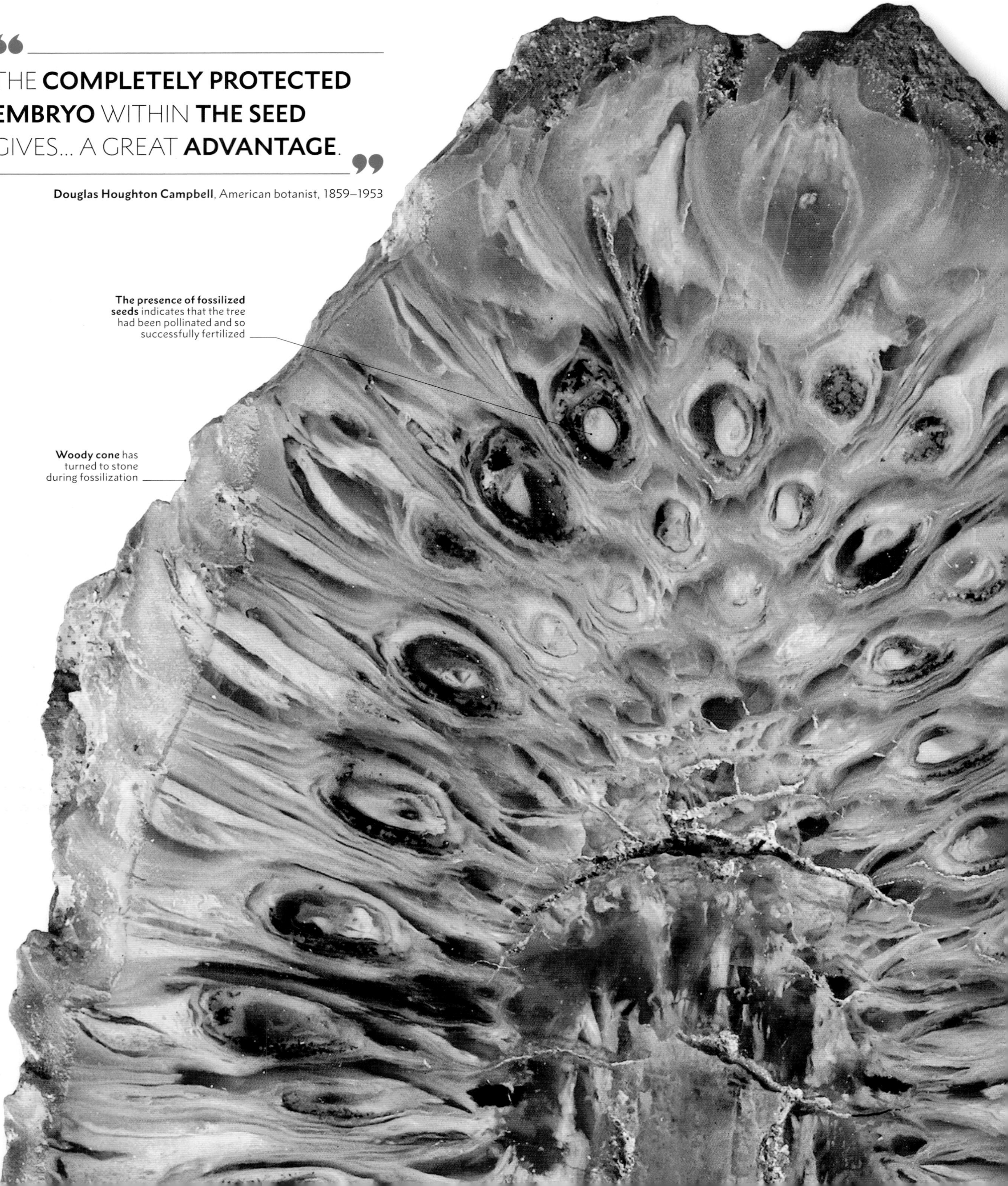
The presence of fossilized seeds indicates that the tree had been pollinated and so successfully fertilized
Woody cone has turned to stone during fossilization

THE **FIRST SEEDS**

About 370 million years ago, a new kind of plant evolved. It produced seeds, which are the ultimate embryo survival kit—packed with nutrients and enveloped in a protective casing. Seeds would shape the history of life and play a key part in our own prehistory.

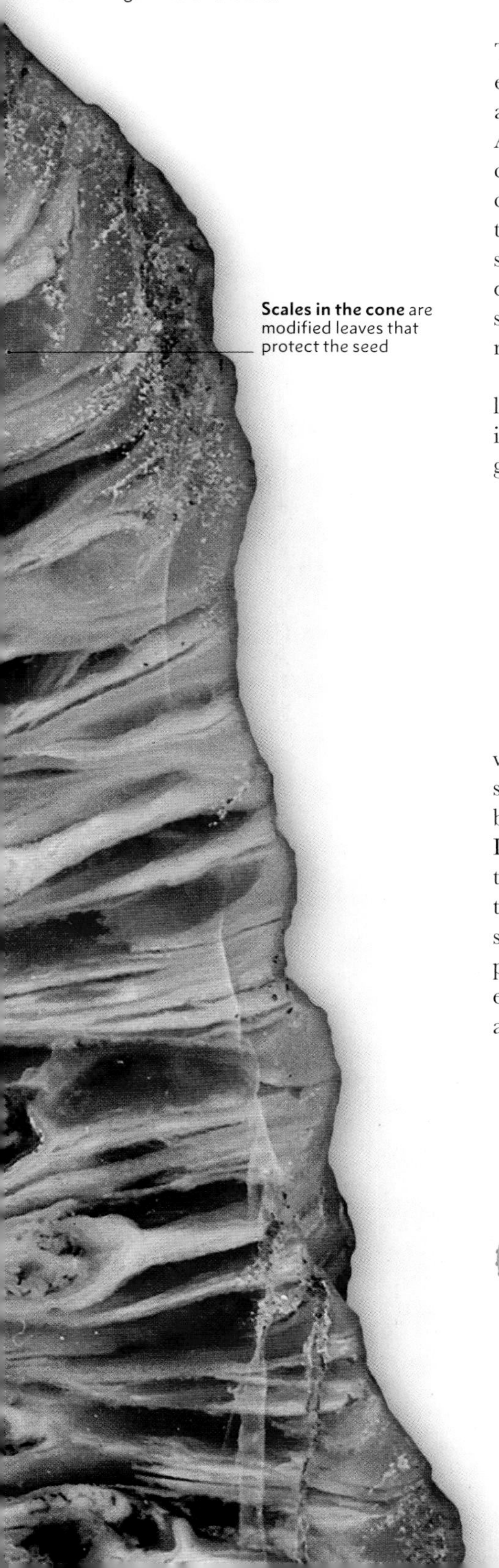

◀ **Monkey puzzle**
This fossilized cone is 160 million years old but is remarkably similar to the cones produced by trees today. This species, *Araucaria araucana*, is known as the monkey puzzle and still thrives in Argentina and Chile.

Scales in the cone are modified leaves that protect the seed

The first algaelike plants completed their entire life cycle—alternating between spores and gametes (eggs or sperm)—under water. As their descendants, mosses and ferns, crept farther inland, more resilient spores could be dispersed into the air. However, their sperm still needed water droplets to swim to the egg: no matter how much their deep roots and tough leaves helped them survive droughts, plants still needed periodic rainfall in order to reproduce.

A new kind of plant broke this restrictive link with water by relocating its fertilization into reproductive shoots away from the ground. Female shoots retained their spores, which grew into eggs. Spores from male shoots became pollen grains that were blown inland to land on female shoots. In the most primitive seed plants, sperm then burst from the pollen grains and swam through the shoot to the egg—something still seen in cycads of today. But in most seed plants, sperm became redundant. Instead, each pollen grain sprouted a tiny thread—a pollen tube—that conveyed a naked male cell nucleus straight to the egg, dispensing with swimming altogether. Pollen allowed plants to spread farther inland than their water-reliant relatives. What is more, these plants completed their break from water by keeping embryos of their next generation in drought-resistant cases—seeds.

MEDULLOSA—A **PRIMITIVE SEED PLANT** THAT LIVED 350–250 MYA—HAD **SEEDS THE SIZE OF EGGS**

HOW SEEDS WORK

Eggs develop inside a thin-walled sac called an ovule. After pollen fertilizes an ovule, its walls thicken, and it becomes a seed. At first, ovules grew exposed on foliage or the scales of cones—reproductive shoots composed of hard scales connected at their base, just like the cones produced by cycads and conifers today. Eventually, most seed plants buried their ovules deeper inside the shoot, beneath a flower (see pp.160–61). When these ovules turn into seeds, the succulent tissue around them becomes fruit. Seed plants had now evolved a method of enticing animals, a different form of complex life, to become part of their life strategy (see pp.164–65).

SEEDS, THEIR SUCCESS, AND US

The pollen method of fertilization and the seed method of dispersal have both been so successful that seed plants now form the basis of all land-based ecosystems and food webs worldwide, including those with humans at the top. Non-seed plants—mosses, ferns, and liverworts—although widespread, no longer dominate any land habitats.

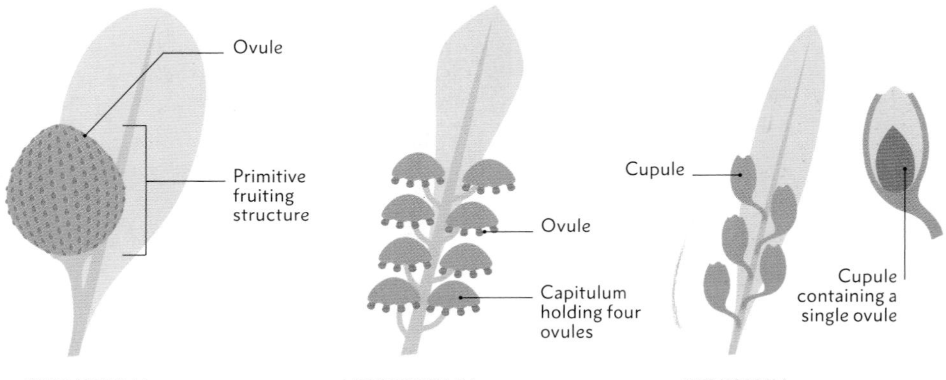

◀ **Primitive seed plants**
The first seed plants are called seed ferns, because of the shape of their leaves, although they are unrelated to the ferns we know today. They grew their ovules in packages attached to the leaves. Cones and flowers eventually evolved in later types of plants.

▼ Life in a shell
Prehistoric reptiles—including dinosaurs—were pioneers of the shelled egg. The embryo could develop inside, safe from dehydration. Its parents may have guarded the egg from predators, just like many reptiles and birds do today.

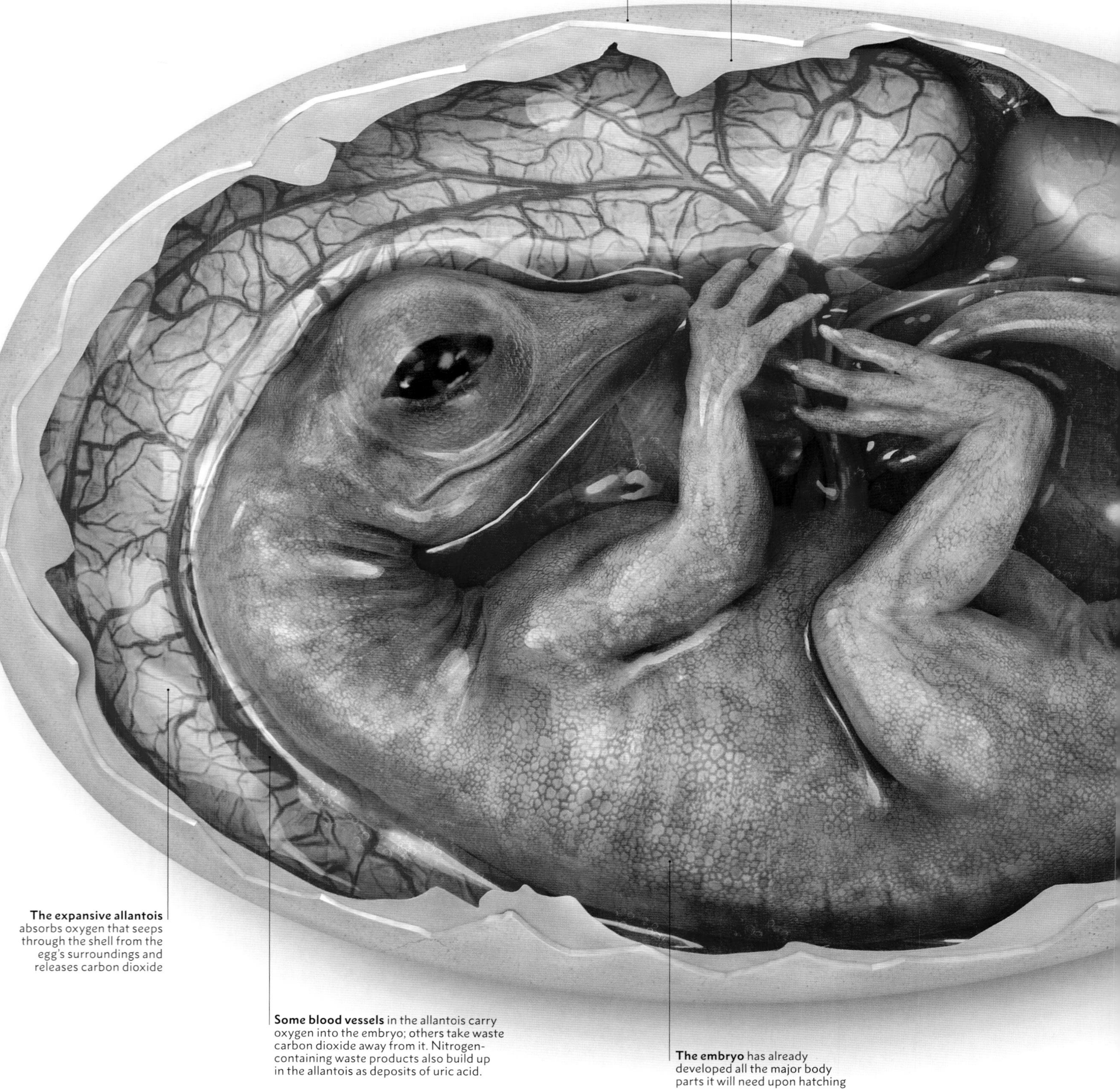
The shell is composed of a chalky material based on calcium carbonate that is hard enough to withstand damage, permeable to allow the exchange of respiratory gases, and sufficiently brittle so the infant can break free upon hatching
White shell membranes conceal the chorion—a transparent embryonic membrane that completely encloses the embryo, amnion, yolk sac, and allantois
The expansive allantois absorbs oxygen that seeps through the shell from the egg's surroundings and releases carbon dioxide
Some blood vessels in the allantois carry oxygen into the embryo; others take waste carbon dioxide away from it. Nitrogen-containing waste products also build up in the allantois as deposits of uric acid.
The embryo has already developed all the major body parts it will need upon hatching

SHELLED EGGS ARE BORN

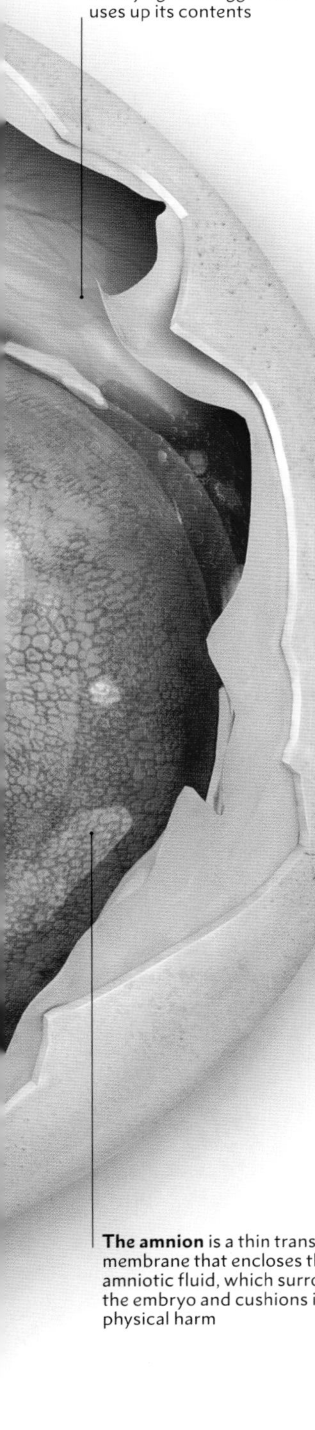
The yolk sac is filled with foods, such as protein and fat, that nourish the developing embryo; it shrinks as the embryo grows bigger and uses up its contents
The amnion is a thin transparent membrane that encloses the amniotic fluid, which surrounds the embryo and cushions it from physical harm

The first backboned animals to live on land could walk, since they had legs, and could breathe air. These early amphibians were still tied to water, however, because they needed a wet place to breed. Reptiles broke this link by producing hard-shelled eggs that could develop on dry land.

Backboned animals originated in water, where fish and amphibians laid their soft eggs encased in nothing but a protective jelly coat. Reptiles not only evolved hard, scaly, waterproof skin as protection from dehydration, but transformed their breeding habits too. They covered their eggs in a shell hard enough to protect and contain the embryo on land, yet permeable enough for it to breathe.

EMBRYO SURVIVAL KIT

The shelled eggs produced by most reptiles and all birds are amazing structures that contain all their embryos need to develop. Until the invention of these eggs, all living embryos developed surrounded by fluid. To reproduce those fluid conditions on land, it was a small and manageable evolutionary step to enclose the fluid within a membrane. The membrane is called the amnion, giving the first animals to possess it the name "amniotes" as well as the more familiar "reptiles." Within the egg, the embryo also has its own larder, the yolk sac, just as fish and amphibians do. But it also has an allantois—a waste-disposal pouch absent in its ancestors. The yolk sac grows smaller and the allantois enlarges as it absorbs oxygen and accumulates waste products while the embryo grows. A final membrane—the chorion—serves to contain the entire embryo "survival kit."

By the time they hatch, reptiles are ready to lead independent lives; on the other hand, most bird chicks need parental care for a time. But, in both cases, hatchlings are ready to eat and breathe as soon as they emerge from the egg.

PREPARED FOR LAND

The shelled egg and its life-supporting membranes enabled the amniotes to complete their life cycle on land. They mated on land and laid eggs in a dry nest. A few living reptiles have abandoned their egg-laying ways and give birth to live young.

THE EARLIEST ANIMAL THOUGHT TO LAY SHELLED EGGS IS ***PALEOTHYRIS***, A REPTILE LIVING **330 MYA**

But one group of amniotes, the mammals, turned live birth into a major asset. They commandeered two membranes—the allantois and chorion—into a placenta, which draws oxygen and nourishment straight from the mother's blood. By nurturing the embryo in the mother's body, mammals improved their offspring's chances of survival beyond those of their larvae-producing ancestors.

◀ **Land colonizer** *Dimetrodon*, a reptile that lived 290–270 MYA, is an example of an early amniote. Its ability to lay eggs with a shell and amnion allowed it to colonize arid habitats where water was not readily available.

HOW **COAL FORMED**

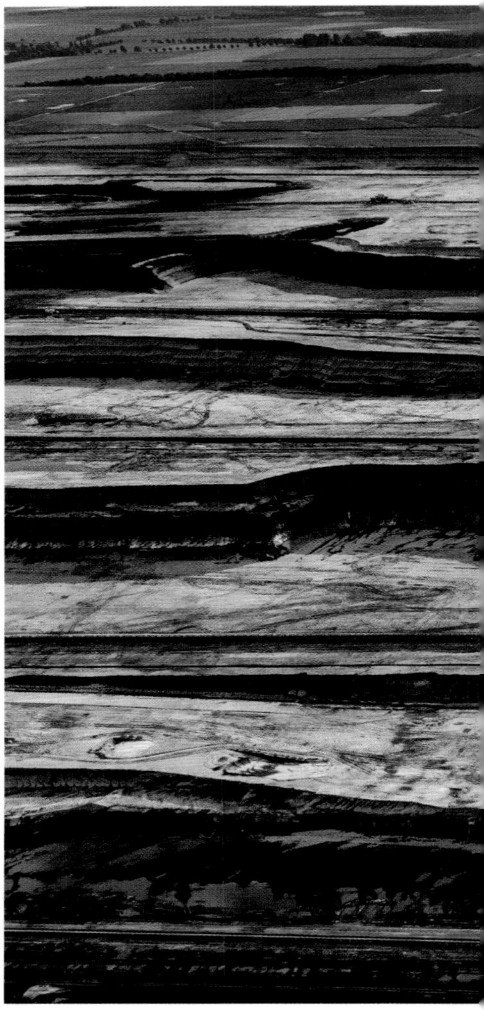

The trees that formed Earth's first forests were giant fernlike plants that resisted decay. Their dead bodies built up, trapping carbon and energy underground. These were the coal forests—and 300 million years later, their compacted remains would fuel an industrial revolution.

The Carboniferous period (359–299 MYA) was a time when life on land prospered more than ever before. Trees grew from mosslike ancestors, insects took to the air in a world already crawling with invertebrates, and giant amphibians were evolving into reptiles. This time in Earth's history would have huge implications for our own history.

THE FIRST FORESTS

For the first time, terrestrial life could live in the trees, imbuing habitats with an extra richness. The first big invasions of land animals, involving millipedes, insects, and arachnids, had already taken place, but now these groups exploded into a multitude of species, including predators such as spiders, scorpions, and centipedes. Carboniferous trees could grow tall because they had evolved a tough supporting material called lignin that formed a protective layer. It would also eventually become the carbon-rich store of energy that formed coal. The trees concentrated lignin in their tissues to more than 10 times the quantity found in today's trees. This not only helped to deter herbivores, but it also resisted decay, because few microbes could digest it. As trees died, their fallen trunks lingered. The lignin, along with the carbon it contained, would be converted to carbon dioxide (CO_2) if it decayed, but it sank into the swampy Earth, locking away its chemical energy. As CO_2 in the atmosphere diminished, oxygen increased, since it would normally be consumed by the same processes of decomposition, which were now suppressed. Oxygen built up in the air to become more than one-third of it by volume. Today, oxygen accounts for only one-fifth of the gas in the atmosphere. The effects of such high oxygen levels would have been bizarre. Ignition would have happened more readily, sparking wildfires. Animals that relied on passive breathing through their skin or body surface became enormous. The biggest insects that ever lived evolved during the Carboniferous, and amphibians grew to the size of crocodiles.

THE ORIGINS OF COAL

Much of the bulk of the Carboniferous trees sank intact beneath the swampy waters, forming layer upon layer of a deposit called peat. In the peat, oxygen was low and acidity high, and instead of decomposing, the carbon-rich remains built up. The peat became compacted under its own weight, squeezing out water and gases, turning first into a form of rock called lignite, and finally

LEPIDODENDRON TREES GREW **UP TO 130FT (40M) TALL** IN THE **CARBONIFEROUS** PERIOD

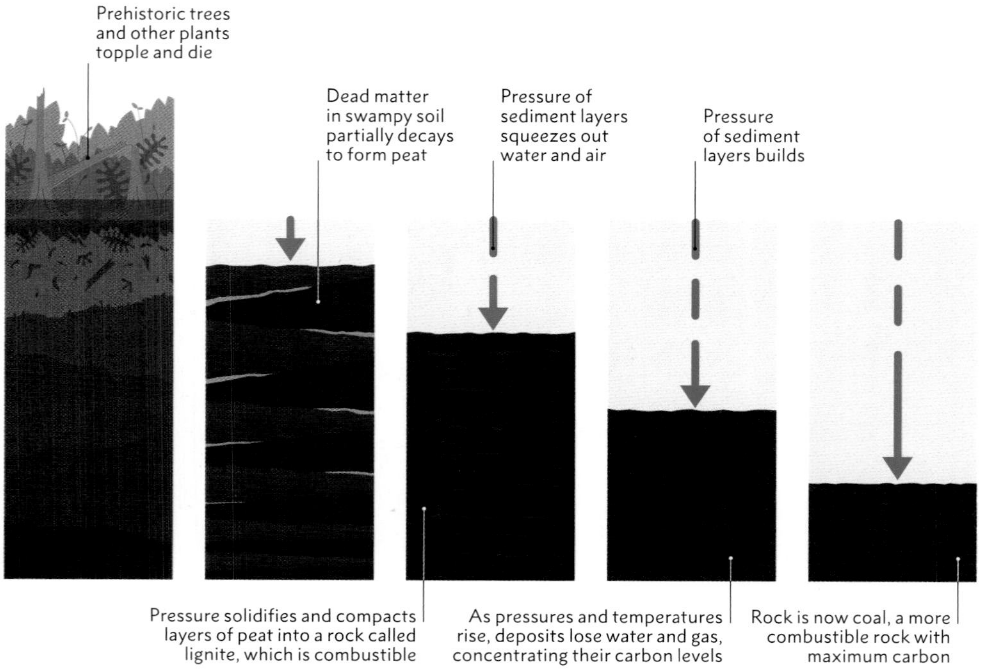
Prehistoric trees and other plants topple and die
Dead matter in swampy soil partially decays to form peat
Pressure of sediment layers squeezes out water and air
Pressure of sediment layers builds
Pressure solidifies and compacts layers of peat into a rock called lignite, which is combustible
As pressures and temperatures rise, deposits lose water and gas, concentrating their carbon levels
Rock is now coal, a more combustible rock with maximum carbon

▶ **Coal formation**
Coal began as undecomposed matter from dead trees. The dead matter was buried as new dead material accumulated on top, and it became compacted under high pressure. Over millions of years of increasing pressure and temperature, the material turned first to the rock lignite, then eventually to coal.

◀ **Prehistoric energy**
Layers of coal can be seen clearly as dark bands packed between the rock at this coal mine in the Lower Rhine region, Germany.

to a harder, denser rock with an especially high concentration of carbon: coal. Coal deposits are found in rocks that date back to before the evolution of land plants—and these probably came from algae. But coal deposits are especially abundant from the Carboniferous period, where conditions were just right for them to form.

Human civilizations, perhaps as far back as 1000 BCE, recognized the potential of using coal as fuel because of its resemblance to charcoal. Both could be burned to release a great deal of heat. The carbon that had been locked away in coal for millions of years was finally released as carbon dioxide. The emergence of large-scale mining (see pp.306–07) could tap deposits in seams far below the surface. Since then, the demand for burning fossil fuels has released so much carbon dioxide in such a short amount of time that it is sparking concerns for humanity. Energy is required for a growing population—however, burning fossils fuels has increased the amount of greenhouse gases and contributed to global warming. Today's civilization must deal with an environmental issue of its own making—and one that affects the entire world.

> “**COAL**, OIL, AND GAS ARE... **FOSSIL FUELS**, BECAUSE THEY ARE MOSTLY MADE OF... FOSSIL REMAINS... THE **CHEMICAL ENERGY** WITHIN THEM IS A KIND OF **STORED SUNLIGHT ORIGINALLY ACCUMULATED** BY **ANCIENT PLANTS**.”
>
> **Carl Sagan**, astronomer and science author, 1934–1996

▶ **Ingredient of coal**
The fossilized trunk of the plant *Lepidodendron*, an abundant tree during the Carboniferous period. Its tall trunk lacked true bark but was thickened by a layer of tough lignin.

Each diamond-shaped scar marks the point where a leaf has broken away from the trunk

LIZARD IN AMBER

Traces—or fossils—that have been left behind in the rocks and stones of Earth are evidence that extinct species were not the same as those living today. Scientists must turn into detectives to work out how they once lived.

More than 99 percent of species that have ever lived on Earth are now extinct. This means that what we know about the history of life is critically dependent on fossil evidence.

Fossils can form in different ways. If dead organisms are buried quickly in sediment before being eaten, they and the sediment turn to rock. When continents shift position over millions of years, rocks that contain these fossils may buckle and rise—exposing the fossils as the surrounding rock is eroded.

The fossilization process is never perfect, and preservation quality varies greatly. Older, soft-bodied species leave frailer traces than younger, harder ones. Hard parts of the body, such as the skeleton, are most likely to be fossilized. Footprints, eggs, and feces can also be fossils. Under the right conditions, the most delicate features, such as skin, feathers, leaves, or even single cells, can be fossilized. Some fossils can be found in amber, such as this lizard. Amber is the solidified resin of trees that has hardened, and animals that get smothered and trapped within it can be exquisitely preserved.

Paleontologists must consider how a fossil formed when interpreting fossil evidence. Clues are studied from multiple disciplines, such as geography and anatomy, to assemble a picture of how different kinds of organisms lived in the past.

Story of the dead body

The study of taphonomy concerns processes that change a dead animal's body as it decays or fossilizes. Tree resin is organic too, so also decays. This lump of resin fossilized well because it was packed under sediment soon after forming. As a result, *Yantarogekko* was preserved perfectly, sealed away from scavengers and erosion.

Prehistoric spider also trapped in amber

Botanical clue

Analysis of this Baltic amber shows that it was produced by a species of conifer, suggesting that *Yantarogekko*'s habitat was coniferous forest. The presence of the fossilized amber indicates that, by this time, these conifer trees had evolved resin (sticky droplets that seal wounds and deter herbivores), perhaps in response to a prehistoric species of herbivore that fed on them.

Coniferous forest, Poland

▼ How fossils form
It takes millions of years for the bodies of living organisms to fossilize. Organic remains decay and harden.

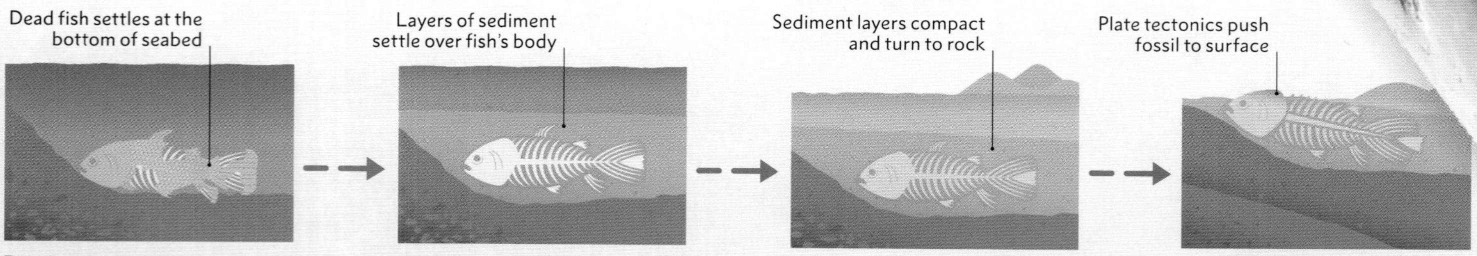

Dead fish settles at the bottom of seabed
Layers of sediment settle over fish's body
Sediment layers compact and turn to rock
Plate tectonics push fossil to surface

Dead animals left uneaten will decay. Left undisturbed the fish's scales settle and are preserved as an outline of its fossilized skeleton.

Buried under sediment, which screens the fish from scavengers, minerals from the water filter into the bones, causing them to crystallize and harden.

Pressure from layers of sediment and rock accumulates over millions of years and the organic parts of the body are entirely replaced by minerals.

Discovery of the fossil may occur when continental drift pushes it to the surface, and erosion may start to wear away at the rock—exposing the fossil.

Locating the habitat

The place where a fossil is found today may differ a great deal from its original habitat. For instance, this amber-entombed lizard was found on the brackish Baltic Sea coastline. When this lizard died 54 MYA, its natural habitat may have been a forest farther inland. The evidence suggests that a river washed lumps of amber from the warm coniferous forest downstream to the coast.

Comparing anatomy

The structure of a fossil's body, or the traces the body leaves behind, can be compared with those of related fossils and species alive today. Only the head, front end of the body, and right forelimb of this lizard are preserved in amber, yet this is enough for paleontologists to recognize it as a species of gecko. This specimen reveals well-developed toe pads and lack of eyelids—features that are preserved in amber but would be lost in a fossilized skeleton preserved in rock.

Banded gecko

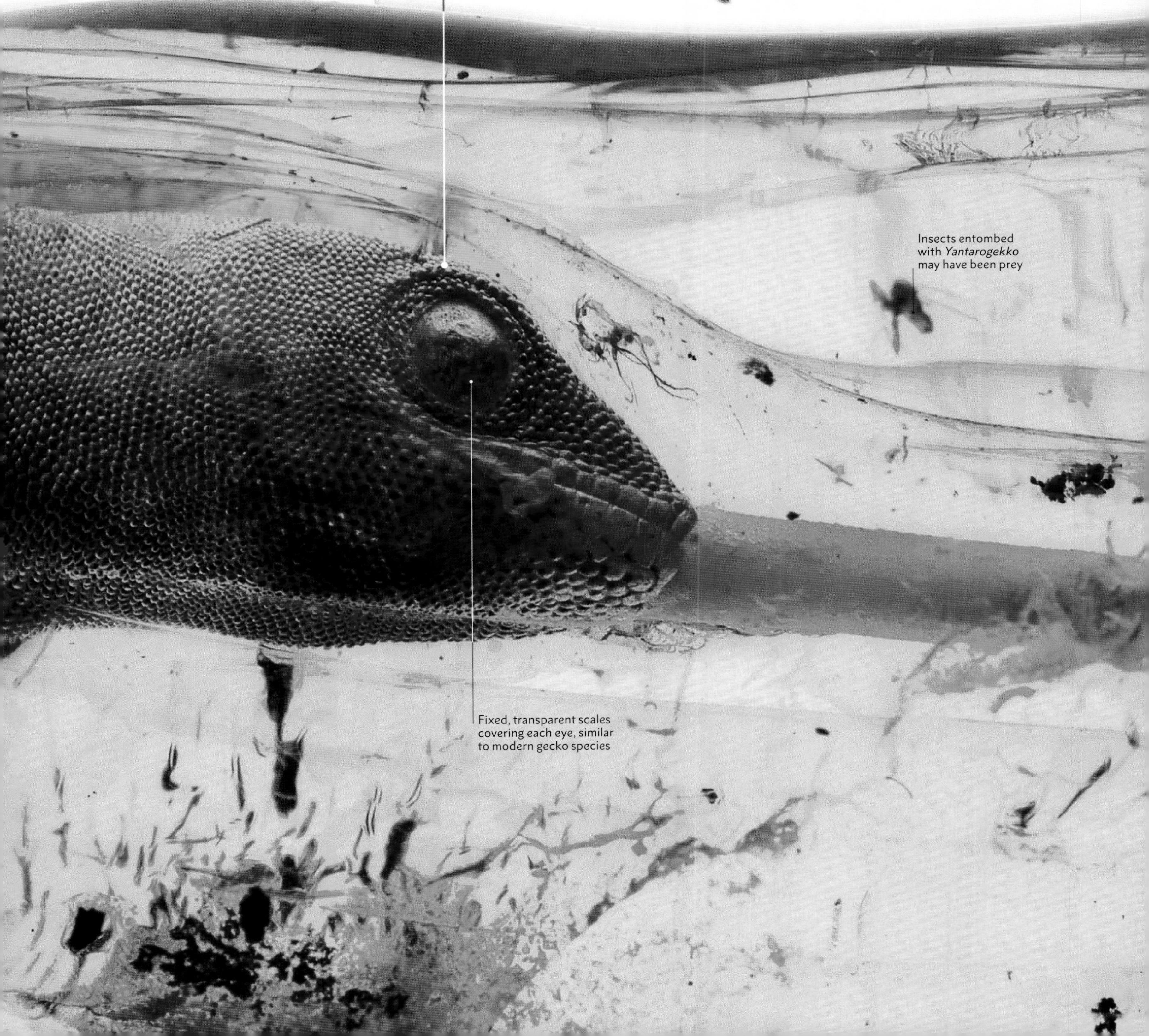
Insects entombed with *Yantarogekko* may have been prey
Fixed, transparent scales covering each eye, similar to modern gecko species

▶ Pangaea
The continents were merged into one over Earth's southern hemisphere between 300–175 MYA. Inland, forests turned to deserts, while diminished coastlines drove many marine species to extinction.

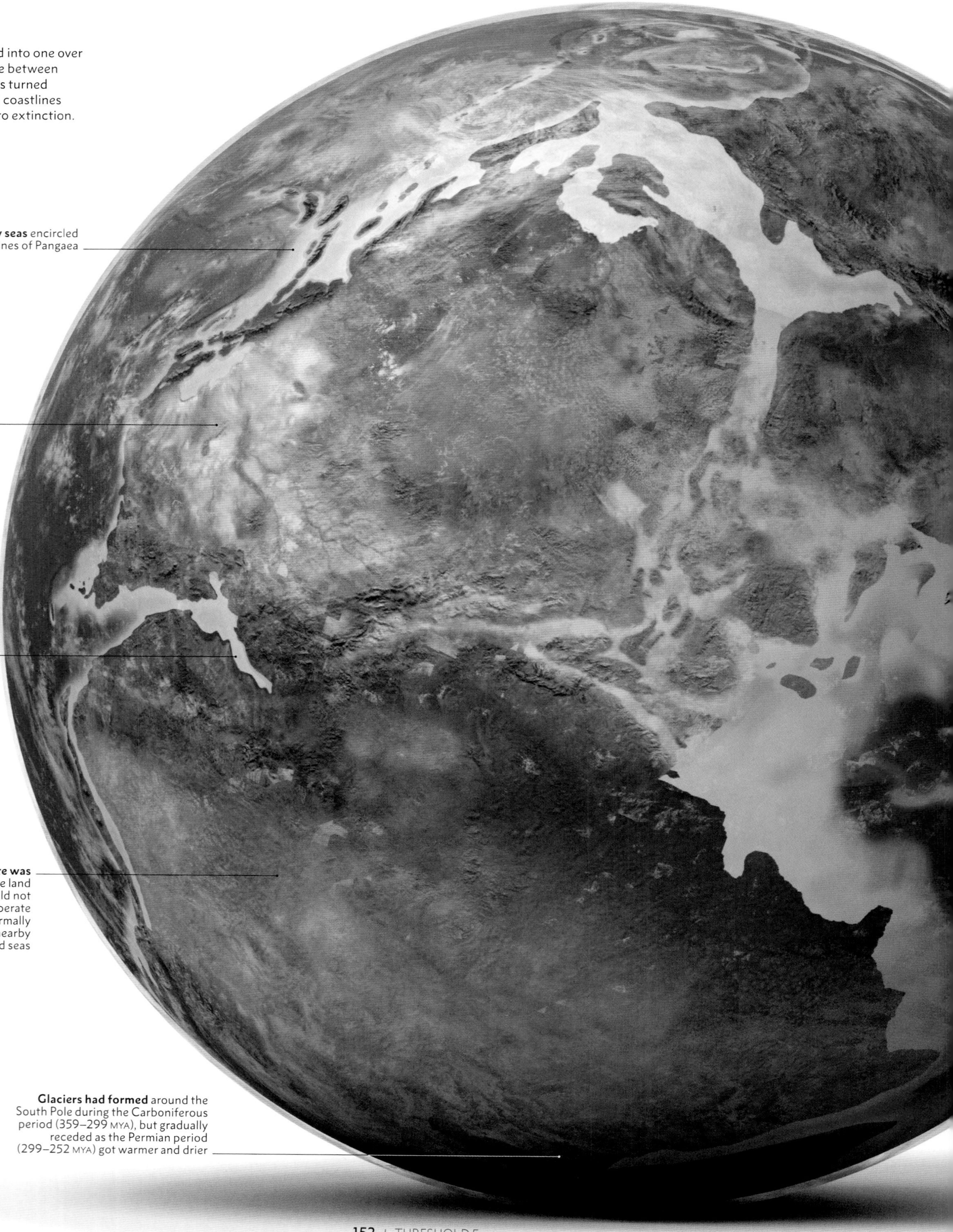
Shallow seas encircled the coastlines of Pangaea
Vast swathes of arid land spread across what would become North America and Europe during the Permian period (299–252 MYA)
Where continents merged, coastlines disappeared—probably resulting in the extinction of marine life
Pangaea's climate was hot and dry, since land in the center would not benefit from temperate climatic effects normally provided by nearby oceans and seas
Glaciers had formed around the South Pole during the Carboniferous period (359–299 MYA), but gradually receded as the Permian period (299–252 MYA) got warmer and drier

THE LAND **DRIES OUT**

After terrestrial life flourished in the swampy coal forests, a global drought that lasted 50 million years changed the direction of life's evolution. As vegetation grew tougher leaves and swamps dried up, some moist-skinned amphibians gave rise to the first scaly reptiles.

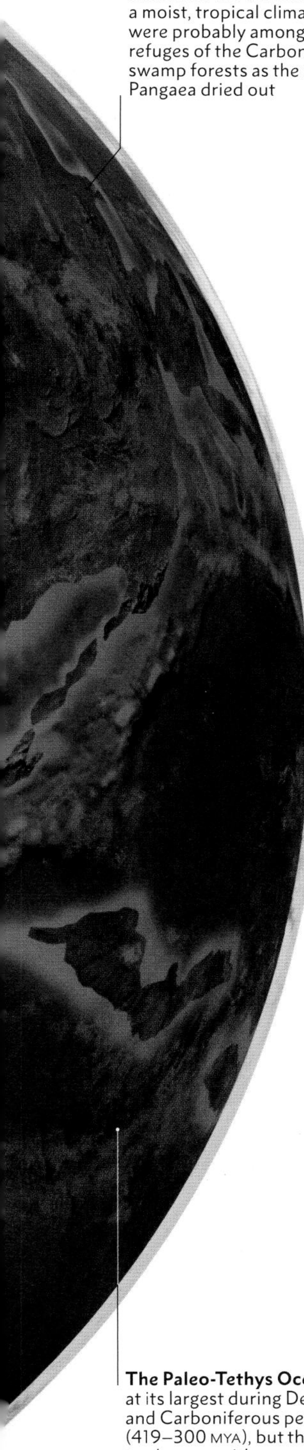

Shores of the coastline basked in a moist, tropical climate, and so were probably among the last refuges of the Carboniferous swamp forests as the rest of Pangaea dried out

The Paleo-Tethys Ocean was at its largest during Devonian and Carboniferous periods (419–300 MYA), but then started to close up with movement of land masses in the Permian period

Around 300 million years ago, all of Earth's landmasses collided to form a single supercontinent called Pangaea. This caused a dramatic change in terrestrial life. Climate change had already triggered a collapse in the great swampy forests of the Carboniferous period (see pp.148–49), but now, at the dawn of the Permian period, much of the landscape of the new supercontinent was about to turn to desert.

> MANY OF THE PERMIAN **REPTILES** POSSESS **FOSSIL CHARACTERISTICS** THAT **FORESHADOW** THE HEAD AND TEETH OF **MAMMALS**.
>
> **R. Will Burnett**, biologist, 1945–

NEW SKIN, LARGER SIZES

Reptiles had evolved in the forests, but now spread across the new parched world. These new vertebrates were better adapted for land than their amphibian ancestors. By evolving hard scales, made from a tough fibrous protein called keratin, they reduced dehydration. Mammals and birds would later use the keratin for their hairs and feathers. The first reptiles to lay hard-shelled eggs (see pp.146–47) also did not need water to breed—unlike their amphibian ancestors. This helped to push vertebrate land colonization like never before.

Two main reptile groups diverged at the start of their reign. One, the diapsids, later went on to produce dinosaurs, birds, and modern lizards. At the time of the Permian it was the second group, the synapsids, that came to rule the arid land. Some evolved to become the biggest land animals of the day. The sail-backed, carnivorous *Dimetrodon* reached the size of a car, and others became the first big herbivores. Later synapsids also included small reptilian ancestors of mammals.

The Permian period closed with violence—a mass extinction so severe that it wiped out more than 70 percent of animal life. With extraordinary volcanic activity releasing noxious gases, the biggest extinction event ever saw many reptiles disappear. But enough descendants of both groups, the synapsids and diapsids, survived to repopulate the land—first with dinosaurs and mammals, and then with birds.

◀ ***Moschops***
With its stocky body, this survivor of the dry Permian world ate tough desert vegetation. It was one of many synapsids—strong-jawed reptiles that would eventually give rise to mammals.

REPTILES **DIVERSIFY**

The coming and going of species define the chapters in the history of life. In the wake of a drying supercontinent, the Age of the Reptiles produced some of Earth's most spectacular animals. Reptilian diversity reached its peak, as giant reptiles conquered sky, land, and oceans.

The great Age of Reptiles spanned more than 200 million years. It began on the parched landscape of Pangaea (see pp.152–53) and ended with an asteroid strike, but even after the demise of the dinosaurs, reptiles prevailed, albeit in a smaller form. Today, lizards and snakes account for nearly one-third of land vertebrate species.

MESOZOIC MONSTERS

During the Mesozoic Era, the stretch of time divided into Triassic, Jurassic, and Cretaceous periods, a group of small, lizardlike reptiles—diapsids—diversified with spectacular results. Some diapsids returned to the ocean habitats of their distant ancestors: the ichthyosaurs and plesiosaurs, such as *Albertonectes*, evolved flippers from limbs and became expert swimmers and hunters of fish.

The most famous diapsids took body size to new extremes. These reptiles—the archosaurs—became crocodilians, flying pterosaurs, dinosaurs, and ultimately birds. They had strong limb muscles that allowed them to walk tall—improving on the lumbering, belly-dragging gait of earlier reptiles.

GIANTS AND MINIATURES

The most successful and diverse archosaurs of the time, dinosaurs evolved into a multitude of predators, grazers, and scavengers. The gigantic, long-necked, herbivorous sauropods, such as *Brachiosaurus*, became about as large as it is possible for a land animal to get. As herbivores evolved into giants, so did their predators. Theropods, the bipedal sprinters of the dinosaur family, were nearly all carnivores. The biggest of these, such as *Tyrannosaurus*, were among the most formidable predators ever to walk on land. Evolution also favored miniaturization among dinosaurs: one group of diminutive theropods grew feathers, turned warm-blooded, and eventually evolved into birds.

MASS EXTINCTION

The reign of the giant reptiles ended with the Cretaceous mass extinction—almost certainly caused by an asteroid or comet striking Earth. Catastrophic conditions followed, including wildfires, acid rain, and a global cloud of debris that blocked the sun's light and brought much of life's food-providing photosynthesis to a temporary halt.

Unable to adapt quickly enough to the rapidly changing conditions, all the giant reptiles—including plesiosaurs, pterosaurs, dinosaurs, mosasaurs, ichthyosaurs, and the giant ancestral crocodilians—became extinct. But lizards, snakes, turtles, and modern crocodiles survived. Surviving along with them were the descendants that would ultimately succeed the reptiles in global domination: birds and mammals.

> “**CREATURES FAR SURPASSING THE LARGEST OF EXISTING REPTILES**... DEEMED SUFFICIENT GROUND FOR ESTABLISHING **A DISTINCT TRIBE**... ***DINOSAURIA***.”
>
> **Richard Owen**, palaeontologist, 1804–1892

Pteranodon

Velociraptor

Edmontosaurus

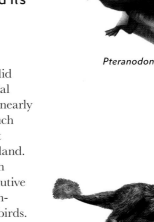

◀ Diversification
The dinosaurs formed one of many reptile groups that dominated Earth for millions of years. Living alongside them, pterosaurs soared in the skies and plesiosaurs and mosasaurs swam in the oceans. In addition, turtles, lizards, snakes, and crocodilians all appeared for the first time.

BIRDS TAKE TO THE **AIR**

Birds are the most varied of the flying vertebrates, and today there are more than 10,000 species. Their origins lie with the dinosaurs, and scientists have been studying fossils for 150 years to better understand this evolutionary transition.

The story of how birds evolved from reptiles provides biologists with a deeper understanding of how evolution works. From one form of life, another can arise so inherently different that at first glance it appears that there is no relationship between the two. Closer inspection of anatomy, the fossil record, and molecular analysis of genomes can lead to surprising connections between seemingly unrelated species.

Superficially, reptiles and birds differ to a large degree. Modern birds look conspicuously distinct from living reptiles, even though they had reptilian ancestors—a group of bipedal, mainly predatory dinosaurs called theropods. Theropods, however, had already evolved to become very unlike the reptiles we know today. Some were not only feathered, but may have been warm-blooded, too.

PREPARING FOR FLIGHT

In some ways, theropods were primed for flying, even if their reasons for doing so are not certain. They walked upright on their hind legs, which meant their front legs were free to become wings. Some small species had hollow bones, which were already lightweight. In some gliding species, long fingers supporting broad, feathered hands provided the lift to sail short distances over ground or from branch to branch. However, genuine wing-flapping flight required at least two more modifications: flight feathers made into stiff blades and stronger muscles capable of sustained flapping.

As birds evolved over time, their breastbones developed a bony protrusion called a keel to which more massive flight muscles attached. Big-keeled birds packed more breast muscle to power their wings. These masters of flight flourished in the forests, grasslands, and wetlands of the post-dinosaur world. They evolved new and better ways to get food, as they caught insects, crushed seeds, or lapped nectar. Others returned to the meat-eating habits of their ancestors and a few, such as ostriches, have abandoned flight altogether and sprint across the ground instead.

150 MYA

Archaeopteryx
This species retained many reptilian features, including a long, bony tail, teeth, and claws on its feathered wings. It lacked the musculature for strong flight, so may have relied heavily on gliding.

125–120 MYA

Confuciusornis
The first bird known to have a toothless beak, it also had a more birdlike tail and a keel on its breastbone. Like *Archaeopteryx*, its shoulder joint was angled lower than in modern birds, and this restricted the depth of its "flap."

0 MYA

Erithacus
The keeled breastbone of modern birds, such as the European robin, supports massive flight muscles (up to 10 percent of the bird's body weight), making flight stronger.

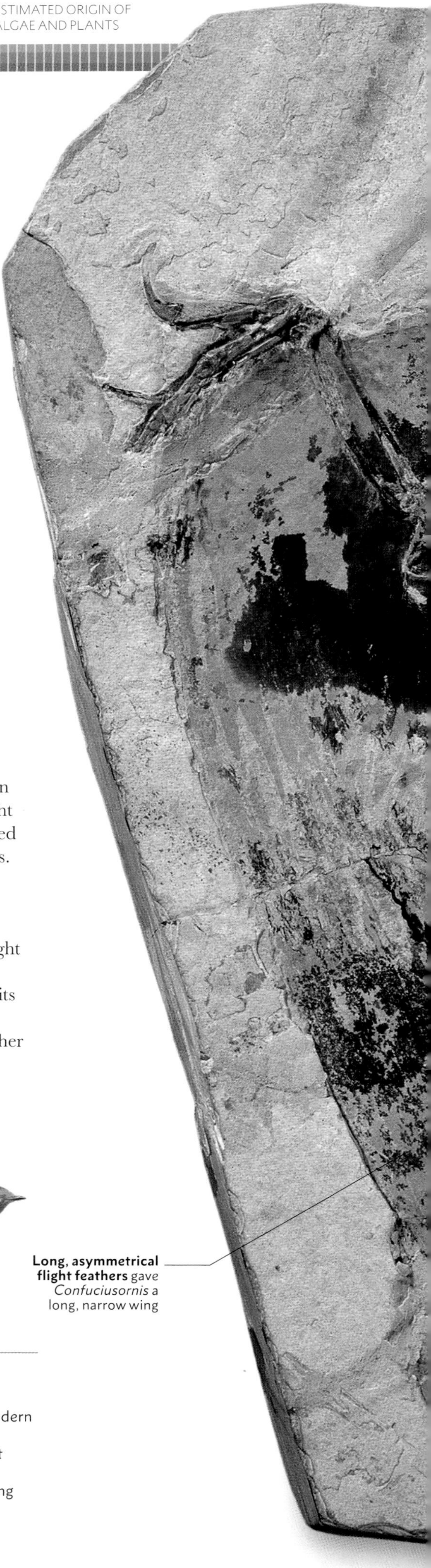

Long, asymmetrical flight feathers gave *Confuciusornis* a long, narrow wing

A toothless beak sets *Confuciusornis* apart from its dinosaur ancestors and from *Archaeopteryx*, which had teeth like a dinosaur
Large claw was probably used to climb trees
The tail vertebrae ended in a bony stump called a pygostyle, typical of all modern birds; like modern birds, if it had a long tail, it was formed of feathers
Backward-facing back toe (hallux) allowed *Confuciusornis* to perch firmly on branches, like many modern birds

◄ Prehistoric flier
A crow-sized primitive bird, *Confuciusornis* lived alongside the dinosaurs 125–120 MYA, during the Cretaceous period. Its fossilized remains have been found in abundance, with many fossils preserving its skeleton and feathers in exquisite detail.

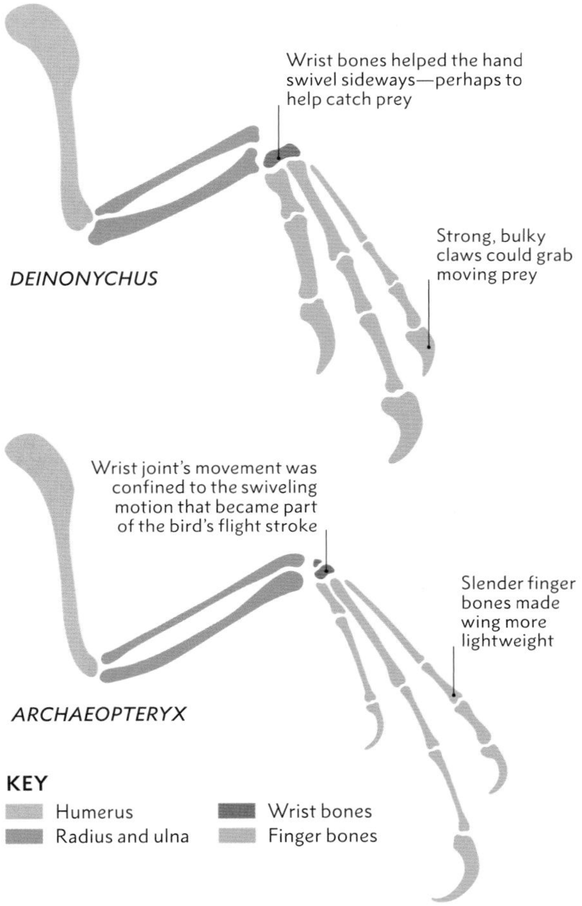
Wrist bones helped the hand swivel sideways—perhaps to help catch prey
Strong, bulky claws could grab moving prey
DEINONYCHUS
Wrist joint's movement was confined to the swiveling motion that became part of the bird's flight stroke
Slender finger bones made wing more lightweight
ARCHAEOPTERYX
KEY
Humerus
Radius and ulna
Wrist bones
Finger bones

▲ Prehistoric hands
Hand and wrist bones of *Deinonychus*, a theropod dinosaur, and *Archaeopteryx*, the earliest known bird, show remarkable similarity in anatomy. However, only *Archaeopteryx* could fly.

> **AVIAN FLIGHT** IS THE **MOST VARIED** AND **SUCCESSFUL** OF ALL FORMS OF VERTEBRATE FLIGHT.

John Ostrom, palaeontologist, 1925–2005

CONTINENTS SHIFT AND **LIFE DIVIDES**

As continents move, they carry with them communities of living things that have evolved over millions of years. Landmasses that split and collide pull species apart and bring others together. As land glides between poles and the equator, climate also affects species.

Land-based life rides on moving continental plates that are pushed and pulled as crust plunges into Earth's interior in some places and is reformed in others (see pp.92–93). Oceans between the crust expand and shrink, while coastal and marine life comes and goes. The shifting surface of Earth helps to explain why fossils found today end up in odd places—such as those of sea-floor animals appearing high in the Himalayas.

CRADLES OF LIFE ON LAND

Relatively early in Earth's history in the Cambrian period (541–485 MYA), giant land masses formed and split, creating the oceans in which life diversified. Once plants and invertebrates had invaded land and diversified, landmasses became centers for evolution. These events happened so long ago that there is scarcely any trace in the distribution of invertebrates and plants alive today. But over 300 MYA—as some amphibians were evolving into reptiles and some spore-bearing plants were evolving into seed plants (see pp.144–45)—the movement of the continents began to have more lasting impacts.

LAND LIFE SPLITS APART

In the Carboniferous period (359–299 MYA), northern and southern land masses collided to form a huge supercontinent called Pangaea (see pp.152–53). It straddled the equator and contained most of Earth's land.

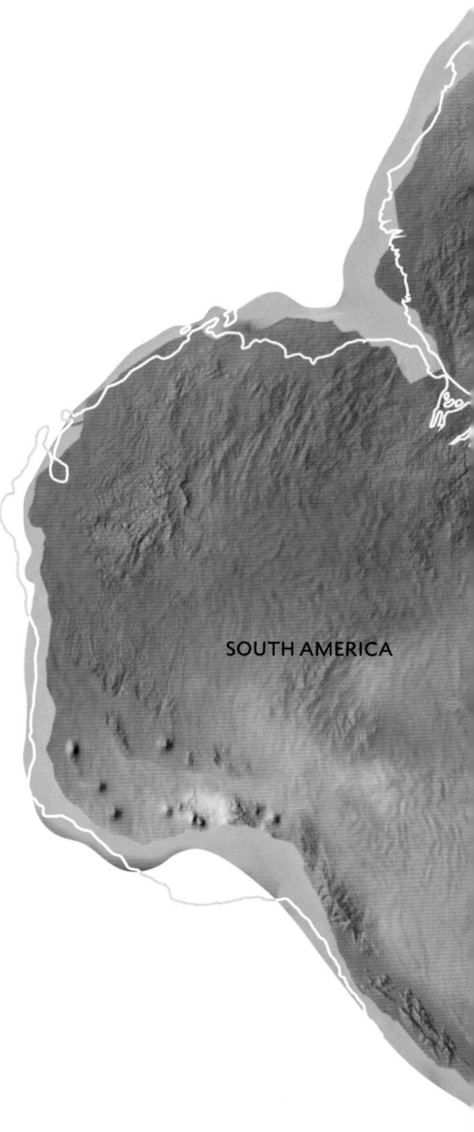

> “**ALL EARTH SCIENCES** MUST CONTRIBUTE **EVIDENCE**... UNVEILING THE STATE OF OUR PLANET IN EARLIER TIMES.”
>
> **Alfred Wegener**, geologist and meteorologist, 1880–1930

▶ **Modern clue**
The African ostrich is a species of flightless ratite bird. Other species of ratites include the South American rhea and Australian emu, providing evidence for a Gondwanan distribution for ratite birds.

Its effect on climate was dramatic—with the dry interior vastly different from the cold, polar extremes. This, coupled with the loss of many coastal habitats, sent many species into extinction, but helped seed plants, reptiles (see pp.154–55), and others diversify.

In the Mesozoic Era 100 million years later, Pangaea began to split. This created a sea barrier for land-based life, and plants and animals were isolated on two supercontinents; Laurasia in the north split from Gondwana in the south. Land-based life could wander across five continents that today are widely separated. Further splitting would produce recognizable landmasses: Laurasia into North America and Eurasia, and Gondwana into South America, Africa, India, Antarctica, and Australia.

We now know that Gondwana was covered in rich rain forests that encouraged diversity. Many groups alive today evolved there first—such as modern marsupial mammals—and spread throughout Gondwana, but could not reach Laurasia. Today, marsupials are restricted to South America and Australia, and have fossils in Antarctica. Flightless ratite birds, such as the Australian emu, also have a remnant Gondwanan distribution. Those evolving in Laurasia, such as salamanders and newts, were restricted to northern continents.

The distribution of fossilized species is evidence for continental drift (see pp.90–91). Certainly, the pattern and movement of continents has had a profound impact on the distribution of all life that followed.

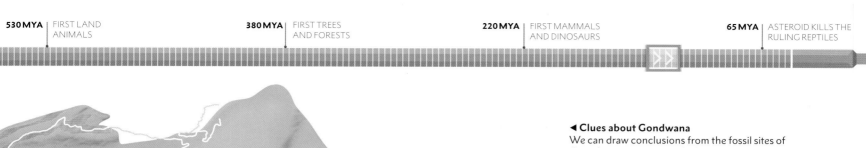

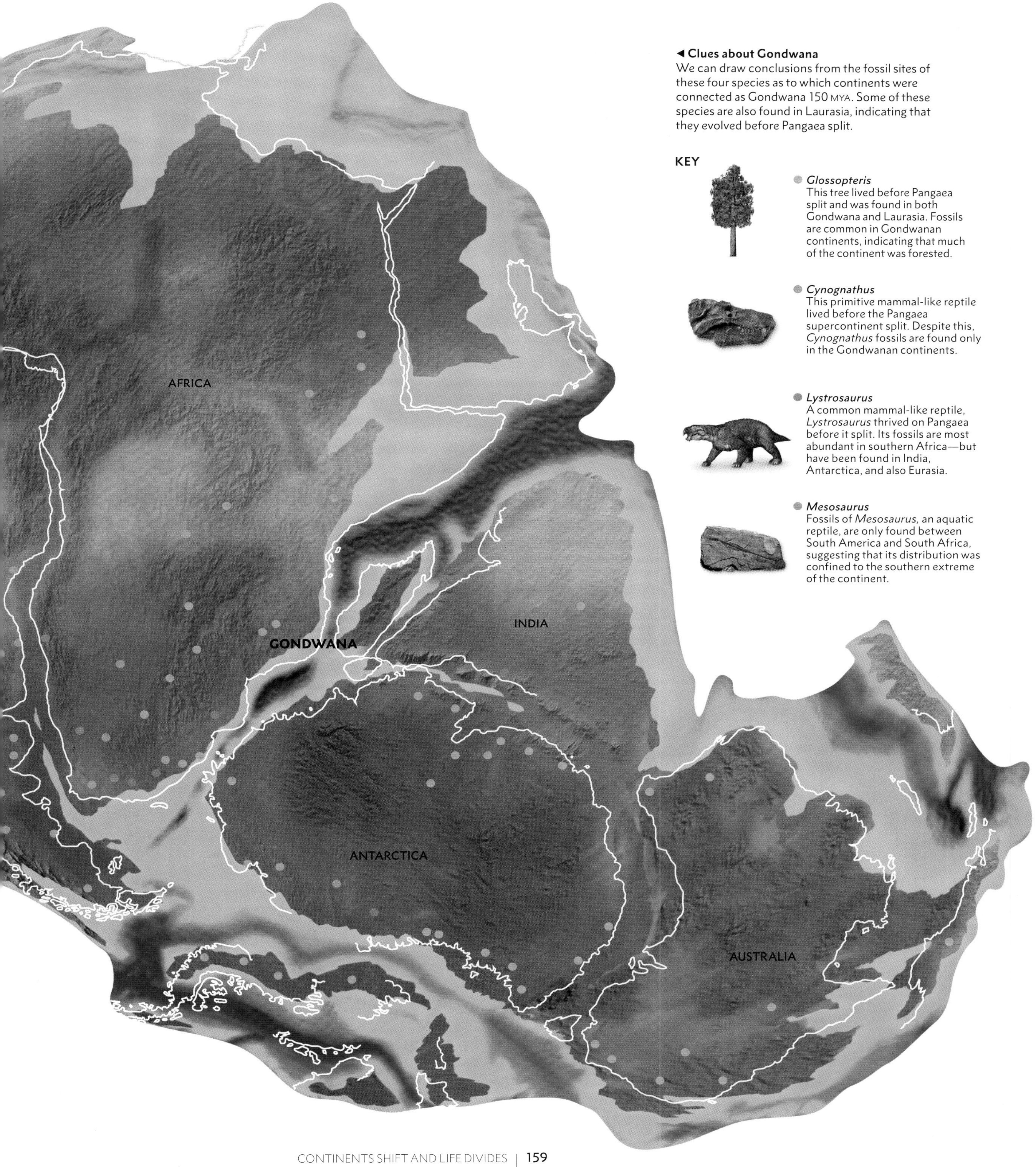

◀ Clues about Gondwana
We can draw conclusions from the fossil sites of these four species as to which continents were connected as Gondwana 150 MYA. Some of these species are also found in Laurasia, indicating that they evolved before Pangaea split.

KEY

● ***Glossopteris***
This tree lived before Pangaea split and was found in both Gondwana and Laurasia. Fossils are common in Gondwanan continents, indicating that much of the continent was forested.

● ***Cynognathus***
This primitive mammal-like reptile lived before the Pangaea supercontinent split. Despite this, *Cynognathus* fossils are found only in the Gondwanan continents.

● ***Lystrosaurus***
A common mammal-like reptile, *Lystrosaurus* thrived on Pangaea before it split. Its fossils are most abundant in southern Africa—but have been found in India, Antarctica, and also Eurasia.

● ***Mesosaurus***
Fossils of *Mesosaurus,* an aquatic reptile, are only found between South America and South Africa, suggesting that its distribution was confined to the southern extreme of the continent.

THE PLANET **BLOSSOMS**

One group of seed plants made the planet burst with color. Flowers gave them more effective ways of spreading their pollen and setting their seeds. Even before the demise of the dinosaurs, forests and other habitats were blooming—and buzzing with pollinators.

Limonium sinuatum

Around 90 percent of all known plant species are flowering plants. As trees, shrubs, and climbers, they dominate rain forests; as grasses, they carpet open ground. Flowering plants thrive in the driest of deserts and cling to rocks on high mountains and Arctic tundra. Some, such as mangroves, even tolerate tidal inundations of salt water along shorelines. While some produce the deadliest of poisons, others supply most of humanity's food. All, in one way or another, provide habitats for animals. Such impressive diversity stems from a uniquely successful reproductive shoot: the flower.

female flowers are receptive. Female parts, the carpels, have special projections, their stigmas, that catch the pollen grains. Many plants rely on wind to disperse pollen, but early in their evolution, some species recruited animal partners to carry it for them. As insects diversified, so did the variety of blossoms (see pp.164–65).

SCATTERING THE SEED

Insects were not the only animals to evolve alongside flowers. Fruit, another innovation of flowering plants, encased the seed and turned fragrant and colorful as it ripened.

Guzmania lingulata

> IT IS DIFFICULT TO CONCEIVE A **GRANDER MASS OF VEGETATION**... ONE MASS OF **BLOSSOMS**... ESPECIALLY THE WHITE ORCHIDS...WHITENING THEIR TRUNKS LIKE SNOW.

Joseph Dalton Hooker, botanist, 1817–1911, *Himalayan journals*

THE FIRST FLOWERS

The first members of the flowering plant group, or angiosperms, evolved around 120 MYA. *Montsechia vidalii*, an aquatic plant with tiny flowers, is thought to have dispersed its pollen in water, similar to its ancestors (see pp.144–45). Angiosperms began to diversify 30 million years later and evolved the flowering structure so integral to their success. Water lilies and magnolias are some of the most primitive species—remaining relatively unchanged today.

MOVING THE POLLEN

Flowers improve the transfer of pollen from male to female parts. Male flower parts, called stamens, split open to release their matured pollen grains at just the right time—when pollinators are active and when

This was perfect for attracting mammals with a nose for scent and birds with an eye for color. Seeds, in turn, became resistant to their digestive processes so they could be dispersed in droppings, readily supplied with a dose of fertilizer.

When plants first used flowers in their reproduction, they were embarking upon an evolutionary pathway with far-reaching repercussions. Tens of millions of years later, animals with a taste for sugar, including humans, would have sweeter foods to plunder, such as fruits, as more seeds scattered and new seedlings grew.

Nymphaea

Anemone pulsatilla

Myrica gale

Globularia alypum

▶ **Bloom of color**
Today, over 250,000 species of flowering plants decorate our planet. Some kinds have specific animal pollinator partners, without whom they would not be able to spread.

Kunzea baxteri

Agapanthus africanus

Austrobaileya scandens

Ostrya japonica

Narcissus pseudonarcissus

Anthericum liliago

Delphinium cardinale

Eriostemon spicata

Potentilla anserine

Rosa rugosa

Choisya ternata

Aesculus hippocastanum

Magnolia campbellii

Paulownia tomentosa

Quercus robur

Xanthoceras sorbifolium

Primula veris

Protea cynaroides

Callistemon viridiflorus

MASS **EXTINCTIONS**

Persistent volcanic activity, glaciation, and climate change can all result in the loss of species. The fossil record preserves five occasions when such mass extinctions were particularly severe.

Life has prevailed on Earth for more than 4 billion years—but individual species come and go. Stable, long-term habitats, such as rain forests or warm coastal seas, that endure for millions of years provide hot spots for evolution that boost the world's diversity of species. Earth is a changeable place and extinction events can act too suddenly for some life to adapt. These events drive multiple species to extinction, but also provide others with fresh opportunities for success.

▲ Rise and fall
The variety of species found at different points in the fossil record provide a wealth of information about the diversity of life on Earth. Diversity levels of marine invertebrates, fish, amphibians, reptiles, birds, and mammals are shown here with rising and falling bandwidths.

KEY

- High diversity
- Medium diversity
- Low diversity

MARINE INVERTEBRATES

FISH

AMPHIBIANS

REPTILES

The Cambrian Explosion, beginning around 542 MYA, gives rise to all modern-day animal groups, but also sees a radiation of many unusual body types, such as *Hallucigenia*, none of which exist today.

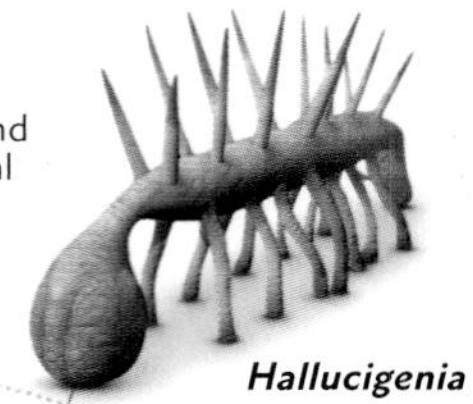

Hallucigenia

500 MYA

The Great Ordovician Biodiversification Event, about 480 MYA, sees marine invertebrates diversify even further.

The Cambrian-Ordovician extinction event 488 MYA affects certain types of brachiopods—small clamlike shellfish.

1 ORDOVICIAN–SILURIAN (445 MYA)

The growth of a huge, thick ice sheet on a continent centered over the South Pole causes sea levels to fall. This obliterates coastal habitats at a time when most life is still ocean-bound. Nearly two-thirds of marine invertebrates become extinct.

450 MYA

Rolfosteus, a species of placoderm fish, lived 383–359 MYA. Placoderms thrived in the Devonian oceans, but the majority were killed off around 365 MYA.

The Lau event 420 MYA sees populations of conodonts (primitive fishlike vertebrates) fall.

End-Silurian extinction event about 416 MYA affects corals; the cause of this event has yet to be determined.

Sponges and corals diversify 405 MYA in the wake of the Ordovician–Silurian mass extinction event. Crinoids such as *Dimerocrinites* inhabit the primitive reefs.

Dimerocrinites

400 MYA

Experimental body types from the Cambrian Explosion go extinct within 100 million years of their evolution.

2 LATE DEVONIAN (365 MYA)

Oxygen levels in the oceans drop during the Late Devonian, the cause of which is uncertain. This decimates coral reefs and many prominent groups of fishes, including placoderms and jawless vertebrates. New kinds of fishes, such as sharks and bony fishes, replace them.

350 MYA

Giant amphibians flourish in lush, tropical swamps 340 MYA.

Pantelosaurus pelycosaurs dating to the Permian (299–252 MYA), were among the victims of the Permian–Triassic mass extincti

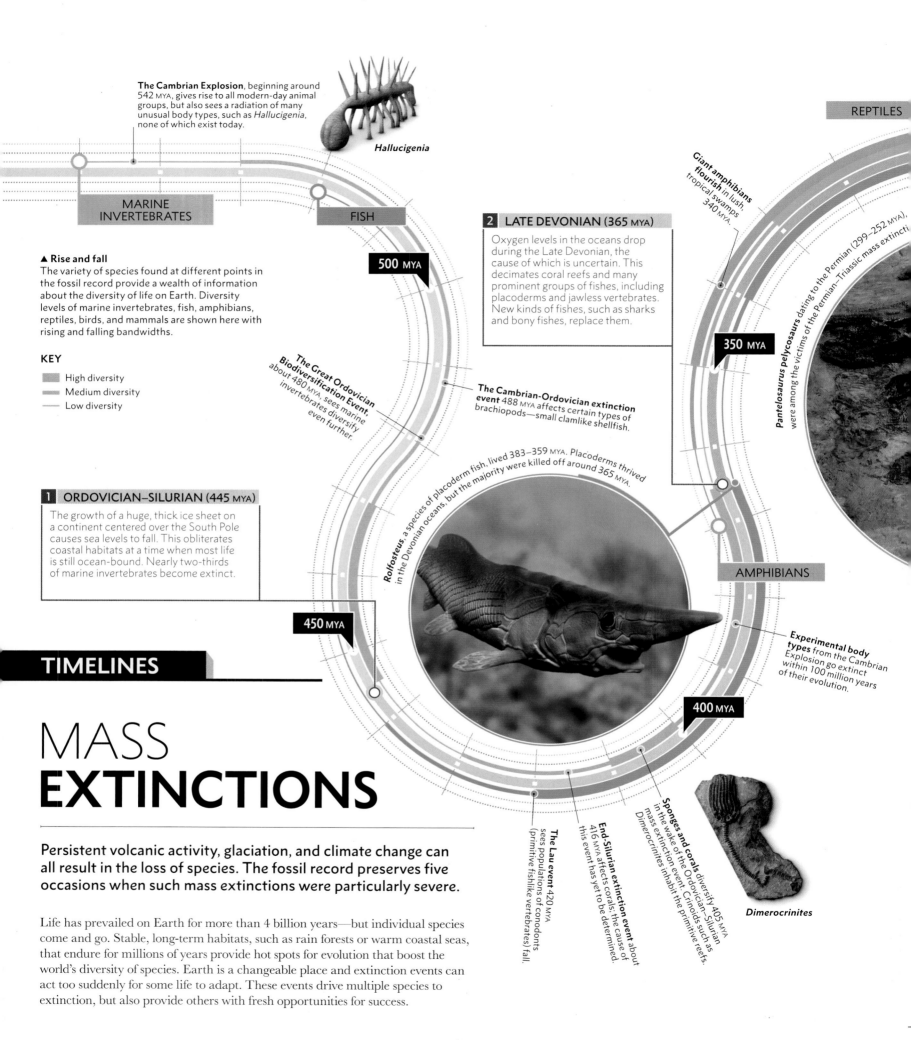

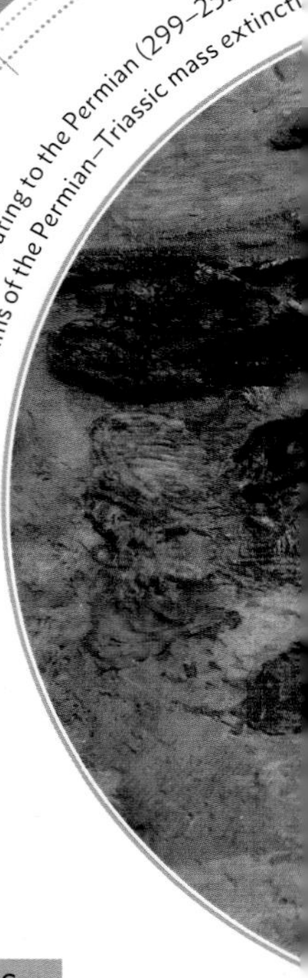

The Carboniferous Rainforest Collapse 305 MYA sees the swampy rainforests recede due to climate change, leading to localized declines in amphibians.

Giant amphibians, such as *Eryops*, are gradually replaced by reptiles 280 MYA in the drying climate of Pangaea.

Eryops

300 MYA

3 PERMIAN–TRIASSIC (250 MYA)

A vigorous period of volcanism erupts copious amounts of greenhouse gases, which cause the biggest mass extinction so far. Global warming creates vast deserts and triggers an event called the "Great Dying." More than 70 percent of all species go extinct.

Amphibian-reptile transitionals, called anthracosaurs, go extinct 260 MYA.

250 MYA

MAMMALS

0 MYA

Large predatory terror birds go extinct about 1 MYA.

Ray-finned fishes diversify 10 MYA and become the biggest class of vertebrates alive today.

Multituberculates, an ancient and diverse group of mammals, goes extinct 30 MYA.

The Grande Coupure 34 MYA sees a change in climate causing the evolution of new mammal species on land and the extinction of some ancient whale ancestors in the sea.

5 CRETACEOUS–PALEOGENE (66 MYA)

An asteroid impact creates acid rain and perpetually dark skies. Plant growth reduces, and the dinosaurs, pterosaurs, ammonites, and mosasaurs such as *Mosasaurus*, among other groups, go extinct. Ancestors of the familiar mammals and birds of today emerge from the aftermath.

Mosasaurus

50 MYA

An asteroid impact 66 MYA spells doom for many of the large reptiles that have dominated Earth for hundreds of millions of years.

100 MYA

200 MYA

Two-legged relatives of crocodiles called ornithosuchids go extinct about 200 MYA.

Toarcian turnover 183 MYA is a minor extinction event as oxygen levels deplete in the oceans, which leads to the extinction of mollusks such as many ammonites.

4 TRIASSIC–JURASSIC (200 MYA)

The breakup of supercontinent Pangaea increases levels of volcanic activity, which warms Earth's climate and causes mass extinction in the oceans and on land. However, the event favors the advancing global domination of dinosaurs.

Dinosaurs grow to enormous sizes; *Brachiosaurus* lived 155 MYA and reached roughly 30 ft (9 m) in height.

150 MYA

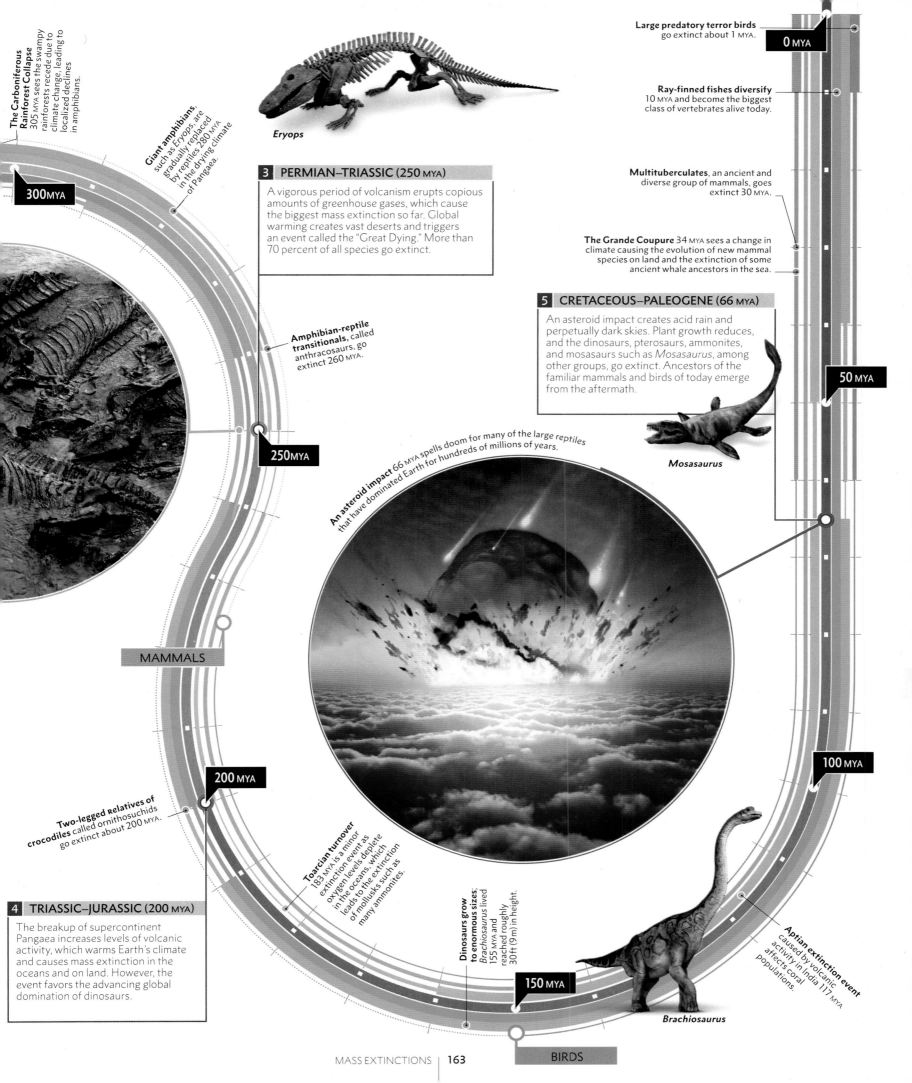

Brachiosaurus

Aptian extinction event caused by volcanic activity in India 117 MYA affects coral populations.

BIRDS

Agent of pollination
The long proboscis of the hummingbird hawk-moth can reach into tubular flowers, such as jasmine and honeysuckle, to feed on their nectar. Pollen easily sticks to the proboscis, making this species an excellent pollinator.

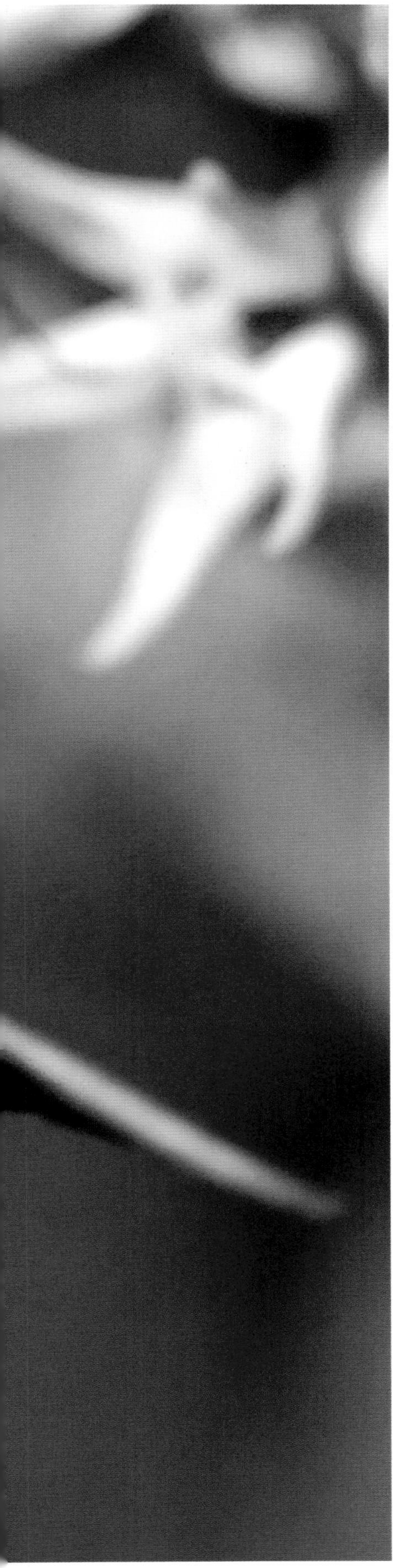

PLANTS **RECRUIT INSECTS**

Species are products of evolution that are shaped, through natural selection, by the environment around them—but species do not evolve in isolation. They interact with each other; some clash when they compete for the same food, but others end up cooperating.

For each species to thrive in its habitat, its members must do whatever it takes to breed. Species that have cooperative relationships with one another are an interesting example of the way life adapts to a changing world.

LIFE AFFECTING LIFE

The relationship between flowering plants and pollinating insects marked an important milestone in evolution. It is no coincidence that flowering plants and insects represent the most diverse groups of plants and animals. There are 250,000 species of flowering plants—while insects number around one million species. Each group diversified together as plants provided insects with nutritious nectar and insects provided the service of pollination. While flowers evolved color and scent to entice pollinators, insects evolved mouthparts that allowed them to extract the reward.

In 1964, American biologists Paul Ehrlich and Peter Raven introduced the term "coevolution" to explain instances of coadaptation. They documented how family trees of butterflies showed a degree of correspondence with those of flowering plants—suggesting closely corresponding pathways of evolution. Coevolution occurs when two species exert selective influences on each other. Both evolve by natural selection, but for each the other species becomes a factor in the selection. This can drive partnerships down increasingly narrow avenues of dependency until two species become entirely reliant on one another. Many plant species have flowers that can only be successfully pollinated by a single kind of insect. A species of Madagascan orchid with an exceptionally long "spur" (hollow tube) is pollinated by a species of hawk-moth with a proboscis (tongue) long enough to reach inside it.

Pollination of flowers by insects is an important example of mutualism—a relationship between two species in which both benefit from each other. One-way benefits, such as where predators or grazers exploit their prey, can also lead to coevolution. Coevolution fashions these kinds of relationships just as it does mutualistic ones.

> "**POLLINATORS**... ARE **KEYSTONE SPECIES**. YOU KNOW HOW AN ARCH HAS A KEYSTONE. IF YOU **REMOVE THE KEYSTONE**, THE **WHOLE ARCH COLLAPSES**."
>
> **May Berenbaum**, zoologist, 1953–

◀ **Pollen collector**
The honeybee is renowned for its nectar-loving diet, and it is an important distributor of pollen for many plant species.

1 SYNAPSIDS

Synapsids are the reptilian ancestors of mammals. They had strong jaw muscles and a powerful bite. Therapsids evolved from synapsids 35 million years later. They had better posture, raising their bodies off the ground.

Eothyris

SYNAPSIDS

Synapsids estimated to diverge from other reptiles 325 MYA.

Earliest known synapsid, *Protoclepsydrops*, is fossilized 312 MYA in Nova Scotia.

300 MYA

Edaphosaurus is fossilized 300 MYA. It is one of the earliest plant-eating sailback reptiles and part of a group of synapsids that will give rise to mammals.

Dimetrodon is fossilized 295 MYA. It is a meat-eating sailback reptile and part of a group of synapsids that will give rise to mammals.

Skull of *Eothyris* is fossilized. Living 285 MYA, this synapsid has two large fangs, indicating it was a carnivore.

THERAPSIDS

Tetraceratops, possibly an early therapsid, is fossilized in Texas 275 MYA.

Raranimus, the first undisputed therapsid, is fossilized 270 MYA in China.

Moschops leaves fossils 265 MYA in South Africa. It is the size of a sheep and one of the largest land animals of its time.

Moschops skull

CYNODONTS

Charassognathus becomes the earliest fossil of a cynodont 259 MYA. Cynodonts were more advanced therapsids.

Primitive synapsids, such as sailbacks *Dimetrodon* and *Edaphosaurus*, go extinct as therapsids replace them.

250 MYA

Cynodonts, such as these two juvenile *Thrinaxodon* that lived 250 MYA, were ancestors to mammals. They were probably nocturnal and sensed their surroundings with whiskers.

This tiny, mouse-sized morganucodont is called *Megazostrodon*. It lived in the shadow of the dinosaurs 199–196 MYA.

MORGANUCODONTS

Massetognathus, a plant-eating cynodont, is fossilized 235 MYA.

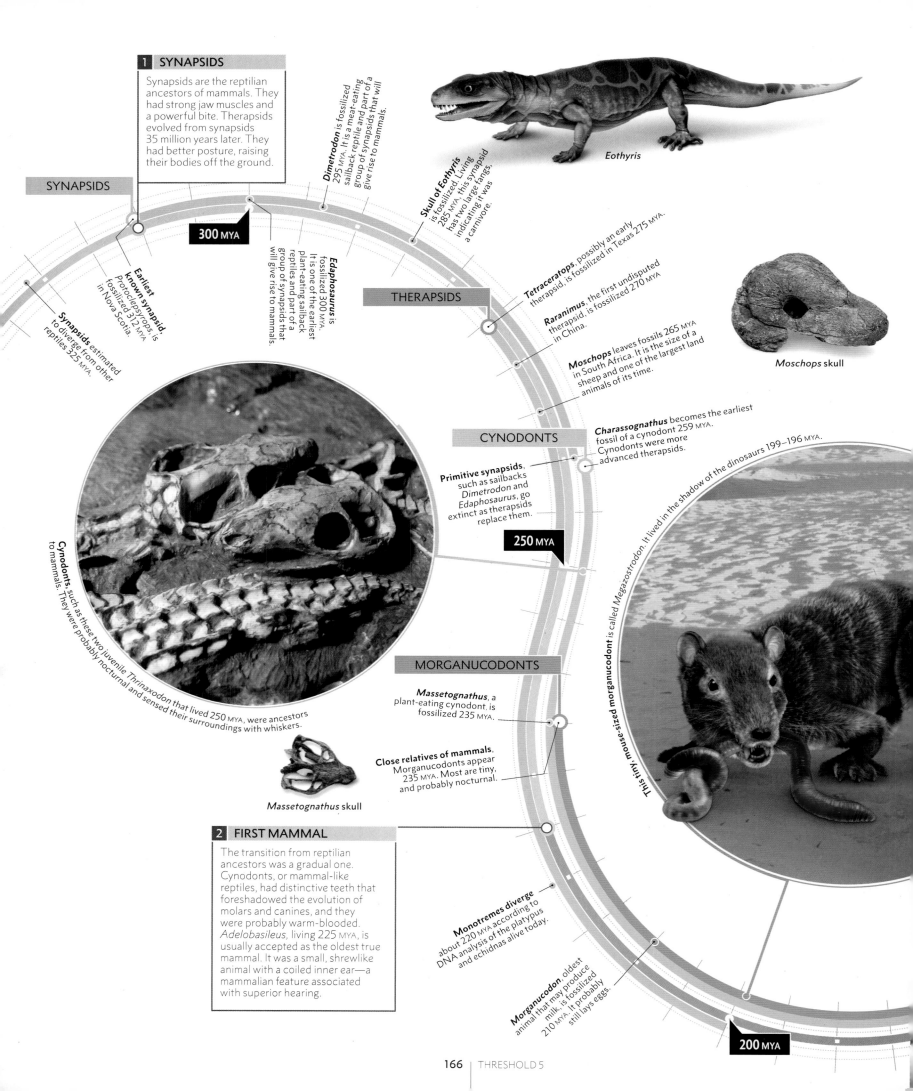

Massetognathus skull

Close relatives of mammals, Morganucodonts appear 235 MYA. Most are tiny, and probably nocturnal.

2 FIRST MAMMAL

The transition from reptilian ancestors was a gradual one. Cynodonts, or mammal-like reptiles, had distinctive teeth that foreshadowed the evolution of molars and canines, and they were probably warm-blooded. *Adelobasileus*, living 225 MYA, is usually accepted as the oldest true mammal. It was a small, shrewlike animal with a coiled inner ear—a mammalian feature associated with superior hearing.

Monotremes diverge about 220 MYA according to DNA analysis of the platypus and echidnas alive today.

Morganucodon, oldest animal that may produce milk, is fossilized 210 MYA. It probably still lays eggs.

200 MYA

Alphadon, a small marsupial, is fossilized 80 MYA along with giant dinosaurs Tyrannosaurus and Triceratops.

Alphadon

Ravenictis, a carnivoran mammal, is fossilized 62 MYA in Canada. Carnivores will later diversify into dogs, cats, bears, and many other groups.

Heomys, a possible first rodent, leaves fossils 60 MYA in China.

Eritherium, the earliest-known elephant, leaves fossils 56 MYA in Morocco.

Diacodexis, the oldest-known even-toed hoofed mammal, leaves fossils 55 MYA. Even-toed hoofed mammals will diversify into deer, pigs, antelope, sheep, goats, cattle, camels, and cetaceans.

Himalayacetus, the oldest cetacean, is fossilized 52 MYA in India. The cetacean group later includes whales and dolphins.

Oldest-known armadillo leaves fossils 48 MYA in Brazil.

50 MYA

Uintatherium

100 MYA

Uintatherium is fossilized 45 MYA. It is part of an extinct group of once-common herbivores called brontotheres.

Andrewsarchus was one of the largest carnivorous mammals ever to have evolved. It lived 45–36 MYA and was a hoofed mammal, unrelated to dogs and cats.

Dogs appear in the fossil record with Hesperocyon, fossilized 37 MYA in Canada.

Arsinotheruim, a distant relative of elephants, is fossilized 30 MYA.

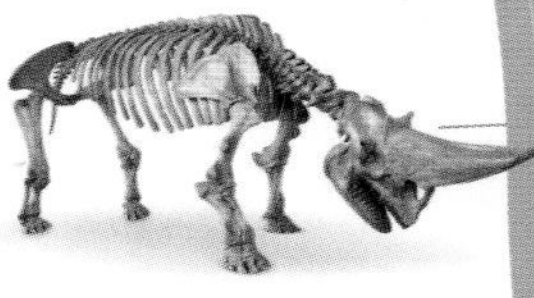

Arsinotherium

Modern-day mammal groups begin to diverge about 120 MYA with ancestors of elephants and manatees, armadillos, and sloths diverging from the rest.

Teinolophus, a platypus-like monotreme, leaves fossils 120 MYA in Australia.

Sinodelphys is the earliest known marsupial, leaving fossils in China 125 MYA. Marsupials produce tiny young that are nurtured in a pouch.

Hoofed mammals diversify about 9 MYA and develop larger body sizes and the ability to run quickly across open grasslands.

MARSUPIALS

0 MYA

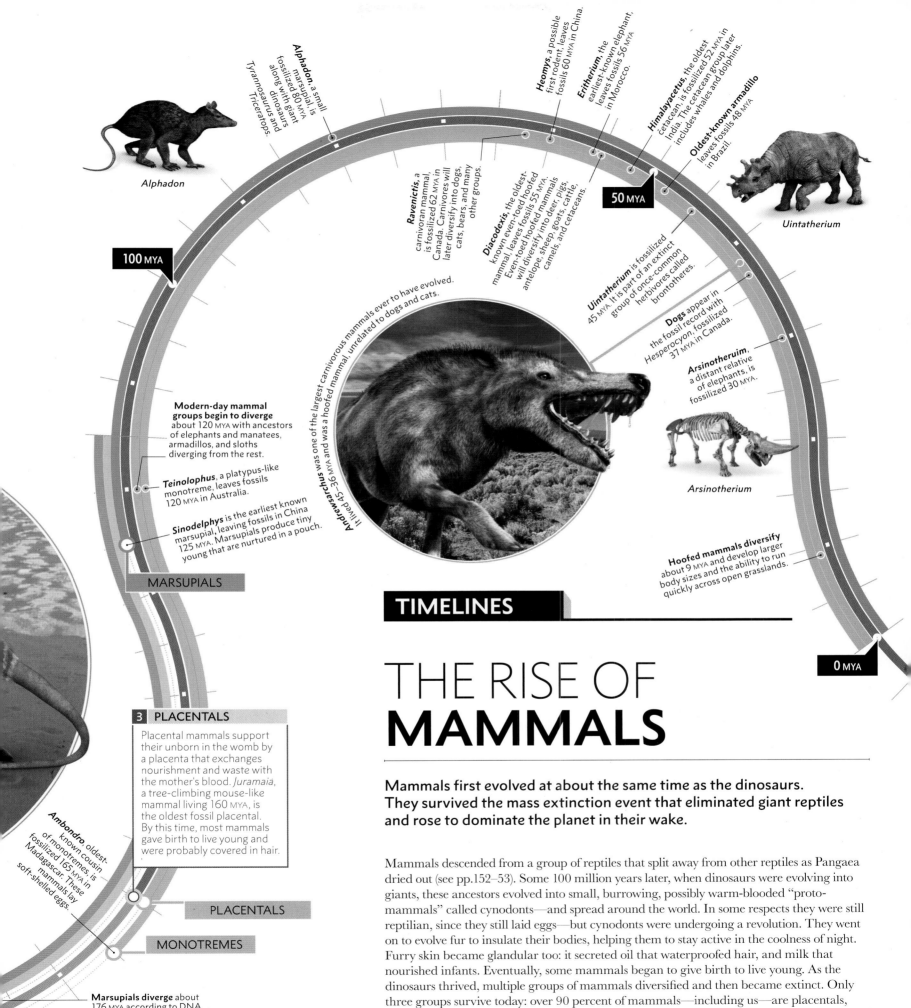

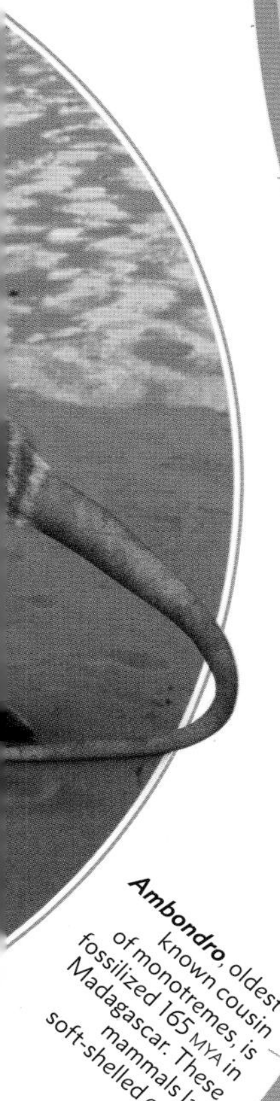

TIMELINES

THE RISE OF MAMMALS

Mammals first evolved at about the same time as the dinosaurs. They survived the mass extinction event that eliminated giant reptiles and rose to dominate the planet in their wake.

3 PLACENTALS

Placental mammals support their unborn in the womb by a placenta that exchanges nourishment and waste with the mother's blood. *Juramaia*, a tree-climbing mouse-like mammal living 160 MYA, is the oldest fossil placental. By this time, most mammals gave birth to live young and were probably covered in hair.

Ambondro, oldest-known cousin of monotremes, is fossilized 165 MYA in Madagascar. These mammals lay soft-shelled eggs.

PLACENTALS

MONOTREMES

Marsupials diverge about 176 MYA according to DNA analysis of species alive today.

Mammals descended from a group of reptiles that split away from other reptiles as Pangaea dried out (see pp.152–53). Some 100 million years later, when dinosaurs were evolving into giants, these ancestors evolved into small, burrowing, possibly warm-blooded "proto-mammals" called cynodonts—and spread around the world. In some respects they were still reptilian, since they still laid eggs—but cynodonts were undergoing a revolution. They went on to evolve fur to insulate their bodies, helping them to stay active in the coolness of night. Furry skin became glandular too: it secreted oil that waterproofed hair, and milk that nourished infants. Eventually, some mammals began to give birth to live young. As the dinosaurs thrived, multiple groups of mammals diversified and then became extinct. Only three groups survive today: over 90 percent of mammals—including us—are placentals, so-called because they carry their young through a long pregnancy nourished by a placenta.

> ...THE **GRASSLANDS** ARE **LARGELY UNDISCOVERED TREASURES** OF AN **IMPORTANT NATIONAL HERITAGE**.
>
> **Francis Moul**, environmental historian, 1940–

Acacia trees dot tropical grasslands in Africa, but do not dominate, offering sparse cover and shade
The lion is an incredibly successful grassland predator, hunting in groups to take down larger, fast-moving prey
Wildebeest graze almost exclusively on short grasses, and in turn are bountiful prey for grassland predators such as lions
Deinotherium was a species of elephant with unusual downward-sloping tusks
Termite mounds produce nitrogen, which promotes lush grass growth
Aardvarks burrow during the day, safe from predators
Dinofelis, a prehistoric cat, possibly ambushed prey from dense undergrowth
Savanna grasses can regrow quickly after heavy grazing
Hyenas use old warthog dens to raise their cubs in hiding, lowering risk of attracting predators on the open savanna
Gazelles are fast and nimble, capable of escaping predators by running away

▲ Life in the savanna
One million years ago, the East African savanna supported an impressive food chain, with herds of grazing hoofed mammals falling prey to meat-eating predators—just as they do today.

GRASSLANDS ADVANCE

In environmental and ecological terms, the grass family is probably the single most important plant group on Earth. Nearly three-quarters of crop species grown by humans are grasses. Remarkably, they only appeared relatively recently—about 55 MYA.

Although grasses evolved about 55 MYA, grassland habitats were not established until 15–10 MYA. Given the right conditions, grasses grow opportunistically in open spaces, spreading quickly by underground stems. A few, such as bamboos, grow tall and woody, but most others stay low before flowering and setting their seed. These are the species that populate the open habitats familiar today, forming vast plains and prairies dominated by a single species. Today, one-fifth of Earth's vegetation cover is grassland.

SURVIVING THE GRAZE

Although grasses can look palatable, most species reinforce their leafy margins with granules of the mineral silica. Some species possess enough silica to make their blades abrasive or even sharp enough to cut skin. This adaptation deterred herbivores, but in response plant eaters evolved stronger jaws or more resilient digestive systems. Grasses evolved yet another tactic: by growing their blades from the base, rather than the tip, they could be grazed close to the ground and still regenerate. Their creeping stems can even send up regenerative shoots after being trampled under heavy hooves. This allows grasses to out-compete other plants in heavily grazed environments.

GRAZERS GROW BIGGER

As grasslands spread across the world, life evolved in turn. Productive growth could support bigger plant eaters—and large bodies were perfect for digesting grass. Big herbivorous mammals evolved digestive systems that worked like fermentation vats, relying on gut microbes to help break down plant fiber. The grassland bounty came at a price: there was no cover from predators. Fleet-footed grazing mammals evolved, gathering in herds for safety.

Today grasslands support some of the biggest concentrations of wildlife on Earth. Two million years ago, the first humans joined the grassland food chain. No terrestrial habitat has been so influential in shaping the evolution of mammals and humankind (see pp.186–87).

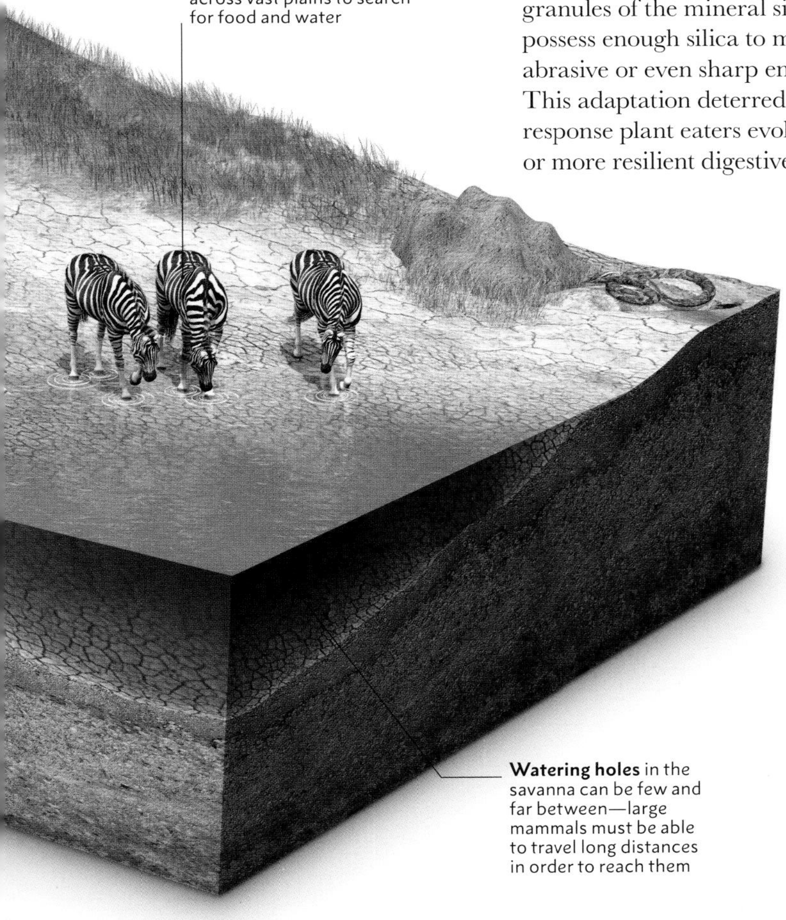
Zebras are perfectly adapted to inhabit grasslands; they can move across vast plains to search for food and water
Watering holes in the savanna can be few and far between—large mammals must be able to travel long distances in order to reach them

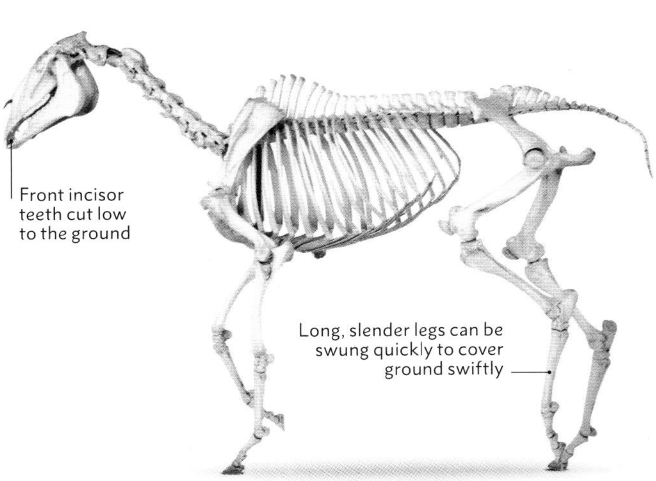
Front incisor teeth cut low to the ground
Long, slender legs can be swung quickly to cover ground swiftly

◀ **Built for grasslands** Grazers such as the horse consume low grass in open places. Their large leg muscles are concentrated at the top of the legs, leaving the slender lower legs free of bulky muscle, so they are light and easily maneuverable for a quick escape.

EVOLUTION **TRANSFORMS LIFE**

Evolution happens by small changes in genes. These changes are inherited from one generation to the next, and over millions of years, these changes can become amplified. Vast stretches of time may pass before new species—with new ways of life—emerge.

Some organisms reproduce so quickly that their evolutionary changes can be observed directly. Resistance to antibiotics, for instance, can spread through bacteria that double their numbers every half hour. But to study changes in living things that breed more slowly and evolve over much longer periods of time, scientists must examine evidence from multiple sources—such as genes, anatomy, and fossils—to work out how evolution has shaped life on Earth through time.

CHANGE AND DIVERGENCE

Natural selection works on the variation created by mutation to bring about adaptation (see pp.108–09). Over many generations of evolution, organisms change so much in their anatomy and behavior that they may become unrecognizable. Populations split as landscapes move and habitats come and go—sending different groups along diverging paths that can result in the evolution of different species. For vertebrate animals this may take a few million years, but for fast-breeding microbes it can happen within our lifetime.

TRACING THE RELATIONSHIPS

Analysis of the chemical sequence of genes helps to uncover the relationships between species (see pp.172–73). This analysis shows, for instance, that humans are closest to chimpanzees—a "sister species"—but more distantly related to gibbons, whose genes have fewer similarities with ours. Genes show that cetaceans—whales, dolphins, and porpoises—share a common ancestor with the hippopotamus, and are therefore derived from the hoofed mammal group. Scientists can estimate the rate of random genetic change that accumulates over time by mutation and devise a "molecular clock" to calculate roughly when species diverged.

HIPPOS GIVE **BIRTH** AND SUCKLE THEIR OFFSPRING UNDER WATER, JUST **LIKE** THEIR CLOSEST LIVING RELATIVES – **WHALES AND DOLPHINS**

▼ From land to sea
The evolution of whales from a land-based ancestor is an example of large-scale genetic change over the course of millions of years.

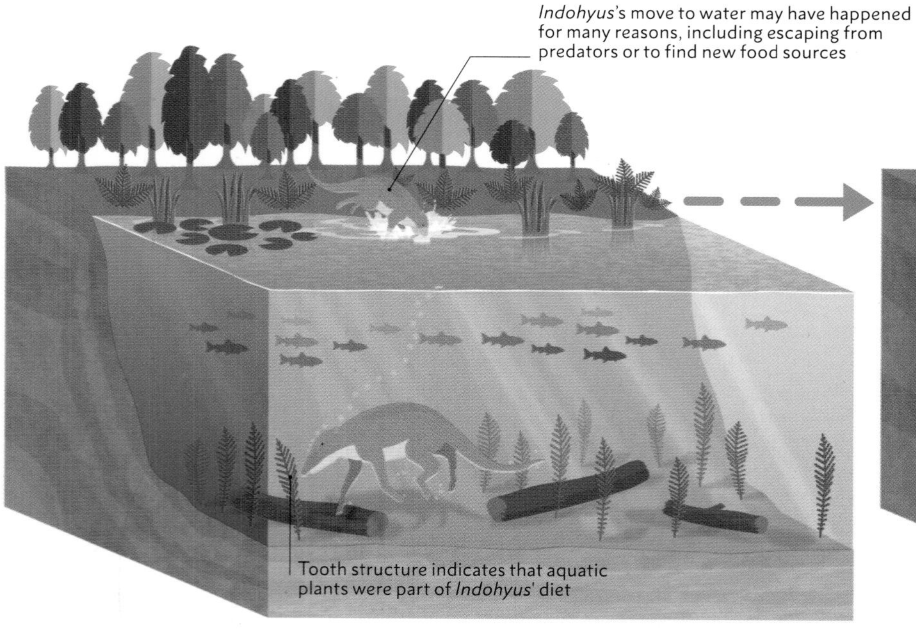
Indohyus's move to water may have happened for many reasons, including escaping from predators or to find new food sources
Tooth structure indicates that aquatic plants were part of *Indohyus*' diet

▲ A small hoofed animal called *Indohyus* was the earliest member of the lineage that led to whales and dolphins. Chemical analysis of its fossils indicates that it spent some time in fresh water. Its skull was thicker in the region of its ear canal, suggesting it had good hearing, perhaps to help it find food under water.

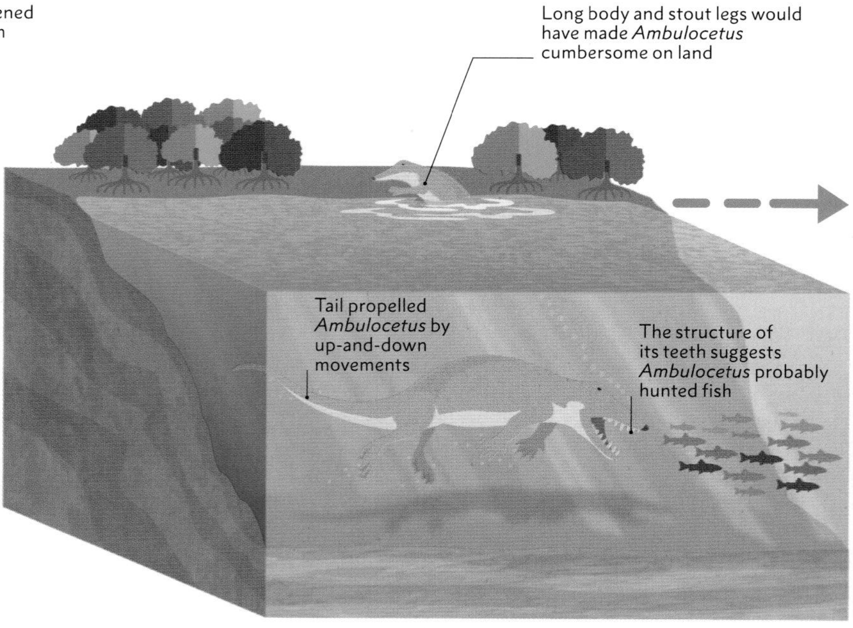
Long body and stout legs would have made *Ambulocetus* cumbersome on land
Tail propelled *Ambulocetus* by up-and-down movements
The structure of its teeth suggests *Ambulocetus* probably hunted fish

▲ *Ambulocetus* was a semi-aquatic animal whose name translates as "walking whale," although it was best suited to life in fresh and salt water habitats. It was less accustomed to movement on land and instead was a better swimmer. Its powerful tail moved up and down—just like the flapping tail of modern whales.

By using this molecular clock, they conclude that the ancestors of whales and hippopotamuses diverged between 50 and 60 million years ago. Genes only provide part of the picture. They can never show what ancestors looked like, and for that scientists rely on fossils.

Fossils show how the anatomy of prehistoric life compares with species alive today. Although their own DNA has degraded, their anatomy—even when fragmentary—can reveal important relationships. Fossils can be dated, which helps to establish when key events took place and support the molecular clock. Scientists can never be sure that fossilized forms of life are the direct ancestors of living ones, but their relative positions in the tree of life can be strongly indicated by the evidence. Dozens of fossil animals are at the base of the cetacean family tree—tens of millions of years before modern whales. They not only help to show how walking limbs evolved into swimming flippers, but even, from chemical analysis, whether the animals lived in fresh or salt water.

After 4 billion years of evolution, Earth is rich with millions of diverse species—and many more have lived and died out in the past. Everything on the great tree of life is connected to the past, and to each other.

▼ Evolutionary pathway
Evidence from anatomy and DNA indicates that whales and dolphins evolved from hoofed animals, and that the hippopotamus is their closest living relative. Numerous fossil species add detail to their cladogram.

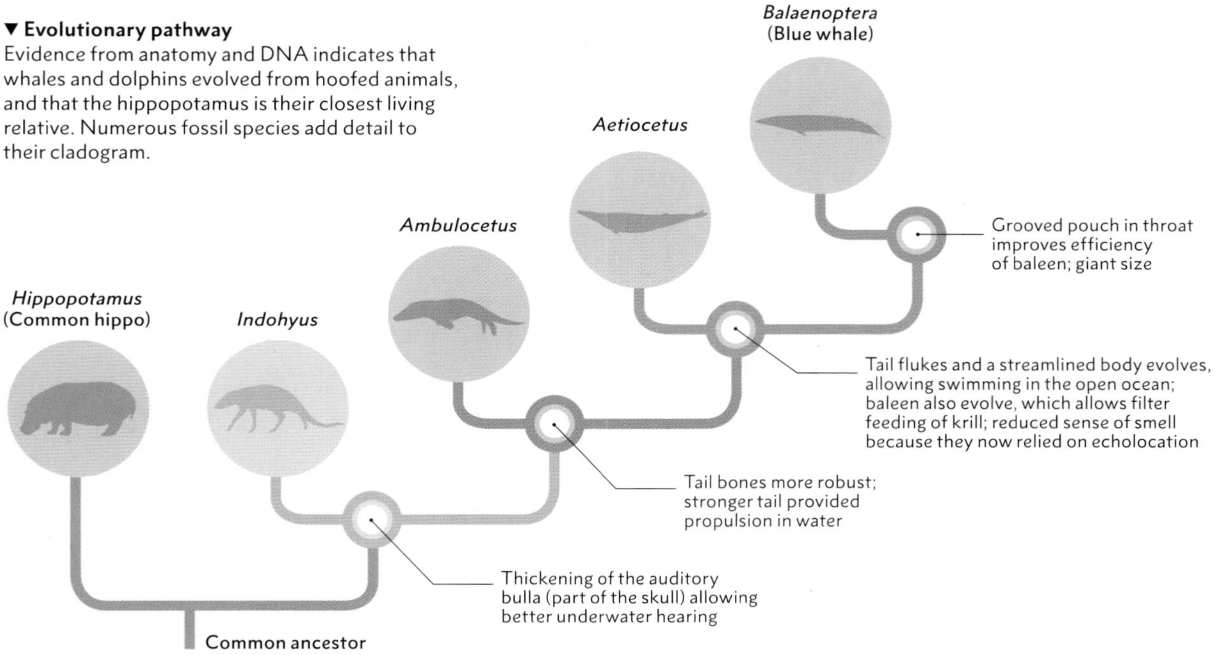
Balaenoptera
(Blue whale)
Aetiocetus
Ambulocetus
Hippopotamus
(Common hippo)
Indohyus
Grooved pouch in throat improves efficiency of baleen; giant size
Tail flukes and a streamlined body evolves, allowing swimming in the open ocean; baleen also evolve, which allows filter feeding of krill; reduced sense of smell because they now relied on echolocation
Tail bones more robust; stronger tail provided propulsion in water
Thickening of the auditory bulla (part of the skull) allowing better underwater hearing
Common ancestor

HUMANS ARE… **A TINY LITTLE TWIG** ON THE ENORMOUSLY ARBORESCENT **BUSH OF LIFE**… **IF REPLANTED** FROM SEED, WOULD ALMOST SURELY **NOT GROW THIS TWIG AGAIN**.

Stephen Jay Gould, palaeontologist, 1941–2002

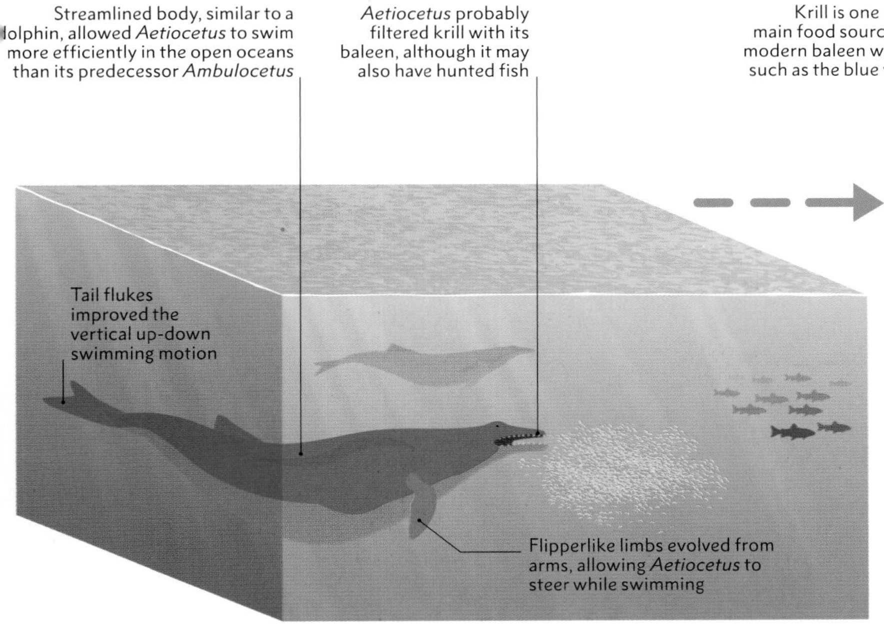
Streamlined body, similar to a dolphin, allowed *Aetiocetus* to swim more efficiently in the open oceans than its predecessor *Ambulocetus*
Aetiocetus probably filtered krill with its baleen, although it may also have hunted fish
Tail flukes improved the vertical up-down swimming motion
Flipperlike limbs evolved from arms, allowing *Aetiocetus* to steer while swimming

▲ *Aetiocetus* was a recognizable whale—no longer capable of moving on land, with a shorter neck, reduced sense of smell, flipperlike limbs, tail flukes, and no external ears. It had a beak, but unlike any living whale, its mouth contained both teeth and baleen—fringes of hornlike material to filter plankton—marking it as a truly transitional animal.

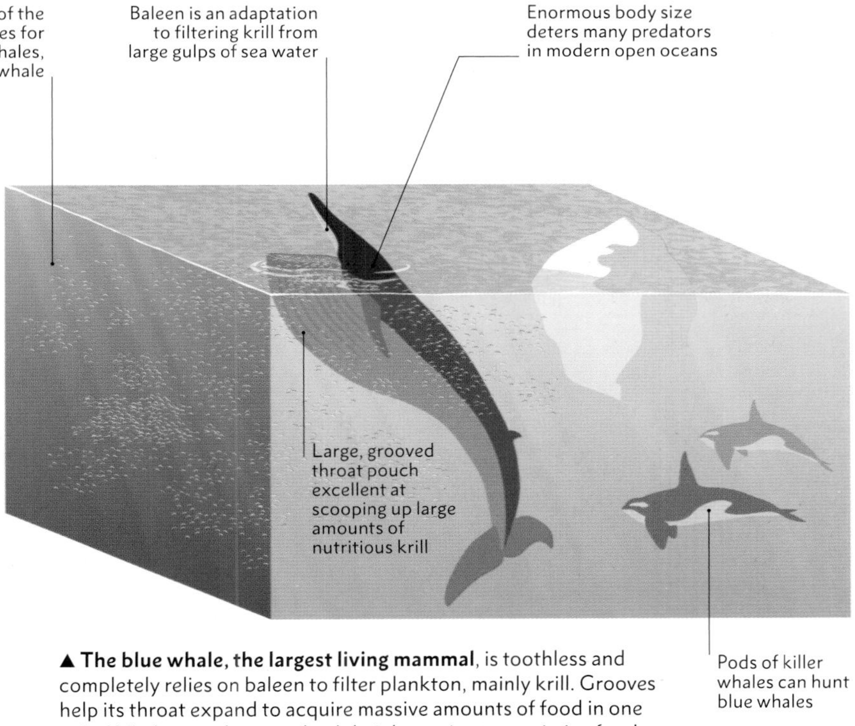
Krill is one of the main food sources for modern baleen whales, such as the blue whale
Baleen is an adaptation to filtering krill from large gulps of sea water
Enormous body size deters many predators in modern open oceans
Large, grooved throat pouch excellent at scooping up large amounts of nutritious krill
Pods of killer whales can hunt blue whales

▲ The blue whale, the largest living mammal, is toothless and completely relies on baleen to filter plankton, mainly krill. Grooves help its throat expand to acquire massive amounts of food in one gulp. Whales may have evolved their large size to maximize food intake—or perhaps to avoid predation from giant prehistoric sharks.

BIG IDEAS

HOW WE **CLASSIFY LIFE**

The classification of living things involves more than unscrambling the order of the natural world. Modern biologists classify species on the basis of their ancestral relationships, and their methods for doing so have been honed over 200 years of studying disciplines as diverse as anatomy, paleontology, and genetics.

Naturalists have been classifying living things for as long as they have been trying to understand them. Early groupings were wholly guided by specific needs. For example, apothecaries classified plants according to their medicinal properties. Ancient Greek thinker Aristotle classified plants and animals along his *scala naturae*, or "ladder of life," imbuing each kind with a "degree of perfection," between base minerals at the bottom and God at the top. Some of Aristotle's categories, such as vertebrates and invertebrates, are still used today, but his belief that each type of organism had an ideal form—an "essence"—pervaded biological thought until the time of Charles Darwin (1809–82), and hampered notions of evolution based on natural variation (see pp.110–11).

THE EARLY NATURALISTS

From the 16th century, botany and zoology moved forward as new researchers made firsthand observations, instead of relying on the received wisdom of ancient philosophers. Renaissance anatomists, such as Andreas Vesalius (1514–64), explored the human body by dissection, and 100 years later, the newly-invented microscope opened up a world of cells and microbes. Naturalists came to devise their own classification systems and made more meaningful comparisons based on accurate knowledge of anatomy. English naturalist John Ray (1627–1705), for instance, recognized that whales were mammals and not fish. He wrote exhaustively on plants and animals and he was the first observer to devise the concept of a biological species: an organism that reproduced always to result in the same form. As more species were being discovered though, they lacked a standard naming system—however, one Swedish botanist was about to change that.

DARWIN SKETCHED A **TREE OF LIFE** IN 1837, 100 YEARS **BEFORE THEY BECAME COMMON**

NAMING LIFE

A botanist named Carl von Linné (1707–78)—later Latinized to Carolus Linnaeus—had been studying the structure of flowers, identifying their parts as reproductive organs and cataloging their diversity. In

▶ Collecting specimens New species are described from preserved specimens—so-called "type specimens"—that are deposited as scientific collections in museums.

> **DARWIN**... SHOWED WHY THERE ARE **NATURAL GROUPS** AND WHY **THEY SHARE 'ESSENTIAL' CHARACTERS**.

Ernst Mayr, 1904–2005
Biologist

1735, he published a pamphlet called *Systema Naturae*, or "Natural System." Initially, it outlined a hierarchical classification system of all known life that was defined by ranks. Classes—such as reptiles, birds, and mammals—were split into orders—such as pigeons, owls, and parrots—and then into genera (singular, genus). The genus rank defined the basic form of an organism, such as bear, cat, or rose. As was the convention of the day, the specific type (equivalent to John Ray's species) was still denoted by a cumbersome Latin description. In 1753, Linnaeus' *Species Plantarum* changed this by substituting one-word names for plants, and his 1758 tenth edition of *Systema Naturae* did the same for animals. For example, the brown bear—which in 1735 was listed in his genus *Ursus*—was now given the specific name of *Ursus arctos*. Linnaeus's 1753 and 1758 publications mark the beginnings of recognized scientific names for plants and animals, respectively. This two-name system became universally adopted in biology: the first name (*Ursus*) denotes the genus, and the second (*arctos*) the species. Linnaeus's taxonomic system is still used today—but with some modifications and additional ranks. As our knowledge about the relationships of species grows, many species move to other genera, changing their two-word scientific name as they go.

CLADISTIC ANALYSIS SHOWS THAT **BIRDS** ARE **CLOSEST TO DINOSAURS**

ORGANIZING LIFE

Even in the 19th century, many still saw variations in individual forms of life as imperfect deviations from an ideal form. Charles Darwin's recognition of the importance of these variations to evolution led to a shift away from this Aristotelian viewpoint. By the early 1900s, species were known to be made up of variable populations and the genetic basis for this variation was better understood (see pp.108–09).

In the 1960s, German biologist Willi Hennig (1913–76) applied more rigorous evolutionary rules to classifying life. Groups at any rank should contain all species descended from a common ancestor. These groups were called clades, the branching diagram showing them called a cladogram, and the new method called cladistics. Cladistics has since been universally adopted as the appropriate way to classify life—because this method clearly shows to what degree one animal is related to another. Classification now reflects evolutionary relationships, and taxonomic groups were redefined on the basis of descent from common ancestors. Knowing how closely related species are is more useful than knowing they are simply similar. If we know that one plant produces a life-saving drug, and we also know which other plants are closely related to it, we can focus our search for new sources for this drug.

Cladistics changed how taxonomists view Linnaean groups. Where once taxonomists understood mammals and birds as groups (classes) of equal rank to reptiles, cladistic groupings have reworked this notion. We now know mammals and birds evolved from reptiles, and reptiles evolved from amphibians, and so on. Therefore, cladistics classifies mammals and birds as two distinct clades within a larger clade that also includes reptiles, because they all share a single common ancestor.

Today, taxonomists have a better tool than anatomy for discovering evolutionary relationships. Biologists have turned to DNA as a source of information ever since they recognized that inherited genes are stored inside it. DNA contains a code—a sequence of chemical components along its chain. Closely related species have similar sequences. Modern analytical techniques, coupled with powerful computer programs, can compare DNA among multiple species, generating the statistical likelihood of a relationship between species. Biologists can even use DNA information to calculate when two organisms diverged from each other (see pp.170–71). They can then create cladograms with time estimates applied to each branching point. These "timetrees" of life can be used to map evolutionary progress over millions, or billions, of years. It means that taxonomic groups are not only defined in terms of descent, but also by their estimated times of origin and divergence.

> **PLANT GROUPS** SHOW **RELATIONSHIPS** ON ALL SIDES... LIKE THE COUNTRIES ON A MAP.

Carolus Linnaeus, botanist, 1707–1778

ICE CORES

Ice cores capture a wealth of clues indicating a vigorous, and largely natural, back-and-forth of climatic conditions. Similar to animals trapped in amber, tiny relics from Earth's past can be held inside ice cores.

Earth's ice sheets are gigantic treasuries of evidence of past climates. These three ice cores, each 3¼ft (1m) long, are samples from a long core drilled from the Greenland ice sheet, which is more than 6,600ft (2,000m) thick. As the ice sheet formed from falling snow, it captured atmospheric gas and airborne particles, which were incorporated into the ice as a record of conditions at the time. Ice builds up year after year, so as scientists drill down, they reach older and older records. This particular core documents 111,000 years of climatic history.

Climatologists analyze ice cores to find clues to Earth's past climate. If dust trapped in the ice contains radioactive elements, radiometric dating (see pp.88–89) can be used to date the sample. Ice cores can reveal what the average temperature was in the past, and can tell us the proportions of gases in the atmosphere. This provides long-term context to the rise in carbon dioxide (CO_2) levels seen in recent decades. Research stations in Earth's polar regions, such as Vostok, Antarctica, have contributed records of CO_2 levels stretching back more than 400,000 years. At Dome C in Antarctica, drillers extracted an even longer ice core. At 10,738ft (3,270m) long, it holds data, such as methane and CO_2 levels, from the last 650,000 years. Ice cores can also capture volcanic ash, dust, sand, and even pollen. These clues can tell us about volcanic activity, the extent of deserts, and the spread of different types of vegetation in the past.

The drivers behind natural climate change include cyclical changes in Earth's orbit and changes to its axis of rotation that are known as Milankovitch cycles. Other natural factors are changes in the sun itself, plate tectonics, and volcanism. Scientists study ice cores to learn about these natural effects on climate and to predict how they might interact with the current human activities that seem to be bringing about rapid climate change (see pp.352–53).

▼ Milankovitch cycles
Long-term changes in Earth's orbit and spin are called Milankovitch cycles. The cycles alter the timing and intensity of our seasons and seem to coincide with regular bouts of climate change, such as ice ages (see pp.176–77).

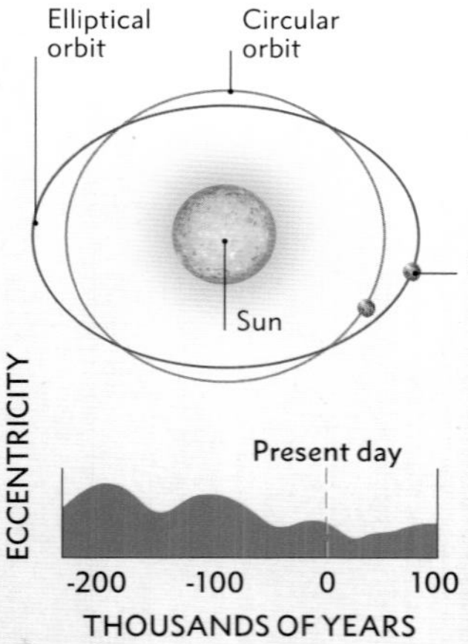

The shape of Earth's orbit changes from circular to elliptical (more "eccentric"), under the influence of Jupiter and Saturn's gravity. This alters the length of our seasons, changing our climatic patterns.

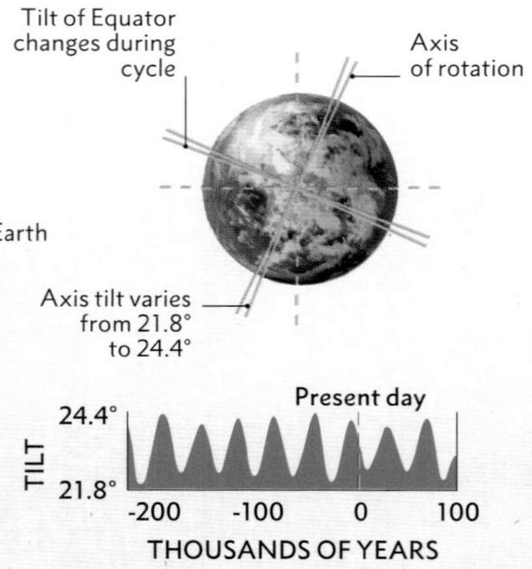

The angle of Earth's axis varies by a few degrees. With a greater tilt, the northern or southern hemisphere is inclined further toward the Sun, which results in more extreme contrasts in our seasons.

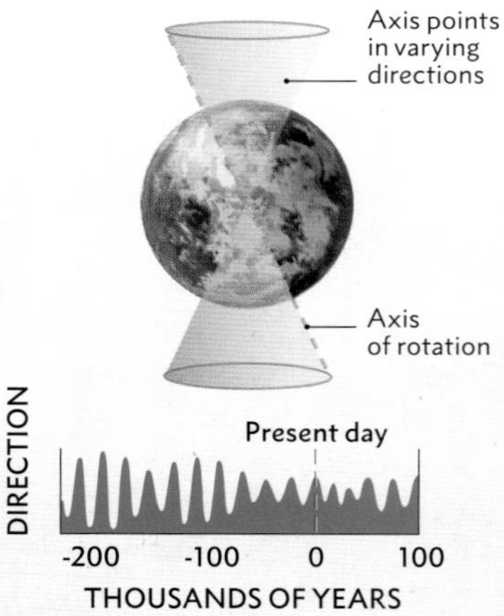

Earth wobbles because it is not a perfect sphere—this causes its axis to trace out imaginary circles over approximately 26,000 years. This alters the timing of midsummer, midwinter, and the solstices.

Atmospheric gases

Each layer of snow that fell on the Greenland ice sheet contains gas from the atmosphere that was trapped as the snow compacted into ice. Climatologists who compare gas levels inside ice cores from varying depths can create a timeline of Earth's climatic past. The level of carbon dioxide in the atmosphere was stable over the last millennium until the early 19th century, when it began to increase. It is now 40 percent higher than before the Industrial Revolution (see pp.304–05).

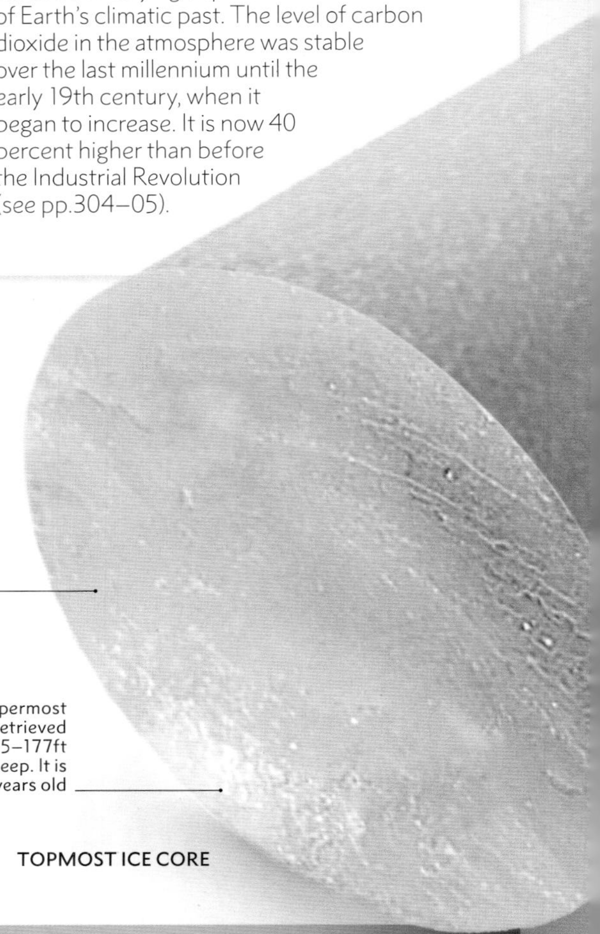

TOPMOST ICE CORE

Extracting ice cores

Ice cores—long columns of ice—have been extracted since the 1950s, largely from the Greenland and Antarctic ice sheets. A large team of scientists is required to drill into an ice sheet and extract a viable ice core. The cores are then stored in temperatures below 5°F (-15°C) to preserve them and prevent cracks.

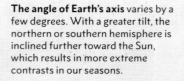

Scientists drill into Antarctic ice

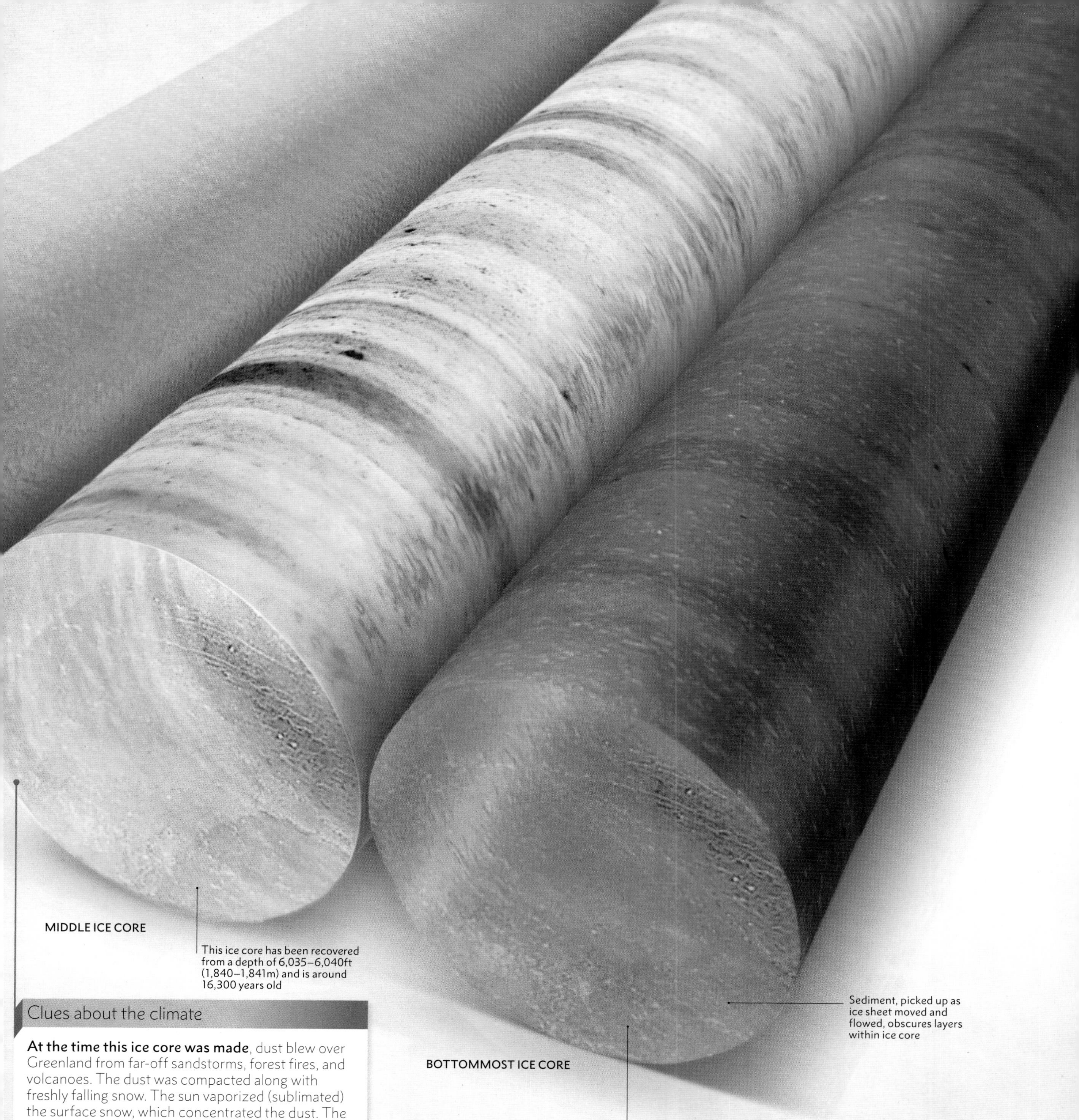
MIDDLE ICE CORE
This ice core has been recovered from a depth of 6,035–6,040ft (1,840–1,841m) and is around 16,300 years old
BOTTOMMOST ICE CORE
Sediment, picked up as ice sheet moved and flowed, obscures layers within ice core
This ice core is from the bottom of the Greenland ice sheet. An exact date is unknown, but it is more than 111,000 years old.

Clues about the climate

At the time this ice core was made, dust blew over Greenland from far-off sandstorms, forest fires, and volcanoes. The dust was compacted along with freshly falling snow. The sun vaporized (sublimated) the surface snow, which concentrated the dust. The dust shows as dark rings within the ice core. Dark rings indicate summer months, and clear rings signify winter months. Thick, dark rings show summers that were particularly long.

EARTH **FREEZES**

Climate change has been a natural part of Earth's history since the planet was formed. At its coldest, at the height of Earth's many ice ages, the world groaned under vast ice sheets that had a massive impact on life—driving some species to extinction and shaping the evolution of others.

Ice ages happen when the temperature of the Earth's surface plunges and extensive sheets of ice start to grow. It is likely that no single cause is responsible: shifts in Earth's orbit or atmospheric change both play their parts. But the effects can go far beyond climate. Freezing temperatures lock ocean water into permanent blocks – ice sheets and glaciers—lowering sea levels and merging lands that were once separated. Populations adapted to a tropical climate may contract toward the equator or even disappear altogether, while cold-adapted species advance.

EARTH ALMOST **COMPLETELY FROZE OVER TWICE IN ITS HISTORY**, WITH **ICE SHEETS ALMOST** 3,300FT (**1,000M**) **THICK**

ICE AGE EVENTS

At least two major ice ages happened before the Cambrian explosion of life, 520 MYA. In each case, our planet turned into a "snowball," almost completely covered in ice. Another ice age took place 460–420 MYA, when fish were filling the oceans. A fourth came as the first forests grew, 360–260 MYA, when the continent of Gondwana drifted over the South Pole and a polar ice cap started to spread. The last ice age—starting just over 2.5 million years ago—is better known, and is ongoing. During this ice age, ice sheets that are currently centered over Greenland in the north and Antarctica in the south have waxed and waned during glacial and interglacial periods. Since the ice sheets have not yet disappeared, Earth is still in this ice age, albeit in a relatively warm, interglacial. The glaciers of the recent past have left their mark in eroded valleys and glacial deposits, while changing temperatures and sea levels have made modern life a product of the glacial age.

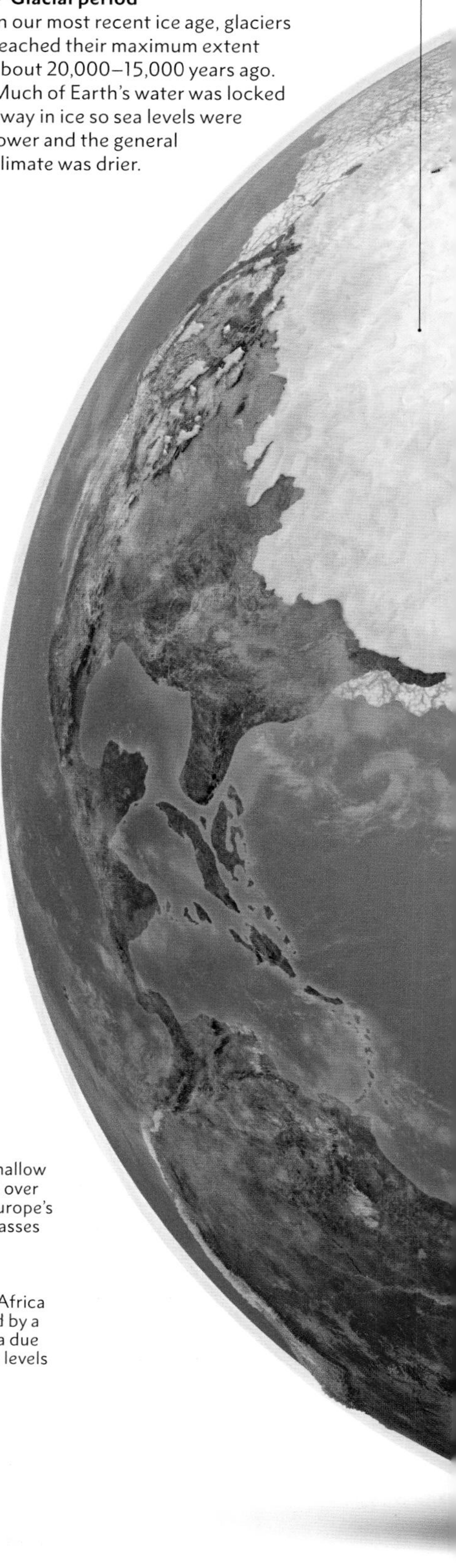
Vast North American ice sheet extended to the center of the continent at its maximum extent

▶ **Glacial period**
In our most recent ice age, glaciers reached their maximum extent about 20,000–15,000 years ago. Much of Earth's water was locked away in ice so sea levels were lower and the general climate was drier.

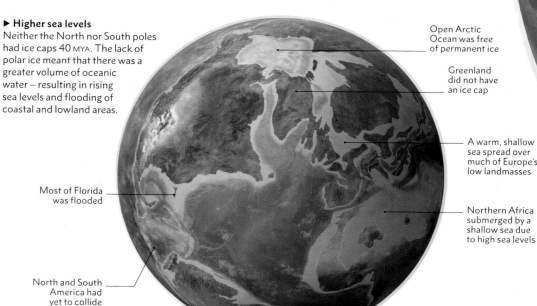

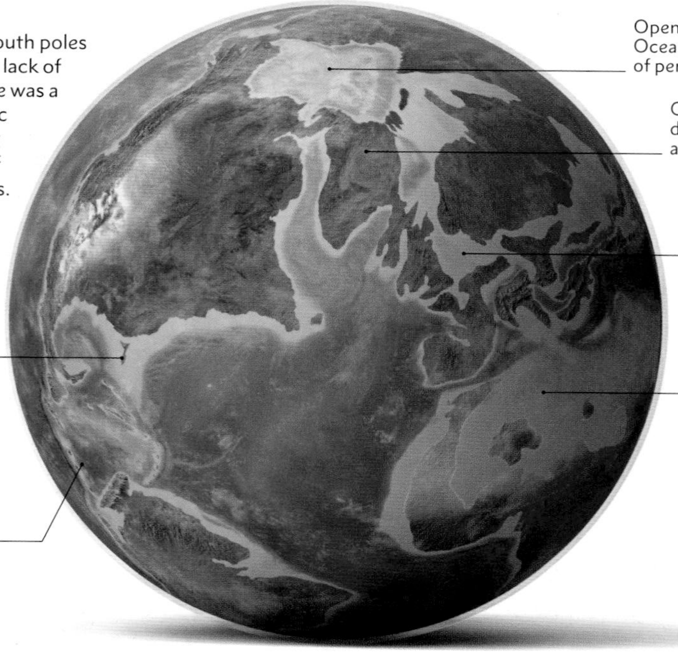
Open Arctic Ocean was free of permanent ice
Greenland did not have an ice cap
A warm, shallow sea spread over much of Europe's low landmasses
Northern Africa submerged by a shallow sea due to high sea levels
Most of Florida was flooded
North and South America had yet to collide

▶ **Higher sea levels**
Neither the North nor South poles had ice caps 40 MYA. The lack of polar ice meant that there was a greater volume of oceanic water – resulting in rising sea levels and flooding of coastal and lowland areas.

40 MYA

Pack ice extended as far south as northern France
Eurasian ice sheet extended across half of Russia and most of northern Europe at its maximum extent
Smaller ice sheets were also present on the Tibetan plateau
Grasslands were the main habitat that encircled the edges of the ice sheet in Europe and Asia
Dry scrub extended across western Asia due to a drier global climate
The British Isles were connected to Europe by land

20,000 YA

▲ Ice age elephant
The origins of woolly mammoths lay with the elephant family in Africa some 5 million years ago. Elephants spread northward around the Earth, evolving shaggy coats that kept them warm during the advancing ice age.

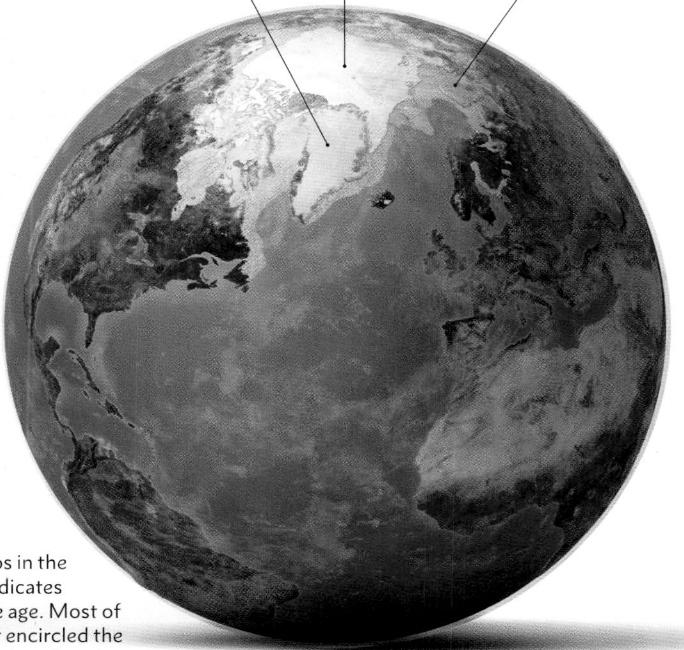
Ice cap over Greenland's landmass
Permanent sea ice over Arctic Ocean
Pack ice around northern Russia

TODAY

► Interglacial period
The presence of ice caps in the Arctic and Antarctic indicates that we are still in an ice age. Most of the vast grasslands that encircled the ice sheets have retreated, replaced with wetter coniferous forests.

THRESHOLD

6

HUMANS **EVOLVE**

With our origins in the stars—like everything else—and sharing a common ancestor with the apes, what makes humans unique? Humans have an ability to innovate, learn, and share experiences like no other species. Through the use of symbolic language, and by sharing and building on knowledge collectively, our human ancestors begin to dominate the landscape.

GOLDILOCKS CONDITIONS

Modern humans evolved relatively recently, around 200,000 years ago. The ability to communicate using symbols, exchange ideas, and build on the knowledge of earlier generations has allowed *Homo sapiens* to create new levels of complexity, and become the single most powerful and influential species on Earth.

Natural selection acting on apes

Broadening hominin diet includes meat and unlocks new energy

Evolution of new genus *Homo* with increased cognitive capacity

Rapidly shifting global climate

What changed?
A new species—*Homo sapiens*—evolves with the capacity for collective learning.

Mammals diversify
Tree-dwelling primates evolve about 65 million years ago. Their large brains, social skills, and manual dexterity allow them to use and develop tools.

Diverse habitats
Primates adapt quickly to life in an unstable climate, surviving in rain forest and savanna environments.

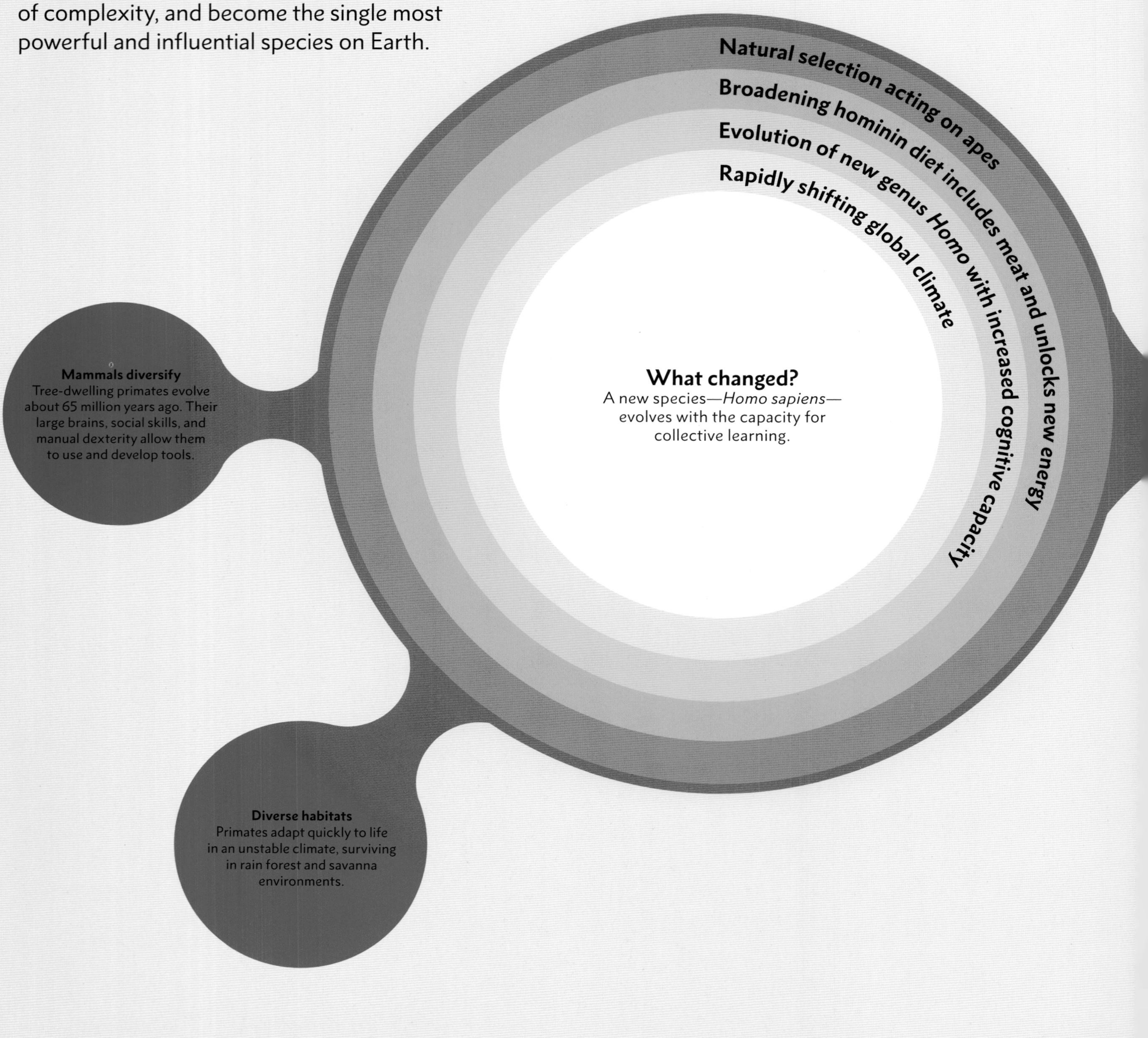

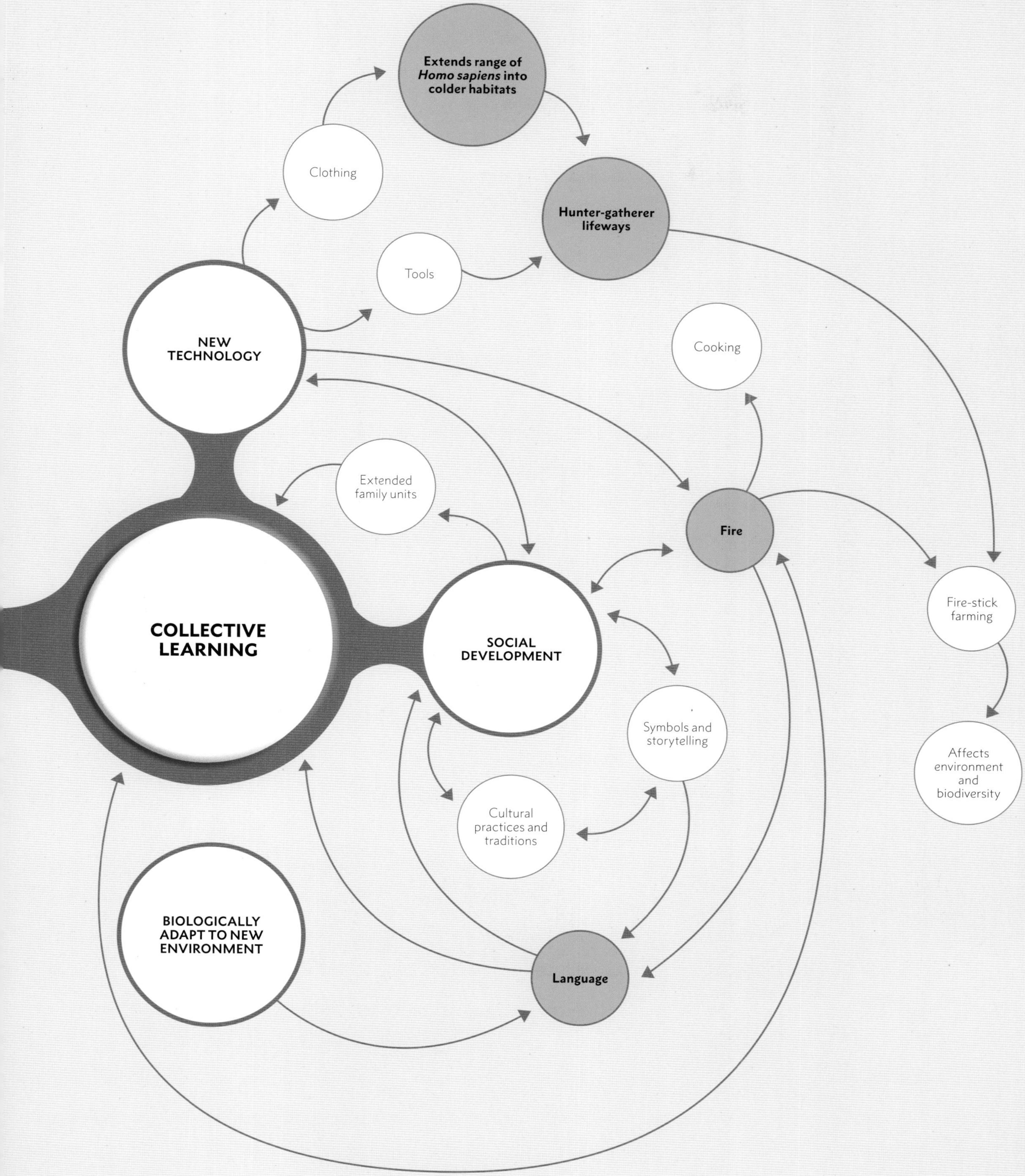

Extends range of *Homo sapiens* into colder habitats
Clothing
Hunter-gatherer lifeways
Tools
NEW TECHNOLOGY
Cooking
Extended family units
Fire
COLLECTIVE LEARNING
SOCIAL DEVELOPMENT
Fire-stick farming
Symbols and storytelling
Affects environment and biodiversity
Cultural practices and traditions
BIOLOGICALLY ADAPT TO NEW ENVIRONMENT
Language

Family connections
Much of human behavior can be seen mirrored in other primates, such as the parental care given to this orangutan baby. Orangutan young are completely dependent on their mothers during their first decade.

THE **PRIMATE FAMILY**

With our large brains, dexterous fingers, and highly complex social structures, it may seem obvious that we are primates. However, the primate order is diverse, and while many species share particular features, it has no single, defining physical characteristic.

Today, about 400 primate species have been identified, ranging from minuscule tarsiers to imposing gorillas. Physically and genetically *Homo sapiens* clearly descends from this order—specifically the line of apes—yet even the apes are only a recent branch of the tree. It took 20 million years for the tiny ratlike proto-primate *Purgatorius* (65 MYA) to evolve into the lemurlike primate *Darwinius*. By this time, two major primate lines had flourished—one leading to lorises and lemurs and another leading to tarsiers. By 40 MYA, the anthropoid line had appeared, and this led to monkeys, apes, and eventually humans. These anthropoids probably emerged in Asia, and their fossils show that the primate face—which had a snout—was already shortening.

ALMOST HUMAN

By 25 MYA, forest environments were filled with a diverse range of monkeys. The tailless *Proconsul*, which lived in East Africa 25–23 MYA, had a mixture of ape and monkey characteristics, and soon, many species of true apes began radiating into Europe and Asia. These were the first of the modern primate species. DNA suggests that the splits leading to orangutans and gorillas happened around 16 MYA and 9 MYA respectively, and each had contemporary relatives, like *Sivapithecus* in Asia and *Chororapithecus* in Ethiopia. From around 9 MYA, a group of huge Asian apes called *Gigantopithecus* evolved, some of which may have existed until very recently. One of the earliest African species thought to have led to the hominin line was *Sahelanthropus tchadensis* (7–6 MYA), which lived around the same time that our ancestors are estimated to have split from chimpanzees.

Behaviorally, early apes probably had the same high degree of dexterity, intelligence, and flexibility as modern primates, and probably lived in similarly diverse communities, featuring strong bonds and complex communication. It is also likely that some of these species used tools, just as various apes and capuchin monkeys do today.

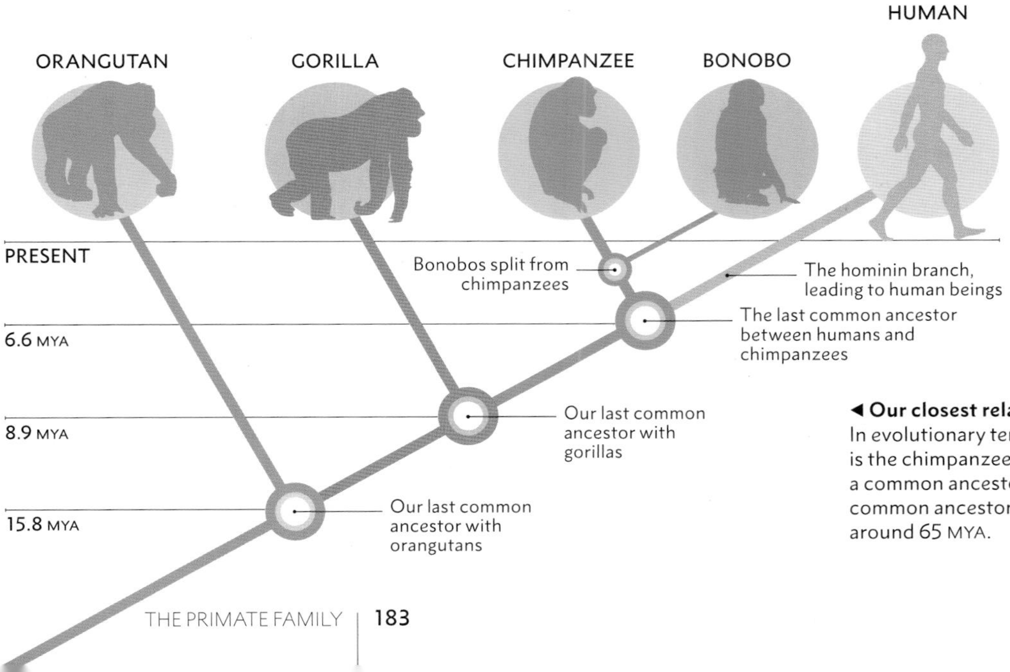

◄ Our closest relatives
In evolutionary terms, our closest relative is the chimpanzee, with whom we shared a common ancestor some 7–6 MYA. The common ancestor of all primates lived around 65 MYA.

HOMININS EVOLVE

Humans belong to the hominin branch of the primate family tree. It is a branch that took over 7 million years to develop and includes all modern humans, extinct human species, and all our recent ancestors.

When tracing our roots, it is tempting to think that our "advanced" characteristics, such as the ability to walk on two feet and use tools, emerged as a result of a single creature becoming ever more complex. But the truth is that early hominins were diverse, and that these traits were shared in various combinations by *Homo habilis*, the earliest *Homo* species, and *Australopithecus*, an earlier hominin genus, and probably evolved independently.

The fossil record is tantalizing. It reveals that slender australopithecines (*A. afarensis* and *A. anamensis*) appeared between 4 and 3 MYA, and later diversified into more robust forms with heavy-duty teeth. However, the earliest *Homo habilis* dates to 2.4 MYA, leaving a considerable gap between the species. A possible bridge was found in Ethiopia in 2015—a fossil jawbone, dating to 2.8–2.75 MYA. The fossil matches the crucial period, and it shows some signature *Homo* features, but without the rest of the skull or any indication of the size of the brain, it is impossible to determine which family its owner belonged to.

In evolutionary terms, a key mark of the *Homo* lineage was its ability to adapt to different environments by changing its diet. The tendency to eat more meat was crucial: this led to a greater reliance on tools for hunting, which in turn favored the larger brains that evolved after 2 MYA (see pp.188–89). This led to shifts in social organization and ranging patterns, culminating in the evolution of *Homo erectus*, probably the first global explorer; *Homo neanderthalis*, our closest hominin relative; and finally *Homo sapiens*.

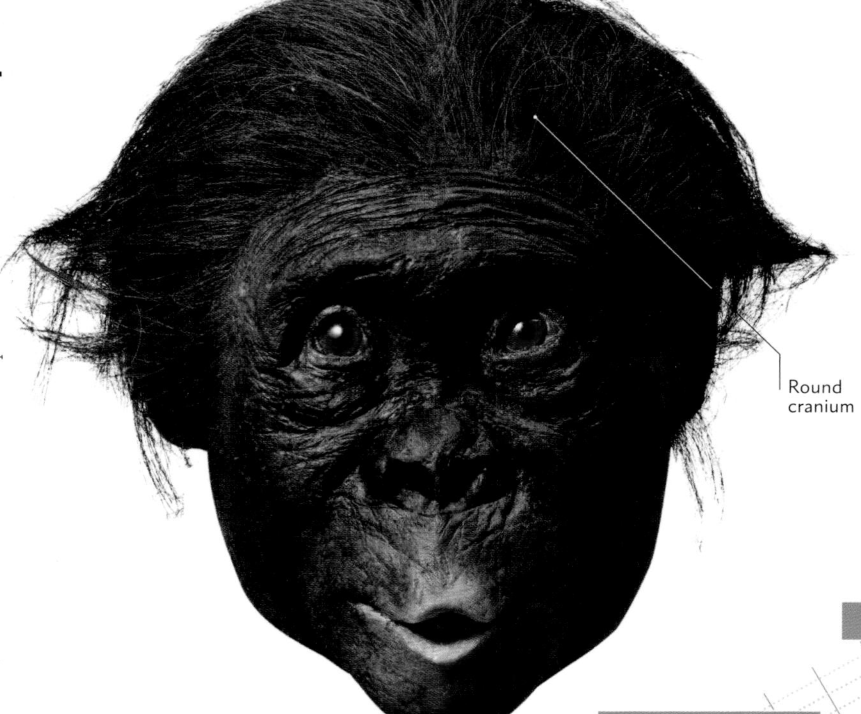

▲ ***Australopithecus africanus***
One of seven known *Australopithecus* species, *A. africanus* was the first early hominin to be discovered in Africa. Dating to around 3–2 MYA, it had a small brain but could walk upright.

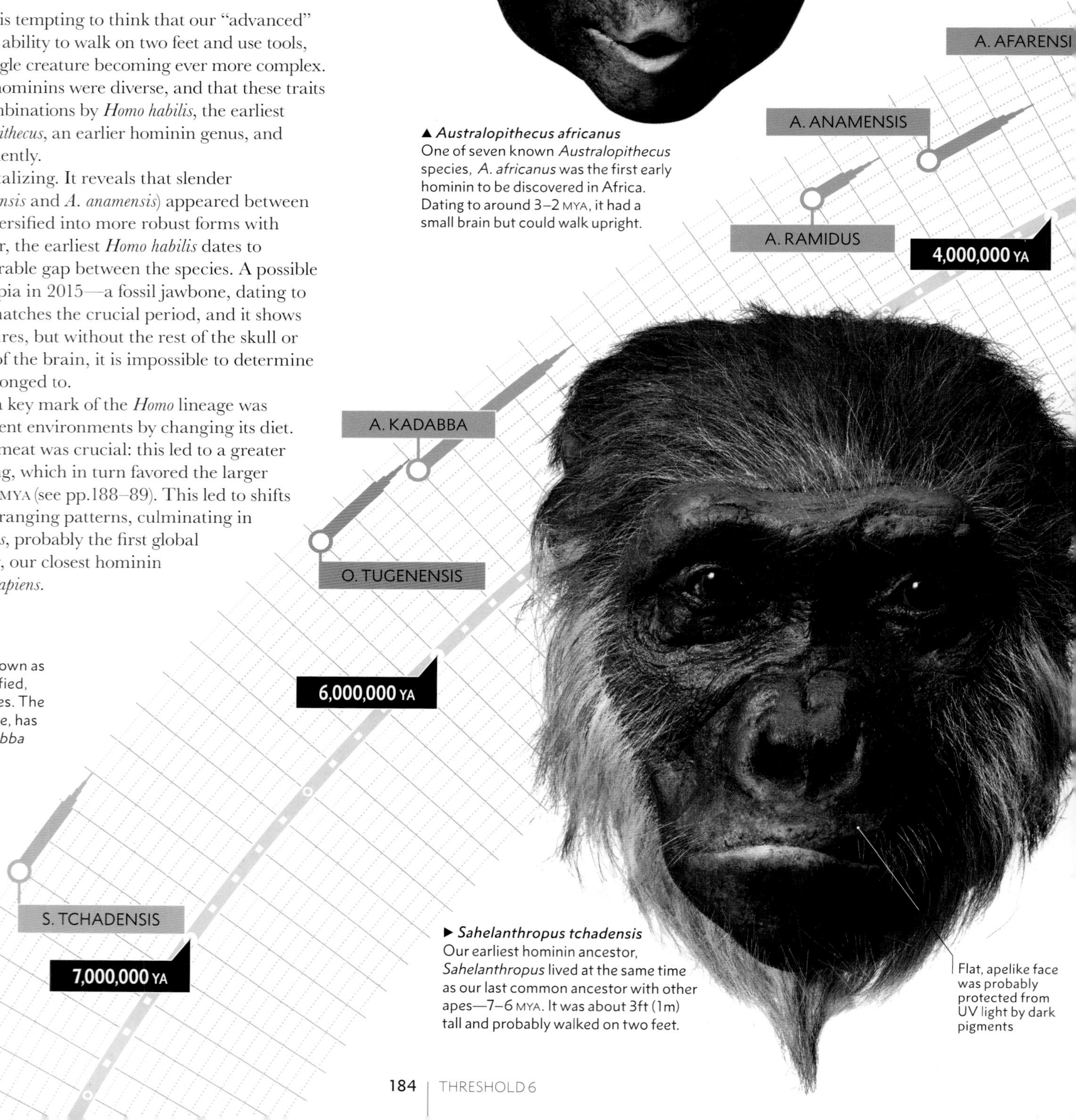

▶ **The hominin family tree**
Seven hominin groups, each known as a genus, have so far been identified, and some contain several species. The genus *Ardipithecus*, for example, has two species, *Ardipithecus kadabba* and *Ardipithecus ramidus*.

KEY
- *Sahelanthropus*
- *Orrorin*
- *Ardipithecus*
- *Kenyanthropus*
- *Paranthropus*
- *Australopithecus*
- *Homo*

▶ ***Sahelanthropus tchadensis***
Our earliest hominin ancestor, *Sahelanthropus* lived at the same time as our last common ancestor with other apes—7–6 MYA. It was about 3ft (1m) tall and probably walked on two feet.

▶ *Homo erectus*
The remains of *H. erectus*, or "upright man," have been found in Africa, China, and Indonesia. It arose just under 2 MYA, and may still have been alive only 50,000 YA.

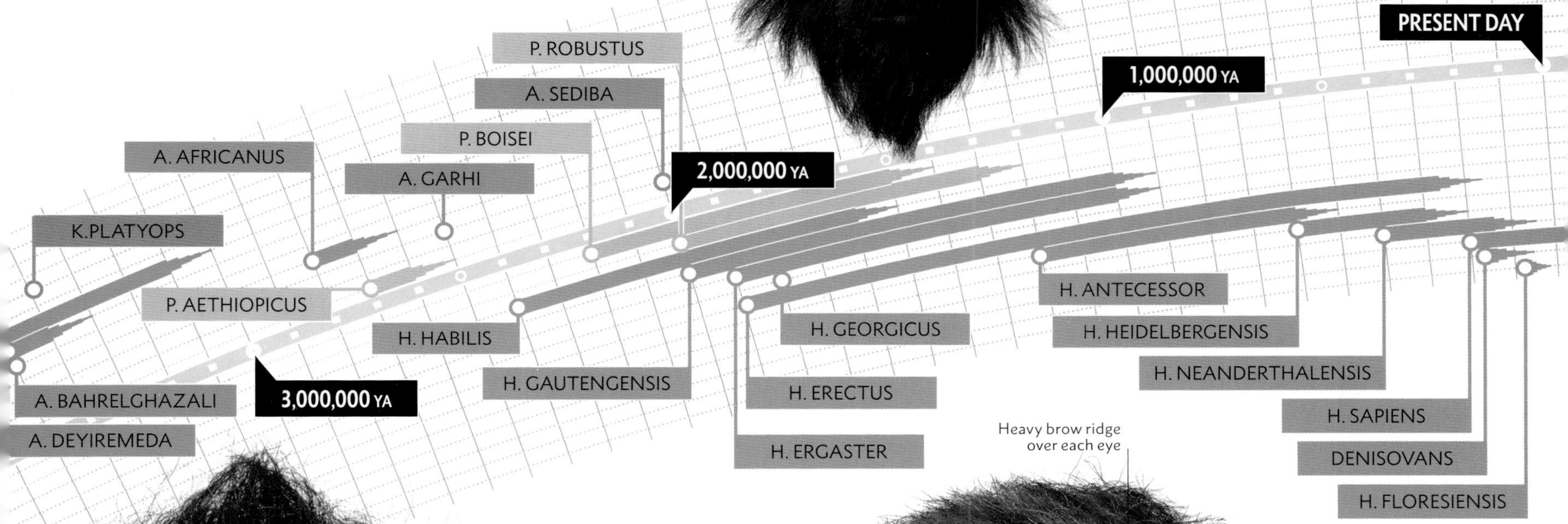

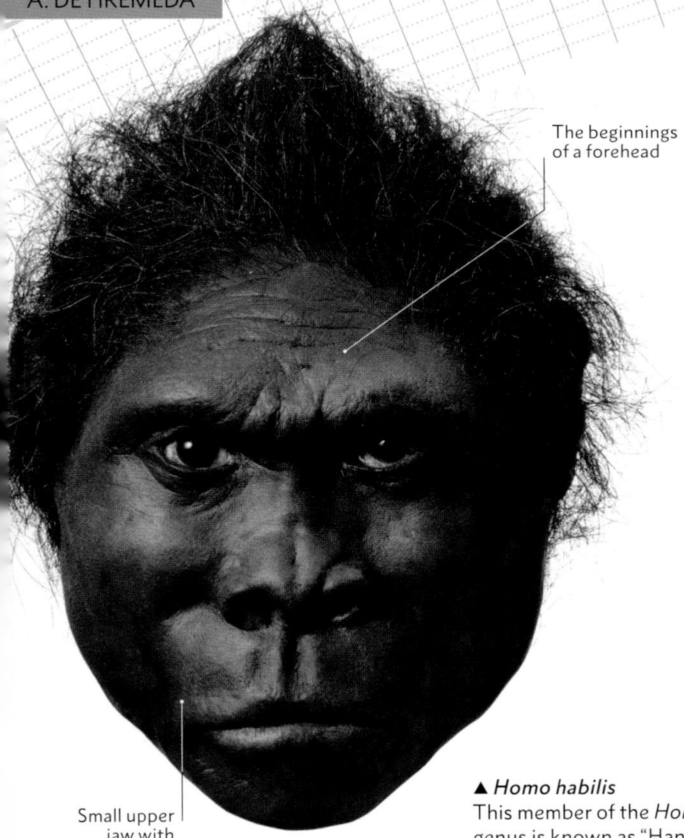

▲ *Homo habilis*
This member of the *Homo* genus is known as "Handy man," referring to its use of stone tools.

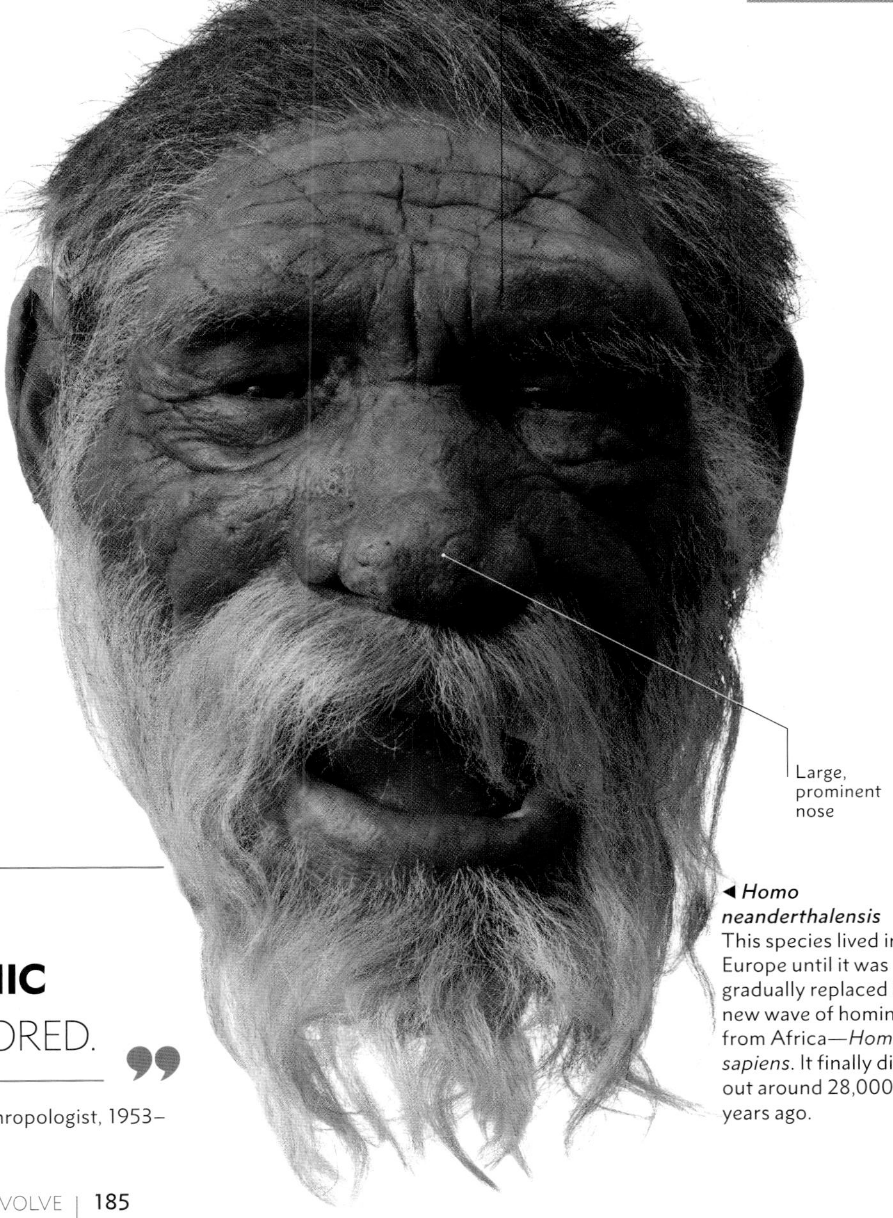

◀ *Homo neanderthalensis*
This species lived in Europe until it was gradually replaced by a new wave of hominins from Africa—*Homo sapiens*. It finally died out around 28,000 years ago.

> "THE **BORDER** BETWEEN **NONHUMAN** AND **HUMAN** IS NOT THE **SHARP ADAMIC EMERGENCE** THAT HAS LONG BEEN FAVORED."

Jean-Jacques Hublin, paleoanthropologist, 1953–

APES BEGIN TO **WALK UPRIGHT**

The journey from tree-climbing apes to ground-walking humans involved major anatomical changes throughout the skeleton. Ancient footprints show that our ancestors already walked like humans 3.7 MYA, but a further 2 million years of refinement were needed to make us into runners.

Colder, drier climates from 35 MYA led to a change from forests to more varied habitats, including open grassland. This has long been seen as the driving force that around 7–4 MYA made some tree-climbing apes change into "bipedal" animals that walked primarily on the ground on two legs. The reality is more complex, since some of the oldest bipedal fossils are from locations that were densely forested. Whatever the reasons, however, a series of fossils offers glimpses of the transition to ground dwelling.

ADAPTING TO THE GROUND

A good model for the starting point of the change is *Proconsul*, an animal close to the base of the ape family tree. It moved by either running along branches or climbing, using hands and feet to grasp tree limbs.

Some fossils from 7 MYA onward show a marked contrast. These are the hominins (see pp.184–85), the group to which humans belong. The oldest, *Sahelanthropus*, already shows evidence of an upright spine, since the entry point of the spinal cord into the skull is on its underside, not the back, as in today's apes. Soon, another hominin evolved with more distinctly ground-dwelling features. This was *Ardipithecus ramidus*, which lived in what is now Ethiopia 4.5–4.3 MYA. It could walk almost upright, but was not fully bipedal, since its feet had opposable toes.

To become fully bipedal, hominins needed feet dedicated to walking on the ground, with in-line big toes and bones and tendons forming a springy arch. Footprints in Africa, left possibly by *Homo ergaster*, suggest these features had evolved 1.5 MYA—2 million years after the famous Laetoli prints (see below). Now, *H. ergaster* and other *Homo* species had become capable runners. They had evolved an S-shaped spine that absorbed vertical shocks, a short, wide pelvis that centered the torso above the hips, and thigh bones angled inward toward the knees, improving balance and gait. By 1 MYA, hominins were striding across most of Africa, Asia, and Europe.

▶ Down from the trees
The transition to bipedal walking on the ground can be summarized by these three key stages.

Proconsul was one of the earliest apes, found in Africa. It lived 23 million years ago in dense tropical forest, used quadrupedal (four-legged) locomotion, and was a good climber. But the lack of a tail showed that living in trees was becoming less important.

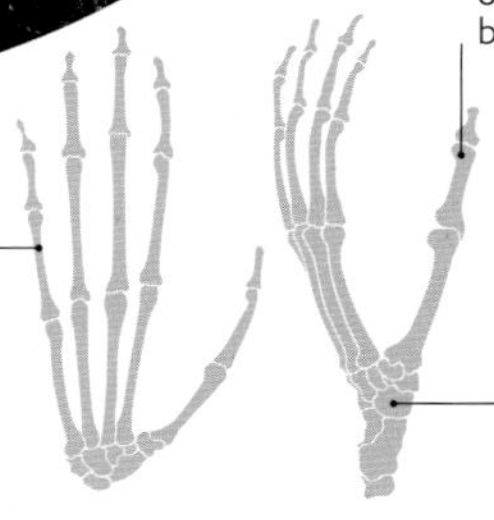

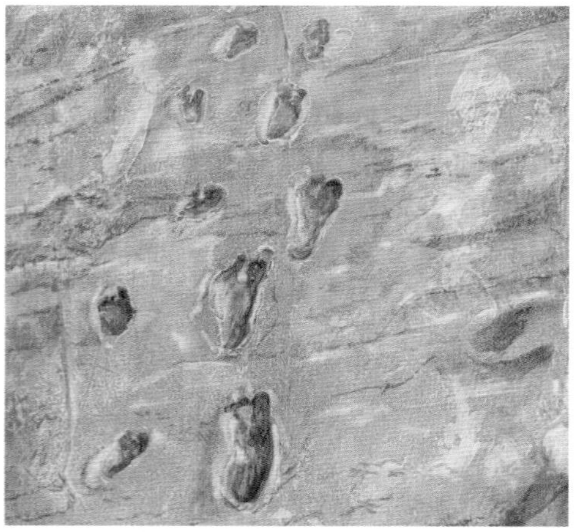

▶ Ancient footsteps
An adult and child *Australopithecus ramidus* made these fossil prints 3.7 million years ago in what is now Laetoli, Tanzania. The 3-D contours of the imprints, compared to those made by modern humans, suggest that they walked with a humanlike gait, not the rocking, bent-knee gait of apes.

► A cooler, less predictable planet
The analysis of core samples from ice sheets (see pp.174–75) and deep-sea sediments have shown that over the last 6 million years, Earth's climate has not only cooled but has also become more variable. The emergence of new hominin species seems to coincide with the rising variability, suggesting that they diversified due to the pressure of environmental change. The adaptability of the hominin skeleton may have enabled individuals to live in a wide range of habitats, whether open or wooded, wet or dry.

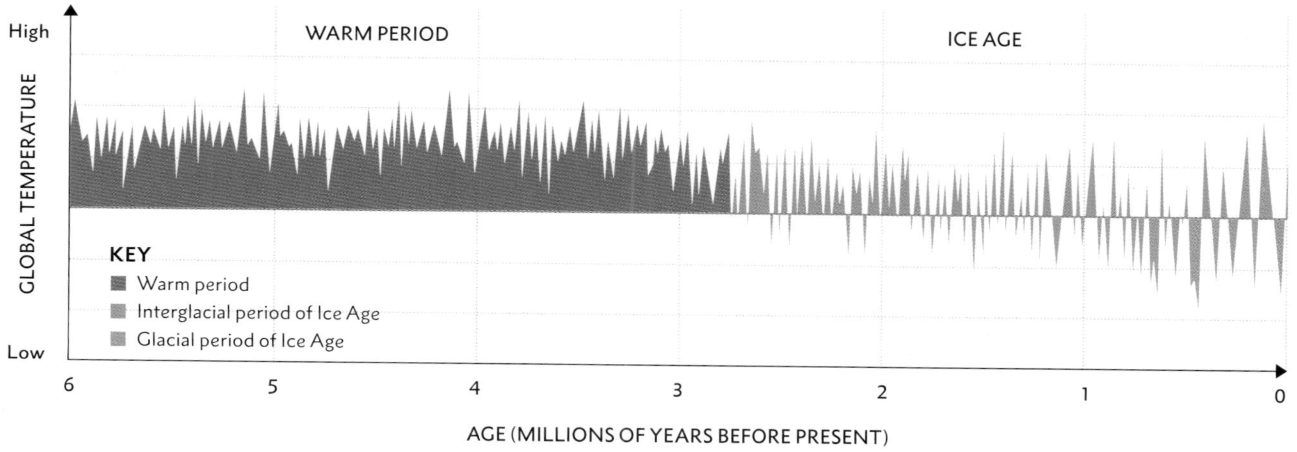

THE ARCHAIC HOMININ ***SAHELANTHROPUS*** MAY HAVE WALKED UPRIGHT **7 MILLION YEARS AGO**

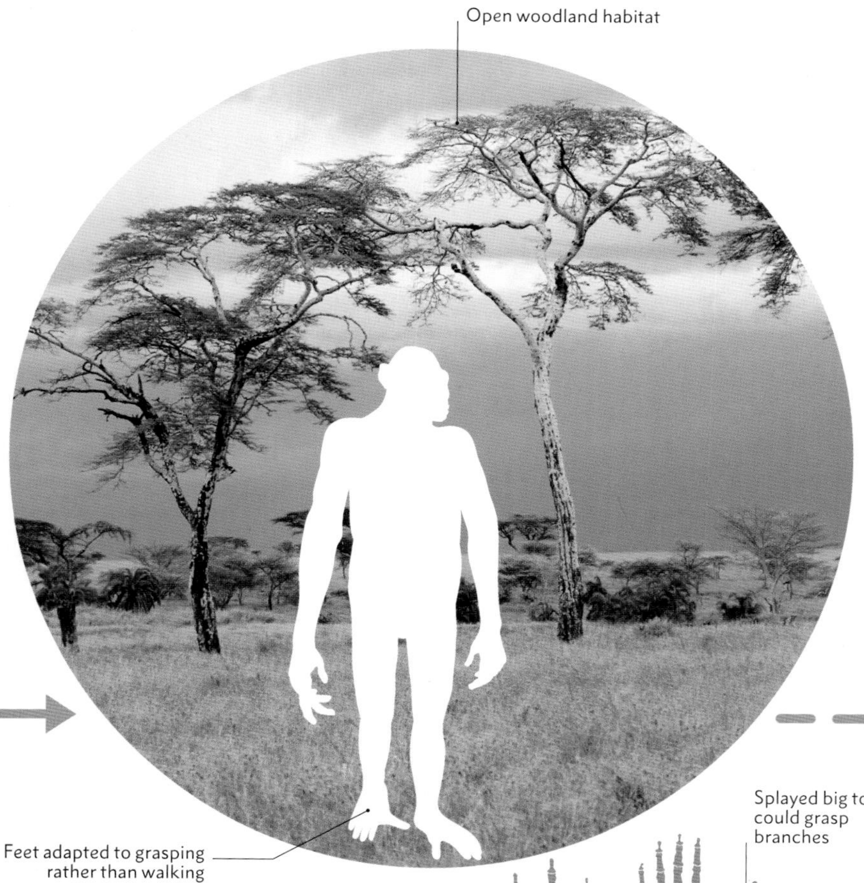

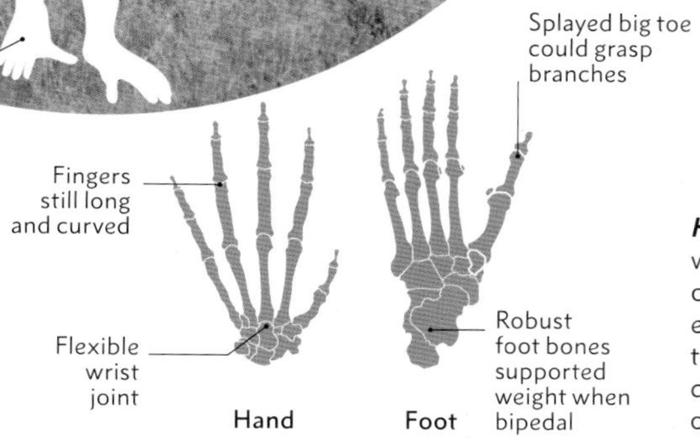

Ardipithecus ramidus was much closer anatomically to a human than *Proconsul*. *Ardipithecus ramidus* had long arms and a grasping foot that enabled it to climb in the forested environments that it seems to have preferred.

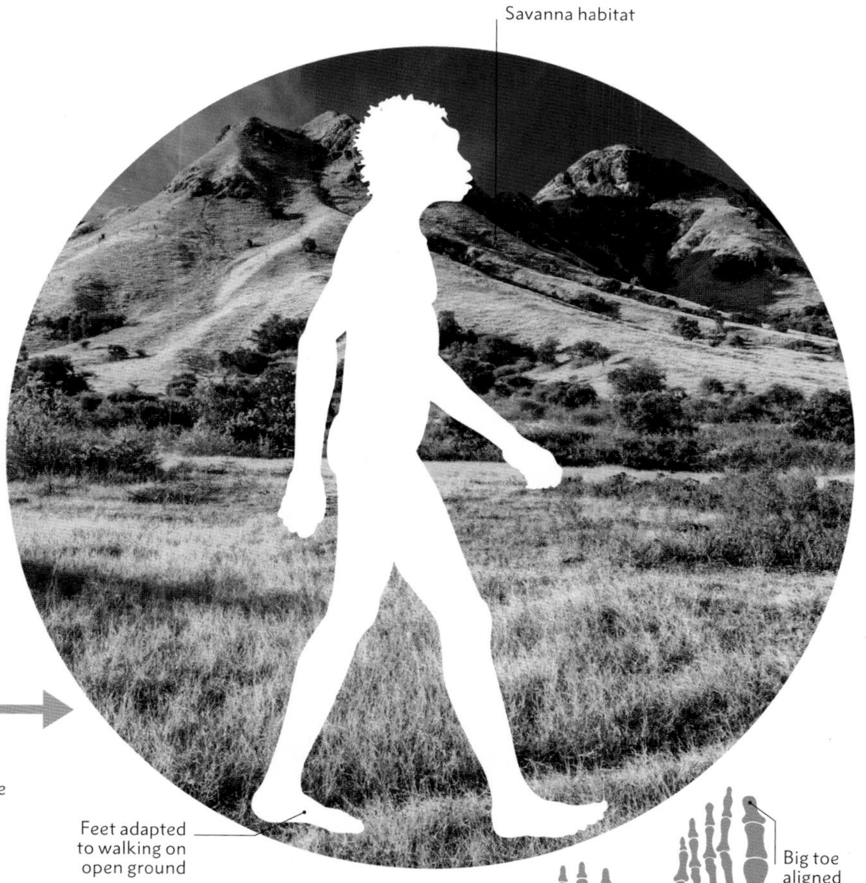

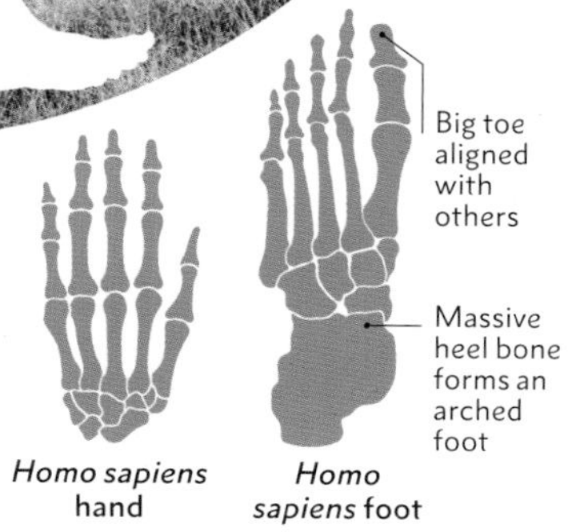

Homo erectus stood as tall as a modern human. It was fully bipedal and its arms were shorter than those of its forest-adapted ancestors. It could cover ground efficiently in open grassland, with hands free to carry tools. A few fossils of its hands and feet have been discovered, and they seem to have approached those of modern humans in form and function.

▲ **Meat-fueled minds?** This Paleolithic cave painting of a bison is from Altamira in Spain. Some theories propose that the switch to a diet including meat was the catalyst for the growth in brain size among hominins.

GROWING A LARGER BRAIN

Biologists have studied differences in brain size and intelligence across the animal kingdom for over a century. The trend toward increased primate encephalization (brain mass relative to body size), most dramatically seen in *Homo sapiens*, is clearly an adaptive feature.

Understanding why and how we developed a large brain—an organ that requires lots of energy to grow and maintain—involves considering many aspects of our evolution. Brain size relative to body size seems to be important: when compared with primates and other mammals, humans stand out with our globelike, inflated skulls enclosing huge brains for our overall bulk.

FOOD FOR THOUGHT

One theory for increasing brain size ratio in hominins relates to changes in diet. While a few primate species, including chimpanzees, regularly consume meat, this is usually in very small amounts. In comparison, the hominin archaeological record shows that the gut shrank over time as eating meat became more common, indicating that fewer hard-to-process plant foods were consumed. Did extra calories and fats from meatier diets, and eventually cooked foods, feed our energy-hungry brains, and even drive their evolution? While there undoubtedly was some impact, the timings don't quite add up. Stone-tool technology, which emerged over 3 million years ago, gave hominins better access to the high-energy foods within animal carcasses. But over the million years between the first australopithecine toolmakers and early *Homo*, the increase in brain size was quite small, only about 6 cu in ($100cm^3$). Not until 500,000 years ago, in *Homo heidelbergensis*, had brain capacity doubled.

200,000 YA | HOMO SAPIENS APPEARS

135,000 YA | FIRST USE OF SYMBOLS

110,000 YA | LAST ICE AGE BEGINS

41,000 YA | EARLIEST PAINTED CAVE ART

12,000 YA | LAST ICE AGE ENDS

THE SOCIAL BRAIN

More recent theories consider not only the brain's overall size, but also how its different parts changed over time, including areas vital for communication, visual processing, planning, and advanced functions such as problem solving. Of particular interest is the link between the size of the neocortex (the outer part of the brain) and social intelligence. The neocortex is involved in many brain functions, ranging from motor control to perception, consciousness, and language. Primates with a proportionately larger neocortex live in bigger social groups, suggesting that the neocortex provides the extra "processing power" the brain needs to keep track of relationships between many individuals. But it's not just about numbers: primate social life involves predicting and even manipulating the behavior of others. When social networks increased in size in hominins, this required even greater investment in the brain.

PRIMATE BRAINS ARE **NEARLY TWICE AS BIG** AS THOSE OF SIMILAR-SIZED MAMMALS

These ideas link to other aspects of brain size noted across different species. Animals with larger eyes, for example, tend to have bigger brains, implying that greater visual acuity needs more processing power. In hominins with increasingly complex social lives, a highly developed visual sense enables individuals to not only find food and detect predators, but also to determine the precise direction of another's gaze and observe subtle gestures.

Bigger-brained species, from mammals to birds, also tend to show greater levels of self-control. They are able to resist impulses and delay satisfaction, and instead reflect on other courses of action, based on previous experiences. While in primates levels of self-control do not necessarily increase with a social group's size, greater self-control may have helped hominins to follow rule-based social strategies for managing status and "getting ahead" in social groups.

COMPLEX ANSWERS

Ultimately, developing larger brains may have been the result of many competing pressures on hominins, which cumulatively demanded greater levels of processing power. Questions about diet are important, but perhaps the gradual broadening of the hominin diet is more crucial than just the introduction of meat. As well as plant foods and meat, "specialized" foods such as fish began to be exploited by early *Homo* nearly 2 million years ago, evidenced by the eating of catfish and turtles at Koobi Fora, Kenya. Wider foraging, and especially increased tool use, required a larger base of motor skills, memory, and overall greater flexibility. In many cases these were probably cooperative activities, relying on the ability to learn, have self-control, and engage in intense social networking.

While there may be diverse reasons behind our increased brain size over the past 20,000 years, human brains have actually started to shrink again. It may be that a better understanding of brain function among *Homo sapiens* will show that intelligence is determined not just by brain size but by smarter wiring too.

▲ The social brain Today the indigenous San people of the Kalahari work in tightly bound social groups, just as other hunter-gatherers usually do. This facility for complex interaction was only made possible by the development of a larger brain.

> **THE BRAIN IS A MONSTROUS, BEAUTIFUL MESS**. ITS BILLIONS OF NERVE CELLS... LIE IN A **TANGLED WEB** THAT DISPLAYS **COGNITIVE POWERS FAR EXCEEDING** ANY OF THE **SILICON MACHINES** WE HAVE BUILT TO MIMIC IT.

William F. Allman, journalist, 1955 –

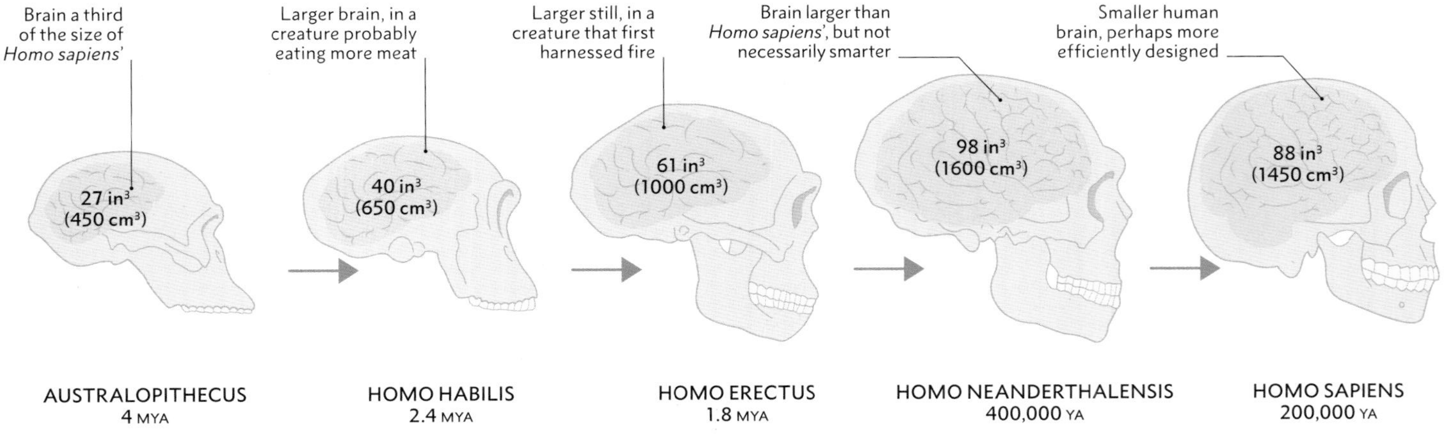

◄ Evolution of the hominin brain Over the last 7 million years, the hominin brain has tripled in size, with most of that growth occurring over the last 2 million years. Measurements of ancient brains are based on the size of skull remains, some of which preserve casts of their interiors.

THE NEANDERTHALS

The Neanderthals are just one of our close hominin relatives, but for centuries they have played a special role in our understanding of human history. Studying these ancient people, who were successful for so long, has transformed our view of ourselves.

▲ Eagle talon jewelry
Eight eagle talons were found in a 300,000-year-old Neanderthal cave in Croatia. Friction marks suggest that they were once strung together.

The branch of the hominin tree that led to the Neanderthals and *Homo sapiens* appeared around 600,000 years ago, and the earliest examples of "Neanderthal-like" features appear nearly 400,000 years ago. These are revealed in a wealth of Neanderthal fossils—one of the largest collections for any hominin species—which includes parts of more than 275 individuals, and some reasonably complete skeletons. Anatomically, they differed from us in subtle ways, having slightly larger skulls, less prominent chins, but bulkier eyebrow ridges. There were also differences in tooth shape. Neanderthals were typically shorter than *Homo sapiens*, and they had more rounded chests, differently proportioned arms and legs, and larger fingertips. When dressed, however, they would have looked very similar to us.

WIDE-RANGING HUNTERS

Neanderthals are often depicted as Ice Age creatures, but their range was far greater than this. They lived through cycles of both glacials and interglacials (some even warmer than today), and were just as much at home in deciduous forests as in open steppe-tundra. Many hundreds of Neanderthal sites are known, in places as far flung as Wales, Israel, Siberia, and Uzbekistan. It is difficult to establish which sites are the most recent due to dating complexities, but it seems that the last Neanderthals lived about 30,000 years ago.

As to their fate, they are no longer considered "extinct," since analysis of nuclear genomes shows that humans and Neanderthals interbred repeatedly at different times and places. There is probably more Neanderthal DNA surviving in the world today—in humans—than there ever was when Neanderthals walked the Earth.

INJURIES ON **NEANDERTHAL SKELETONS** FOLLOW A PATTERN SIMILAR TO THAT OF **MODERN-DAY RODEO RIDERS**

Another transformation has been in our view of the culture and cognitive capacities of Neanderthals. Their stone tools were far from crude or unchanging, instead showing regional diversity and development over time. They made blades, the earliest multipart tools, the earliest synthetic material (birch bark adhesive), and various wooden utensils. They were undoubtedly top hunters too, with a diet that varied according to where they lived and included many plants and small game such as tortoises.

The fact that humans repeatedly had relationships with Neanderthals, and that the resulting children survived, suggests that cognitively they cannot have been alien. They used red and black pigments, collected shells, and had a unique interest in the feathers and claws of birds, especially large raptors. On the other hand, there is no Neanderthal art that matches the work of later Ice Age human populations, and this could point to a difference in cognitive ability. As for their disappearance, the reasons are likely to have been myriad and complex, including competition for food, climatic stress, and disease.

▶ The Neander Valley
The Neanderthals take their name from the Neander Valley, near Düsseldorf, Germany, where some of the earliest fossil remains of the species were found in a cave in 1856.

▶ Another kind of human
The Neanderthals were remarkably similar to *Homo sapiens*, with whom they bred for thousands of years. Up to 20 percent of their DNA may survive in humans today.

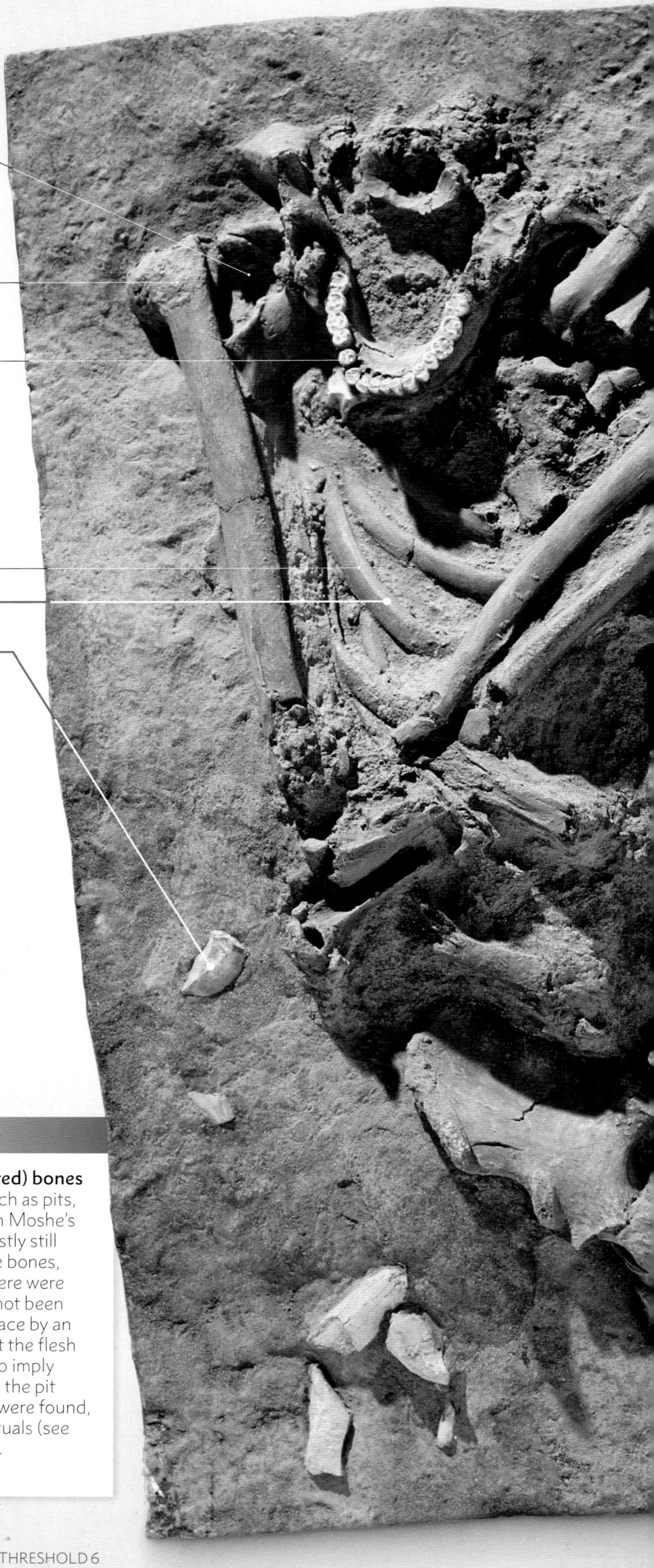

Other than the lower jaw, the skull was missing. No fragments were found, which suggests that it was probably removed by erosion
Thick bones and large joints show that the arms and hands were muscular and powerful
The teeth are heavily worn; Neanderthals may have used their teeth like a vice to help them hold animal skins or other objects as they worked
The relatively complete rib cage enabled scientists to reconstruct the shape of the thorax (chest area) from the curvature of the ribs

Neanderthal anatomy

The rib cage shows that Moshe had a barrel-shaped chest and large lungs. It was thought that European Neanderthals had developed big lungs as an adaptation to the cold. Living in cold climates consumes a lot of energy, requiring more oxygen to fuel energy-releasing reactions in the body; large lungs also help to warm and moisten inhaled air. But since Moshe lived in the more temperate eastern Mediterranean, some scientists now discount this theory. They suggest that the large lung size was an existing anatomical feature, inherited from earlier African hominins, that equipped Neanderthals for a high-energy hunting lifestyle. It probably did, however, help them to colonize the cooler parts of Europe.

Dating techniques

Archaeologists employ a range of techniques to date remains. Two of these, thermoluminescence (TL) and electron-spin resonance (ESR), measure the amount of radiation damage, in the form of electrons, that accumulates in a material over time from background sources and cosmic rays. While TL is used on stone tools, ESR is applied to human and animal teeth. Tests on burnt flints and gazelle teeth found at Kebara indicate that the skeleton is around 60,000 years old.

A technician conducts TL analysis of a specimen

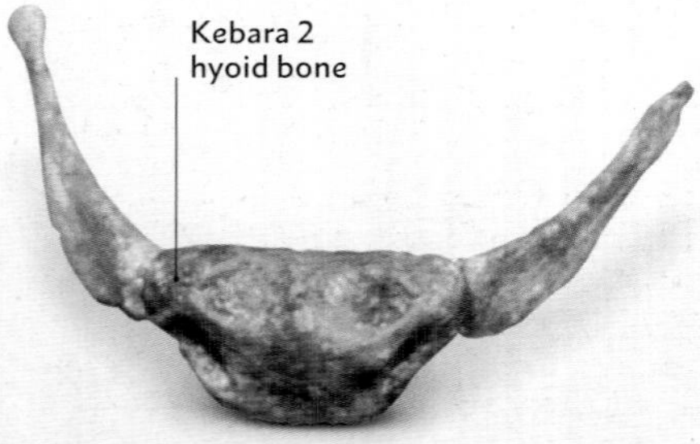

Kebara 2 hyoid bone

▲ Unique hyoid
Moshe's hyoid bone is virtually identical to that of *Homo sapiens*. In modern humans, this bone, which is rooted in the cartilage surrounding the larynx, anchors the throat muscles that facilitate speech. The Kebara hyoid raises the possibility that Neanderthals may also have had language capabilities (see pp.202–03).

Laid to rest

Skeletons with articulated (connected) bones that are found in distinct contexts, such as pits, are suggestive of intentional burials. In Moshe's case, the body parts present were mostly still correctly joined together, and delicate bones, such as the hyoid, were unbroken. There were no carnivore marks, so the body had not been scavenged or dragged to its resting place by an animal. Body posture and the fact that the flesh seems to have decomposed in situ also imply that Moshe was deliberately placed in the pit after his death. Since no grave goods were found, we cannot infer that there were any rituals (see pp.218–19) associated with the burial.

HARD EVIDENCE

KEBARA NEANDERTHAL

In 1983, a well-preserved skeleton of an adult Neanderthal was uncovered in Kebara Cave on Mount Carmel, Israel. Such physical remains, whether fossilized or not, are treasure troves of information about our hominin relatives.

Remains of up to 17 individuals were found at Kebara. They included an infant, known as KMH1 or Kebara 1, discovered near a wall in what may have been a midden. The adult, called KMH2 or Kebara 2, was lying on its back in a pit, with one arm across its chest and the other across its abdomen. Bone growth, dental wear, and the shape of the pelvis showed that it was a male aged 25–35. Nicknamed "Moshe," he was about 5ft 7in (1.7m) tall—slightly taller than the average Neanderthal. Although the skull and most of the legs were missing, the skeleton provided the first full sets of Neanderthal ribs and vertebrae, the first complete pelvis, and the only Neanderthal hyoid bone, which enables speech in modern humans.

A clue to diet comes from chemical analysis of the ratio of carbon to nitrogen in bones. Neanderthal bones have a higher proportion of carbon, indicating that they ate a lot of meat (high nitrogen levels signify a more herbivorous diet). This is supported by the many gazelle and deer bones at Kebara that bear the cut marks of butchery and signs of burning.

Recent studies of Neanderthal teeth reveal different information to analysis of bones, showing that plants may have been consumed more often than scientists once thought. Plant remains in Kebara cave, including charred peas in hearths, suggest these Neanderthals consumed a range of wild legumes, grasses, seeds, fruits, and nuts, though in what quantities we cannot be sure.

While Moshe's bones show no evidence of injury, many Neanderthals had healed fractures, possibly sustained when hunting large animals at close quarters. As well as being an indication of health, signs of disease and injury can sometimes suggest some level of care between group members. Shanidar 1, a male Neanderthal from Shanidar Cave, Iraq, had received a blow to the skull that probably blinded him and perhaps caused brain damage; he also had one withered arm and had lost his other forearm entirely. He could only have survived to his estimated age of 40–45 years with the help of others in his community.

Burial site

Moshe's body lay in the cave's main living area, which had the greatest concentration of hearths and animal bones. It was found in a shallow grave cut into the thick black hearth deposits. The grave contained a yellow sediment that differed from the surrounding hearth layer. This is evidence that the pit had been filled in after the body was placed inside it.

Kebara Cave, where Moshe was found

8 MYA | HOMININS APPEAR

2.6 MYA | STONE TECHNOLOGY IS DEVELOPED

2.5 MYA | GENUS HOMO APPEARS

300,000 YA | FIRST WEAPONS WITH HANDLES

▼ Spreading around the world
Archaeologists use the distribution of hominin skeletons and artifacts such as tools to reconstruct routes of dispersal. The routes and timings are constantly being refined as more evidence comes to light.

Red Deer Cave people Fossils found at Maludong Cave in southwest China are remarkable because they seem to be from a species of human that is found nowhere else and yet they are relatively recent—dating from just 14,500–11,500 years ago, long after modern humans had already reached China.

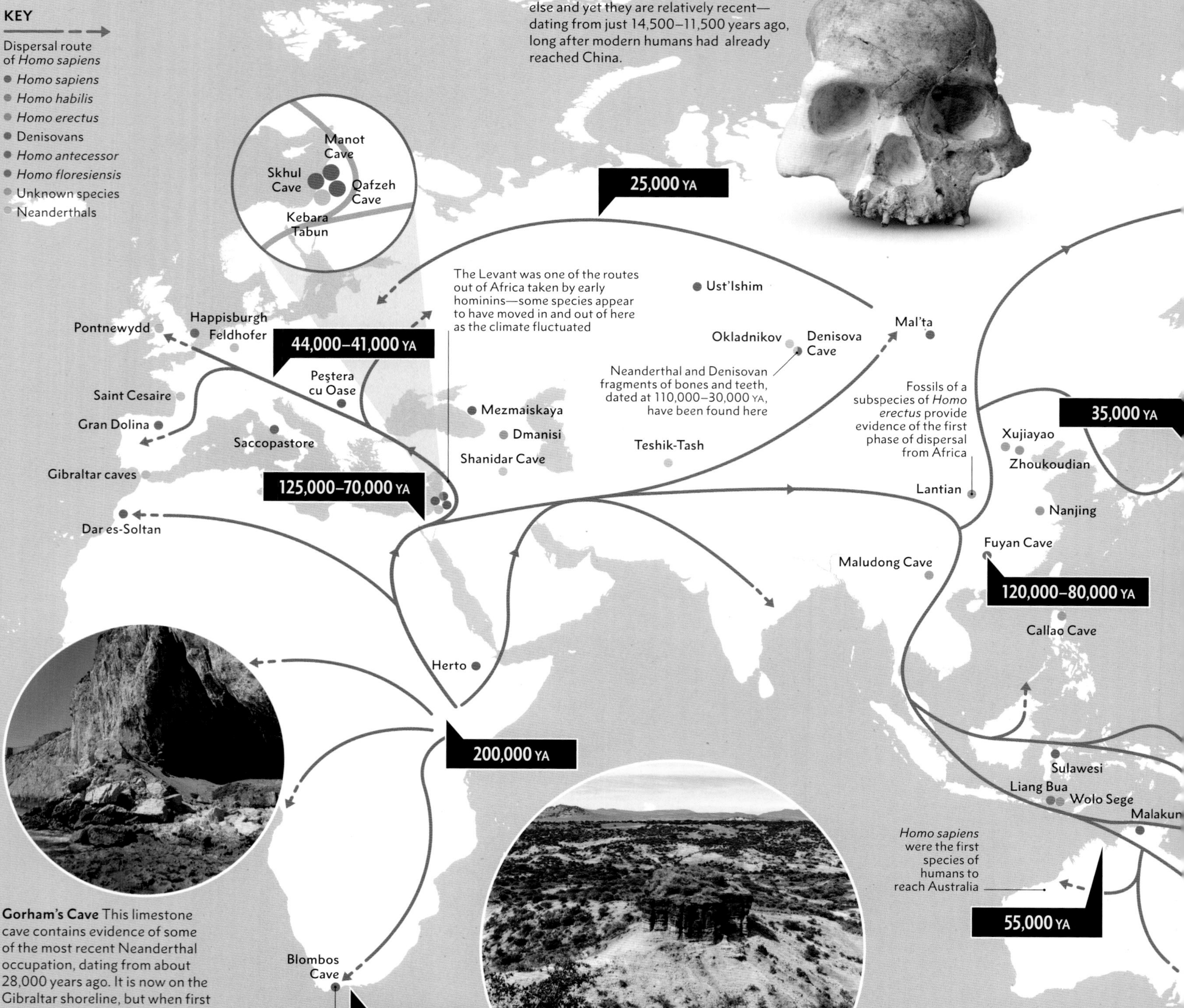

Gorham's Cave This limestone cave contains evidence of some of the most recent Neanderthal occupation, dating from about 28,000 years ago. It is now on the Gibraltar shoreline, but when first inhabited, 55,000 years ago, it was about 3 miles (5km) inland.

Olduvai Gorge This vast ravine in northern Tanzania was formed as a stream cut downward through lake deposits, volcanic ash, and lava flows. Not only do the layers contain the remains of several hominin species, they can also be accurately dated, providing a valuable record of human evolution spanning a period from about 1.75 MYA to 15,000 years ago.

200,000 YA | HOMO SAPIENS APPEARS
135,000 YA | FIRST USE OF SYMBOLS
110,000 YA | LAST ICE AGE BEGINS
41,000 YA | EARLIEST PAINTED CAVE ART
12,000 YA | LAST ICE AGE ENDS

AT ABOUT **3FT 3IN (1M) TALL**, THE HOMININS DISCOVERED AT LIANG BUA CAVE IN **FLORES, INDONESIA**, ARE THE SMALLEST EVER FOUND

EARLY HUMANS **DISPERSE**

The first hominins were found only in Africa. Helped by the ability to adapt to new environments, the various species of the genus *Homo* dispersed around the world and inhabited almost all parts of Earth's land surface.

Early humans probably dispersed from their African savanna habitat in at least two phases. The first of these may have begun about 2 million years ago, resulting in fossil finds of a species similar to *Homo habilis* at Dmanisi, Georgia, dated at 1.8 million years old. The same dispersal may also account for fossil finds in China and Indonesia dated at 1.6–1.1 million years old, although these are more similar to *Homo erectus*. A later phase of dispersal followed. This led to the occurrence in Europe of *Homo antecessor* in Spain and Britain at least 900,000 years ago.

These two phases of dispersal placed hominin species in Africa, Asia, and Europe. The populations diversified and new hominin species developed. For example, between 500,000 and 400,000 years ago, Neanderthals originated in Europe and, simultaneously, other species, such as the Denisovans, were emerging in Asia.

At some time between 150,000 and 120,000 years ago, groups of modern humans (*Homo sapiens*) left Africa, moving first into Asia and later into Europe. The demanding sea crossings to New Guinea and Australia were made by 55,000 years ago, although colonization of North, South, and Central America had to wait for the traversal of the Bering Strait after the peak of the last Ice Age, about 18,000 years ago.

Compared with earlier hominins, modern humans dispersed relatively quickly. Adapting to new environments required them to exploit new sources of food, adjust to colder, more seasonal climates, and withstand climate change. Crucial to their survival were the abilities to invent new technologies, learn new skills, and exchange resources and information.

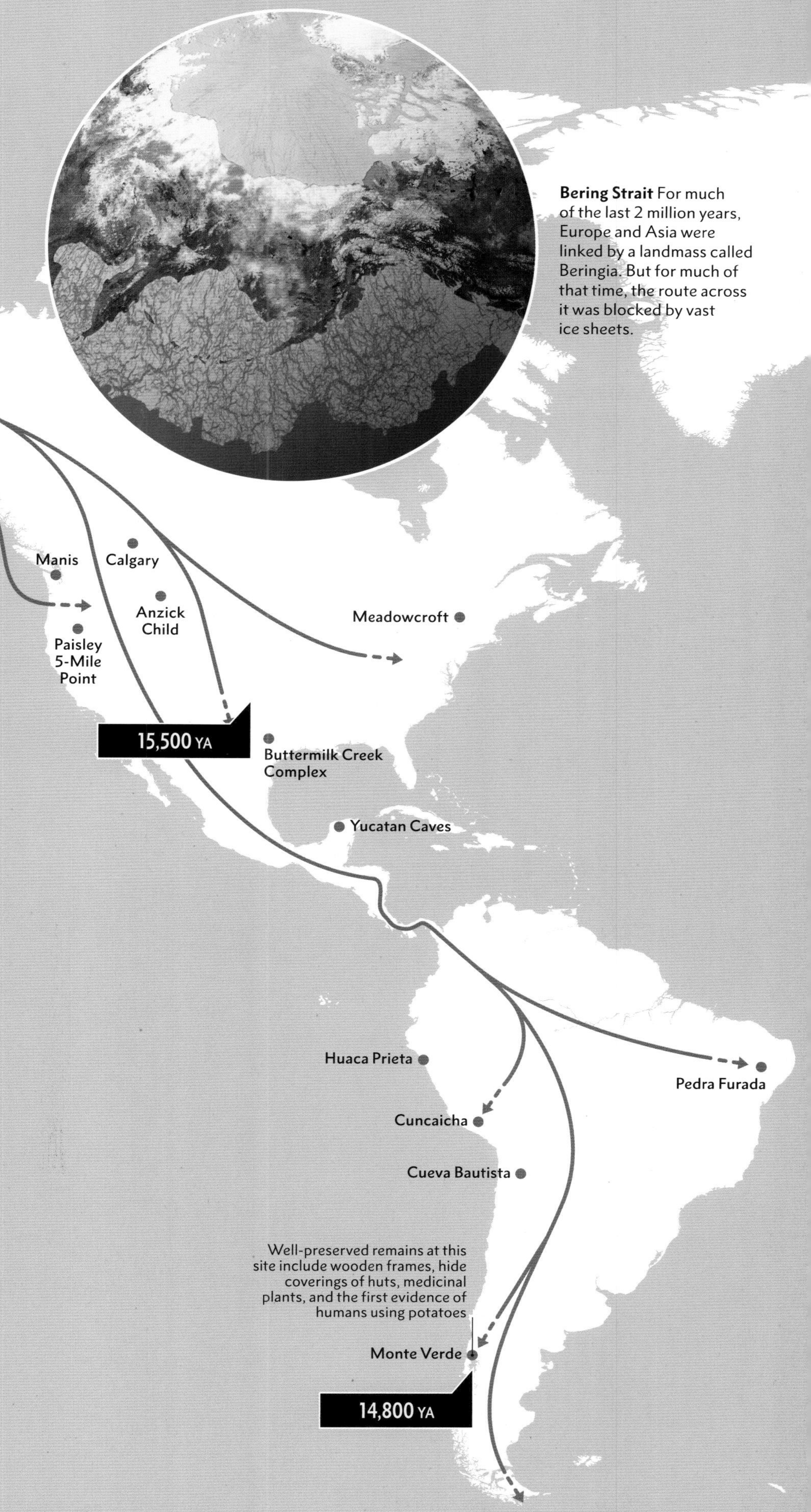

Bering Strait For much of the last 2 million years, Europe and Asia were linked by a landmass called Beringia. But for much of that time, the route across it was blocked by vast ice sheets.

ANCIENT **DNA**

Over the past decade, advances in analyzing ancient DNA—the genetic material found in cells—have revolutionized our understanding of human evolution and led to some surprising discoveries.

DNA (deoxyribonucleic acid) is a very long molecule made up of small individual units. DNA is found in the cells of all living things. The order of the small units is like a set of coded instructions, genes, that determine the characteristics of an individual.

The oldest DNA so far obtained is from 400,000-year-old Neanderthals at Sima de Los Huesos, Spain, and suggests *Homo sapiens* split from other ancient hominins between 760,000 and 550,000 years ago. This and other samples show that Eurasia was always a melting pot, and that globally there was more interaction and breeding between ancient groups and with *Homo sapiens* than we previously suspected based on evidence from fossils and archaeology.

One 40,000-year-old human from Oase, Romania, may be as few as four generations removed from a Neanderthal ancestor. Other branches of our family were genetic "dead-ends": an individual from Ust'-Ishim, Siberia, dated to 45,000 years ago, had Neanderthal ancestry but did not contribute genetically to later *Homo sapiens* populations. Similarly, there were at least four large population replacements in Europe between the earliest *Homo sapiens* colonizers and modern times.

We have only just begun to decipher the details of this ancient DNA and understand how genetic differences between species impacted on their—and our—success. As techniques advance and early DNA is decoded, especially from African and Asian remains, we can expect to unlock more secrets about our origins, migrations, and unique genetic adaptations, and also uncover further links between different branches of the hominin tree.

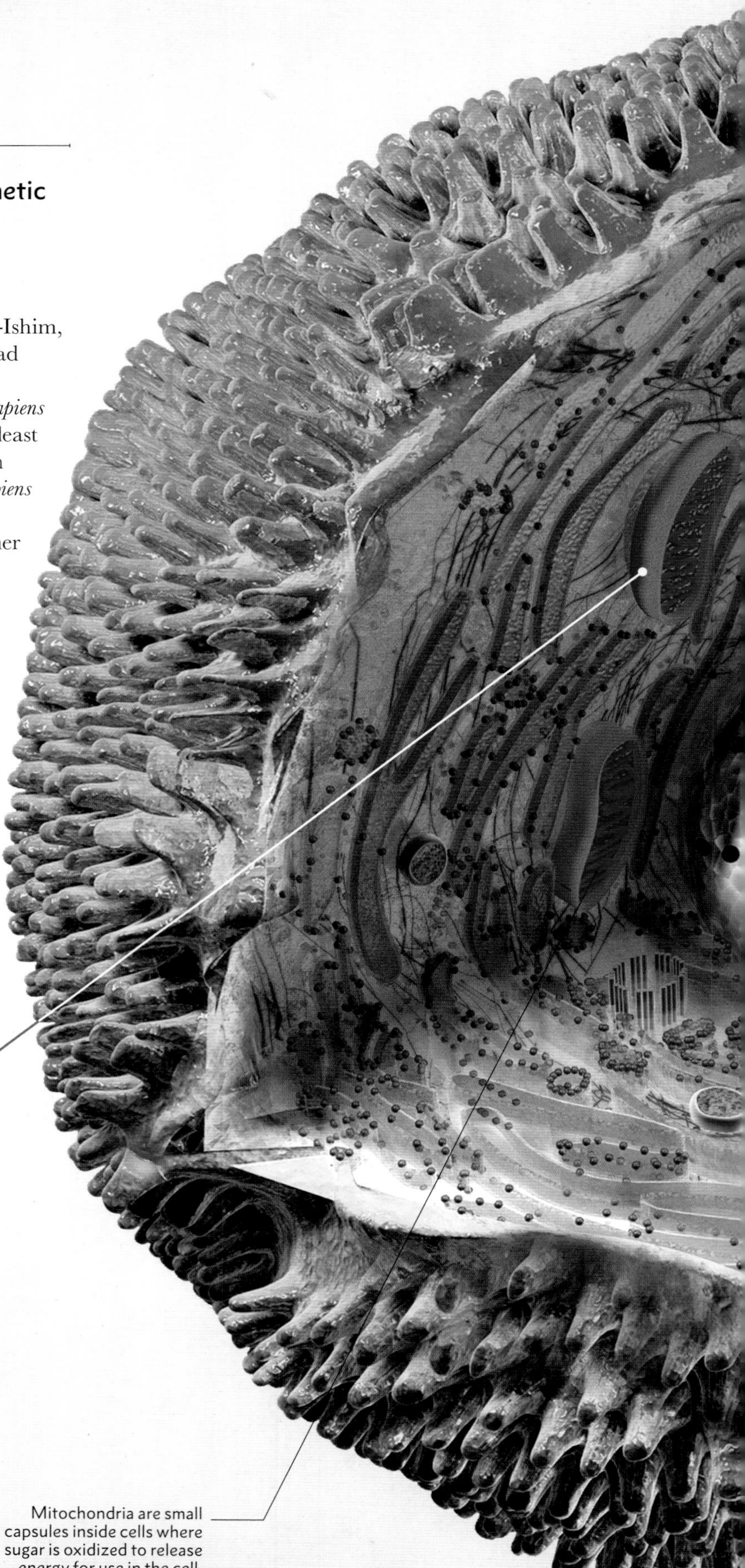

Mitochondria are small capsules inside cells where sugar is oxidized to release energy for use in the cell. Each one has its own DNA containing 37 genes that allow it to function

Mitochondrial DNA

We inherit mitochondrial DNA (mtDNA) from our mothers. This type of DNA is found not in the cell nucleus but in other cell structures called mitochondria. Since mtDNA only traces the maternal lineage, studying samples from many thousands of people has enabled scientists to construct a genetic "family tree" that indicates a common female ancestor for everyone alive today. This "Mitochondrial Eve" had many contemporaries, but they did not contribute to our mtDNA. She lived between 200,000 and 100,000 years ago, and was probably African or one of the earliest *Homo sapiens* to colonize Eurasia.

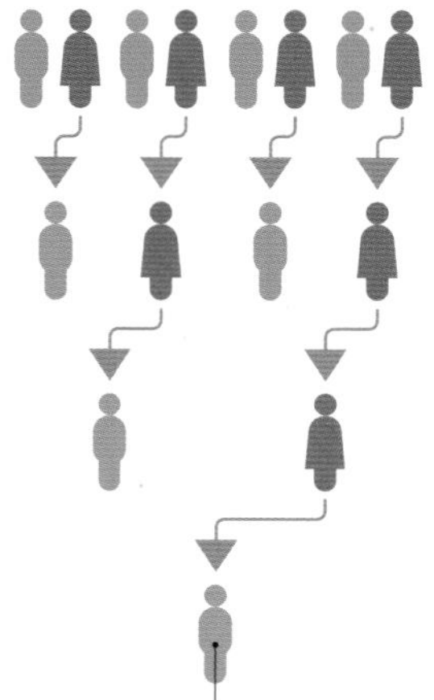
Maternal lineage of mtDNA
Only one female mtDNA lineage is present in people today

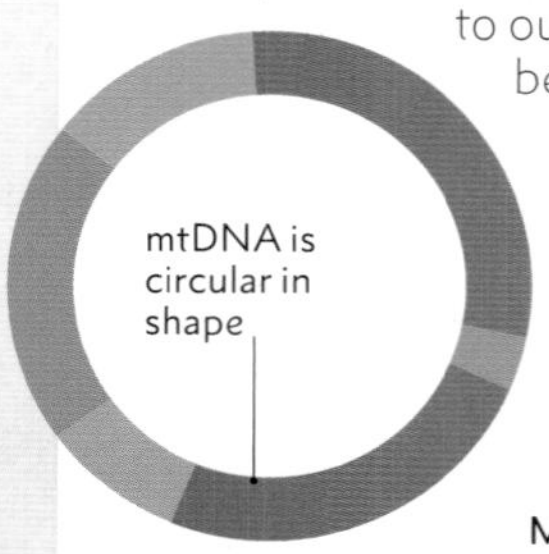
mtDNA is circular in shape
MTDNA

Nuclear DNA

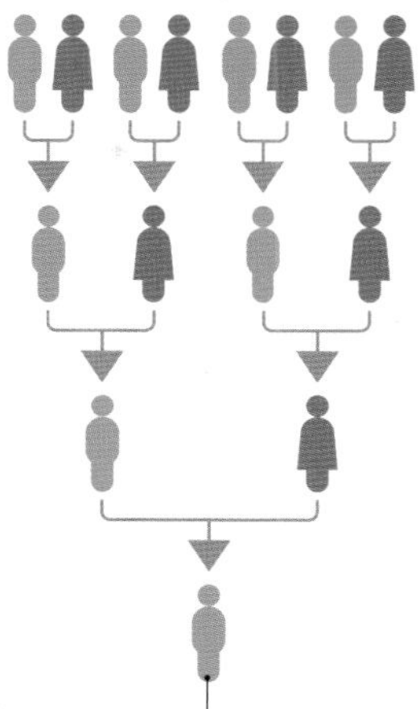

Most DNA is located within the cell nucleus. Both parents pass on nuclear DNA to their offspring, so this type of DNA reveals much more about the relatedness of species, genetic differences, and adaptive traits. Recent studies have shown that during early dispersals from Africa, *Homo sapiens* populations interbred with hominins already living in Eurasia, but at different times and places, leaving living humans with varying amounts of their DNA. Genes from Neanderthals and other hominins probably aided our survival and eventual global success, such as by improving our immunity and metabolism.

NUCLEAR DNA

Extracting DNA

Archaeologists extract DNA from teeth, bones, and mummified tissues. Mitochondrial DNA is easiest to recover intact: there are up to 1,000 mitochondria in every cell, each with 5–10 copies of the short mtDNA strands. The much longer strands of DNA in a cell's single nucleus are more likely to degrade over time and with changing soil temperature. Often the best chance of recovering nuclear DNA is from dental cementum—the mineralized outer layer of the tooth root. This is because the hard mineral matrix helps to preserve any cellular material trapped within it.

Discovering the Denisovans

In 2010, DNA analysis of a 50,000-year-old fragment of a girl's finger bone from Denisova Cave, Siberia, revealed a mystery hominin population. The "Denisovans" had brown eyes, hair, and skin, and showed an ancient relationship to Neanderthals, whom they lived alongside in Eurasia. By 2015, two further individuals had been identified at the site, one of whom lived 60,000 years earlier. Studies show that non-Africans alive today have varying proportions of Denisovan DNA, up to 4 percent in the island populations of Melanesia. This indicates that only some early *Homo sapiens* migrants interbred with the Denisovans, but where this occurred remains uncertain.

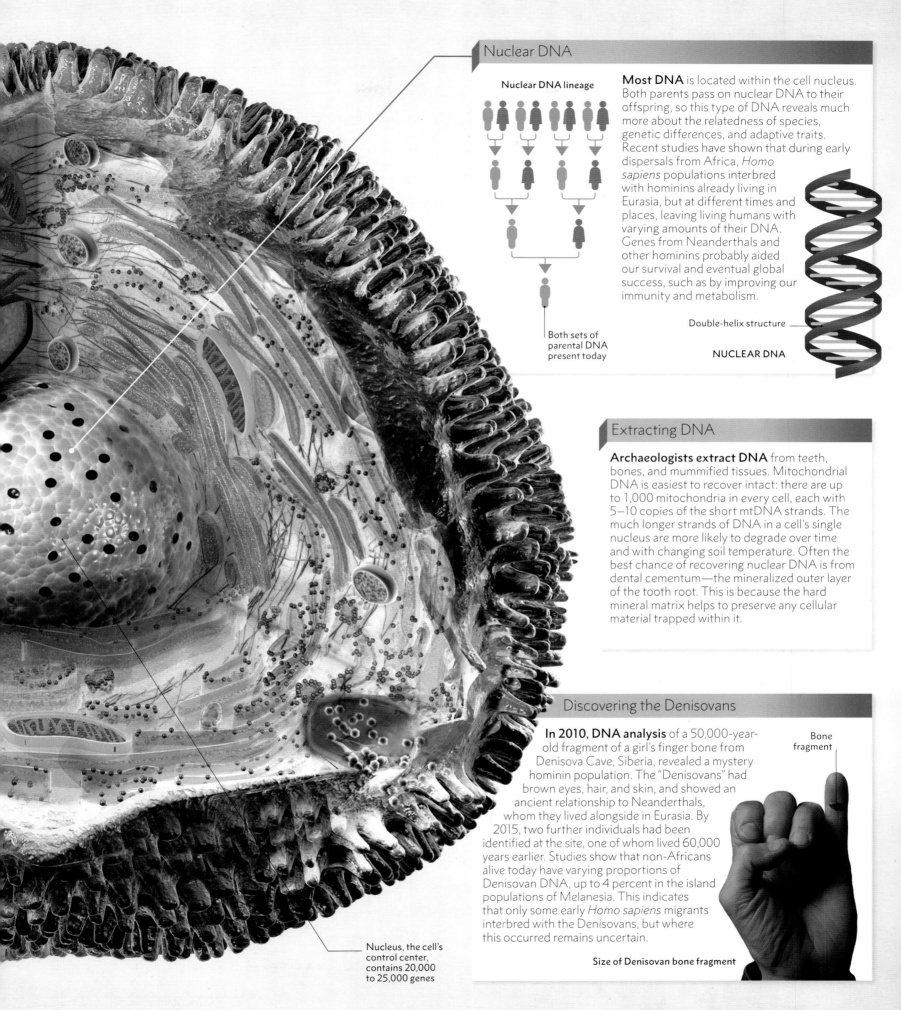

Size of Denisovan bone fragment

► Herto skull
This skull from Herto, Ethiopia, shows slight differences to others from early *Homo sapiens*. Some anthropologists suggest it represents a subspecies, *Homo sapiens idaltu*.

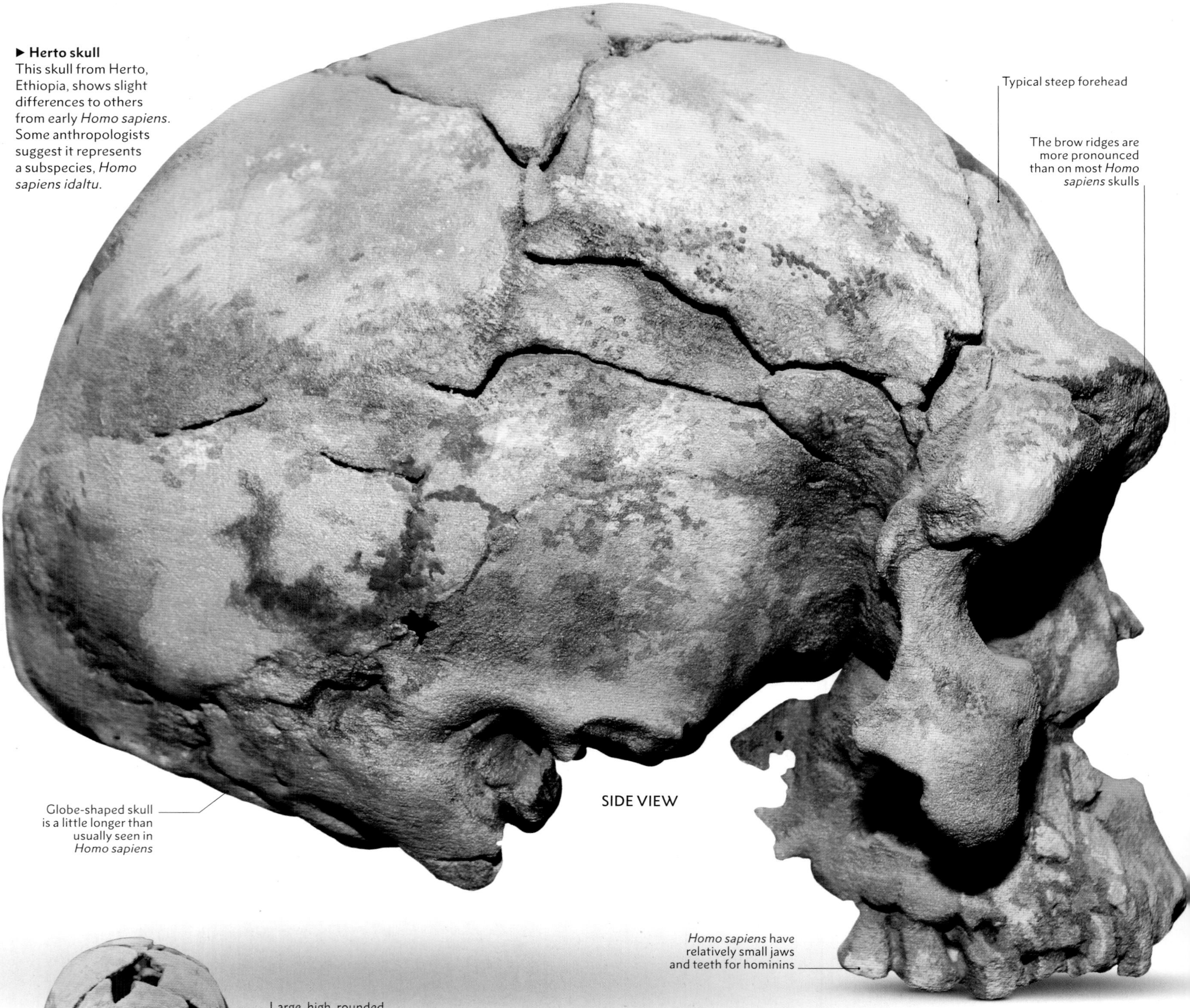

Typical steep forehead
The brow ridges are more pronounced than on most *Homo sapiens* skulls
Globe-shaped skull is a little longer than usually seen in *Homo sapiens*
Homo sapiens have relatively small jaws and teeth for hominins

SIDE VIEW

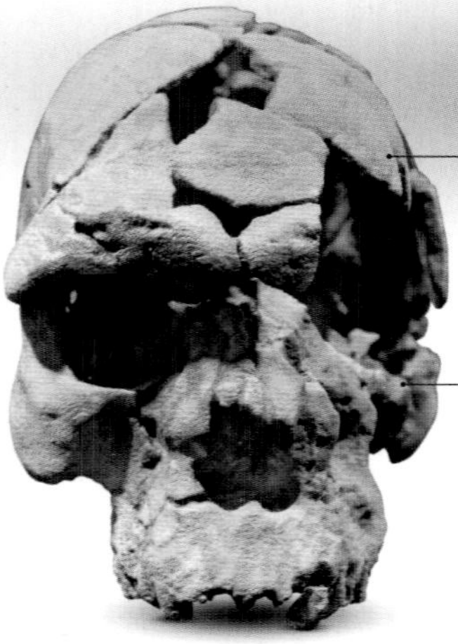

Large, high, rounded skull to accommodate increased brain capacity
Short, flat face with narrow cheekbones

FRONT VIEW

> WE CAN SEE THE FOCUS, THE **CENTER OF EVOLUTION,** FOR MODERN **HUMANS IN AFRICA** APPARENTLY **MOVING AROUND** FROM ONE PLACE TO ANOTHER, **DRIVEN BY CLIMATE CHANGES**.

Chris Stringer, anthropologist, 1947–

200,000 YA | HOMO SAPIENS APPEARS
135,000 YA | FIRST USE OF SYMBOLS
110,000 YA | LAST ICE AGE BEGINS
41,000 YA | EARLIEST PAINTED CAVE ART

THE FIRST **HOMO SAPIENS**

Of all the hominin species, and all the variants of genus *Homo*, only *Homo sapiens* remains today, having survived the challenges of the last ice age. It did so thanks to its unique anatomy, which came together in Africa nearly 200,000 years ago.

The distinctive "package" of anatomical features that identify living people today as *Homo sapiens* developed gradually, beginning around 500,000 years ago. Some key characteristics include: globular skulls, very large brains, shorter tucked-under faces, and smaller teeth, together with a more slender, lighter skeleton, smaller arm-to-lower-limb proportions, and narrower ribs. The appearance of these modifications was complex, occurring at different times and places, and in different combinations, but brain size continued to increase everywhere.

The oldest *Homo sapiens* fossils come from Omo Kibish, Ethiopia. Dated to around 195,000 years ago, the fragmented skulls and skeletons of two individuals show a modern morphology (form and structure), but one has less modern features than the other. Other early modern fossils have been discovered at Herto in Ethiopia, Singa in Sudan, Laetoli in Tanzania, Jebel Irhoud in Morocco, and Border Cave and Klasies River Mouth in South Africa. All of these are between 200,000 and 100,000 years old and display modern characteristics, albeit with variation in morphology.

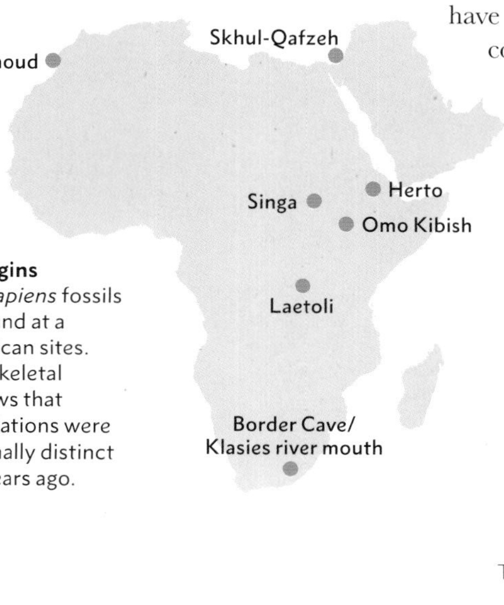

▶ **African origins**
Early *Homo sapiens* fossils have been found at a variety of African sites. Genetic and skeletal evidence shows that African populations were already regionally distinct by 120,000 years ago.

BEGINNING THE LONG WALK

By 120,000 to 80,000 years ago, early *Homo sapiens* had moved into the Middle East and western Asia. The remains of over 20 individuals recovered from the caves at Skhul and Qafzeh in Israel still show some morphological differences. However, 47 distinctly modern teeth, with flat crowns and thin roots, have been found thousands of miles farther east in Fuyan Cave, Daoxian, China. It is clear from this that we are missing fossils from large swathes of *Homo sapiens*' long journey into Asia, and that at least some of the stone tools found along their route in this period, such as in India, were made by anatomically modern people. It also makes it very likely that the oldest stone tools in Australia, dating back 55,000 years, were made by *Homo sapiens*, since they had already been in Asia a long time.

We cannot be sure what stimulated *Homo sapiens*' dispersal from Africa, leading eventually to a single global human species. It is unlikely to have been technological progress, since their stone tools were little more advanced than those made 100,000 years previously. Populations may have been increasing and climate change may have played a part, but important cognitive and social changes were also taking place at the time. Increased symbolic expression after 150,000 years ago point to innovations that probably coincided with *Homo sapiens* acquiring a brain size similar to that of people living today.

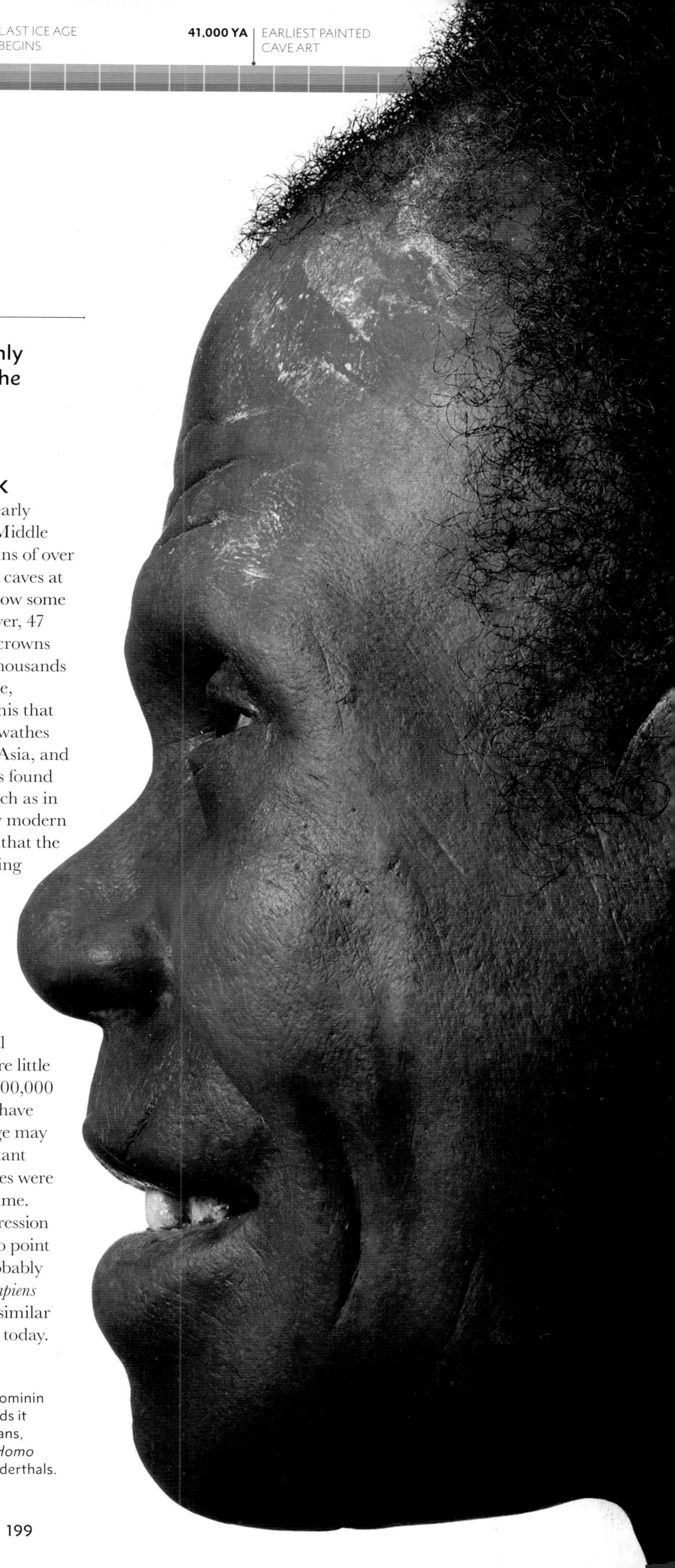

▶ **Sole survivor**
Homo sapiens is the last hominin species, but for long periods it coexisted with other humans, including *Homo erectus*, *Homo floresiensis*, and the Neanderthals.

A family affair
Unlike the young of other primates, human children spend years being cared for by parents, grandparents, and friends of the family. These prolonged childhoods provide ample time for learning the ways of the world.

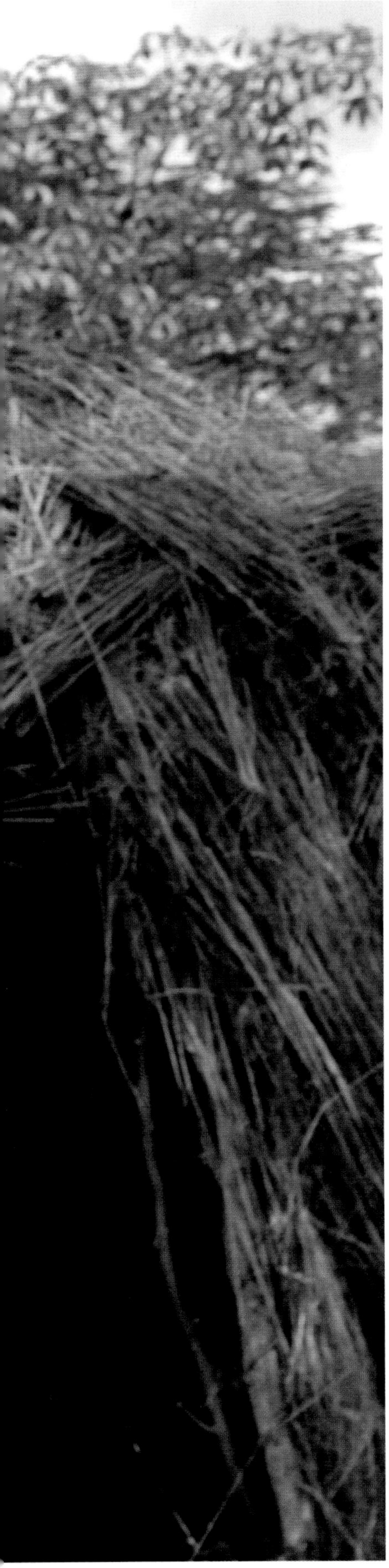

BRINGING UP **BABIES**

Changes in the human reproductive cycle played an important part in the success of *Homo sapiens*. Increasing brain size probably made childbirth harder, but it also enabled us to evolve the very culture we need to rear our relatively undeveloped young.

The labors of *Homo sapiens* are long, painful, and risky. Our infants are large, have big heads, are mostly helpless, and are born with only 30 percent of their adult brain size. Pregnancies would need to be 16 months long to attain the same development as newborn chimpanzees. Our childhood development is also extended, demanding high levels of care, not just by parents, but by other family members and friends.

To explain these complications, it is often said that greater brain size (see pp.188–89) coupled with bipedalism—which gave us narrower pelvises—created a biological trade-off. Potentially fatal births were avoided by limiting the length of pregnancies, forcing babies to be born early. It is certainly possible that by about 500,000 years ago, hominins were already experiencing tricky births, and that women may have had some level of assistance, or at least company, during labor. Other social primates, such as bonobos, exhibit similar behavior. However, it is also true that nonbipedal primates have a tight fit in the birth canal, that capuchins and chimpanzee babies have relatively undeveloped brains, and that human gestation is actually longer than expected given our body size. It may be that the upper limit on pregnancy length is actually metabolic—the point at which mothers can no longer biologically support a growing baby.

COOPERATIVE BREEDING

Anatomical changes also affected how we bring up our young. As australopithecine feet lost the "big toe" associated with tree climbing, infants were less able to cling to their mothers, and required greater care. It is possible that the exploitation of animal skins may have been driven more by the need to make baby slings and wraps than a need for warm clothing.

Although the length of time spent breastfeeding was probably comparable to that of other apes—lasting several years, as it does today—the greater demands of a hominin infant may have promoted the evolution of cooperative breeding, by which several adults bring up a child. The role of nonrelated adults and older generations in caring for children probably became important too, creating a rich environment in which experienced individuals could be observed finding food and making tools—vital skills that were then passed on to the next generation.

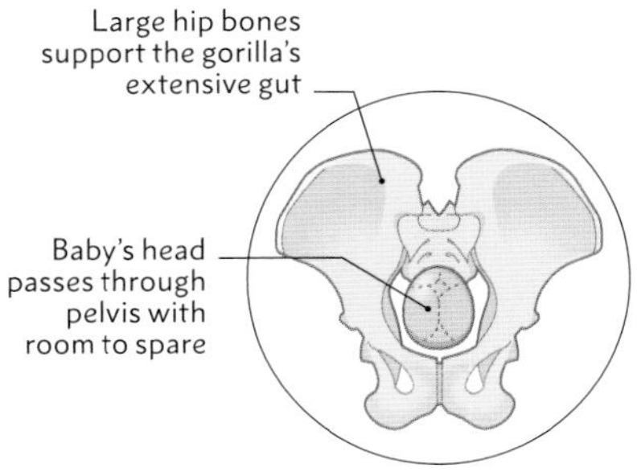
Large hip bones support the gorilla's extensive gut
Baby's head passes through pelvis with room to spare

▲ Gorilla birth
Due to its small brain, a baby gorilla's head passes through its mother's birth canal with room to spare, making labor shorter and less risky.

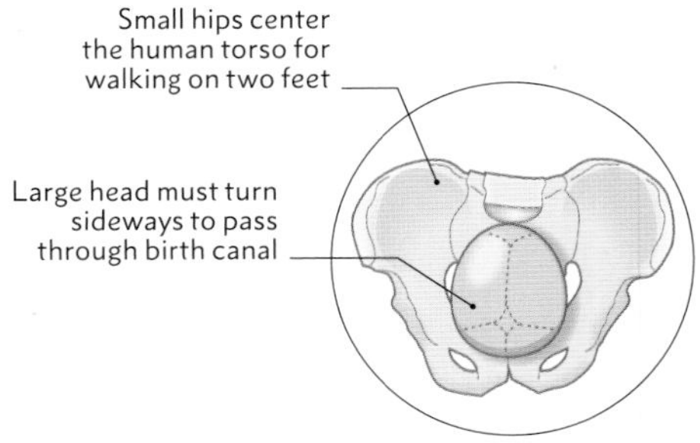
Small hips center the human torso for walking on two feet
Large head must turn sideways to pass through birth canal

▲ Human birth
The head of a human baby must rotate to descend through its mother's birth canal, making childbirth longer and more painful.

HOW **LANGUAGE EVOLVED**

Many animals call to each other with sounds that stand for "Danger!," "Food!," or "Here!," but only humans can think conceptually—can talk, for example, about the nature of food or danger. For this, language had to evolve, and with it came storytelling, information-sharing, and our first attempts at understanding the world.

In evolutionary terms, the ability to speak emerged as a result of the hominin larynx lowering in the throat, enabling our ancestors to produce more diverse sounds than those of any other primate. The biological price of this was high, as an elevated larynx had enabled us to breathe and swallow simultaneously; now, we ran the risk of choking when we ate. At the same time, the hyoid bone, which connects the larynx to the root of the tongue, also changed position in a way that helped facilitate vocalization. Judging from the fossil record, this happened between 700,000 and 600,000 years ago, as Neanderthals and probably our common ancestor both had a "modern" hyoid bone. Our exceptional breath control, essential when speaking, also seems to date from this time.

Casts of fossil skulls show that Neanderthals had structures in the brain that were equivalent to our own "Broca's area." This area is vital to speaking and understanding language, and to perceiving meaningful gestures. Indeed, gestures may have been key: studies show that chimpanzees repeatedly use hand signs when vocalizing, indicating that early language may not have been purely vocal. However, the functions performed by different parts of the brain can change over time, so even if other hominins had brain structures similar to ours, they may not have been used for language.

TODAY THERE ARE NEARLY **7,000 LANGUAGES**, BUT EACH USES **ONLY A SMALL NUMBER** OF THE SOUNDS THAT A **HUMAN BEING** CAN MAKE

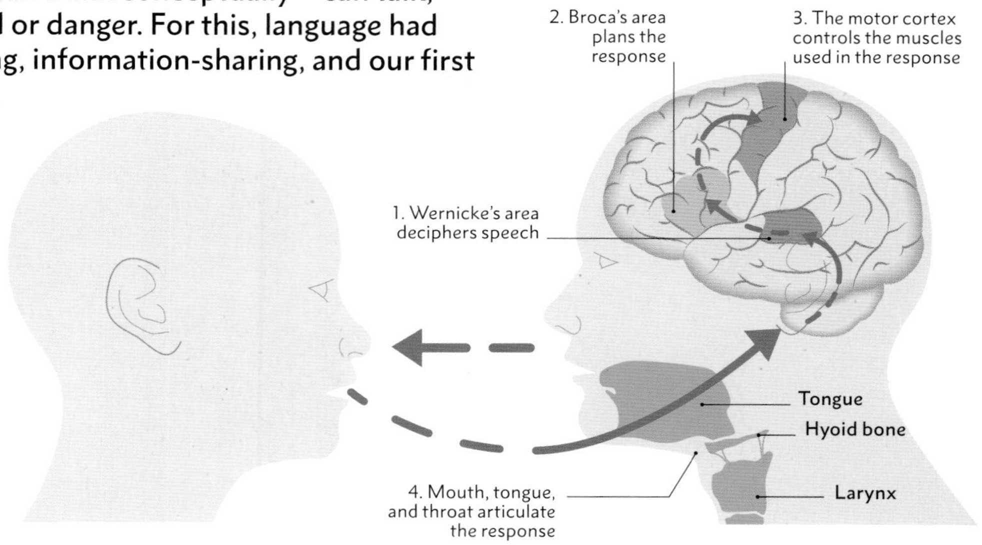
2. Broca's area plans the response
3. The motor cortex controls the muscles used in the response
1. Wernicke's area deciphers speech
Tongue
Hyoid bone
Larynx
4. Mouth, tongue, and throat articulate the response

▲ **How humans process speech**
The emergence of speech required the evolution of several key structures in the throat and brain. This included the hyoid bone, which is vital in producing varied vocal sounds.

SYMBOLS AS EVIDENCE

The artifacts left by our ancestors are better forms of evidence. Among the most striking are those created by early *Homo sapiens* in South Africa between 100,000 and 50,000 years ago. At Blombos Cave, for example, red ocher blocks were shaped and carefully covered with delicate crosshatch designs (see p.207). Even more impressive are the ostrich eggshells found at Diepkloof Cave, also in South Africa (see p.208). These were engraved with complex geometric patterns that show changes over time, hinting at shifts in meaning. Very much older than these, however, is a seashell from Trinil, Indonesia, which bears the incised zigzag markings of a *Homo erectus* who lived some 540,000–430,000 years ago (see p.206). It reveals that the common ancestor of several hominins used graphic symbols, and so had probably developed language—a fact that is supported by anatomical evidence.

Another type of symbolic evidence comes from personal ornaments, which often communicate social meanings—for instance,

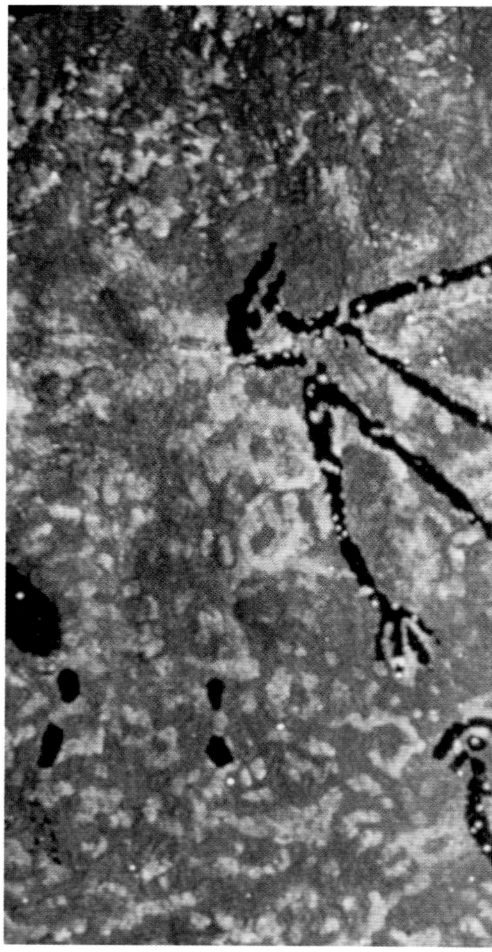

about personal status or group affiliation—which can only be established through language. For example, the first use of shell beads occurs at the same time as engravings become more common; beads from Skhul Cave in Israel date from 135,000–100,000 years ago, while those from Grotte des Pigeons in Morocco date from 80,000 years ago. At Blombos Cave, too, groups of beads were excavated from layers dating from around 80,000 years ago, many showing areas of polish that suggest they were strung together, possibly as necklaces. The markings also show that the arrangement of the beads changed over time, suggesting not only that they were symbolic, but that their meanings evolved, like those of the Diepkloof eggshells.

FROM SYMBOLS TO STORIES

Taken together, the evidence shows that *Homo sapiens* had evolved symbolic culture and language by 70,000 years ago—and that Neanderthals did this independently. However, the evidence for language being used in narrative, storytelling senses comes much later, after 45,000 years ago. For example, the famous Lion-man ivory statue from Hohlenstein-Stadel, Germany (see p.208) was carved around 40,000 years ago. It merges a lion's head with a human body, indicating both an imaginative leap by the artist, and a narrative to give it meaning.

The most striking examples of Paleolithic narrative come from later European cave art. One scene painted at Lascaux, France, around 17,000 years ago, features a wounded bison charging a male figure who lies above some fallen spears and a line topped by a bird. There are many interpretations of the scene, but all of them agree that the man, the bison, and the bird only make sense in a storytelling context. This and other examples indicate that rich oral traditions, full of meaning and symbolism, were part of Paleolithic life, and likely had been for many thousands of years. They were our first attempts at fathoming the world around us—of giving it a narrative shape.

◀ **Almost talking**
Campbell's monkeys from Ivory Coast seem to be on the verge of speaking. They have a "proto-syntax" composed of alert calls, which they use to communicate detailed information—such as what type of predator is coming and how it was detected.

> "A COMPLEX **TRAIN OF THOUGHT** CAN BE NO MORE **CARRIED OUT** WITHOUT **WORDS**... THAN A **CALCULATION** WITHOUT THE USE OF **FIGURES**."
>
> **Charles Darwin,** *The Descent of Man*, 1871

◀ **The birdman of Lascaux**
Dating from around 17,000 years ago, this strange image of a man—apparently dressed as a bird—being charged by a bison is probably evidence of story-telling. It may also show a shamanic experience.

COLLECTIVE **LEARNING**

The emergence of language set *Homo sapiens* apart from other species, for with language came the ability to share and store information across generations. This ensured that new generations could know more than the last, and so be more effective in the world.

The practice of storing and sharing information is called "collective learning." At its simplest, this means that we only need to invent the wheel once, for that knowledge can then be stored and shared publicly. The alternative is to imagine us as a group of networked computers. Without the network—without connectivity—how could human history unfold?

SURVIVING COOPERATIVELY

Humans appear to be predisposed to work together to a far greater degree than other animals. The roots of this tendency can be seen in primates, the majority of which live in social groups, with strong kin relationships and friendships. However, humans live in unusually diverse societies, and our high level of cooperation is a unifying characteristic. Hunter-gatherer groups, for example, typically number between 25 and 50 individuals, but they are usually part of extended social networks, consisting of blood relations and other types of kinship. Within and between these groups, food, labor, and childcare are shared—as is vital information about water, predators, and the availability of food.

The evolution of this ability to cooperate can be seen in the archaeological record. Stone tools began to be transported increasingly long distances around 200,000 years ago, pointing to expanding social networks. By then, multipart tools, such as spears, were being made, probably collaboratively. More spectacular examples of these, such as atlatls and bows, came later, and after 40,000 years ago many of these were lavishly ornamented. The Mas d'Azil atlatl, for example, is one of five almost identical objects found at different sites in the Pyrenees. Each is carved into the shape of an ibex, demonstrating a common artistic tradition, and probably some level of apprenticeship. Moreover, an atlatl, like a bow, is a "tool for using a tool"—in this case a tool for propelling spears—which is of a whole new order of complexity. It shows that by 17,000 years ago we were adapting ourselves ever more cleverly to our environment—alone of all creatures through cultural rather than genetic change. Thanks to collective learning, human history could begin.

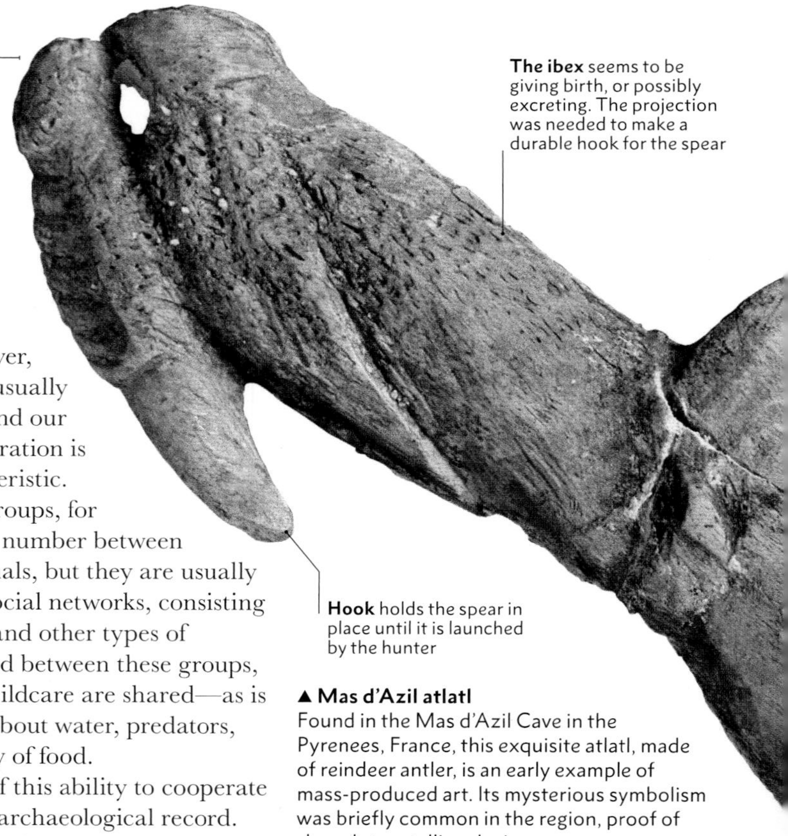
The ibex seems to be giving birth, or possibly excreting. The projection was needed to make a durable hook for the spear
Hook holds the spear in place until it is launched by the hunter

▲ **Mas d'Azil atlatl**
Found in the Mas d'Azil Cave in the Pyrenees, France, this exquisite atlatl, made of reindeer antler, is an early example of mass-produced art. Its mysterious symbolism was briefly common in the region, proof of shared storytelling devices.

▶ **Sharing information**
Today, the San people of the Kalahari make fire using knowledge passed down for tens of thousands of years by their forebears.

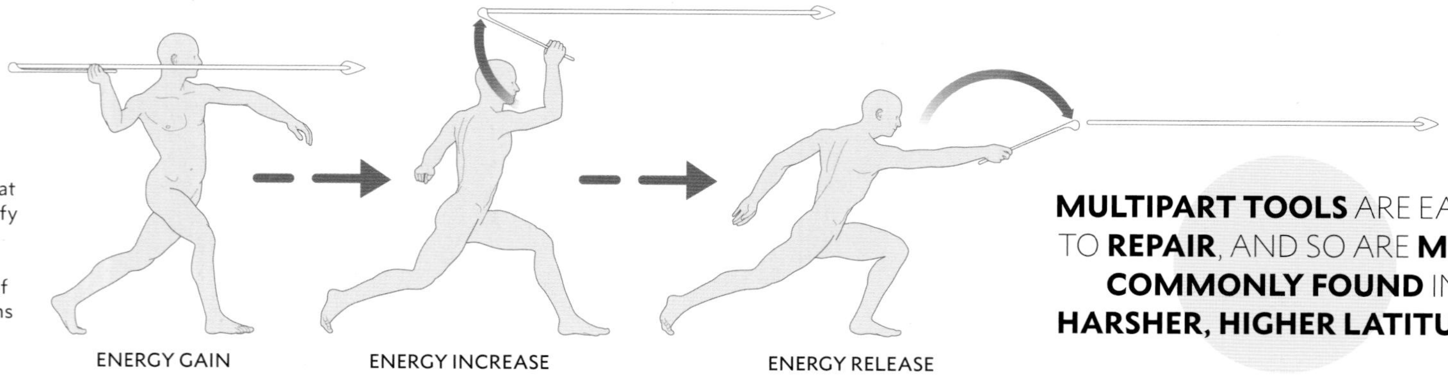
ENERGY GAIN
ENERGY INCREASE
ENERGY RELEASE

▶ **Throwing power**
An atlatl, or spear-thrower, is a device that uses leverage to amplify throwing power. The spear is kept in place by a hook at the rear of the atlatl, and this gains energy as the hunter throws the spear.

MULTIPART TOOLS ARE EASIER TO **REPAIR**, AND SO ARE **MORE COMMONLY FOUND** IN **HARSHER, HIGHER LATITUDES**

> **A GROUP** CAN POOL THE **HARD-WON DISCOVERIES** OF **MEMBERS**, PRESENT AND PAST, AND END UP **FAR SMARTER** THAN A **RACE OF HERMITS**
>
> **Steven Pinker**, cognitive scientist, 1954 –

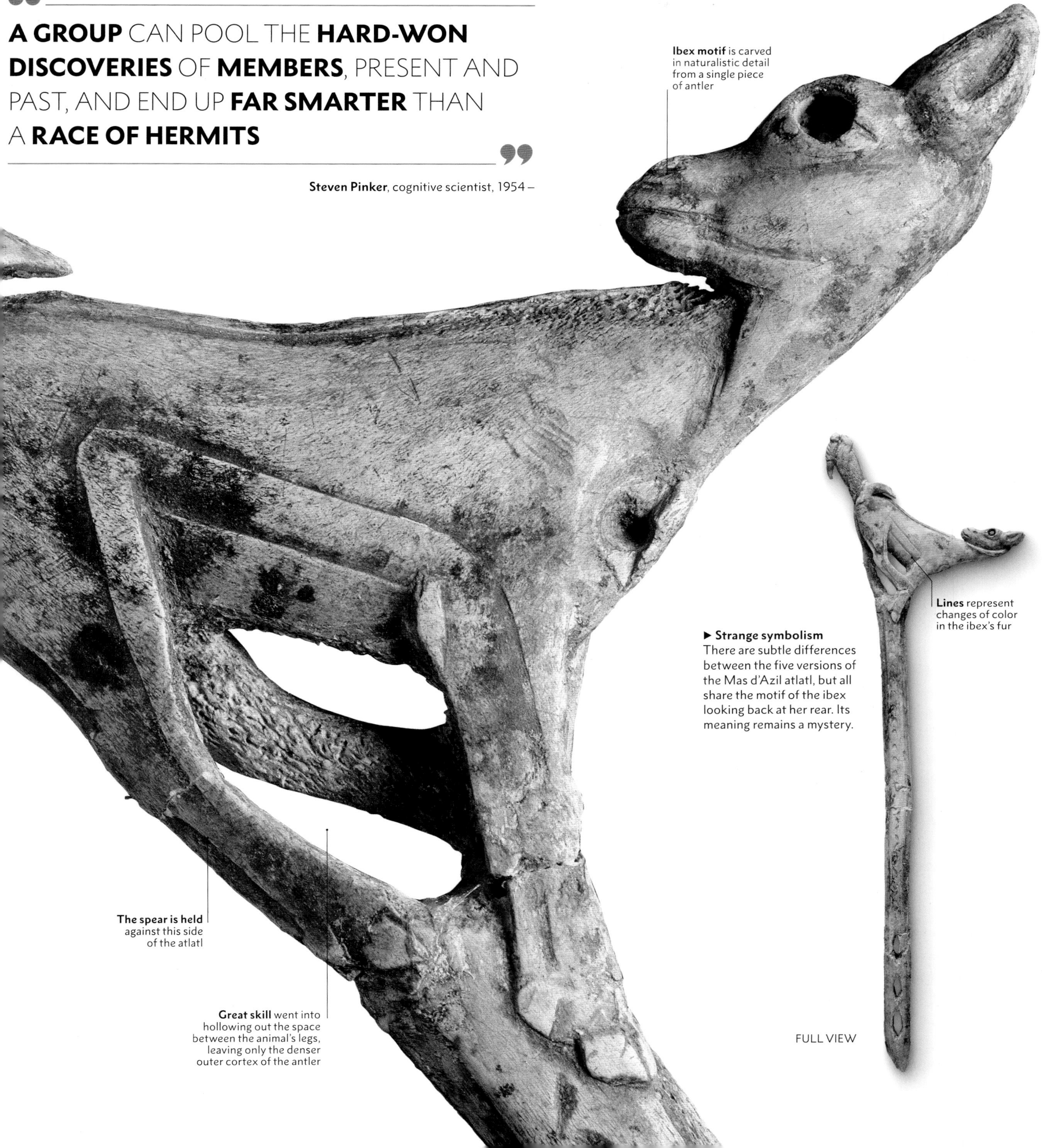
Ibex motif is carved in naturalistic detail from a single piece of antler
The spear is held against this side of the atlatl
Great skill went into hollowing out the space between the animal's legs, leaving only the denser outer cortex of the antler
Lines represent changes of color in the ibex's fur

FULL VIEW

► **Strange symbolism**
There are subtle differences between the five versions of the Mas d'Azil atlatl, but all share the motif of the ibex looking back at her rear. Its meaning remains a mystery.

TIMELINES

THE BIRTH OF **CREATIVITY**

With the evolution of genus *Homo* an entirely new form of complexity appeared—the first to be shaped not by natural selection, but by intelligence and design—namely culture.

This uniquely creative ability can be traced through the archaeological record, which itself has been through a revolution recently. Many "first appearances" have shifted back in time, and the popular idea that all major innovations were made over the last 50,000 years—and only by *Homo sapiens*—can no longer be supported. The first hints of symbolic communication can be seen in *Homo erectus*, and the oldest synthetic material, birch bark pitch, was made by Neanderthals.

Overall, however, there is a clear trend toward increasing complexity over time, albeit punctuated by long periods in which creativity seems to pause. Innovations appear, flourish, and then disappear again for tens of thousands of years, suggesting that the complexity of a culture is determined as much by social conditions as by innate cognitive capacity.

AUSTRALOPITHECINES

3 MYA

H. HABILIS

Stone tool cut marks are left on animal bones in Dikika, Ethiopia, c.3.4 MYA, possibly made by *Australopithecus africanus*. If genuine, they are the oldest evidence of meat-eating by hominins.

The earliest flaked stone artifacts, dating to 3.3 MYA, are made at Lomekwi, Kenya. Pre-dating genus *Homo*, they are the work of an australopithecine species.

Verified stone tool cut marks appear on bones at Gona and Bouri, Ethiopia, around 2.5 MYA.

More advanced stone tools are made c.2.6 MYA. This "Oldowan" technology (named after the Olduvai Gorge, Tanzania) consists of simple flakes and "choppers," both of which are used to cut plants and animal tissues.

Oldowan chopper

2 MYA

H. ERECTUS

Double-sided "bifaces" or hand axes appear in Kenya and Ethiopia around 1.75 MYA. Their *Homo erectus* makers show an expert grasp of stone fracture mechanics.

Oldowan choppers are still the most complex technology found at many sites—nearly a million years after their invention.

1 MYA

The earliest controlled use of fire, 790,000 YA, at Gesher Benot Ya'aqov, Israel.

The first extended sea crossing, between 1 MYA and 700,000 YA, as *Homo erectus* colonizes Indonesia. The crossing was made by raft or by swimming.

H. HEIDELBERGENSIS

500,000 YA

Prepared-core technology emerges in Kenya, 550,000–500,000 YA. With this technique, predictably shaped flakes or blades are struck from a prepared stone "core." Bifacial tools also remain prevalent.

Hand ax

Symbols are engraved on a sea shell at Trinil, Java, c.540,000–430,000 YA. They are made by *Homo erectus*, long before the appearance of *Homo sapiens*.

Engraved shell from Trinil, java

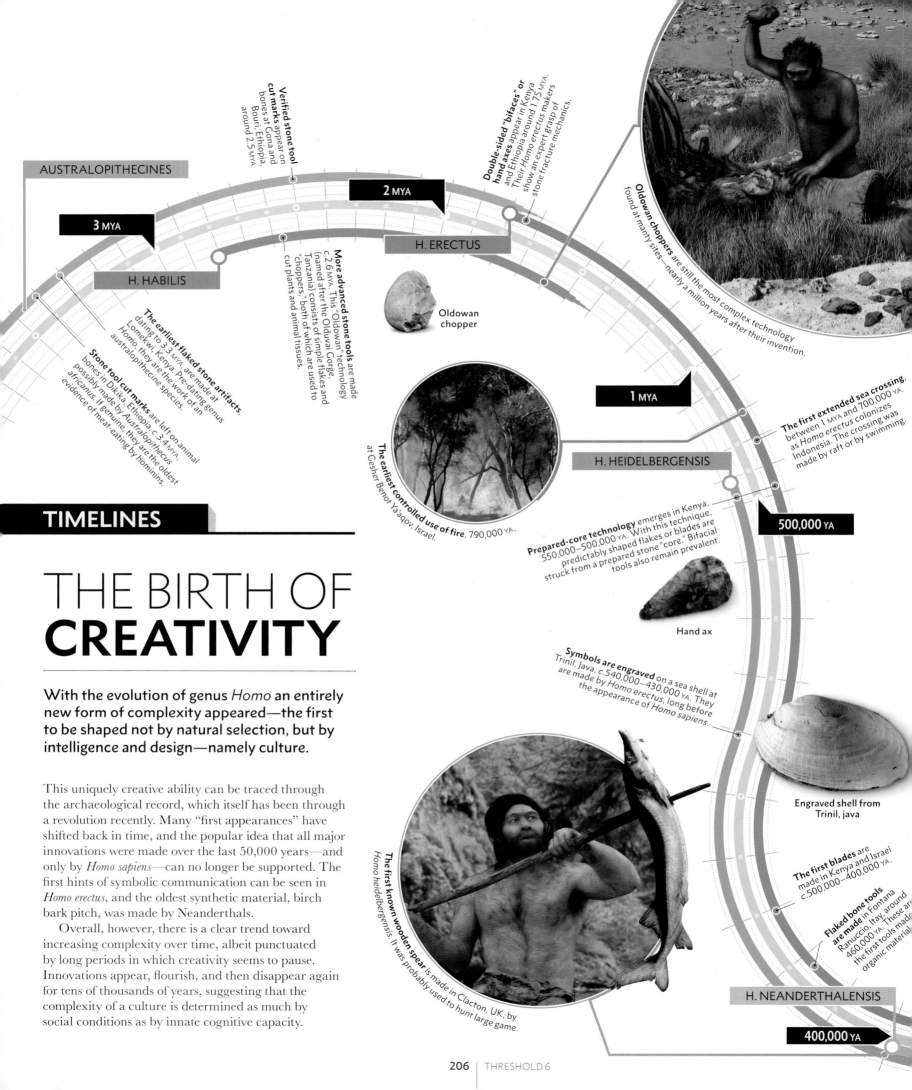

The first known wooden spear is made in Clacton, UK, by *Homo heidelbergensis*. It was probably used to hunt large game.

The first blades are made in Kenya and Israel c.500,000–400,000 YA.

Flaked bone tools are made in Fontana Ranuccio, Italy, around 460,000 YA. These ar the first tools mad organic materia

H. NEANDERTHALENSIS

400,000 YA

The first synthetic material, birch bark pitch, and the first multipart tool are produced by Neanderthals at Campitello, Italy, 300,000–250,000 YA.

300,000 YA

Pigments are extensively collected, possibly for symbolic expression, at Twin Rivers, Zambia, c.300,000 YA.

Levallois points appear in Europe 250,000–200,000 YA. They are a refinement of the prepared-core technology invented by Homo erectus.

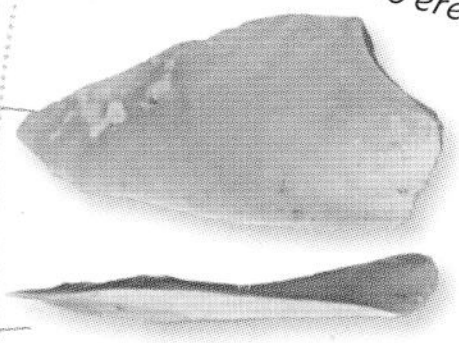

Levallois points

Two-part hafted throwing weapons are made by Neanderthals at Biache-Saint-Vaast, France, c.250,000 YA.

H. SAPIENS

200,000 YA

Clothing is invented, at leastt 170,000 YA, based on the genetic analysis of head and body lice.

Heat-treating stone tools is under way around 165,000 YA, at Pinnacle Point, South Africa. The process enables crafsmen to split stone with greater accuracy.

The first ornamental shell beads are made at Skhul Cave, Israel, 135,000–100,000 YA.

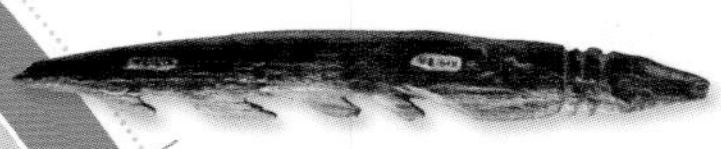

Katanda Harpoon

Bone harpoons are fashioned at Katanda, Zaire, 90,000 YA.

100,000 YA

Leather is being tanned in Neumark, Germany, 100,000 YA.

The earliest known burial takes place at Qafzeh Cave, Israel, 92,000 YA.

String is being used by Neanderthals in France and the Netherlands by 90,000 YA. It is made of twisted plant fibers.

Pigments used as paint in Blombos Cave, South Africa, 100,000 YA, suggesting a flowering of artistic talent. Sticks of ocher pigment are engraved with symbols.

▶ **Prepared-core technology**
The ability to visualize a rock as a material volume in 3D is a milestone in cognitive development. The first such "prepared core" method appeared about 800,000 YA. A refinement was the Levallois technique, in which tools were struck from hierarchically organized core surfaces using a harder stone as a hammer.

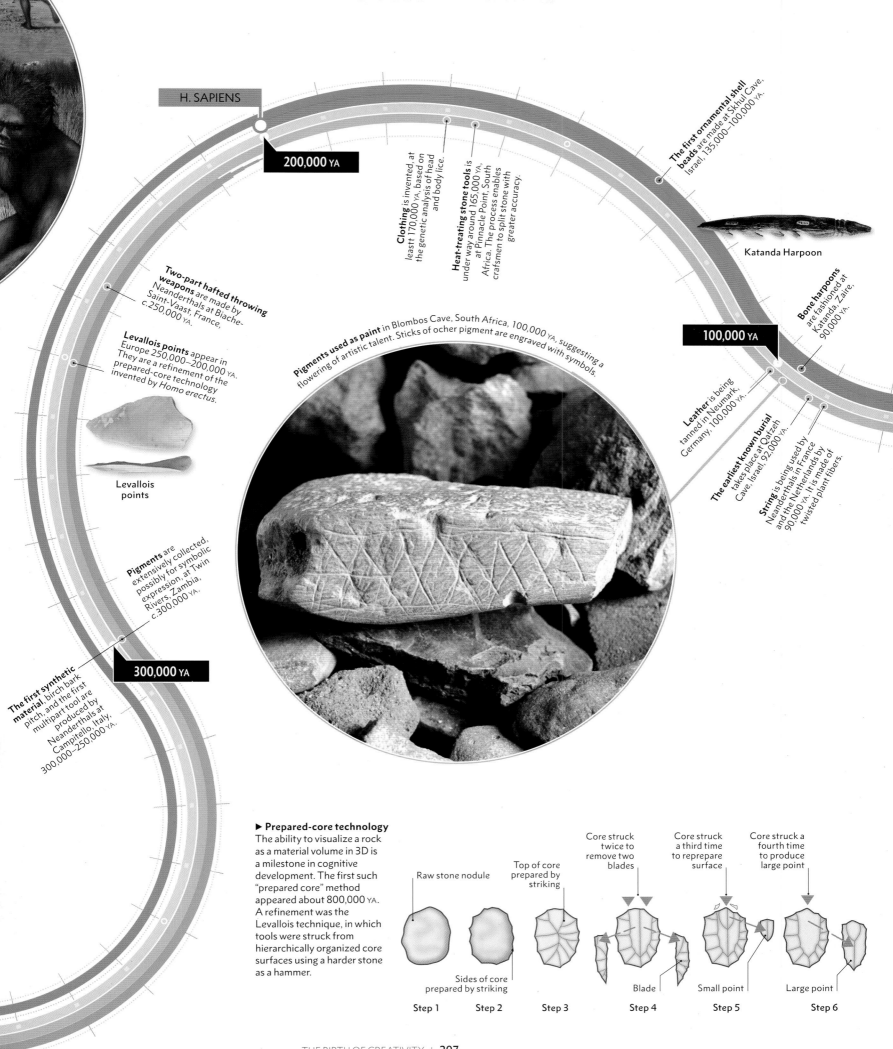

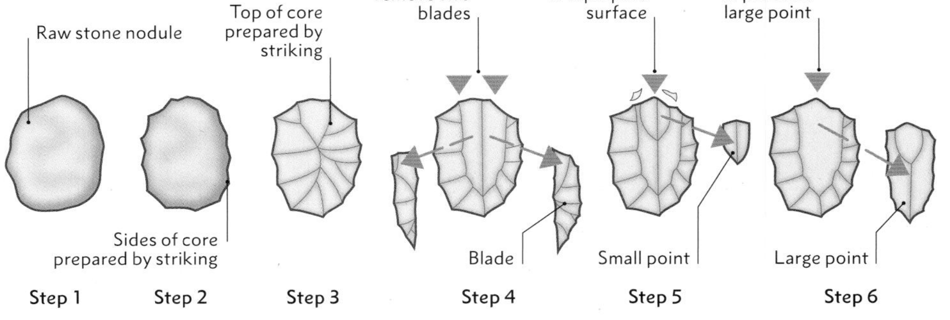

SCULPTURE AND CAVE ART, WHICH EMERGED ABOUT **40,000 YEARS AGO**, PROVIDE EVIDENCE OF THE USE OF **SYMBOLS** TO REPRESENT **IDEAS**

80,000 YA

Formally worked bone tools, possibly spear tips, are produced at Blombos Cave, South Africa, 84,000–76,000 YA.

Plants are used to create bedding areas in Sibidu, South Africa, 77,000 YA.

Pressure flaking of stone begins at Blombos Cave, 75,000 YA. Small flakes are prized out of the stone rather than struck, giving the craftsman greater control over the final shape of the tool.

Bead necklaces are made at Blombos Cave, South Africa, and the Grotte des Pigeons, Morocco, 82,000–75,000 YA.

Decorated ostrich eggshell

70,000 YA

Symbols are engraved on eggshells at the Diepkloof Rock Shelter, South Africa, 100,000–60,000 YA.

Projectile and microblade technologies emerge at Pinnacle Point, South Africa, 71,000–64,000 YA. Geometric points are also produced at Sibudu.

60,000 YA

Australia is colonized by 55,000 YA, requiring ocean-worthy canoes.

Flutes made of bird bone and mammoth ivory are produced at Geissenklösterle, Germany, 43,000–42,000 YA. These are the oldest known musical instruments.

40,000 YA

Figurative sculpture emerges in Europe, c.40,000 YA. The Lion-man of the Hohlenstein-Stadel is the oldest known work; the contemporaneous Hohle fels Venus (see p.213) is the oldest human figure.

Lion-man sculpture

The oldest painted cave art—a red dot—is made in El Castillo, Spain, c.40,000 YA. At almost the same time, handprints are produced on a cave wall in Sulawesi, Indonesia.

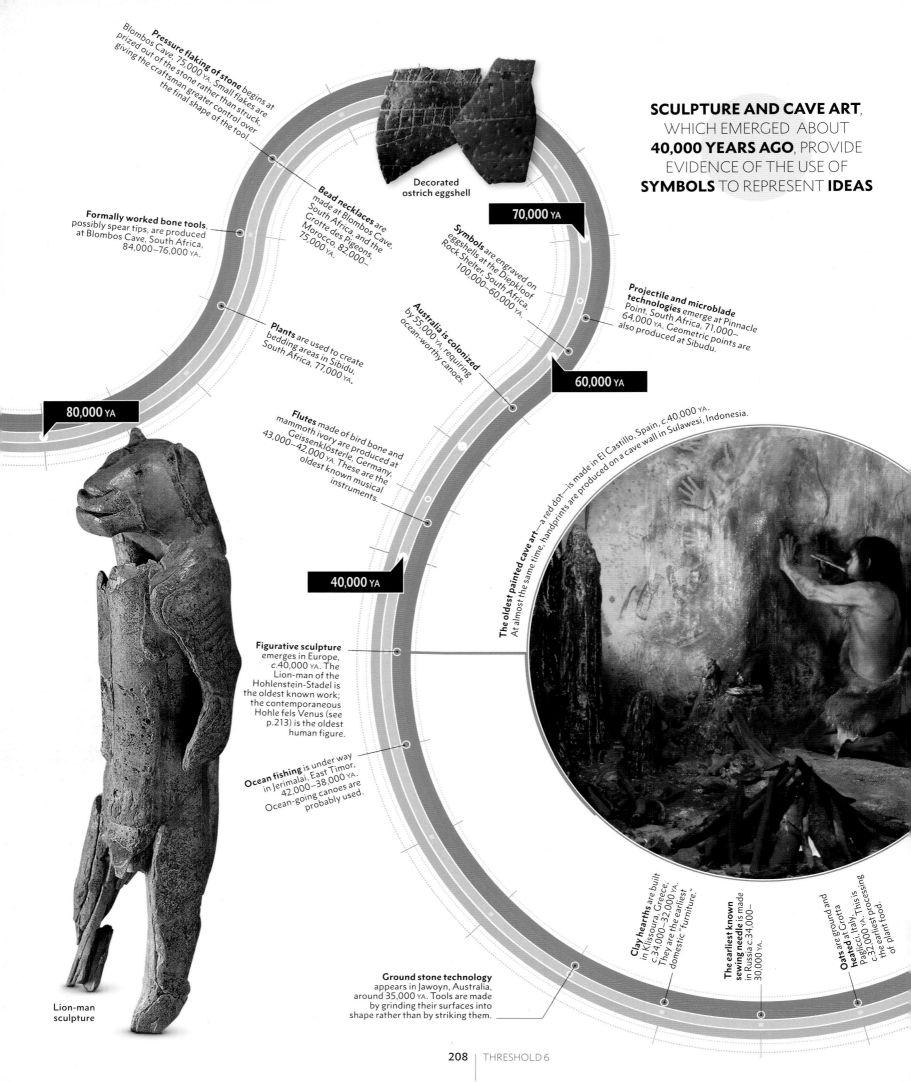

Ocean fishing is under way in Jerimalai, East Timor, 42,000–38,000 YA. Ocean-going canoes are probably used.

Ground stone technology appears in Jawoyn, Australia, around 35,000 YA. Tools are made by grinding their surfaces into shape rather than by striking them.

Clay hearths are built in Klissoura, Greece, c.34,000–32,000 YA. They are the earliest domestic "furniture."

The earliest known sewing needle is made in Russia c.34,000–30,000 YA.

Oats are ground and **heated** at Grotta Paglicci, Italy, c.32,000 YA. This is the earliest processing of plant food.

Shamanism is possibly practiced in Europe by 26,000 YA. This is suggested by individuals buried with typically shamanic effects, such as ocher, beads, and a staff.

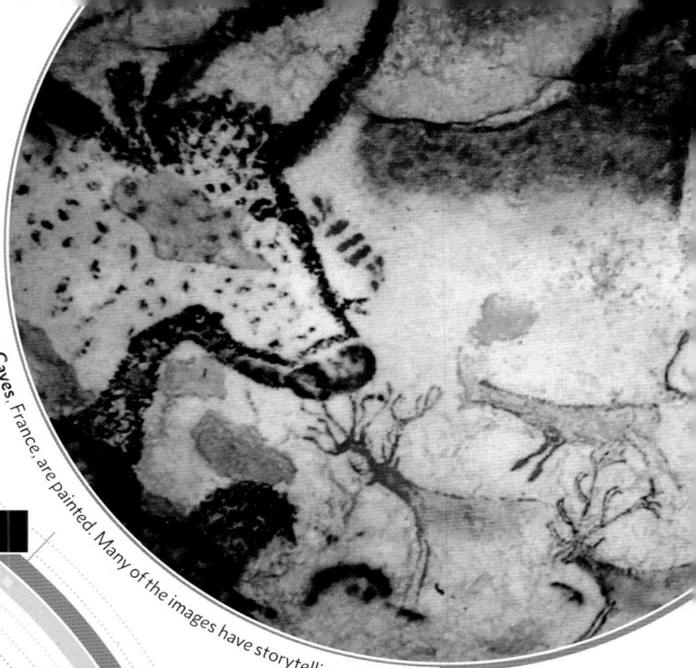

The Lascaux Caves, France, are painted. Many of the images have storytelling elements.

20,000 YA

The first clay pots are produced in Xianrendong, China c.20,000 YA.

Fish hooks for large-scale ocean fishing are made in Jerimalai, East Timor, 23,000–16,000 YA.

Woven textiles, basketry, nets, and string are produced in Pavlov and Dolni Vestonice, the Czech Republic, 28,000–25,000 YA. The linenlike textiles were probably made on a loom.

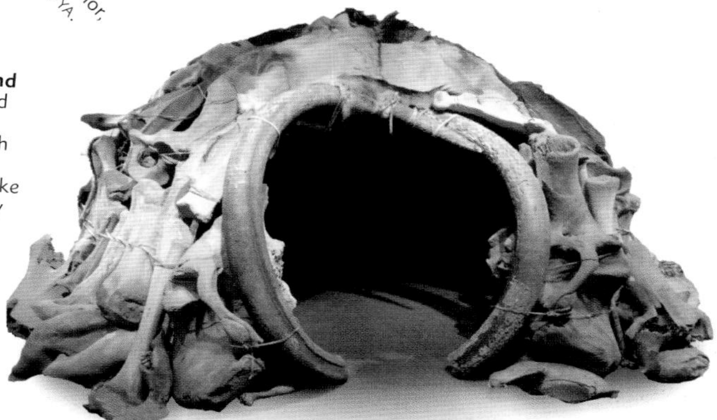

Replica of house made from mammoth bones and hide

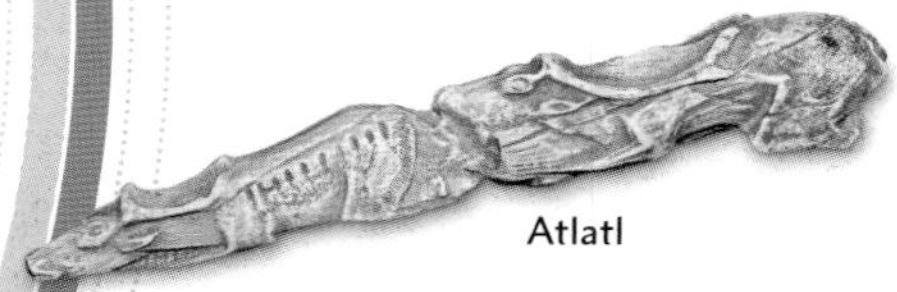

Atlatl

The first houses are built in Dolni Vestonice, the Czech Republic, c.26,000 YA. They are made of mammoth bones and possibly coated with daub.

Ornately carved atlatls (spear-throwers) are widely used in Western Europe 17,000–12,000 YA. Their decoration is proof of story-telling, and their use proof of apprenticeship.

Wild cereals and starchy plant foods are used intensively, such as at Grotte des Pigeons, Morocco, 15,000–14,000 YA. Plants are becoming a major part of *Homo sapiens'* diet.

Microblades are produced in Mongolia c.28,000 YA, using a prepared-core technique.

Clovis point arrowheads appear in New Mexico c.13,500 YA. These are struck from a stone core, then pressure-flaked around the edges.

The dog is domesticated c.30,000 YA.

The Chauvet Cave, France, is painted, beginning 32,000 YA. The images are largely of local animals, many of which are shown in perspective, and some with a sense of movement.

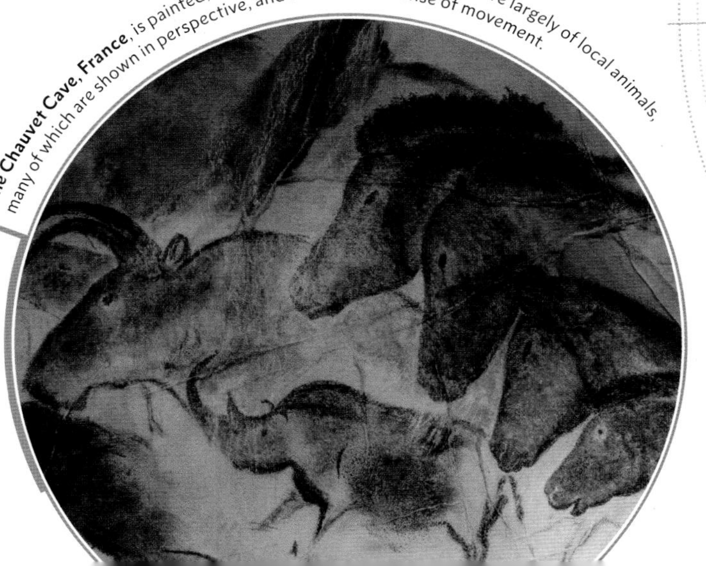

The bow and arrow is widely used in Europe, particularly among the Ahrensburgian people of Germany, 15,000–10,000 YA.

30,000 YA

The earliest known fish trap is made in Dublin, Ireland, 10,000–9,000 YA. Fish resources are increasingly exploited.

Clovis point

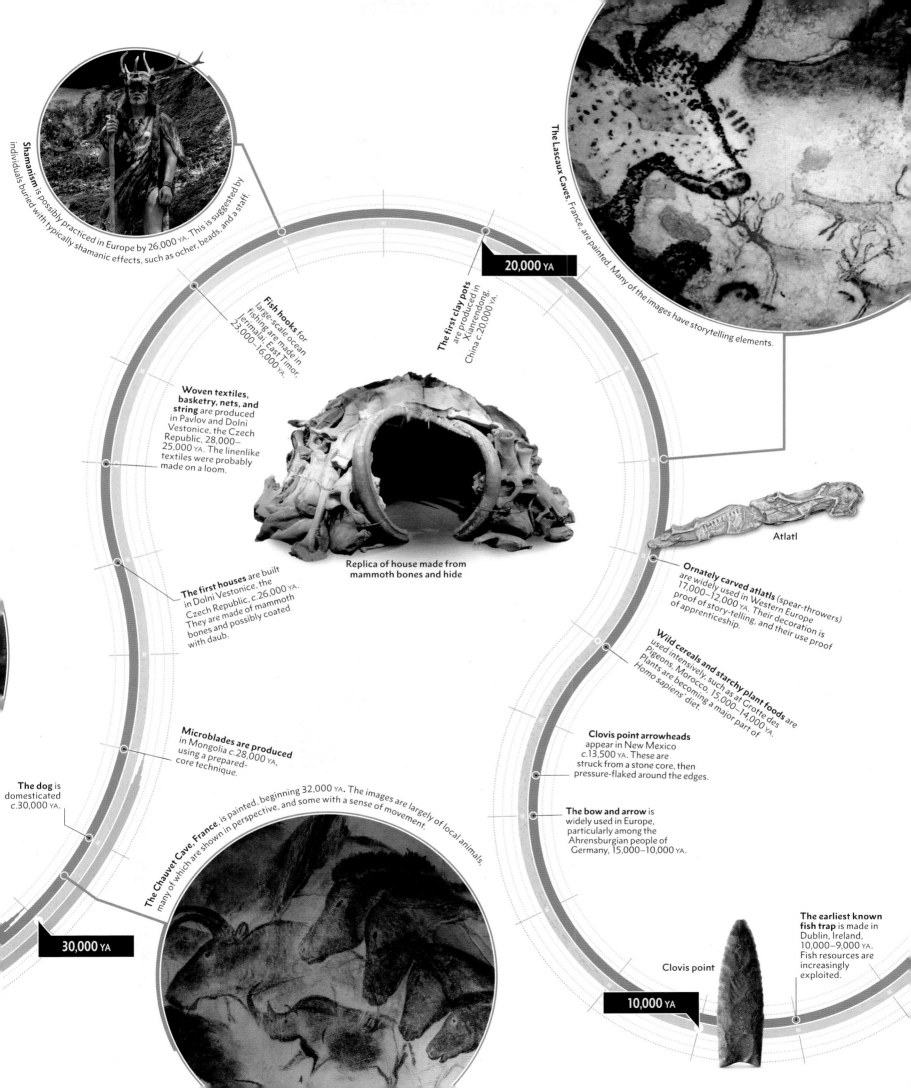

10,000 YA

Ancient practices
The San people have been hunting the landscapes of the Kalahari for thousands of years. Large game accounts for about 20 percent of their diet—the remainder is made up of plants and smaller animals caught in traps.

HUNTER-GATHERERS **EMERGE**

From the earliest times, most hominins survived by gleaning what they could from the world around them, rather than producing their own food. The range of items eaten and the ways of sourcing them varied according to the environment, and demanded high levels of social organization.

Early members of the hominin family had diverse diets—primarily of fruit, leaves, and insects—and some probably used cobbles to crack nuts, as primates do today. The first stone tools made food processing easier, and while these were being made by pre-*Homo* species at least 3.3 MYA, the earliest evidence of their function comes from around a million years later. Analysis of the surfaces of tools found at Kanjera South, Kenya, shows that plants and meat were being processed, probably by *Homo habilis*. Dating to about 2 MYA, the tools were made using an early technology known as Oldowan. At the same site, there is also evidence of hunting—or at least of scavenging kills made by other animals. Whole carcasses of small gazelles were brought in and cut up; because tooth marks from carnivores overlie them, it is clear that hominins had first access.

Around 1.8 MYA, with the emergence of *Homo erectus* and an improved way of making handaxes, called Achelean technology, hunting seems to have increased. Hundreds of footprints found at Ileret, Kenya, dating to 1.5 MYA, reveal that small groups of adults circled the lake shore—just as carnivores do. At the very least, it shows cooperative foraging was under way.

By around 700,000 years ago, diets had diversified. At Gesher Benot Ya'aqov, in Israel, there is evidence of nut-cracking as well as the exploitation of large animals, including elephants, although it is unclear whether the elephants were hunted or scavenged. Various plant resources are also thought to have been important during the European ice ages, both for Neanderthals and, later, *Homo sapiens*, but large amounts of fat and meat were still vital for survival.

LEARNED ADAPTABILITY

Making the most of varied food resources across different environments required an investment in complicated technologies and a dedication to preserving knowledge. The ability to hunt large animals suggests that hominins from *Homo erectus* onward were learning how to track, probably from early childhood. From 200,000 YA, Neanderthals were hunting birds, and at least 120,000 YA *Homo sapiens* were exploiting shellfish. Our species colonized the harshest environments, including the Arctic, suggesting that we had especially flexible skills.

Foragers typically lived in mobile communities, which were broadly egalitarian. However, abundant and predictable resources, such as fish, could encourage people to stay in the same locations, and even to become semi-sedentary—eventually, an alternative to the hunter-gatherer way of life would emerge.

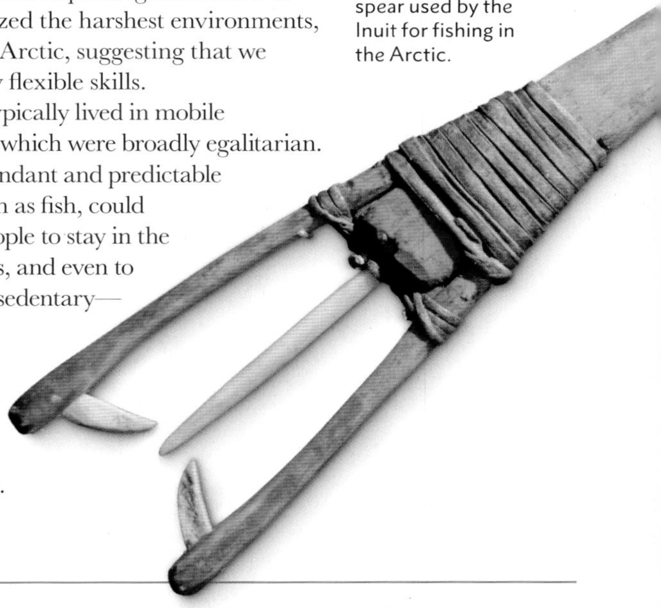

▼ **Evolving technologies** *Homo sapiens* spread across the globe by inventing new technologies, such as this three-pronged spear used by the Inuit for fishing in the Arctic.

> "THERE WAS **NOTHING** THAT THEY COULD NOT **ASSEMBLE** IN **ONE MINUTE**, WRAP UP IN THEIR **BLANKETS**, AND CARRY ON THEIR **SHOULDERS** FOR A JOURNEY OF **A THOUSAND MILES**."

Laurens van der Post, writer and conservationist, 1906–1996, on the San bushmen of the Kalahari

PALEOLITHIC **ART**

For many, the word "art" means representational imagery, and "Paleolithic art" is shorthand for a purely European tradition. However, Paleolithic art is much more diverse than this, and can be traced back to symbolic graphic creations produced over 100,000 years ago.

Early stirrings of artistic expression can be seen in the engraved eggshells found at the Diepkloof Rock Shelter in South Africa (see p.208), dating to over 100,000 years ago. However, we have no clear depictions of recognizable figures before 50,000 years ago. Currently, the two oldest paintings in the world (both *c.*40,000 YA) are a single red dot found in El Castillo cave, Spain, and a hand stencil found in Leang Timpuseng cave in Sulawesi, Indonesia. This proves that art was being practiced far beyond Europe at the time, even though most of the surviving dated examples are European. The Chauvet Cave in France, for instance, preserves some of the most stunning images of the era, including representations of nearly 450 animals. They were painted in two phases, the first starting nearly 37,000 years ago, the second over 2,000 years later. The walls of the cave were carefully prepared by the artists, and the images show a profound understanding of movement and perspective.

Around the same time, a piglike animal was painted in Leang Timpuseng. The very first Australian Aboriginal cave painting to have been firmly dated appeared soon after (*c.*28,000 YA), and from 20,000 years ago multiple traditions began flowering across the world. Throughout this time, "portable" art was also produced, including a female carved pendant discovered at Hohle Fels, Germany (*c.*40,000 YA). Known as the Hohle Fels "Venus," it is the oldest known depiction of a human being. Other traditions included the carving of ivory, bone, and antlers, and, in Eastern Europe, firing clay to make animal and human figurines. The meaning of these works can only be guessed at, but their growing significance to the people of the time is beyond doubt.

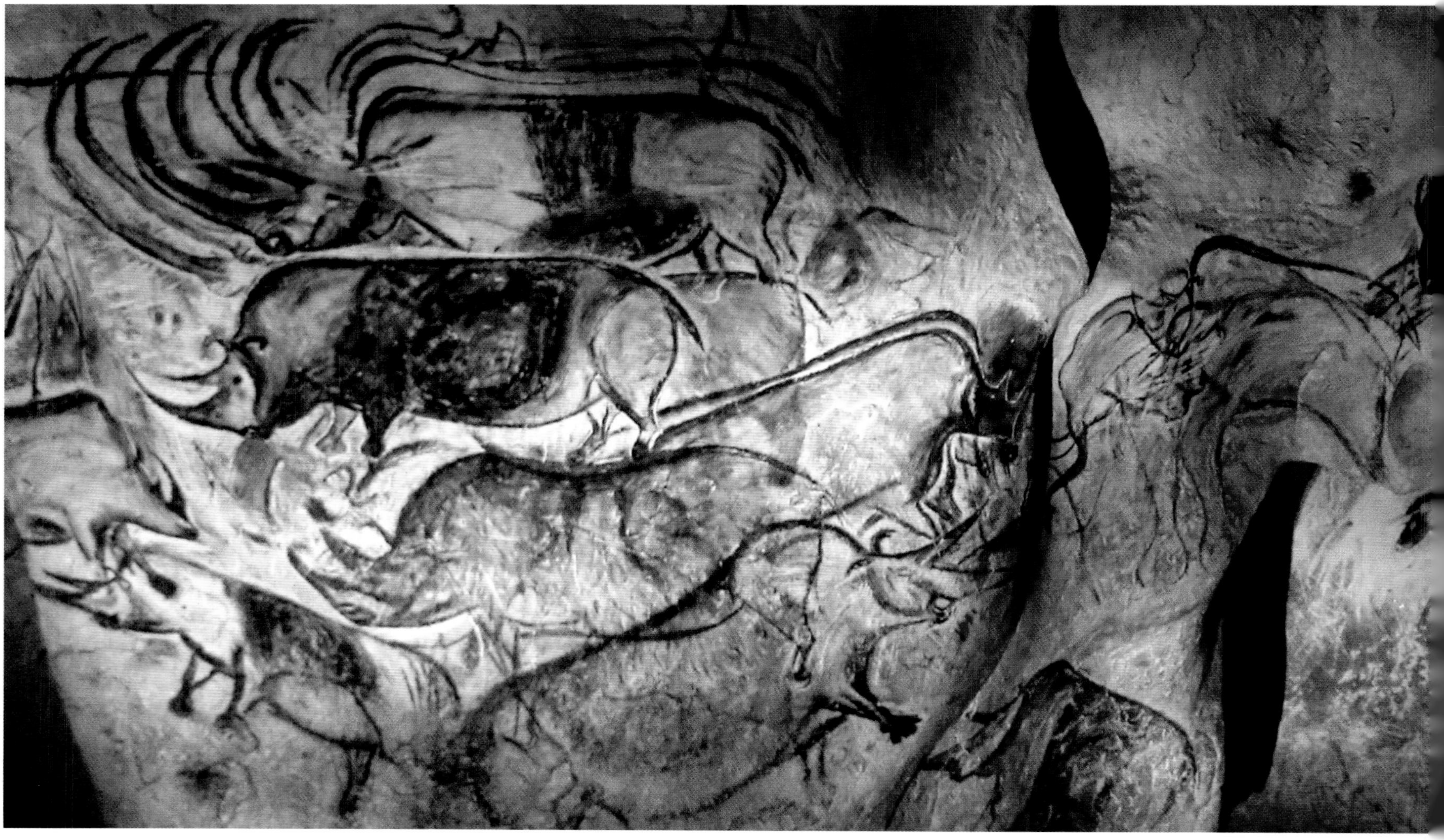

▼ Painted cave
The Chauvet Cave, in France, features huge panels of animals, including bison, horses, and lions—but not one single complete human being.

◄ **The Zaraysk bison**
This reconstructed figurine from Russia is a masterpiece of naturalistic carving. Made from ivory and rubbed with red pigment, it was smashed before being buried in a pit.

► **Hohle Fels Venus**
This ivory pendant is the oldest known female figurine.

> [ANCIENT ART] MAY BE AN **ATTEMPT TO NEGOTIATE**... WITH THE **HUMAN INTELLECT** AND ITS CAPACITY TO **OCCUPY OFTEN ILLUSORY REALMS DISTINCT** FROM **THE REALITY** OF THE **REST OF NATURE**
>
> **Jill Cook**, archaeologist, 1960 –

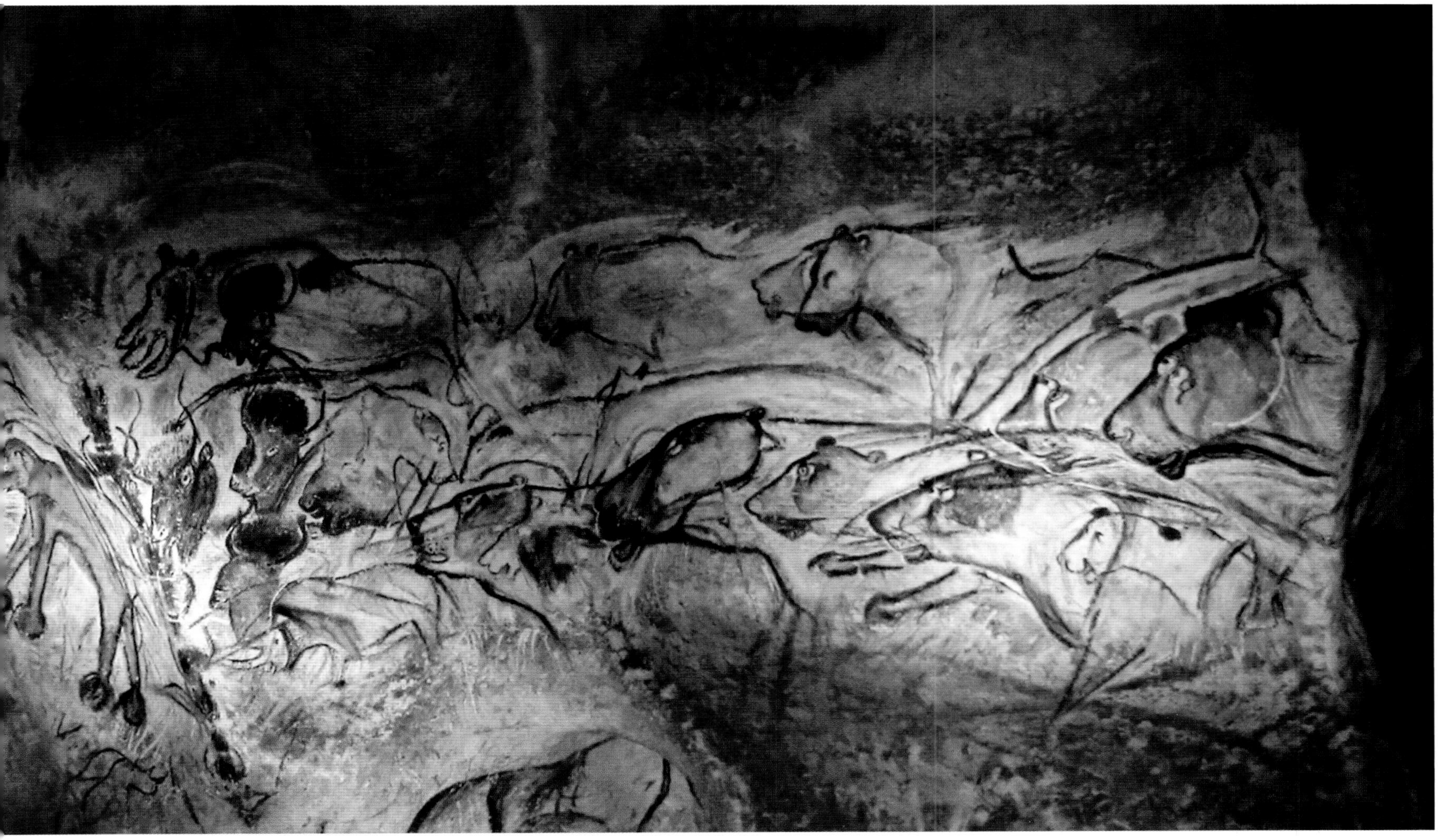

THE INVENTION OF **CLOTHING**

Clothing protects us from the cold, from sunburn, from insect bites, and even from certain weapons. In short, it makes us more adaptable. In Paleolithic times, it allowed us to live in a range of hostile environments, and to begin our spread across the globe.

We know from physical evidence that early *Homo sapiens* and Neanderthals used pigments and may have worn jewellery, but the earliest evidence for clothing is mostly indirect as clothing does not survive well in the ground. Biological studies suggest that during very cold glacial phases, hominin species in the northern hemisphere needed tailored body coverings. Even the Neanderthals, who are thought to have been physically cold-adapted, needed to cover at least 80 percent of their body, especially their hands and feet. Another clue comes from the study of parasites. Body lice are adapted to living in clothes, and the estimated age for their split from head lice, based on DNA studies, is at least 170,000 years ago. That long ago there were numerous types of humans—including Neanderthals, Denisovans, *Homo floresiensis*, and ourselves—and it is possible that exchanges between the different types of humans led to both the habit of wearing clothes and the spread of these parasites.

“[THE DISTINCTION BETWEEN **HEAD** AND **BODY LICE**] PROBABLY AROSE WHEN **HUMANS** BEGAN TO MAKE **FREQUENT USE OF CLOTHING**.”

Mark Stoneking, American geneticist, 1956–

▶ Buried prince
Only the shells remain of the clothes worn by the “Young Prince” found in the Arene Candide cave, Italy. He was buried over 23,000 years ago.

THE FIRST FABRICS

The earliest clothes were probably animal-based. Tiny scraps of tannin-soaked organic material were found stuck to a stone tool in Neumark-Nord, Germany, suggesting that over 100,000 years ago Neanderthals were tanning skins. They didn’t have needles, but could have sewn pieces of leather and fur using existing tools designed for piercing and threading. Bone tools with rounded ends have been found in 40,000-year-old Neanderthal sites, and these were probably “lissoirs”—leather-softening tools very like the ones still used today. The oldest bone needles date to 20,000 years ago, but these were probably used for bead embroidery as much as sewing other materials.

The use of plants for producing fabrics seems to begin with *Homo sapiens*. Dyed plant fibers have been found at Dzudzuana Cave in Georgia, dating to 30,000 years ago. Other sites show that from at least 28,000 years ago, fabric was being woven.

Tiny impressions in baked clay fragments from the sites of Pavlov and Dolni Vestonice, in the Czech Republic, show fine textiles comparable to linen, possibly made from flax or nettle, alongside netting and basketry. We cannot be sure that these fabrics were used for clothing, but some of the carved human figurines from the same region and period seem to show that plaited or woven caps and belts were worn. Other carvings from the Siberian site of Mal’ta, a few thousand years later, may represent full-body outfits with hoods, possibly made from fur.

The production of plant-fiber textiles continued through Mesolithic times, when bast (from tree bark) was spun into clothes. However, there is no evidence for their replacement with softer animal fibers, such as wool, until the advent of farming.

200,000 YA | HOMO SAPIENS APPEARS

135,000 YA | FIRST USE OF SYMBOLS

110,000 YA | LAST ICE AGE BEGINS

41,000 YA | EARLIEST PAINTED CAVE ART

12,000 YA | LAST ICE AGE ENDS

▶ Prehistoric dress

This reconstruction of a prehistoric person is based on remains found in the Abri Pataud site in Aquitaine, France. The site contained human remains, figurines, tools, and cave paintings from between 47,000 and 17,000 years ago, a period during which archaeologists believe clothing had become relatively sophisticated.

Hair may have been twined together to form dreadlocks, to keep it easy to clean and to avoid the potential illnesses associated with matted hair. Evidence for caps or simple hats has also been found, as have bandeaux—thin strips of fiber that hold the hair in place

EVIDENCE SUGGESTS THAT **HUMANS** MAY HAVE BEEN WEARING **JEWELRY** AS FAR BACK AS **75,000 YEARS AGO**

Snoods made from fur would have provided warmth during winter and at night

Tunics may have been made from woven nettle and hemp stalks

Clothes may have been dyed, with the dyes obtained from berries, roots, and leaves of plants

Elaborate jewelry made from stone, shell, bone, ivory, and antler was worn around the wrists and neck, and sewn onto clothing

Long string skirts and simple belts may have been common, as were boots made from animal skins laced together

HUMANS **HARNESS FIRE**

The ability to use fire is uniquely human, and may have been a significant impetus for the evolution of genus *Homo*. However, we may not have fully controlled it until relatively late in hominin evolution.

The earliest evidence of fire comes from Wonderwerk Cave, South Africa, where careful analysis of sediments nearly a million years old reveals that bones and plants were deliberately burned deep inside the cave. However, it is possible that early hominins took advantage of fires started by natural events, such as lightning strikes. The first repeated, controlled use of fire dates from just after 800,000 years ago, at Gesher Benot Ya'aqov, Israel, where burned materials recur over a 100,000-year period, shows that *Homo erectus* was both making and maintaining fire.

TECHNOLOGY AND SOCIAL LIFE

After 400,000 years ago, the increased frequency of sites with deep layers of ashes, charcoal, and burned bone reveals a habitual control of fire. In Europe, this coincides with the appearance of the Neanderthals, who seem to have been the first to use it for manufacturing purposes. At the Italian site of Campitello (*c.*300,000 YA), stone tools were found covered in birch bark tar, which was used to make multipart tools.

Familiarity with fire also changed our social life. Domestic spaces centered on fires appear by 200,000 years ago, and may have played a key role in the development of language. The campfire increased the amount of light to work by, but not sufficiently to perform difficult tasks, thereby creating opportunities for conversation and storytelling. It was also where the first experiments in cooking took place, nearly 800,000 years ago.

From around 35,000 years ago, in eastern Europe, people experimented with fire and clay for some 5,000 years, producing animal and human figurines; by 20,000 years ago, the first pottery was produced in China. From then on, fire drove many new technologies, especially as people abandoned hunter-gatherer lifestyles.

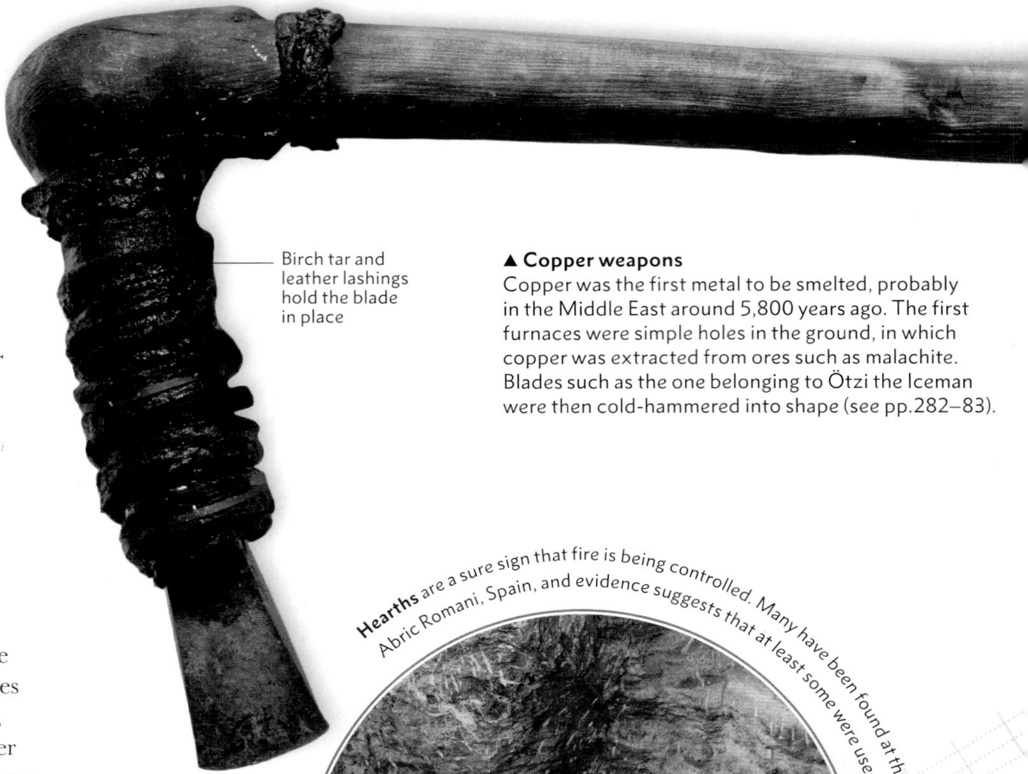

Birch tar and leather lashings hold the blade in place

▲ Copper weapons
Copper was the first metal to be smelted, probably in the Middle East around 5,800 years ago. The first furnaces were simple holes in the ground, in which copper was extracted from ores such as malachite. Blades such as the one belonging to Ötzi the Iceman were then cold-hammered into shape (see pp.282–83).

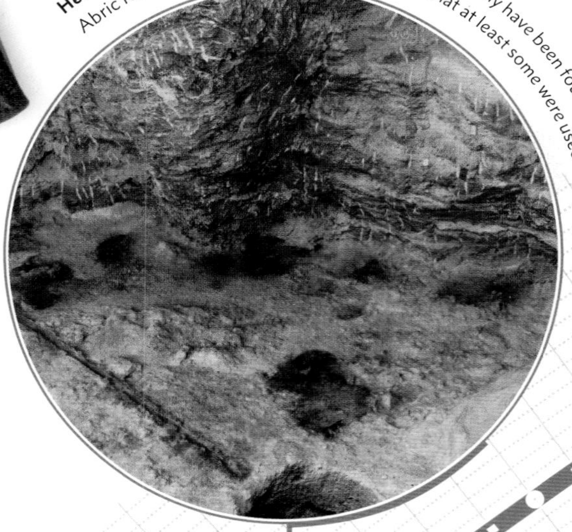

Hearths are a sure sign that fire is being controlled. Many have been found at the Neanderthal site of Abric Romani, Spain, and evidence suggests that at least some were used at the same time.

1 MYA
Early hominins use fire occasionally, perhaps sourced from bush blazes

780,000 YA
Habitual use of fire by late *Homo erectus*. Evidence of burned seeds suggests cooking

500,000 YA

400,000–300,000 YA
Evidence of the controlled use of fire, such as burned bones, becomes more common

300,000–250,000 YA
Neanderthals use fire to make birch bark pitch, the first synthetic material

200,000 YA
Activity areas in Paleolithic sites become centered on hearth areas

165,000 YA
Heat treatment to improve stone tool-making by early *Homo sapiens* in South Africa

50,000 YA
Neanderthals cook plant foods, as proved by starch grains found on teeth

Impression made by wooden haft, probably for a knife

▲ Birch bark tar
This 80,000-year-old piece of birch bark tar comes from Königsaue, Germany. On its reverse side, it bears the thumbprint of a Neanderthal—possibly its maker—who used it as an adhesive to attach a piece of flint to a wooden shaft.

The first campfires were probably sourced from bush fires that had started naturally. Typically, a fire would be kept alive in a cave, where it was sheltered from the elements.

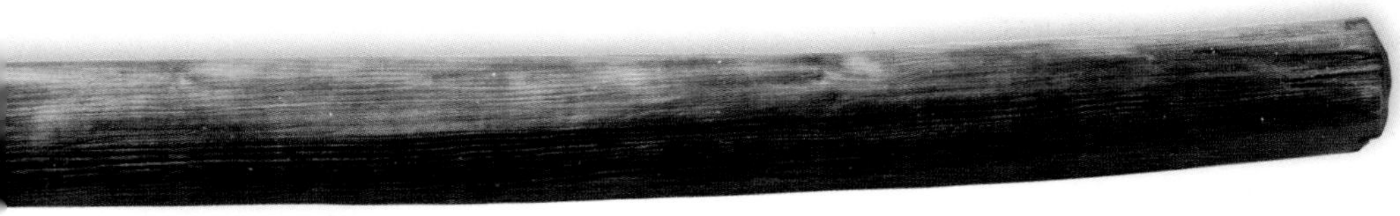

◀ Clay vessels
Clay vessels, such as this pot made by the Jomon people from Japan, greatly enhanced our ability to cook and store food. Their production became widespread with the adoption of the agricultural lifestyle, when strong containers for grain and other foodstuffs were needed.

40,000 YA
35,000–30,000 YA Clay-lined hearths, clay lamps, and the oldest fired ceramic figurines
20,000 YA The first known pottery food containers, made in China
13,500 YA
5,800 YA The earliest crucibles, evidence of copper smelting
5,500 YA
5,000 YA Pottery vessels found worldwide, with varied techniques of manufacture
3,800–3,200 YA The earliest evidence of iron smelting
3,000 YA
2,000–1,800 YA Glass manufacture begins in Egypt and possibly India

◀ Clay figurines
The oldest fired figurine is the "Venus of Dohlni Vestonice," from Croatia. Made some 29,000–25,000 years ago, she is proof that her makers were experimenting with fire.

▲ Bronze armor
Bronze was produced by adding tin to copper during the smelting process. It was considerably harder than copper, giving bronze-clad soldiers a distinct advantage in battle. This helmet was worn by a Greek soldier around 2,650 years ago.

> **HEARTHS** ARE PLACES TO SIT AND **EXCHANGE NEWS**... **MAKE AND REPAIR STONE TOOLS**, DISCUSS THE DAY'S **HUNT**, AND **MAKE PLANS** FOR THE **FUTURE**.
>
> **John McNabb**, archaeologist, 1960–

▼ Making birch bark pitch
Birch bark pitch, the first synthetic substance, was used as an adhesive from Palaeolithic times onward. First made by the Neanderthals, it was produced by "cooking" birch bark in a fire for several hours under controlled temperatures. The liquefied tar was then collected and allowed to cool. As it was hardening, it was applied like putty to the joints of multipart tools.

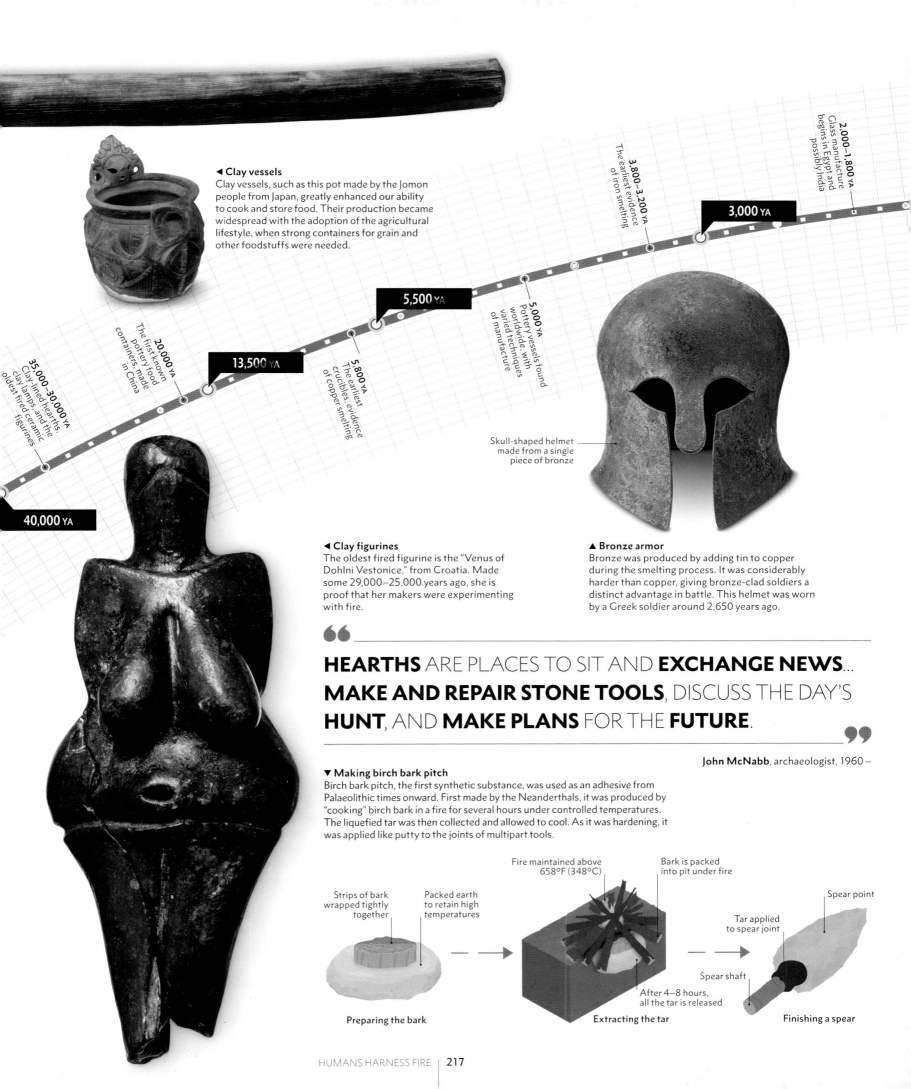

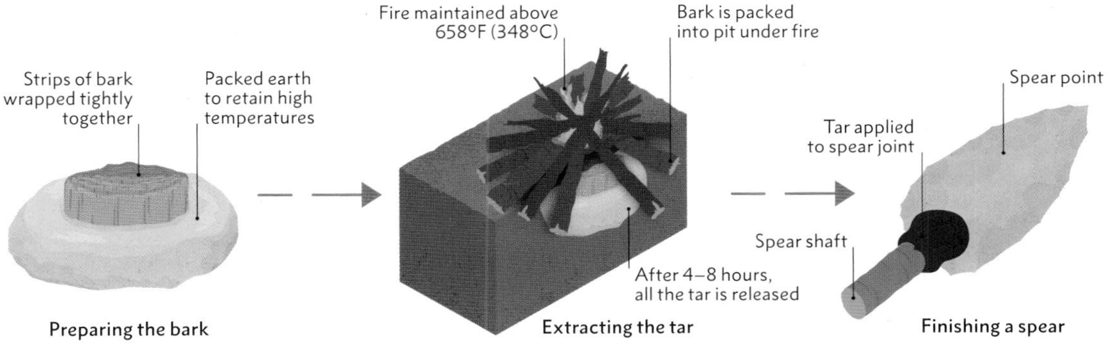

BURIAL PRACTICES

The Paleolithic period saw the emergence of that most human of characteristics—having respect, even concern, for the dead. The rituals of the day were simple, but they foreshadowed a time when tombs would be built for entire generations of ancestors.

Practices relating to death are important because they point to key intellectual capacities, such as an understanding of time. The ability to comprehend that an individual has moved from a living state to one of death seems uniquely human, but other species show hints of understanding. Elephants, for example, can be reluctant to leave the bodies of dead group members, and chimpanzees show an extraordinary range of reactions, from extreme agitation to quietly staying by a body for hours, and sometimes carrying infants' corpses for weeks. However, it is impossible to know if these reactions are simply the effects of confusion and distress or true expressions of loss and sadness.

THE FIRST **OPEN-AIR BURIALS** TOOK PLACE AROUND **40,000 YEARS AGO**. BEFORE THEN, **ALL FUNERARY PRACTICES** TOOK PLACE IN **CAVES**

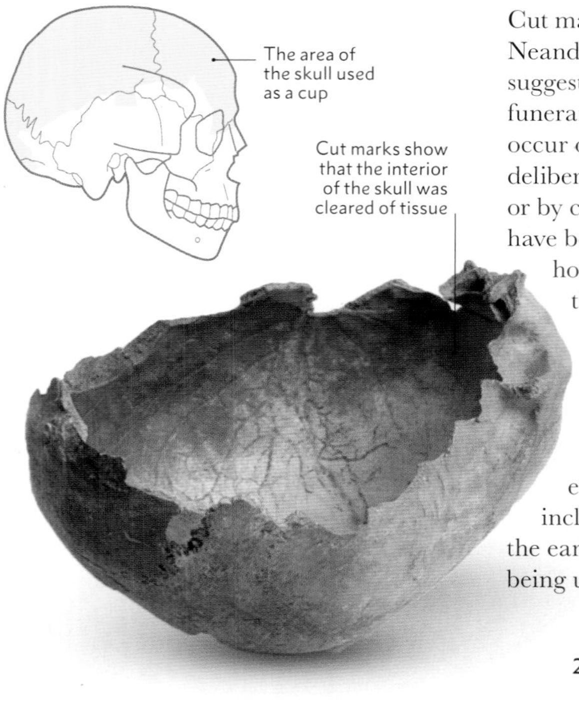
The area of the skull used as a cup
Cut marks show that the interior of the skull was cleared of tissue

▶ Skull cup
The skull remains found at Gough's Cave show all the signs of manufacture. The meticulous cutting and cleaning of the bone suggests that the skull was used for ritualistic purposes.

THE FIRST BURIALS

The earliest evidence of hominins acknowledging the dead is the practice of "caching," or collecting, bodies. Around 430,000 years ago, at Sima de Los Huesos, Spain, at least 28 hominins were deliberately placed in a deep pit, accompanied by a single strikingly colored stone tool. The complete interment of bodies began much later. Some 92,000 years ago, a number of *Homo sapiens* were buried at Qafzeh and Skhul, in Israel. These included a young adult and a child who were buried together, and a teenager who was interred with antlers covering his chest.

After 40,000 years ago, the frequency of burials increases, as does the number of bodies buried with objects. In Sungir, Russia, around 25,000 years ago, two children were buried head-to-head in a single grave, accompanied by spears, thousands of beads, and a single adult femur filled with red pigment. Such rich burials were rare, however. Simpler burials were common, and isolated body parts were sometimes interred separately.

EATING THE DEAD

Cut marks from stone tools found on both Neanderthal and *Homo sapiens* remains suggest another facet of Paleolithic funerary practices. The marks, which occur on various bones, were caused by the deliberate removal of flesh from the body or by cutting the body to pieces. This may have been a means to interact with or honor the dead, but may also indicate that the population had to resort to cannibalism for nutritional purposes. A number of such bones was found at Gough's Cave, in the UK, in 1898. Dating to 14,000 years ago, and almost certainly evidence of cannibalism, the bones included several carved "skull cups"—the earliest examples of human skulls being used as drinking vessels.

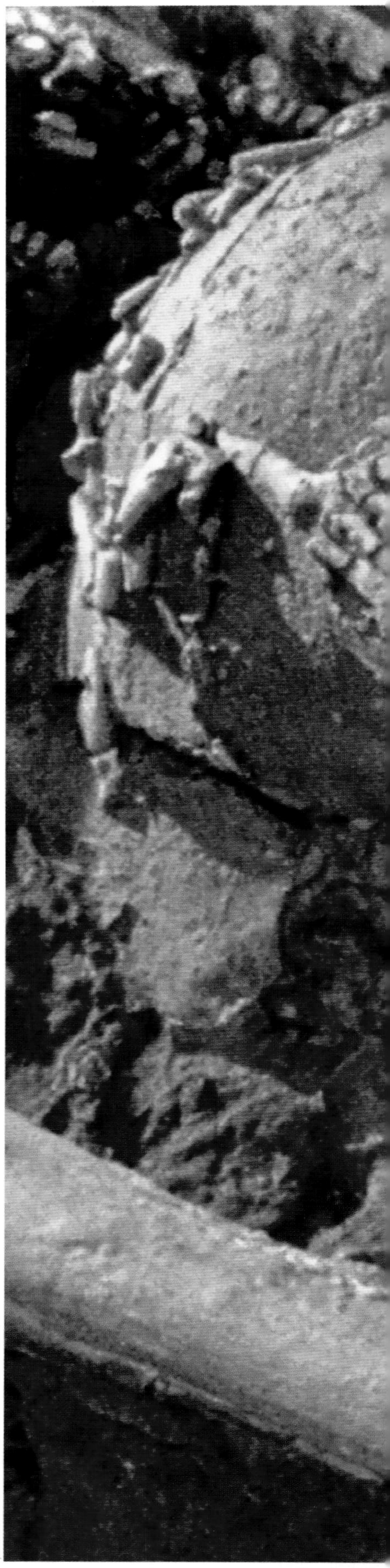

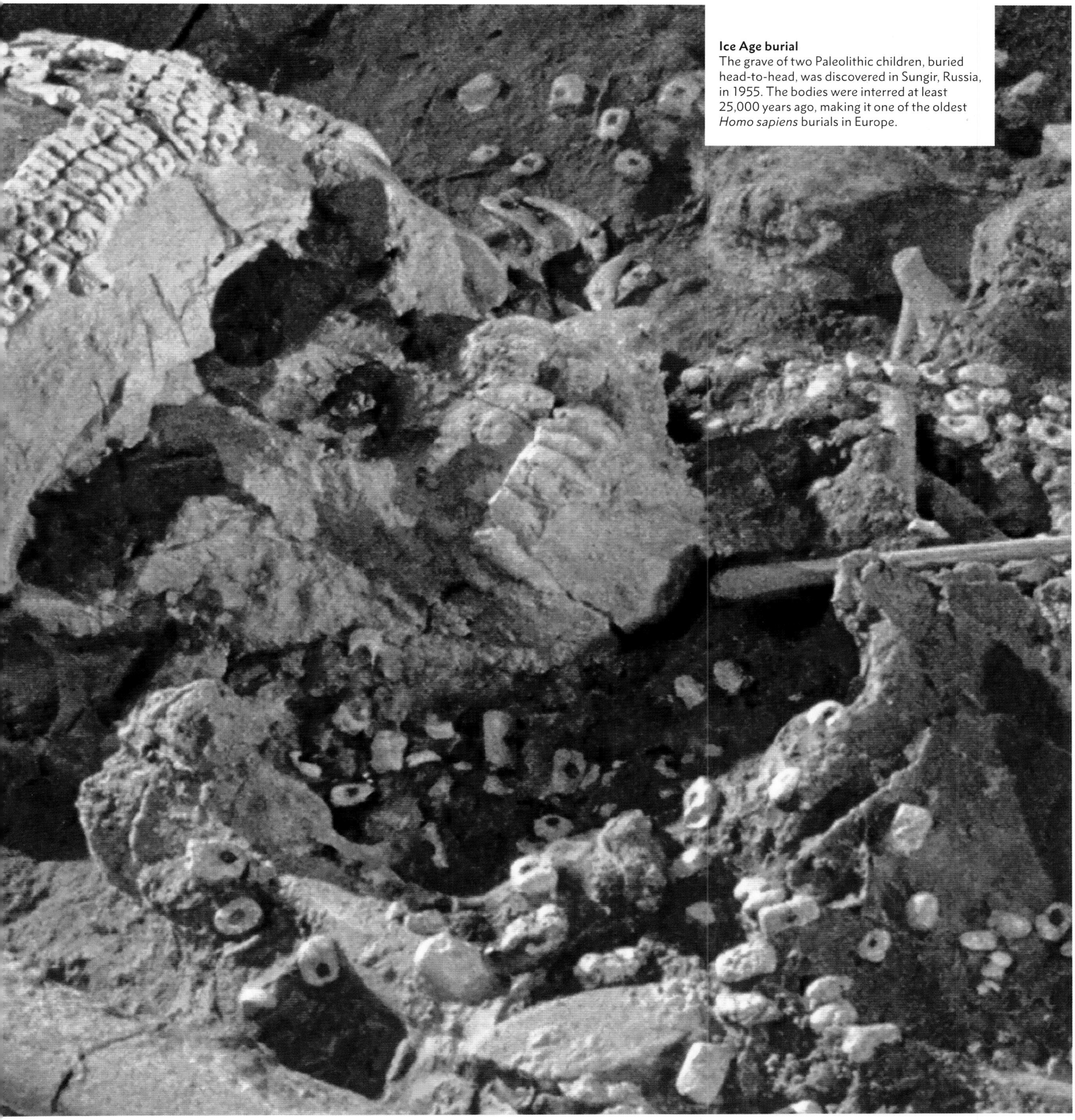

Ice Age burial
The grave of two Paleolithic children, buried head-to-head, was discovered in Sungir, Russia, in 1955. The bodies were interred at least 25,000 years ago, making it one of the oldest *Homo sapiens* burials in Europe.

HUMANS BECOME **DOMINANT**

The end of the last Ice Age nearly 12,000 years ago marked the beginning of the Holocene—our current geological epoch. Climatic change was nothing new for the hominin family. However, two things were different: there was now only one surviving human species, and we had already begun to alter the habitats and landscapes around us.

The end of the Pleistocene brought warmer, wetter conditions to most of the planet. In many areas, grasslands gave way to mixed deciduous forests, and desert areas grew increasingly moist. Over tens of thousands of years, *Homo sapiens*' dispersal had resulted in settlement all the way from the South African coast, through Eurasia, into Australia and up to the tip of South America. The unfamiliar, often harsh, surroundings they found, along with climatic changes, required innovative survival strategies. Humans exploited their ability to problem-solve and learn skills—including forging new relationships with flora and fauna. These new relationships, in turn, started to shape the local environments in which *Homo sapiens* lived.

EFFECT ON ANIMAL LIFE

The earliest known examples of hominins exploiting marine resources are the gathering of shellfish by *Homo sapiens* at Pinnacle Point, South Africa, around 160,000 years ago, and by Neanderthals at Bajondillo Cave, Spain, some 150,000 years ago. Such small-scale activity made little impact on shellfish populations, but as harvesting escalated over time, it began to have a negative effect. In South Africa, the average size of some shellfish species decreased around 50,000 years ago, which indicates that collecting had become more intense. This may, perhaps, be due to changing settlement patterns, with more humans moving to coastal areas, or may have been caused by an increase in the existing local human population. There were similar reductions in shellfish size following the human colonization of Papua New Guinea 30,000 years ago and southern California 10,000 years ago.

Most animal populations no doubt recovered from temporary local pressure exerted by humans, but our species may have a deep history of more permanent, catastrophic impacts on biodiversity. The so-called megafaunal "overkill" hypothesis correlates reductions in the diversity of large animal species with evidence of increasing *Homo sapiens* occupation toward the end of the last Ice Age. This is most obvious in Australia and North America, where the arrival of humans, 55,000 years ago and 15,000 years ago respectively,

MELTING GLACIAL ICE AT THE BEGINNING OF THE **HOLOCENE** EPOCH CAUSED **WORLD SEA LEVELS** TO **RISE 115FT (35M)**

▼ Fire-stick farming
The practice of burning vegetation to create grassland habitats that suit the animals humans wish to hunt may have been in use in Australia for 50,000 years. It can radically change the landscape and even a region's climate.

corresponded with the disappearance of a great number of animal species, including the spectacular giant ground sloths. However, climate changes around the same time may also have played a part, and certainly *Homo sapiens* had been present in Europe since before 40,000 years ago with no clear associated mass extinctions.

even ventured beyond. But not so far back in time—within the blink of an eye in geological terms—humans were few in number, scattered, and surviving on what they could find or catch. Yet even during these early stages of our history, we were making an impact on the world around us through our daily lives.

was conducted so systematically that it could justifiably be described as quarrying. Even if this occurred over a long period of time, the mounds of waste material dramatically changed the local landscape.

More subtle open-air artistic traditions begin to appear around 40,000 years ago, sometimes involving the transformation of

> THE FACT THAT ***HOMO SAPIENS*** IS THE **ONLY HOMINID SPECIES** ON THE EARTH TODAY MAKES IT EASY TO ASSUME THAT OUR **LONELY EMINENCE** IS HISTORICALLY A NATURAL STATE OF AFFAIRS—**WHICH IT CLEARLY IS NOT**.
>
> **Ian Tattersall**, British paleoanthropologist, 1945–

It may be that in especially challenging environments the arrival of a new, skilled predator, *Homo sapiens*, was just enough to push particular species into extinction. One of the strongest cases to support the overkill hypothesis is the Caribbean ground sloth, which went extinct less than 5,000 years after the arrival of humans; even then, however, the process seems to have taken about 1,000 years.

While there is no evidence of *Homo sapiens* having an extinction-scale impact on plant communities at this time, we may have been significantly altering some environments. Charcoal from sediment cores may indicate that people were burning forests in Southeast Asia around 50,000 years ago and also in Australia between 60,000 and 50,000 years ago. Although natural causes of forest fires cannot be entirely dismissed, "fire-stick farming," where forest is burned to increase ecological productivity and attract animals, is known to have a long history in North America and Australia, and there is evidence that Mesolithic communities may also have developed similar practices in some areas.

CULTURAL LANDSCAPES

Homo sapiens' world-spanning civilization is today easily visible from space; our robotic crafts have explored the solar system and

By persistent activity at a particular place, organisms begin to change their surroundings. For hominins, this can be seen in the accumulations of detritus within caves. Thousands of caves around the world show deep sediments formed from the waste of countless generations. These unintentional creations were not limited to caves, but also occurred where people continually inhabited the same open-air sites, and they provide evidence of how people lived. For example, some shell middens (refuse heaps) may have had symbolic significance. At some sites they contain human remains, as well as discarded mollusk shells. One such place is Klasies River Mouth, South Africa, in a region where very few burials have been found.

Hominins interacted with cultural deposits in other ways, too, including the digging of burial pits through older occupational layers left by Neanderthals and *Homo sapiens*. They often found cultural deposits to be a useful resource, and it became common practice to recycle old stone tools made centuries earlier.

The overall effect on the landscape of using vast amounts of rock for stone tools over millions of years is hard to calculate, but some sites show intense activity. The exploitation of flint in Israel over 500,000 years ago, for example,

entire valleys into outdoor symbolic arenas, such as the 5,000 engravings at Côa, Portugal. The alteration of stone on such a large scale foreshadows the oldest megalithic structures, built at Gobekli Tepe, Turkey. They were made by hunter-gatherers some 11,000 years ago, within a few centuries of the beginnings of early agriculture.

> WE ARE **PROBABLY** THE MOST **ADAPTABLE MAMMAL** THAT HAS **EVER EVOLVED** ON **EARTH**.
>
> **Rick Potts**, American paleoanthropologist, 1953–

THRESHOLD

CIVILIZATIONS **DEVELOP**

As our highly adaptable and ingenious species starts to modify nature in order to sustain itself, we turn from hunter-gatherers into farmers. This is a pivotal point in the story of our species. Farming sets us on a path of expansion. The population grows, and small nomadic communities turn into permanent cities, states, and—eventually—empires with complex new power structures.

GOLDILOCKS CONDITIONS

Agriculture developed after years of collective learning enabled humans to extract more resources from their environment. The ability to innovate and manipulate nature altered both the biosphere and society itself: larger populations required organization to function effectively, and new, more complex power structures began to emerge.

Warmer climate
Population and resource pressures of denser human communities
Buildup of collective learning
Hunter-gatherers
Armed with information accrued through generations of collective learning, foragers band together to collect a variety of seasonal foods from large areas of land. Populations remain small and highly mobile, but teamwork is important, particularly when hunting or trapping large animals for their meat.
What changed?
The warmer climate transforms the landscape, and the food and energy sources available to humans, reducing the need to move on. With the elderly, infirm, and very young no longer left behind, communities settle, grow, and learn to cultivate their own food, and extract more from the environment.
Fire-stick farmers
Hunters with knowledge of local flora and fauna use fire to clear land to create favourable grassland habitats for hunting and gathering.
Affluent foragers
Foraging communities in areas with abundant natural resources devise methods of storing food to consume out of season, and enjoy a more settled way of life.

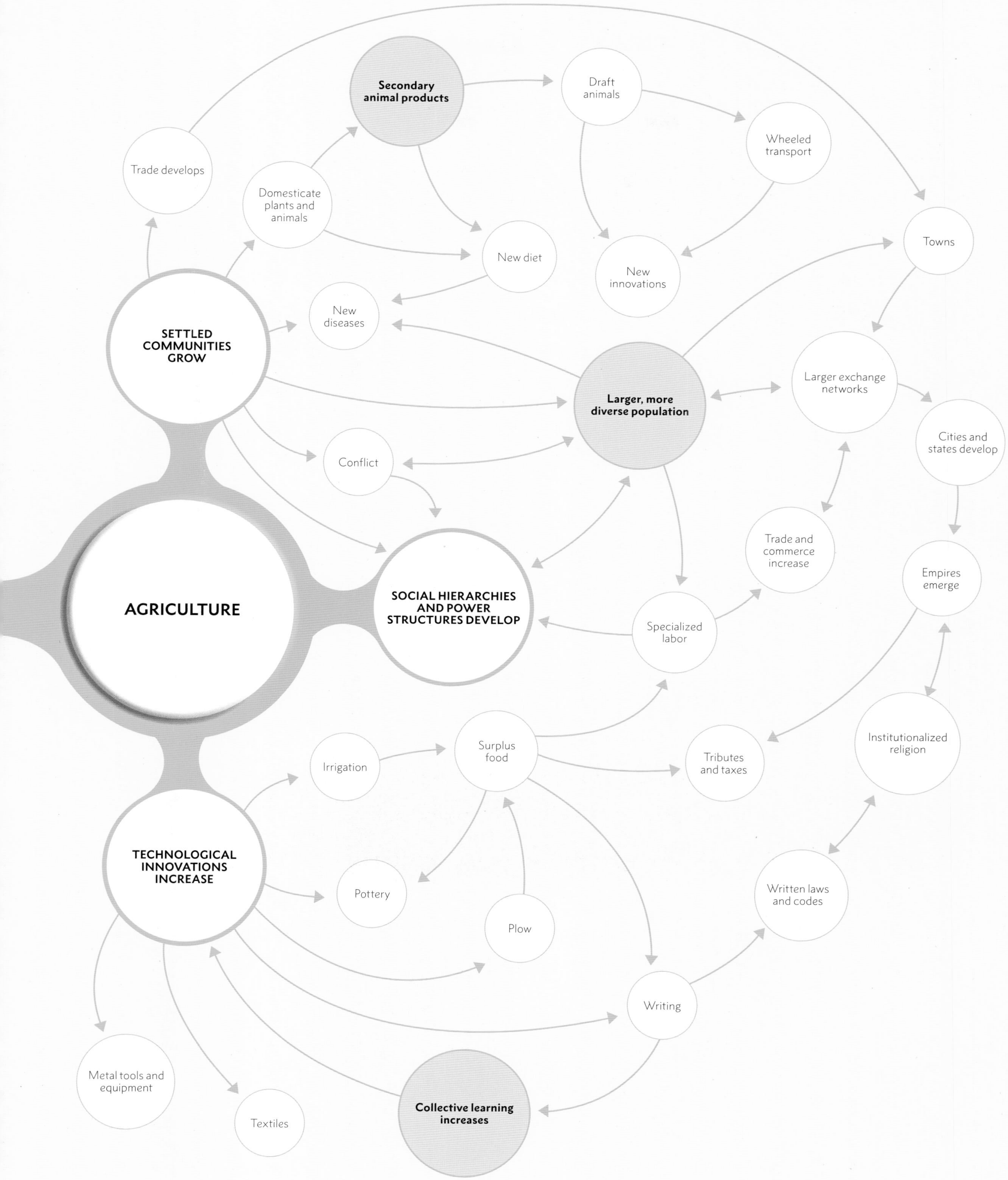
AGRICULTURE
SETTLED COMMUNITIES GROW
Trade develops
Domesticate plants and animals
Secondary animal products
Draft animals
Wheeled transport
New diet
New innovations
Towns
New diseases
Larger, more diverse population
Larger exchange networks
Cities and states develop
Conflict
SOCIAL HIERARCHIES AND POWER STRUCTURES DEVELOP
Trade and commerce increase
Empires emerge
Specialized labor
Irrigation
Surplus food
Tributes and taxes
Institutionalized religion
TECHNOLOGICAL INNOVATIONS INCREASE
Pottery
Plow
Written laws and codes
Writing
Metal tools and equipment
Textiles
Collective learning increases

CLIMATE CHANGES THE **LANDSCAPE**

From around 9600 BCE, global temperatures rose rapidly, beginning the current geological period known as the Holocene ("wholly recent"). Humans were forced to find new ways to hunt and gather. Eventually, they discovered a very different way of life—one based on farming.

As the climate warmed up, the ice sheets melted. This raised sea levels and released more water into the atmosphere, which increased rainfall. Asia was cut off from America, and Britain and Japan became islands. The wetter climate produced forests, grasslands, and new lakes and rivers. There was a mass extinction of Ice Age big game, such as the mammoth, the woolly rhino, and the giant elk.

MESOLITHIC ABUNDANCE

During a transitional period called the Mesolithic, or Middle Stone Age (280,000–25,000 BCE), people adapted to the new conditions by hunting smaller animals, such as deer, using the bow and arrow—a new invention, ideal for stalking animals in woodland. They also caught more fish and learned to eat a wider range of plant foods, including grasses and acorns, which required processing or cooking. By trial and error, people discovered which plants were poisonous and which could be made edible. In fact, some coastal regions were so rich in food resources that hunters and gatherers were able to settle down, for the first time, in permanent villages.

Throughout this period, people were building up and sharing their knowledge about plants and animals, which would contribute toward the new way of life.

At the same time, there was a dramatic rise in the human population, which had by now spread to every inhabitable part of the world. Rising sea levels also meant that huge areas of land that were once rich hunting grounds were now lost under water. Our planet may have reached its carrying capacity for the hunter-gatherer way of life. Climate change, and the pressure of competition for resources, eventually led some people, in different parts of the world, to begin farming.

> IN MANY PARTS OF THE HABITABLE GLOBE, **THE CHANGE IN CLIMATE** MUST HAVE BEEN **REMARKABLE** EVEN IN **ONE** LONG **LIFETIME**.

Geoffrey Blainey, Australian historian, 1930–, *A Short History of the World* (2000)

▸ Climatic ups and downs
Although the Holocene era has been generally warm, climatic fluctuations have occurred within it, as shown on this chart. The first farmers may have begun to grow crops in response to cool and dry periods, when there was a decline in the availability of wild food plants.

KEY Warmer Cooler

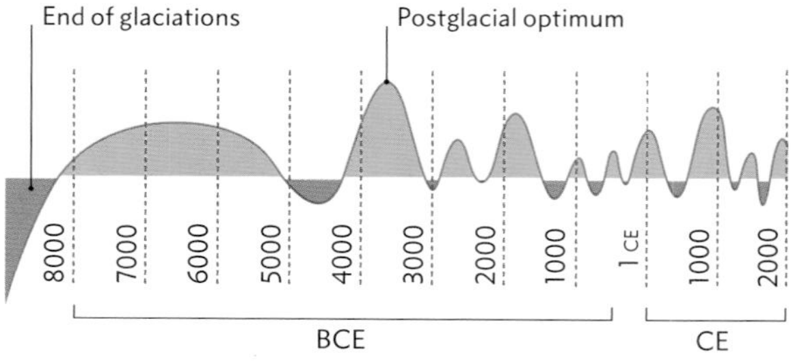

A good day's hunting
This African rock painting shows hunters using bows and arrows to kill deer. During the Mesolithic period, Africa's Sahara—which is now desert—was a grassland habitat, and a plentiful source of game.

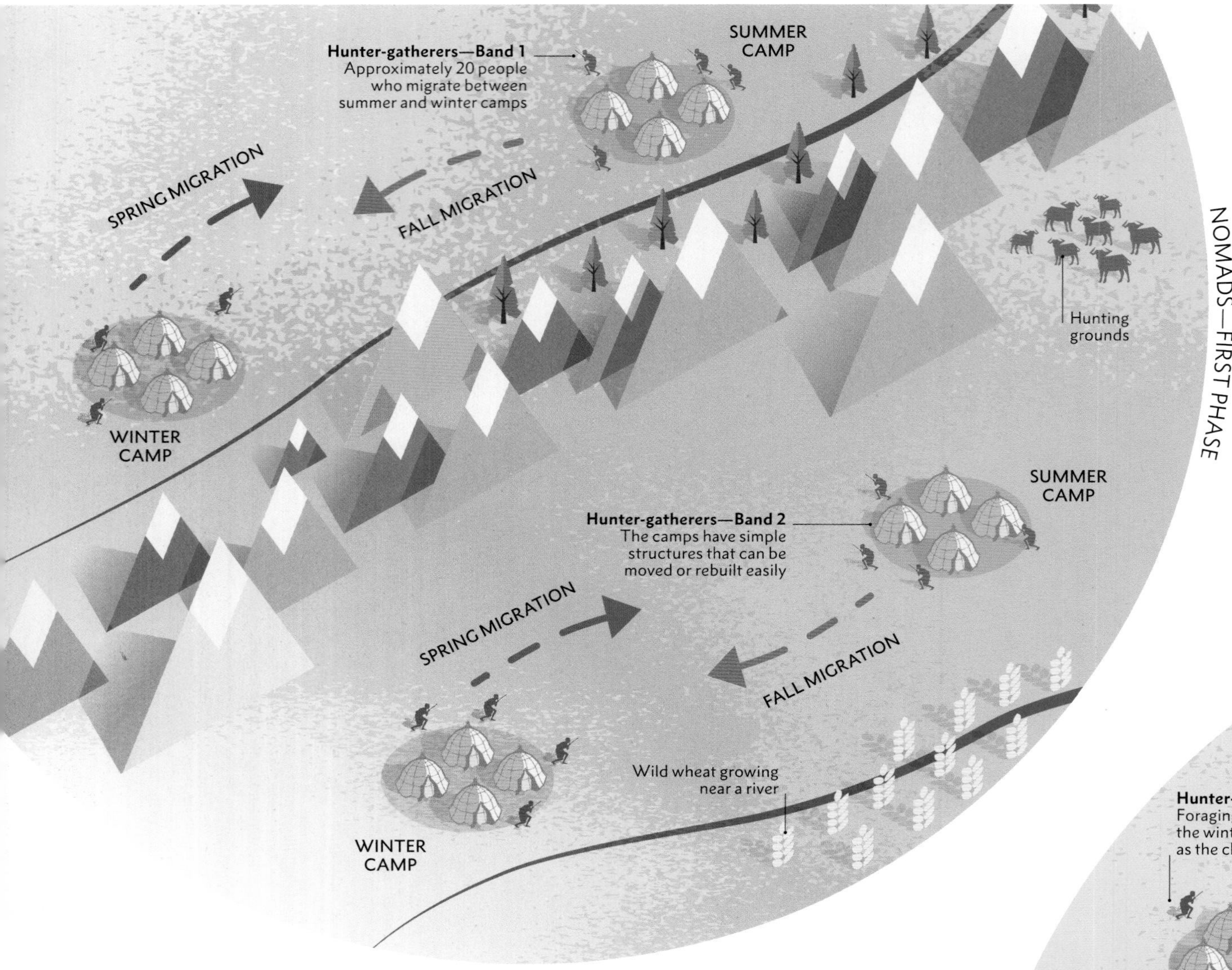
SUMMER CAMP
Hunter-gatherers—Band 1
Approximately 20 people who migrate between summer and winter camps
SPRING MIGRATION
FALL MIGRATION
WINTER CAMP
Hunting grounds
NOMADS—FIRST PHASE
SUMMER CAMP
Hunter-gatherers—Band 2
The camps have simple structures that can be moved or rebuilt easily
SPRING MIGRATION
FALL MIGRATION
Wild wheat growing near a river
WINTER CAMP

◄ **Nomads**
23,000–13,000 BCE
People are organized into small family groups (bands) that rely on hunting and gathering for food. They live a nomadic lifestyle, moving to new sites as the seasons and resources change. A nomadic lifestyle puts natural restrictions on population growth.

▼ **Early settlers and affluent foragers**
13,000 BCE
The climate becomes warmer and wetter. Rivers swell, grasslands and forests spread, creating a richer landscape. Some bands continue their nomadic lifestyle, but others settle in one prime spot.

FORAGERS **BECOME FARMERS**

As the warming climate transformed the landscape, hunter-gatherers and foragers across the world discovered new ways to boost their food supplies, most dramatically by farming. Instead of continually moving to find food, they could now settle permanently in one place.

Settling down had many unforeseen consequences. Without the need to move on, technology became heavier and more complicated. This led to quern stones for grinding grain, looms for weaving, and pottery. More permanent settlements meant children did not have to be carried over long distances on the annual migrations, and the elderly and infirm were no longer left behind to fend for themselves until the band returned. As a result, the birth rate went up and people lived longer, but there were now more mouths to feed.

Gradually, these settled populations came to depend on the limited number of crops they could grow, rather than the wide but seasonal range of wild foods obtained by foraging. In many ways, settling down was a trap. Although farming could support significantly larger populations than foraging, people had to work much harder for their food.

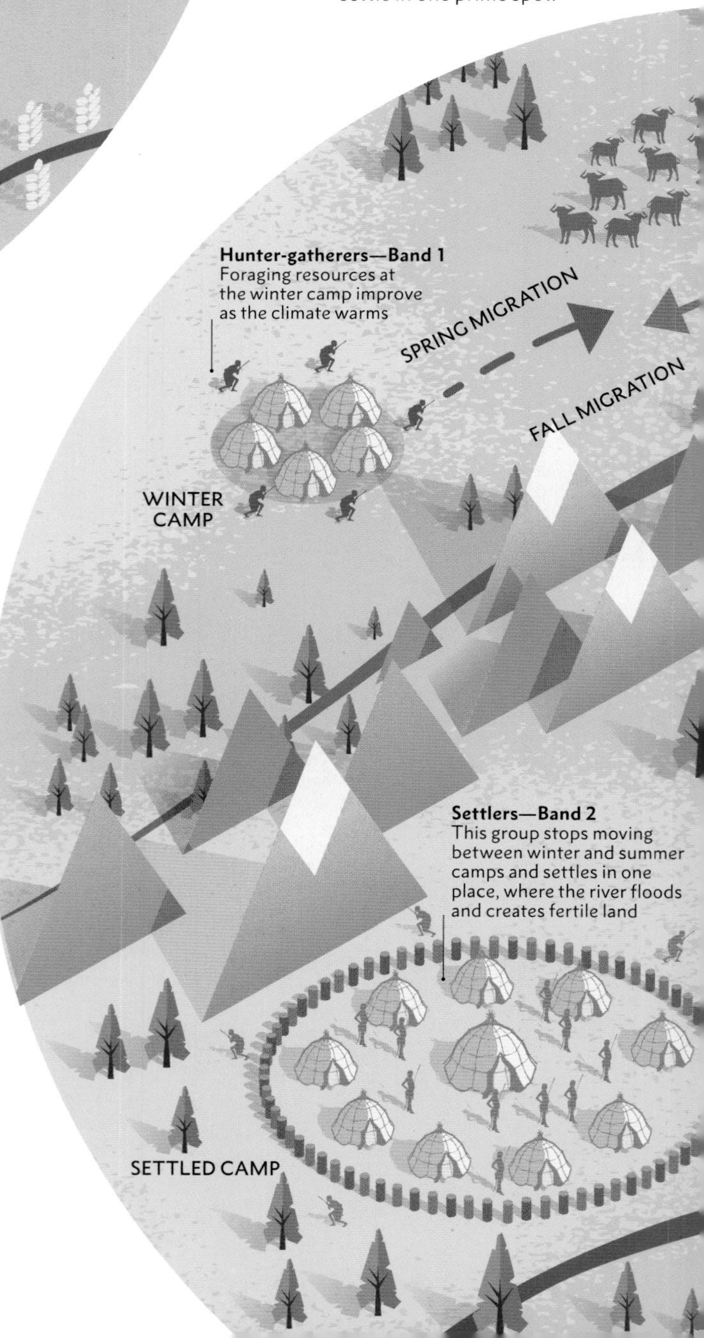
Hunter-gatherers—Band 1
Foraging resources at the winter camp improve as the climate warms
SPRING MIGRATION
FALL MIGRATION
WINTER CAMP
Settlers—Band 2
This group stops moving between winter and summer camps and settles in one place, where the river floods and creates fertile land
SETTLED CAMP

1 BCE | 1 CE

1300 CE | RENAISSANCE BEGINS

1439 CE | PRINTING PRESS SPARKS THE INFORMATION REVOLUTION

1600 CE | COLUMBIAN EXCHANGE

EARLY FARMING COULD SUPPORT **50 TO 100 TIMES** AS MANY PEOPLE **AS FORAGING** IN A SIMILAR AREA

▶ **Growing settlements**
6000 BCE
The population has continued to rise and people are more firmly tied to their land: buildings are permanent and villages are defended. With more mouths to feed, wild cereal crops are deliberately cultivated, and animals are penned to supply the community with meat.

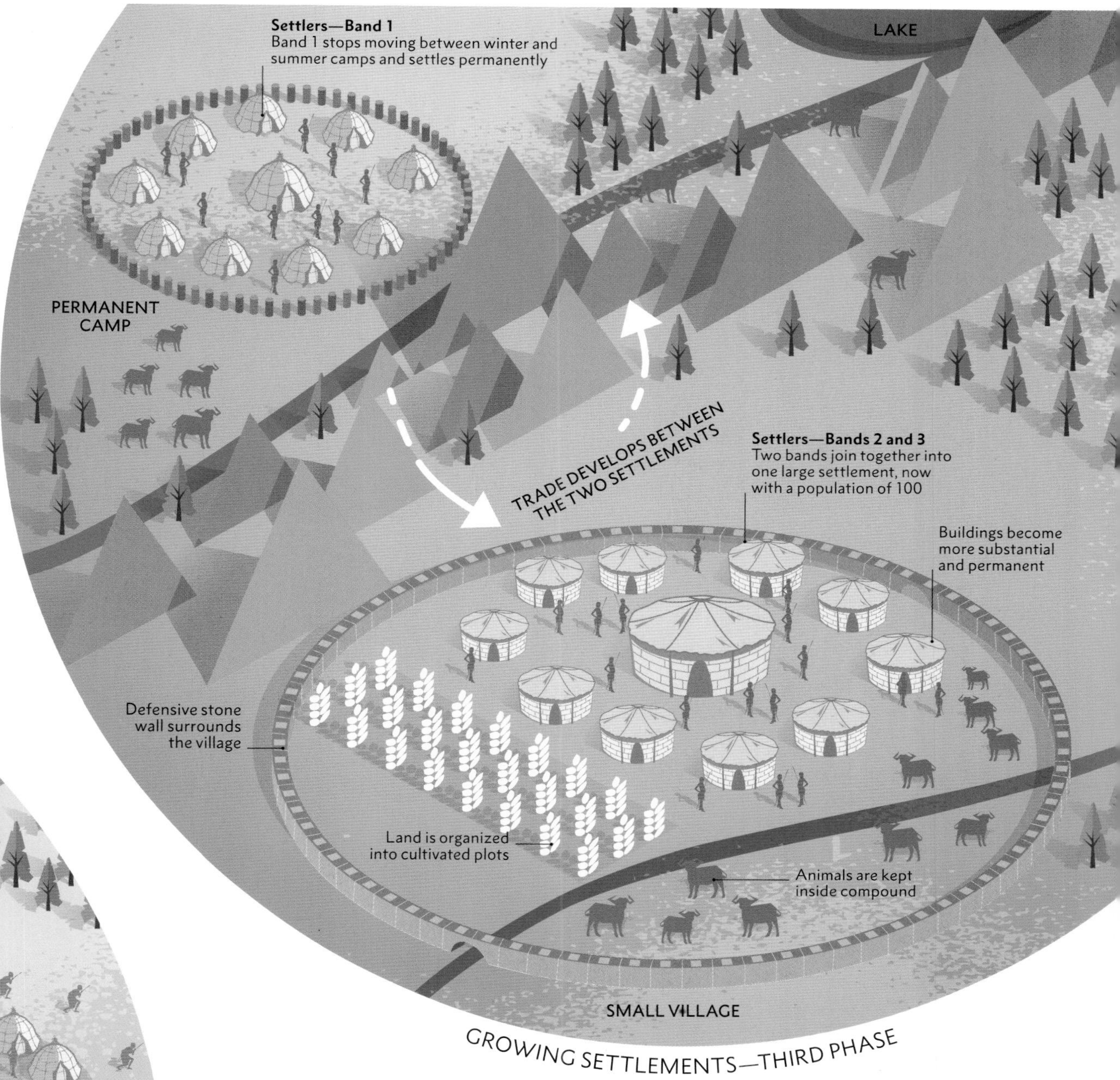

Settlers—Band 1
Band 1 stops moving between winter and summer camps and settles permanently
LAKE
PERMANENT CAMP
TRADE DEVELOPS BETWEEN THE TWO SETTLEMENTS
Settlers—Bands 2 and 3
Two bands join together into one large settlement, now with a population of 100
Buildings become more substantial and permanent
Defensive stone wall surrounds the village
Land is organized into cultivated plots
Animals are kept inside compound
SMALL VILLAGE
GROWING SETTLEMENTS—THIRD PHASE

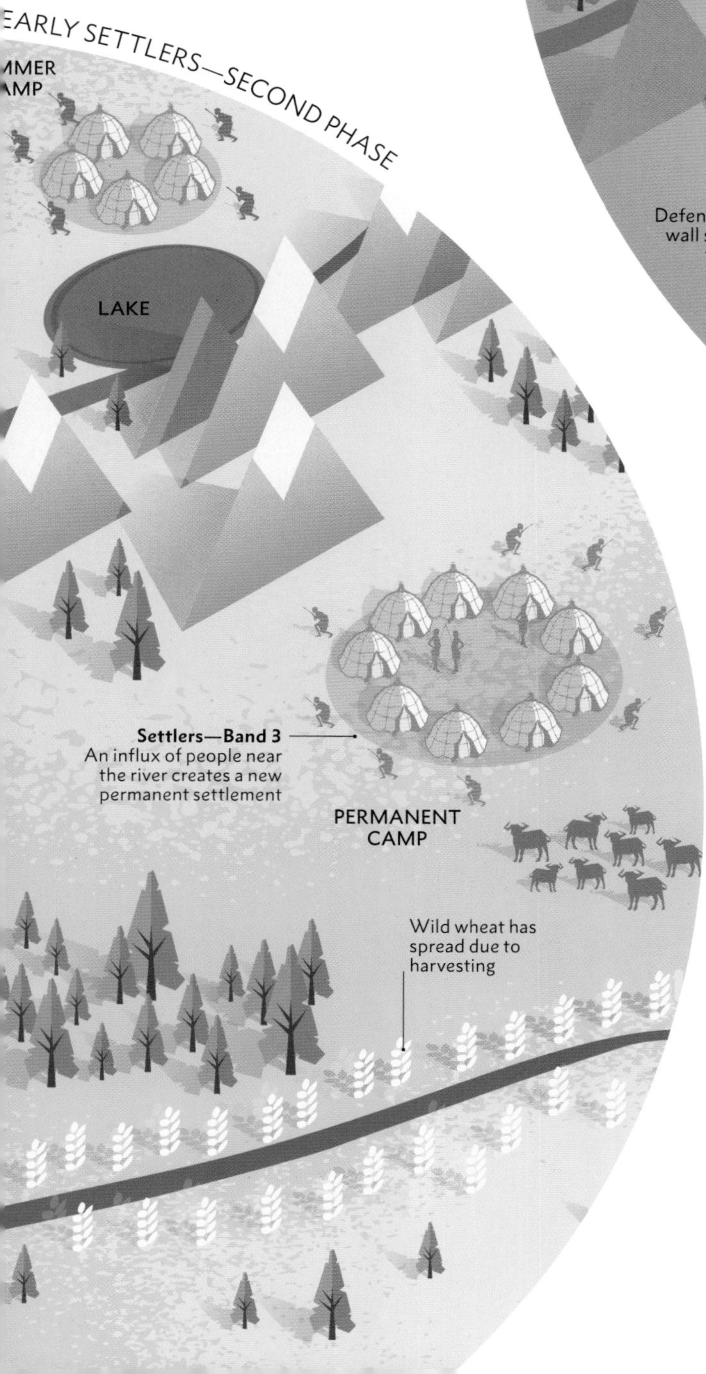

EARLY SETTLERS—SECOND PHASE
MMER
MP
LAKE
Settlers—Band 3
An influx of people near the river creates a new permanent settlement
PERMANENT CAMP
Wild wheat has spread due to harvesting

◀ **Always on the move**
Modern pastoralists continue to follow the nomadic way of life, moving with their animals to find better pasture and water if climatic conditions change. This gives them a substantial advantage over settled farmers, who can lose their crops and animals in periods of drought.

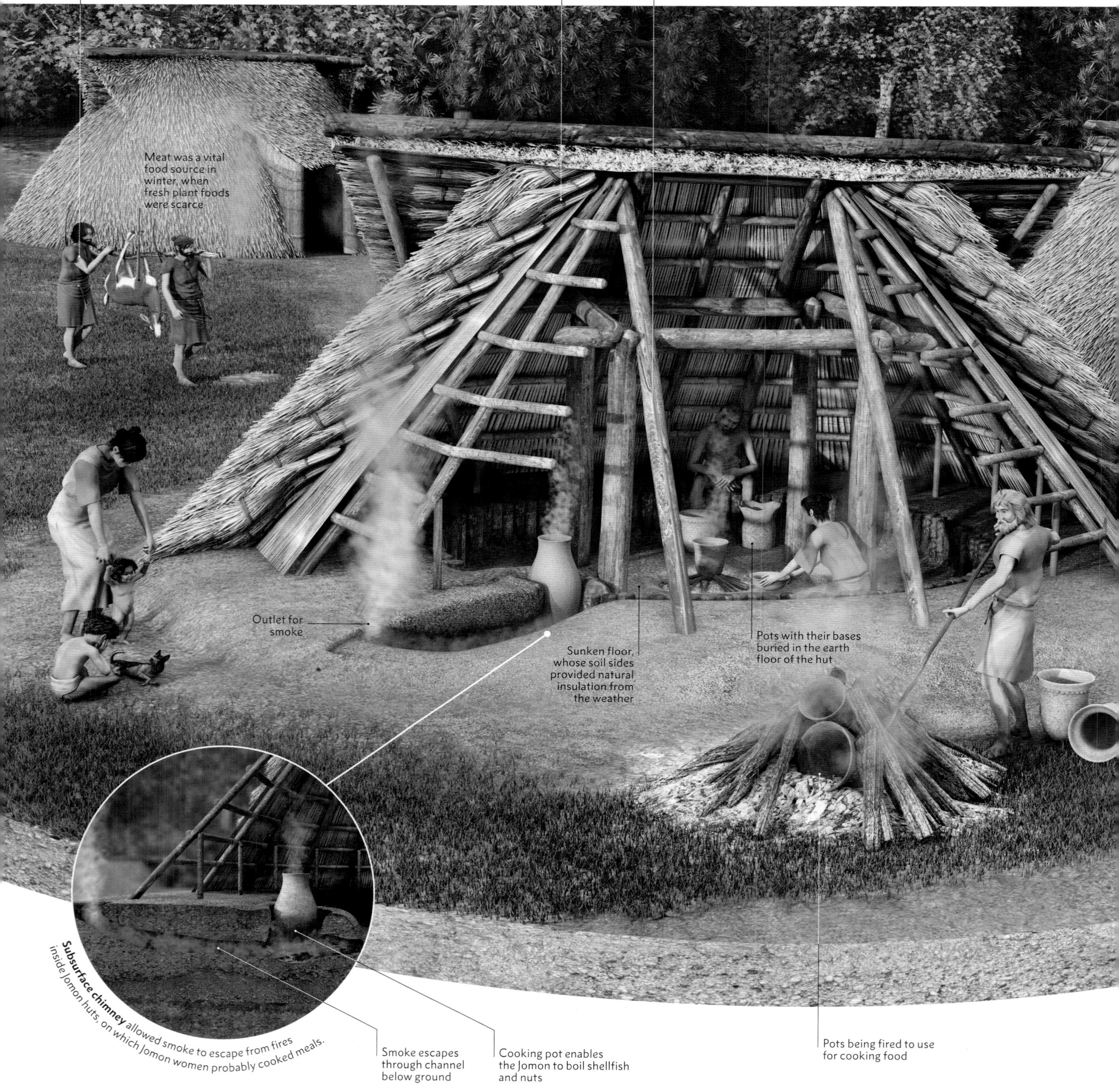
Jomon hunters caught game, including wild boar, deer, and bears, using pit traps and bows and arrows
Jomon houses were usually 10–13ft (3–4m) across
This cross-section shows how the huts may have been constructed. The main evidence comes from sunken floors and post-holes for timber
Meat was a vital food source in winter, when fresh plant foods were scarce
Outlet for smoke
Sunken floor, whose soil sides provided natural insulation from the weather
Pots with their bases buried in the earth floor of the hut
Subsurface chimney allowed smoke to escape from fires inside Jomon huts, on which Jomon women probably cooked meals.
Smoke escapes through channel below ground
Cooking pot enables the Jomon to boil shellfish and nuts
Pots being fired to use for cooking food

It is likely that the roofs and sides were thatched, helping to ventilate the interior
The forest was rich in plant foods, such as berries, walnuts, chestnuts, and acorns, which the women gathered in fall
Salmon drying on a wooden frame. This process involved many people, and is evidence of community cooperation
The Jomon fished using specialized tools: spears, nets, basket traps, and lines
Boat made from a hollowed-out log
Rivers and lakes yielded salmon and other freshwater fish, while tuna, mackerel, turtles, and shellfish were harvested from the sea
ns and other plant foods were kept in ts and storage pits
Grinding grain collected from wild plants

Life in the village
This is a typical Japanese Jomon village of *c.*13,000 BCE. At the time, villages were small, consisting of around five pit-houses. Settlements gradually grew larger until, by 9000 BCE, some contained as many as 50 or 60 houses.

AFFLUENT FORAGERS

At the end of the last ice age, climates became warmer and wetter, which enabled human communities to stay in the same place for longer, while still living as hunter-gatherers. They are described as "affluent foragers."

Affluent foragers settled in areas of natural abundance and were able to live off the fruits of the land. Among the most successful affluent foragers were the Jomon of Japan, who first settled in villages around 14,000 BCE. They lived in small communities—without adopting farming—for more than 13,000 years. The Jomon lived beside forests, but also stayed close to the coasts, river estuaries, and lakes. Their mixed environment provided a rich, varied diet of seasonally available plant foods, fish, and wild animals. This, combined with their more sedentary lifestyle, allowed affluent foragers to invest more energy in larger specialized tools and technology rather than just portable objects. The Jomon were the first people to invent pottery, in around 13,000 BCE, which they used to cook fish and store food to consume out of season.

▶ Flame-rimmed vessel
Jomon pottery was fired in the open air, in bonfires. From simple beginnings, Jomon pots grew more elaborate. This richly decorated vessel dates from the late Jomon period.

HUNTERS BEGIN TO **GROW FOOD**

The first farmers worked the land with wooden digging sticks and stone-bladed hoes and adzes. This method, called horticulture, was not productive enough to create a surplus. It was subsistence farming, in which people grew only enough crops to feed their own families.

The simplest agricultural tool is a digging stick—a strong, straight, pointed stick, often hardened in a fire. To remove weeds, farmers used a hoe, which had a blade made of stone or antler set at an angle to the handle. Without plows or draft animals, people could grow crops only in light, easily worked soils, such as loess, a fertile topsoil formed by windblown dust.

▲ **Wooden adze**
Flint-bladed tools are remarkably strong. An adze can cut a large hardwood tree in a matter of hours.

FARMING WITH FIRE

Long before farming, hunter-gatherers had burned forests to create open areas where they could hunt grazing animals, and encourage the growth of useful plants such as hazel and willow for making baskets. The first farmers used fire in a similar way. After cutting down an area of forest with stone-bladed adzes, they left the vegetation to dry and then burned it. The ash made the soil fertile for planting seeds. But after two years, the fertility of a field dropped, and farmers had to move on to create a new one.

Using fire to create fields is called slash-and-burn or swidden farming, from an old Norse word for "burned ground." It is still practiced by between 200 million and 500 million people worldwide, mostly in the tropical rain forests of South America, Southeast Asia, and Melanesia. Slash-and-burn is sustainable in these regions because high rainfall and a warm climate permit a year-round growing season. But it is only practical where there are relatively few people and the area of forest is large enough to support the population's size.

Slash-and-burn proved unsustainable in the cooler, drier latitudes of Eurasia, where farming began. The short growing season meant that vegetation took much longer to recover after a fire. As the population grew, people were forced to invent new ways to increase the yield from their fields. Their challenge was to find better tools than the hoe and digging stick, and new ways to fertilize the soil.

Despite this, we know that slash-and-burn was once practiced across large areas of Eurasia. Studies of ancient peat bogs in northern Europe show the disappearance of pollen from oak trees, accompanied by a rise in pollen from cereal crops along with layers of powdered charcoal—clear evidence of slash-and-burn farming.

FOREST GARDENING

Human interaction with the forest was not always quite so devastating. As people living beside rain forest rivers and on wet foothills in monsoon regions began to adapt to their immediate surroundings, they learned which species were helpful to the growth of food plants, and which were a hindrance. They protected useful plants and removed unwanted species. Later, they introduced beneficial plants from elsewhere to these "forest gardens."

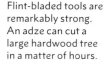

"THE **PROUD GETAE** LIVE HAPPILY, GROWING FREE FOOD FOR THEMSELVES ON LAND THEY **DO NOT WANT TO CULTIVATE FOR MORE THAN A YEAR**."

Horace, Roman poet, 1st century BCE

Destructive harvest
Areas of Laos still follow a tradition of slash-and-burn cultivation. However, it is highly destructive to the rain forest—crops are only grown for one year as they quickly deplete the soils, and harvests are poor. The area then has to be left for between four and six years to regenerate.

8000 BCE | AGRICULTURE EMERGES
6000 BCE | FIRST CITIES FORM
4000 BCE | WRITING DEVELOPS
3100 BCE | FIRST CIVILIZATIONS EMERGE

FARMING BEGINS

Farming began once people started to store and plant seeds and tubers. Archaeology shows that within a few millennia, agriculture had emerged separately in different parts of the world that had no contact with each other.

Reasons for becoming farmers may have varied. In some places, people responded to a shortage of wild foods, due to climate change or a rising population. In other areas, they may simply have preferred one food crop over others. They would not have made a conscious decision to become farmers—they had no idea what the new way of life would be like. However, food production could only begin where there was a source of animals and plants suitable for domestication. The range varied from region to region and as a result, farming had different impacts in each world zone. The crops of eastern North America and New Guinea were much less nutritious than those of other farming areas, so people continued to depend on wild foods, and farmers lived alongside hunter-gatherers. It was very different in the Fertile Crescent and China, where agriculture offered such a complete food production package that farmers were able to outcompete their hunter-gatherer neighbors.

> **EACH** OF THE **FOUR WORLD ZONES** WAS ITS **OWN WORLD FOR A TIME**.
>
> **Cynthia Stokes Brown**, American historian, 1938–

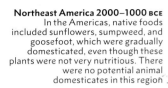

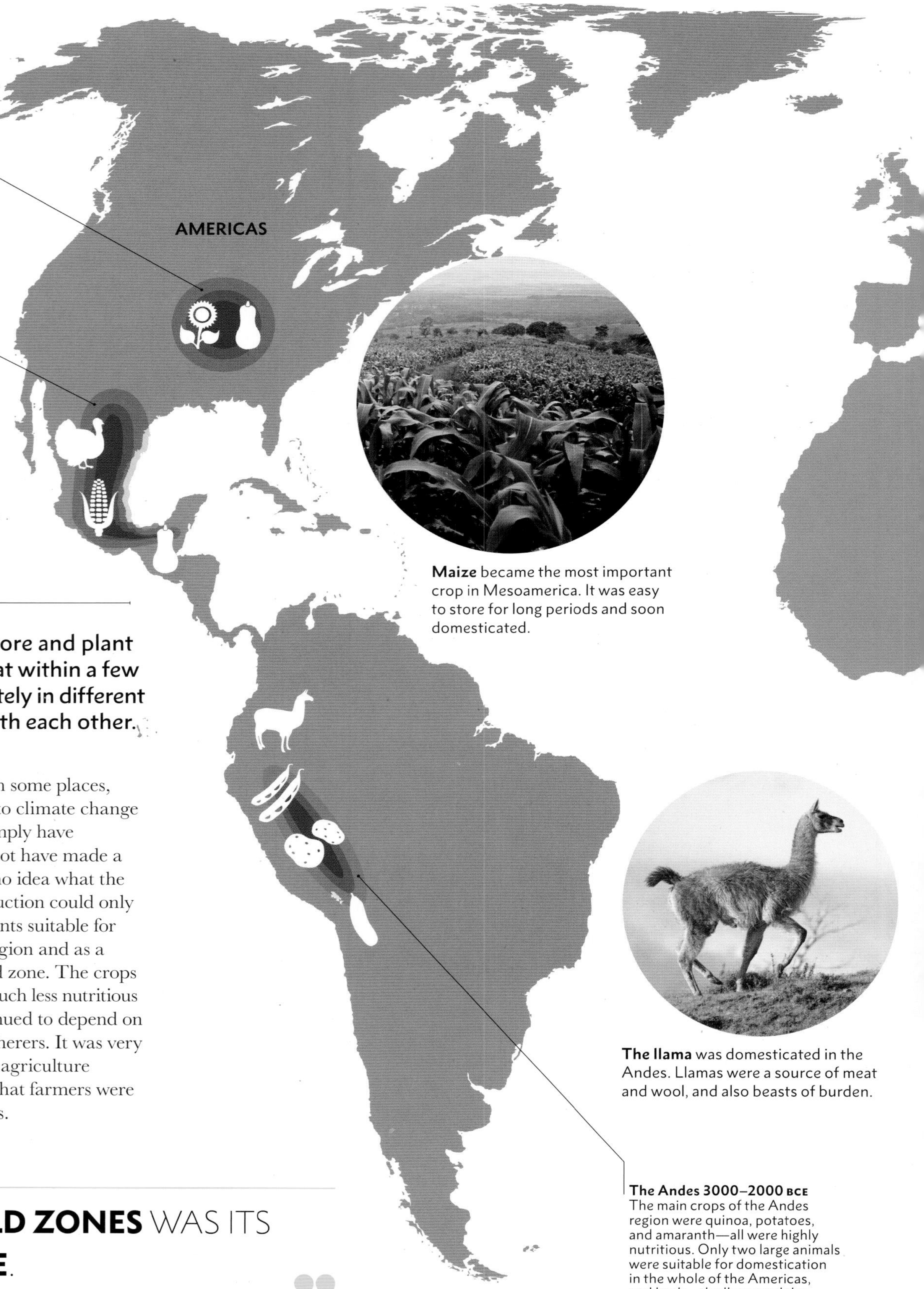
AMERICAS
Northeast America 2000–1000 BCE
In the Americas, native foods included sunflowers, sumpweed, and goosefoot, which were gradually domesticated, even though these plants were not very nutritious. There were no potential animal domesticates in this region
Mesoamerica 3000–2000 BCE
Mesoamerican farmers had an ideal combination of crops, with maize and beans grown alongside each other. Turkeys and dogs were the only domesticable animals, and were raised for meat
The Andes 3000–2000 BCE
The main crops of the Andes region were quinoa, potatoes, and amaranth—all were highly nutritious. Only two large animals were suitable for domestication in the whole of the Americas, and both—the llama and the alpaca—were found in the Andes

Maize became the most important crop in Mesoamerica. It was easy to store for long periods and soon domesticated.

The llama was domesticated in the Andes. Llamas were a source of meat and wool, and also beasts of burden.

Sheep were first domesticated in the Fertile Crescent and herding remains a way of life in Egypt today.

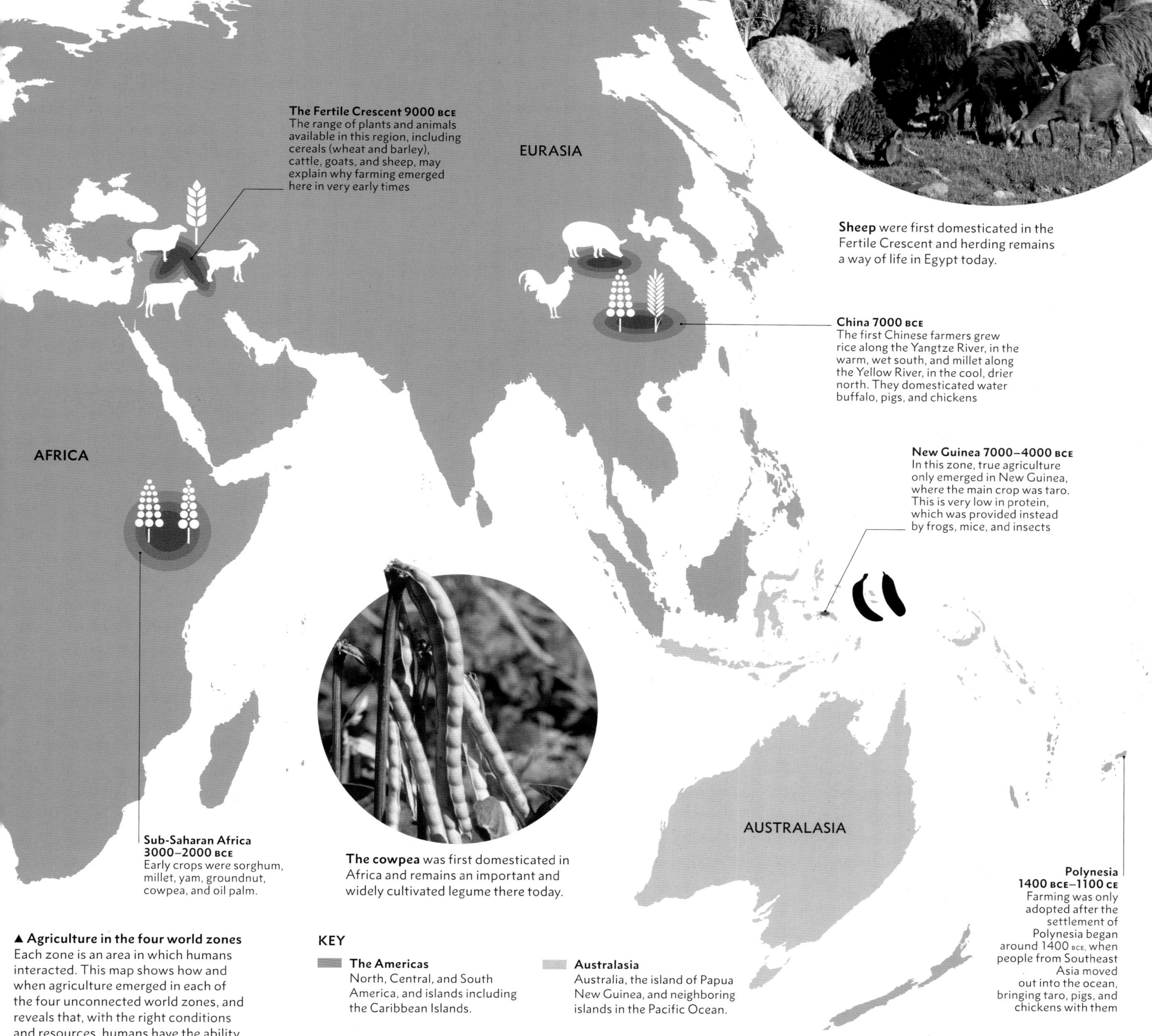

The Fertile Crescent 9000 BCE
The range of plants and animals available in this region, including cereals (wheat and barley), cattle, goats, and sheep, may explain why farming emerged here in very early times

China 7000 BCE
The first Chinese farmers grew rice along the Yangtze River, in the warm, wet south, and millet along the Yellow River, in the cool, drier north. They domesticated water buffalo, pigs, and chickens

New Guinea 7000–4000 BCE
In this zone, true agriculture only emerged in New Guinea, where the main crop was taro. This is very low in protein, which was provided instead by frogs, mice, and insects

Sub-Saharan Africa 3000–2000 BCE
Early crops were sorghum, millet, yam, groundnut, cowpea, and oil palm.

The cowpea was first domesticated in Africa and remains an important and widely cultivated legume there today.

Polynesia 1400 BCE–1100 CE
Farming was only adopted after the settlement of Polynesia began around 1400 BCE, when people from Southeast Asia moved out into the ocean, bringing taro, pigs, and chickens with them

▲ Agriculture in the four world zones
Each zone is an area in which humans interacted. This map shows how and when agriculture emerged in each of the four unconnected world zones, and reveals that, with the right conditions and resources, humans have the ability to innovate and often find similar solutions to similar problems.

KEY

The Americas
North, Central, and South America, and islands including the Caribbean Islands.

Afro-Eurasia
Africa and the Eurasian landmass, including islands such as Britain and Japan.

Australasia
Australia, the island of Papua New Guinea, and neighboring islands in the Pacific Ocean.

The Pacific Islands
Societies such as New Zealand, Micronesia, Melanesia, and Hawaii.

WILD PLANTS **BECOME CROPS**

Domestication is a process through which plants are brought under human control. As a result of human selection, plants changed until they were unable to reproduce successfully in the wild. Domestication was a two-way process, which benefited plants as well as people.

The most important domesticated plants are the grasslike grain crops, which offer little nutrition individually but can be gathered in bulk. The heads of wild grasses shatter when ripe, so that the grains can spread in the wind. However, it was easier for early foragers to harvest grains that stayed longer on the plants. Eventually, a new plant developed with heads that no longer needed to shatter: domesticated plants wait to be harvested.

The growing season was also changed by domestication. Wild seeds tend to germinate piecemeal over a long period, which ensures that in a changing climate some plants will survive. Humans created plants that all germinated at the same time. Domesticated plants also grew to around the same height, which made them easier to harvest. The grains themselves grew bigger and became easier to remove from their husks.

These changes were not consciously planned by farmers. They occurred as a natural result of selecting seeds from the most desirable plants to harvest and sow in the following year. Yet the more plants were brought under human control, the more human lives revolved around the needs of domesticated plants. As farming developed, people found themselves forced to spend long days caring for wheat, rice, and corn.

▼ Held for harvest
The difference between wild and domesticated wheat grains is subtle, but significant. The change from an easily breakable rachis (shaft) on wild plants to one that needed to be threshed meant that more grain could be collected—but it took a lot more effort.

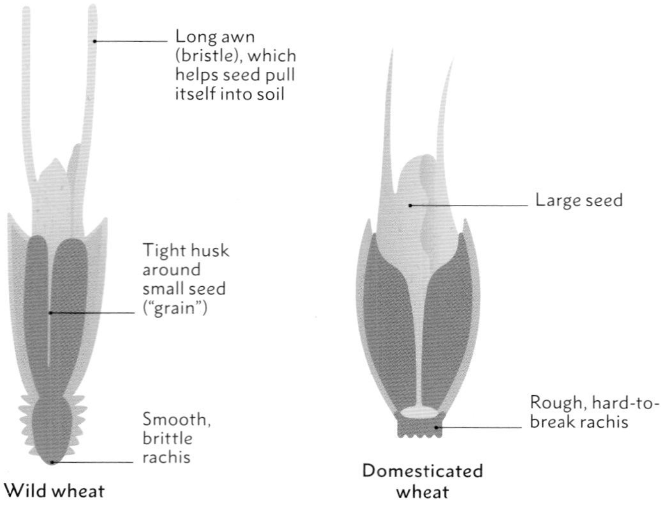

▲ Vital commodity
Rice now provides one-fifth of all the calories consumed by humans worldwide. It can be grown even on steep hillsides, using terraces.

FIRST CROPS

Domestication of wheat began in the region known as the Fertile Crescent, in the Middle East. Here, between 11,000 and 9000 BCE, early farmers domesticated two types of wheat—wild emmer and einkorn. Then, in Iran around 7000 BCE, domesticated emmer wheat crossed with a wild goatgrass to become bread wheat; this has larger grains, easily removable husks, and higher gluten levels, which creates an elastic dough that rises to form soft bread.

Unlike other cereals, rice is a marsh plant, suitable for growing in water. It was domesticated between 4900 and 4600 BCE in southern China, south of the Yangtze River. Wild rice has a long awn, a hard husk, a tiny grain, and a strong stem, allowing the plant to regenerate itself. Domesticated rice developed a bigger grain, and lost its awn, hard husk, and ability to regenerate itself.

In the 5th millennium BCE, farmers in southwest Mexico transformed the wild teosinte plant into maize, or corn. Teosinte yields less than 12 kernels, while corn produces up to 600. Teosinte kernels are protected by a hard outer covering, but corn kernels are naked. The plants look so different that the relationship between them was only discovered in the 20th century.

Beans were domesticated 6,000 years ago in Mesoamerica and also in the Andes. The plants selected produced bigger beans or better yields and were easier to harvest. In the Andes, the plant changed from a tall vine to a more productive bush.

▶ More gatherable grains
Over time, wheat evolved from wild shattering varieties with small grains, to a plant with nonshattering heads and bigger grains. Farmers also selected for head size, plant height, growing season, and grains that were easy to remove from their husk. Scientists have recently begun making hybrids by crossing modern varieties with their wild relatives to reintroduce old characteristics such as resistance to drought, heat, and pests.

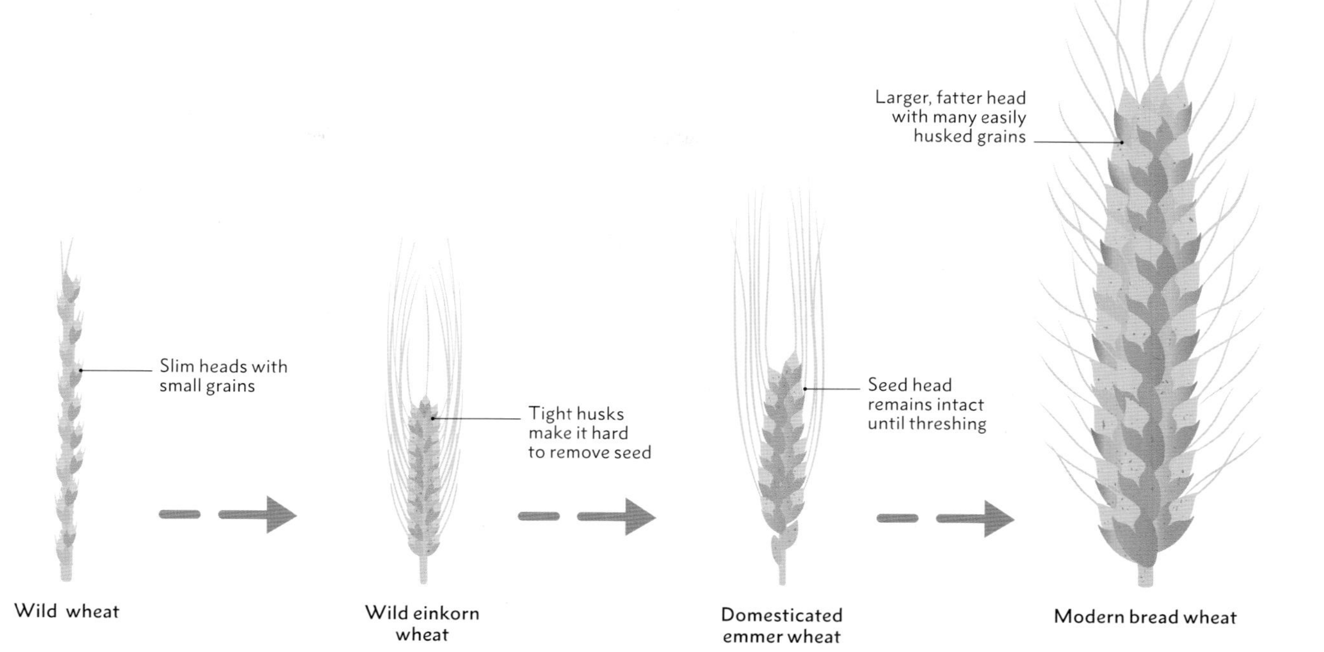

▶ Bigger beans
Wild beans were a staple of the Mesoamerican diet because they contain amino acids that corn does not have. Wild plants have small pods that twist when ripe, splitting open to release their seeds (the beans). Domesticated species have more beans in bigger pods, but the beans stay in their pods until humans split them open. In Mesoamerica, beans were planted alongside corn, which acted as a support, and squashes, to suppress weeds, in what is known as the "three sisters" planting scheme.

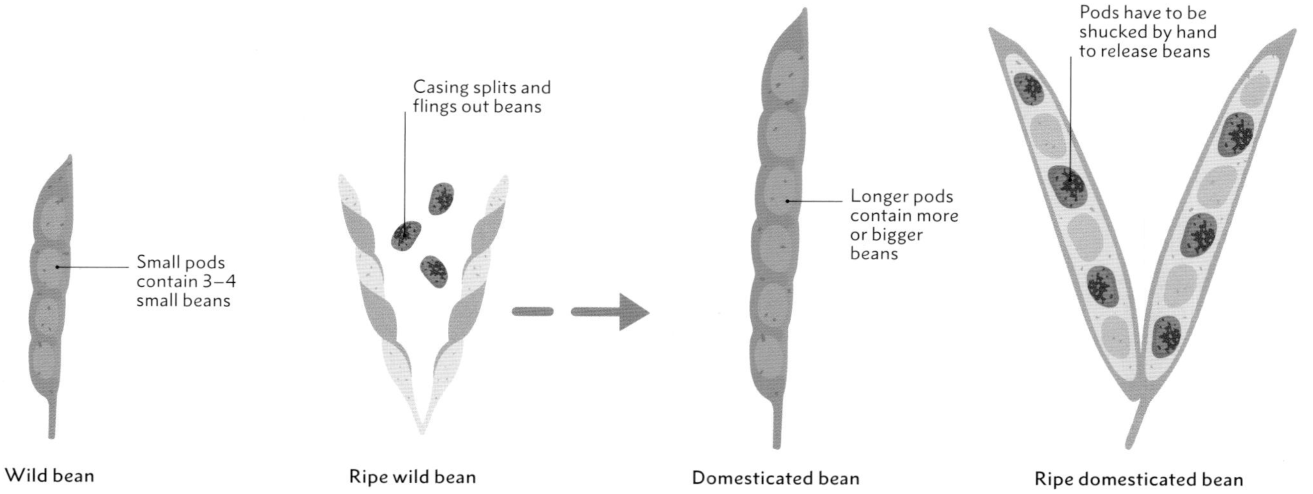

> **PLANTS DOMESTICATED** ***HOMO SAPIENS***, RATHER THAN **VICE VERSA**.

Yuval Noah Harari, Israeli historian, 1976–, *Sapiens: A Brief History of Humankind*

▶ Gigantic improvement
Teosinte, a wild form of maize (corn), has only a few kernels on a head less than 1in (2.5cm) long. A modern domesticated corn cob, packed with kernels, can measure more than 12in (30cm). The discovery of phytoliths (plant microfossils) and starch grains from a number of sites in Mexico suggests that domesticated forms of corn may have existed much earlier than previously thought.

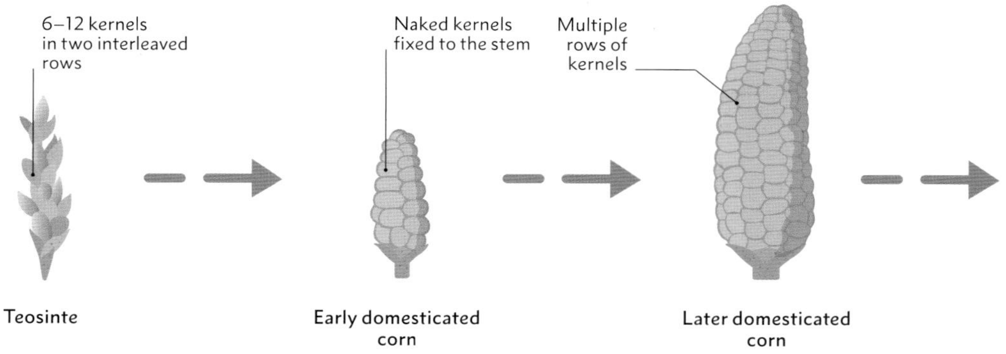

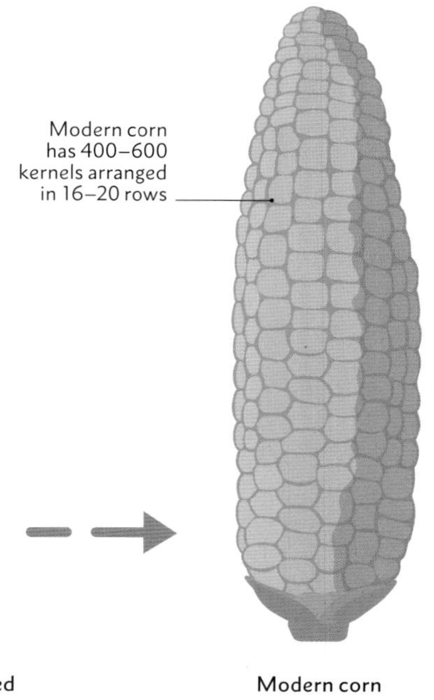

POLLEN **GRAINS**

Small amounts of plant residue can reveal a wealth of information about climatic conditions, the history of agriculture, and the lives of our ancestors thanks to forensic techniques like pollen analysis.

The study of pollen, plant spores, and microscopic plant organisms is known as palynology. Pollen grains, which are the male reproductive bodies of flowering plants, are produced in vast quantities in nature. Thanks to its hard outer shell, a pollen grain can survive for millions of years in favorable conditions. Different plants have distinctively shaped pollen grains, which makes it possible to identify the plants that produced them.

Pollen survives best in peat bogs, lake beds, and cave sediments. Ancient pollen associated with humans is also found in mud bricks, storage pits, boats, pottery vessels, tombs, preserved bodies, and coprolites (fossil feces). It can also be detected on the surfaces of grinding stones and stone tools.

Palynologists use an electron microscope to identify individual pollen grains, counting the grains of each type. Using this data, they recreate a picture of the climate and environment in one area at a particular time. By repeating the study with different depths of soil deposits, they build up a pollen chronology, which shows how the range of plants changed over time. Archaeological sites can be dated by matching the range of pollen collected with the known chronology.

Palynology has revealed the huge impact that early farming had on the environment. Wherever it was practiced, agriculture was marked by a decline in tree pollen and a rise in pollen from cereals and opportunistic weeds, such as darnel, that are associated with their growth.

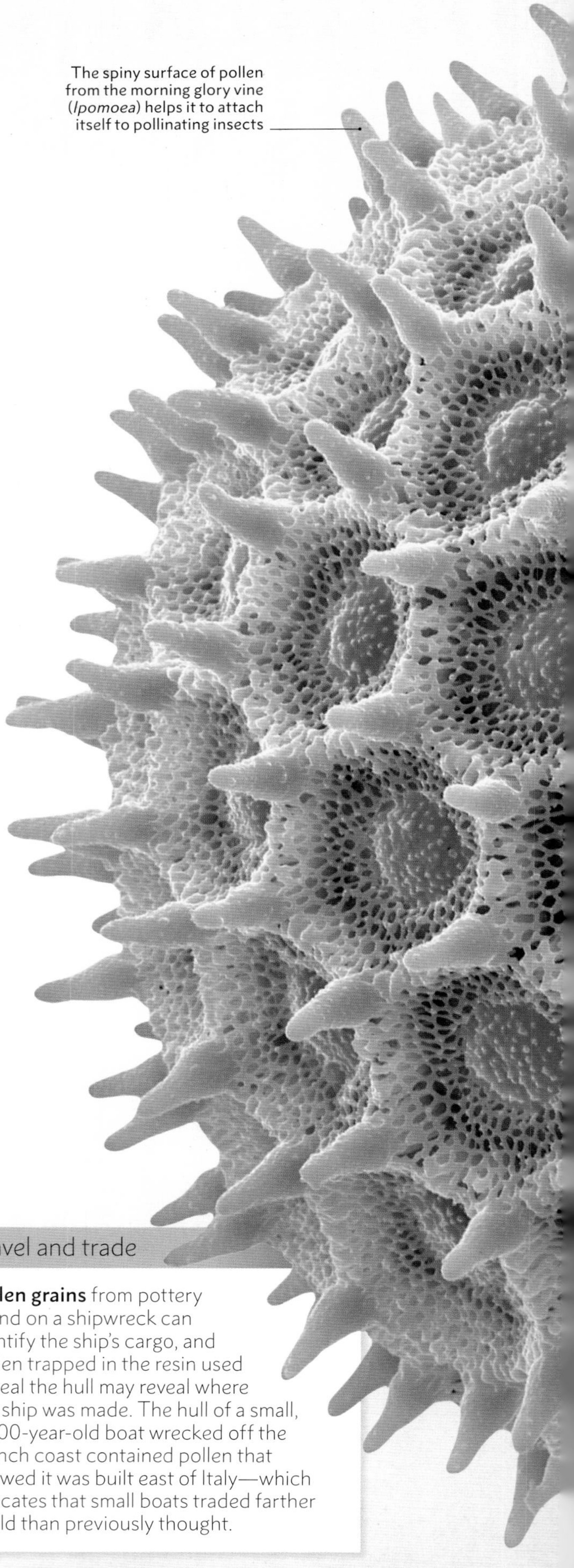

The spiny surface of pollen from the morning glory vine (*Ipomoea*) helps it to attach itself to pollinating insects

Orange
(*Citrus sinensis*)

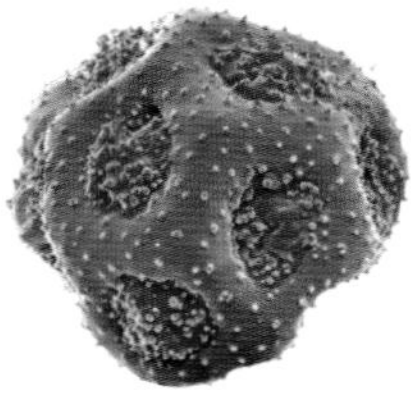

Primula
(*Primula* sp.)

Geranium
(*Geranium* sp.)

Scots pine
(*Pinus sylvestris*)

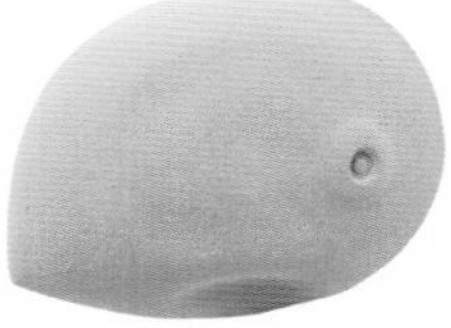

Maize
(*Zea mays*)

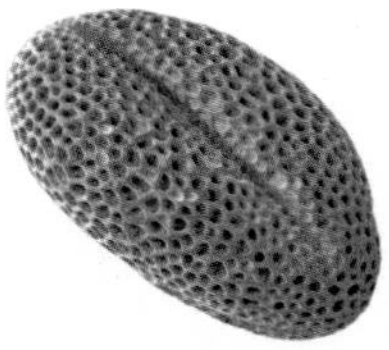

Rapeseed
(*Brassica napus*)

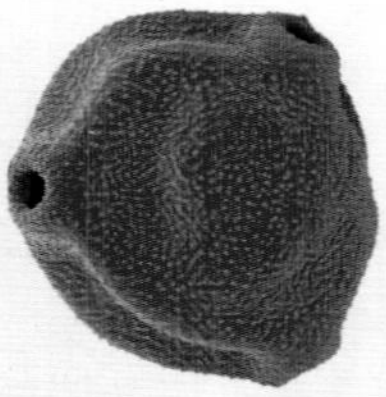

Silver birch
(*Betula pendula*)

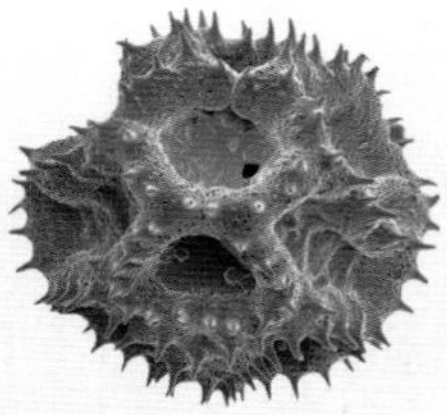

Narrow-leaved hawksbeard
(*Crepis tectorum*)

Wheat
(*Triticum spp.*)

▲ Pollen gallery
This selection of pollen shows how distinctly shaped the pollen from different plants is. Pollen also ranges widely in size, from 5 to 500 microns (1 micron is 0.001 mm).

Travel and trade

Pollen grains from pottery found on a shipwreck can identify the ship's cargo, and pollen trapped in the resin used to seal the hull may reveal where the ship was made. The hull of a small, 2,000-year-old boat wrecked off the French coast contained pollen that showed it was built east of Italy—which indicates that small boats traded farther afield than previously thought.

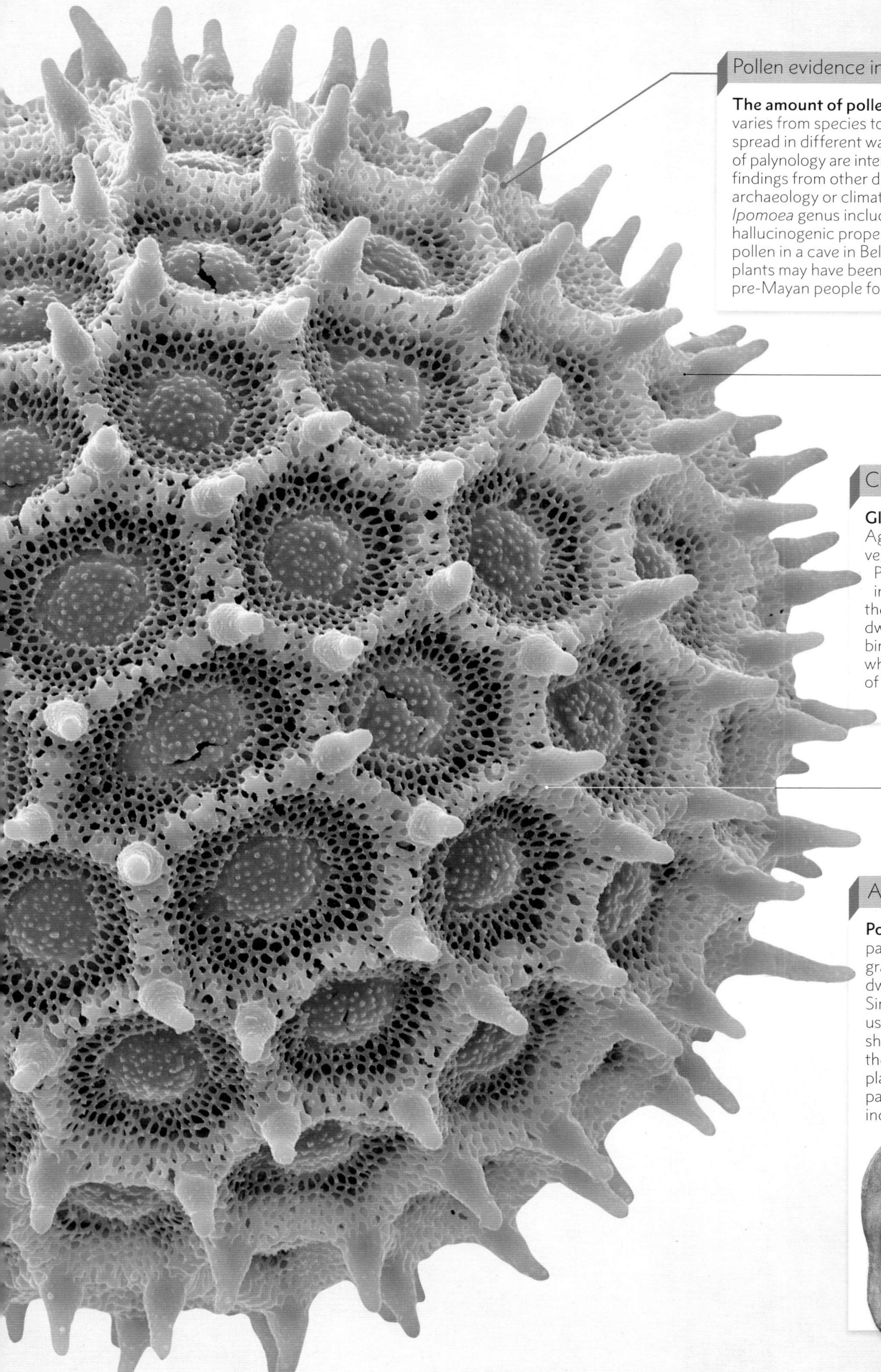
The tough, rigid, waterproof shell prevents the pollen grain from rotting or drying out.
This pollen grain can be identified as *Ipomoea purpurea* from its size, shape, and surface features

Pollen evidence in context

The amount of pollen plants produce varies from species to species, and is spread in different ways, so the results of palynology are interpreted alongside findings from other disciplines such as archaeology or climate science. The *Ipomoea* genus includes plants with hallucinogenic properties; *Ipomoea* pollen in a cave in Belize suggests the plants may have been taken there by pre-Mayan people for ritual purposes.

Morning glory vine

Climate change

Global warming at the end of the Ice Age caused a dramatic change in vegetation across northern latitudes. Pollen collected from lake sediments in Britain shows that before 9600 BCE the only trees there were cold-hardy dwarf birches. As the climate warmed, birches were replaced by Scots pines, which in turn gave way to a wider variety of trees, including hazels, elms, and oaks.

Agriculture and food

Pollen can help us understand what past peoples farmed and ate. Pollen from grass and other fodder plants stored in dwellings tells us how livestock were fed. Similarly, pollen clinging to grinding stones used by New Mexico's Anasazi people shows that alongside domesticated maize they also harvested a wide variety of wild plants. In the American Southwest, palynologists recreated a prehistoric individual's diet from pollen found in coprolites (fossilized feces), while in Scotland pollen on 5,000-year-old pottery shards was used to recreate the recipe for heather ale, drunk by early Celtic farmers.

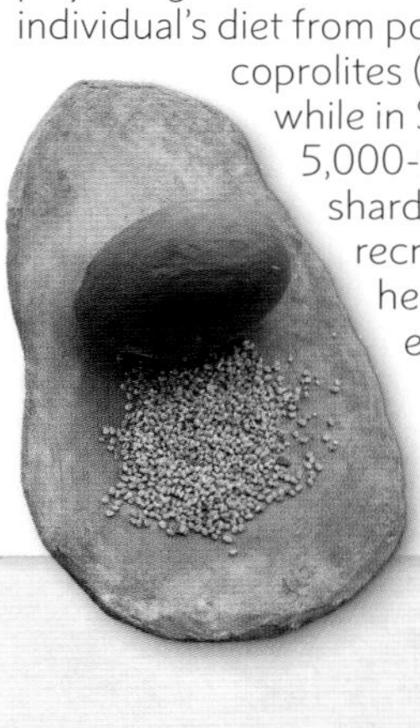

Stone quern with sandstone rubber

FARMERS **DOMESTICATE ANIMALS**

The domestication of animals began at roughly the same time and in the same areas as the domestication of plants. The process probably began with men guarding a local herd of animals as it moved, assisted by dogs. Eventually the herd was enclosed, fed, and protected.

A domesticated animal is one that has been bred in captivity and has become modified from its wild ancestor. Some animals, such as elephants and bears, can be tamed, but this is not the same as domestication. Tamed elephants remain wild animals, and never adapt completely to their new conditions.

Animals needed certain characteristics to be suitable for domestication. They had to be a manageable size and relatively docile with social structures, early sexual maturity, and a high reproductive rate. Herbivores were better than carnivores because they would survive on local plants. Just 14 large mammals met all these requirements, almost all of them in Eurasia.

Attempts to domesticate other animals failed: bison are related to cattle, but they are more aggressive, faster, and can leap 6ft (1.8m) into the air. Similarly, zebras are more aggressive than horses, and have better peripheral vision, which makes them almost impossible to catch with a rope. Gazelles have a tendency to panic, and are likely to batter themselves to death when placed in an enclosure.

HOW ANIMALS CHANGED

Animals separated from their natural environment began to change as farmers bred from specimens that met their needs. Because people selected smaller animals that were easier to manage, domesticated cattle became smaller than their wild ancestor, the aurochs. Evolution by natural selection also played a part—adaptations for survival in the wild, such as intelligence and long horns, were no longer necessary. Domesticated animals did not have to fear predators or search for new sources of food, and so their brains reduced in size.

In the wild, male mammals are much larger than females because they have to compete with other males for mates. This competition ended in captivity, because breeding was controlled by humans. As a result, male cattle, sheep, and goats became the same size as the females, as well as losing their long horns.

The willingness of these animals to become domesticated ultimately ensured their evolutionary success. There are now 1.4 billion cattle on the planet—but their wild ancestor, the aurochs, became extinct during the 17th century.

> **DOMESTICABLE** ANIMALS ARE **ALL ALIKE**; EVERY UNDOMESTICABLE ANIMAL IS **UNDOMESTICABLE IN ITS OWN WAY**
>
> **Jared Diamond**, American scientist, 1937–, *Guns, Germs, and Steel*

▶ **Wild at heart**
Bees are semi-domesticated. Through selective breeding, humans modified bee behavior, making them less likely to sting and swarm than wild bees. Although managed by humans, bees still forage for their food and retain the ability to survive in the wild.

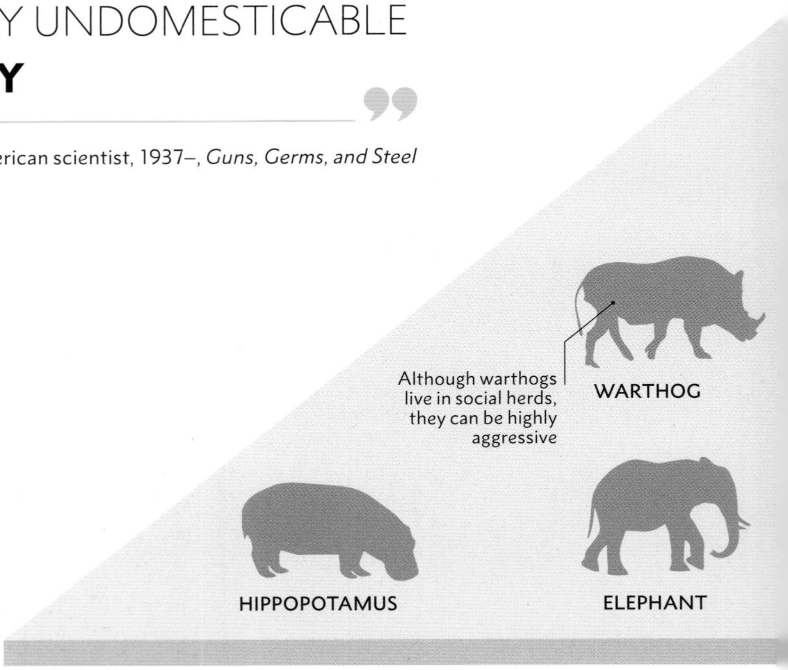
Although warthogs live in social herds, they can be highly aggressive
WARTHOG
HIPPOPOTAMUS
ELEPHANT

DOGS WERE DOMESTICATED FROM WOLVES **35,000 YEARS AGO**

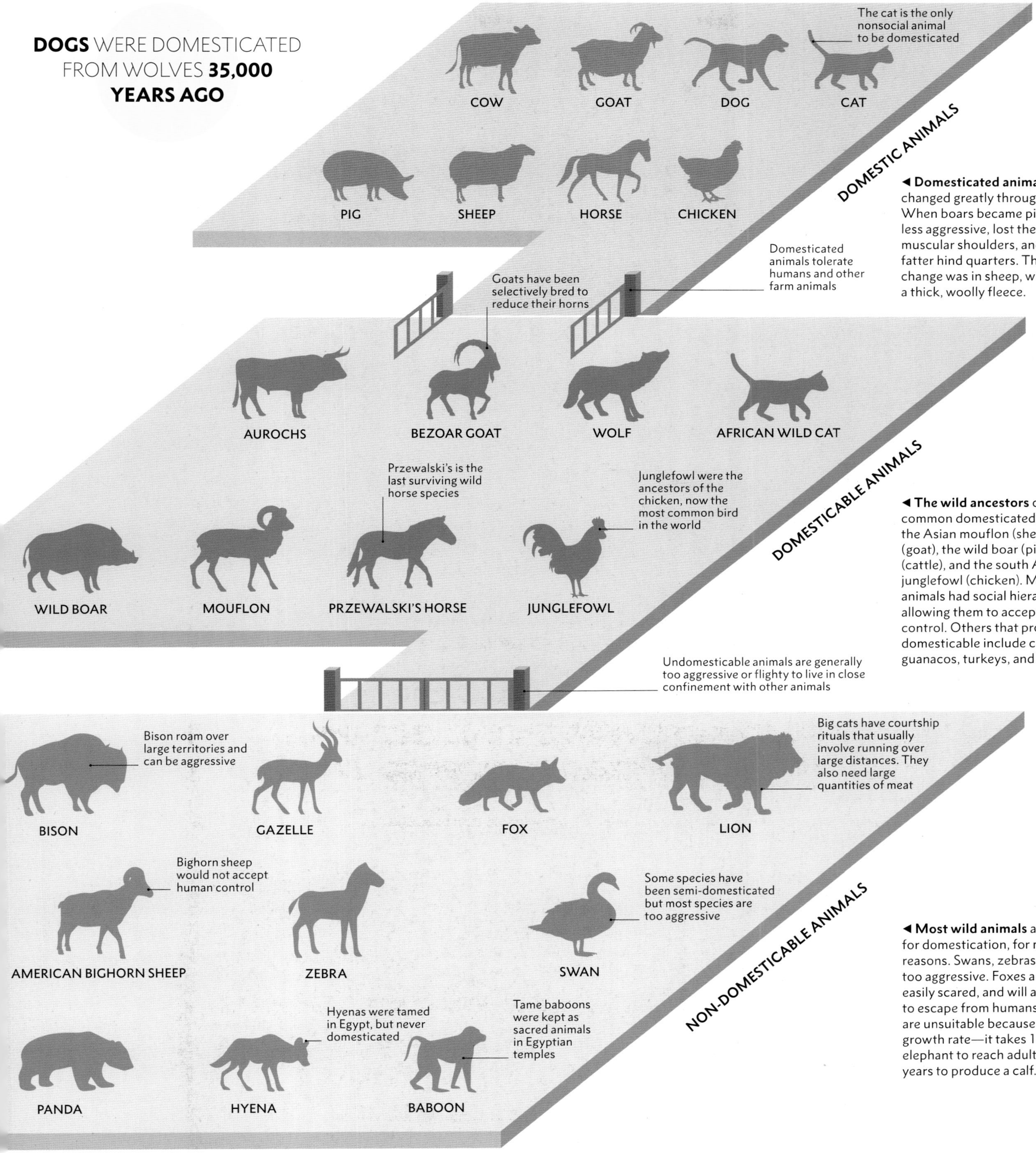
DOMESTIC ANIMALS
COW
GOAT
DOG
CAT
The cat is the only nonsocial animal to be domesticated
PIG
SHEEP
HORSE
CHICKEN
Domesticated animals tolerate humans and other farm animals
Goats have been selectively bred to reduce their horns
DOMESTICABLE ANIMALS
AUROCHS
BEZOAR GOAT
WOLF
AFRICAN WILD CAT
Przewalski's is the last surviving wild horse species
Junglefowl were the ancestors of the chicken, now the most common bird in the world
WILD BOAR
MOUFLON
PRZEWALSKI'S HORSE
JUNGLEFOWL
Undomesticable animals are generally too aggressive or flighty to live in close confinement with other animals
NON-DOMESTICABLE ANIMALS
Bison roam over large territories and can be aggressive
Big cats have courtship rituals that usually involve running over large distances. They also need large quantities of meat
BISON
GAZELLE
FOX
LION
Bighorn sheep would not accept human control
Some species have been semi-domesticated but most species are too aggressive
AMERICAN BIGHORN SHEEP
ZEBRA
SWAN
Hyenas were tamed in Egypt, but never domesticated
Tame baboons were kept as sacred animals in Egyptian temples
PANDA
HYENA
BABOON

◀ **Domesticated animals** almost all changed greatly through human control. When boars became pigs, they became less aggressive, lost their tusks and muscular shoulders, and developed fatter hind quarters. The most extreme change was in sheep, which developed a thick, woolly fleece.

◀ **The wild ancestors** of the most common domesticated animals are the Asian mouflon (sheep), the bezoar (goat), the wild boar (pig), the aurochs (cattle), and the south Asian junglefowl (chicken). Most of these animals had social hierarchies, allowing them to accept human control. Others that proved domesticable include camels, yaks, guanacos, turkeys, and donkeys.

◀ **Most wild animals** are unsuitable for domestication, for many different reasons. Swans, zebras, and bison are too aggressive. Foxes and gazelles are easily scared, and will always attempt to escape from humans. Elephants are unsuitable because of their slow growth rate—it takes 15 years for an elephant to reach adult size, and two years to produce a calf.

Corn, or maize, transformed farming in eastern North America when it arrived from Mesoamerica in c.2000 BCE.

FARMING SPREADS

After its original adoption in several areas of the planet, farming spread in all directions. The new way of life expanded much more rapidly in Eurasia than in the Americas, due to the different shapes of each landmass.

Agriculture spread in two ways. The most common was where farmers were forced to leave their homeland due to the pressures of a rising population and competition over land. Moving on, they took their animals and crops with them. The second, less usual, was for hunter-gatherers to partly adopt the new way of life. As they came into contact with farmers, some hunter-gatherers acquired domesticated cattle, sheep, and goats, and so became herders.

The differing rates of spread were determined by the continents' axes. While Eurasia stretches east to west, the Americas stretch north to south. It was much easier for farmers to move their crops and livestock within the same latitude, because they found similar climactic conditions, seasonality, day lengths, pests, and diseases. But for crops to move from one latitude to another—as corn did in the Americas—the plant had to evolve to suit different conditions.

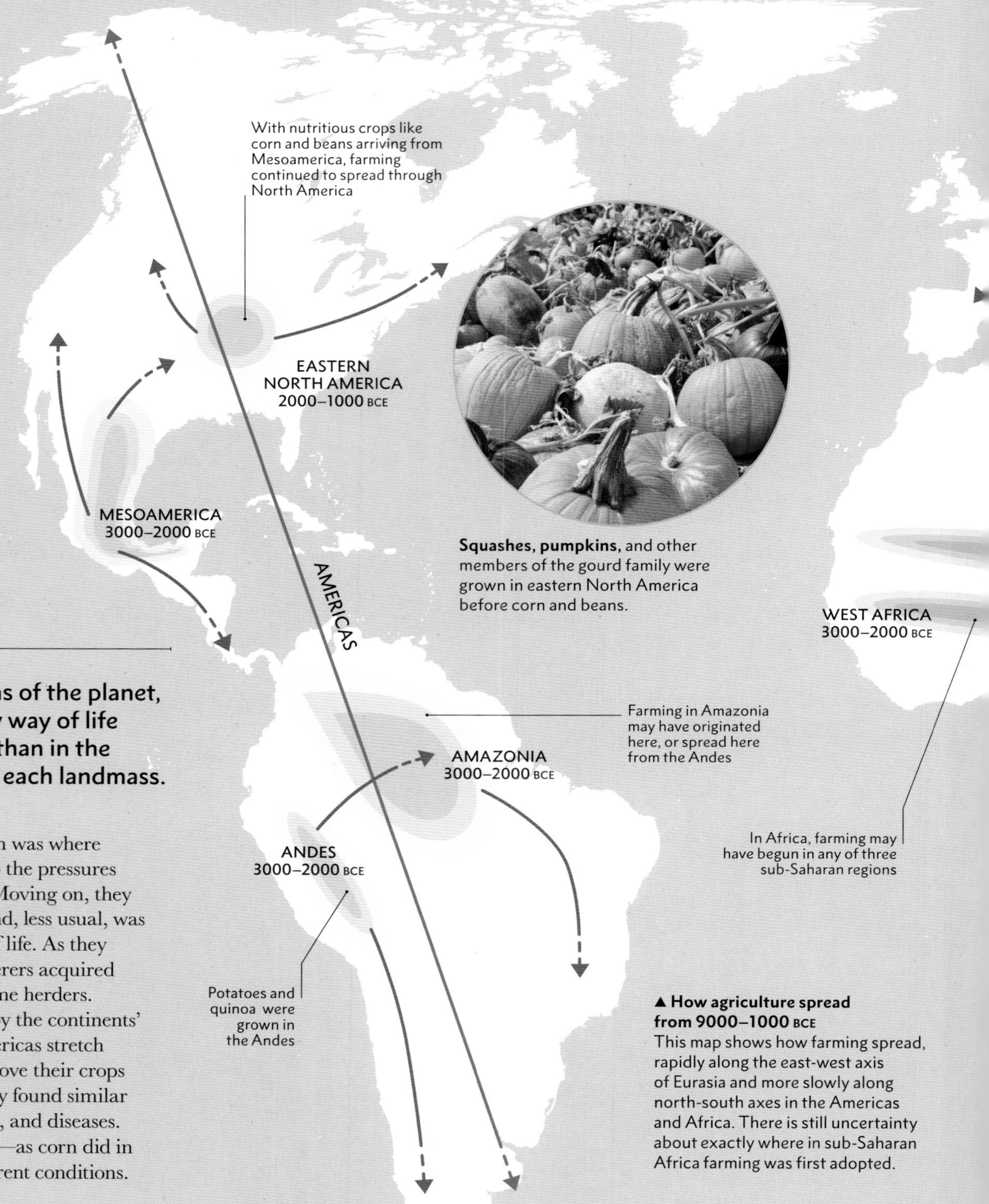

Squashes, pumpkins, and other members of the gourd family were grown in eastern North America before corn and beans.

▲ How agriculture spread from 9000–1000 BCE
This map shows how farming spread, rapidly along the east-west axis of Eurasia and more slowly along north-south axes in the Americas and Africa. There is still uncertainty about exactly where in sub-Saharan Africa farming was first adopted.

> **WHY SHOULD WE PLANT** WHEN THERE ARE SO MANY **MONGONGO NUTS** IN THE WORLD?

African Kalahari bushman, quoted by Richard Lee, *What Hunters Do for a Living*

1 BCE | 1 CE

1439 CE | PRINTING PRESS SPARKS THE INFORMATION REVOLUTION

1600 CE | COLUMBIAN EXCHANGE

Wheat, evolved from wild grasses in the Fertile Crescent over 11,000 years ago, is the earliest crop to be domesticated (see p.237).

Rice, today one of the world's most common staples, spread throughout Asia, reaching Europe during antiquity and the Americas at the time of European colonization.

Taro spread from Southeast Asia to Oceania and New Guinea, where both its root and its leaves remain popular food staples.

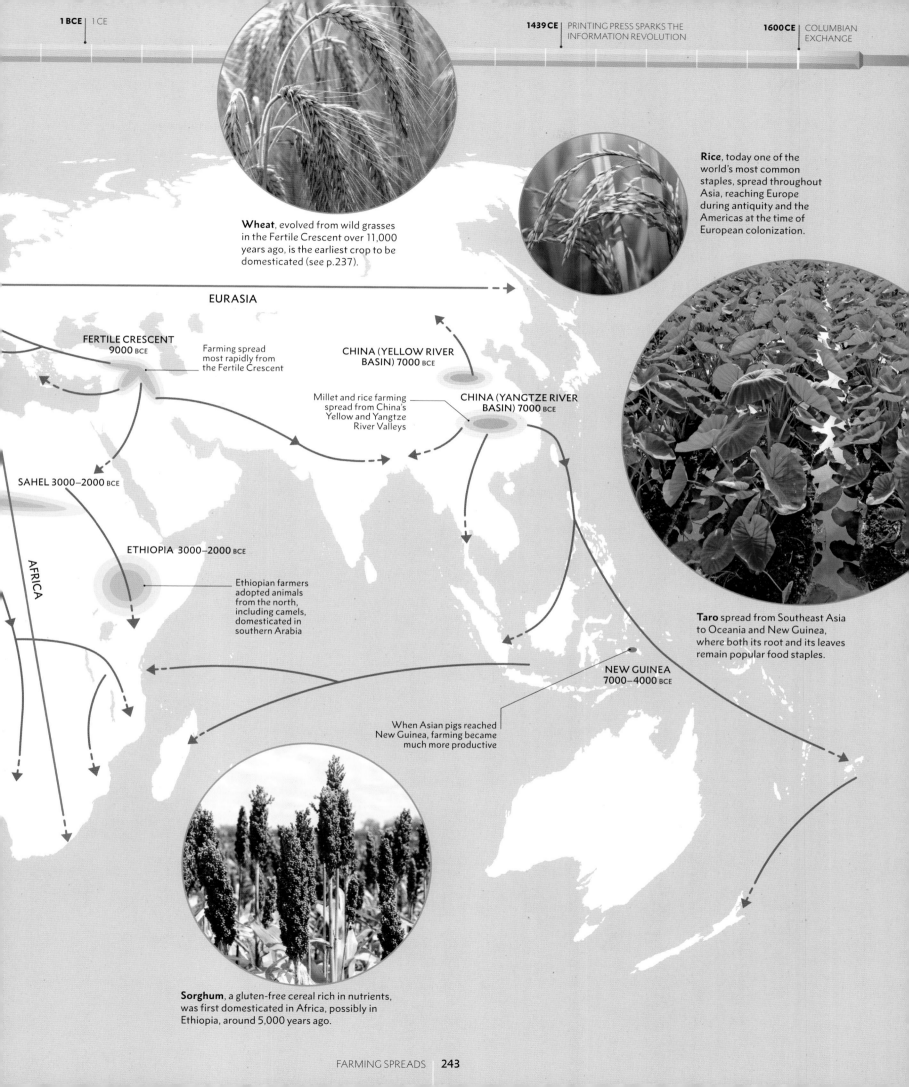

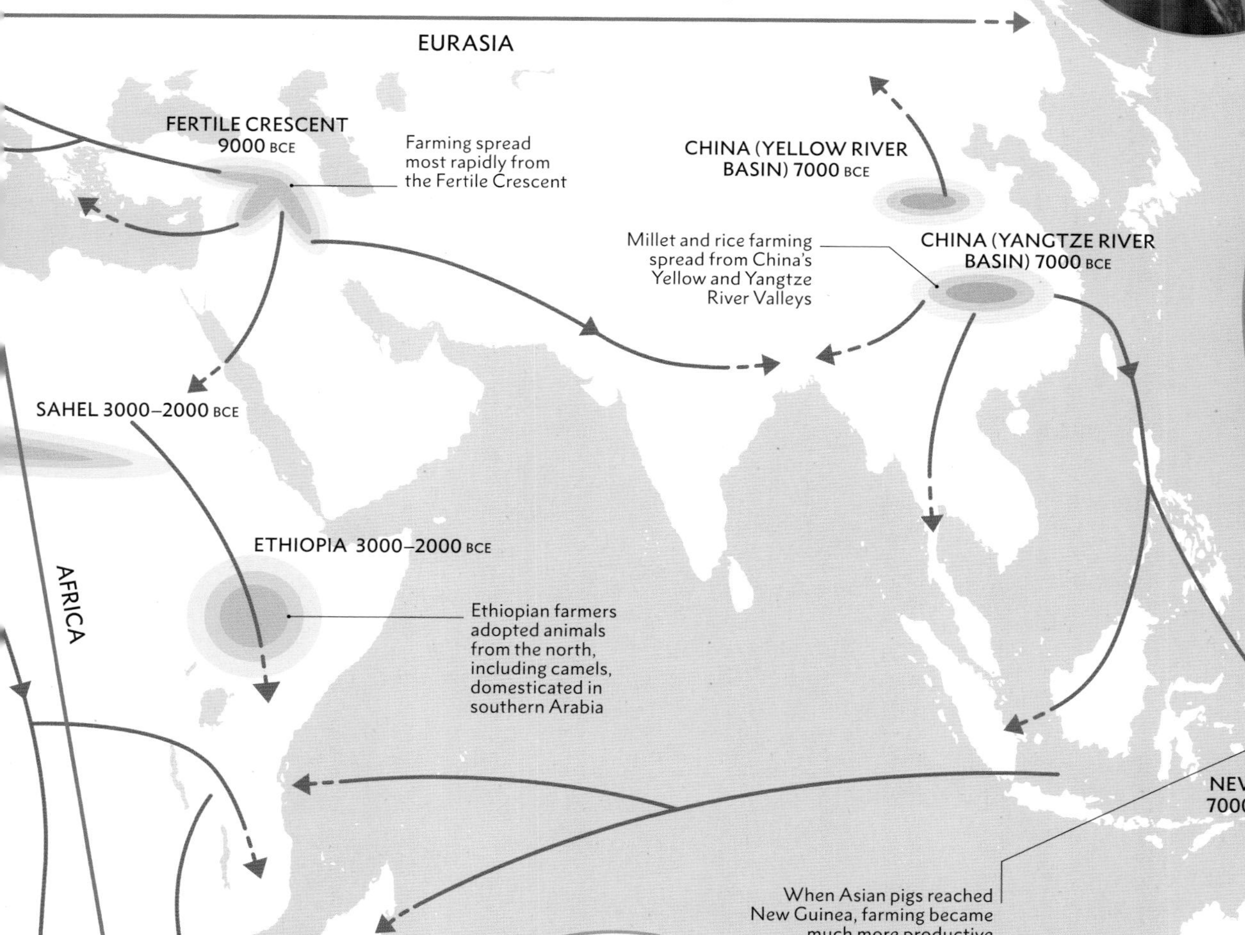

EURASIA
FERTILE CRESCENT 9000 BCE
Farming spread most rapidly from the Fertile Crescent
CHINA (YELLOW RIVER BASIN) 7000 BCE
Millet and rice farming spread from China's Yellow and Yangtze River Valleys
CHINA (YANGTZE RIVER BASIN) 7000 BCE
SAHEL 3000–2000 BCE
ETHIOPIA 3000–2000 BCE
AFRICA
Ethiopian farmers adopted animals from the north, including camels, domesticated in southern Arabia
NEW GUINEA 7000–4000 BCE
When Asian pigs reached New Guinea, farming became much more productive

Sorghum, a gluten-free cereal rich in nutrients, was first domesticated in Africa, possibly in Ethiopia, around 5,000 years ago.

MEASURING **TIME**

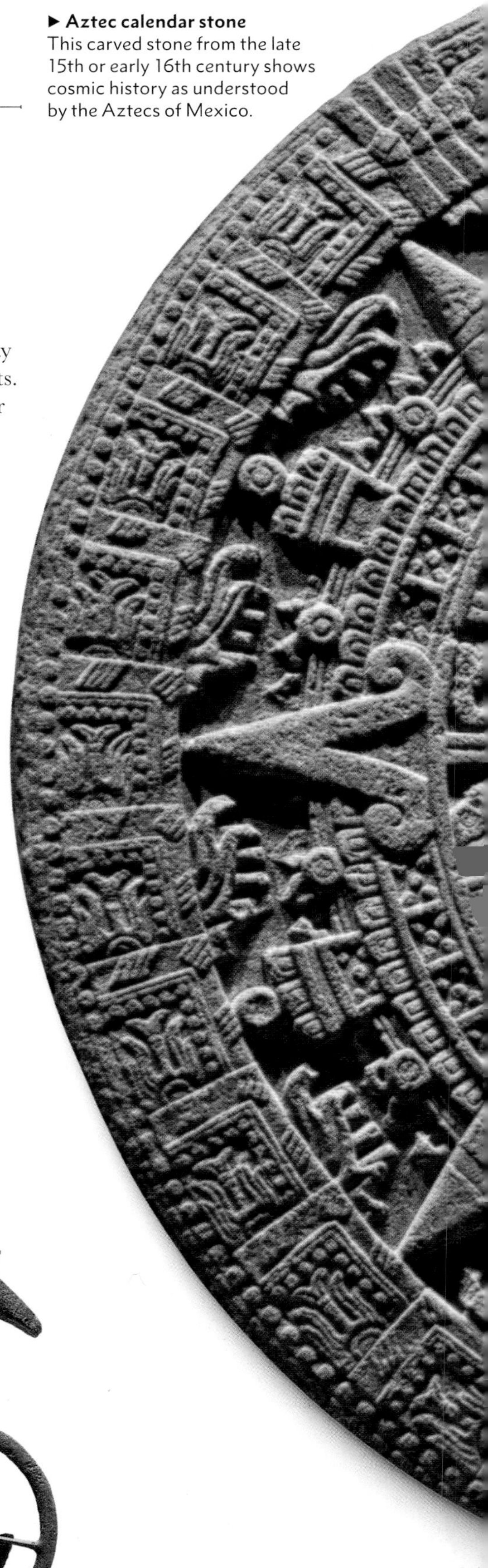

► **Aztec calendar stone**
This carved stone from the late 15th or early 16th century shows cosmic history as understood by the Aztecs of Mexico.

Agriculture gave a new importance to keeping track of time, since farmers needed to know when to plow, sow, and gather the harvest. With the rise of states, calendars became a means of social control, regulating work and coordinating the activities of large populations.

Hunter-gatherers knew about time passing because of seasonal changes, including the migrations of animals, birds, and fish, and the fall appearance of fruits and nuts. They could see the passage of time in the sky, evidenced by the moon's phases, the sun's daily journey, and the regular reappearance of constellations, such as the Pleiades and Orion, throughout the year.

CONTROL BY CALENDAR

Agriculture requires long-term planning, so early farmers built on their astronomical knowledge to invent the first calendars. In the northern hemisphere, where people were especially aware of the sun's seasonal movements, standing stones were used to track the progress of the year from where it rose and set on the horizon. Stonehenge in England, for example, was aligned with the midwinter sun.

There was also a religious motivation in the creation of written calendars. Often the work of priests—who had the time and skills to make astronomical observations—such calendars were made for the regulation of festivals and for divination. The ability to predict eclipses was a particularly good way to keep the populace in line at key moments. Written calendars later came to be used for more mundane things—when to collect taxes, when to go to war, and to establish the sailing season for merchant ships.

THE WORKING WEEK

Different cultures developed differing understandings of the passage of time. Mesoamericans, such as the Aztecs, saw time as a cyclical pattern of recurring events, in which the world was regularly destroyed and recreated.

Early societies devised a cycle of work days and rest days. The week was ten days long in China and Egypt, and seven in Mesopotamia. The day was divided into hours measured by clocks, the earliest types being water clocks and sundials. As societies grew more complex, people's lives were increasingly ruled by calendars and clocks, which measured human, social time rather than the cycles of nature.

Solar disk decorated with motifs and patterns on both sides
The model is made of bronze; only one side of the disk is gilded

▼ **Sun chariot**
This Danish model, from around 1400 BCE, imagines the sun's journey through the sky as made by horse and chariot. Markings on the sun's disk have led one archaeologist to suggest that it may have functioned as a calendar.

Symbols around the edge of the stone represent aspects of the heavens, including stars, the sun's rays, and the planet Venus

In the center is the face of Tonatiuh, the fifth and present sun god

Each square around the face represents a previous era and sun, named after Jaguar, Wind, Rain, and Water

The fifth and current era and sun are represented by the shape of the frame enclosing the central signs

This circle shows the 20 signs used to name the Aztec days

▲ Observing the heavens
This curved structure was built in the 1420s as part of Sultan Ulugh Beg's Samarkand observatory. It allowed his astronomers to calculate when sunrise and sunset would fall each day, as well as the length of a year.

“THE GODS **CONFOUND THE MAN** WHO FIRST FOUND OUT **HOW TO DISTINGUISH HOURS**.”

Aulus Gellius, Roman author, *c.*125–185 CE, *Attic Nights*

NEW USES FOR **ANIMALS**

Animals were first domesticated to provide a ready source of meat and hides. Later, farmers discovered that animals could also be used as a renewable resource, to provide milk, wool, and power. This new way of using animals is known as the secondary products revolution.

The first secondary product was milk. The earliest evidence, from the 7th millennium BCE, is pottery found in Turkey containing traces of milk. At the time, adults—unlike babies—lacked the enzyme needed to break down lactose, the main sugar in milk. But early farmers were able to reduce lactose levels by fermenting heated milk, making yogurt and cheese. Fermentation was also the best way to preserve and store milk. Around 5500 BCE, people in Central Europe developed lactose tolerance. They were able to digest milk, giving them a rich new source of protein. Lactose tolerance spread across Europe and also appeared later in West Africa and parts of Asia. Today, about a third of humanity can drink milk.

Another new product that came into use around this time was sheep's wool, which was spun and woven into textiles. Farmers in western Asia selected animals with the best quality hair for breeding. As a result, sheep developed thick woolly fleeces between 7000 BCE and 5000 BCE.

POWER AND MOTION

The most important secondary product was animal power, which gave humans their first new source of energy since the control of fire. Around 4500 BCE, donkeys were domesticated as pack animals. Later, people in western Asia harnessed oxen to pull loads, at first on simple sleds. Then, in about 3500 BCE, the plow was invented and wheels—devised for turning clay pots—were fitted to sleds to make carts. Horses were also domesticated around this time. Riding horses gave humans their first fast mode of transportation. Horses and carts enabled people to move with their grazing animals and survive on Eurasia's grassy steppes—an unsuitable environment for growing crops.

▶ **Pulling power**
Wheeled carts spread so quickly across Eurasia that it is difficult to know exactly where they originated. This 4,000-year-old pottery model of an ox cart comes from the Indus civilization of India.

> THE... REVOLUTION TURNED **DOMESTICATED HERBIVORES** INTO **EFFICIENT MACHINES** FOR TRANSFORMING **GRASS INTO ENERGY** USABLE BY HUMANS.

David Christian, Big History historian, 1946–, *Maps of Time*

Milking time
Early milking scenes often show calves. In the early days of dairying, the calf's presence was needed to make the cow release her milk. This 7th-century CE carving is from a cave temple in Tamil Nadu, India.

INNOVATIONS **INCREASE YIELDS**

Larger, settled populations inevitably needed to produce more food. Farmers began to innovate and develop new agricultural methods, such as plowing and the use of fertilizers. These new technologies enabled farmers to intensify production and increase yields.

With a pair of oxen and a plow, one man could prepare a whole field for planting in much less time than it took a team of workers with digging sticks. Plows made it possible to farm in heavier soils, greatly increasing the area of land available for cultivation. Plowing is also an efficient way of removing weeds.

The plow was an adaptation of the digging stick, allowing it to be dragged continuously through the ground. It may have been invented in Mesopotamia, where images of plows have been found dating from the 4th millennium BCE. The earliest type was the scratch plow, or ard, which had a wooden tip (share) that could cut only a shallow furrow. To plow efficiently with an ard, farmers had to cross-plow, going over the field twice, with the second plowing at right angles to the first. Later improvements included metal-tipped shares and a blade called the coulter, which sliced the soil in front of the share.

In the 1st century BCE, the Chinese further refined the plow with the addition of the moldboard—a curved blade that turned over the soil, burying weeds and bringing nutrients to the surface. Use of this plow was carried west across Eurasia, reaching Europe by the 7th century CE. Thanks to the moldboard, farmers no longer had to cross-plow. This doubled the amount of land a plow team could prepare.

Plows could only be used where there were suitable draft animals, such as oxen, water buffalo, horses, mules, and camels. The plow was never invented in the Americas, where there were no domesticated animals strong enough to pull such a device.

IMPROVING THE SOIL

One great advantage of the use of draft animals is that their dung enriches the soil. American farmers, who did not have draft animals, found other kinds of fertilizer. The Incas of Peru collected vast amounts of seabird droppings (guano) that they spread on their fields. Guano is an ideal fertilizer because it is rich in nitrogen, potassium, and phosphate—all vital nutrients for growing plants. In ancient China, farmers used human manure (nightsoil), collected from towns at night.

UNEXPECTED CONSEQUENCES

Agricultural intensification had its problems as well as its benefits. Despite better harvests, which sparked increases in the population, food remained scarce for most people. Intensive irrigation and farming fields without a fallow period eventually impoverished the soil. Communities regularly faced shortages and periodic famines, which led to malnourishment, disease, and shorter lifespans. Scarcity also brought social disorder and led to war, mass migration, and cultural disruption.

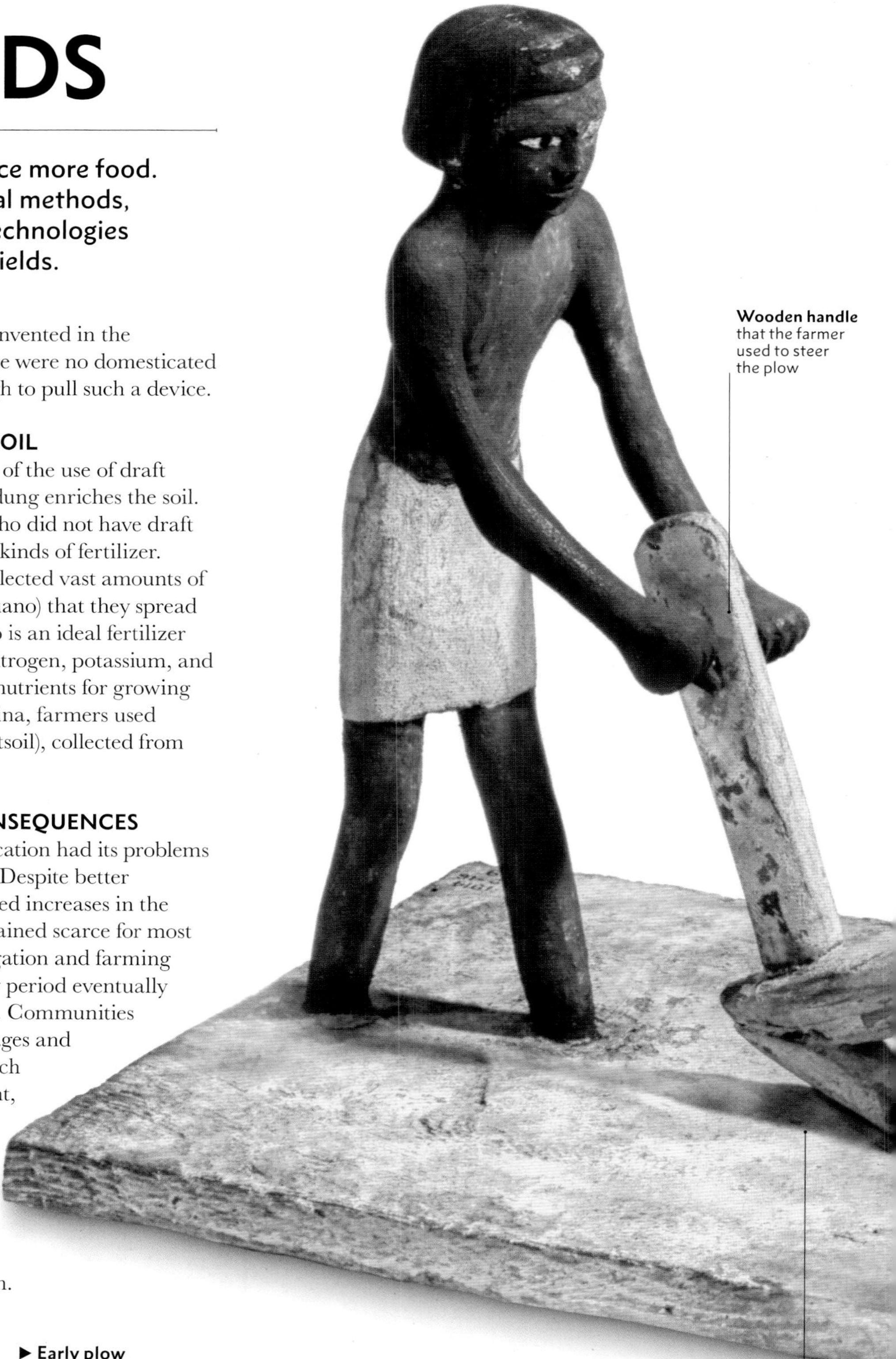

▶ Early plow
This model of a farmer using an ard, or scratch plow, comes from an Egyptian tomb of c.2000 BCE. The Nile floods deposited nutrients on the surface, so the plow did not need to turn over the soil, merely break it up for sowing.

OX PLOWS COULD NOT BE USED IN **SUB-SAHARAN AFRICA**, BECAUSE CATTLE WERE VULNERABLE TO **TRYPANOSOMIASIS**, A DEADLY DISEASE PASSED ON BY THE **TSETSE FLY**

▲ **Cutting-edge technology**
The Egyptians harvested grain using wooden scythes set with flint teeth, cutting off the ears and leaving the stalks standing for livestock to feed on. The quest for higher yields led to the need for more manpower and, in some places, slave labor.

> **GET TWO OXEN**, BULLS OF NINE YEARS. THEIR STRENGTH IS UNSPENT AND THEY ARE **BEST FOR WORK**. THEY **WILL NOT FIGHT** IN THE FURROW AND **BREAK THE PLOW**.

Hesiod, Greek poet, *c.*700 BCE, *Works and Days*

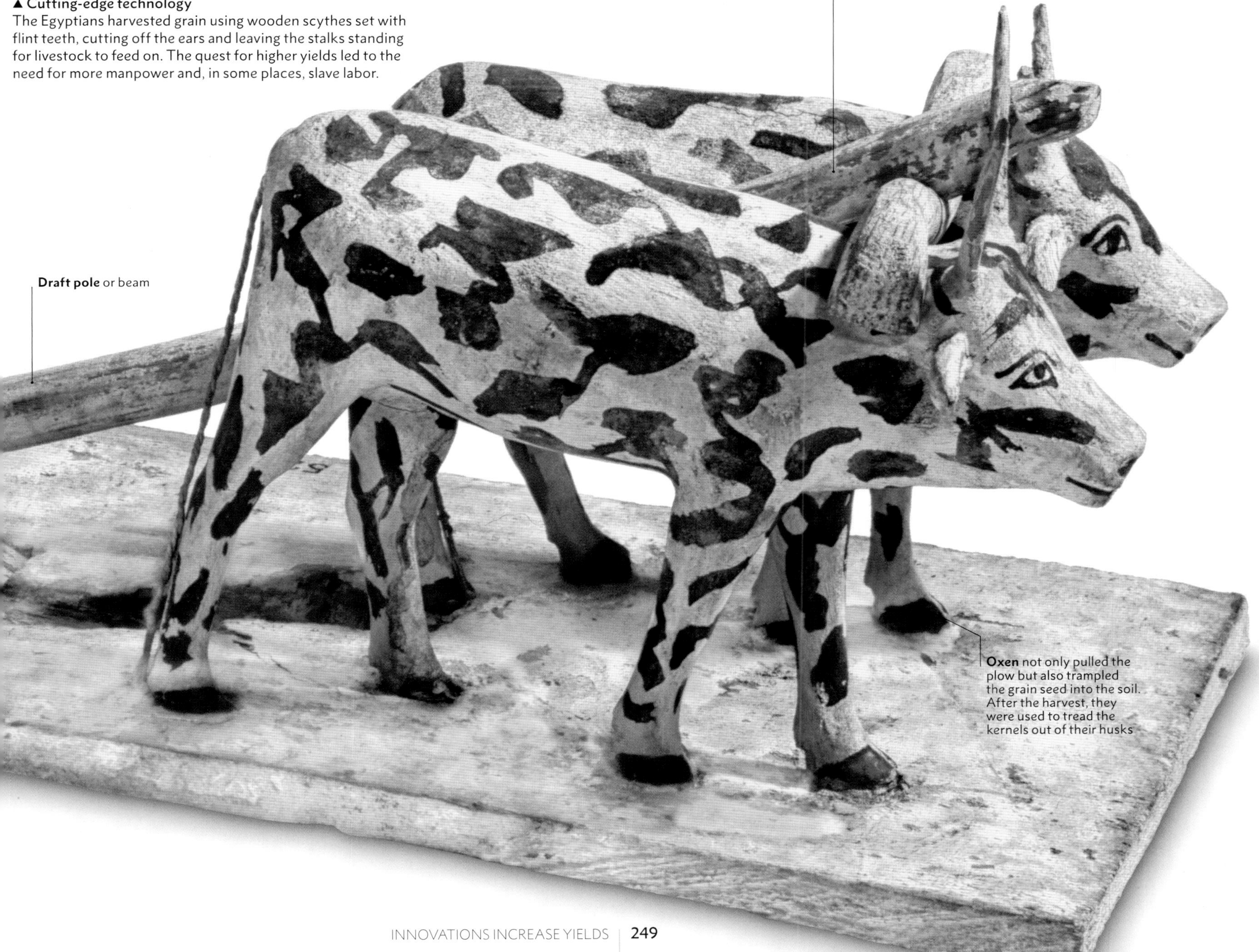

A wooden cross-piece called a yoke linked the oxen to the plow

Draft pole or beam

Oxen not only pulled the plow but also trampled the grain seed into the soil. After the harvest, they were used to tread the kernels out of their husks

SURPLUS BECOMES POWER

Once farmers learned to grow surplus food, they needed ways to store it for future use. Granaries built to store surplus grain were central to the creation of early states: these surpluses became a form of wealth that were taxed by rulers and used for trade or to reward loyal subjects.

To store grain, it must be protected from rodents and pests, and kept dry so that it does not rot or germinate. Many societies across Africa and Eurasia built granaries with raised floors, which deterred rodents and let air circulate underneath. Egypt's arid climate meant that raised floors were not necessary there. The Inca of Peru sited granaries on steep hillsides, exposed to the drying effects of mountain winds.

Large states needed ways to measure and record their grain supplies, which required an unprecedented level of organization and central control. In Egypt, this was done by measuring grain by volume, based on the *hekat*, a small barrel holding 1.1 gallons (4.8 liters). The *hekat* was the standard measuring unit used throughout the Eastern Mediterranean from 1500 BCE to 700 BCE.

The Chinese measured grain by weight, with the basic unit being the amount one man could carry on a shoulder pole. In China, archaeologists have found hundreds of vast underground grain silos dating from the Sui and Tang dynasties (581–907 CE). The walls of these state granaries bear inscriptions recording the variety, quantity, and source of the stored grain, and the date of its storage.

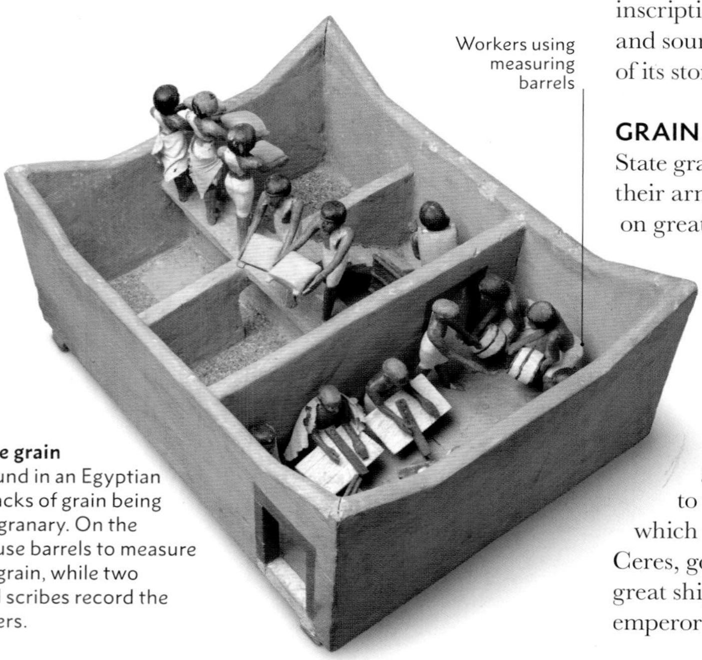

▶ Counting the grain
This model, found in an Egyptian tomb, shows sacks of grain being brought into a granary. On the right, workers use barrels to measure the amount of grain, while two shaven-headed scribes record the harvest in ledgers.

GRAIN AND STATE POWER

State granaries enabled rulers to feed not only their armies but also the workers who toiled on great building projects, such as the pyramids of Egypt and the Great Wall of China. Granaries also provided vital famine relief in years with a bad harvest. Rulers knew that the grain supply was vital to maintain the good will of the people. Roman emperors gave a monthly ration of free grain to the citizens of their capital city, which was distributed from the Temple of Ceres, goddess of grain. It was imported in great ships from Sicily and Egypt, which the emperor maintained as his personal estate.

> “I HAVE **HEAPED GRAIN** IN THE GRANARIES FOR THE PEOPLE. **IN ORDER THAT THEY MIGHT EAT** IN THE SEVEN YEARS OF EMPTY HUSKS, I HAVE COLLECTED GRAIN **FOR THE PEOPLE**.”

Epic of Gilgamesh, *c.*2000 BCE

Traditional granaries
The Dogon people of Mali still live in an agricultural society. They store millet in tall granaries, built from clay and raised on rocks, with thatched roofs that protect them during the rainy season.

8000 BCE | AGRICULTURE EMERGES
6000 BCE | FIRST CITIES FORM
4000 BCE | WRITING DEVELOPS
3100 BCE | FIRST CIVILIZATIONS EMERGE

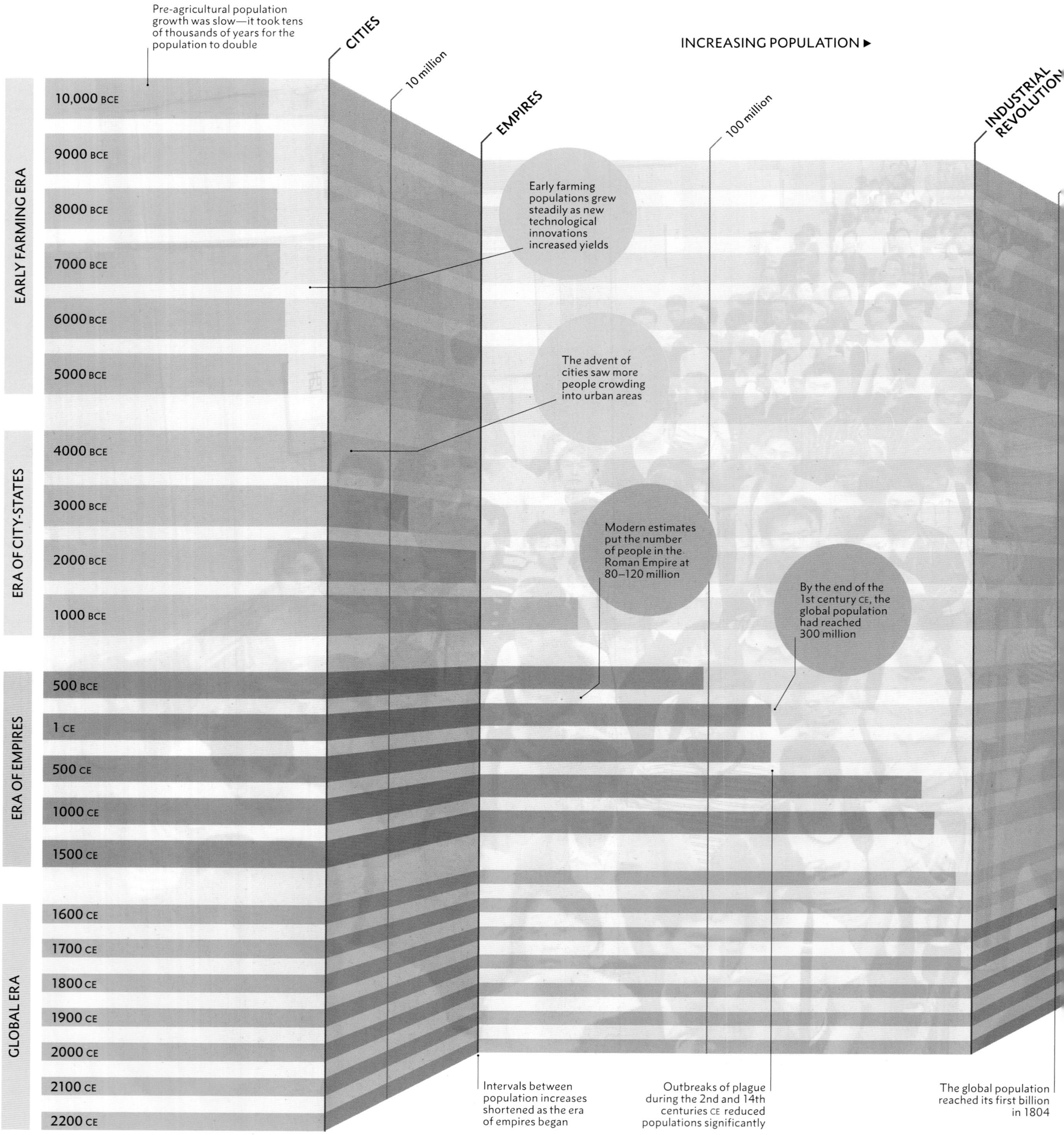

Pre-agricultural population growth was slow—it took tens of thousands of years for the population to double
CITIES
10 million
EMPIRES
100 million
INCREASING POPULATION ▶
INDUSTRIAL REVOLUTION
EARLY FARMING ERA
10,000 BCE
9000 BCE
8000 BCE
7000 BCE
6000 BCE
5000 BCE
ERA OF CITY-STATES
4000 BCE
3000 BCE
2000 BCE
1000 BCE
ERA OF EMPIRES
500 BCE
1 CE
500 CE
1000 CE
1500 CE
GLOBAL ERA
1600 CE
1700 CE
1800 CE
1900 CE
2000 CE
2100 CE
2200 CE
Early farming populations grew steadily as new technological innovations increased yields
The advent of cities saw more people crowding into urban areas
Modern estimates put the number of people in the Roman Empire at 80–120 million
By the end of the 1st century CE, the global population had reached 300 million
Intervals between population increases shortened as the era of empires began
Outbreaks of plague during the 2nd and 14th centuries CE reduced populations significantly
The global population reached its first billion in 1804

IN 2015, THE **ANNUAL** HUMAN **BIRTH RATE** WAS MORE THAN **TWICE THE DEATH RATE**

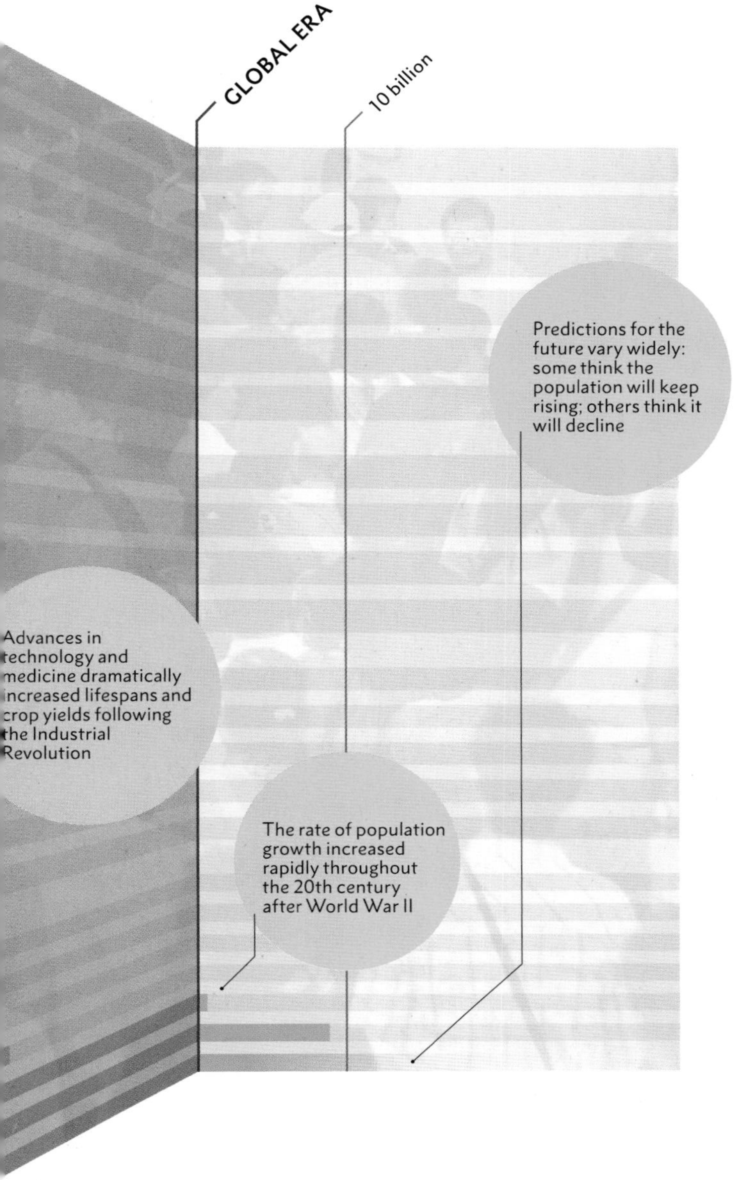
GLOBAL ERA
10 billion
Predictions for the future vary widely: some think the population will keep rising; others think it will decline
Advances in technology and medicine dramatically increased lifespans and crop yields following the Industrial Revolution
The rate of population growth increased rapidly throughout the 20th century after World War II

▲ Population growth
Human numbers grew slowly up to 1700 CE. From around 1750 through to present day, population growth has been rapid, thanks to farming innovations, industrial production, and the spread of more productive food crops, such as manioc and corn, following the Columbian Exchange (pp.296–297).

POPULATION **STARTS TO RISE**

The switch to agriculture and the creation of food surpluses led to population growth: even early farming could support 50 to 100 times more people than hunting and gathering. Agricultural innovations, such as the plow and irrigation, accelerated the increase in population.

There are many different estimates for early world populations, ranging from 2–10 million in 10,000 BCE to 50–115 million in 1000 BCE. Whatever the true figures, there is consensus that the period saw a dramatic increase in the global population as a result of farming. As large human populations spread, living in ever denser settlements, they became vulnerable to disease and periodic famine. In two periods populations fell significantly as a result of famine and plague: in the Roman Empire during the 2nd century CE and in 14th-century Eurasia.

Changes in population growth are often attributed to "Malthusian cycles." In the 18th century, economist Thomas Malthus argued that human populations always rise faster than the food supply, which results in famine and decline. Malthusian cycles often began with a new innovation: for example, improved horse collars in Europe allowed animals to pull plows that cut deeper soils, thus improving productivity. As agricultural innovations spread, the population rose, which led to larger areas being farmed. Periods of growth stimulated commercial activity and encouraged towns to expand—and their populations needed to be supplied with food. Larger populations exchanged more ideas and innovations, but ultimately in the agrarian era population growth would outpace the rate of change and was followed by a Malthusian crash.

◀ Around one in every five people on Earth is Chinese. There are as many people in China today as there were in the whole world just 150 years ago. India is predicted to displace China as the most populous country in around 2050.

NEW FOODS

The spread of new food crops could also stimulate population growth. In the 11th century CE, China adopted a new variety of early-ripening rice from Champa, Vietnam, which could produce up to three harvests a year. This drought-resistant crop could be grown on higher ground, doubling the area available for rice cultivation. This enabled the population of China to double from the 10th–11th centuries. During the 16th century, the introduction of American corn and potatoes—which could be grown at even higher altitudes than rice—led to further population growth in China.

> THE RAGING **MONSTER UPON THE LAND** IS **POPULATION GROWTH**.

E.O. Wilson, American biologist, 1929–, *The Diversity of Life*

THE FENTON **VASE**

Pottery is one of archaeology's greatest resources, as it survives in the ground when organic materials decay, providing invaluable clues to the cultures and technologies of ancient civilizations.

This beautifully decorated ceramic pot, discovered in Guatemala in 1904, provides a fascinating glimpse into the life and times of the Maya, a pre-Columbian civilization of Mesoamerica. The Maya occupied much of southeastern Mexico and northern Central America, and this vase dates from 600–800 CE. Like many Mayan vases, it was placed in the burial of a noble and depicts a scene from court life—here, the offering of a tribute—providing invaluable evidence of ritual, belief, and the daily life of the elite.

Dating pottery has become increasingly sophisticated. In the late 19th century, archaeologist Flinders Petrie used different styles of Egyptian pottery to invent sequence dating. He recorded the various styles of pots and arranged them in order according to the depths at which they were discovered. Sequence dating is still used to date archaeological sites.

Pots can now be dated scientifically using a technique that exploits the property of clay to absorb and trap electrons. If the clay is heated in a lab, the electrons are released as light. Measuring how much light is released indicates when the pot was fired. The Maya probably sourced the clays for their pots from river valleys, as their descendants do today. Chemical analysis of the clays used provides a "chemical fingerprint" that helps to identify where the clay was sourced.

The distribution of a particular style of pot can also provide clues to trade or migration. One group of Neolithic peoples, who moved into western Europe between 2800 and 1800 BCE, made a distinctive style of pot known as the bell beaker, so archaeologists refer to them as the Beaker people. The beakers found at burial sites around Europe have revealed how extensively the Beaker people traveled.

The lord's name and titles written in glyphs

Ancient writing

Mayan vases often include vital information in the form of hieroglyphs, a sophisticated writing system that was unique in Mesoamerica. Sets of glyphs used within a scene on a pot record the names and titles of the key individuals portrayed. Some pots also have text around the rims, to dedicate the vessels and list their contents.

Kneeling noble presents a Spondylus seashell
Basket piled high with maize cakes
Elaborate headdress of lord marks his rank
Glyphs in panels identify figures shown in scene
Scribe records the exchange in a screenfold book
Figures wear jewelry, elaborate clothing, and turbans decorated with flowers

The entire scene depicted on the Fenton Vase reveals a lord seated in a palace throne room receiving tribute from Mayan nobles, whose status is indicated by their ornate turbans. The five figures are individually named by the glyphs in the panels beside them. The lord points at a basket filled with tamales (maize pancakes) on top of bolts of cloth. Behind him, a scribe records the details of the tribute.

Red slip used to paint details

How was it made?

The first pots were made by coiling strips of clay, or beating clay into slabs. Then, around 3400 BCE, the potter's wheel was invented in Mesopotamia. Early wheels were turned slowly by hand, but later, the foot-operated kick wheel made it possible to "throw" pots quickly, enabling potters to produce ceramics on a large scale. Thereafter, pottery became a specialized craft, usually practiced by men. The Fenton Vase would have been made by hand, probably using the coil technique, as the potter's wheel was unknown in pre-Columbian America.

Ancient Egyptian at potter's wheel

Making a mark

The Maya decorated their pots with colored clay slips, fine mixtures of clays and minerals that fuse to a pot when it is fired. Black and red slips were used on the Fenton Vase. The earliest designs on pots such as this European bell beaker were made using incised marks.

Bell beaker

Food from the past

Pots often contain microscopic traces of the food kept within them, providing information about what people ate in the past. Scrapings from Mayan pots such as the Fenton Vase show that they were used to hold chocolate. The 4,000-year-old bowl of noodles shown below was found in China in 2005. Analysis of the noodles' starch grains revealed that they were made of millet.

The world's oldest noodles

EARLY SETTLEMENTS

As agriculture became more productive, people began to live in more dense, permanent settlements. Alongside farming, they developed impressive crafts, and created regional trade networks.

The oldest and largest early settlement was Catal Höyük in Central Turkey, which lasted from around 7300 to 5600 BCE. It covered 32 acres (13 hectares) and had a population of several thousand people. Catal Höyük was the world's first true town. Another early town was 'Ain Ghazal in Jordan, founded around 7200 BCE. 'Ain Ghazal was slightly smaller than Catal Höyük.

LIFE IN THE FIRST TOWNS

The people of these early towns were farmers, who kept large herds—sheep at Catal Höyük, and goats at 'Ain Ghazal. Both towns grew wheat, barley, peas, and lentils. They also hunted local wild animals, including aurochs (wild cattle), deer, and gazelles.

These first towns may have been quite isolated except for their trading routes. We do not know if they had any contact with neighboring hunter-gatherer bands, if these existed. As trade developed, it prompted the development of new crafts and skills. Vital new technology—ploughs, wheels, bronze tools—would later emerge from the specialist artisans living in towns.

The first experiments in urban living, these towns developed in different ways. Buildings in each were rectangular, and densely packed together. 'Ain Ghazal had courtyards and narrow lanes between the houses, which were entered through doorways. By contrast, at Catal Höyük, the houses were built against each other without passageways. They were entered through rooftop openings, reached by ladders.

Houses at 'Ain Ghazal vary considerably in size, which suggests that some of its inhabitants were wealthier than others. However, at Catal Höyük, there is no evidence of different classes: there were no high-status homes, public buildings, or even public open spaces. People here seem to have lived lives of equality.

> “THE **QUALITY AND REFINEMENT** OF NEARLY EVERYTHING MADE HERE **IS WITHOUT PARALLEL** IN THE CONTEMPORARY NEAR EAST.”
>
> **James Mellaart**, British archaeologist, 1925–2012

This 6½in-high (16cm) clay figure, found at Catal Höyük, depicts a woman who is flanked by two leopards
Figure is thought to represent a mother goddess, who controlled the fertility of the earth
STONE GODDESS
Strong, colorful textiles were woven on looms
Wild plants, such as fruit trees, provided an additional source of food
Graves for the dead under house floors. Bodies were exposed to vultures and then the skeletons were buried
DECORATED INTERIORS
House walls were plastered with white clay and then painted with geometric patterns or images of hunting scenes

▶ **Bone tools** used for sewing and weaving provide evidence of textile making at Catal Höyük. Patterned pottery seals have also been found, which may have been used to print patterns on fabrics or on people's skin for decoration.

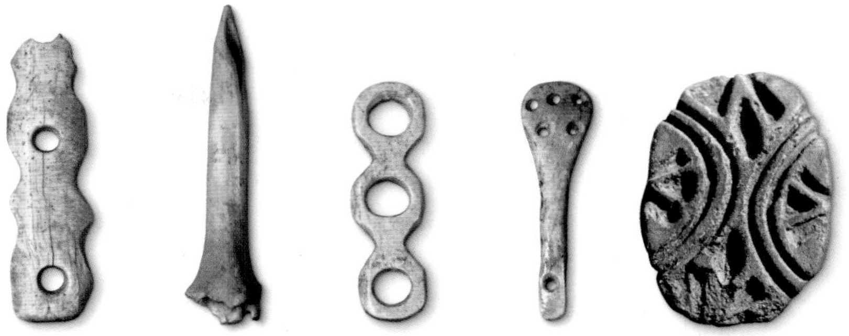

BONE TOOLS USED FOR MAKING TEXTILES

POTTERY SEAL

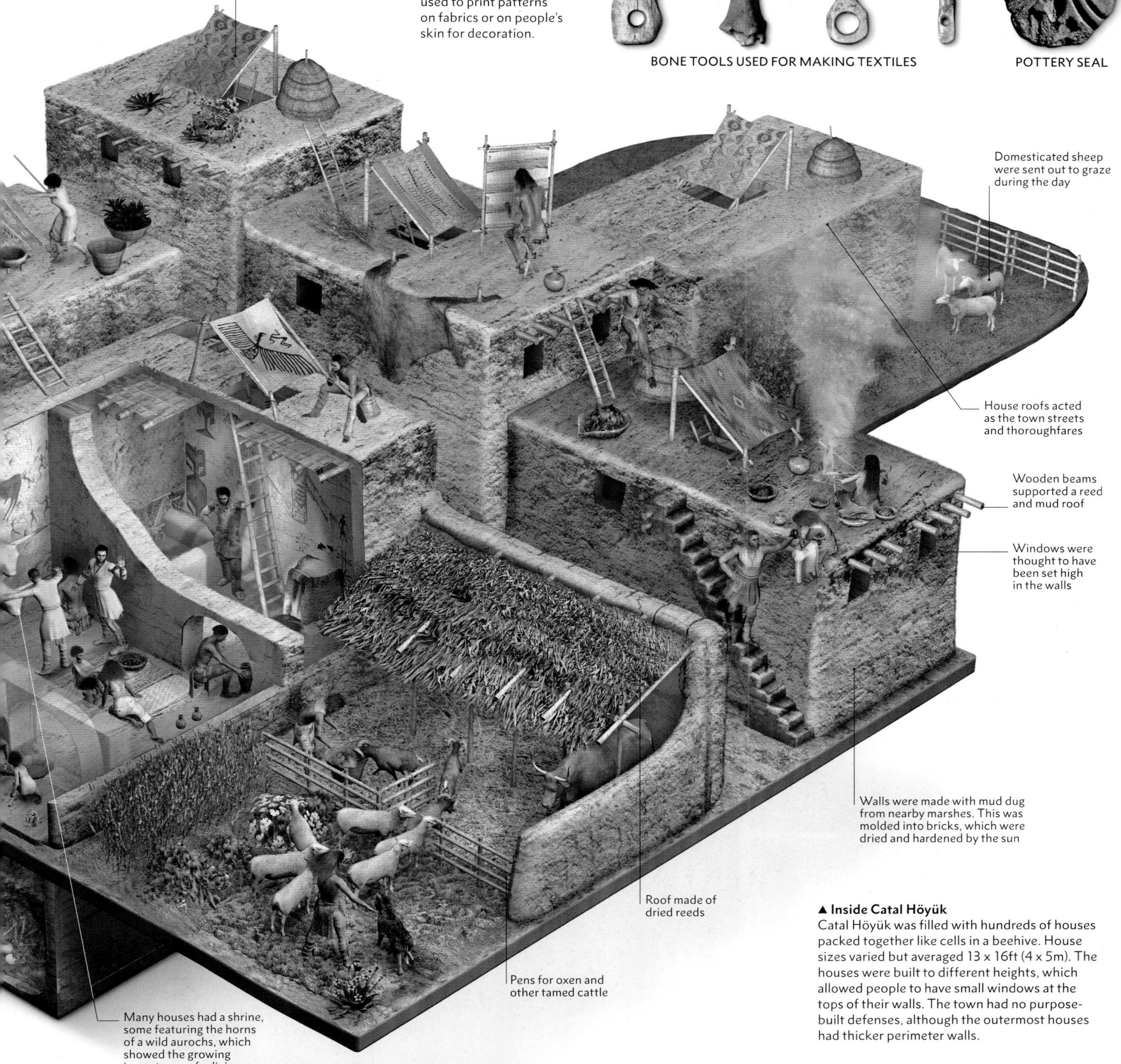
Entrances to houses offered ventilation. They may have been covered by woven awnings to block the sun
Domesticated sheep were sent out to graze during the day
House roofs acted as the town streets and thoroughfares
Wooden beams supported a reed and mud roof
Windows were thought to have been set high in the walls
Walls were made with mud dug from nearby marshes. This was molded into bricks, which were dried and hardened by the sun
Roof made of dried reeds
Pens for oxen and other tamed cattle
Many houses had a shrine, some featuring the horns of a wild aurochs, which showed the growing importance of religion

▲ **Inside Catal Höyük**
Catal Höyük was filled with hundreds of houses packed together like cells in a beehive. House sizes varied but averaged 13 x 16ft (4 x 5m). The houses were built to different heights, which allowed people to have small windows at the tops of their walls. The town had no purpose-built defenses, although the outermost houses had thicker perimeter walls.

► **Social hierarchy in ancient Egypt**
Egyptian society, as in many states, resembled a pyramid, with the king at the top and different ranks beneath. There have been many other types of social pyramids. In the states of medieval Europe and Japan, warriors became the dominant class. Merchants were powerful in the Indus civilization, but had a lowly position in Imperial China. Some societies, such as the Roman Empire, depended on the mass use of slaves, who were people with no rights and were regarded as property.

SLAVES WERE AT THE **BOTTOM** OF ANY **SOCIETY** IN WHICH THEY WERE FOUND

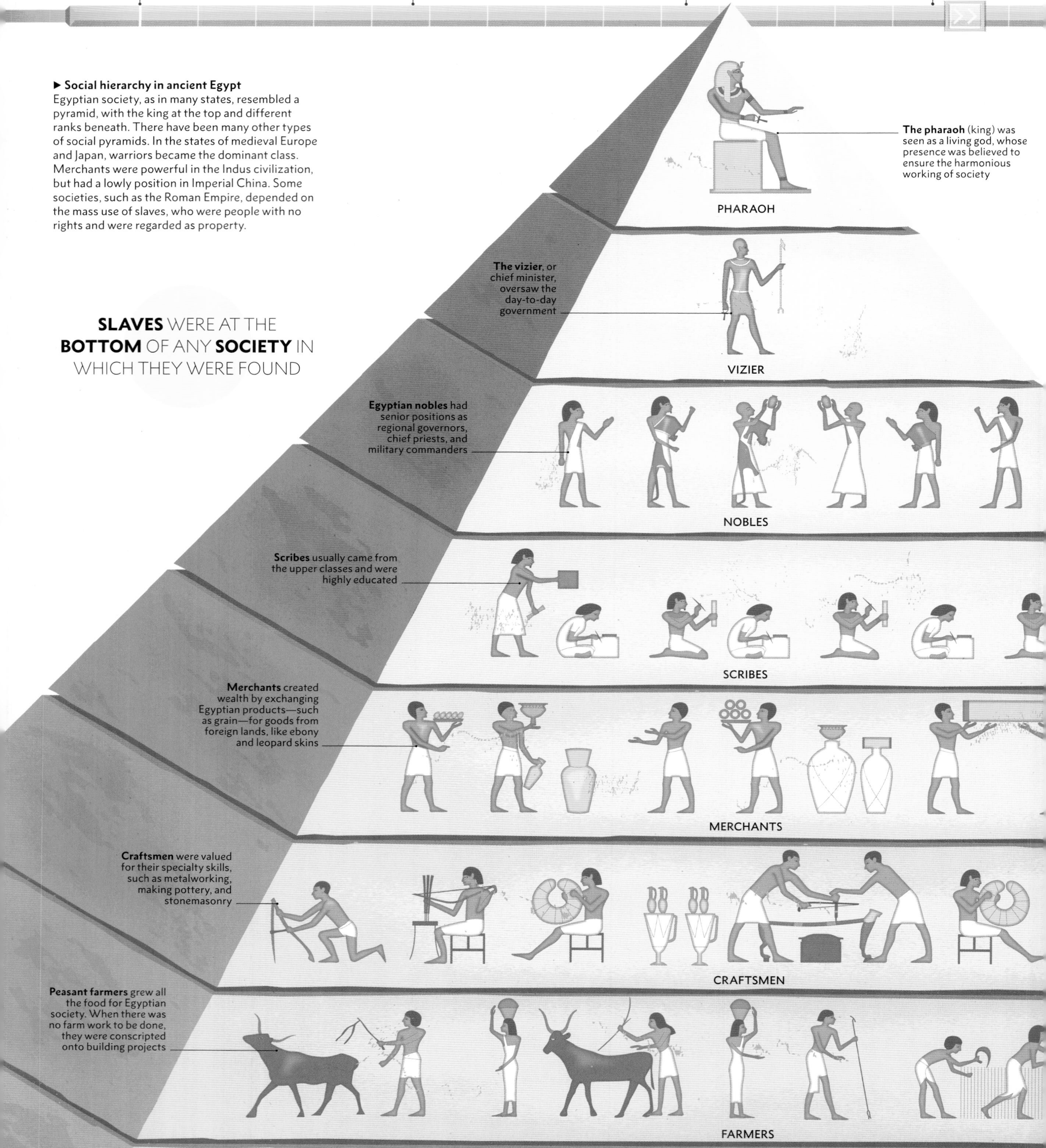
The pharaoh (king) was seen as a living god, whose presence was believed to ensure the harmonious working of society
PHARAOH
The vizier, or chief minister, oversaw the day-to-day government
VIZIER
Egyptian nobles had senior positions as regional governors, chief priests, and military commanders
NOBLES
Scribes usually came from the upper classes and were highly educated
SCRIBES
Merchants created wealth by exchanging Egyptian products—such as grain—for goods from foreign lands, like ebony and leopard skins
MERCHANTS
Craftsmen were valued for their specialty skills, such as metalworking, making pottery, and stonemasonry
CRAFTSMEN
Peasant farmers grew all the food for Egyptian society. When there was no farm work to be done, they were conscripted onto building projects
FARMERS

▲ **Mother and child**
Between 100 BCE and 250 CE, the Jalisco people of Mexico made many pottery figures of mothers with babies, reflecting women's primary role in their society.

SOCIETY GETS ORGANIZED

As the population increased, humans had to learn, for the first time, to live in peace alongside large numbers of strangers. There were new forms of social organization, ultimately leading to the creation of the state, with a king at the top presiding over a hierarchy of different classes.

Hunter-gatherers lived in small bands of 25 to 60 individuals who were related through family and marriage ties. Bands were egalitarian: there were no leaders, although certain members were highly respected because of their wisdom or skill at hunting or gathering. Men and women were also equal, with each contributing food supplies, the men hunting and the women gathering.

With the advent of agriculture, people settled down in larger groups, coming together as tribes. A tribe is a group of up to a few thousand, often united by a belief in their descent from a shared ancestor. Early tribal societies remained egalitarian and decisions were made communally. Many tribes had a "big man" whose opinion was valued, but his status came through force of personality rather than inheritance.

Once a population reached several thousand, people had to live alongside others to whom they were not related. Powerful chieftains kept the peace, claiming a monopoly on the right to use force. Tribal members paid tribute to the chief, who redistributed it to his followers. This led to the emergence of different classes. Kinship was still important, but the chieftain's own lineage came to be seen as superior.

THE FIRST STATES

States emerged once populations exceeded 20,000 people—too great a number for kinship to play a role. State organization resembled a pyramid, with an all-powerful ruler at the top and a hierarchy of classes below, including priests and administrators. The largest class of all was made up of peasant farmers. They were at the bottom of the pyramid, even though it was their hard work that created the surpluses on which the whole system was based.

The meticulous record keeping of the scribes was essential for the state to function. Royal scribes were rewarded with wealth and power
Egyptian craftsmen had their own hierarchies, with royal artisans having a much higher social status than ordinary craftsmen

> **SOCIETY HAS ARISEN** OUT OF THE **WORKS OF PEACE**: THE ESSENCE OF **SOCIETY IS PEACEMAKING**.

Ludwig von Mises, Austrian economist, 1881–1973

PATRIARCHY EMERGES

After people switched from hunting and gathering to farming, women gradually lost their equality within the tribe and came under male control—a system known as patriarchy. Men now supplied the food or income, while women were tied to the home, giving birth and caring for children. Many states prevented women from owning property and placed them under the legal control of husbands or fathers. In some societies, men were allowed to take multiple wives. Sons were preferred over daughters, and there was infanticide of female babies.

War captives
This Mayan wall painting of *c.*790 CE shows King Chan Muwan of Bonampak, in the center, triumphing over captured warriors from a rival city. The captives, stripped of their high-status clothing, have had their fingernails torn out as a demonstration of his superiority and power.

RULERS EMERGE

As societies grew larger, power began to shift from consensual kinship relationships to top-down, coercive rule. The new rulers, called chieftains or kings, backed up their position with armed force, which they used to exact tribute from their subjects.

Rulers were able to achieve their positions of power by redistributing the tribute they received. They armed and rewarded elite groups, creating a class of warriors or nobles, while disarming the mass of people.

Why did the majority of people allow a small minority to rule over them? To begin with, there may have been a consensual element, as people willingly gave up power in exchange for organization, security, and protection. Alternatively, the process may simply have been imposed on them from above by forceful and ruthless individuals.

All over the world, rulers found similar ways to display their power. They sat on raised seats (thrones), wore tall headdresses, and held ornamental staffs called scepters. The Egyptian pharaohs carried a shepherd's crook and a flail, symbolizing the king's protective and coercive role as the "shepherd" of his people.

Success in war was also a sign that rulers had the support of the gods. In public art, kings had themselves depicted triumphing over enemies, who were often shown naked to emphasize their powerlessness.

▼ King's coffin
The coffin of Pharaoh Tutankhamun (*c.*1327 BCE) is covered with symbols of the king's royal authority and divine status. It was made of gold, which was seen as the flesh of the gods, and inlaid with blue enamel.

DIVINE BACKING

Royal authority was usually justified by supernatural claims, in which the ruler's well-being was portrayed as essential to society. Egyptian pharaohs, for example, were said be the earthly embodiment of the sky god Horus, Chinese emperors claimed to have the "Mandate of Heaven," and Mayan kings claimed descent from divine ancestors, who were believed to retain power over the living. Subjects who approached kings were expected to adopt submissive postures, such as bowing or prostrating themselves.

Polynesian chieftains were surrounded by religious taboos that forbade their subjects from even touching their shadow. To do this would be to damage the chieftain's sacred power, or *mana*. As the chief's *mana* was vital to maintain the ritual security of the community, such actions were thought to place the entire population at risk.

The cobra and the vulture represent the pharaoh's supreme power and authority over the Upper and Lower Kingdoms of Egypt
Striped linen headdress (*nemes*) was only worn by the pharaoh
Crook signifies pharaoh as a shepherd, or protector
Ceremonial false beard was a symbol of divinity
Flail, a whip used to goad livestock, shows the pharaoh's power to punish

LAW, ORDER, AND JUSTICE

Large, complex societies need an objective set of rules to govern conduct and resolve disputes peacefully. The earliest law codes were compiled by rulers as a means of social control. Later, an ethical sense developed, based on the idea that justice should be equally available to everyone.

The rise in populations following the introduction of agriculture led to many more opportunities for disputes. Unlike hunter-gatherers, who had no sense of private ownership, farmers quarreled over land, property, water rights, inheritance, and many other matters.

Before the rule of law developed, it was the family or kinship group's responsibility to avenge wrongs against individual members. Failure to avenge a wrong, such as a killing, brought dishonor on the whole kinship group. This could set in motion a cycle of violence, a blood feud, that might last for generations. Blood feuds have been common in societies throughout history, and they form the subject of Greek myths, Icelandic sagas, and Japanese samurai tales.

ROYAL CODES

As states emerged, rulers were quick to assume a monopoly of the right to use violence. To resolve disputes peacefully and prevent feuds, they compiled lists of punishments for crimes, or compensations to be paid by perpetrators to victims.

The earliest surviving law code is that of the Sumerian city of Ur-Nammu, of *c.*2100 BCE. It lists various compensation sums for a wide range of specific injuries. For example, "If a man has cut off another man's foot, he is to pay ten shekels of silver."

The most famous early law code of all is that of Hammurabi, king of Babylon from 1792–50 BCE. He had 282 decrees inscribed on a 7ft 5in (2.25m) high cone-shaped stele—a stone pillar—set up in the center of Babylon for all to see. Hammurabi's Law Code is best known for "If a man put out the eye of another man, his eye shall be put out."

At the top of the stele, Hammurabi declared that he had been commanded by the gods "to bring about the rule of righteousness in the land, to destroy the wicked and the evildoers; so that the strong should not harm the weak." He suggested that any man who felt wronged should go to the stele and have its laws read out: "Let him see the law that applies to him, and let his heart be at rest."

For kings like Hammurabi, dispensing justice was a way of winning popularity. When they were not fighting wars or performing religious ceremonies, many ancient rulers spent much of their time listening to appeals and judging disputes.

According to his biographer, Plutarch, King Demetrius I of Macedon was once on a journey when an old woman approached him and asked for an audience. The king

▶ Mark of proof
Evidence has become important to provide a basis for provable fact. Today's evidence law is influenced by Roman legal practices. In early times, evidence was primarily oral, occasionally written, and only rarely physical.

> **LAW IS THE KING** OF ALL THINGS, BOTH **DIVINE AND HUMAN**.
>
> **Chrysippus**, Greek philosopher, *c.*279–206 BCE, *On Law*

answered that he was too busy, at which she shouted, "Then don't be king!" Stung by the rebuke, he stopped and spent the next few days giving audiences to all who asked for them, beginning with the old woman. Plutarch concludes, "And indeed there is nothing that becomes a king so much as the task of dispensing justice."

DIVINE LAWS

The emergence of moral religions brought a new attitude to law, with many crimes or transgressions now being seen as offenses against God rather than against society or individuals. The Hebrew Torah (Law) is a collection of instructions for every aspect of life, which Jews believe were handed to Moses by God. The most important of these instructions was the Ten Commandments, which were inscribed on stone tablets and kept in the central shrine of the Jewish Temple in Jerusalem.

Islamic Sharia law is a similar set of commandments for every aspect of life. Sharia is based on the Koran, traditions about the Prophet Muhammad, and *fatwas*—rulings—by Islamic scholars. Sharia means "the clear path" in Arabic. In some Muslim countries, Sharia Law has continued the ancient tradition of "an eye for an eye." In 2009, an Iranian Sharia court offered a woman, blinded in an acid attack, the opportunity to pour acid into the eyes of her attacker. She chose to pardon him, saying, "I knew I would have suffered and burned twice had I done that."

CHINESE PHILOSOPHIES

In China, from the 6th century BCE, two very different approaches to law developed, based on contrasting views of human nature. The philosopher Confucius argued that people will behave well if they are set a good example of proper behavior by those in authority. He said, "To govern simply by law, and to create order by means of punishments, will make people try to avoid the punishment but have no sense of shame. To govern by virtue, and create order by the rules of propriety, will not only give them the sense of shame, but moreover they will become good."

ANGLO-SAXON LAW CODES **LIST MONIES TO BE PAID** FOR EVERY KIND OF **INJURY**, DOWN TO **A LOST FINGERNAIL**

The Legalists rejected Confucianism. They viewed people as innately greedy, self-interested, and lazy, and they advocated controlling behavior through strict laws and harsh punishments. Legalism was adopted by the state of Qin in the 4th century BCE. Lord Shang, the chief minister of Qin, wrote, "Those who do not carry out the king's law are guilty of death and should not be pardoned, but their punishment should be extended to their family for three generations." Lord Shang eventually fell out of favor and suffered under his own harsh laws. In 338 BCE, he was torn apart by five chariots and his whole family was killed. Legalism enabled the kings of Qin to create an authoritarian state and then conquer the other kingdoms. In 221 BCE, the unification of China was completed by the First Emperor, who imposed Legalism on the whole country. All Chinese families were organized into mutual responsibility groups, in which each member would be punished for crimes committed by another. Confucian books were banned. The First Emperor's rule proved so harsh that the Qin dynasty survived for only four years after his death in 210 BCE.

The Han dynasty succeeded the Qin. The Han Emperor Wu (ruled 141–87 BCE) combined Confucianism and Legalism. Confucianism, with its emphasis on moral behavior and filial duty, became the state philosophy. Yet it was backed up by strict Legalist punishments. This was summed up by the saying "Confucian on the outside, Legalism within." Legalism has been at the core of the Chinese system ever since.

ROMAN LAW

The Romans were the first people to treat law as a science, with jurists analyzing the principles underlying laws and their application. Roman jurists argued that the spirit or intent behind a law was more important than its precise wording. Another principle was that the accused should be given the benefit of the doubt.

Over centuries, a mass of Roman laws and legal commentaries, often contradictory, built up, which lawyers and magistrates were expected to study. This was reduced to a manageable form in 528–33 CE by the Emperor Justinian, who commissioned a team of experts to collect all the existing Roman laws in one volume—the *Corpus Juris Civilis* (Body of Civil Law). They created a second work, the *Digest*, by editing the legal commentaries to remove repetitions and contradictions. Justinian's Law Code spread to the West where, from the 11th century, the *Digest* was used to educate generations of lawyers. The Code itself influenced many later ones, including the French Napoleonic Code of 1804. In his 1951 book, *Natural Law: An Introduction to Legal Philosophy*, the Italian author Alessandro d'Entreves declared, "Next to the Bible, no book has left a deeper mark upon the history of mankind than the *Corpus Juris Civilis*."

"**JUSTICE** IS A **CONSTANT**, UNFAILING DISPOSITION TO GIVE **EVERYONE HIS LEGAL DUE**."

Ulpian, jurist quoted in Justinian's *Digest*, c.533 CE

THE **WRITTEN WORD**

With the spread of farming and trade, the need to keep accurate records led several early civilizations to invent writing systems. Writing was soon put to other uses, including setting down laws, composing religious texts, chronicling events, spreading scientific ideas, and creating literature.

Writing began around 3300 BCE in Egypt and Mesopotamia as a way to store vital information. Initially it only benefited the ruling classes, as the first systems used so many signs that only a small elite group, the scribes, could master them.

The Phoenician invention of an alphabet, using less than 30 signs to represent sounds, helped to extend literacy beyond just the scribal classes. In the 1st millennium BCE, the alphabet was spread throughout the Mediterranean by Phoenician traders, and then adapted by the Greeks and Romans. Writing was increasingly used for everyday purposes, such as composing letters, making shopping lists, and labeling possessions to indicate ownership.

Books were a valuable tool for collective learning: knowledge could be shared between cultures, and passed down to future generations. They were collected in ancient libraries, the most renowned being the Great Library of Alexandria in Egypt, which was a major center of Greek learning from the 3rd century BCE. One of its chief librarians was the mathematician Eratosthenes, who accurately calculated Earth's circumference in around 200 BCE.

That we know about Eratosthenes today is due to the preservation of Greek and Latin books through the Middle Ages by the Roman Catholic Church and the Byzantine Empire, and also by Islamic scholars, who translated them into Arabic.

Printing with moveable type, which allowed books to be cheaply mass produced, marked the next great advance in the rise of literacy. The first European printed book was Johannes Gutenberg's Bible, printed in 1455 CE. By 1500, presses in Europe were turning out 10–20 million volumes a year, and 35,000 different books were in print.

> “WE MUST... THANK **OUR PREDECESSORS**, BECAUSE THEY DID NOT LET ALL GO IN JEALOUS SILENCE, BUT **PROVIDED A RECORD** IN WRITING **OF THEIR IDEAS OF EVERY KIND**.”
>
> **Vitruvius**, Roman architect, *c.*80–15 BCE, *On Architecture*

▶ In loving memory
The invention of writing allowed people's names to live on after their death. This funeral stele (memorial stone) from Yemen bears an inscription in the Old South Arabian alphabet, which was used from the 9th century BCE to the 6th century CE.

Medieval books were prized for their decoration, so they survived long after the Latin language in which they were written had fallen out of use

▶ A beautiful read
Until the coming of printing, only the wealthy could afford books, which were designed to be admired as objects of beauty as well as read. This 15th-century handwritten prayer book, or "book of hours," is in Latin, which limited its readership.

Ornate initial capitals announced text divisions or highlighted important sections of a work

Pictures were an aid to literacy, helping the reader to understand the text

Upper and lowercase lettering only emerged in the 8th century CE

The text was hand written before the page was illustrated

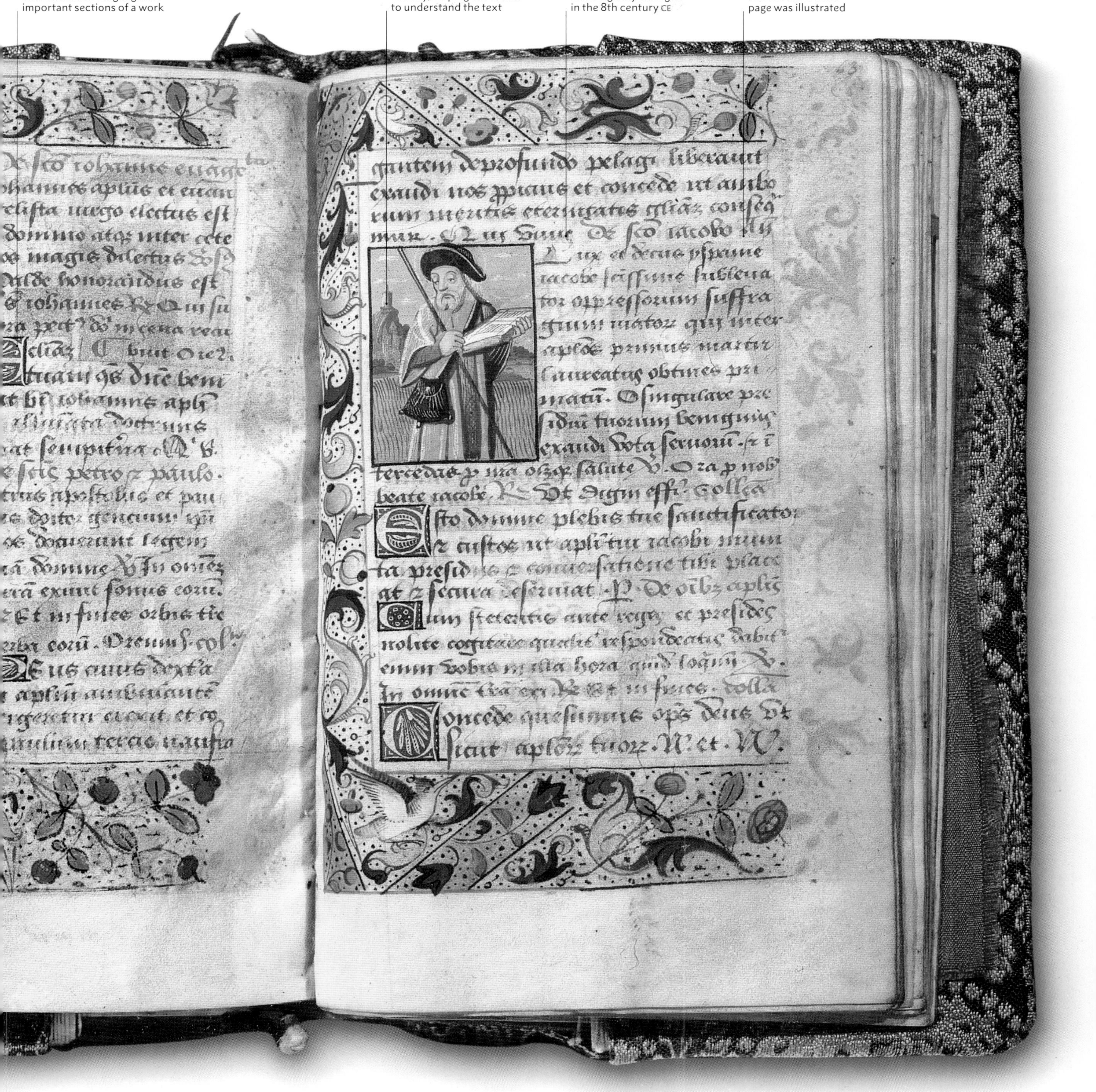

Egyptian Hieroglyphs, a system based on picture signs, were first used about 3300 BCE.

AFRICA

3300 BCE

Sumerian cuneiform is developed about 3200 BCE. It is another system based on picture signs.

Papyrus, an early form of paper, was developed about 3000 BCE in Egypt.

WEST ASIA

The Indus people invent a script in 2600 BCE that is still undeciphered. There are around 400 signs, which are read right to left.

EAST ASIA

The earliest-surviving Chinese writing dates from 2500 BCE.

The Akkadians of Mesopotamia in 2350 BCE simplified cuneiform, reducing the number of signs to about 600.

2220 BCE

EUROPE

The Minoans in Crete invent the "Linear A" system in 1900 BCE. It is inspired by trade with Egypt and is mainly used to keep accounts.

The Proto-Sinaitic (also called Canaanite) alphabet is developed in Canaan or Sinai between 1850 and 1650 BCE.

1500 BCE

The Mycenaeans adapt Linear A in 1450 BCE to create Linear B, the earliest form of Greek.

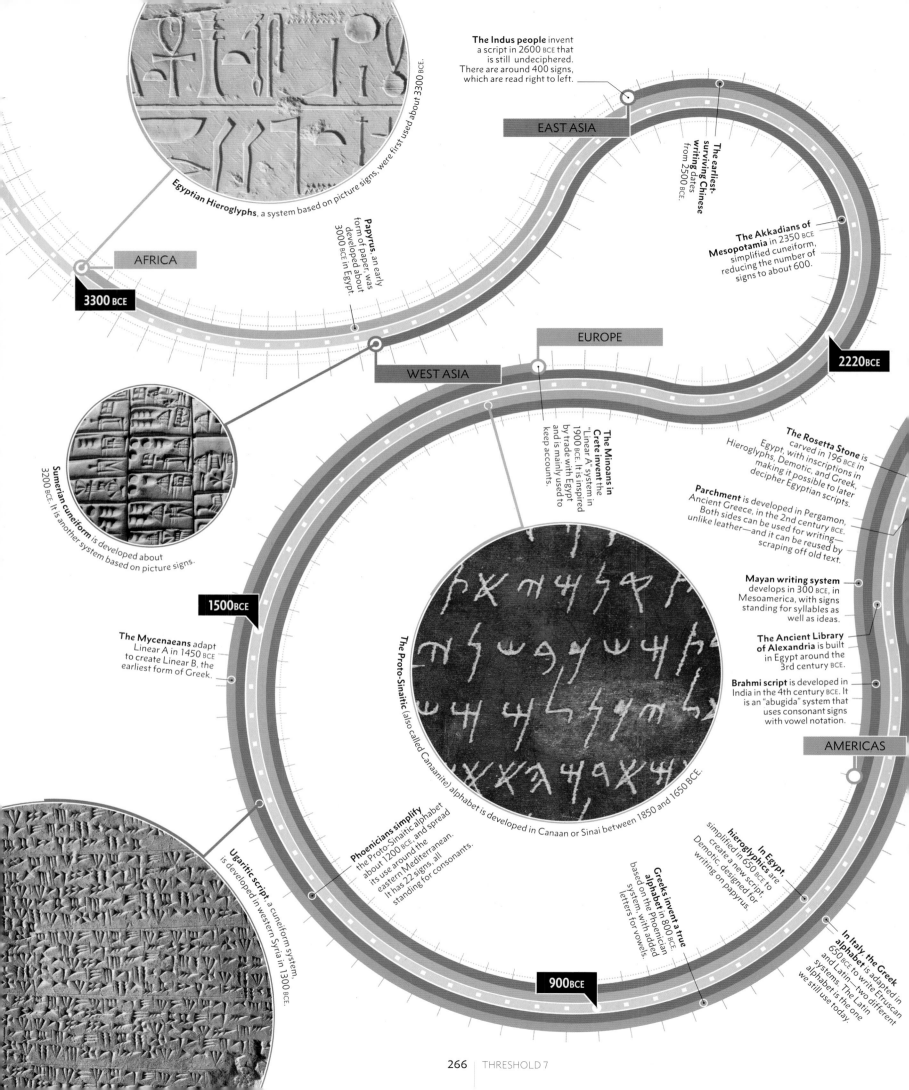

Ugaritic script, a cuneiform system, is developed in western Syria in 1300 BCE.

Phoenicians simplify the Proto-Sinaitic alphabet about 1200 BCE, and spread its use around the eastern Mediterranean. It has 22 signs, all standing for consonants.

900 BCE

Greeks invent a true alphabet in 800 BCE, based on the Phoenician system, with added letters for vowels.

In Egypt, hieroglyphics are simplified in 650 BCE to create a new script, Demotic, designed for writing on papyrus.

In Italy, the Greek alphabet is adapted in 650 BCE to write Etruscan and Latin—two different systems. The Latin alphabet is the one we still use today.

Brahmi script is developed in India in the 4th century BCE. It is an "abugida" system that uses consonant signs with vowel notation.

The Ancient Library of Alexandria is built in Egypt around the 3rd century BCE.

Mayan writing system develops in 300 BCE, in Mesoamerica, with signs standing for syllables as well as ideas.

AMERICAS

Parchment is developed in Pergamon, Ancient Greece, in the 2nd century BCE. Both sides can be used for writing—unlike leather—and it can be reused by scraping off old text.

The Rosetta Stone is carved in 196 BCE in Egypt, with inscriptions in Hieroglyphs, Demotic, and Greek, making it possible to later decipher Egyptian scripts.

WRITING DEVELOPS

Writing emerged when the earliest civilizations began to use pictographs to keep economic records. Sumerian and Egyptian pictographs could stand for words, ideas, and sounds.

Different styles of writing evolved to suit the materials used. Sumerian cuneiform used simple wedge shapes, because it was written by pushing a pointed stylus into soft clay. The flowing appearance of Chinese writing arose because it was originally painted on bamboo strips with a brush. Alphabets were created when the eastern neighbors of the Egyptians adapted around 30 hieroglyphs, using them to represent sounds. The earliest alphabets only contained signs for consonants. Later, the Greeks added vowel signs, too.

Writing systems can be difficult to date because it depends on the accidental survival of ancient texts. While Sumerian clay tablets have survived for millennia, early Chinese writing, on bamboo, has been lost.

1CE

The Chinese invent printing on paper using carved woodblocks in the 1st century CE.

The Romans create the codex in the 1st century CE. It is the first book with separate pages, which is more convenient to read than a scroll.

In Ireland, Ogham writing—featuring straight marks made in stone—develops around the 4th century CE.

The last-known hieroglyphic inscriptions are made in 394 CE, in an Egyptian temple.

Japanese scribes adapt the Chinese script in the 5th century CE, and also invent two syllabic systems (hiragana and katakana)—giving Japan three writing systems.

Literacy starts to be encouraged in Benedictine monasteries in the 5th century CE, leading to the creation of scriptoriums to copy books. Monasteries begin lending to each other, which increases collective learning.

The Cathach of St. Columba is the first document to show a large capital letter at the start of each psalm, followed by decreasing-size letters to regular small capitals, in 610 CE. This is the start of writing in uppercase and lowercase.

Monasteries in England and Ireland begin producing illuminated manuscripts—featuring elaborate decorations and illustrations—from the 7th century CE.

Carolingian script is developed in western Europe in the 8th century. It introduces spacing between words and produces some of the most clear and legible documents in the history of writing.

The Islamic Golden Age begins around 786 CE, when scholars gather in the court of Harun al-Rashid to translate classical texts into Arabic, thus preserving them.

1000CE

The Yongle Encyclopedia is completed in China in 1408 CE. It is the world's largest encyclopedia until Wikipedia, comprising 11,095 volumes and copies of 7,000 texts.

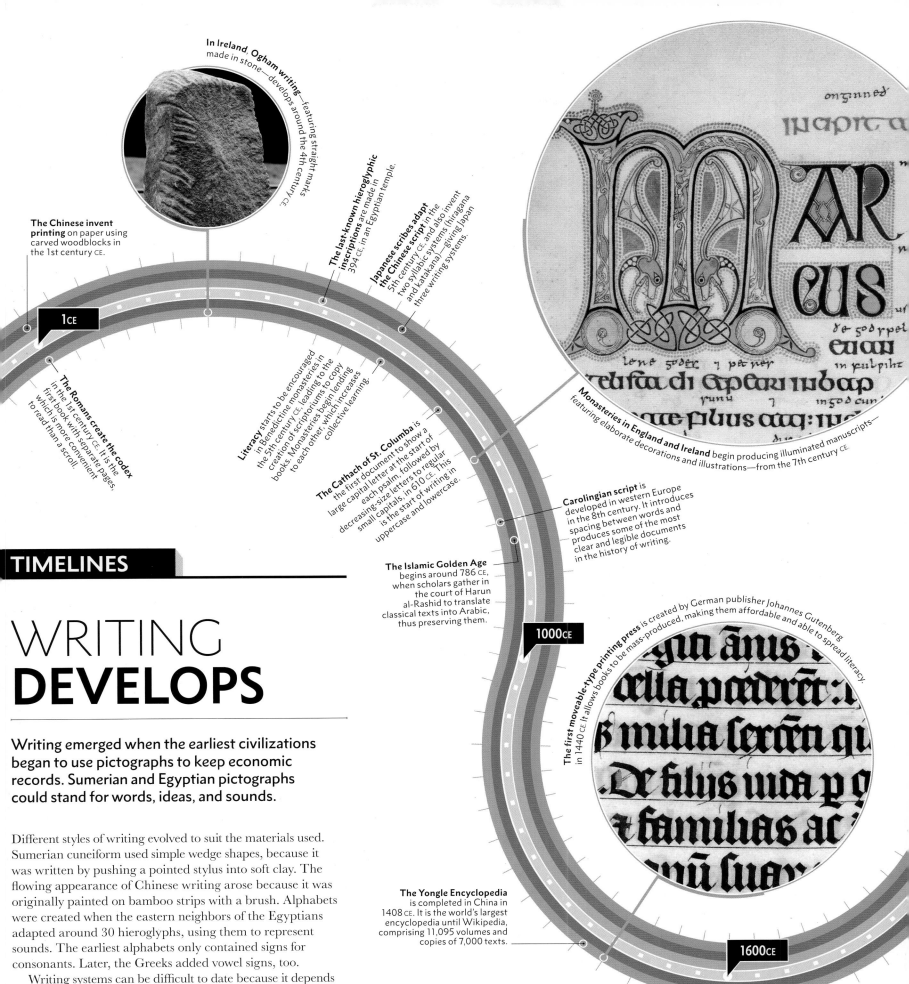

The first moveable-type printing press is created by German publisher Johannes Gutenberg in 1440 CE. It allows books to be mass-produced, making them affordable and able to spread literacy.

1600CE

▼ Irrigating the fields
This reconstruction shows a typical farming village, with its irrigation system, in southern Mesopotamia. It is based on archaeological evidence, such as dried up irrigation canals, and Mesopotamian texts, including instructions for irrigating fields.

Sluice gates
Canal at a higher level
Water regulators controlled the water supply from the canal
Tapered bank to control water flow
WATER REGULATOR
Reeds were harvested for roofing and weaving into baskets
Reed fishing boat
Mesopotamian rivers carried a lot of silt and often changed course
Vegetable and salad crops required plenty of fresh water
Well provided groundwater in times of drought
Fruit trees, such as apples, olives, date palms, and pomegranates, were grown closest to the canal
Trees provi shade crops
Dyke held back floodwaters and prevented deposition of silt into canal
Peas and chickpeas fixed nitrogen in the soil
Shaduf
Small footbridge
Reservoir stored water for emergencies
Weir maintained upstream water level of canal
Animals provided a secondary income from their products
Date palm
Fields were allowed to fall fallow in alternate years to reduce salinization
Cattle plowing field ready for planting
Livestock fertilized fallow fields and acted as insurance against drought—farmers could revert to nomadism
Pigs in village compound were fed on scraps
VILLAGE

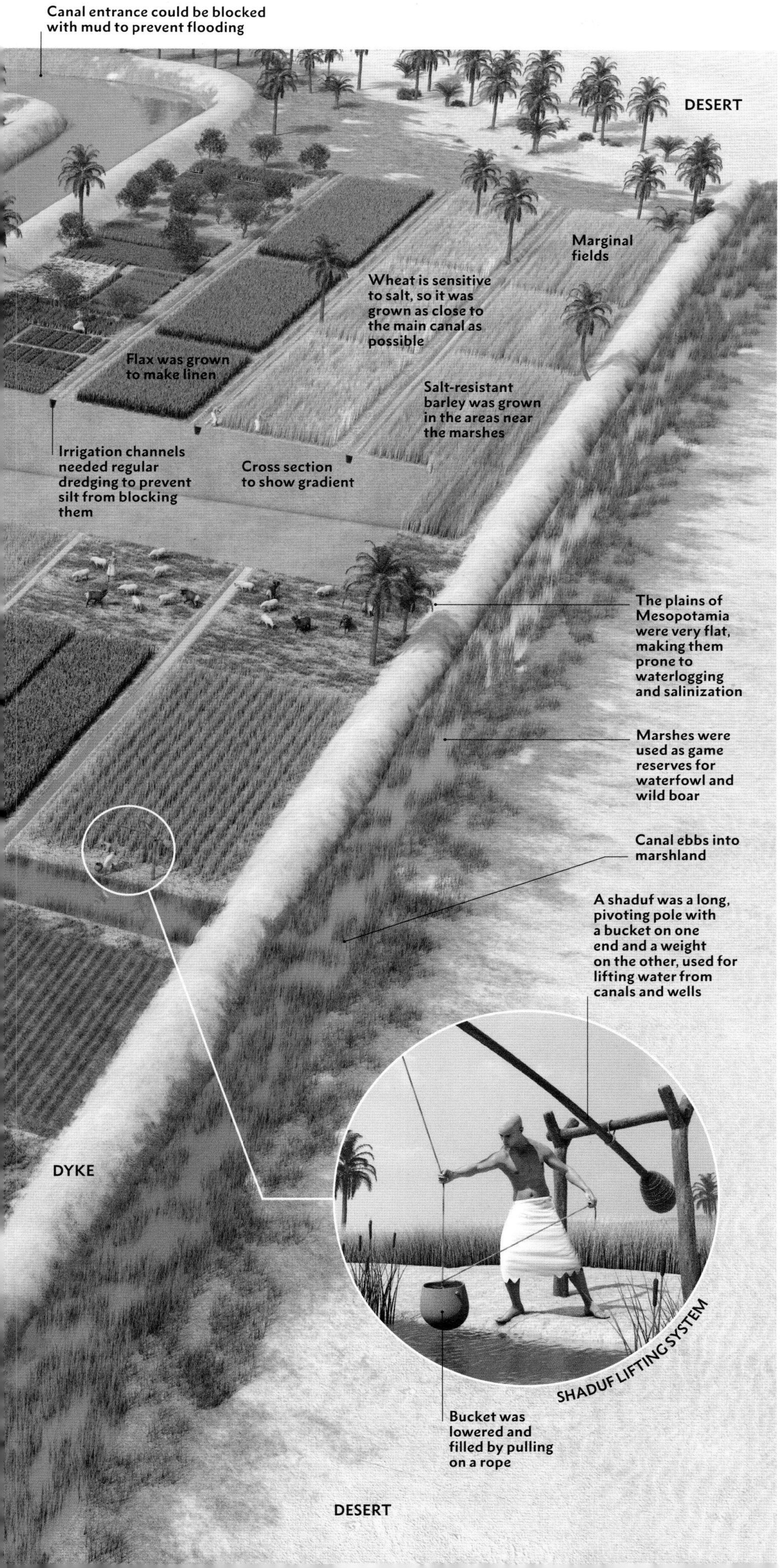
Canal entrance could be blocked with mud to prevent flooding
DESERT
Marginal fields
Wheat is sensitive to salt, so it was grown as close to the main canal as possible
Flax was grown to make linen
Salt-resistant barley was grown in the areas near the marshes
Irrigation channels needed regular dredging to prevent silt from blocking them
Cross section to show gradient
The plains of Mesopotamia were very flat, making them prone to waterlogging and salinization
Marshes were used as game reserves for waterfowl and wild boar
Canal ebbs into marshland
A shaduf was a long, pivoting pole with a bucket on one end and a weight on the other, used for lifting water from canals and wells
DYKE
SHADUF LIFTING SYSTEM
Bucket was lowered and filled by pulling on a rope
DESERT

WATERING THE DESERT

The ability to transfer water from rivers to fields and store it in reservoirs for later use allowed farmers to grow crops beyond the limits of rain-fed agriculture, and even transform desert into fertile land.

Irrigation was very labor intensive, and called for large-scale social cooperation. The first civilizations—Egypt, Mesopotamia, the Indus, and China—all developed extensive irrigation systems. Egypt and Mesopotamia had low rainfall, but benefited from major rivers that flooded every year, depositing nutrient-rich silt on the surrounding fields. In Mesopotamia, where the river flooded at the wrong time of year to grow crops, the water had to be diverted and stored for later use.

DYKES AND CANALS

To divert and control the water, people dug wide canals alongside the rivers. They used the excavated soil to build dykes, which protected their fields and villages from flooding. From the larger canals, smaller channels ran downhill into reservoirs and fields. Weirs and regulators allowed them to adjust the flow from the canals into the channels.

One problem with irrigation is that when water evaporates it leaves behind salt, which builds up in the soil, reducing its fertility. The Mesopotamians dealt with this by leaving fields fallow to recover, and by growing barley, which is more salt resistant than other crops, but overly salty fields were eventually abandoned.

Irrigation demanded a huge amount of work, maintaining dykes and removing silt from the canals. Despite this, the system proved so productive that, in the 4th millennium BCE, the first city-states grew out of these busy and prosperous agricultural towns.

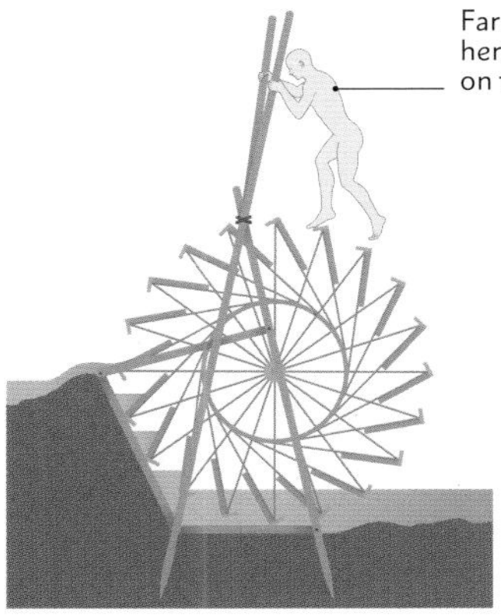
Farmer stood here and walked on the rim

▲ Paddle wheel
Farmers in China lifted water onto their fields using the paddle wheel. The operator stood on the wheel and used the tread of his feet to make it turn and scoop up water.

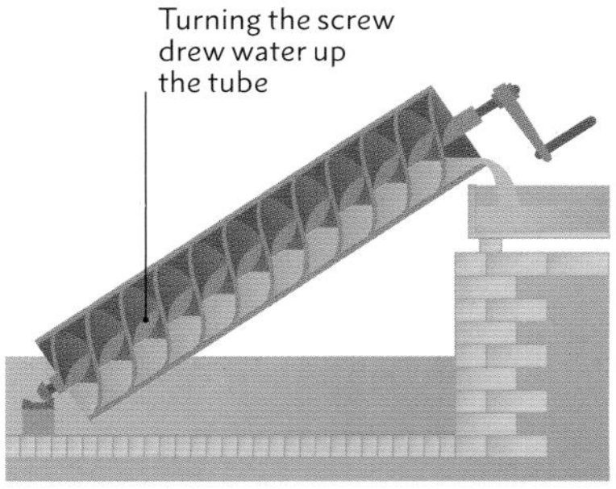
Turning the screw drew water up the tube

▲ Archimedes' screw
This hand-operated pump consisted of a rotating metal screw inside an angled tube. It was said to have been invented by the Greek scientist Archimedes in the 3rd century BCE.

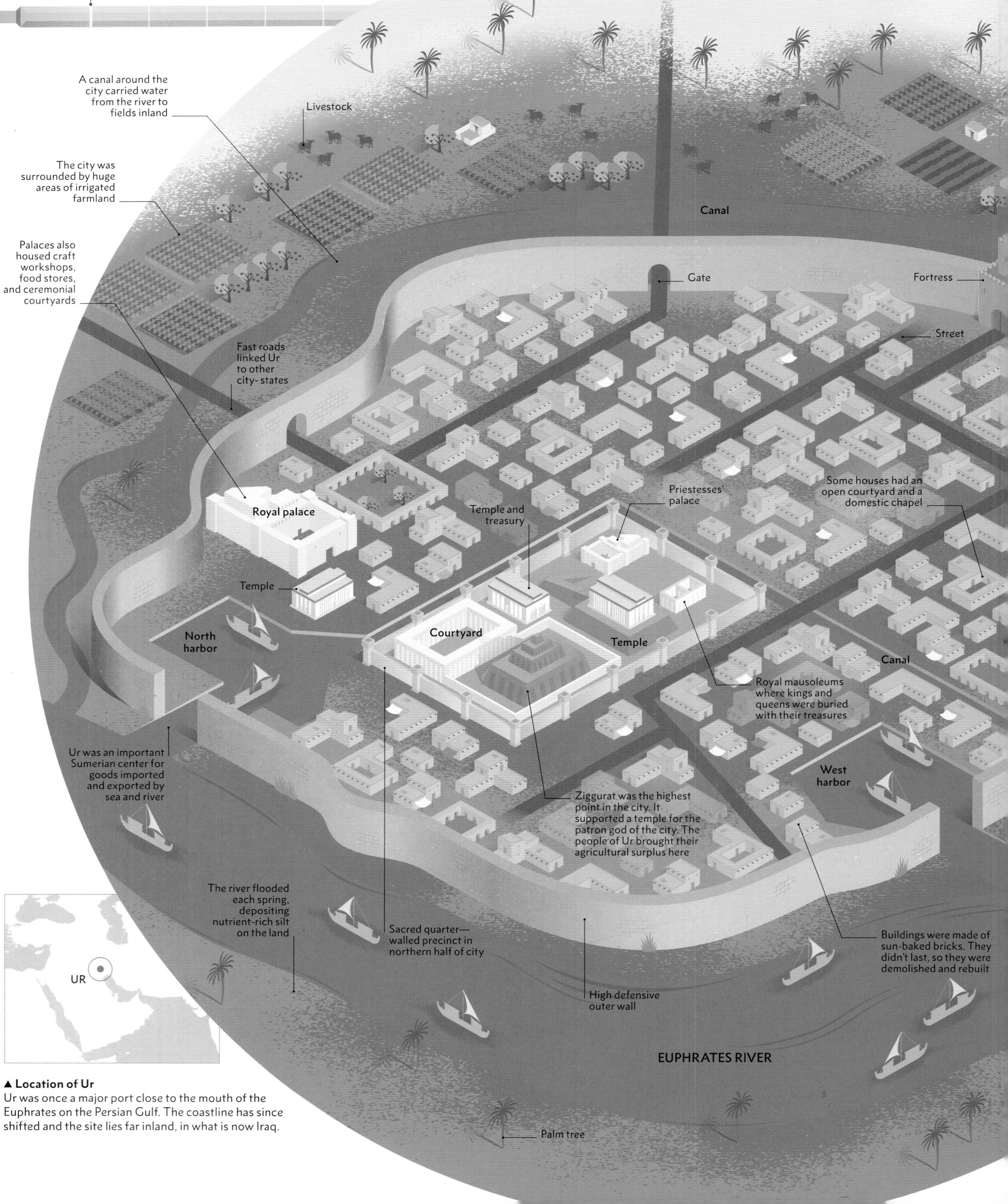

A canal around the city carried water from the river to fields inland
Livestock
The city was surrounded by huge areas of irrigated farmland
Palaces also housed craft workshops, food stores, and ceremonial courtyards
Canal
Gate
Fortress
Street
Fast roads linked Ur to other city- states
Royal palace
Priestesses' palace
Some houses had an open courtyard and a domestic chapel
Temple and treasury
Temple
North harbor
Courtyard
Temple
Canal
Royal mausoleums where kings and queens were buried with their treasures
West harbor
Ur was an important Sumerian center for goods imported and exported by sea and river
Ziggurat was the highest point in the city. It supported a temple for the patron god of the city. The people of Ur brought their agricultural surplus here
The river flooded each spring, depositing nutrient-rich silt on the land
Sacred quarter—walled precinct in northern half of city
Buildings were made of sun-baked bricks. They didn't last, so they were demolished and rebuilt
High defensive outer wall
EUPHRATES RIVER
Palm tree
UR

▲ Location of Ur
Ur was once a major port close to the mouth of the Euphrates on the Persian Gulf. The coastline has since shifted and the site lies far inland, in what is now Iraq.

CITY-STATES EMERGE

Around 3500 BCE, farming villages and towns along the Tigris and Euphrates rivers, in southern Mesopotamia, were transformed into the world's first cities. In seven other places worldwide, cities emerged independently and human history entered a new era: an age of agrarian civilizations.

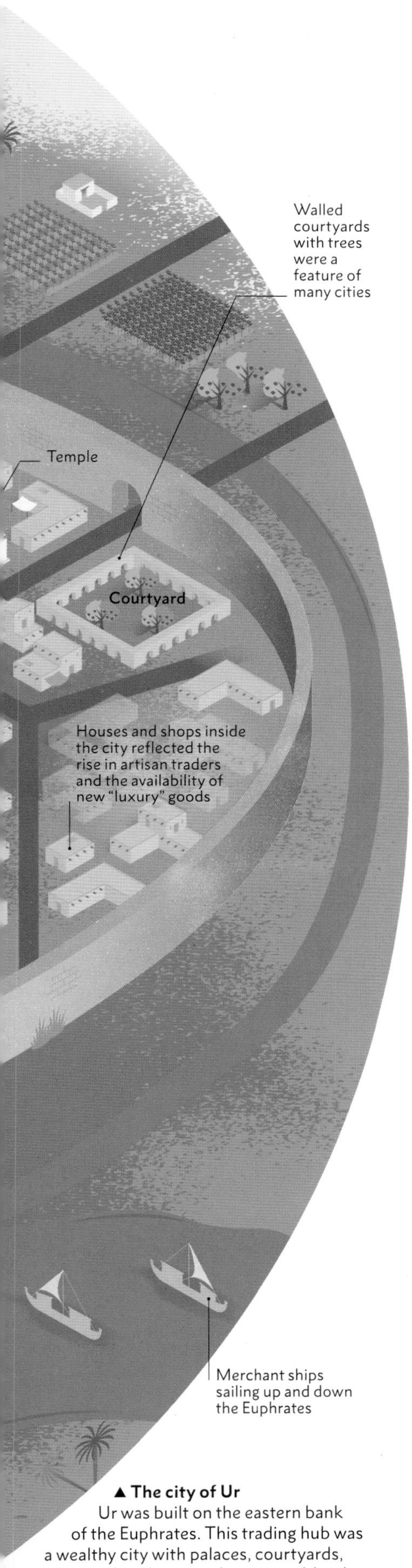
Walled courtyards with trees were a feature of many cities
Temple
Courtyard
Houses and shops inside the city reflected the rise in artisan traders and the availability of new "luxury" goods
Merchant ships sailing up and down the Euphrates

▲ The city of Ur
Ur was built on the eastern bank of the Euphrates. This trading hub was a wealthy city with palaces, courtyards, temples, marketplaces, and many mud-brick houses, where ordinary people lived.

The first cities were more than just the large villages of the early agrarian era, which consisted of similar, self-sufficient households. These cities saw humanized environments emerge with new forms of hierarchy and complexity. One factor that led to the emergence of cities was rapid population growth, the result of increases in productivity after collective learning led to the invention of new technologies.

Uruk was the first of several cities that appeared in southern Mesopotamia, or Sumer. The area was surrounded by desert, which led to the development of settlements with irrigation systems. This innovation made it possible to support a larger population: these cities attracted settlers from more arid parts of the region, and became important centers of exchange. The region of Sumer lacked raw materials, and the need for resources led to the development of long-distance trading networks. Sumerian cities exchanged pottery and grain for tin and copper from Anatolia and gold from Egypt.

By 3000 BCE, there were a dozen Sumerian cities including Uruk, Ur, and Lagash, each with a population of between 50,000 and 80,000 people. Cities were complex economic structures that required new forms of social organization: kings and priestly elites emerged and specialized occupations developed. This led to the creation of states with political, social, and economic hierarchies. During a period of extraordinary invention, the elements of what we call civilization were born: kingship, social hierarchy, monumental architecture, tax collecting, law codes, and literature.

> “THIS IS THE **WALL OF URUK**, WHICH **NO CITY** ON EARTH **CAN EQUAL**. SEE HOW **ITS RAMPARTS GLEAM LIKE COPPER** IN THE SUN.”
>
> *Epic of Gilgamesh*, *c.*2000 BCE

◀ Center point
Sumerian cities were dominated by tall mud-brick temples called ziggurats, which could be seen for miles around. The size of the temple displayed the importance of the local god and the wealth and power of the city that built it. This ziggurat, at Ur, has been partially reconstructed.

FARMING IMPACTS THE **ENVIRONMENT**

When farmers reshaped the landscape to make it favorable for growing food, there were unforeseen consequences. Deforestation, the removal of tree cover, caused soil erosion and the loss of woodland species, while irrigation gradually turned the soil so salty that it could not sustain crops.

The pollen record shows a massive loss of forests across Eurasia as a result of farming. Forests were cut down to provide timber, charcoal for iron working, and arable and grazing land. The Mediterranean lost its deciduous forests, leaving thin soils only suitable for olive trees. In China, felling the trees of the Loess Plateau allowed mineral-rich soil to be washed into the Yellow River, giving its waters their distinctive hue.

Deforestation has a disastrous impact in arid lands, where trees have adapted to the low rainfall by growing deep roots. Between 200 and 400 CE in southern Peru, the Nazca people removed all the local huarango trees. The huarango has the deepest root system of any tree, which helps to maintain the soil's fertility and moisture levels. Pollen samples reveal that the trees were replaced by cotton and corn. Without the anchoring huarango roots, Nazca fields were devastated by soil erosion from high desert winds and seasonal flooding. The land became unsuitable for agriculture, much of it turning to desert. Salinization—the deposition of mineral salts when irrigation water evaporates from fields—also helped to hasten the end of the Nazca culture. The salts accumulate at the soil's surface, making it toxic to most plants. By 500 CE, only salt-tolerant weeds grew on what was once productive Nazca farmland.

Other South American cultures induced similar crises. The Maya, for example, were forced to abandon their cities and pyramids after over-intensive use of water and land.

EASTER ISLAND

When Polynesians arrived at Easter Island (Rapa Nui) in the Pacific, in about 1200, it was covered with a thick palm forest. Pollen studies tell us that by 1650 the last trees had been cleared by slash-and-burn farming. Without wood, the islanders could no longer build boats to fish. They managed to survive the loss of the trees by scattering rocks over half of their island. Called lithic mulching, this system reduces evaporation and soil erosion, and helps replace lost nutrients.

▶ **Planting techniques** The deforestation of Easter Island by the mid-17th century resulted in wind-lashed, infertile fields. The islanders responded by building thousands of planting enclosures called *manavai*. These circular stone walls preserve moisture in the soil and protect young plants from high winds as well as grazing cattle.

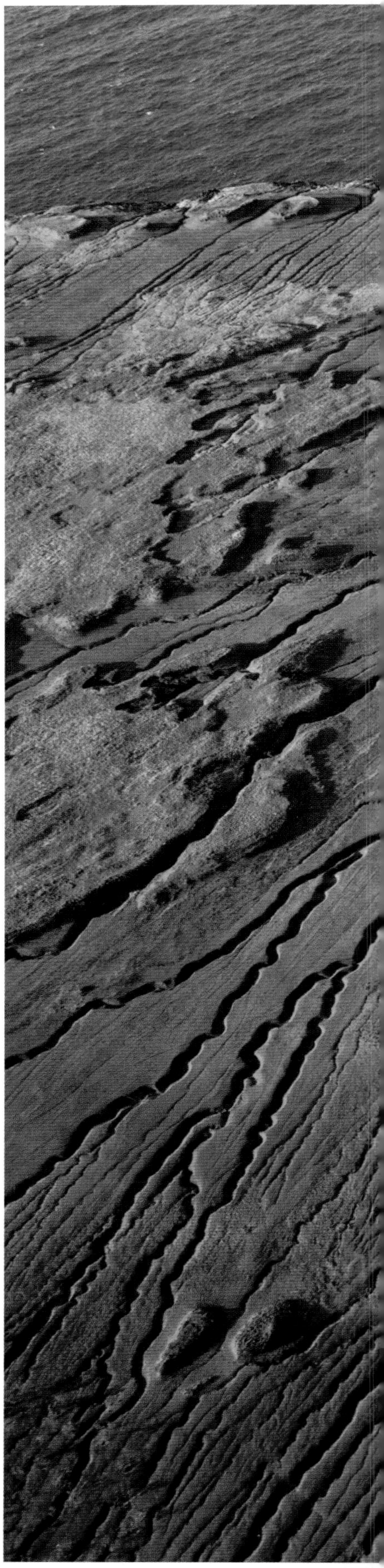

Stripped bare
An aerial view of part of Easter Island shows signs of the massive erosion caused by the loss of its palm trees over three centuries ago. The nutrients in the soil were washed away by heavy rainfall and not replaced, which led to a loss of plant and animal diversity.

BELIEF SYSTEMS

Humans have long believed in the supernatural, but these beliefs have altered over time in response to changing lifestyles. As hunter-gatherers became farmers, beliefs shifted from animism to the worship of ancestors and new gods. Later, as societies grew larger and more complex, universal faiths were established, most of them monotheistic.

The earliest religion we know of is animism or shamanism, which is still practiced by modern hunter-gatherers. This is based on the belief that people, animals, and forces of nature all have spirits, which can be contacted through ceremonies. Bad weather, sickness, or an unsuccessful hunt can all be explained by displeased spirits. Religious specialists, called shamans, enter a trance state to contact the spirits, and then perform rituals to appease them.

With the shift to farming and settled communities, there was a new focus on the worship of the ancestors—the spirits of the dead, who were thought to watch over the living. In many farming communities, people even kept the bodies of the dead in their houses and made offerings to them. The earliest religious structures are great tombs, megaliths, and passage graves, often built on hilltops. The local people's claim to the land they farmed would have been strengthened by the visible presence of their ancestors in the landscape.

> I BELIEVE IN THE **FUNDAMENTAL TRUTH OF ALL GREAT RELIGIONS** OF THE WORLD.
>
> **Mahatma Gandhi**, Indian independence leader, 1869–1948

Farmers also worshipped the Earth, or Great Mother, because it produced new life, and the sun, on which they depended for a good harvest. The Incas of Peru called their sun god Inti and the Earth goddess Pachamama, meaning "World Mother." Farmers in the Andes still perform rituals for Pachamama before the sowing season.

It was widely believed that the favor of supernatural forces could be won by offering gifts, called sacrifices. People in Bronze Age Europe threw precious bronze swords and shields into lakes and rivers, which were seen as portals to the spirit world. The more precious the offering, the more effective it would be. Humans were killed as sacrifices in many cultures, including Bronze and Iron Age Europe and Mesoamerica.

A FAMILY OF GODS

Over time, natural forces and abstract ideas were personified, and families of gods emerged. The Indo-Europeans were pastoralists who, from around 4000 BCE, migrated across western Eurasia, spreading the family of languages. They carried with them the worship of a sky and thunder god, called Dyaus Pita in India, Zeus in Greece, and Jupiter in the Roman Empire. He was the head and king of a family of gods.

The rise of states went hand-in-hand with organized religions, and with temples and priests dedicated to local patron gods. State religions provided a new common bond, uniting large numbers of people who were not tied by kinship. This benefited rulers by creating an ideological framework for the transfer of wealth from the masses to elites. Farmers were expected to bring tribute to offer to the gods at their local temple.

Just as hierarchical state systems emerged, gods also came to be ranked in terms of seniority. Kings justified their rule by claiming to have a unique relationship to the gods, and would intercede on behalf of the people to obtain successful harvests.

Polytheistic religion was inclusive and always open to new gods. The Romans thought that the more gods they could call on, the safer their empire would be. Visitors to other cities were happy to take part in ceremonies honoring local gods without feeling disloyal to their own deities. Polytheistic gods also had no concern with morality. The gods in Homer's *Iliad*, which was the closest thing that the Greeks had to a sacred text, behave just as badly as the human protagonists. The question of what people believed was unimportant; some Greek philosophers even questioned whether gods existed. Around 580 BCE, the philosopher Xenophanes stated that humans create gods in their own image: "Ethiopians say the their gods are flat-nosed and dark, Thracians that theirs are blue-eyed and red-haired. If oxen and horses had hands and were able to draw, horses would draw the shapes of gods to look like horses and oxen to look like oxen."

UNIVERSAL RELIGIONS

A major shift took place with the rise of universal religions offering moral teaching, emotional fulfillment, and salvation. The most important were Zoroastrianism in India, Buddhism in India, Confucianism in China, and Judaism, Christianity, and Islam in the Mediterranean world. These were all founded by male teachers, who were thought by their followers to be divinely inspired.

Universal religions first appeared in the 1st millennium BCE, after the emergence of great empires and the rise of urban life. They were a response to the human need to find meaning in a world of increasing social complexity. Historians of religion call this period the Axial Age, because it was the time when most of today's religions and philosophies emerged.

> THE **CHRISTIAN BIBLE** IS THE WORLD'S **BEST-SELLING BOOK**

In the Americas, there was no Axial Age and no universal religion, perhaps because urban living developed much later than in Eurasia and there was no long-distance trade network that allowed ideas to spread.

CONCERNING **THE GODS**, I HAVE **NO MEANS OF KNOWING** WHETHER THEY **EXIST OR NOT**.

Protagoras, Greek philosopher, *c.*485–415 BCE
On the Gods

ONE GOD

Most universal religions were monotheistic, based on the worship of a single, all-powerful God whose primary concern was human behavior. Religions that addressed moral actions were of use to states in enforcing conformity, enabling rulers to claim that the social order was divinely inspired. Religion offered those who suffered in this life the consolation of an afterlife, and a promised reward in paradise made people willing to sacrifice their own lives for the greater good. This willingness among individuals to sacrifice themselves made the state more successful in warfare.

Universal religions flourished when they were adopted by empires. Christianity and Zoroastrianism became the state religions of the Roman Empire and Persian Empire respectively, and Confucianism became the state philosophy of China. The new religions spread widely thanks to the Eurasian trade networks. From its Indian birthplace, Buddhism was carried east along the Silk Road to China, Japan, and Southeast Asia. Islam spread even further, thanks to its control of the Mediterranean hub region. In the century after the Prophet Muhammad's death, in 632 CE, Muslim armies conquered many lands and established an empire that stretched from Spain to India. Missionaries and merchants went on to carry Islam around the Indian Ocean.

STRONG BELIEFS

Unlike the polytheistic religions, the universal monotheistic faiths placed great importance on beliefs. The problem was that they offered different interpretations of what people should believe. This clash of belief systems caused tensions between nations and cultures. For the first time, people went to war over religion.

The major conflict was between Islam and Christianity. As a result of interfaith wars, the Eurasian trade network became divided into rival blocs, with Christian Europe cut off by the Islamic Ottoman Empire from the Silk Route to China. This led, in the 15th century, to the Age of Exploration, when Christopher Columbus and other European explorers set off to discover new maritime routes to the East.

In this way, religion acted as a major trigger for globalization—the linking up of the entire world by European Christian nations as they traveled, traded, and conquered in the name of faith.

IN 2010, **ISLAM** HAD **1.6 BILLION** FOLLOWERS, A **QUARTER OF THE WORLD'S POPULATION**

◄ **Face of the god**
The elephant-headed Ganesh is one of the best-known and most-popular deities in the Hindu pantheon. Known as the Remover of Obstacles, he is the god of wisdom and learning.

Lord of the dead
This reconstruction of the Lord of Sipán's tomb shows his richly dressed body in the center, with four people around him. His male attendants had had their feet cut off, perhaps to prevent them from deserting their posts.

GRAVE **GOODS**

People have long believed that death is followed by an afterlife: the practice of burying the dead with items that would be useful in the next life goes back more than 30,000 years. The coming of agriculture and the rise of civilization saw a huge increase in grave goods.

Through grave offerings, we can trace the rise of different social classes. The graves of the first farmers, who were buried with simple pots or cuts of meat, show no signs of social distinction. By the Bronze Age (*c.*3000 BCE), chieftains had emerged, buried under large grave mounds with rich treasures.

Grave goods tell us a lot about daily life and beliefs in the past because they include items considered important or valuable at the time. High-status grave goods—evidence of technology—include Iron Age British and Chinese chariots and complete Anglo-Saxon and Viking ships. They also provide evidence of long-distance trade. The 7th century Anglo-Saxon king buried in his ship at Sutton Hoo in England had silver bowls and spoons that had been brought all the way from Constantinople in the Roman Empire (now Istanbul, Turkey).

The absence of grave goods is also significant. It provides evidence of a changed view of the afterlife, spread by new religions. The change is most obvious in late Roman cemeteries, which pagans—buried with grave goods—shared with Christians, who were buried without offerings and with their feet pointing east, toward Jerusalem.

ROYAL GRAVES

The most elaborate offerings come from royal graves, such as that of the Moche Lord of Sipán, on the north coast of Peru. He was buried in around 300 CE with 451 precious objects, made from gold, silver, and feathers. Sharing his tomb were three women, two men, a child, two llamas, and a dog. They were probably sacrificed to accompany their lord in the afterlife.

Human sacrifice was also practiced in the royal tombs of early China, Egypt, and Mesopotamia. As the custom died out, models were used as substitutes for real humans. In Egypt, wooden servants performed work on behalf of the living, while in China, the First Emperor, Qin Shi Huang (259–210 BCE), was buried with a complete terra-cotta army to defend him from the angry ghosts of the people he had killed during his reign.

◀ **Terra-cotta guardian**
This kneeling warrior is one of 7,000 life-sized figures buried to guard the tomb of China's First Emperor. The position of his hands suggests that he held a crossbow.

> “MEMBERS OF THE **KING'S HOUSEHOLD** ARE **BURIED BESIDE HIM**... ALL OF THEM STRANGLED. **HORSES** ARE BURIED TOO, AND **GOLD CUPS** AND **OTHER TREASURES**.”

Herodotus, Greek historian, describing the funeral of a Scythian king, *c.*484–425 CE

CLOTHING SHOWS STATUS

Clouds represent the celestial realm, signifying rain, luck, and never-ending fortune

The production of textiles dates back to the early days of agriculture, when skills from basket weaving were first applied to plant and animal fibers. As textile production developed, fabric became a highly tradeable commodity, and clothing became a new way to demonstrate social rank.

Textiles were invented independently in several parts of the world, using various materials. The earliest textiles, from about 7000 BCE, were linen made from fibers of the flax plant, which was domesticated in Southwest Asia, and cotton, domesticated in India. Later there was wool, which came from sheep in Eurasia and from alpacas and llamas in South America. The main fabrics in Mesoamerica were cotton and ayaté, made from the maguey plant.

MAKING FABRICS

Weaving began with the development of the loom, a device designed to keep warp (lengthwise) threads tight while weft (cross) threads are woven between them. In the Americas, this was achieved by attaching the loom to the weaver's back. Eurasian weavers used an upright wooden frame with weights tied to the warp threads.

Textiles were colored with dyes from plants, minerals, insects, and shellfish. The ancient world's most expensive dye was purple, produced from the *Murex* sea snail in the eastern Mediterranean. This dye was so highly prized that the people who traded in it came to be called Phoenicians, meaning "purple people" in Greek.

STATUS AND SILK

Clothing became an important way for people to display status. In both Egypt and Mesopotamia, linen, which is lighter and smoother than wool, was a high-status material worn by the wealthy. Many societies had laws governing the clothes people were allowed to wear. In Tudor England, members of the royal family alone could wear cloth of gold. In China, only the emperor and his closest relatives were allowed to wear bright yellow.

Silk was the most sought-after textile because of its luster, softness, smoothness, and isothermal properties, which made it cool in summer and warm in winter. It was made in China before 4000 BCE from cocoons of the *Bombyx mori* moth, the world's only fully domesticated insect. Through selective breeding, the moths lost their ability to fly and the legs of the larvae shrank so that they could not crawl away from the trays on which they were kept.

> "**EVEN MEN** HAVE NOT BEEN ASHAMED TO ADOPT **SILK CLOTHING IN SUMMER** BECAUSE OF ITS LIGHTNESS."
>
> **Pliny the Elder**, Roman scholar, 23–79 CE, *Natural History*

◄ Chinese silk
This early 12th-century Chinese painting shows women ironing silk. This fabric was so valued that the overland route from Asia to Europe along which it was traded became known as the Silk Road. Until the 6th century CE, China maintained a monopoly in silk production by making it a capital crime to export silkworms or cocoons. The painting itself was made on a sheet of silk.

Red and blue are lucky colors

◀ Dragon lord
This embroidered yellow silk robe from the 18th century was worn by the Chinese emperor on festive occasions. The color and symbols shown on it were reserved for imperial use.

Flaming pearl is one of the Eight Treasures (pearls of wisdom), which stand for perfection and enlightenment

Dragon is a symbol of good fortune and an emblem of rank and high power. Five-clawed dragons show that wearer is an emperor—lower ranks have three or four claws

Nine dragons appear in total, as nine is the number reserved for the emperor

Dragon flies from the waves to the heavens, bringing rain and fertility

Hem of robe represents the sea

TIMELINES

USING **METALS**

The invention of metallurgy was one of the most important technological advances in history. Metal tools can be molded, hammered into new shapes, and resharpened when they grow blunt.

Metallurgy developed in stages, as Eurasian people gradually learned how to work harder metals. The earliest metal was copper, which is a soft metal, so tools needed regular resharpening. Later, people learned how to make bronze by adding a small amount of tin to copper. This produced a harder metal suitable for swords, spears, and shields. Copper and tin are both scarce metals, so bronze was mainly used by the elite.

Iron was the last metal to be worked, because it requires very high temperatures to smelt. Yet the use of iron, to make high-status weapons and low-status tools and nails, would change the lives of everyone across Afro-Eurasia.

1 FIRST METALS

Metalworking began in the Fertile Crescent, in the Middle East, *c.*7000 BCE, with people in farming communities making jewelry from naturally occurring nuggets of gold, copper, and lead. These soft metals could be worked without using heat, although too much hammering would make them brittle.

7000 BCE

PURE METALS

6000 BCE

Copper is first extracted from its ores by heating over a fire (smelting) in Western Europe and East Asia *c.*5000 BCE. The metal is poured into molds to make tools.

5000 BCE

Metal foundries process gold, copper, lead, zinc, tin, and iron in Metsamor, Armenia, from 5000 BCE.

Death mask of Tutankhamun is made using techniques for purifying gold from ores, which emerged in Egypt *c.*1327 BCE.

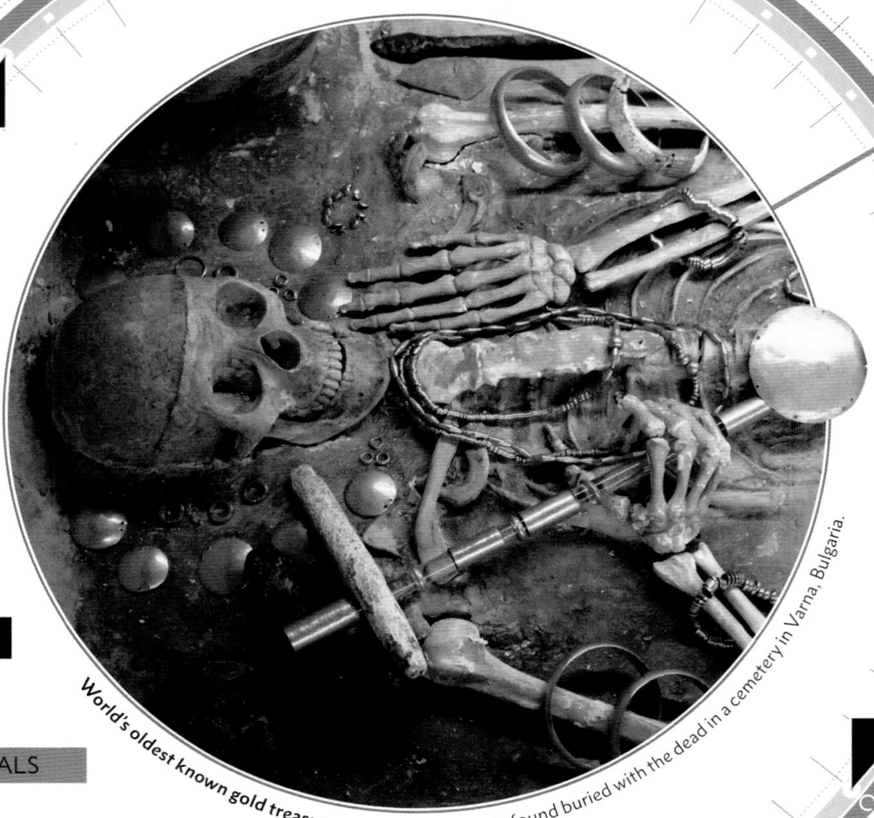

World's oldest known gold treasure, *c.*4600–4200 BCE, is found buried with the dead in a cemetery in Varna, Bulgaria.

Tin bronzes are made in Pločnik, Serbia, *c.*4500 BCE. Knowledge of the process is lost when the Vinča culture from the region dies out.

4000 BCE

Tuyere pipes, depicted here in a 15th century BCE painting, are used to increase the temperature in a crucible while smelting copper, from *c.*4000 BCE.

Copper is cast, using the lost-wax process, in what is now Israel *c.*3700 BCE.

BRONZE

Bronze is made by the Sumerians in Western Asia *c.*3500 BCE. Gold and silver, as native metals, are also exploited in the region.

2 BRONZE BEGINS

Bronze is a mixture of copper and another metal, usually tin. Since copper and tin ores are rarely found together, tin bronzes are evidence of trade. Most archaeologists date the start of the Bronze Age to *c.*3500 BCE, but there is evidence that tin bronze was being worked in Serbia 1,000 years before that.

Bronze casts are used to make weapons *c.*3100 BCE, although these are too expensive to be used by anyone but the elite.

Two forms of bronze are in use: "classic bronze" for casting, and "mild bronze" for sheet metalwork and armor (*c.*3000–2100 BCE).

3000 BCE

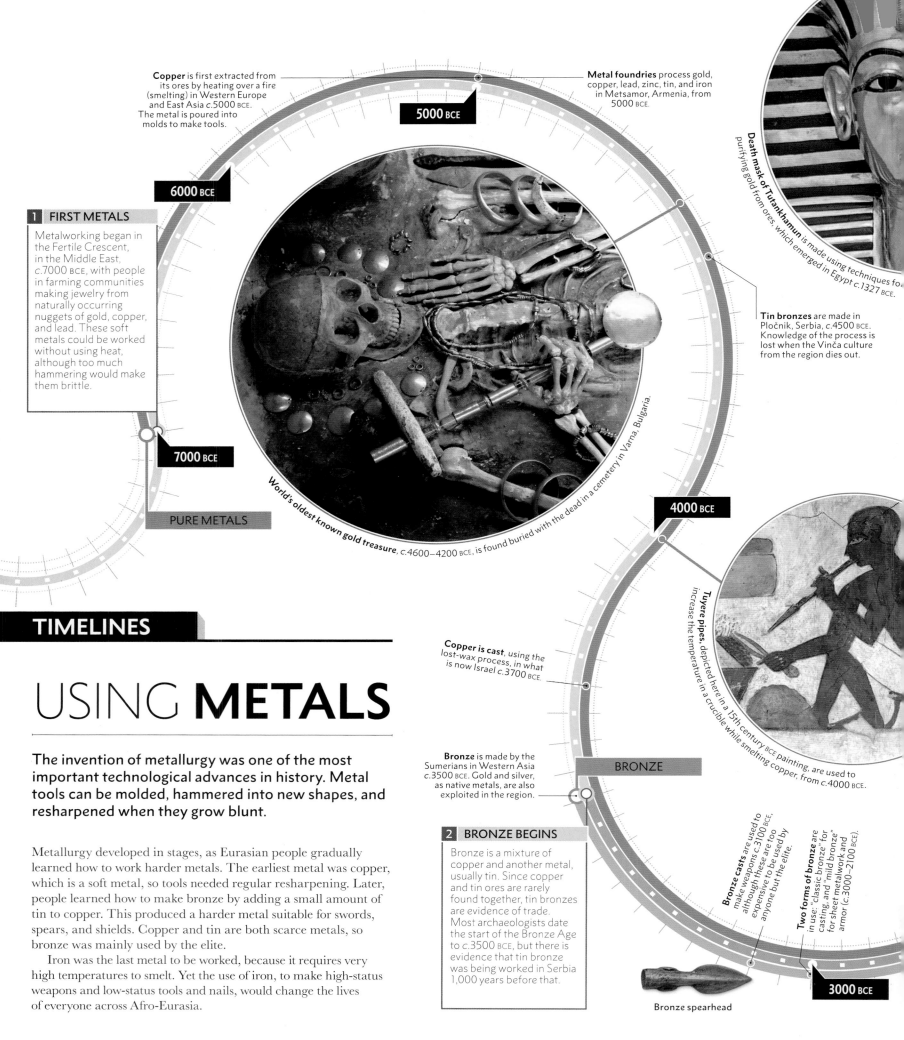

Bronze spearhead

IRON

Bronze objects from the Chinese Shang dynasty become more decorative (*c.*1500 BCE).

Shang dynasty taotie (bronze animal mask)

IRON SHIFT

Dates given for the start of the Iron Age vary, but iron objects found in India and evidence of steel manufacturing in Anatolia, Turkey, date back to 1800 BCE. Iron is an abundant metal, but it requires high temperatures to smelt. It is possible that disruption in the tin trade forced the shift in use from bronze to cheaper iron.

1000 BCE

Iron working reaches western Europe *c.*800 BCE. The European Iron Age brings increased warfare, reflected in the building of hill forts and defenses.

Wootz steel, an alloy of iron with carbon, is invented in southern India *c.*550 BCE and is exported to the West.

The blast furnace, for making cast iron, is invented in China *c.*500 BCE. It is centuries before the technology is matched in Europe.

Lead smelting takes place in the Roman Empire (*c.*366 BCE–36 CE), contributing to early global pollution.

Damascus steel swords are made by eastern Mediterranean metalworkers *c.*330 BCE, using imported Wootz steel from India. Knowledge of their sword-making technique is later lost.

Soldering is invented by Chavin metalworkers in Peru *c.*330 BCE, along with lost-wax casting. They also invent tumbaga, an alloy of gold with copper, which is used to make beautiful artworks.

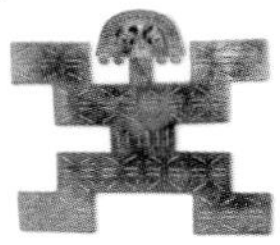

Gold tumbaga pectoral

100CE

Carbon steel is first created by the Haya people of Tanzania *c.*100 CE, centuries before its production in Central Europe.

Ironwork spreads across western Asia and the Mediterranean, 1200–1100 BCE, where this Greek vase depicting a smith at a forge was later painted in the 6th century BCE.

Worked gold necklace, *c.*2000 BCE, is found in a grave near Lake Titicaca, Peru. The metal is made from a naturally occurring nugget, cold-hammered into shape.

European sword-makers develop stronger swords by welding together successive layers of iron with added carbon, or by beating out thin iron strips and welding them together, 700–800 CE.

European iron sword

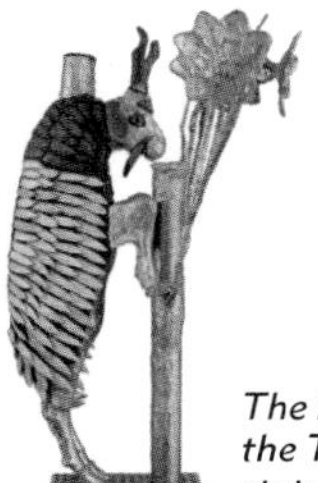

The Ram in the Thicket statuette

Silver starts to be mined and smelted in Anatolia, Turkey, 3000–2500 BCE.

Gold and silver are among the materials used to make *The Ram in the Thicket* statuette, from Ur, Iraq, *c.*2550 BCE.

Western Mediterranean countries initially make bronze with arsenic, but poisoning forces them to switch to using tin, which is more expensive.

Bronze knife, *c.*2700 BCE, found at Dongxiang, provides the earliest evidence for bronze-making in China.

English iron foundries convert cheap coal to coke and use it instead of charcoal in the 1600s, to produce cast iron.

Bessemer process for making steel, invented in the 19th century, has its roots in East Asia *c.*1200 BCE.

1000 CE

Gunpowder weapons, including the "flying-cloud thunderclap eruptor," a cast-iron cannon, are invented in China *c.*1200 CE.

Cast iron is developed in Europe during the 1400s. Because it is strong and can be cast into tube shapes, it finds an immediate use in the manufacture of artillery.

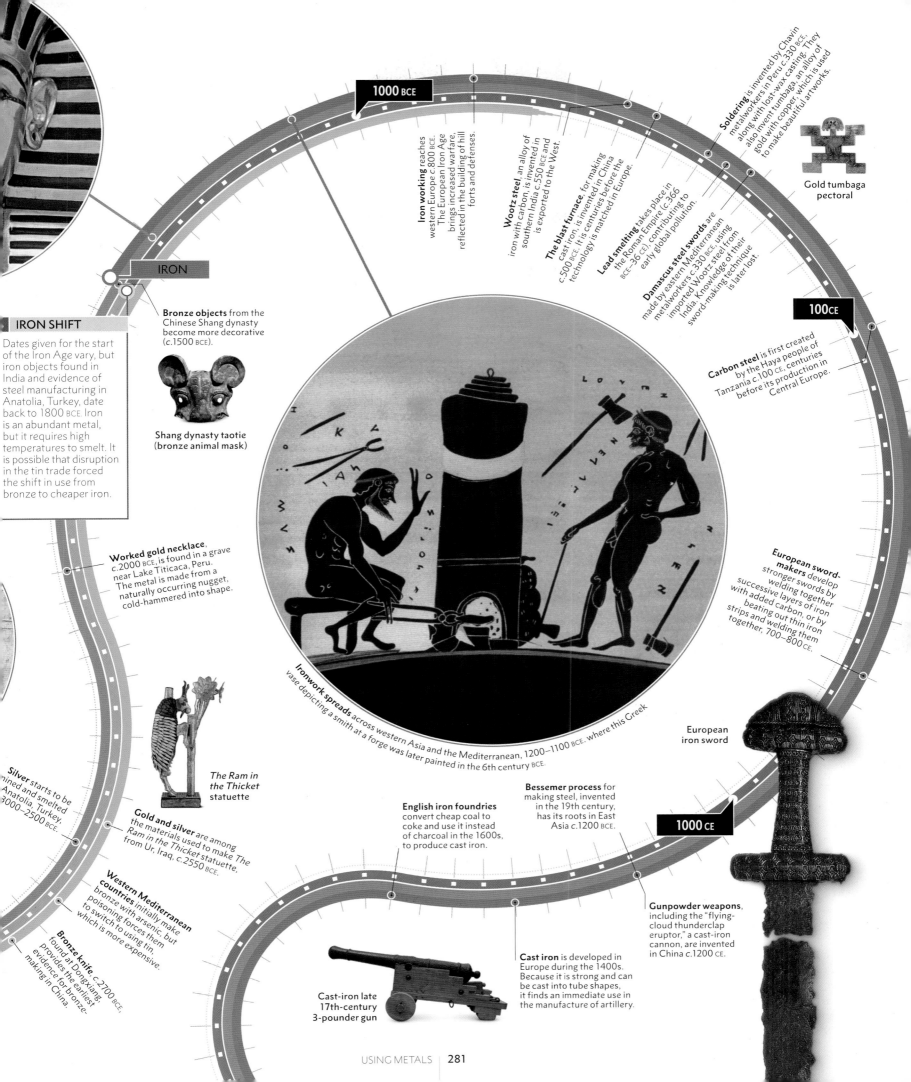

Cast-iron late 17th-century 3-pounder gun

Health issues

Ötzi suffered with arthritis, from a life of hard physical work. He was also infested with intestinal whipworms, from drinking dirty water, which would have given him stomach pains and diarrhea, and he may have had Lyme disease—a bacterial infection caused by tick bites. Growth patterns in his one surviving fingernail show that Ötzi had been seriously ill three times during the last year of his life.

Comfortable shoes

The outer covering of Ötzi's shoes was made of deerskin. Inside was a woven-grass netting that held an insulating layer of hay in place. Both parts were fastened by leather straps to a bearskin sole. The shoes would have been warm and comfortable, but they were not waterproof.

Leather straps
Inner shoe

Goat leather loincloth was fastened with a belt

► **A museum reconstruction** allows us to visualize what Ötzi may have looked like. He was short in stature and had a wiry, but strong, frame. He lacked a twelfth pair of ribs and had no wisdom teeth.

Ötzi reconstructed

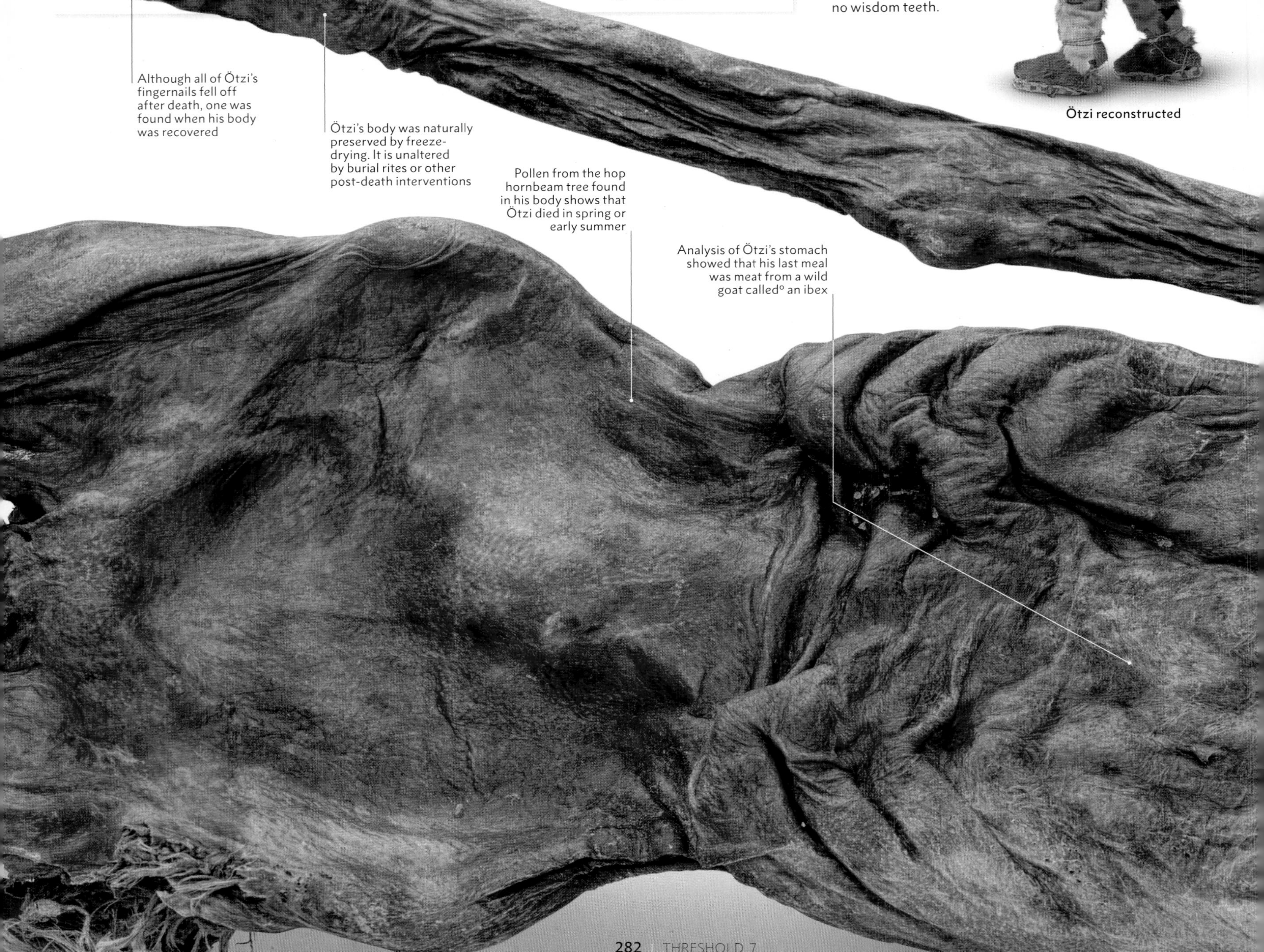

Unhealed knife wound on right hand, between thumb and index finger
Although all of Ötzi's fingernails fell off after death, one was found when his body was recovered
Ötzi's body was naturally preserved by freeze-drying. It is unaltered by burial rites or other post-death interventions
Pollen from the hop hornbeam tree found in his body shows that Ötzi died in spring or early summer
Analysis of Ötzi's stomach showed that his last meal was meat from a wild goat called an ibex

ÖTZI THE **ICEMAN**

In 1991, a naturally mummified man was found in the Ötzal Alps, between Austria and Italy. Nicknamed Ötzi, items found with him tell us that he lived and died around 5,300 years ago.

Ötzi's body, discovered with 70 items of clothing and equipment, gives us a unique and detailed snapshot of one individual who lived and died during the Copper Age (*c.*4500–3500 BCE), when metal tools were first used in Europe.

Although he belonged to a farming community, Ötzi was also a hunter. The copper axe he carried was a symbol of the status he held in his community. Ötzi had the typical health problems of early farming peoples, including bad teeth and arthritis.

Ötzi's clothing, made from the hides of domesticated goats alongside wild deerskin and bearskin, consisted of a loincloth, a belt with a tool pouch, leggings, shoes, a coat, and a cap. The clothes were infested with fleas. He may have used a piece of grass matting to shelter from the rain.

Ötzi died violently. Not long before his death he had fought off an attacker who had wounded him in the hand with a knife. Ötzi escaped but was later killed when an arrow struck him in the back. His body was quickly covered by snow and ice, which protected it from decomposition.

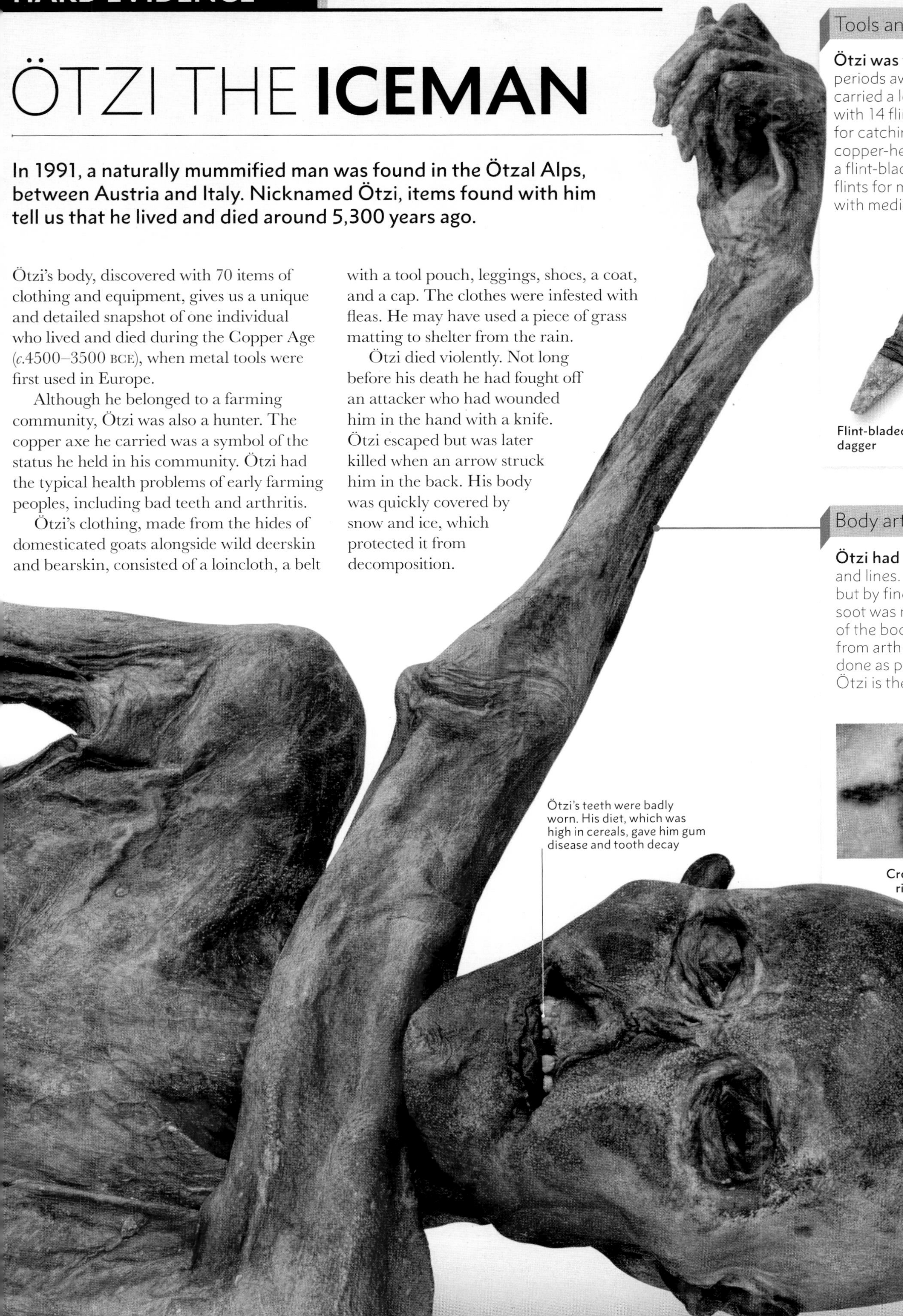

Ötzi's teeth were badly worn. His diet, which was high in cereals, gave him gum disease and tooth decay

Ötzi originally had brown hair, but it all fell out while he was in the ice. Particles of copper in his hair suggest he may have been a coppersmith

A wound to the back of the head was caused by a fall or an assault

As well as long head hairs, shorter, curly hairs were also found at the site, indicating that Ötzi probably had a beard

Tools and equipment

Ötzi was well equipped to survive long periods away from home. For hunting he carried a longbow made from springy yew, with 14 flint-headed arrows, and a string net for catching birds and rabbits. He had a copper-headed axe, for felling trees, and a flint-bladed dagger. His gear also included flints for making fire and fungi with medicinal properties.

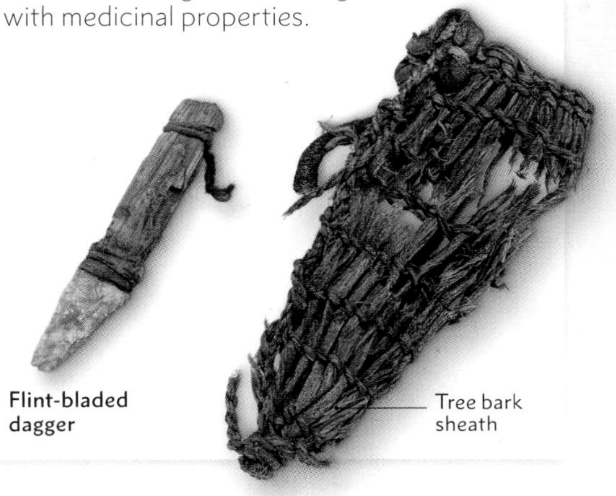

Flint-bladed dagger

Tree bark sheath

Body art or pain relief?

Ötzi had 61 tattoos, mostly crosses and lines. They were made not by needles but by fine cuts to the skin, into which soot was rubbed. The tattoos are on areas of the body where Ötzi would have suffered from arthritic pain. They may have been done as pain relief, like acupuncture. Ötzi is the world's oldest tattooed mummy.

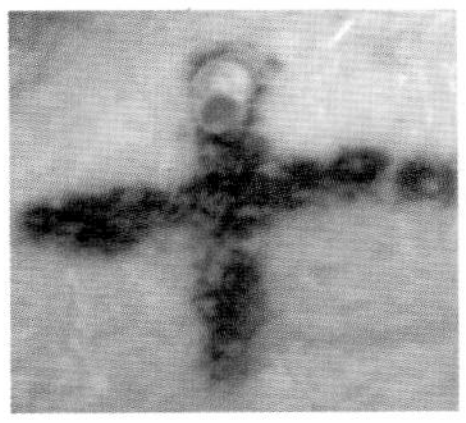

Cross behind right knee

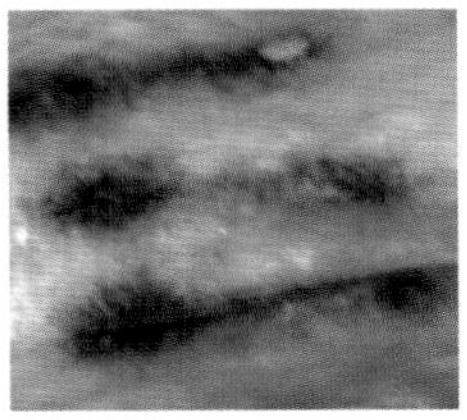

Three lines on inner right ankle

CONFLICT LEADS TO WAR

For most of human history, the population was small enough to avoid intercommunal violence on any great scale. Warfare began as populations rose and demand increased for land and resources. As communities grew larger, conflicts became ever more deadly.

The earliest evidence of targeted collective violence comes from a cemetery in Egypt, where archaeologists discovered 24 skeletons of hunter-gatherers who had been killed by flint arrowheads around 13,000 years ago.

The birth of agriculture led to a steep rise in violent conflict. Farmers had land, goods, and livestock to protect, and they were vulnerable to attack. Groups competed over resources, with conflict intensifying when harvests were poor. Evidence of early massacres comes from three mass graves found in Germany, from around 5000 BCE, where the dead were slain with stone adzes.

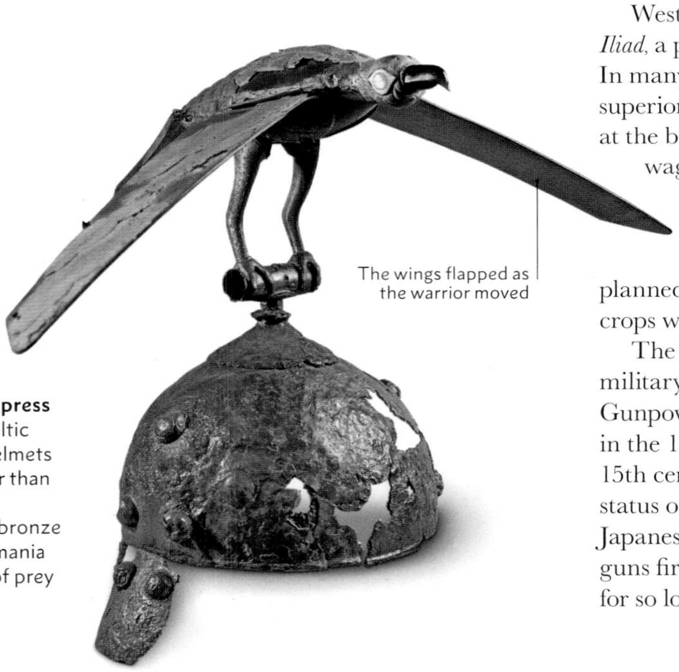

The wings flapped as the warrior moved

► **Dressed to impress**
High-ranking Celtic warriors wore helmets for display rather than protection. This 4th-century BCE bronze helmet from Romania has a huge bird of prey as its crest.

THE FIRST ARMIES

The formation of states led to the creation of armies and the development of new military technologies. These technologies included the chariot, used by elite warriors across Bronze Age Eurasia, and the composite bow, which combined horn and wood to make a small weapon of great power. After the domestication of the horse had opened up the steppes of Asia to nomadic pastoralists, swift-moving tribes of mounted nomads armed with these bows became a constant threat to the settled civilizations of China and western Eurasia.

Western literature begins with Homer's *Iliad*, a poem glorifying heroic warriors. In many cultures, warriors were considered superior to all other classes, with farmers at the bottom of the social hierarchy. Yet waging war was only possible thanks to the work of farmers, who grew the crops that armies depended on. Military campaigns had to be planned to coincide with the period when crops were available to feed the troops.

The Eurasian trade network allowed military innovations to spread widely. Gunpowder weapons, invented in China in the 13th century, reached the west in the 15th century. Gunpowder ended the elite status of warriors. European knights and Japanese samurai were both vulnerable to guns fired by conscripted peasant soldiers—for so long their social inferiors.

"**WAR—I KNOW IT WELL**, AND THE **BUTCHERY OF MEN**... IN CLOSE FIGHTING, I KNOW **ALL THE STEPS** OF THE **WAR GOD'S DEADLY DANCE**."

Homer, Greek poet, c.800–700 BCE, *Iliad*

Battlefield technology
This Persian painting depicts a cavalry battle between Persian and Turk forces in 589 CE. Both sides are armed with small, powerful composite bows. The Turkish ruler Bagha Qaghan (right) is killed by an arrow fired by the Persian general, Bahram Chobin (left).

Edge of the empire
Hadrian's Wall, built by the Romans across northern Britain in 122 CE, was both a defensive barrier and a means of controlling the population on either side. It split the territory of the local Brigantes tribe in two, and was used to monitor, and tax, movement from one side to the other.

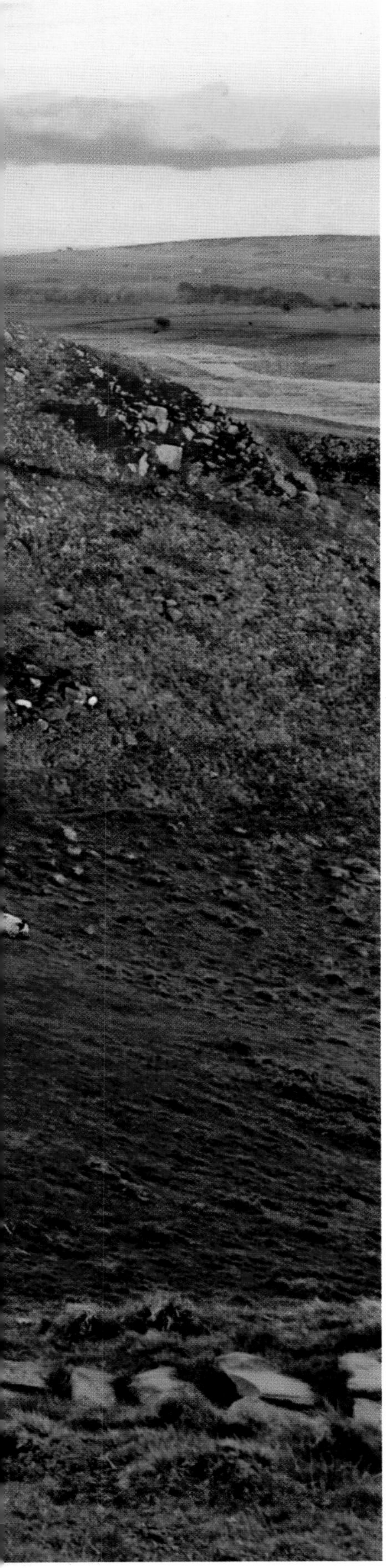

AGE OF **EMPIRES**

Empires arose as states expanded out of their own regions, conquering other areas to acquire more resources. In the process, rulers had to work out how to keep a diverse range of conquered subjects under control, exact tribute, and govern far-flung lands.

The simplest form of empire is one based on indirect rule. In the 15th century CE, the Aztecs conquered a huge empire that stretched from the Pacific to the Gulf of Mexico, but they did not directly rule any of its peoples. Instead, the conquered cities were expected to send annual tributes of luxury goods—including textiles, jade, and feathers—to the Aztec capital, Tenochtitlan. The disadvantage was that their subjects resented Aztec rule and, when the chance came, rebelled against it.

Other empires were able to enforce direct rule by installing governors in conquered cities. In the 540s BCE, Cyrus the Great, founder of the Persian Empire, created 26 satrapies—local governorships. The Persian Empire was diverse and multicultural: stone reliefs show people from all over the empire, in their distinctive dress, bringing tribute to the Great King. The weakness of this system was that the conquered had no reason to remain loyal to Persia, and satraps were able to create independent power bases.

THE ROMAN EMPIRE

The most effective and long-lasting empire was that of the Romans, whose innovation was to open up citizenship to new conquests. Elites were offered the chance to become Roman, with all the rights and privileges that entailed. Unlike the Persian Empire, Rome offered a shared culture, with the same languages (Latin and Greek), clothing, and gods throughout its territories. Men in places as distant as Egypt and northern Britain wore the Roman toga.

The Romans also offered stable rule, known as the *Pax Romana* (Roman Peace), which encouraged trade. They linked the lands of their empire with a vast network of roads and rid the Mediterranean of piracy. The rich empire was also a market for goods from distant lands, including silk from China, Baltic amber, and Indian spices.

Although the Roman Empire finally fell, it left a lasting legacy in the form of roads, towns, literature, architecture, and a template for effective imperial governance that would inspire nations and rulers for millennia.

> **LET US PRAY** THAT ALL **THE GODS** AND THEIR CHILDREN GRANT THAT **THIS EMPIRE** AND THIS CITY **FLOURISH FOREVER**.
>
> **Aelius Aristides**, Greek rhetorician and Roman citizen, 117–181 CE, *The Roman Oration*

▼ The Oxus Chariot
The Persian Empire was the first to use a road system as a means of governance and communication. Satraps and messengers could travel quickly on the royal roads, in chariots similar to the one depicted by this tiny gold model.

HOW EMPIRES **RISE AND FALL**

Throughout history, hundreds of empires have risen and fallen, often following a similar lifespan—a period of vigorous growth, followed by a decline. Some empires fragmented into smaller states. Others were conquered by new rising empires.

Empires were hard to sustain. Armies had to be funded and maintained. As long as an empire was expanding, the expense could be met by new conquests. However, once it reached its largest size, this had to be done by taxing the population. Empires were vulnerable to external enemies and internal conflict, as well as environmental factors, such as famine and disease.

The earliest empire we know of is that of Sargon of Akkad, who conquered all of Mesopotamia around 2300 BCE. He pulled down the fortifications of conquered cities and installed his sons as governors. The Akkadian Empire broke up around 2150 BCE after a series of rebellions and foreign invasions. Although Sargon's empire fell, he had set an example that many later Mesopotamian rulers attempted to match.

◄ **Early emperor**
This bronze head is thought to be Sargon of Akkad. He was admired by the Mesopotamian conquerors who followed him.

LASTING LEGACIES

The most successful conqueror was Alexander the Great (356–323 BCE), whose empire stretched from Egypt to Afghanistan. Although his empire did not survive his death, Alexander's astonishing conquests inspired both the Romans and Chandragupta Maurya, founder of India's first empire. Greek ideas, art, and culture greatly influenced the Romans.

Meanwhile, more than two hundred theories have been put forward to explain why the Roman Empire "fell." Today, historians tend to describe its end as a gradual transformation rather than a sudden collapse. What is more interesting, perhaps, is the fact that while central rule ended, the Roman Empire, like Sargon's and Alexander's, left a lasting legacy through collective learning. By 1300 CE, universities that were founded in many European cities introduced Greco-Roman ideas to European intellectual life. And the Roman legal system, reorganized by Emperor Justinian, is still the basis of legal systems in most of Europe today.

Conquers other states with power vacuums and valuable economic assets
Well-governed, strong city-state reaches the limits of its growth and resources

► **Rise and fall of empires**
Throughout history, empires across every part of the globe have grown and then collapsed, all folllowing a similar process with these common elements contributing to their rise and decline.

> ANY **KING** WHO WANTS TO **CALL HIMSELF MY EQUAL, WHEREVER I WENT** [CONQUERED], **LET HIM GO**.

Sargon of Akkad, Emperor of the Akkadian Empire, d.2215 BCE

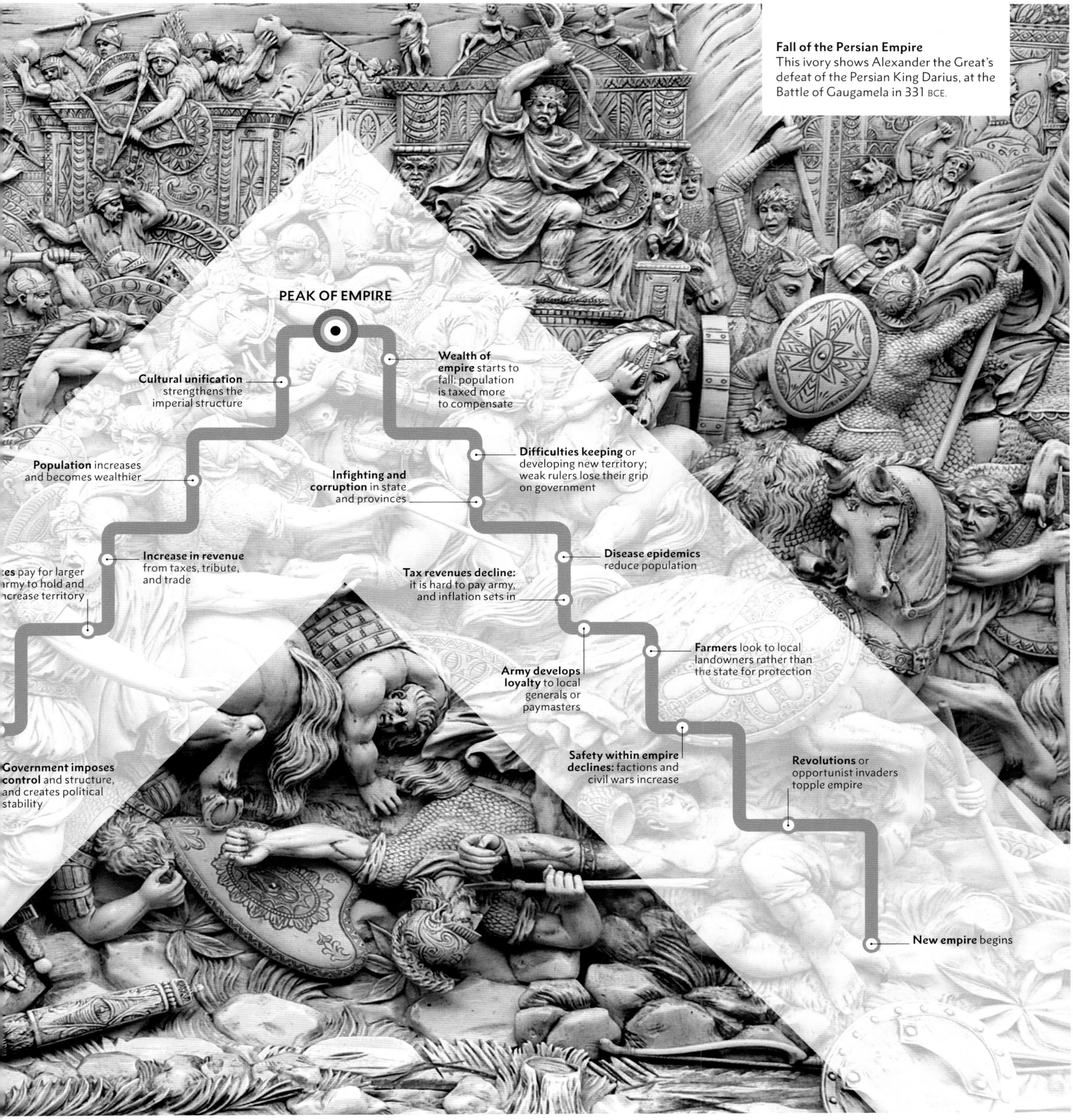
PEAK OF EMPIRE
Cultural unification strengthens the imperial structure
Population increases and becomes wealthier
Increase in revenue from taxes, tribute, and trade
es pay for larger army to hold and crease territory
Government imposes control and structure, and creates political stability
Wealth of empire starts to fall: population is taxed more to compensate
Difficulties keeping or developing new territory; weak rulers lose their grip on government
Infighting and corruption in state and provinces
Disease epidemics reduce population
Tax revenues decline: it is hard to pay army, and inflation sets in
Army develops loyalty to local generals or paymasters
Farmers look to local landowners rather than the state for protection
Safety within empire declines: factions and civil wars increase
Revolutions or opportunist invaders topple empire
New empire begins

Fall of the Persian Empire
This ivory shows Alexander the Great's defeat of the Persian King Darius, at the Battle of Gaugamela in 331 BCE.

▶ Qianlong coin
This coin of China's Qianlong Emperor (ruled 1736–95) follows the model of the First Emperor's coinage. It has a powerful symbolic design, asserting the emperor's universal authority. The coin was minted in denominations of 1 and 10.

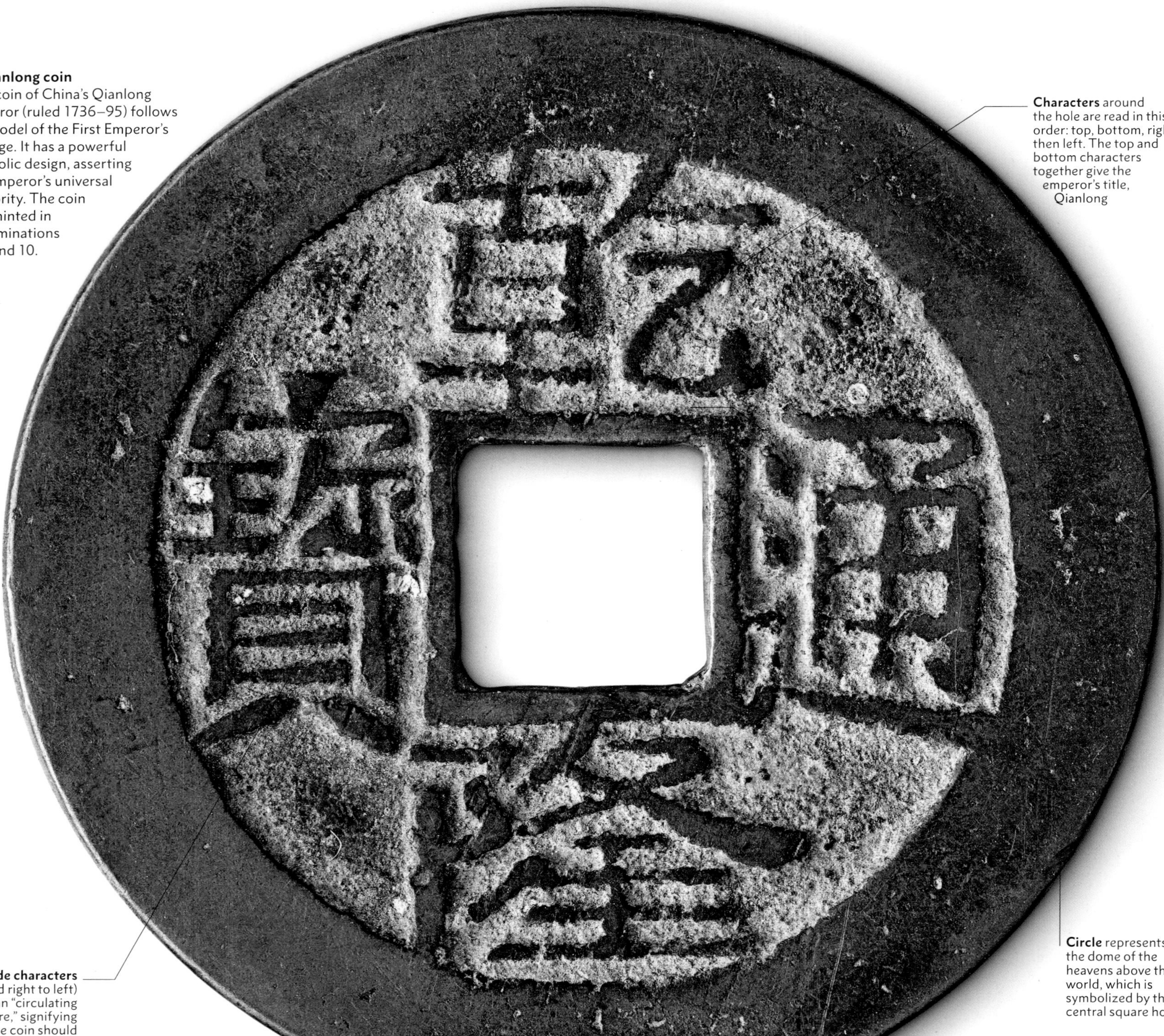
Characters around the hole are read in this order: top, bottom, right, then left. The top and bottom characters together give the emperor's title, Qianlong
Side characters (read right to left) mean "circulating treasure," signifying that the coin should circulate freely
Circle represents the dome of the heavens above the world, which is symbolized by the central square hole
Coin is made from copper alloy cast in a mold

▶ Replicated designs
These coins show how the idea of money spread across Europe. At left is a gold Greek coin issued by Philip of Macedon (ruled 359–336 BCE). Philip's coin was copied by the Parisii, a Celtic tribe of northwest Europe. On later Parisii coin designs, the imagery became less realistic.

Head of Apollo
Stylized horse
Abstract design

4th century BCE Greek coin (front) | 4th century BCE Greek coin (back) | 1st century BCE Parisii coin (front) | 1st century BCE Parisii coin (back) | Later Parisii coin (front) | Later Parisii coin (back)

MAKING MONEY

Money is a symbolic token of value, used as a means of exchange. At first, items that had local significance, such as cowrie shells, feathers, textiles, or cacao beans, were used as tokens. These were replaced by more valuable metals, which greatly improved trade between regions.

The earliest form of trade was bartering. The problem with bartering is that both sides in the exchange must have something of equivalent value that the other wants. To solve the problem, the earliest civilizations invented money.

Currencies that were used for trade over wider areas used metals, especially gold, silver, and bronze. Gold and silver are most valued because of their scarcity, beauty, durability, and the effort needed to extract them. At first, weighed silver was used as a currency. Then, in the 1st millennium BCE, as the Eurasian trade network expanded, states began to issue coins—metal tokens stamped with their values.

The first true coins were made in Lydia, in what is now Turkey, around 600 BCE. From Lydia, coinage spread to Greece. Each Greek state minted its own coins, usually decorated with an image of a patron god or the god's sacred animal.

The act of issuing coins was an assertion of political authority and the right to rule. Rulers realized that they could use coins to promote their public image and spread ideas or information widely and quickly. Roman coins combined a portrait of the reigning emperor with news of his achievements—for example, a military victory or the building of a new temple. Similarly, Islamic caliphs issued coins bearing religious inscriptions, such as: "In the name of God, Muhammad is the messenger of God."

COINS AS EVIDENCE

The distribution of coinage is evidence of the new trade networks, and the spread of ideas, across Eurasia. Roman coins found as far away as Afghanistan and India bear witness to the trade in spices from the East.

A decline in the quality of coinage is an indication of an empire in economic trouble. The Roman Antonianus was a silver coin, first issued in 215 CE. Over time, its silver content was reduced; by the 270s, it was only silver-coated copper. This led to inflation as traders raised prices in response to what they perceived as a less valuable currency.

CHINESE COINS

Coins, in the form of miniature cast-bronze tools, became widespread in China during the Warring States period (475–221 BCE). The northern and eastern states shaped their coins like knives, while the central states modeled theirs on spades.

After uniting China in 221 BCE, the First Emperor introduced a uniform circular copper coin. It had a square hole in the center so that coins could be strung together. Copper is not as valuable as bronze, but the intrinsic value of the material from which the coins were made no longer mattered, because everyone in China was using the same monetary system. The important factor was that the right to mint coins was a monopoly held by the imperial government.

As trade increased, so did the demand for money. Around 900 CE, Chinese merchants, who wanted to avoid carrying around thousands of coins, started trading receipts from shops where they had left money or goods. The government then granted a monopoly to certain shops, giving them the right to issue the receipts. In the 1120s, the government took over the system, and issued the world's first paper money.

MESOPOTAMIAN COUNTING TOKENS

◀ Symbolic worth
These clay tokens were used by early Mesopotamian merchants to keep their accounts. Different-shaped tokens stood for different goods. The tokens were often passed between merchants as bills of trade in clay "envelopes" that recorded how many tokens were inside.

> "WITH THIS **PAPER-MONEY** THEY CAN **BUY WHAT THEY LIKE** ANYWHERE **OVER THE EMPIRE**, WHILST IT IS ALSO VASTLY **LIGHTER TO CARRY** ABOUT ON THEIR **JOURNEYS**."
>
> **Marco Polo**, Venetian merchant, c.1254–1324, *The Travels*

◀ Stone money
On the island of Yap in Micronesia, huge disks carved from limestone are a traditional form of currency (*rai*). The disks were quarried on the islands of Pulau and Guam and towed on rafts to Yap. A stone's value depends on its size, workmanship, and history—especially how difficult or dangerous it was to transport to Yap. Ownership is recorded orally, and the stones often remain in situ despite changing hands.

The Triumph of Death
Following the Black Death, The Triumph of Death became a popular subject in European art. This wall painting from Sicily, painted in the 1440s, shows Death as a skeleton riding a horse, shooting down all classes, including emperors, nobles, and churchmen, with a bow and arrow.

UNHEALTHY DEVELOPMENTS

Farming could support many more people than hunting and gathering, but the move to a limited diet proved to be a less healthy way to live. As the population rose, and communities became denser and more widely connected, diseases spread rapidly and with devastating effect.

The skeletons of early farmers reveal problems caused by the new way of life. Grain-based diets caused scurvy and rickets from a lack of vitamins C and D. Farmers also suffered injuries caused by hard, repetitive work. Female skeletons from the first farming site, Abu Hureyra in Syria, show damaged lower backs and knees, and deformed big toes, all caused by long hours kneeling to grind grain.

Periodic famine was an inadvertent consequence of agriculture. People had replaced their broad hunter-gatherer diet with a smaller number of crops and animals, all of which could fail due to climate, disease, or pests. In Egypt, farming depended on the annual flooding of the Nile, which usually reached 26ft (8m) high. A 23ft (7m) flood would result in a poor harvest, but anything less would lead to famine. Repeated failures led to the collapse of some civilizations.

DEADLY DISEASES

Close proximity made it easier for bacteria and viruses to change their host species from domesticated animals to humans. Measles, for example, evolved from the rinderpest virus, a deadly disease in cattle. Diseases could be passed on by direct contact with animals, or transmitted by blood-sucking insects, such as fleas and lice. The most devastating was bubonic plague, caused by the *Yersinia pestis* bacterium, passed from rats to fleas to humans. The worst outbreak—the 14th-century Black Death—began in Asia and was then carried west along trade routes, killing one-third of Europe's population.

Hunter-gatherers rarely had contact with rats, but human settlements, with all their garbage, made an ideal habitat for rodents. Drinking water sources were often contaminated with human and animal feces. Roundworm infections, and two deadly bacterial diseases—cholera and typhoid, both caused by sewage-polluted water—were common occurrences. Even something as simple as an infected cut could prove fatal before the advent of modern medicine.

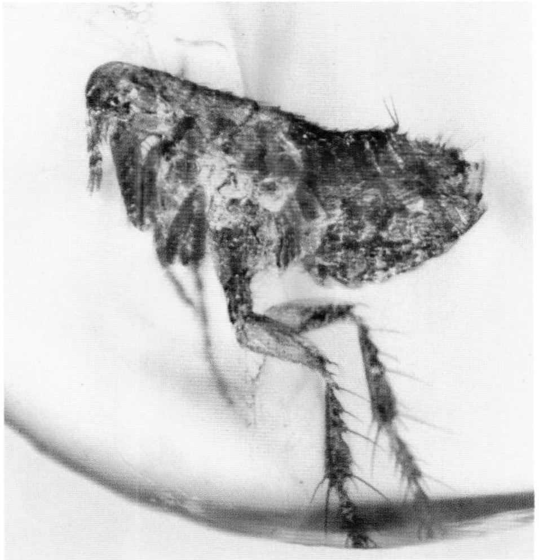

◀ **Plague carrier**
Bubonic plague is an ancient disease of rodents, but humans caught it only after they began to settle in large communities. This 20 million-year-old flea, preserved in amber, carries plague bacteria in its mouthparts.

> **GREAT PITS WERE DUG** AND PILED DEEP WITH **HUGE HEAPS** OF THE **DEAD**... AND I, AGNOLO DI TURA, **BURIED** MY FIVE CHILDREN **WITH MY OWN HANDS**.

Agnolo di Tura, Italian merchant and chronicler, c.1347

TRADE NETWORKS DEVELOP

As agrarian civilizations grew, they were linked together in vast interconnected networks, where goods, languages, technology, microbes, and genes were all exchanged. The most important exchange network of the Agrarian Era is known today as the Silk Roads.

▼ In search of pasture Modern Kazakh nomads, riding horses and using camels to carry their belongings, herd their flocks on the Altai Plain of China, which was part of the Silk Roads. Their way of life has changed very little in 6,000 years.

The treeless steppes stretch for 3,000 miles (4,800km) from eastern Europe to the borders of China. For the last 6,000 years, the steppes have been home to nomadic pastoralists. Mounted on horses or camels, these people were constantly on the move in search of fresh pastures for their animal herds. The extreme mobility of the steppe nomads enabled the creation of the Silk Roads. This collection of routes spanned the steppes of Eurasia. During the Agrarian era, they connected the entire Afro-Eurasian world zone. Other world zones had early exchange networks, such as the American trade networks of the Andes mountains and Mesoamerica, but they were smaller and less varied than the Silk Roads. While warfare played a role in connecting different civilizations, the most influential networks were built through trade.

THE SILK ROADS

The Silk Roads included land routes across China, Central Asia, and the Mediterranean and also trade that took place by sea. By the first major period of Silk Roads trade, from 50 BCE to 250 CE, small early agrarian civilizations had been consolidated into vast and powerful empires, enabling large-scale exchanges. The four ruling dynasties—the Roman, Parthian, Kushan, and Han empires—constructed road networks that connected their territories. Technological advances in metallurgy and transportation, intensified agricultural production, and the emergence of coinage all contributed to conditions in Afro-Eurasia that allowed for unprecedented levels of material and cultural exchange. Meanwhile, large and powerful

nomadic communities had appeared across the harsh interior of Inner Eurasia. They helped to link up the different civilizations, and travelers relied on these nomadic people once the Silk Roads formed.

Long-distance trade between China and the Mediterranean flourished from around 200 BCE, following the Han dynasty's expansion into Central Asia. Merchants crossed the steppes and deserts, carrying Chinese silk, jade, and bronze, Roman glass, Arabian incense, and Indian spices. Control of the trade brought great wealth to oasis towns in the deserts, and to the cities of Northern Persia and Afghanistan.

Even more important were the ideas and religions, including Buddhism and Islam, that were carried along the Silk Roads. In the 550s CE, monks from the Byzantine Empire reached China, where they managed to smuggle silkworm eggs back to the West, allowing the Byzantines to begin silk manufacture and breaking China's long-held monopoly of this sought-after fabric.

The Silk Roads also made it easy for disease to spread. During the 2nd and 3rd centuries CE, there were deadly epidemics of the same diseases in both Han China and the Roman Empire. Over time, this exchange of microbes allowed the peoples of Afro-Eurasia to build up resistance to diseases.

All these different types of exchange resulted in Afro-Eurasia having common technologies, artistic styles, cultures, and religions. Through these exchanges, the Silk Roads encouraged higher levels of collective learning, which contributed to growth and innovation.

◀ **On horseback**
Polo, invented in Central or South Asia, spread all the way to China along the Silk Road. This pottery Tang dynasty burial figure (618–907 CE) features one of the much-prized "heavenly horses," which were traded along the Silk Road.

> “THEY HAD **BROUGHT THE EGGS** TO **BYZANTIUM**... THE METHOD HAVING BEEN LEARNED, THUS **BEGAN** THE ART OF **MAKING SILK**... IN THE **ROMAN EMPIRE**.”

Procopius of Caesarea, Roman historian (*c.*500–560), on the spread of silk production

EAST MEETS WEST

Until 1492, people in the "Old World" of Afro-Eurasia and the "New World" of the Americas were each unaware that the other existed. European explorers brought the two worlds together, leading to the "Columbian Exchange": a transfer of people, animals, crops, diseases, and technology.

BETWEEN 1492 AND 1650, UP TO **90 PERCENT** OF THE **AMERICAN INDIAN POPULATION** WAS **WIPED OUT** BY **EPIDEMICS**

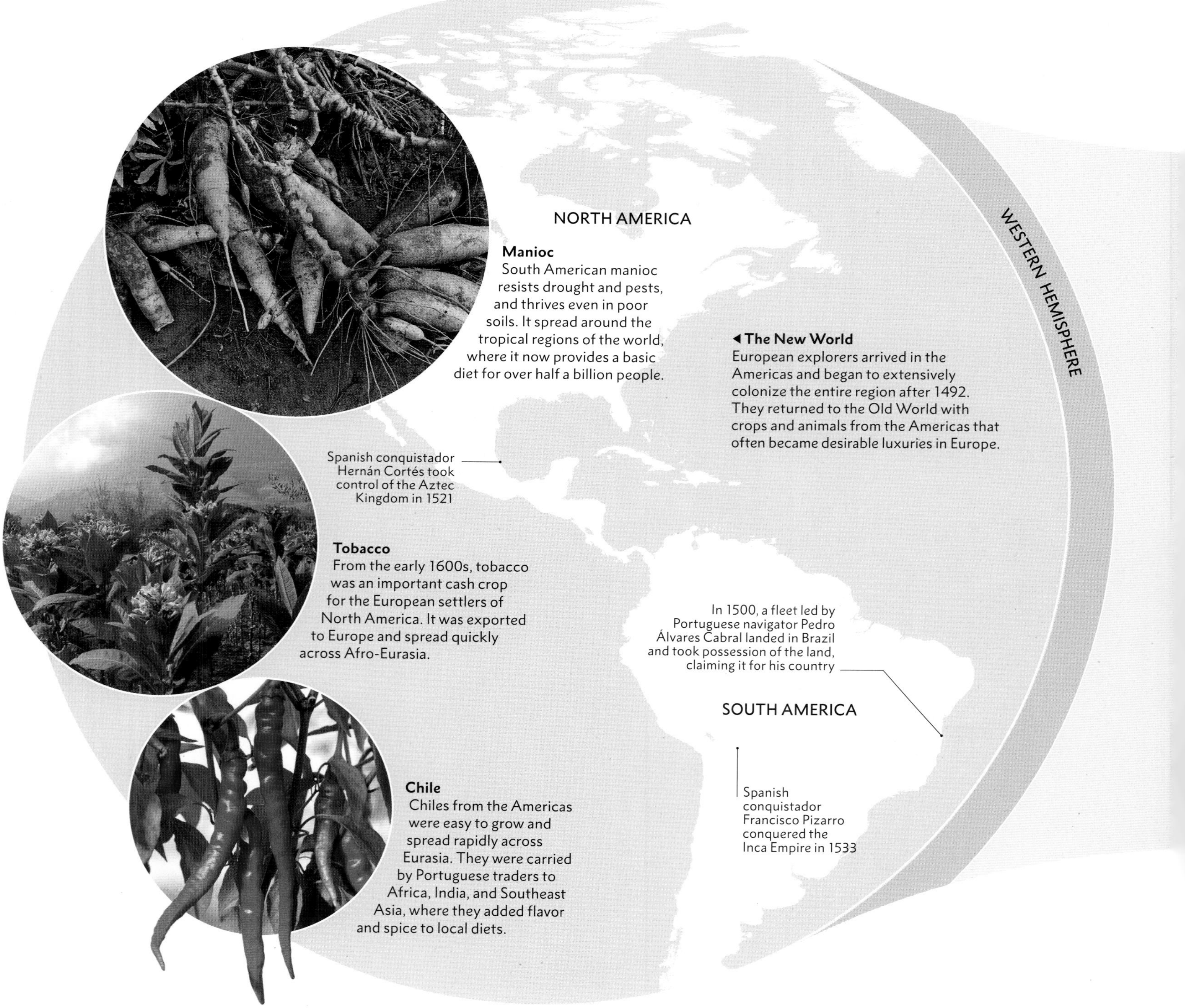
NORTH AMERICA
SOUTH AMERICA
WESTERN HEMISPHERE
Spanish conquistador Hernán Cortés took control of the Aztec Kingdom in 1521
In 1500, a fleet led by Portuguese navigator Pedro Álvares Cabral landed in Brazil and took possession of the land, claiming it for his country
Spanish conquistador Francisco Pizarro conquered the Inca Empire in 1533

Manioc
South American manioc resists drought and pests, and thrives even in poor soils. It spread around the tropical regions of the world, where it now provides a basic diet for over half a billion people.

◀ The New World
European explorers arrived in the Americas and began to extensively colonize the entire region after 1492. They returned to the Old World with crops and animals from the Americas that often became desirable luxuries in Europe.

Tobacco
From the early 1600s, tobacco was an important cash crop for the European settlers of North America. It was exported to Europe and spread quickly across Afro-Eurasia.

Chile
Chiles from the Americas were easy to grow and spread rapidly across Eurasia. They were carried by Portuguese traders to Africa, India, and Southeast Asia, where they added flavor and spice to local diets.

European explorers made full use of their superior technology—horse-riding, guns, and steel weapons—to conquer the peoples of the New World, and the diseases they carried with them also helped. The Columbian Exchange transformed life across the world. Everywhere, people benefited from new foods, resulting in global population growth for the next two centuries. Crops and animals spread, along with improved agrarian techniques and new organizational methods. The power of governments was enhanced and they began to expand their territories to increase their populations and revenues, resulting in an increase in human control of the land.

New global exchange networks emerged, and the cultural impact of the Columbian Exchange was felt most profoundly in two regions: the Americas and Europe. In the Americas, it devastated cultural and political traditions: American languages died out, as people learned to speak European languages. But Europe lay at the heart of the newly created global exchange network, and flows of new information had the greatest impact here. Surprisingly, this did little to increase rates of innovation. In 1700, the world was still traditional, but the scale at which existing ideas, goods, people, crops, and diseases were exchanged had increased, paving the way for a spectacular burst of innovation in the late 18th century.

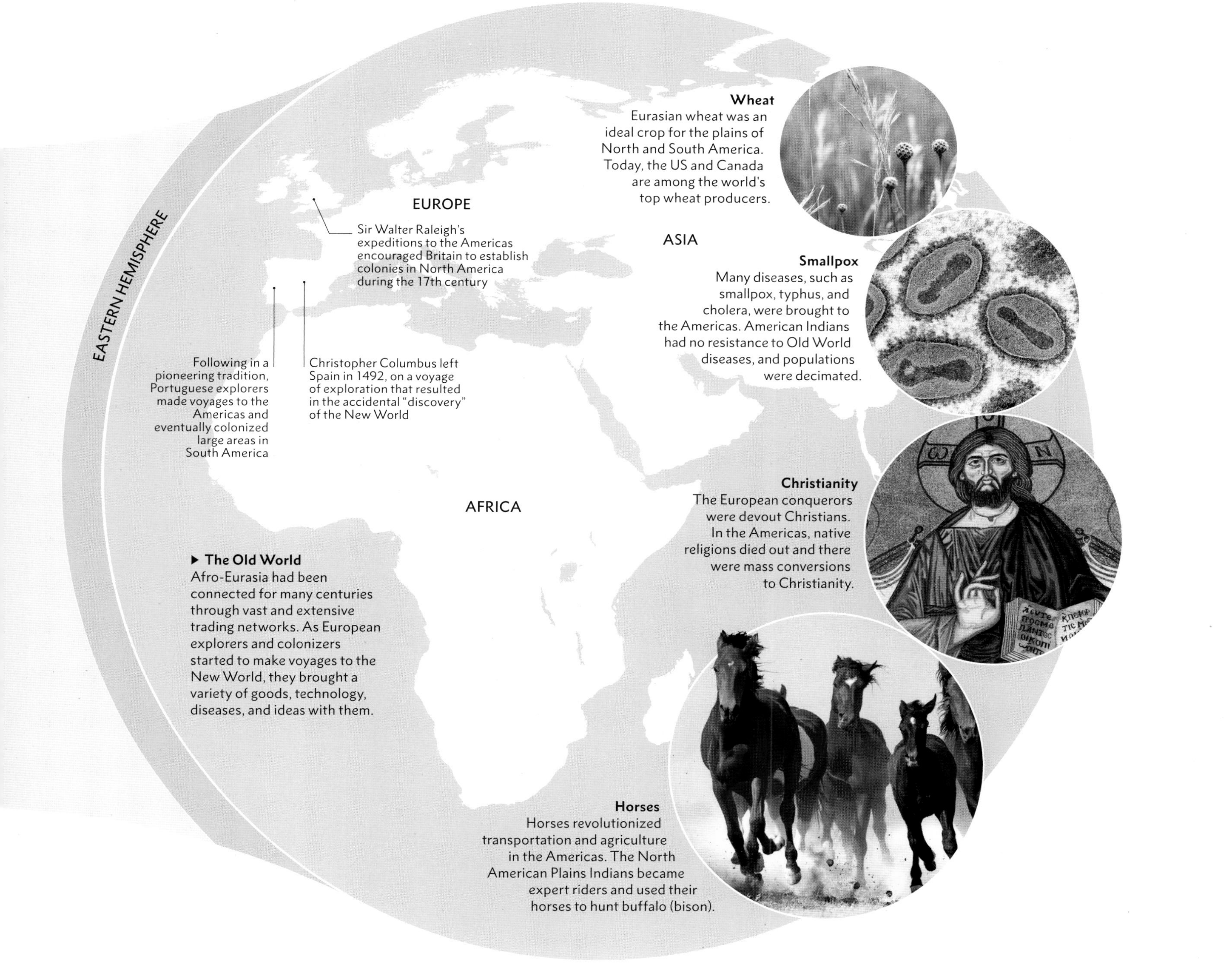
EASTERN HEMISPHERE
EUROPE
ASIA
AFRICA
Sir Walter Raleigh's expeditions to the Americas encouraged Britain to establish colonies in North America during the 17th century
Following in a pioneering tradition, Portuguese explorers made voyages to the Americas and eventually colonized large areas in South America
Christopher Columbus left Spain in 1492, on a voyage of exploration that resulted in the accidental "discovery" of the New World
Wheat
Eurasian wheat was an ideal crop for the plains of North and South America. Today, the US and Canada are among the world's top wheat producers.
Smallpox
Many diseases, such as smallpox, typhus, and cholera, were brought to the Americas. American Indians had no resistance to Old World diseases, and populations were decimated.
Christianity
The European conquerors were devout Christians. In the Americas, native religions died out and there were mass conversions to Christianity.
Horses
Horses revolutionized transportation and agriculture in the Americas. The North American Plains Indians became expert riders and used their horses to hunt buffalo (bison).

▶ **The Old World**
Afro-Eurasia had been connected for many centuries through vast and extensive trading networks. As European explorers and colonizers started to make voyages to the New World, they brought a variety of goods, technology, diseases, and ideas with them.

TRADE GOES **GLOBAL**

From the late 15th century, the world became globally connected for the first time, as European ships traversed the oceans, creating a worldwide system of maritime trade. Most important was the linking of Eurasia and the Americas, but the effects of globalization were felt worldwide.

Globalization began in 1492, when Christopher Columbus sailed west across the Atlantic, hoping to reach Asia. Instead, he found the Americas, "a New World" whose existence had not even been suspected in Eurasia. Six years later, a Portuguese fleet under Vasco da Gama sailed south and east to India. Then, in 1519–22, the Spanish expedition of Ferdinand Magellan sailed all the way around the world. Soon the English, French, and Dutch were also making long-distance voyages.

EUROPEAN MOTIVATION

Why was it Europeans rather than other peoples who connected the globe? Europe was at the wrong end of the Eurasian trade network, far from the source of spices and silks, and cut off from the overland route by the rise of the hostile Ottoman Empire. So Europeans, all too aware of their exclusion, set about creating technology—including ships, navigational devices, and maps—to bring the spices within reach. In this, the countries of northwest Europe had an advantage over Mediterranean nations, since their coasts faced out into the Atlantic.

Europe was then a continent divided by rivalry and conflict. This spurred European countries to conquer lands overseas in search of riches to fund their frequent wars.

While China, too, had the technology to explore new lands, the country was unified and there was no incentive to investigate the wider world. There was one brief period of exploration in the early 1400s, when fleets of junks sailed as far as Africa, but the purpose was to display Chinese power rather than to discover new sources of wealth. After 1433, when the emperor called an end to these expeditions, China became inward looking.

There were no long-distance trade routes in the Americas; the Aztecs of Mexico and Incas of Peru were not even aware of each other's existence. As a result, the peoples of the Americas had no idea that other lands were worth exploring and no reason to build ocean-going ships.

> **WORLD TRADE** [DATES] FROM THE 16TH CENTURY ... FROM THEN ON THE **MODERN HISTORY OF CAPITAL** STARTS TO UNFOLD.
>
> **Karl Marx**, German scholar, 1818–1883, *Das Kapital*

▼ The world on an egg
Made in Europe around 1500, this is the earliest known globe to depict the New World. It was carved on two half ostrich eggs from Africa—further evidence of global connections.

Asia with the Indian Ocean
There are 71 place names. On the east coast of Asia (not visible here) is written *Hic sunt dracones*, meaning "Here are dragons"
Africa
Madagascar
OLD WORLD

"Isabel" is La Isabella, Columbus's settlement in what is now the Dominican Republic
South America, labeled "Mundus Novus" (New World)
Terra Sanctae Crucis ("Land of the Holy Cross")
NEW WORLD

THE NEW GLOBAL NETWORK

As a result of new global connections, the focus of trade networks shifted. Northwest Europe, formerly at the margins of the Eurasian network, became the center of a rapidly expanding new global network. This is why four of the most widely spoken languages today are English, Spanish, Portuguese, and French. The previously important trading hubs of southern Europe, such as Venice, went into long-term decline.

European economies changed as wealth poured in from the Americas and other lands. Power shifted from landholding aristocrats to merchants, marking the birth of what would become modern capitalism.

◀ **Portuguese trade**
In 1543, the first Portuguese ships, sailing from Goa, India, reached Japan. They exchanged Chinese silks and porcelain and Indian cloth for Japanese metalwork and artwork. This Japanese painting shows a Portuguese carrack, a type of large merchant ship.

SOUTH AMERICAN SILVER

In 1545, the Spaniards discovered a mountain of silver ore at Potosi in Bolivia. This was the biggest source of silver ever found. By 1660, about 60,000 tons of silver had been shipped to Spain, tripling the amount of the metal in Europe.

Silver, sought after by Asian merchants, soon became the foundation of the world economy. Much of it found its way to China, where it was used to buy silks and porcelain. Spanish galleons, sailing from Mexico, carried the silver across the Pacific to the Philippines. Portuguese ships also went east, using New World silver to buy cotton and spices in India, and porcelain and silks in China, which they then traded in Japan.

The flood of silver from America caused widespread inflation in Europe and beyond. Through trade, Spanish silver coins reached the Ottoman Empire, rendering the local coinage, with its lower silver content, less valuable. Government officials and soldiers found they could no longer live on their pay.

Despite the flow of silver from the Americas, the Spanish crown, constantly engaged in wars, was always in debt. The wealth ended up in the hands of foreign bankers who serviced the royal debt.

DESTRUCTIVE IMPACT

Globalization also spread Eurasian diseases throughout the world. Their impact on the indigenous peoples of the Americas, Australia, and the Pacific Islands was especially devastating.

At first, mines and farms in the Americas were worked by American Indians, but so many perished as a result of ill treatment and introduced diseases that a new source of labor was required. From 1534, Europeans began to transport African slaves—who had a resistance to Old World diseases—to the Americas. Over the next 350 years, 12–25 million Africans crossed the Atlantic, chained in the holds of slave ships.

The impact of globalization on the environment was also catastrophic. The introduction of sheep to Australia and goats to the Pacific Islands, for example, resulted in widespread deforestation and the extinction of many species of native wildlife.

◀ **Spanish silver**
Famous for their consistent weight and purity, Spanish silver coins set a standard against which other coins were measured.

THRESHOLD

8

INDUSTRY **RISES**

Spurred on by the need to feed and care for a growing population, humans unlock a new source of energy from the Earth—fossil fuels. These power the rise of industry and consumerism, creating a new world order in which humans become a dominant force for change on Earth.

GOLDILOCKS CONDITIONS

In large, diverse, interconnected societies, collective learning is a powerful force. The journey to our highly complex modern world began in the 18th century, when new global connections enriched existing networks of exchange. The pace of change started to accelerate—as did the human capacity to control the biosphere.

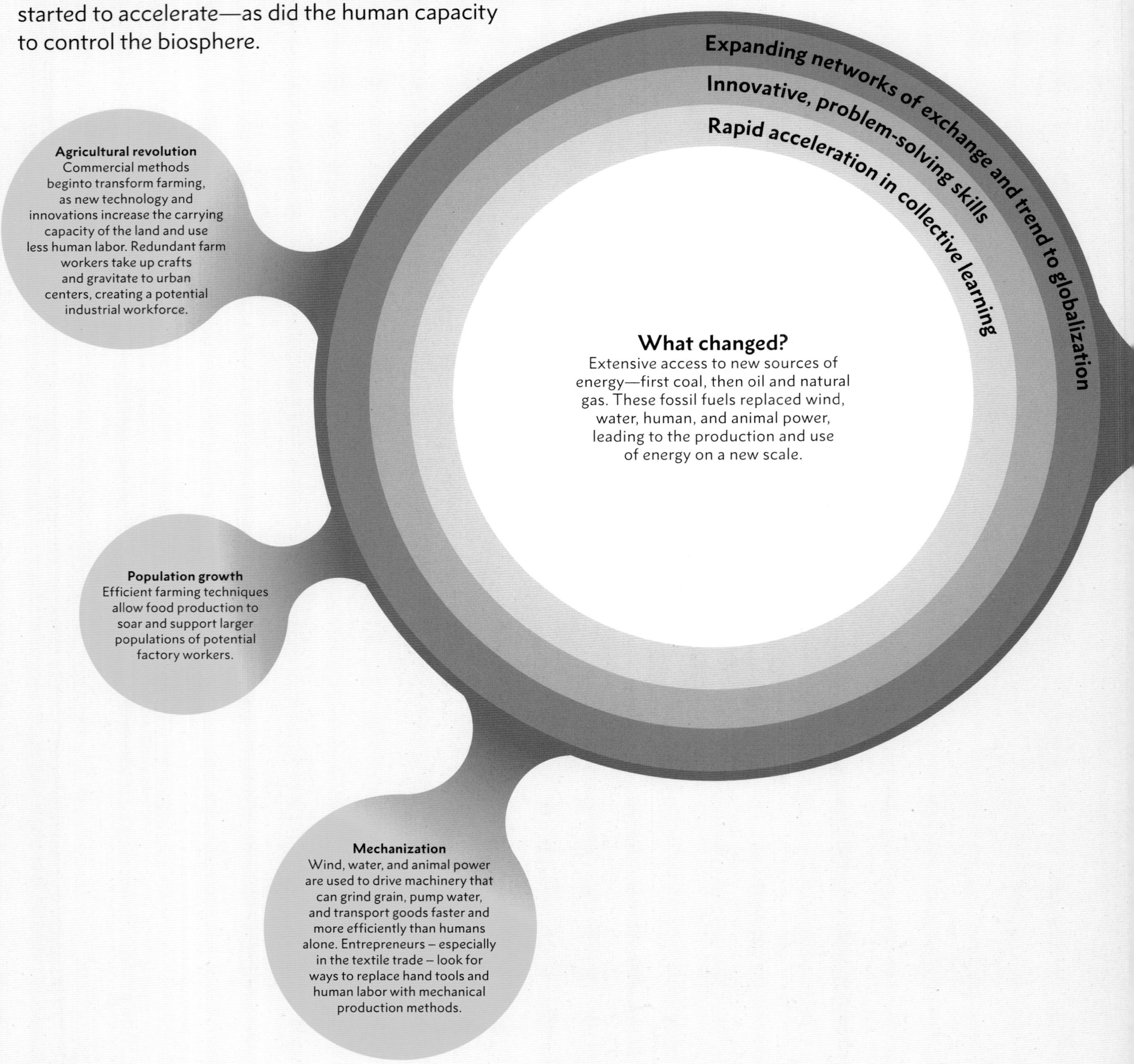

Expanding networks of exchange and trend to globalization
Innovative, problem-solving skills
Rapid acceleration in collective learning
What changed?
Extensive access to new sources of energy—first coal, then oil and natural gas. These fossil fuels replaced wind, water, human, and animal power, leading to the production and use of energy on a new scale.
Agricultural revolution
Commercial methods beginto transform farming, as new technology and innovations increase the carrying capacity of the land and use less human labor. Redundant farm workers take up crafts and gravitate to urban centers, creating a potential industrial workforce.
Population growth
Efficient farming techniques allow food production to soar and support larger populations of potential factory workers.
Mechanization
Wind, water, and animal power are used to drive machinery that can grind grain, pump water, and transport goods faster and more efficiently than humans alone. Entrepreneurs – especially in the textile trade – look for ways to replace hand tools and human labor with mechanical production methods.

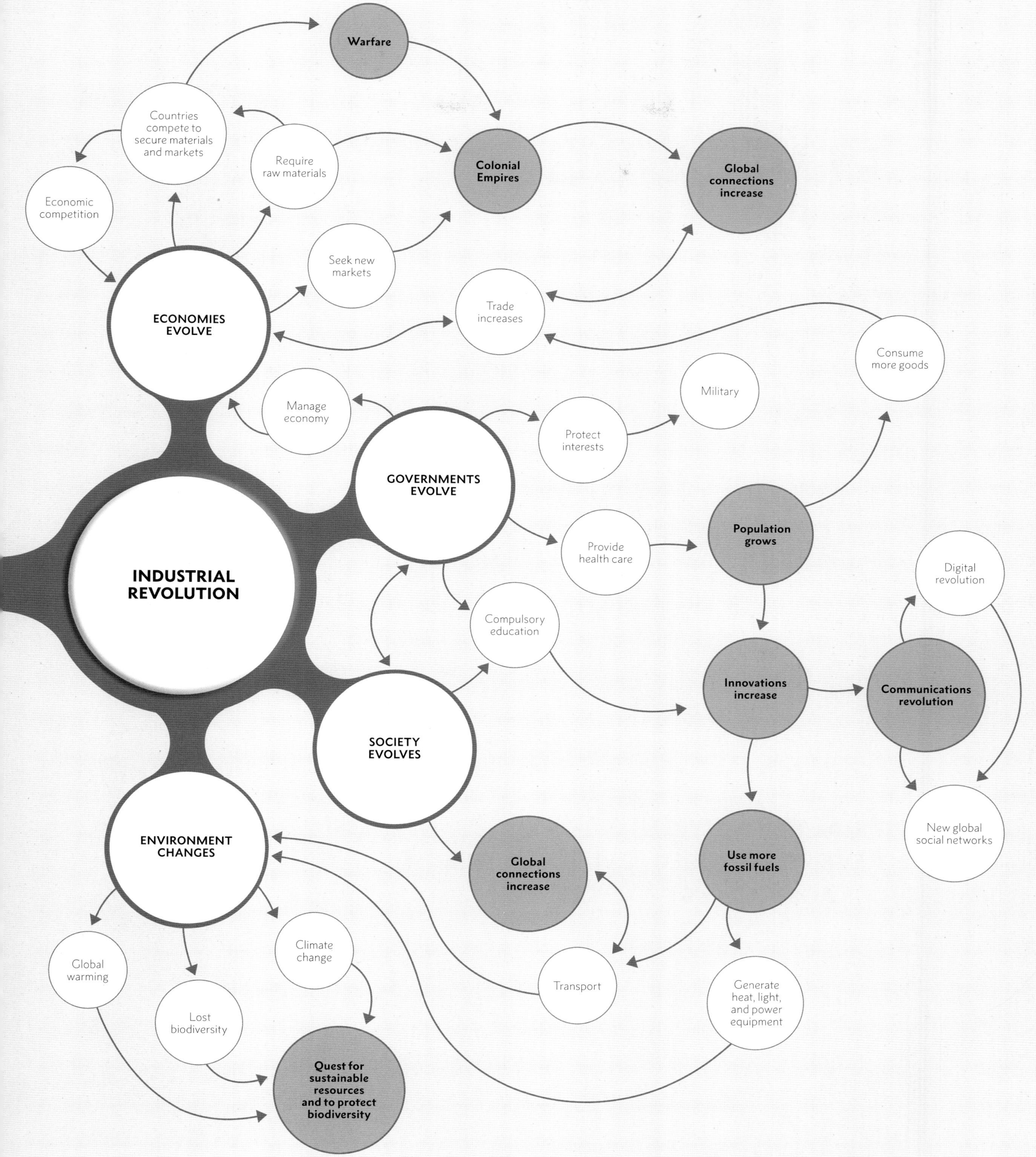
INDUSTRIAL REVOLUTION
ECONOMIES EVOLVE
Economic competition
Countries compete to secure materials and markets
Warfare
Require raw materials
Colonial Empires
Global connections increase
Seek new markets
Trade increases
Consume more goods
Manage economy
GOVERNMENTS EVOLVE
Protect interests
Military
Provide health care
Population grows
Compulsory education
Digital revolution
Innovations increase
Communications revolution
SOCIETY EVOLVES
New global social networks
ENVIRONMENT CHANGES
Global connections increase
Use more fossil fuels
Global warming
Climate change
Lost biodiversity
Transport
Generate heat, light, and power equipment
Quest for sustainable resources and to protect biodiversity

THE INDUSTRIAL **REVOLUTION**

In the mid-18th century, after hundreds of years of slow development, a series of innovations in Britain began a process that would change the world forever. This process is now known as the Industrial Revolution.

The Industrial Revolution transformed agrarian societies that discovered how to use fossil fuels like coal to replace human and animal power in manufacturing, communication, and transportation. It began in Britain, when several factors—both global and local—ushered in an era of relatively fast technological change.

WHY THERE, WHY THEN?

Britain's industrialization followed a period of rapid population growth in Europe. Innovations in agriculture, such as the horse-drawn seed drill and the adoption of modern farming methods, had combined to increase the carrying capacity of the land, fueling population growth (see pp.252–53). It was also a time of social change: with landowners able to produce twice as much food using less labor, many agricultural workers moved to the cities or took up crafts. Landowners no longer took tribute from their peasants, who became wage earners. For the first time, the structure of society began to change from an agrarian society to a commercialized one.

This was a significant change. Rates of innovation are slower when social and ideological conditions offer no incentives to innovate, and the political climate played a part in that too. During the 18th century, Europe's absolute monarchies stifled innovation, but Britain had a parliamentary monarchy with a government that supported commercial projects and rewarded innovation, as part of an intellectual climate known as the Enlightenment.

Most of Britain's national income came from commerce, which was protected by a strong army and navy, and provided the essential capital needed for industrialization. London was an important trade hub; at the center of an international trade network connecting Europe and the Americas, Britain was perfectly placed to benefit from new inventions as a result of collective learning.

In theory, with its large population, China could have industrialized at any point from the 11th century, when it developed an iron and steel industry, powered by coal. But its coalfields were located in the unstable north of the country, far from the economic centers, which had moved south after the Mongol invasions in the 13th century. The political climate was also unfavorable: the Confucian ideals promoted by the government emphasized stability, and industrialization was seen as disruptive.

> **[THE INDUSTRIAL REVOLUTION]** WAS PROBABLY THE MOST **IMPORTANT EVENT** IN **WORLD HISTORY... SINCE THE INVENTION** OF **AGRICULTURE** AND **CITIES**.
>
> **Eric Hobsbawm**, British historian, 1917–2012

A GROWING PROBLEM

Britain's population doubled between 1750 and 1800. This led to a shortage of wood, so coal was increasingly used as a source of fuel. As the shortage became acute, demand for coal rose. Britain had abundant reserves, but they were underground and difficult to access. This created a need to innovate; Britain had all the necessary conditions for innovation to thrive—and thrive it did.

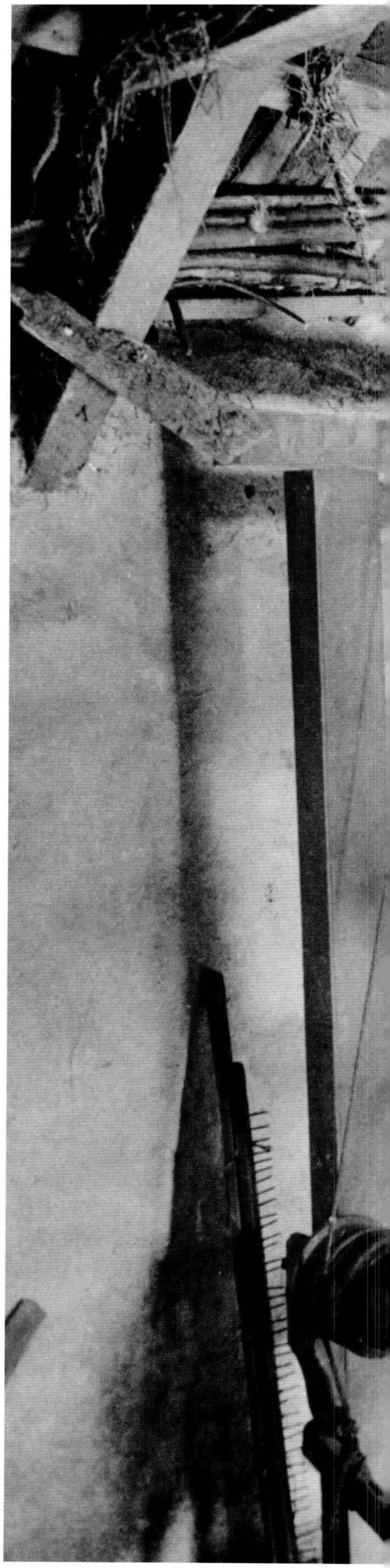

Working from home
Agricultural workers freed up to work in cottage industries, such as textiles, boosted the economy, trade, and exports, helping to pave the way for industrialization.

▼ Workings of a coal mine
As Britain began to industrialize, more coal was needed to fuel steam engines and furnaces. As a result, coal production increased, mines got deeper, and the industry became more dangerous.

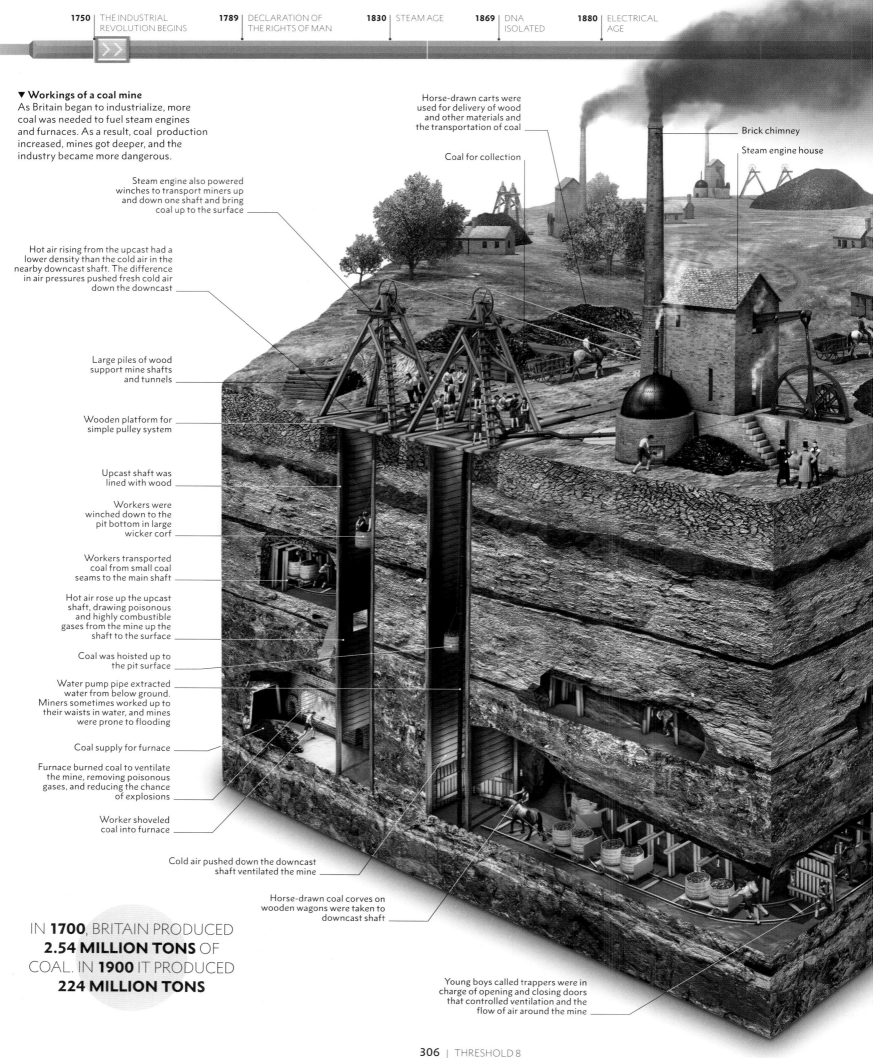

Horse-drawn carts were used for delivery of wood and other materials and the transportation of coal
Brick chimney
Steam engine house
Coal for collection
Steam engine also powered winches to transport miners up and down one shaft and bring coal up to the surface
Hot air rising from the upcast had a lower density than the cold air in the nearby downcast shaft. The difference in air pressures pushed fresh cold air down the downcast
Large piles of wood support mine shafts and tunnels
Wooden platform for simple pulley system
Upcast shaft was lined with wood
Workers were winched down to the pit bottom in large wicker corf
Workers transported coal from small coal seams to the main shaft
Hot air rose up the upcast shaft, drawing poisonous and highly combustible gases from the mine up the shaft to the surface
Coal was hoisted up to the pit surface
Water pump pipe extracted water from below ground. Miners sometimes worked up to their waists in water, and mines were prone to flooding
Coal supply for furnace
Furnace burned coal to ventilate the mine, removing poisonous gases, and reducing the chance of explosions
Worker shoveled coal into furnace
Cold air pushed down the downcast shaft ventilated the mine
Horse-drawn coal corves on wooden wagons were taken to downcast shaft
Young boys called trappers were in charge of opening and closing doors that controlled ventilation and the flow of air around the mine

IN **1700**, BRITAIN PRODUCED **2.54 MILLION TONS** OF COAL. IN **1900** IT PRODUCED **224 MILLION TONS**

COAL FUELS INDUSTRY

Access to large reserves of coal was the breakthrough that fueled the machines of industry and set the modern age alight. Coal was the first of several fossil fuels used to power the industrial world.

The history of coal is far older than the mines of 18th-century Europe. It was used in China as early as 1000 BCE to heat homes, smelt copper, and fuel blast furnaces to create iron; by the 11th century CE, the Song Dynasty relied on coal to produce the iron needed to make weapons and armor. In Britain, coal was used as fuel from the 2nd century CE, when the Romans mined coal near the surface to heat their forts, fuel furnaces, and burn sacrifices at altars in honor of their gods. After the Romans departed in the 5th century, the use of coal declined. For most people, wood was a far more accessible source of fuel, but from the 13th century sea coal—an abundant resource that washed up on the beaches of northeast England—was collected and distributed by boat.

Fuel was needed for industrialization, and in Britain coal deposits were fortuitously located in thick seams, albeit deep underground. However, early mining was hazardous: pits continually filled with water, and horsepower removed it too slowly. The steam engine, invented by Thomas Newcomen and developed by James Watt, was the breakthrough that made it possible to effectively pump water out of mines and access more coal at greater depths.

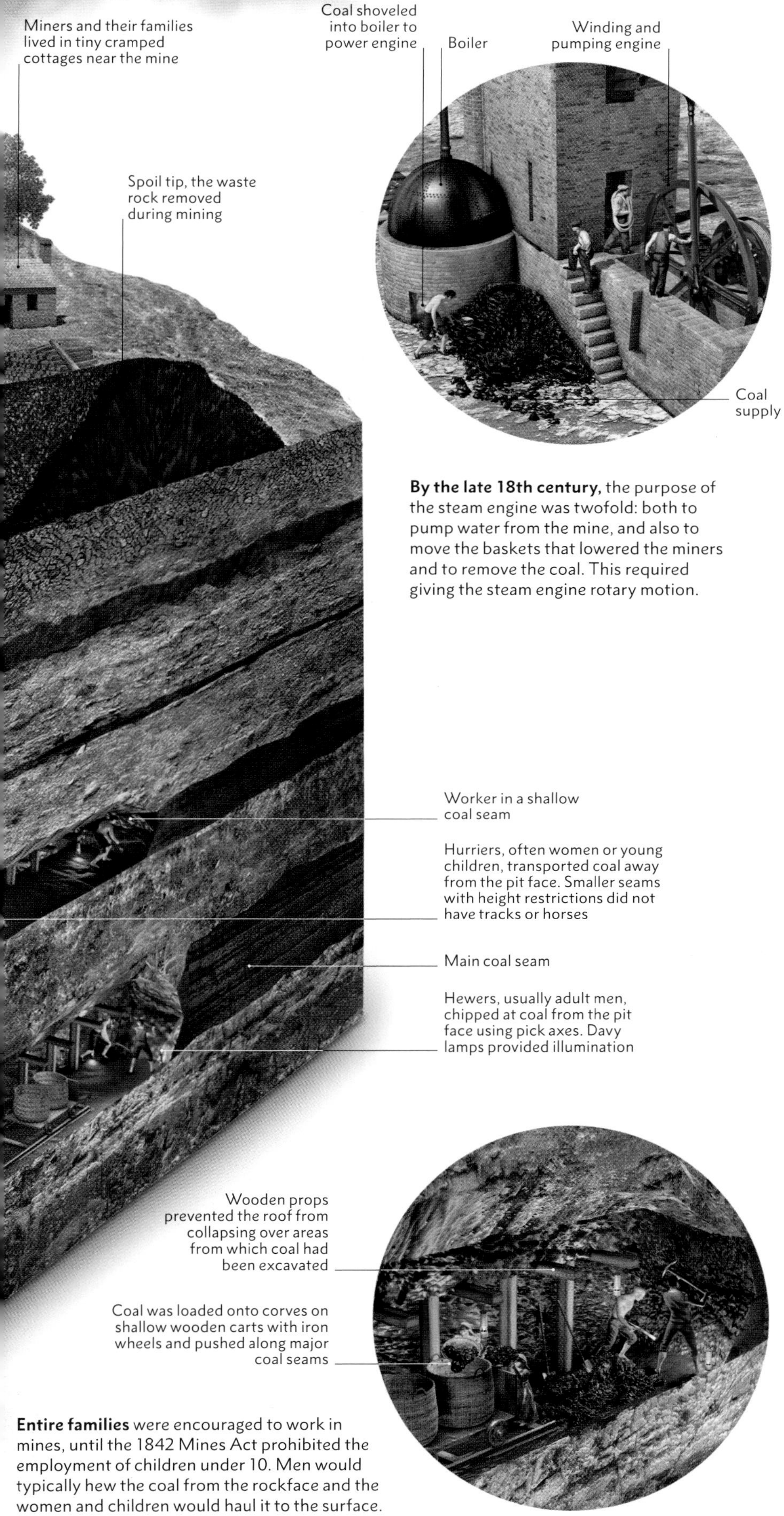

Miners and their families lived in tiny cramped cottages near the mine
Coal shoveled into boiler to power engine
Boiler
Winding and pumping engine
Spoil tip, the waste rock removed during mining
Coal supply
Worker in a shallow coal seam
Hurriers, often women or young children, transported coal away from the pit face. Smaller seams with height restrictions did not have tracks or horses
Main coal seam
Hewers, usually adult men, chipped at coal from the pit face using pick axes. Davy lamps provided illumination
Wooden props prevented the roof from collapsing over areas from which coal had been excavated
Coal was loaded onto corves on shallow wooden carts with iron wheels and pushed along major coal seams

By the late 18th century, the purpose of the steam engine was twofold: both to pump water from the mine, and also to move the baskets that lowered the miners and to remove the coal. This required giving the steam engine rotary motion.

Entire families were encouraged to work in mines, until the 1842 Mines Act prohibited the employment of children under 10. Men would typically hew the coal from the rockface and the women and children would haul it to the surface.

◀ **Screening coal** Women and children sorted the coal and separated it into different groups based on size. The sorted coal was washed and dried, before being transported from the mine.

STEAM POWER **DRIVES CHANGE**

Developed in the 18th century to pump floodwater from mines, the steam engine was the defining invention of the Industrial Age. Fueled by the newly available coal, steam engines replaced human, animal, and water power and led to the rise of factories, railways, and steamships.

▲ Powering industry
Watt's improvements to the Newcomen engine enabled it to power factory machinery, leading to the rise of mass-production.

In 1712, British ironmonger and engineer Thomas Newcomen invented a steam engine that could pump water with the power of twenty horses from mines deep underground. This made it possible to mine to greater depths and unlock the seemingly endless supply of British coal. Newcomen's machine became so popular that by 1755 his engines were installed in France, Belgium, Germany, Hungary, Sweden, and the United States. However, Newcomen's engine was large, inefficient, and consumed enormous quantities of coal: without improvements it could operate only in coal mines. In 1765, inventor James Watt realized a lot of coal and steam was going to waste in Newcomen's machine and built an engine with a separate condenser to eliminate this wastage.

RISE OF THE FACTORY SYSTEM

Although mining engines relied on an up-and-down motion, the industrialist Matthew Boulton recognized the potential of Watt's improved design to be adapted to the rotary motion needed to drive factory machinery. Boulton was the owner of Birmingham's Soho manufactory, which produced small metal trinkets and toys. Like many industrialists of the time, Boulton relied on a waterwheel to power his machinery, and when a drought left the river bed dry, production came to a halt.

▼ Driving change
Railways carried passengers, raw materials, and manufactured goods. Steam locomotives provided cheap transportation that encouraged further industrialization.

Boulton gave Watt the tools and engineers to build a prototype, and in 1776, Watt's steam engine was invented, releasing manufacturing from the constraints of natural power. It produced the same amount of power as the Newcomen engine on a quarter of the fuel, and could be installed anywhere. Soho became the first steam-powered factory in the world and its employees began toiling on production lines in the new mass-production of goods. Steam engines enabled the growth of a new mode of production: the factory system.

The shift to machine-based manufacturing began with the textile industries in Britain, the United States, and Japan. Steam power transformed the industry and the mass-production of textiles transformed the British economy. Attaching a steam engine to spinning and weaving machines allowed cotton textiles to be produced at unprecedented speeds. By 1850, Britain was using 10 times more cotton than in 1800, and textiles became cheap and widely available. The demand for more American cotton kept the country's slave plantations in business.

THE SPREAD OF STEAM

Steam engines made it possible to work and produce goods without being reliant on proximity to waterways. Towns sprang up around steam-powered factories at the turn of the century. To supply these towns with the necessary amounts of coal, raw materials, and goods for market, new transport links were created: turnpike roads, canals, and then railways.

▲ Women weavers
The power loom was gradually adopted in textile factories. When it became more efficient, women and children could operate the equipment and replaced men as weavers.

> “THOSE WHO ADMIRE **MODERN CIVILIZATION** USUALLY **IDENTIFY IT** WITH THE **STEAM ENGINE** AND THE ELECTRIC TELEGRAPH.”
>
> **George Bernard Shaw,** Irish playwright and political activist, 1856–1950

Railways were part of a second wave of industrialization that was made possible through the mass-manufacture of iron. British engineer Abraham Darby learned how to smelt iron by burning coke in the early 18th century, and Britain's iron production rapidly increased thanks to the new availability of coal. The marriage of iron and smaller, high-pressure steam engines allowed for the manufacture of steam locomotives and tracks to run them along. During the 19th century, new railway lines joined up other industrializing nations too. Iron, coal, and railways became the central symbols of the industrial revolutions in Germany, Belgium, France, and the United States. Railways were another example of the incessant drive in the industrial age to improve existing technologies.

Introducing a turbine system to the steam engine allowed the technology to power ships. The introduction of a screw-propeller, which was more efficient than the earlier paddle wheels, enabled a more consistent propulsion. By 1840, steamships were making trips across the Atlantic to transport goods and people. By the end of the 19th century, the ironclad warship, a steam-propelled vessel protected by iron or steel plates, showed that the power of steam could also be used as a weapon.

▲ Shipping lines
The Royal Netherland Steamship company transported goods, passengers, and mail between Europe and the Dutch East Indies.

1750 | THE INDUSTRIAL REVOLUTION BEGINS

1789 | DECLARATION OF THE RIGHTS OF MAN

1830 | STEAM AGE

THE PROCESS OF **INDUSTRIALIZATION**

As the first country to undergo an industrial revolution, Britain provided a template that other nations could follow. Each country took a unique path, but they all shared common factors.

Industrialization transformed agrarian economies. It produced a series of technological innovations that increased the use of natural resources and led to the mass production of manufactured goods. Access to new energy unlocked a chain of innovation; the invention of machines that increased production but required less human energy to operate allowed work to be organized differently in factories, which led to increased specialization and division of labor. As science was increasingly applied to industry, new materials like iron contributed to developments in transportation and communication infrastructures.

Eventually, industrialization resulted in political, social, and economic change as trade expanded, economies grew, governments responded to the needs of the new industrialized societies, and new cities and empires emerged.

INGREDIENTS FOR INDUSTRIALIZATION

Coal

SOURCE OF ENERGY
Countries industrialized by harnessing an energy source: water, coal, oil, or gas. Coal was the main energy of the Industrial Revolution, used in steam engines, iron-producing blast-furnaces, and as fuel.

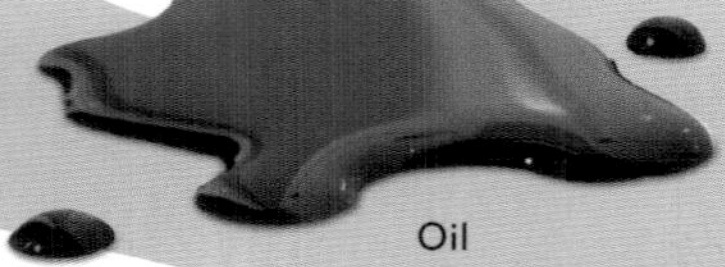

Oil

A WORKFORCE
Population growth created by innovations in agriculture led to the specialization of labor: artisans, craftsmen, weavers, and wage workers were no longer tied to rural areas and could migrate to cities to find work in factories.

EFFICIENT TRANSPORTATION NETWORKS
Getting raw materials to factories and finished goods to market was an essential aspect of industrialization. Turnpike roads were followed by canals and later railways. Steamships enabled fast transportation across the Atlantic.

TECHNOLOGICAL ADVANCES
Improvements in steam power technology were constantly made, leading to locomotives and steamships. Coal-burning steam engines still power the world by producing much of its electricity.

Steam engine

INNOVATIVE MINDSET
New machinery, such as the water frame, cotton gin, and spinning jenny, allowed goods to be mass-produced. Large machines powered by steam engines led to the rise of the factory system.

FREEDOM TO EXCHANGE IDEAS
The exchange of ideas between innovators and industrialists led to the creation of new technologies, such as the steam engine. Industrial espionage and expanding trade routes enabled these technologies to spread.

Iron

STRONG TRADE LINKS
Industry created wealth: governments and industrialists provided the capital. New domestic and international markets were opened to provide raw materials and buyers for the finished products.

◄ **How industrialization works**
Industrialization was a process that transformed agricultural societies and economies. Spurred on by new inventions and technologies, it resulted in vast social and political changes, new economic doctrines, and the creation of powerful industrial empires.

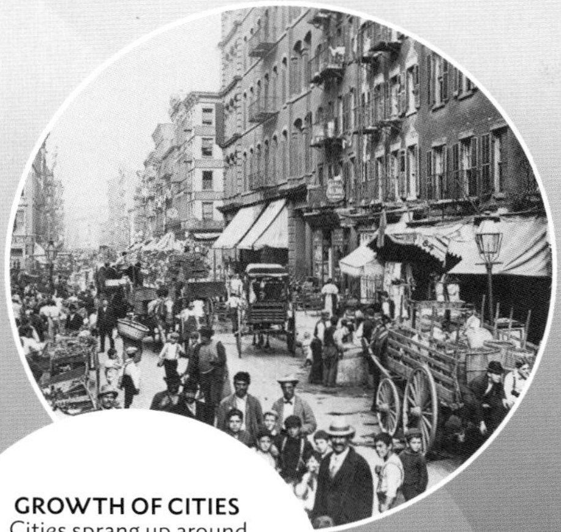

GROWTH OF CITIES
Cities sprang up around industrial centers. Mass-urbanization often led to overcrowding, squalor, and the spread of disease. Industrial cities were dirty, and provided little sanitation or running water for working-class inhabitants.

NEW PRODUCTION METHODS
Factories housing the new industrial machines mass-produced goods. There were social consequences, as workers toiled for long hours in terrible conditions and with very little pay.

CONSUMER CULTURE
The availability of luxury products at low cost, an influx of foreign goods through new trade networks, and higher wages led to the rise of the middle classes. The consumer revolution created capital that could be reinvested.

MONEY MANAGERS
Industrial governments began to manage markets. Financial institutions were created to control and accumulate wealth, including banks, stock markets, and insurance agencies.

MILITARY TECHNOLOGY
Building strong militaries was a major concern of industrial powers. Military technology, such as machine guns, gave governments control over markets and encouraged some unindustrialized nations to open to trade.

POLITICAL PARTNERSHIPS
Revolution, the rise of the middle classes, and political and social reforms led to new social contracts between governments and their citizens, the creation of the modern state, and the rise of democracy.

SOCIAL REFORMS
In the 19th century, governments began to act to improve the lives of their citizens, introducing laws to control working hours and child labor; compulsory public education; health systems; and instigating sanitation projects to clean up cities.

COLONIAL STRENGTH
Industrial powers used their strong armies and navies to colonize parts of the world rich in the raw materials needed for their factory-made products, in a practice known as imperialism.

POWERFUL MILITARIES
Industrial wealth enabled governments to create militaries large enough to compete with other industrialized nations. These militaries were sometimes also used to control vast colonial empires.

NEW IDEOLOGIES
As governments of industrial countries adopted the institutions of the modern state, concepts of nationalism and imperialism developed. Inherent in imperialist doctrine were the ideals of supremacy over peoples and nations of the unindustrialized world.

ECONOMIC STRENGTH
Industrialization drove consumer capitalism, which created wealth. It resulted in a growing divide between rich and poor citizens, and an even larger division between industrialized and unindustrialized countries.

→ REASONS TO CONTINUE INNOVATING → NEW INFRASTRUCTURE AND INSTITUTIONS → SOCIAL, POLITICAL, AND ECONOMIC CHANGE

TIMELINES

INDUSTRY **GOES GLOBAL**

Britain's early industrialization gave the country a new economic might and an ability to exploit its position on a global scale. Other countries soon tried to emulate its success.

Industrialization arose in Britain as the result of a combination of unique and unplanned forces, but it could be replicated by other countries, where it was implemented by strong governments or by entrepreneurs. The new industrial societies developed in their own way, each with distinctive features, but they all descended from the British predecessor and shared common elements, such as the importance of coal, iron, steel, and the textiles industry.

Britain tried to protect her advantage by preventing new technology and skilled workers from leaving, but countries determined to compete smuggled out machinery, sent spies to learn British secrets, and bribed entrepreneurs to set up factories abroad. The first countries to industrialize were the ones that were geographically or culturally similar to Britain: Belgium, France, and Prussia were next to develop railways and factories essential for industrialization.

1 CHANGE BEGINS

British inventors and innovators lead the way, mechanizing the textile industry and introducing the factory system. Using raw materials from the overseas colonies and the newly mechanized factory processes to mass-produce cheaper goods, Britain begins to dominate world trade.

1750

GREAT BRITAIN

Flying Shuttle (patented 1733) is more widely adopted by textile manufacturers c.1753, doubling the output of individual weavers.

Bridgewater Canal opens, 1761. It is entirely man-made and revolutionizes transport in Britain.

Spinning jenny, invented in 1765, lowers the cost of yarn and helps meet demand from weavers.

Steam engine is developed by James Watt, 1765. It is more efficient than the Newcomen engine (1712) and able to power machines, leading to the rise of the factory system.

Soho Manufactory opens, 1766. It pioneers mass production of a range of metal and glass goods using steam power.

Water frame, a spinning machine powered by water and invented by Richard Arkwright, in 1769, requires less skill to operate.

Cromford Mill, in Derbyshire, built in 1771, is the first water-powered cotton spinning mill. The system is adopted by other industrializing countries.

Adam Smith's *The Wealth of Nations* is published in 1776. It expounds a new economic theory for the industrial age, advocating free trade (trade without restrictions).

World's first cast iron bridge is constructed by Abraham Darby III over the Severn River, Shropshire in 1779. It becomes known across the world as a symbol of the Industrial Revolution.

Spinning mule, 1779, combines the spinning jenny with the water frame and fully automates the weaving process, which leads to expansion of the textiles industry.

Puddling furnace, invented by Henry Cort in 1783, increases wrought iron production by 400 percent in 20 years.

Threshing machine, invented in 1786, increases efficiency in farming.

Coal gas first used by William Murdock to light his home in 1792. Eventually coal gas replaces candles and oil lamps and is used to light up streets, houses, and factories.

US

Cotton gin, invented by Eli Whitney, 1793, efficiently separates cotton seeds from fiber; production—and demand for slave labor—soars.

Cotton gin

BELGIUM

European Industrial Revolution begins in 1799 when Englishman William Cockerill contravenes British law to build spinning machines in Belgium.

1800

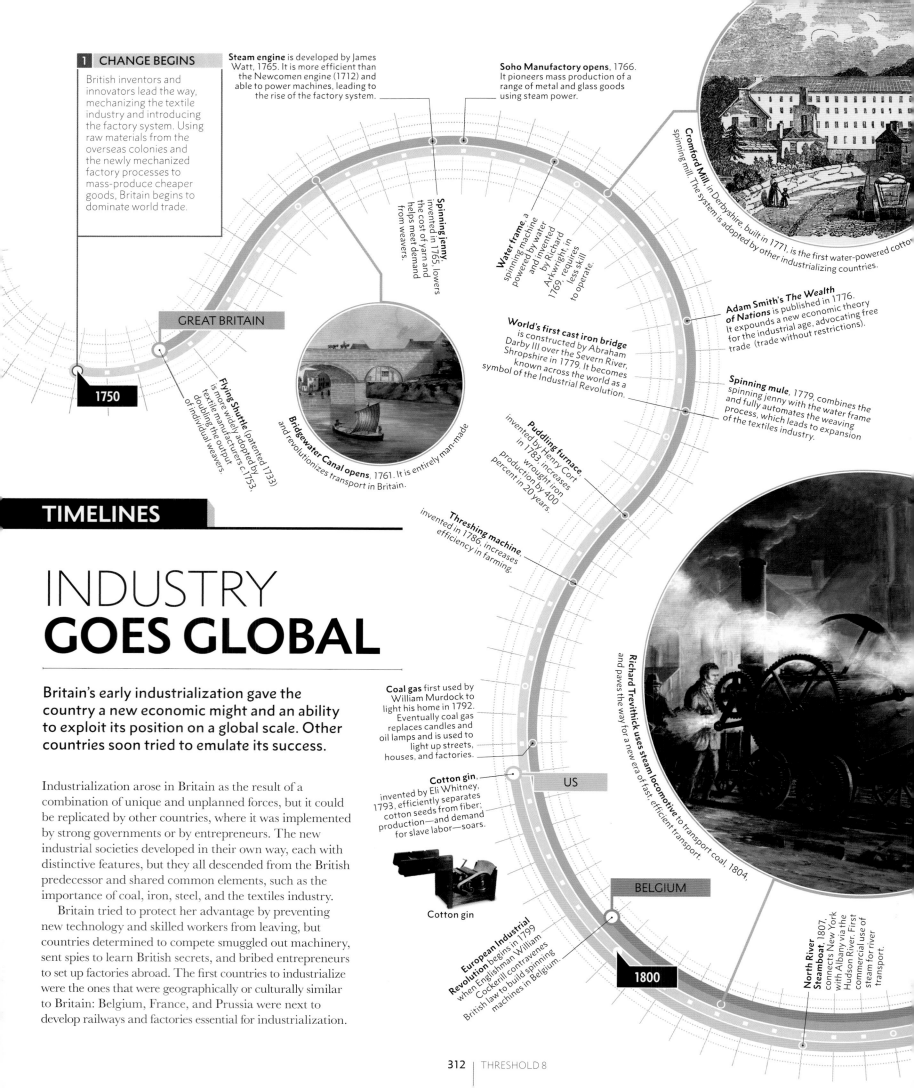

Richard Trevithick uses steam locomotive to transport coal, 1804, and paves the way for a new era of fast, efficient transport.

North River Steamboat, 1807, connects New York with Albany via the Hudson River. First commercial use of steam for river transport.

William Cockerill imports a Watt steam engine to Liege, 1813. Steam power transforms the Belgian coal, iron, and textile industries.

2 THE REVOLUTION SPREADS

Political revolutions in 1830 and 1848 cause social upheaval but spread liberal ideas and new innovations across Europe. The US also starts to industrialize, capitalizing on the rich natural resources of her newly acquired territories, and the pace of change accelerates after the American Civil War ends in 1865.

First passenger train runs from Stockton to Darlington, 1825.

FRANCE

Rail engineer Horatio Allen ships America's first steam locomotive from England, 1829.

French textile factories open in Rouen, Lille, and Mühlhausen, 1830, and the textile industry drives industrialization.

PRUSSIA

First rotating electric motor created by Moritz von Jacobi, 1834. Four years later, his improved design was strong enough to power a boat.

Launch of SS Great Western, 1837, the first paddle steamship to cross the Atlantic. Steamships revolutionize the transportation of goods and people.

Four railways built in France by British engineer Thomas Brassey in 1841. France develops a railway system, which is essential for industrialization.

USS Princeton, 1843, was the first screw-propelled warship powered by steam.

Pneumatic tire created by Robert Thompson, 1845, makes travel more comfortable.

Mass manufacture of weapons is introduced to Europe by German steelworker Alfred Krupp in 1847 after he constructs the first steel-cast cannon.

James Young begins distilling oil in 1848 to produce petroleum, paraffin, and kerosene for use in lamps and for lubricating machinery.

1850

American steam-powered warships arrive in Tokyo Bay in 1853 and use force to open Japan to trade with industrial nations.

Bessemer process, 1856 enables the mass-production of steel, which reduces the cost and allows it to be used for large-scale engineering work.

The commercial American oil industry is born when Edwin Drake strikes oil in Titusville, Pennsylvania in 1859. By the end of the 19th century oil is a major fossil fuel used to power engines.

Gatling gun first used in the American Civil War, in 1860, is intended to reduce deaths from wartime disease.

Siemens-Martin process improves steel production and makes it cheaper, 1865.

JAPAN

Rapid industrialization of Japan instituted by the Meiji government in 1868.

3 INDUSTRIAL WORLD

Driven by the desire to compete, Japan and Russia industrialize against a background of domestic social and political upheaval. European powers also compete to secure raw materials in the scramble for Africa.

Pennsylvania steelworks built by Scotsman Andrew Carnegie in 1875. It utilizes the Bessemer system, and the American steel industry greatly expands.

Benz automobile

First automobile is patented by Karl Benz in 1886. It is powered by the internal combustion engine, which runs on the newly available fossil fuel—oil.

Richmond Union Passenger Railway opens in Virginia, in 1888. It is the first successful large electric street railway system.

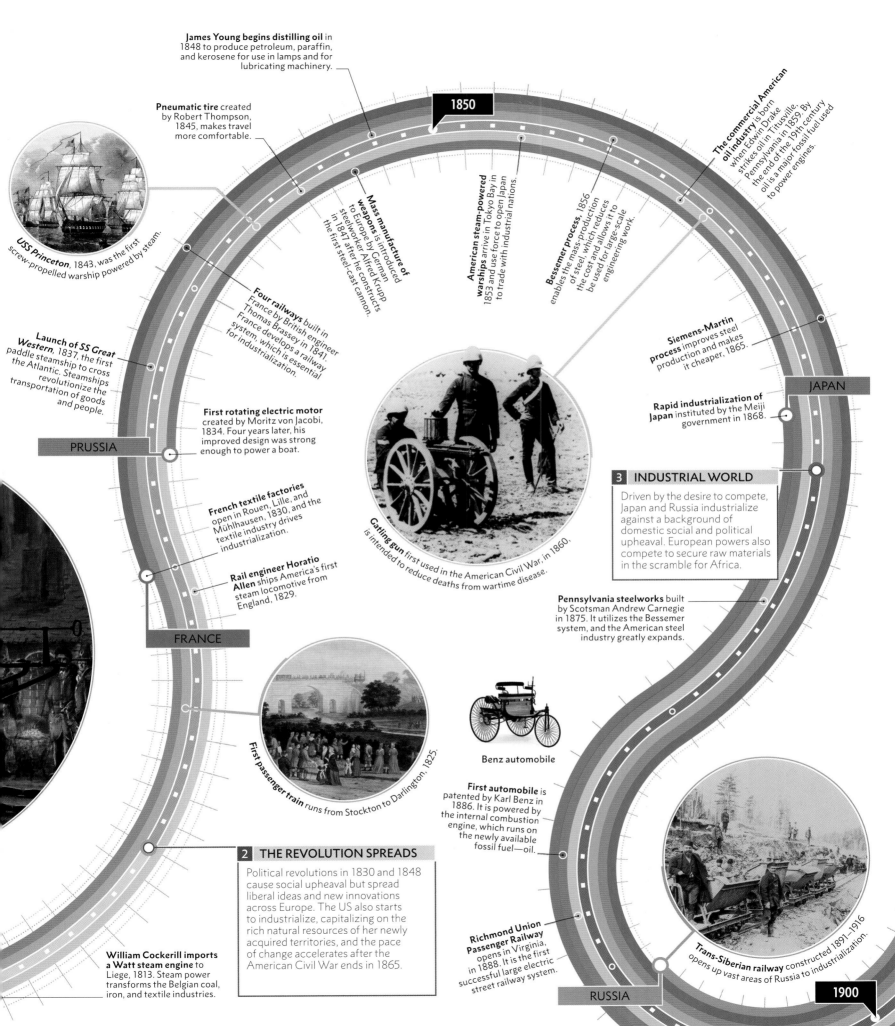

Trans-Siberian railway constructed 1891–1916 opens up vast areas of Russia to industrialization.

RUSSIA

1900

GOVERNMENTS EVOLVE

Governments soon realized that industrialization could increase their country's wealth—and it changed the way they ruled. They began to work in partnership with industry, and this ushered in a new balance of power, as governments became managers of markets and citizens.

The process of industrialization saw the nature of government change. Structures that served agrarian civilizations either evolved or were replaced by new institutions that were developed to manage the wealth and power of industrial economies. As the first country to industrialize, the British government led the way in facilitating the creation of more wealth, cooperating with merchants and using its navy to protect their overseas interests. Successful commerce led to larger markets and even greater wealth, so the government began to encourage innovations to meet demand and increase output. Other countries observed that industrialization produced revenues that could be used to fund their militaries, and their governments also became increasingly concerned with trying to support industrialists, control the new economies, and manage the growing numbers of wage workers—all of which led to more bureaucracy and the creation of the modern state.

The way in which modern states evolved varied dramatically. France created a completely new bureaucracy after the social and political revolution of 1789 swept away the institutions associated with its *ancien régime*. Britain already had an established representative assembly and gradually developed other institutions over time. To ensure the loyalty of their citizens, leaders began to develop national ideologies, and, by 1914, the modern state had begun to shape the politics of countries around the world.

“THE **SUBJECTS OF EVERY STATE** OUGHT TO **CONTRIBUTE TOWARD** THE SUPPORT OF THE **GOVERNMENT** IN PROPORTION TO THEIR RESPECTIVE ABILITIES.”

Adam Smith, Scottish philosopher and pioneer of the political economy, 1723–1790

▼ Pressures of power
Industrialization transformed society, distributing wealth far more broadly, and different groups began to make demands on governments.

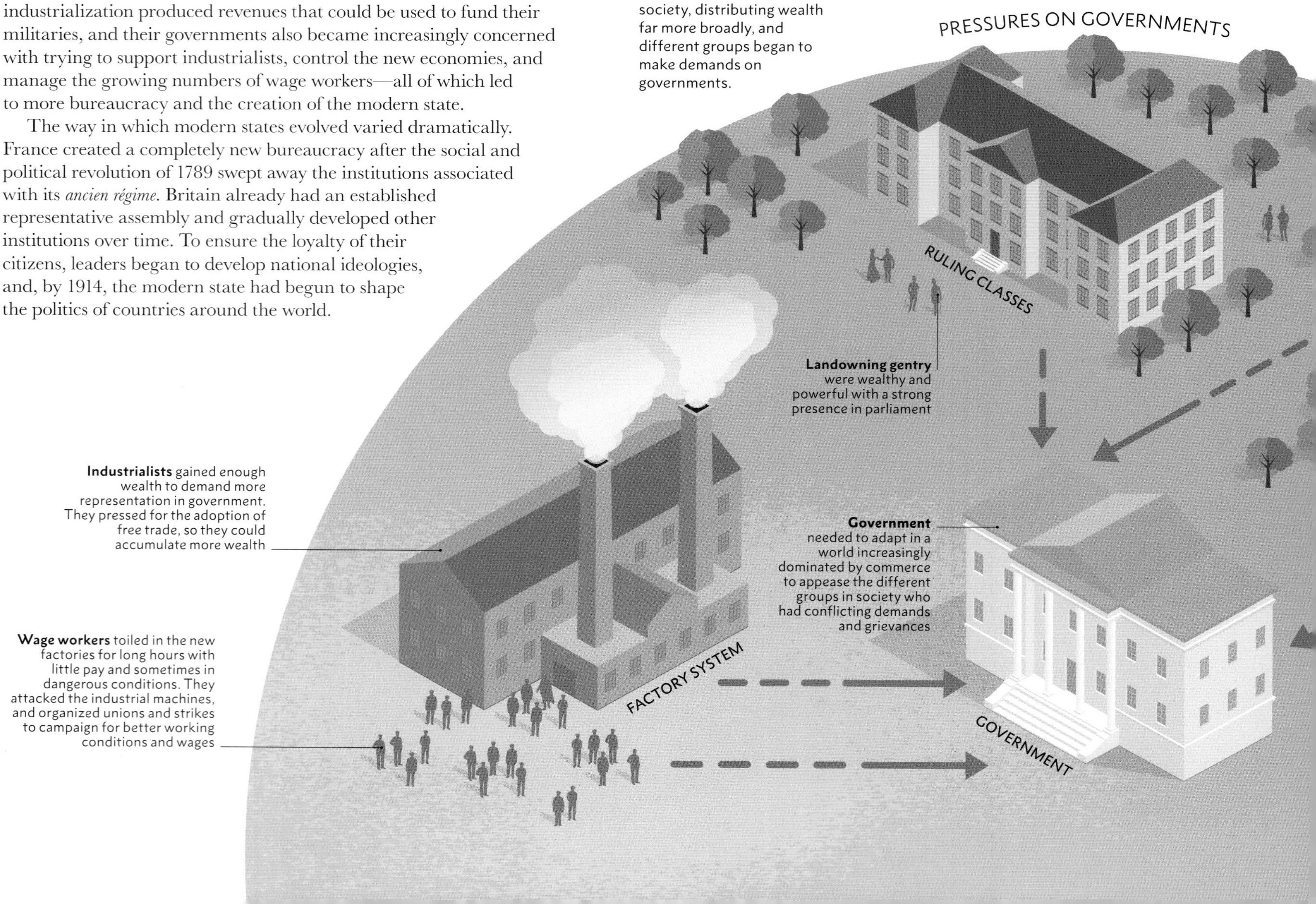
PRESSURES ON GOVERNMENTS
RULING CLASSES
Landowning gentry were wealthy and powerful with a strong presence in parliament
Industrialists gained enough wealth to demand more representation in government. They pressed for the adoption of free trade, so they could accumulate more wealth
Government needed to adapt in a world increasingly dominated by commerce to appease the different groups in society who had conflicting demands and grievances
FACTORY SYSTEM
Wage workers toiled in the new factories for long hours with little pay and sometimes in dangerous conditions. They attacked the industrial machines, and organized unions and strikes to campaign for better working conditions and wages
GOVERNMENT

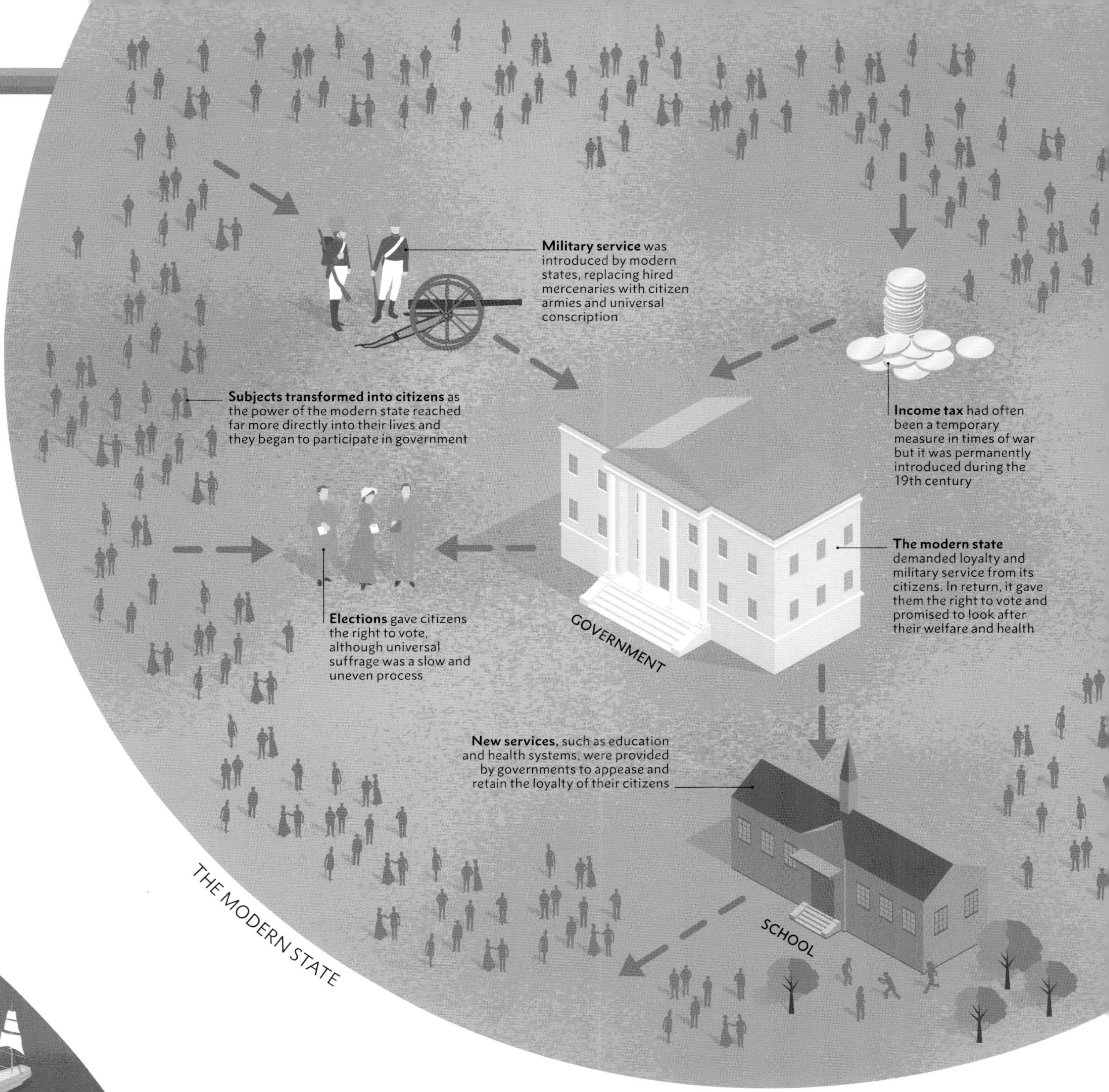
Military service was introduced by modern states, replacing hired mercenaries with citizen armies and universal conscription
Subjects transformed into citizens as the power of the modern state reached far more directly into their lives and they began to participate in government
Income tax had often been a temporary measure in times of war but it was permanently introduced during the 19th century
The modern state demanded loyalty and military service from its citizens. In return, it gave them the right to vote and promised to look after their welfare and health
GOVERNMENT
Elections gave citizens the right to vote, although universal suffrage was a slow and uneven process
New services, such as education and health systems, were provided by governments to appease and retain the loyalty of their citizens
SCHOOL
THE MODERN STATE

▶ The new establishment Governments were forced to adapt to manage the growing wealth and power of industrial economies. To do so, they increasingly developed the institutions of modern states: citizen armies, taxation, and services that included infrastructure, protection, education, and hospitals.

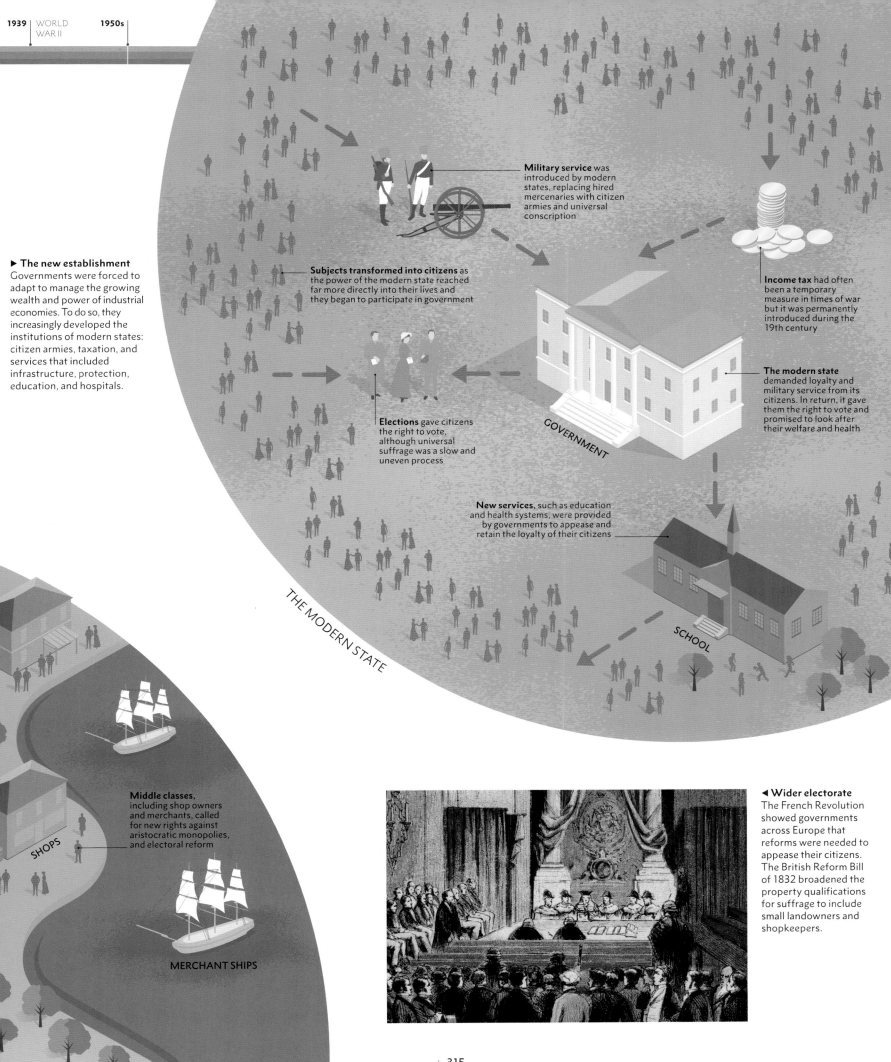

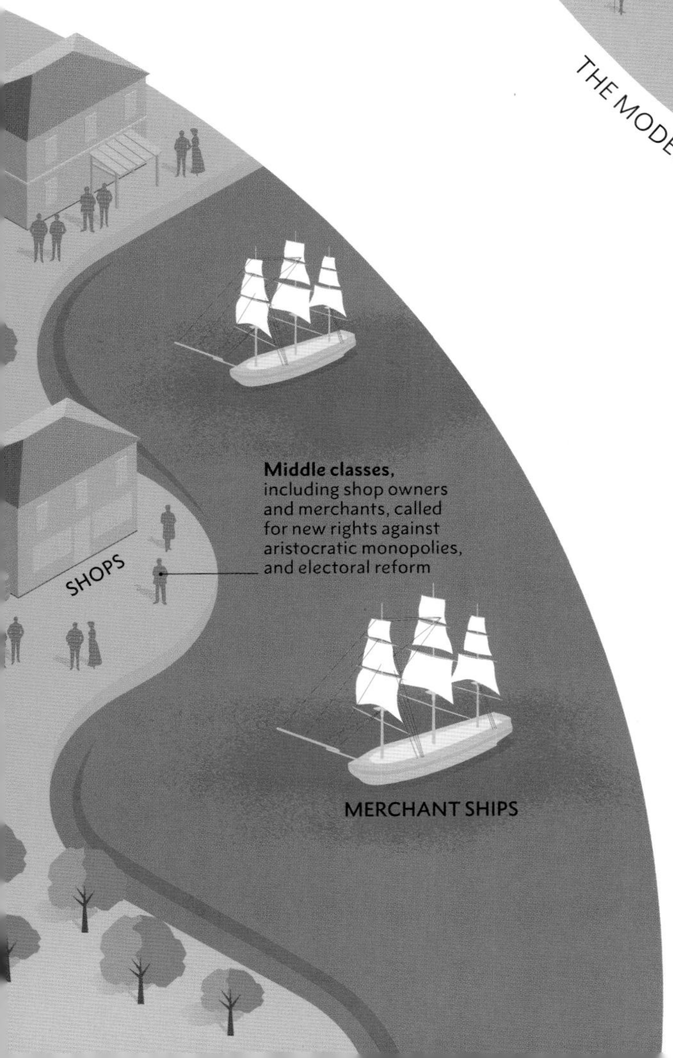
Middle classes, including shop owners and merchants, called for new rights against aristocratic monopolies, and electoral reform
SHOPS
MERCHANT SHIPS

◀ Wider electorate The French Revolution showed governments across Europe that reforms were needed to appease their citizens. The British Reform Bill of 1832 broadened the property qualifications for suffrage to include small landowners and shopkeepers.

CONSUMERISM TAKES OFF

Industrialization meant that land was no longer the only source of wealth. It became possible to generate wealth through manufacturing and trading goods. During the late 18th century, the middle classes grew, leading to a new emphasis on upward mobility and consumption.

▲ Household items for everyone
Expansion in the pottery industry increased consumer choice, and laborers who once ate from metal platters dined from Wedgwood porcelain.

Industrialization brought improvements in transport and manufacturing technology, which increased the availability of consumer products. This, combined with increased international trade, brought an unprecedented array of new goods to the domestic market. Rising prosperity and social mobility allowed the middle classes to increase their ranks and more people had a disposable income to spend. The middle classes were not a homogeneous group but a broad band of the population, which fell between the aristocracy and the workers. At the lower end were the storekeepers and at the top were the capitalists who owned companies. They included businessmen and entrepreneurs, doctors, lawyers, and teachers. The emergent middle classes all shared a common interest in the expansion of the economy and held specific ideas about the role the government should play in its management. They wanted an economy unfettered by government restrictions, as they thought this was the best way to foster individual achievement. They also shared

▼ Consumer culture
With the advent of the department store, customers could buy an astonishing array of goods all in one place and shopping became a leisure activity.

common values: they believed that through hard work and self-reliance it was possible to achieve economic success.

The notion of self-improvement was a key part of middle-class culture. As they rose up the social ladder, and in order to ensure the aristocracy no longer had an unfair advantage, the middle classes campaigned for electoral reform and free trade. These were seen as the necessary conditions to make it possible for anyone to succeed through their own efforts.

In Britain, the middle classes converted economic success into political power with the 1832 Reform Act, as a more aspirational society began to demand and expect more from the government.

CONSPICUOUS CONSUMPTION

The middle classes often aspired to the consumption patterns of the aristocracy. Clothing and household possessions were a way of communicating one's social position and by the end of the 18th century these status symbols were within the financial reach of many. The dizzying array of goods on offer included textiles, furniture, clothes, hats, china, books, jewelry, lace, perfume, and food. Middle-class wives filled their homes with new material possessions and purchased fashionable clothing to display their husbands' financial success.

Wages were high in 18th-century Western Europe, especially in Britain, which meant that even members of the lower classes could afford some of these consumer goods. Most 18th-century towns had taverns offering cheap meals, and coffee houses where coffee and chocolate could be consumed and ideas could be exchanged.

Greater purchasing power and a gradual fall in prices led to rising demand for new consumer products, which in turn fueled the economies of industrializing countries. These commodities were made affordable by slave labor, with over 11 million slaves producing the goods that flowed into Europe's ports. These slaves were part of a system known as the "triangular trade": European merchants transported slaves from Africa to work on plantations in the Americas and the Caribbean, and then transported the commodities produced by the slaves back to Europe.

ADVERTISING AND ASPIRATIONS

English entrepreneur Josiah Wedgwood noticed the way aristocratic fashions slowly filtered down through society. He sold tea services to the British Queen, and his "Queen's Ware" became a must-have item among the middle classes. Wedgwood realized that he needed to convince consumers they wanted to buy his wares, and that consumers were primarily women. He opened a showroom where women were encouraged to meet, drink tea, and be shown his new ranges of china. His pottery reached every industrial market in Europe and North America. He is often considered the father of modern advertising. Wedgwood's marketing genius had a knock-on effect in London and then abroad, as retail outlets made products more easily available to the consumer. This was manifest in the growth of department stores, which opened in Paris in the 1830s, Russia in the 1850s, and Japan in the 1890s.

With the rapid growth of towns and cities, by the 19th century shopping had become an important cultural activity, as a shift in behavior meant people began buying for fashion rather than necessity. Storefronts displayed mirrors, bright lights, colorful signs, and advertisements, and all of their products to entice shoppers inside. Many stores tried to appeal to the wealthier end of the market, but cheaper mass-manufactured goods and an abundance of food markets made shopping a cultural activity open to every class.

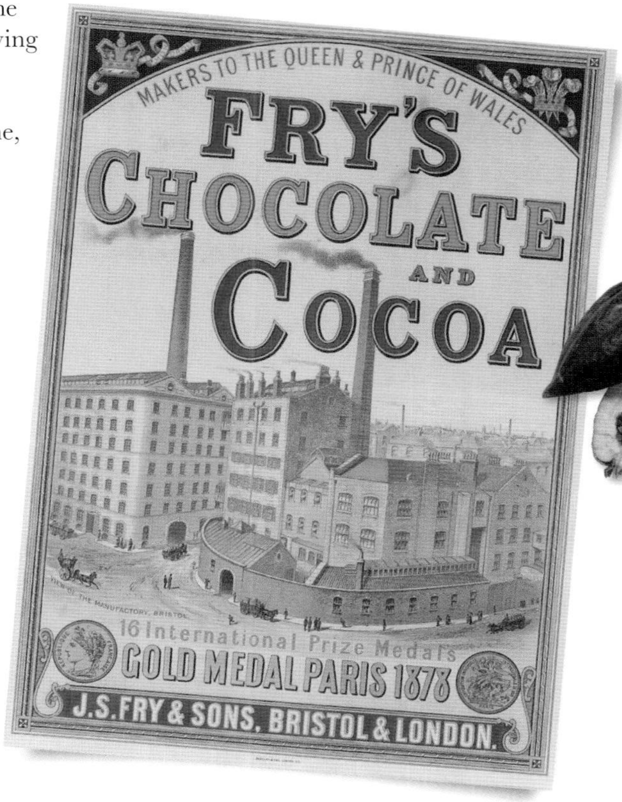

Cocoa bean

◀ **Chocolate temptations**
Once the favored drink of the aristocracy, chocolate became accessible to the general public, and manufacturers targeted women and children in advertising campaigns.

◀ **Luxury and slavery**
Imports of raw cotton, sugar, rum, and tobacco came from slave plantations in the Caribbean, where African slaves were the primary source of labor.

Popular catchphrases of the American and French revolutions, the concepts of liberty, equality, and fraternity were drawn from 17th and 18th century Enlightenment ideals of reason, knowledge, and the freedom of people to improve their condition. The marriage between the philosophies of the Enlightenment and the actions of the revolutionaries launched a sea change in Western politics. People began to demand freedom from the oppression of absolute monarchies and imperial overlords and wanted a new social contract in its place. On a practical level, this included greater representation in government and the right to own land, but it also brought about a general shift in consciousness. The existence of universal natural rights became part of a new, more empathetic world view, which fed into the development of the modern state.

BIG IDEAS

EQUALITY AND **FREEDOM**

Revolutionary ideas promoting liberty, equality, and fraternity were introduced to the industrializing world in the late 18th century, after revolutions in France and America dismantled established aristocratic regimes. These ideas echoed through the politics of the 19th century and became central to modern beliefs about human rights.

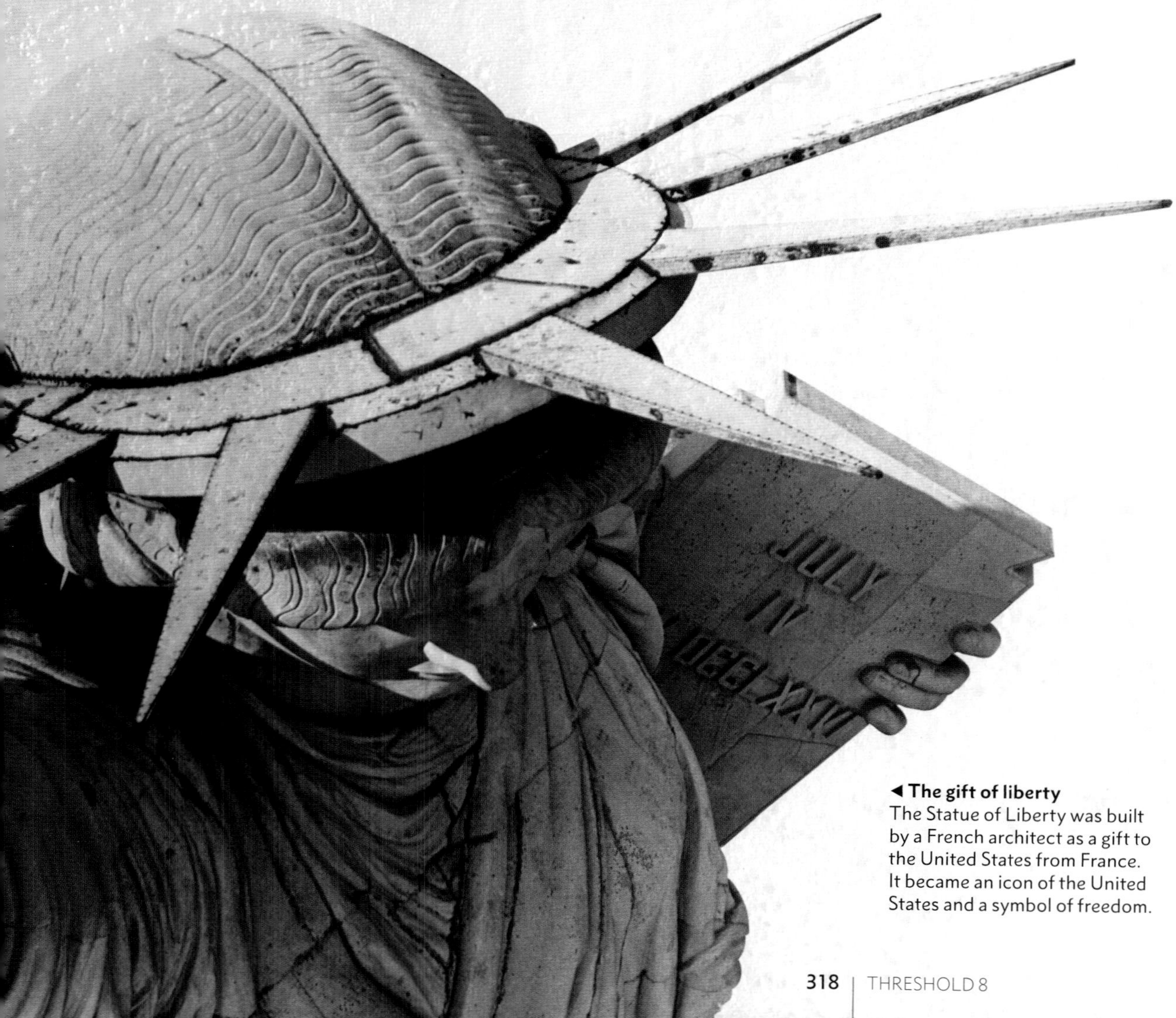

◀ The gift of liberty
The Statue of Liberty was built by a French architect as a gift to the United States from France. It became an icon of the United States and a symbol of freedom.

INTERNATIONAL EXCHANGE

These principles were first asserted by Thomas Jefferson, a key figure in the American War of Independence and author of the 1776 Declaration of Independence. The Declaration stated that all men are born free, are equal before the law, and have natural rights to property, life, and liberty—ideas that remain central tenets of democracy today. Democracy itself was not a new concept: it had been established in ancient Athens around the 5th century BCE, and was rediscovered during the Renaissance. The Athenian experience helped inspire revolutions against absolute monarchies, such as in France.

The Declaration of Independence—and the American revolution itself—was heavily influenced by international figures: English philosopher John Locke argued that legitimate governments needed the consent of the governed; writer and activist Thomas Paine advocated for the right to revolt against a government that did not protect its citizens' needs. They published their arguments in polemical pamphlets, distributed through a revolutionary exchange network, including men who had participated in both the American and the French revolutions, such as the Marquis de Lafayette—a French hero of the American war. This ensured that ideas, such as Paine's *Rights of Man* (1791), reached an international audience. The spread of ideas between America and France was the most important political exchange network of the time. America showed the world what was possible: many Frenchmen helped in its liberation from British rule and returned home influenced by what they had seen. After France's own uprising, the Marquis de Lafayette enlisted help from Thomas Jefferson—in Paris at the time—to pen the *Declaration of the Rights*

200 COPIES OF THE AMERICAN **DECLARATION** OF INDEPENDENCE WERE **PRINTED AND DISTRIBUTED**

of Man and of the Citizen. The American and French revolutions revealed how powerful uncensored ideas could be.

The exchange of Enlightenment ideas was encouraged among the bourgeoisie, the middle class who led the French revolution. They were ambitious and well-read, schooled in the theories of Montesquieu, Rousseau, and Voltaire, the thinkers known as the "philosophes" who advocated the uncensored exchange of ideas and freedom of the press. The philosophes spread their views through the Republic of Letters, a community of European and American intellectuals who communicated through letters, essays, and published papers.

In the 17th and 18th centuries, Enlightenment thought led to a shift away from religious dogma toward scientific experiment and empirical modes of thought. Scientific progress and technological innovation helped incubate the industrial revolution in Britain. The exchange of broader Enlightenment ideas was encouraged among the middle classes of Europe through "societies of thought" such as reading rooms, coffee houses, Masonic lodges, and scientific academies. Coffee houses became famous meeting places for later revolutionaries such as Karl Marx and Friedrich Engels, key figures of the 1848 revolutions in Europe. They harnessed the power of the rotary press, invented in 1843, which enabled the mass-production of print books and newspapers. Marx's own newspaper, the *Rheinische Zeitung*, reported on the events of the 1848 uprisings, and helped to spread the revolutionary message to the masses.

THE LEGACY

The American, French, and other revolutions of the 19th century were all based on the Enlightenment idea that humans have certain inalienable rights. The government's role would be to recognize and secure the rights and property of its citizens, and it would be formed by elected, tax-paying citizens. Women, slaves, and foreigners were not included. However, in the aftermath of the French revolution, a new consciousness began to spread through Europe. Many people developed an empathy with the plight of others—progressive thinkers called for the reform of prisons, an end to harsh sentences, and the abolition of slavery. France was first to abolish slavery in 1794; Britain and America followed in 1807 and 1808, respectively. By 1842, the Atlantic slave trade was over.

The human rights ideal played an important role in Europe in 1820, 1830, and 1848 when revolutionary activity broke out across the continent. Thinkers from both the right and the left, the two sides that defined modern politics, echoed the principles of the *Declaration of the Rights of Man and of the Citizen* and argued that the ideals of universal rights justified their political action. Crucially, the Declaration's clause that "the source of all sovereignty resides essentially in the nation" was evoked constantly during the rise of nationalism and the formation of the modern nation states of Europe.

A key principle of the *Declaration of the Rights of Man and of the Citizen*, that "all human beings are born free and equal in dignity and rights," spread widely throughout the 19th century. Progressives across the world believed that the universal, equal, and natural human rights espoused in the Declaration would overturn all undemocratic forms of rule. Simon Bolivar (1783–1830), the liberator of Spanish Venezuela, Ecuador, Bolivia, Peru, and Columbia, openly admired the French Revolution. Hindu reformer Ram Mohun Roy (1772–1833) argued for freedom of speech and religion as natural rights in condemning India's caste system. And during the late 19th and 20th centuries, educated Asian and African leaders argued that European colonization contravened the human rights of the indigenous people. Eventually the principle became enshrined in the first article of the United Nations 1948 Universal Declaration of Human Rights, which set out to protect the fundamental rights to which all peoples—no matter where they come from—are inherently entitled.

10,000 AFRICAN SLAVES WERE **FREED** AFTER **THE FRENCH REVOLUTION**

“TO **DENY PEOPLE** THEIR **HUMAN RIGHTS** IS TO **CHALLENGE** THEIR VERY **HUMANITY.**”

Nelson Mandela, South African civil rights activist, 1918–2013

“WE HOLD **THESE TRUTHS** TO BE SELF-EVIDENT, THAT **ALL MEN** ARE **CREATED EQUAL...**”

Thomas Jefferson, 1743–1826, *Declaration of Independence*

NATIONALISM EMERGES

The second half of the 18th century was a period of immense revolution, in both social and political terms. These profound changes in the world order led to the formation of new nation states and a growing sense of nationalism, as countries began to assert their individuality.

The roots of modern nationalism can be traced back to the political philosophy of John Locke in 17th-century England, with its emphasis on the individual and his rights, and the human community. It was also influenced by the unprecedented social changes brought about by the industrial revolution and by the liberal ideals of the Enlightenment philosophers. Essentially, modern nationalism demanded loyalty to one's country and embodied a sense of common identity and history shared by rulers and citizens alike.

Unity under a free and equal democracy was central both to the liberal nationalism of the American Revolution of 1776, and to the outbreak, in 1789, of the French Revolution, which paved the way for the modern nation state as a united community enjoying equal rights under a Constitution. The French revolutionaries introduced a centralized administrative bureaucracy with uniform laws, and established French as the common language of the land.

▼ Unifying force
In 1871, Chancellor Otto von Bismarck finally achieved his aim of bringing 300 small kingdoms and principalities together to form a unified Germany.

NEW NATION STATES

The growing sense of nationalism in Europe sparked struggles for independence in Greece and Belgium (where there was a successful revolution against Dutch rule). In 1848, revolution once again erupted across Europe, as huge swathes of the populace vented their dissatisfaction, demanding national unification and constitutional reform. The Kingdom of Italy was finally created in 1861 and Germany in 1871, but these two unifications came at a cost. Absolute monarchies were re-established and liberal institutions such as the popular press were persecuted. A misplaced blend of nationalism and beliefs about racial superiority led European nations to colonize many countries in the late 19th century.

Culturally, nationalism often took the form of a celebration of a nation's history, culture, and achievements. Proud of their rapid modernization, the great industrial nations of the world hosted impressive international trade exhibitions to show off their latest manufacturing—the supreme expression of confidence in their nation.

“**PATRIOTISM** IS WHEN **LOVE OF YOUR OWN PEOPLE** COMES FIRST; **NATIONALISM**, WHEN HATE FOR PEOPLE OTHER THAN YOUR OWN COMES FIRST.”

Charles de Gaulle, former President of France, 1890–1970

Patriotic display
The Great Exhibition of 1851, in Britain, was the first international exhibition of manufactured products, as well as a display of national pride.

1750 | THE INDUSTRIAL REVOLUTION BEGINS **1789** | DECLARATION OF THE RIGHTS OF MAN **1830** | STEAM AGE **1869** | DNA ISOLATED **1880** | ELECTRICAL AGE

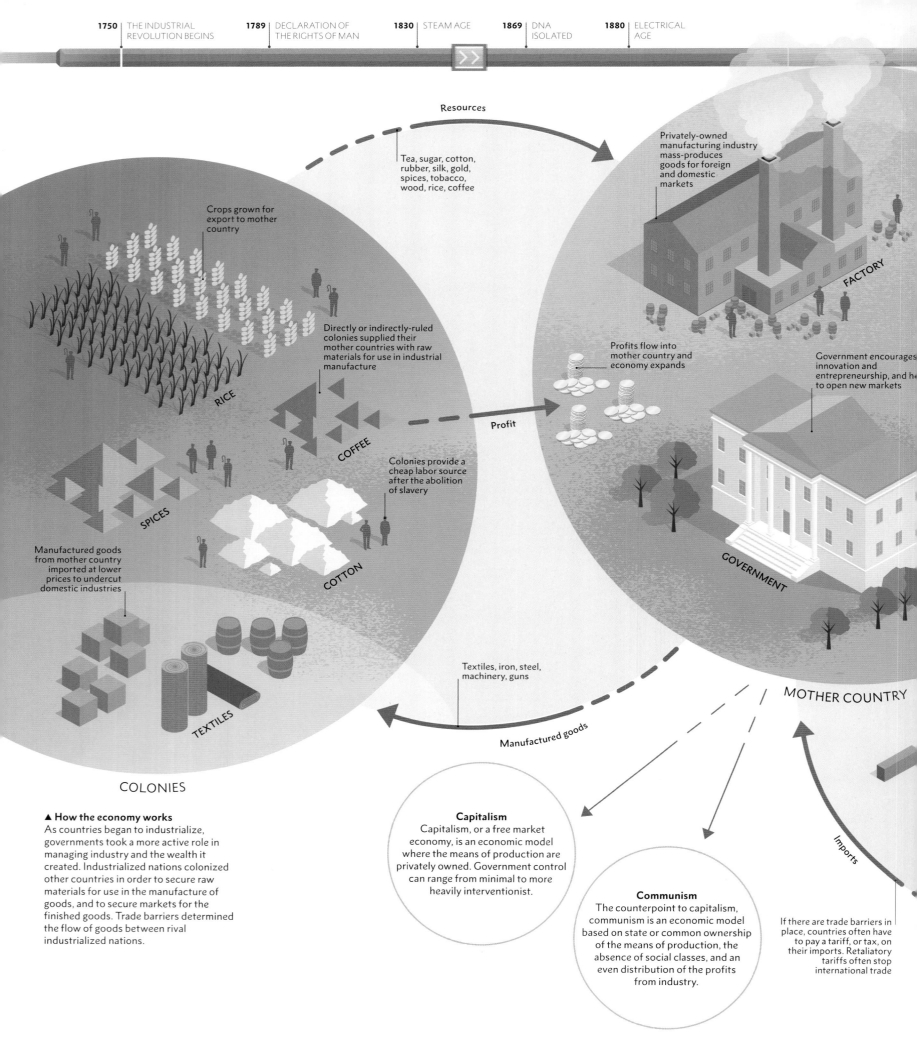

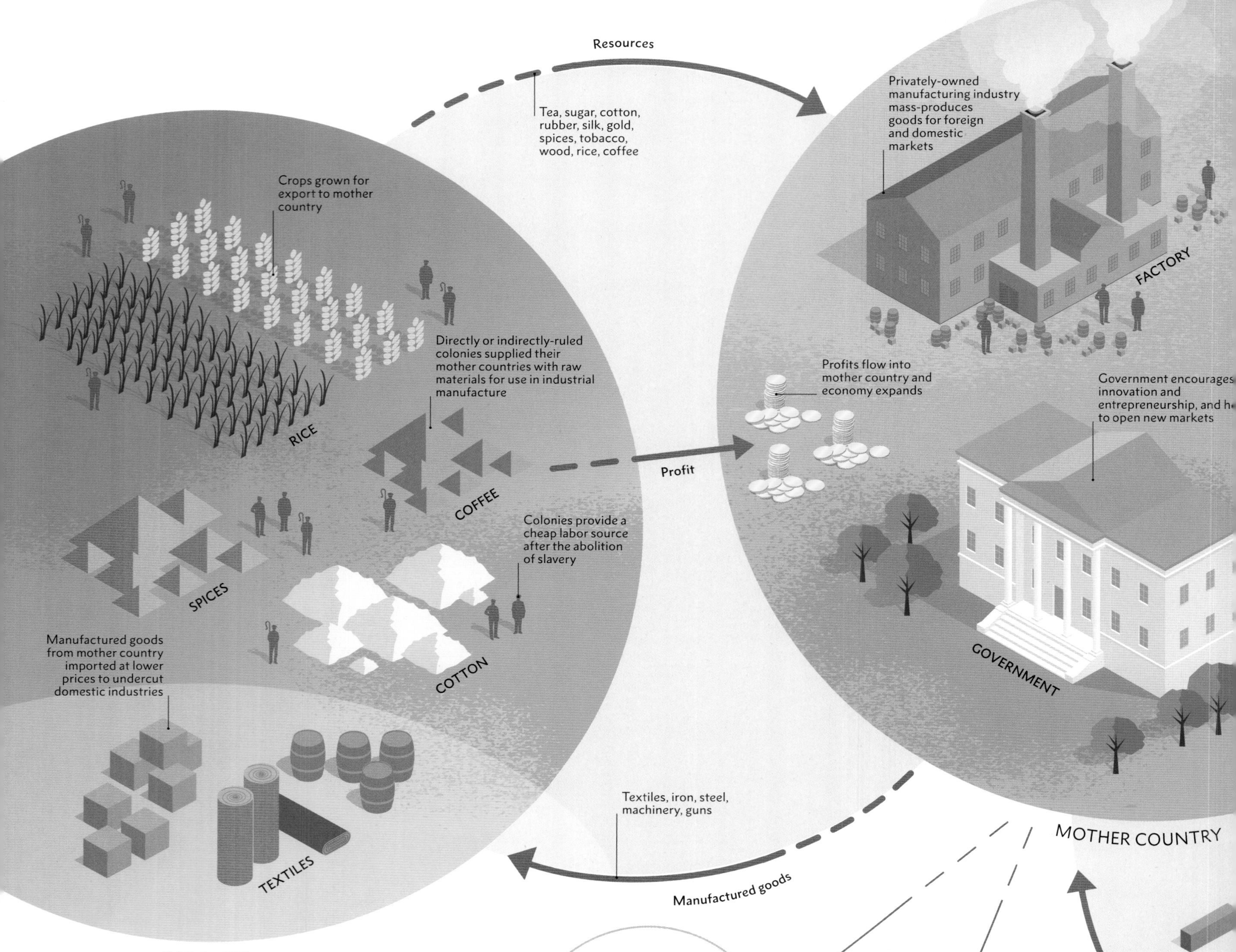

Resources
Tea, sugar, cotton, rubber, silk, gold, spices, tobacco, wood, rice, coffee
Privately-owned manufacturing industry mass-produces goods for foreign and domestic markets
FACTORY
Crops grown for export to mother country
RICE
Directly or indirectly-ruled colonies supplied their mother countries with raw materials for use in industrial manufacture
COFFEE
Profit
Profits flow into mother country and economy expands
Government encourages innovation and entrepreneurship, and h to open new markets
SPICES
Colonies provide a cheap labor source after the abolition of slavery
COTTON
GOVERNMENT
Manufactured goods from mother country imported at lower prices to undercut domestic industries
TEXTILES
Textiles, iron, steel, machinery, guns
Manufactured goods
MOTHER COUNTRY
COLONIES

▲ How the economy works
As countries began to industrialize, governments took a more active role in managing industry and the wealth it created. Industrialized nations colonized other countries in order to secure raw materials for use in the manufacture of goods, and to secure markets for the finished goods. Trade barriers determined the flow of goods between rival industrialized nations.

Capitalism
Capitalism, or a free market economy, is an economic model where the means of production are privately owned. Government control can range from minimal to more heavily interventionist.

Communism
The counterpoint to capitalism, communism is an economic model based on state or common ownership of the means of production, the absence of social classes, and an even distribution of the profits from industry.

Imports

If there are trade barriers in place, countries often have to pay a tariff, or tax, on their imports. Retaliatory tariffs often stop international trade

1939 | WORLD WAR II
1950s | ANTHROPOCENE EPOCH BEGINS
1970 | DIGITAL AGE
1973 | OIL CRISIS
1989 | WORLD WIDE WEB INVENTED
2001 | 9/11
2008 | GLOBAL FINANCIAL CRISIS

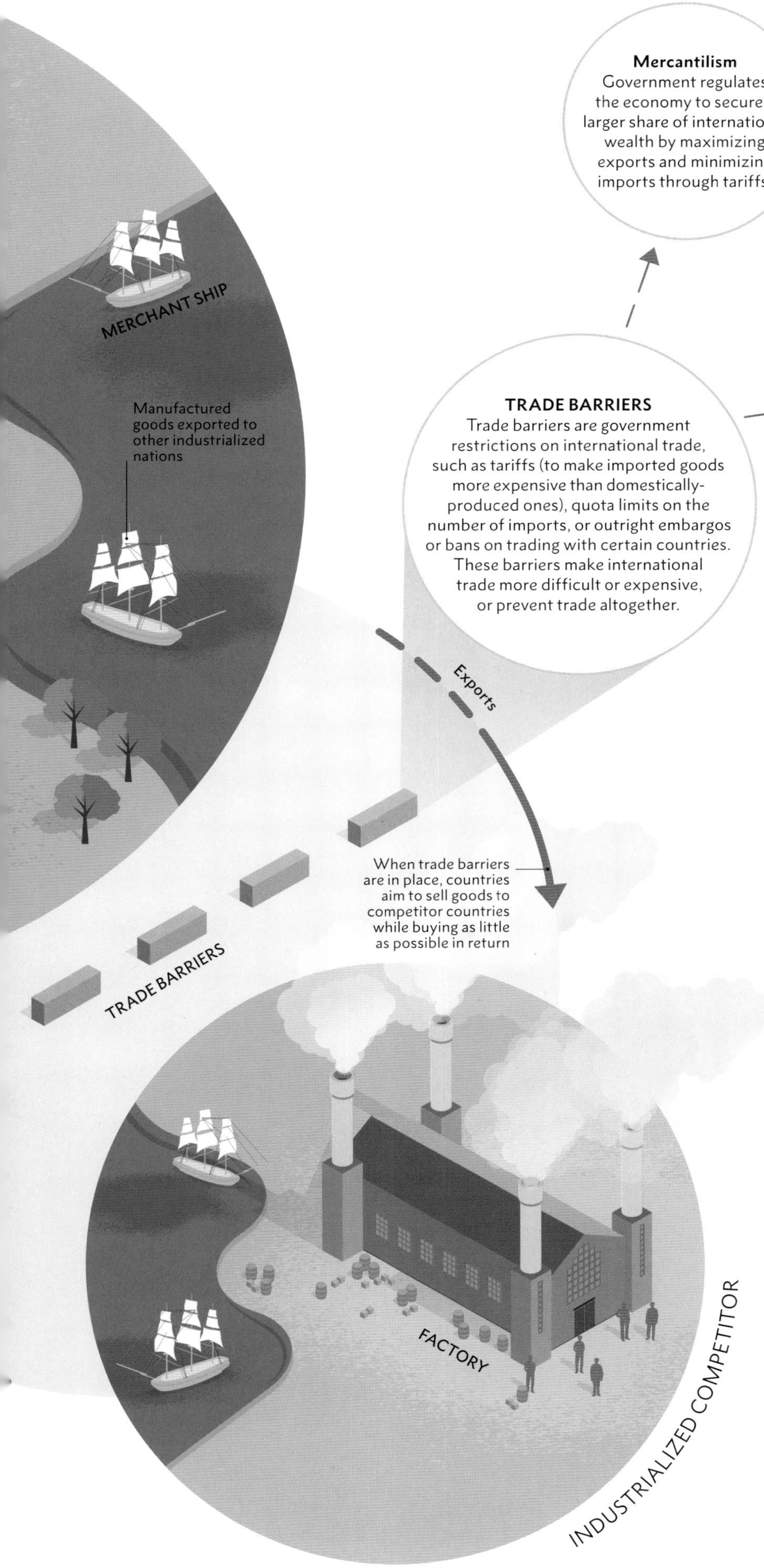

Mercantilism
Government regulates the economy to secure a larger share of international wealth by maximizing exports and minimizing imports through tariffs.
TRADE BARRIERS
Trade barriers are government restrictions on international trade, such as tariffs (to make imported goods more expensive than domestically-produced ones), quota limits on the number of imports, or outright embargos or bans on trading with certain countries. These barriers make international trade more difficult or expensive, or prevent trade altogether.
Protectionism
Government restricts international trade to protect domestic industries from potential rivals—tariffs, subsidies and import quotas, and exclusion from the market.
MERCHANT SHIP
Manufactured goods exported to other industrialized nations
Exports
When trade barriers are in place, countries aim to sell goods to competitor countries while buying as little as possible in return
TRADE BARRIERS
FACTORY
INDUSTRIALIZED COMPETITOR

BY 1913, THE UNITED STATES, GERMANY, THE UNITED KINGDOM, FRANCE, AND RUSSIA PRODUCED 77 PERCENT OF THE WORLD'S MANUFACTURED GOODS

THE INDUSTRIAL ECONOMY

The industrial revolution created new possibilities for countries to increase their wealth. Nations adapted to cope with the corresponding increase in international trade, and the pitfalls that came with it.

Before the industrial revolution, mercantilism had been the dominant European economic model. In this approach, a nation encouraged exports and discouraged imports, with the belief that the world's wealth was finite. However, the introduction of mass production on an industrial scale increased economic output, and showed it was possible to create new wealth. Industrialists realized they could make more money by importing cheap raw materials from unindustrialized countries and manufacturing them into goods they could sell to both foreign and domestic markets. With this increase in both production and profit, for a time it became beneficial for countries to trade freely with each other. Industrialists pressured governments to adopt a policy of free trade—trade without barriers or government interference, where imports are tariff-free and exports are not subsidized. This was the start of a period of great wealth accumulation, the founding of new financial institutions, and the birth of capitalism—a term coined by economist Adam Smith—which remains the dominant economic model for industrializing countries today.

However, as well as increasing the flow of goods and wealth between nations, the consequences of free trade can include economic instability, exploitation, and clashes over the sources of wealth—colonies, markets, and raw materials. To counteract this, governments create trade barriers to try and protect their interests; this can result in cyclical periods in which trade increases or decreases between nations.

Gunboat diplomacy
Samurai warriors row out to meet an American "black ship," which introduced Japan to gunboat diplomacy and made its 17th-century weaponry instantly ineffective.

THE WORLD OPENS **TO TRADE**

The 19th century marked a major turning point for world trade, as industrialized nations sought to expand their commercial reach. It was not always a peaceful process, but it laid the foundations of the modern international economy.

In the 1840s, the policy of free trade—trade without government interference or tarrifs on imports or exports—led to a period in which industrializing nations were able to accumulate great wealth. Factories enabled the cheap and rapid production of an unprecedented array of products for domestic and foreign markets, as a rising demand for consumer goods, in turn, fueled more economic growth. This expansion of world trade by the industrializing powers of Britain, and later western European countries and North America, eventually resulted in the need for each world power to protect their own economic position.

CONTROLLING MARKETS

The most rewarding and efficient form of free trade was to control both the raw materials and the markets. This was often achieved by force, as the growing disparity in technologies between industrialized nations and the rest of the world soon showed. Historically, countries such as Japan and China had been largely unwilling to import European goods: they did not need or want them. Britain imported tea from China but—other than silver acquired by selling slaves from Africa to Spanish colonists in the Americas—had nothing to offer the Chinese in exchange. With the abolition of slavery the supply of silver dried up, so Britain began selling Chinese opium instead. China's resistance to the exploitation of its people sparked two Opium Wars in the mid-19th century.

America also adopted a policy of armed intervention in the East, regarding Japan as a backwater where its traders could open up new markets. In 1853, four American gunboats entered the prohibited waters of Edo Bay, Japan. Bristling with modern weaponry, the black ships intimidated the Japanese into opening their borders to trade with America and Europe. The arrival of the more technologically advanced Americans encouraged Japan's own industrialization and path to modernity.

BETWEEN **1809** AND **1839**, **BRITISH IMPORTS DOUBLED** AND **EXPORTS TRIPLED**

Following their victories in the Opium Wars, the British government imposed a series of treaties on China that gave Britain favored and unequal trading privileges. Japan signed a similar treaty with the United States, and other industrialized European powers also followed suit, imposing unequal treaties on trade with Latin America and the Middle East. Countries wishing to trade had to set low tariffs on European imports and adopt legal measures favorable to European interests.

◀ Opium pipe
Imports of British opium led to widespread addiction in China, resulting in the Opium Wars of 1839–42 and 1850–60. After being defeated, China was forced to open more ports to foreign trade.

▼ The Gatling gun
Richard Gatling justified his 1861 invention of the rapid-fire gun on humanitarian grounds. He claimed it would save lives by reducing the carnage on the battlefield and shortening the length of wars.

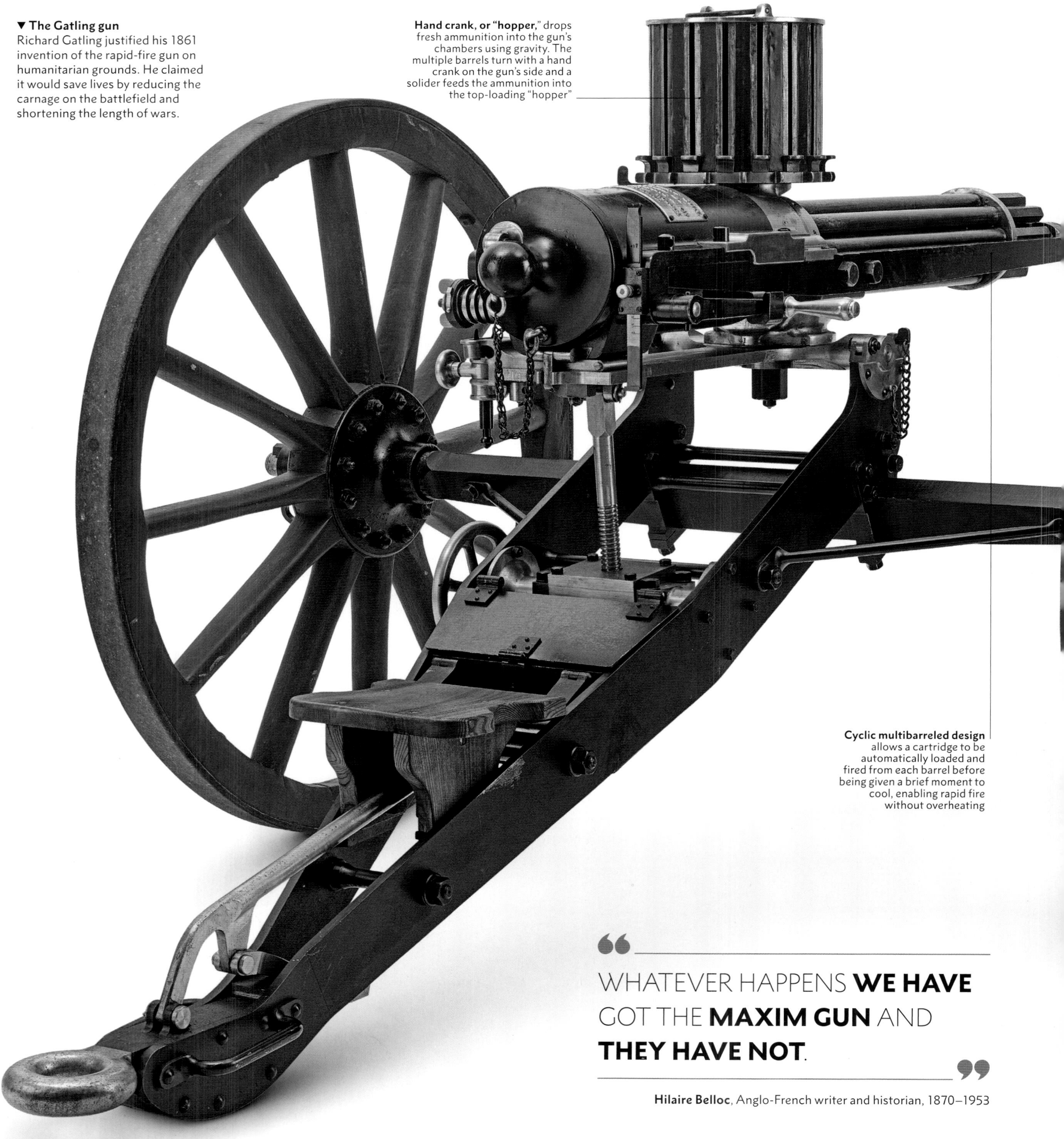

Hand crank, or "hopper," drops fresh ammunition into the gun's chambers using gravity. The multiple barrels turn with a hand crank on the gun's side and a solider feeds the ammunition into the top-loading "hopper"

Cyclic multibarreled design allows a cartridge to be automatically loaded and fired from each barrel before being given a brief moment to cool, enabling rapid fire without overheating

> “WHATEVER HAPPENS **WE HAVE** GOT THE **MAXIM GUN** AND **THEY HAVE NOT**.”
>
> **Hilaire Belloc**, Anglo-French writer and historian, 1870–1953

WAR DRIVES **INNOVATION**

New innovations that increased the effectiveness of war machines made the "scramble for Africa" possible. This rapid colonization of the continent by the powers of Europe was justified by racist ideology, and had the search for raw materials at its heart.

Industrialization and the need for raw materials and new markets were important drivers of imperialism. Notions of cultural and racial superiority were also used to justify it; many 19th century Europeans believed they were duty-bound to bring civilization to the nonwhite world. As they increased their power and productivity at home and abroad, European perceptions of the world begun to change. Racist thinking came to be expressed in scientific terms and the Darwinian concept of the survival of the fittest was applied to society. Europeans argued it was natural for them to displace those they considered "inferior" or "backward" races.

BATTLEFIELD BREAKTHROUGHS

Industrialization also provided the means for colonization: technological innovations were key. The steamboat and quinine, which helped prevent malaria, allowed European traders unprecedented access to the interior of sub-Saharan Africa. This opened up a treasure trove of raw materials, but trading with local economies led to crises, which prompted European powers to annex territory. International rivalries became a factor in the land-grab that followed and, in the ensuing scramble, the borders of the continent were drawn up on maps in European boardrooms. Strong armies were a crucial factor in European empire-building, and the development of new military technologies was a consequence of industrial innovation.

The machine gun, developed by Richard Gatling and perfected by Hiram Maxim, showed that in modern warfare, military technology was paramount. British soldiers used the Maxim gun to slaughter over 10,0000 Sudanese Mahdists in the 1898 Battle of Omdurman, in which the British suffered less than 50 casualties of their own. The machine gun also provided a reminder that the people of Africa did not simply acquiesce to imperial rule. When Ethiopia successfully repelled Italian attempts to colonize it in 1896, it was the first time a European power had been defeated in Africa and a wounding blow to the notions of racial superiority of Europeans.

▲ Colonial control
Askari soldiers were African troops hired and trained by European powers. Native troops were crucial in keeping colonies under control. In Africa there were often up to 200 Askari soldiers to every seven European officers.

◀ Winning weaponry
Ethiopian Emperor Menilek II decimated Italian troops in 1896 using modern guns he had bought in Europe.

▶ The 20th century imperial world
In 1800, European powers and their colonies covered one third of the globe; by the dawn of the 20th century, they had carved up much of the world. The British Empire was by far the largest empire in terms of land and population.

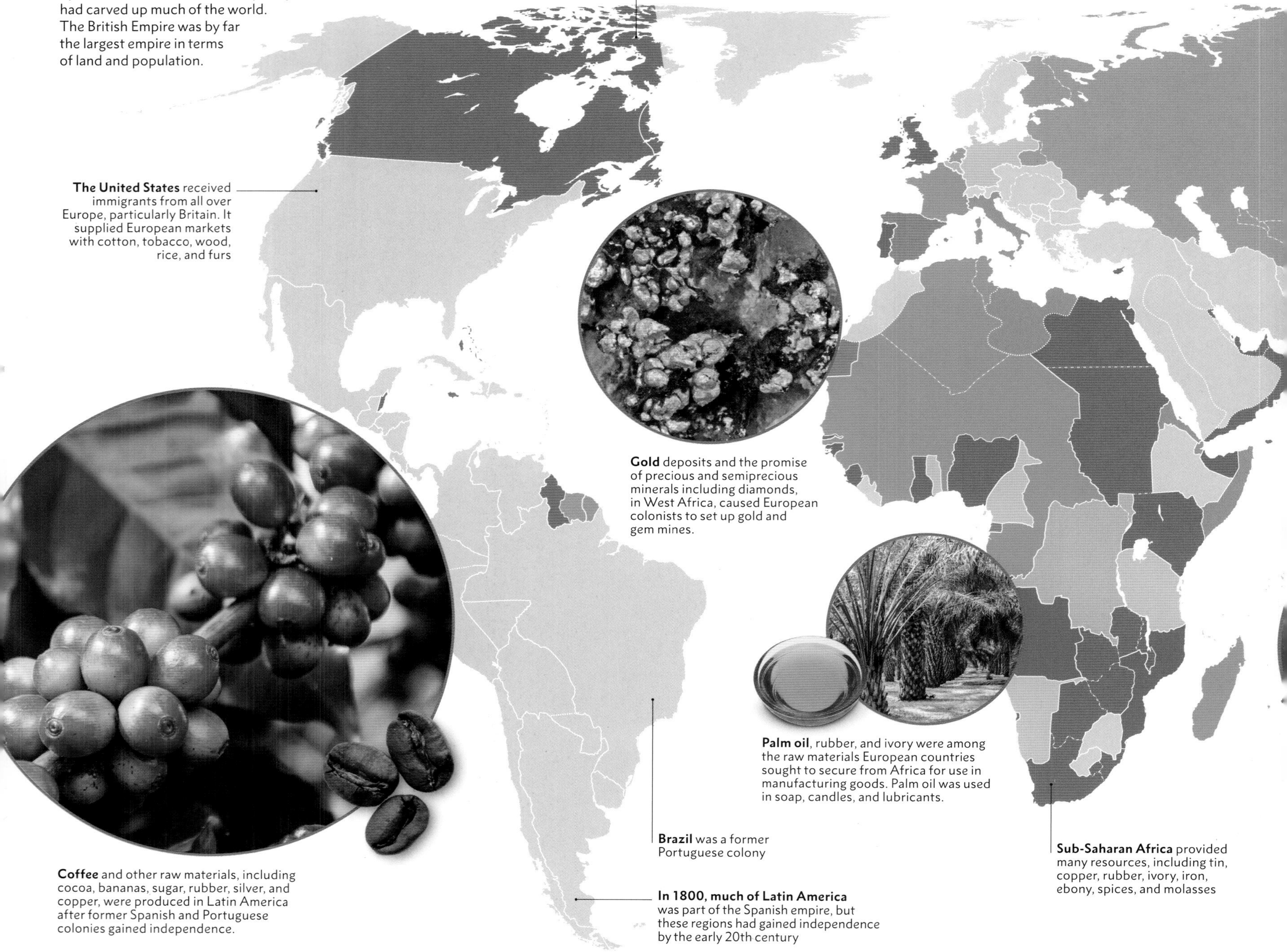
British settlers immigrated to Canada during the 19th century
The United States received immigrants from all over Europe, particularly Britain. It supplied European markets with cotton, tobacco, wood, rice, and furs
Gold deposits and the promise of precious and semiprecious minerals including diamonds, in West Africa, caused European colonists to set up gold and gem mines.
Palm oil, rubber, and ivory were among the raw materials European countries sought to secure from Africa for use in manufacturing goods. Palm oil was used in soap, candles, and lubricants.
Coffee and other raw materials, including cocoa, bananas, sugar, rubber, silver, and copper, were produced in Latin America after former Spanish and Portuguese colonies gained independence.
Brazil was a former Portuguese colony
In 1800, much of Latin America was part of the Spanish empire, but these regions had gained independence by the early 20th century
Sub-Saharan Africa provided many resources, including tin, copper, rubber, ivory, iron, ebony, spices, and molasses

BRITAIN
Duration: 1603–1949
Britain began its process of overseas control through trading posts, which led to colonial expansion across the world and the largest empire in world history.

RUSSIA
Duration: 1721–1917
At its peak in 1866, Russia had the second-largest empire in world history, with territories in its control extending from eastern Europe right across Asia.

BELGIUM
Duration: 1885–1962
Belgium gained independence from the Netherlands in 1830. The Congo was its largest colony and was over 75 times as big as Belgium itself.

GERMANY
Duration: 1871–1918
Germany used its new navy, built to compete with Britain, to colonize parts of West Africa and the South Pacific during the late 19th century.

FRANCE
Duration: 1870–1946
Bruised by defeat in the Franco-Prussian war of 1870, France acquired colonial possessions in Africa, the Pacific, and Southeast Asia from 1871.

1939 | WORLD WAR II — **1950s** | ANTHROPOCENE EPOCH BEGINS — **1970** | DIGITAL AGE — **1973** | OIL CRISIS — **1989** | WORLD WIDE WEB INVENTED — **2001** | 9/11 — **2008** | GLOBAL FINANCIAL CRISIS

COLONIAL EMPIRES GROW

Raw materials for industry, land for settlers, and markets for surplus goods were all factors that drove the imperial expansion of the 19th century, as European countries began to dominate the world.

Competition between imperial powers was fierce and colonies became a symbol of prestige. Large areas of arid, sparsely populated land were often annexed simply to prevent a rival from doing so. As political rivalries and mistrust grew in Europe, colonial wealth was used to control the empires and build up arms.

Once colonies were established, the mother country had to work out how to keep control of its new territory. Often this took the form of indirect control—collaboration with indigenous leaders in Asia and Africa was a vital part of European rule. Imperial military intervention only occurred in unstable regions or places with no pre-existing central control. However, people living in the Americas, Africa, India, and Southeast Asia often experienced racial prejudice, political oppression, and violence at the hands of imperial powers. In the Belgian Congo, the families of workers were held hostage and raped and murdered if the rubber quota was too low. Some indigenous peoples, including the New Zealand Maori and Australian Aborigines, were killed, displaced, or fell victim to European diseases.

From the early 20th century, colonized countries began to gain independence. This process picked up momentum after World War II, when European nations no longer had the wealth, means, or inclination to control faraway territories. The newly independent nations inherited none of the wealth of their past rulers, and were left to create their own institutions. Some have been highly successful; others were plagued by corruption and poverty.

Sugar from Indian plantations became an important export for the British Empire. Once a luxury item, as sugar became available to ordinary people in Europe, demand for it increased.

Nutmeg and cloves were among the spices produced in Indonesia for the Dutch empire, along with sugar and coffee.

Cotton produced in India was shipped to Britain where it was used to make textiles. Britain exported its cloth back to India, undercutting the prices of locally produced textiles.

British settlers immigrated to Australia during the 19th century, easing overcrowding and social unrest at home

EUROPEAN POWERS CONTROLLED AROUND **85 PERCENT** OF THE **WORLD'S LAND** BY **1914**

ITALY
Duration: 1861–1947

Italy colonized Eritrea, Libya, and part of Somalia. The empire ended in 1947 as Italy was forced to abandon its colonies in the aftermath of World War II.

PORTUGAL
Duration: 1415–2002

The first global empire, with territories across several continents, Portugal's was the longest-lasting European colonial empire, spanning almost six centuries.

NETHERLANDS
Duration: 1543–1975

Building up indirect colonial control via the Dutch East and West India Companies before 1800, the Dutch empire reached its height during the 19th century.

JAPAN
Duration: 1868–1945

Japan demonstrated a growing military strength by defeating Russia in the 1904 Russo-Japanese war and winning Korea in the process.

SPAIN
Duration: 1402–1975

Spain gained control of large parts of Latin America by the 18th century, but by the 20th century had lost almost all of its territories.

Factory life
The working class labored on the factory floor, supervised by their middle-class bosses, and surrounded by new machines they were often forced to clean during their lunch break.

SOCIETY **TRANSFORMS**

Industrialization changed every aspect of life for working people. Dangerous and unregulated working conditions existed in factories and workers often lived in overcrowded slum towns, until widespread government reforms improved the plight of this new working class.

As factories replaced farms and fields, the men, women, and children of the peasantry seeking employment were exposed to an unprecedented level of social and technological change.

THE PLIGHT OF THE POOR

The middle classes were the real winners of industrialization; in Britain, the 1832 Reform Bill even gave middle-class men the vote. The laboring classes suffered most. Workers toiled for at least 13 hours a day in factories, and hearing loss, lung disease, and severe injury were common. There was no legal protection for workers: the middle-class factory overseers and owners were king. It was these brutal economic inequalities that stoked the wave of revolutionary activity that broke out across Europe in 1848, mobilized in part by German philosopher Friedrich Engels, who described the misery of factory workers in *The Condition of the Working Poor.*

THE NEW CITIES

Workers lived in slum towns that grew up around factories. The rise in urbanization was everywhere: by 1850, 50 percent of England's population were living in cities; Germany reached this level by 1900, America by 1920, and Japan by 1930. Industrial cities in every country suffered similar problems: overcrowding, pollution, a lack of running water, no waste disposal, and unhealthy housing. These conditions all encouraged the spread of disease, and waves of cholera broke out across India, Europe, and North America. An 1832 French study into cholera showed the link between slums, poverty, and poor health, and English physician John Snow demonstrated that cholera was spread via contaminated drinking water in 1849.

Armed with this knowledge, governments began to take action, introducing sewage systems, running water, and trash collection into cities. Across Europe and North America, other social and political reforms were enacted. Labor laws provided protection for workers, with improved safety regulations, and education became compulsory for children.

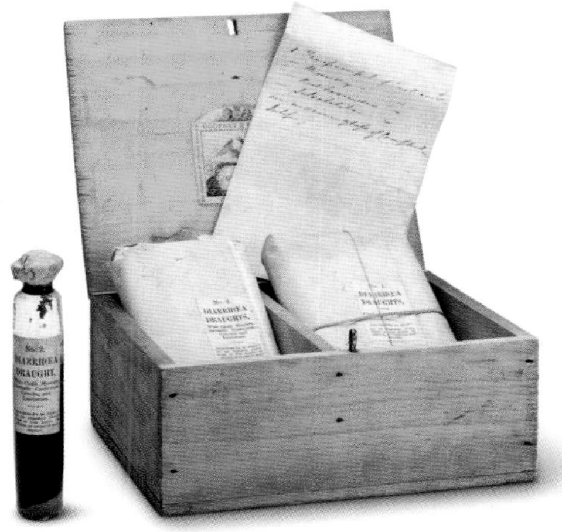

◀ Cholera medicine
By the end of the 19th century, cholera epidemics no longer appeared in Europe and North America. Standards of living rose, sanitation practices improved, and permanent boards of health were established.

> THE WATER... **IN FRONT OF THE HOUSES** IS COVERED WITH A SCUM... ALONG THE BANKS ARE HEAPS OF **INDESCRIBABLE FILTH**...THE AIR HAS... THE **SMELL OF A GRAVEYARD**.

Henry Mayhew, journalist and campaigner for better housing, 1812–1887

EDUCATION EXPANDS

Education plays an essential part in collective learning and innovation. Realizing its importance, many governments introduced widespread reforms to make education compulsory after the mid-19th century. By 2000, 80 percent of the world population could read and write.

The importance of literacy has a long history. European literacy levels had risen steadily from the 16th century, especially in France, Germany, and Britain. A society that valued knowledge and ideas fitted with the Age of Enlightenment beliefs that drove industrialization. In Britain, hundreds of schools were opened in the early 18th century to cope with the rising population. However, there was great disparity between people with access to education and those without. Education had to be paid for during the 18th century and therefore it was not available to the working class. Neither was education considered important for women—working-class women were expected to work from childhood, while middle-class women were only schooled until they married.

EDUCATED NATIONS

In the 19th century, ideas about education began to change. This arose partly from Enlightenment ideals about the value of reason, knowledge, and the free exchange of ideas. Individual reformers worked to rouse popular support for new government intervention in slavery, public health, and education. Intervention was needed to appease citizens after the 1848 revolutions; the middle classes demanded reform and the working classes seemed poised to revolt. Governments realized an educated nation would keep the military strong, encourage patriotism, and reduce the desire for rebellion. From 1870, compulsory state-run school education spread across western Europe and into the northeastern states of America. Other countries outside Europe set up education systems after 1900, including China, Egypt, and Japan. These were partly created to encourage patriotism and also to imitate the institutions that had helped make the western empires so powerful.

Improved access to education has allowed world literacy to rise steadily over the last 150 years. More people than ever before can read and write—and contribute to growing networks of exchange and collective learning. However, even today, access to education is not equal; illiteracy is highest in the some of the poorest parts of the world, and also among women. In 2011, three-quarters of all illiterate adults lived in southern Asia, the Middle East, and sub-Saharan Africa. Furthermore, two-thirds of the world's 774 million illiterate adults were women.

▲ Textbook learning
The school system in America was largely private until reforms in the 1840s began to introduce public schools and standardized textbooks.

THE INFORMATION AGE

Education is an important tool in disseminating information and knowledge at an individual level. Throughout human history, as collective learning has increased, networks of exchange have expanded and their power has grown at faster rates, enabling information to accumulate more and more rapidly. Today, we are living in what could be described as the Information Age. The Digital Revolution has led to a shift from an economy driven by traditional industry to one based on computerized information. In this information society and knowledge-based economy, it is flows of information that

MORE THAN **83 PERCENT** OF THE **GLOBAL POPULATION** WAS **LITERATE** IN 2016

▶ Improving child welfare
In Britain, industrialization led to young working-class children toiling in factories and mines, until government reforms prohibited this. The Education Act of 1880 introduced compulsory schooling up to the age of ten.

◀ **Primary education**
In 1994, the government of Malawi, Africa, introduced free primary school education, but dropout rates remain high, especially for girls. This is often the case in many poor countries in sub-Saharan Africa, where children work to supplement the family income.

drive profit making. Globalization saw the world zones become connected, and now the control and movement of information and wealth has started to blur national boundaries even more. Of the top 100 economies of the world ranked by Gross Domestic Product (GDP) in 2009, 60 are countries and the rest are companies, many of them multinational oil and gas companies such as Sinopec and Shell, and technology and communications companies such as Apple and Samsung. Never has information been so important.

In recent years, the growth of the software and biotechnology industries has placed a new emphasis on the need for highly skilled labor. Industrialization created a system like a pyramid, with large amounts of unskilled labor at the bottom and a small number of capitalist business leaders and creative classes at the top. Education may be the key to moving toward societies in which the pyramid is inverted: if more people have access to a good level of education, they can participate in high-value jobs at the top of the pyramid, while automation reduces the need for large numbers of people to perform unskilled tasks.

THE PURSUIT OF INNOVATION

Education, as a form of collective learning, is crucial for innovation. During the 20th century, for many industrialized societies one of the main drivers of innovation was the pursuit of innovation itself—often, as in the past, with the support of governments, business, and educational institutions. During the 17th century, when the first scientific societies were founded in Europe, the British government offered incentives for innovation, and in the first century of industrialization, it profited from major scientific and engineering breakthroughs. In the 19th century, governments and businesses realized science was a crucial source of innovation, wealth, and power, and began to take an active role in promoting and organizing scientific research. By the 20th century, innovation in science and technology had proved to be fundamental components of military, political, and economic power for industrialized nations.

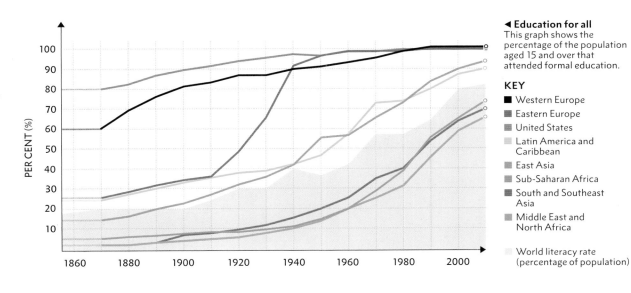

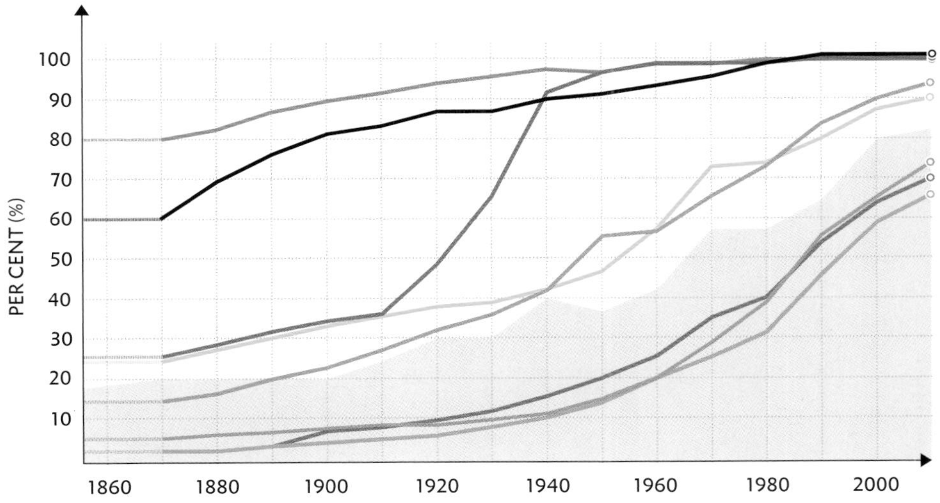

◀ **Education for all**
This graph shows the percentage of the population aged 15 and over that attended formal education.

KEY
- Western Europe
- Eastern Europe
- United States
- Latin America and Caribbean
- East Asia
- Sub-Saharan Africa
- South and Southeast Asia
- Middle East and North Africa
- World literacy rate (percentage of population)

MEDICAL **ADVANCES**

From the late 18th century, there was a great acceleration in medical knowledge in industrializing countries, as scientific research, innovation, and disease prevention allowed people to live longer and healthier lives.

As expanding trade networks and urbanization brought people into closer contact than ever before, diseases spread. Edward Jenner's breakthrough smallpox vaccination, in 1796, was warily considered a medical miracle. During the 19th century, germ theory and the discovery of bacteria eventually led to safer surgery and an understanding of the importance of sanitation in public areas. New innovation gave physicians practical tools to help them diagnose ailments. Medical innovation and improved knowledge had a positive impact on health, especially for the very young and very old.

The 20th century was marked by an explosion in medical technology, as health systems tried to keep up with the epidemics, famines, and wars of the modern age. Scientific research of the new millennium led to stem cell research, the sequencing of the human genome, and the ability to create new life. The internet made details of these medical breakthroughs widely accessible, as well as providing an ever-growing resource for the sharing of medical knowledge—for both practitioners and patients.

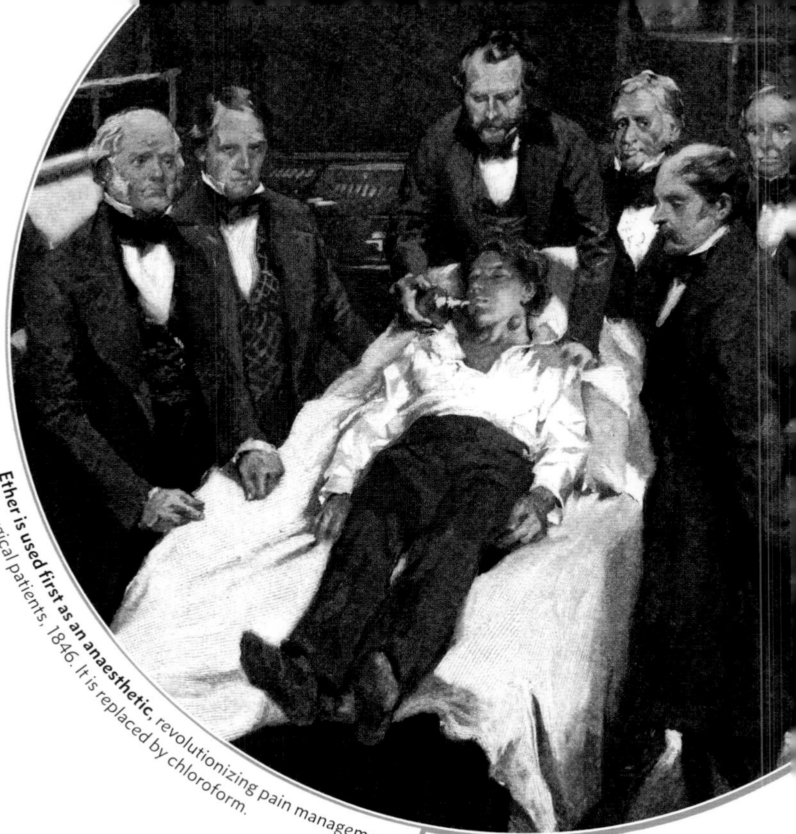

Ether is used first as an anaesthetic, revolutionizing pain management for surgical patients, 1846. It is replaced by chloroform.

Chloroform inhaler

Stethoscope

DRUGS AND ANAESTHESIA

INSTRUMENTS, INNOVATIONS, AND TECHNOLOGY

DISCOVERIES AND BREAKTHROUGHS

GERMS, DISEASES, AND VACCINES

1800

1775

Germ theory, which states that a specific disease is caused by a specific organism, is demonstrated by Louis Pasteur, in the 1850s; Pasteur goes on to produce vaccinations for rabies and anthrax.

Reforms in hygiene and patient care introduced by Florence Nightingale in 1854, during the Crimean War, lead to modern nursing and sanitation in hospitals.

Ignaz Semmelweis proves surgeons with unwashed hands spread infection, 1847.

Aspirin synthesized, 1852, a new medicine that relieves suffering caused by disease.

Stethoscope, invented by Rene Laennec in 1816, enables doctors to listen to the sounds from a patient's chest. It remains a vital diagnostic tool.

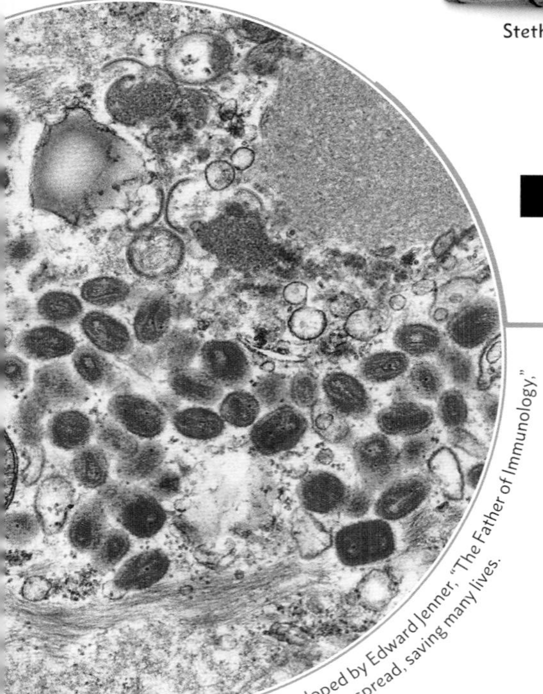

Smallpox vaccination is developed by Edward Jenner, "The Father of Immunology," 1796, and vaccination becomes widespread, saving many lives.

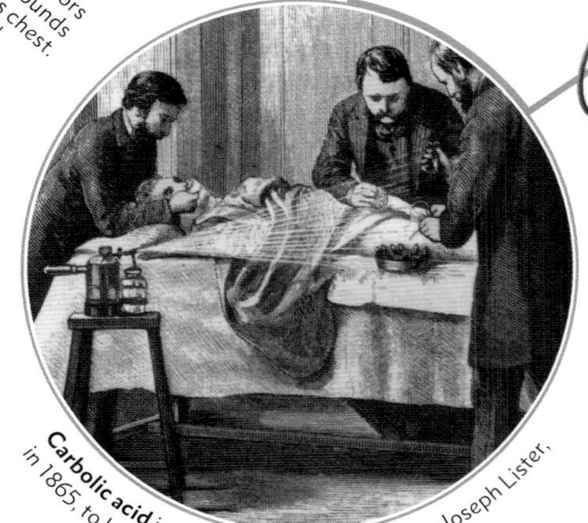

Carbolic acid is first used as an antiseptic by Joseph Lister, in 1865, to kill germs and reduce the risk of infection.

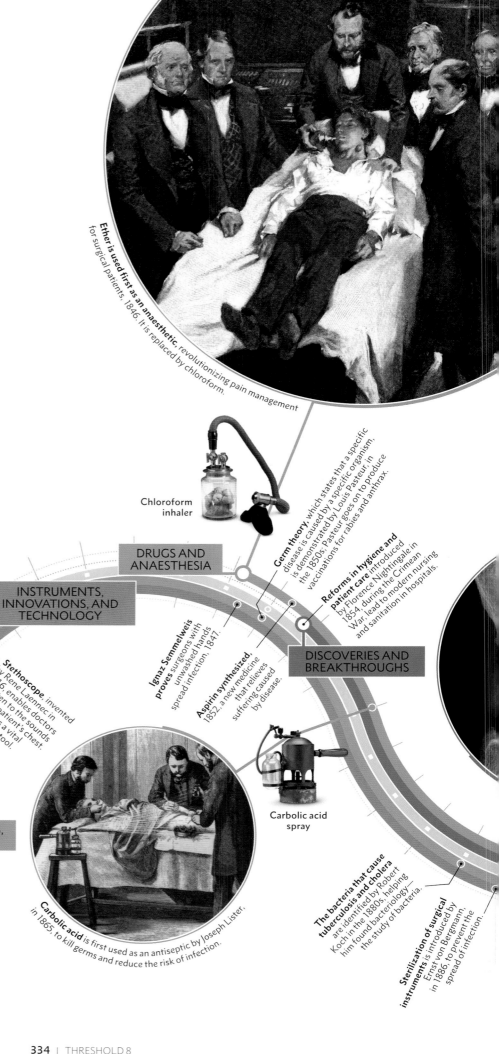

Carbolic acid spray

The bacteria that cause tuberculosis and cholera are identified by Robert Koch in the 1880s, helping him found bacteriology—the study of bacteria.

Sterilization of surgical instruments is introduced by Ernst von Bergmann, in 1886, to prevent the spread of infection.

Ronald Ross discovers that mosquitoes carry malaria and publishes his findings in 1897. This allows him to develop "vector control" to help eradicate malaria-carrying mosquitoes.

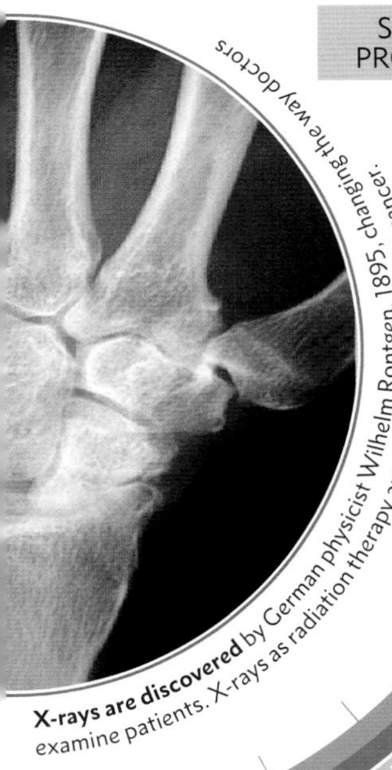

X-rays are discovered by German physicist Wilhelm Rontgen, 1895, changing the way doctors examine patients. X-rays as radiation therapy are eventually used to treat cancer.

1900

The "ABO" blood groups are described by Karl Landsteiner, 1900, leading to safe blood transfusions.

German Paul Ehrlich releases drug treatment for syphilis in 1910—a major breakthrough in chemotherapeutic medicine.

The antibiotic penicillin, accidentally discovered by Alexander Fleming in 1928, leads to the mass production of antibiotics, improving the health of millions.

Electron microscope

Electron microscope is invented in 1931, allowing doctors to see bacteria and viruses for the first time.

SURGICAL PROCEDURES

DNA is described by James Watson and Francis Crick, 1953, leading to a greater ability to diagnose diseases early on.

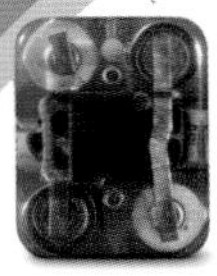

Cardiac pacemaker

Cardiac pacemaker produced, 1958. It is an important innovation that uses electrical pulses to regulate the heart beat.

Birth control pill for women released to the American market, 1960, leading to greater female control over family planning and the "sexual revolution."

First successful heart transplant carried out by South African physician Christian Barnard, 1967. Today over 3,500 heart transplants are performed each year.

Computerized axial tomography (CAT) scanner developed by Godfrey Hounsfield, 1970. It allows doctors to locate abnormalities such as tumors in patients.

First "test tube" baby is born, 1978, after the in vitro fertilization of an egg outside the body, enabling people who could not conceive to have children.

Commercial magnetic resonance imaging (MRI) scanners, which use magnets and radio waves to form an image of a patient's body and organs, aid diagnoses of internal abnormalities, 1981.

Highly active antiretroviral therapy (HAART) to control HIV infection introduced, 1996. By 2010 it has led to a marked drop in the AIDS-related mortality rate in many Western countries.

Dolly the Sheep, the first mammal to be cloned from an adult cell, 1996, brings scientists closer to the theoretical capability of cloning human life.

2000

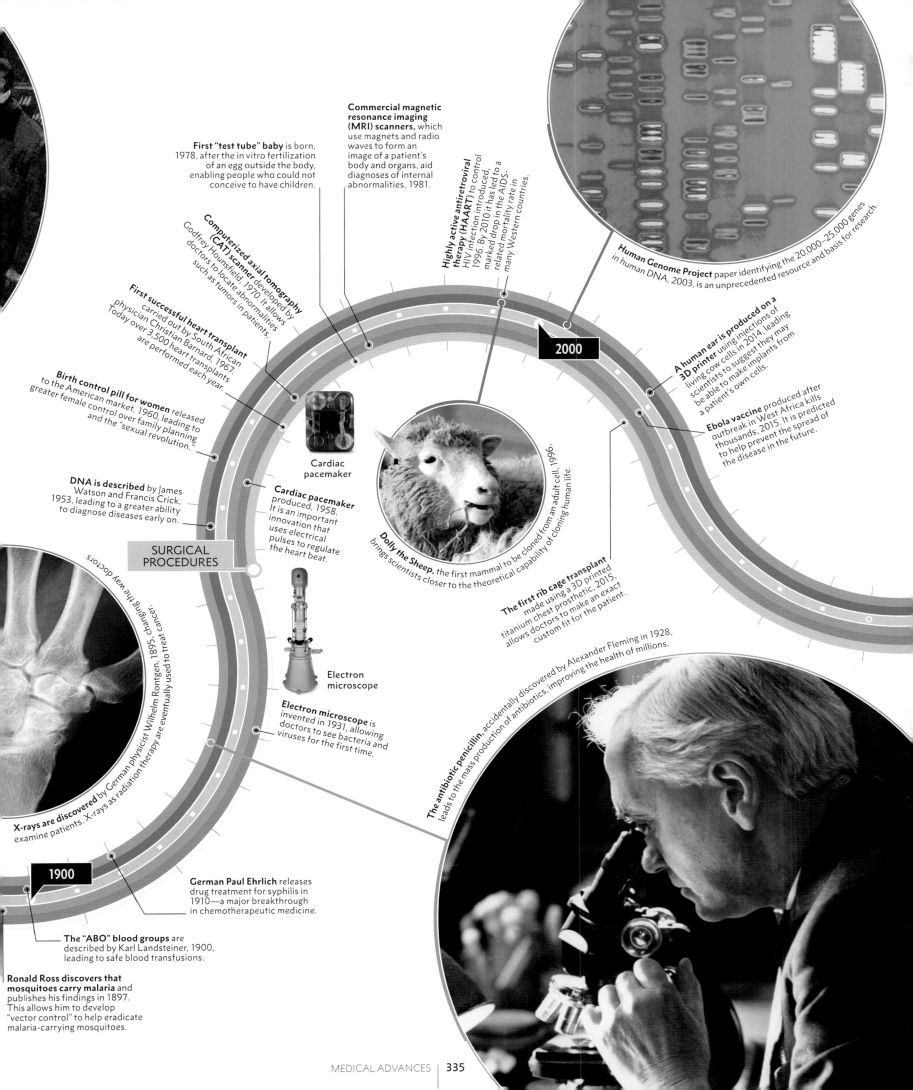

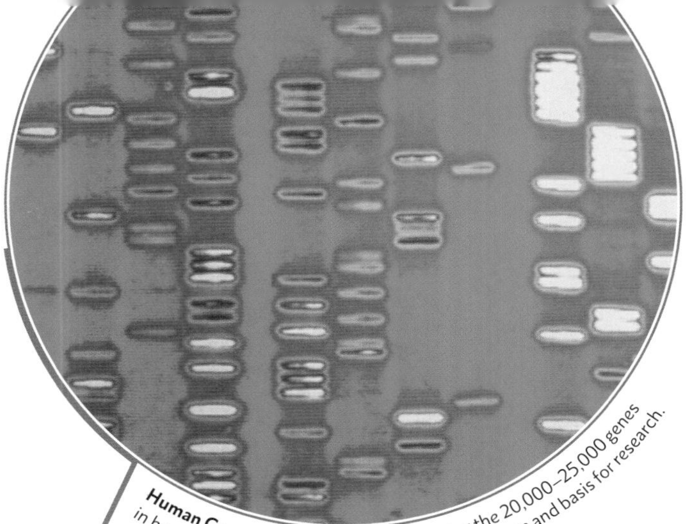

Human Genome Project paper identifying the 20,000–25,000 genes in human DNA, 2003, is an unprecedented resource and basis for research.

A human ear is produced on a 3D printer using injections of living cow cells in 2014, leading scientists to suggest they may be able to make implants from a patient's own cells.

Ebola vaccine produced after outbreak in West Africa kills thousands, 2015. It is predicted to help prevent the spread of the disease in the future.

The first rib cage transplant made using a 3D printed titanium chest prosthetic, 2015, allows doctors to make an exact custom fit for the patient.

1750 | THE INDUSTRIAL REVOLUTION BEGINS
1789 | DECLARATION OF THE RIGHTS OF MAN
1830 | STEAM AGE
1869 | DNA ISOLATED
1880 | ELECTRICAL AGE
1914 | WORLD WAR I
1930S | GREAT DEPRESSION

THE AMERICAS

PACIFIC ISLAND SOCIETIES

FOUR WORLD ZONES
By the start of industrialization in the 18th century, European explorers had already connected the four world zones.

AFRO-EURASIA

AUSTRALASIA

INDUSTRIALIZED ZONE
UNINDUSTRIALIZED ZONE

1900
The four world zones had become two zones: an industrialized zone, made rich through mass-production, and a poorer unindustrialized zone that was exploited for raw materials, labor, and land.

Transportation
Falling transportation and communications costs underpin international trade from 1820 onward. The price of inland transportation drops by 90 percent between 1800 and 1910; transatlantic transportation costs fall by 60 percent between 1870 and 1900.

ROAD TO **GLOBALIZATION**

The merging of the world into two zones resulted in a worldwide exchange network of trade, capital, migration, culture, and knowledge in a process known as globalization.

Globalization is not a modern concept: global exchange networks expanded greatly after the Age of Discovery, which spanned the 15th–18th centuries and opened the new world to the old. During this period, money, people, crops, ideas, and diseases traveled between the two worlds, mostly to the benefit of the countries of Western Europe. This model of globalization sped up during 19th-century imperialism, and, by the end of the century, large colonial empires connected specialized regions of industry and agriculture within a new world economy focused on accumulating capital.

Alongside industrial technologies, including the telegraph and railways, the new organizational structures of the modern state were introduced to the unindustrialized world, including legal systems and state bureaucracy. As countries in the unindustrialized zone developed distinctive specializations—like the growing and exporting of tea—they did so under a system that came with its own rules, regulations, and language. After decolonization in the 20th century, those colonies able to grow their own economies did so with the guidebook left behind by empire.

By the 21st century, innovation in communications technology had become as important as transportation in creating modern globalization: cheap and efficient containerization contributed to the rise of China as an economic superpower, and fiber optics and broadband helped establish India as a global services hub. The innovations continue today, as ever more advanced smartphones connect the world's population, and a new global culture begins to emerge.

Trade agreements
Decolonized countries begin to make trade agreements based on mutual advantage. Many of these countries adopt a free-trade model, which leads to a new transnational economic dynamic.

LATE 20TH/EARLY 21ST CENTURY
Many unindustrialized nations become cogs in a global manufacturing machine, assembling products from raw materials shipped in from around the world.

> “THE **COUNTRY** THAT IS **MORE DEVELOPED INDUSTRIALLY** ONLY **SHOWS** TO THE **LESS-DEVELOPED** THE IMAGE OF ITS OWN **FUTURE**.”

Karl Marx, German philosopher, economist, and sociologist, 1818–83

1939 | WORLD WAR II

1950s | ANTHROPOCENE EPOCH BEGINS

1970 | DIGITAL AGE

2008 | GLOBAL FINANCIAL CRISIS

Migration
Innovative new forms of transportation help to increase the migration and movement of peoples around the world. The Irish potato famine and overcrowding in Britain lead to mass immigration to the colonies, along with imperial bureaucrats and migrant workers.

Resources
In the 19th century, industrial powers introduce free-market capitalism to the world. Empires seize resources, subordinate labor, and turn the globe into a vast agricultural resource for western Europe.

Cultural exchange
The movement of people creates opportunities for cultural exchange in all areas of life, including social customs; academic and business culture; religious and political ideologies; literature, music, and art; clothes and beauty; eating customs and food.

Financial institutions
The rise of powerful financial institutions, such as the International Monetary Fund, leads to investment deals for industrializing countries, which come with obligations attached. This creates a more integrated global financial system.

New players
The fall of the Soviet Union and the opening of China bring new economic players to the global capital market, resulting in a surge of international transactions and investment in postcommunist economies.

Foreign investment
Trade agreements encourage multinational corporations from industrialized nations to invest directly in unindustrialized economies. This prompts increased privatization and greater foreign ownership of assets in unindustrialized countries.

AFTER WORLD WAR II

Modern globalization begins as capitalism and liberalizing of trade create new world economies increasingly controlled by multinational corporations and powerful financial institutions.

Movement
The removal of trade barriers and even cheaper transportation expands human migration for work purposes. This leads to a larger cultural exchange and the rising economy of remittances—money sent from a foreign worker back to their home country.

Cultural homogeneity
The rise of a global services economy, improvements in communications technology, and the spread of multinational corporations all help promote cultural homogeneity, where brands, music, television, and food are found and recognized all over the world.

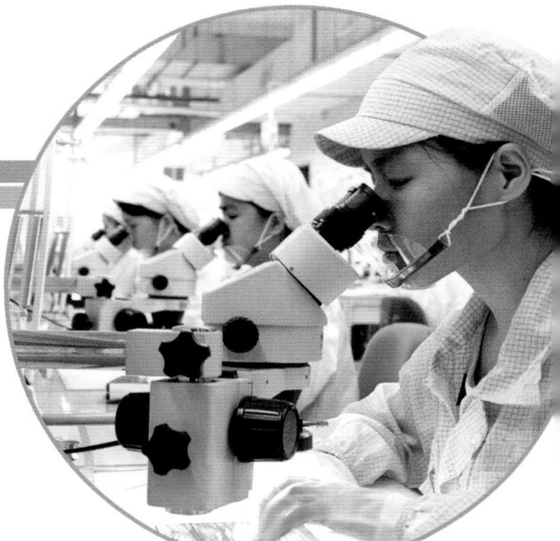

Industrial development
Global capitalism enables the industrialization of many countries in the unindustrialized zone and the creation of wealth through the manufacture of cheap consumables for the market. This leads to more employment opportunities and a reduction of people living in poverty.

▼ Engine of change
A gas-fueled combustion engine was built by Karl Benz to power his "horseless carriage" in 1885. The following year, he created the Benz "Patent-Motorwagen"—the world's first automobile. It shared many features with cars today.

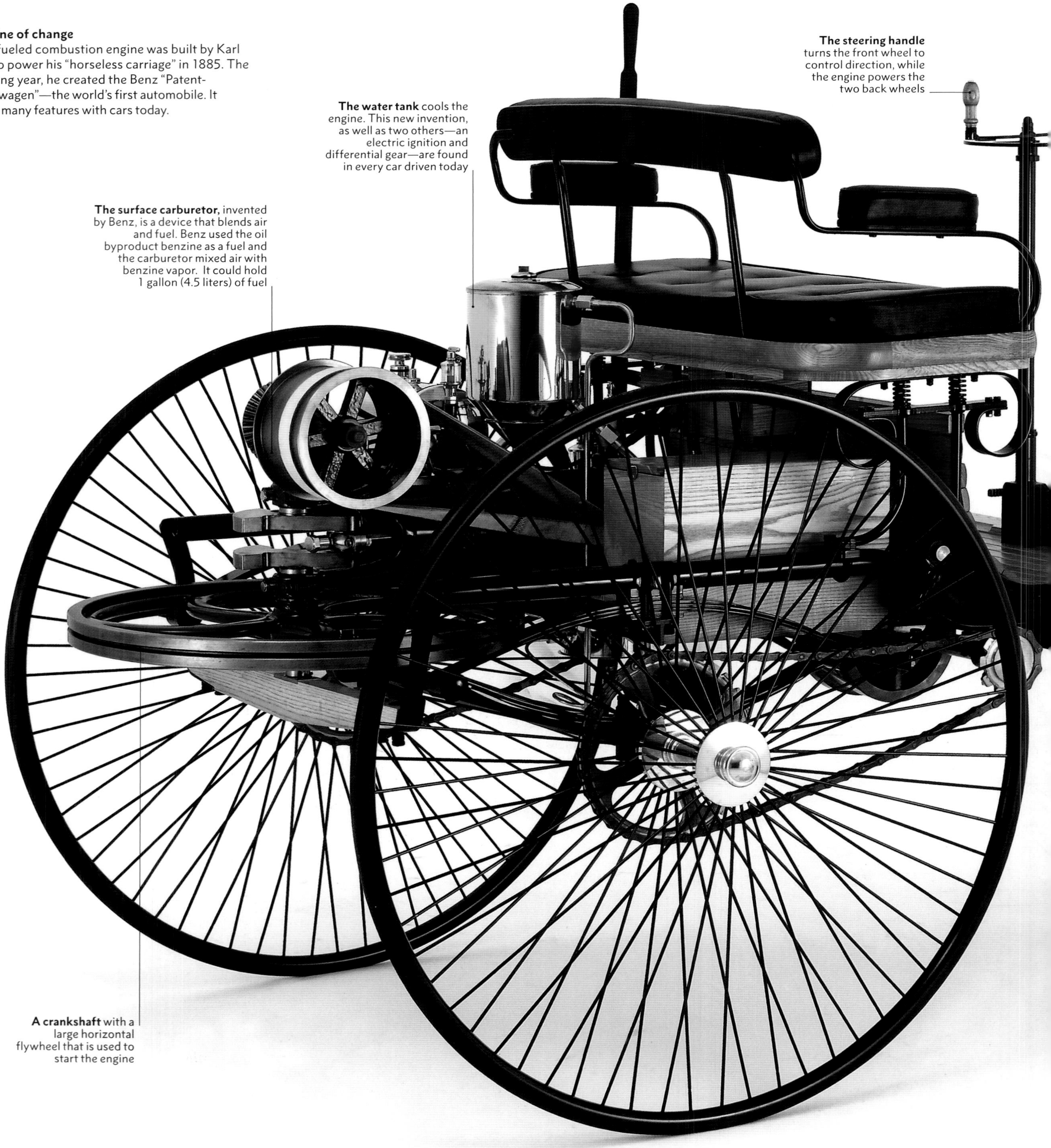

The steering handle turns the front wheel to control direction, while the engine powers the two back wheels

The water tank cools the engine. This new invention, as well as two others—an electric ignition and differential gear—are found in every car driven today

The surface carburetor, invented by Benz, is a device that blends air and fuel. Benz used the oil byproduct benzine as a fuel and the carburetor mixed air with benzine vapor. It could hold 1 gallon (4.5 liters) of fuel

A crankshaft with a large horizontal flywheel that is used to start the engine

ENGINES **SHRINK THE WORLD**

Transport played a key part in the spread of industrialization. In the last two centuries, railways, steamships, and airplanes—and innovations in communication—have vastly increased the rate and speed at which people exchange goods, ideas, information, and technology.

By the late 19th century, railway tracks crisscrossed Europe and America, greatly accelerating the exchange of goods, people, and ideas, as well as making travel more widely available. Rail networks lowered the cost of moving goods between the manufacturer, retailer, and buyer, which, in turn, reduced the cost of consumer goods. The ability to move raw materials and manufactured goods across land and sea at relatively rapid speeds and low costs was as significant to the success of early industrial economies as it is today.

NEW TRANSPORT CONNECTIONS

Just as coal fueled 19th-century railways, the transport revolution of the early 20th century would not have been possible without the wide-scale availability of fossil fuels. Innovative new uses of oil and gas included the invention of the internal combustion engine—which burned oil—and led to the development of automobiles and jet planes. In 1913, entrepreneur Henry Ford devised an assembly line to mass-produce the first affordable motorcar. This was the start of consumer capitalism, as workers became the target market for goods they were making, which would once have seemed like luxuries.

Ford's vision to put a car in every driveway transformed modern Western society as governments built roads and traffic systems to accommodate cars. In the 1950s, oil-fueled cars, buses, and trucks became vital to the transportation of goods and people. Commercial air travel took off after World War II, as wartime aviation experts turned their attentions to creating a peacetime aviation industry. This sped up the transport of people and mail. People began to travel more for a variety of reasons, including business and leisure, which increased networks of exchange. Innovations in transportation drove growth and in turn led to more innovation.

In the early 1960s, humans invented rockets that could carry them into space. The Soviet Union was first to launch a human into space, and in 1969 the United States landed a man on the moon. As the world became increasingly accessible—with more goods, people, and ideas being exchanged than ever before—it also began to appear much smaller.

▲ **Cars for the masses**
The introduction of fast and efficient assembly lines in factories reduced production costs and enabled goods, such as the Ford Model T, to be sold at affordable prices.

> IF I HAD ASKED PEOPLE **WHAT THEY WANTED**, THEY WOULD HAVE SAID **'FASTER HORSES'**.
>
> **Henry Ford**, American industrialist and founder of the Ford Motor Company, 1863–1947

NEWS **TRAVELS FASTER**

The desire to communicate and connect with those around us is an important part of the human story. The technology we use to do this has changed vastly since our ancestors began painting their stories on cave walls, largely due to innovations that date from the 18th century.

The cornerstone of any form of communication is its ability to bring people closer together, and in the 21st century the World Wide Web has completely reshaped the way billions of people create and share information on every topic imaginable. And where early forms of telecommunication, such as the telephone, allowed for one-to-one correspondence, today's online world is geared toward wide—often global—dissemination. This can include anything from a concise political comment on Twitter to a lengthy news article updated in real time.

Perhaps more than anything, it is speed that now defines communication. News that once took days or even weeks to deliver by letter on a ship or train can now be transmitted via email or a Facebook post in a matter of seconds. Alongside this rapid exchange of data comes the sheer volume of information: with 24-hour television news coverage now ubiquitous, and social media in the hands of billions of smartphone users, global communication networks today are more complex and varied than ever before.

British postal delivery service, made available from 1635, uses horse-mounted postmen to deliver mail, allowing people who are living in different places to relay information at a reasonable speed and with some degree of predictability.

1780

POSTAL SERVICE

Royal Mail stage coaches are protected by guards from 1784, creating a more secure British postal service.

VISUAL SIGNALING

Semaphore is invented by Claude Chappe, 1792. It allows messages to be conveyed by waving flags, and is the first telecommunications system of the industrial age.

1800

TELEGRAPH AND TELEPHONE

Telegraph system developed by Samuel Morse, 1837, sends messages via electrical telegraph lines using Morse code, revolutionizing long-distance communication.

Uniform Penny Post, 1840, allows letters to be sent for one penny, bringing an affordable postal service to people in the United Kingdom and Ireland.

NEWS AND BROADCASTING

Typewriter invented by Charles Thurber, 1843, becomes widely used in offices and business communication.

Foreign post offices are opened in Chinese ports in 1844 following the Opium Wars. This lays the foundation for China's first national mail service.

Transatlantic cable, 1866, enables eight words a minute to be transmitted between America and Europe. Messages had previously taken 10 days by ship.

Signal lamps utilizing a form of Morse code, 1867, allow British naval ships to transmit messages across long distances.

Telephone

Telephone, patented by Alexander Graham Bell in 1876, becomes the most widespread communication system of the modern age.

Phonograph invented by Thomas Edison, 1877, is the first device to both record and play back sound, transforming the music industry.

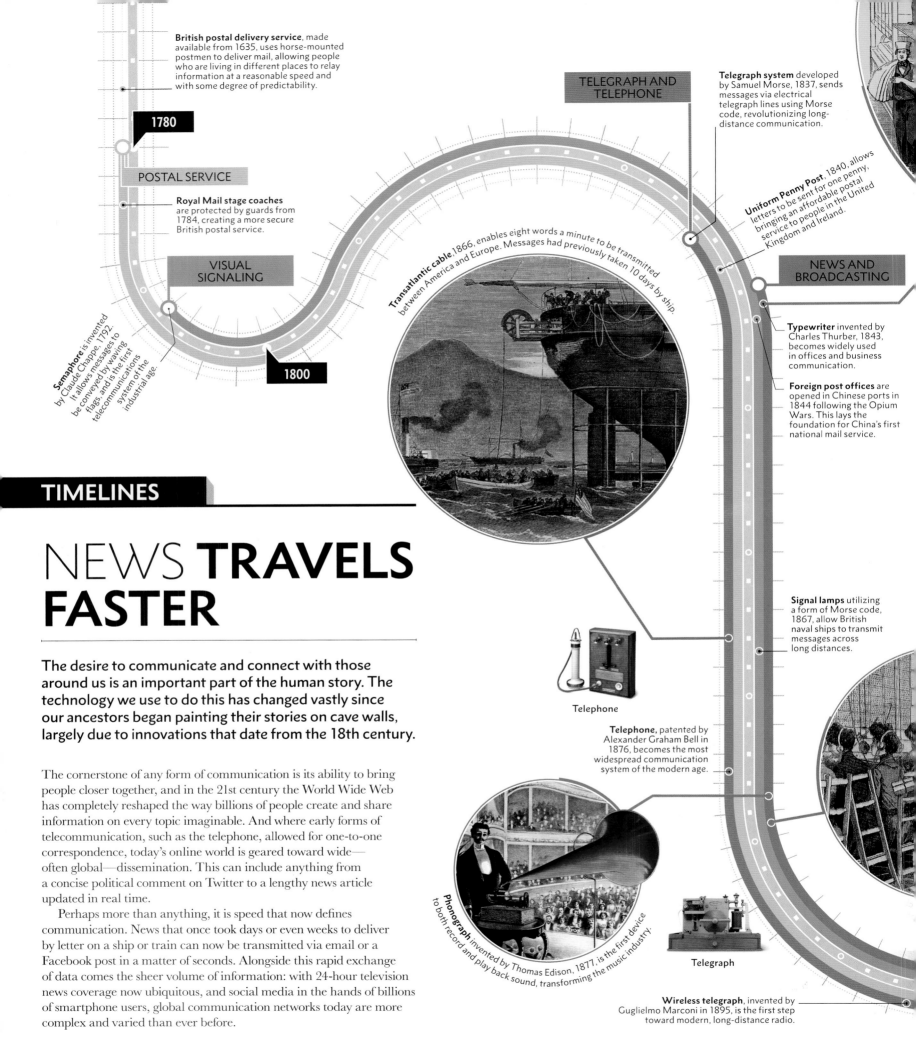

Telegraph

Wireless telegraph, invented by Guglielmo Marconi in 1895, is the first step toward modern, long-distance radio.

Rotary printing press becomes widely available, 1843, and paves the way for industrial-scale printing, especially at newspapers.

Telephone exchange set up in Great Britain in 1879 allows phone calls to be connected between different users.

1900

First radio broadcast of voice and music is transmitted by inventor Reginald Fessenden in the US, 1906.

First transcontinental telephone call made in 1915 between Alexander Graham Bell in New York and Thomas A. Watson in San Francisco.

First true television images are projected in 1926 by inventor John Logie Baird.

Baird television

World's first national broadcasting organization, the British Broadcasting Corporation (BBC), is founded in 1927, to provide a television and radio service.

BBC offers first regularly scheduled television service in 1936, which includes sports, dramas, and cartoons.

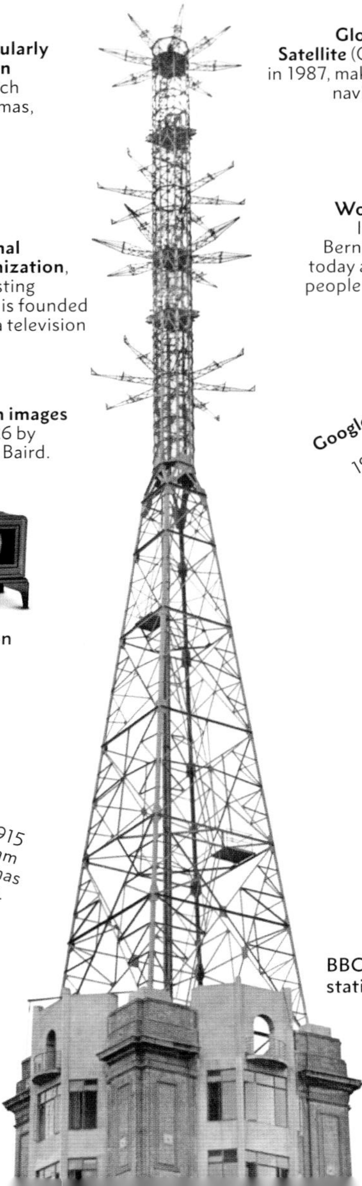

BBC television station, 1936

Radio broadcasts are made in over 15 countries in 1939, from Vatican City to the Soviet Union, leading to a growing worldwide audience of radio listeners.

Pre-cellular mobile telephone service is launched in the US in 1946; unlike mobile phones today, the phones were usually built inside cars and trucks.

Fifteen-minute television newscasts begin in the US in 1948, making news more widely available across the country.

ONLINE WORLD

Telstar 1 satellite beams television, phone calls, and fax messages across the Atlantic Ocean for the first time in 1962, opening up new and faster methods of global communication.

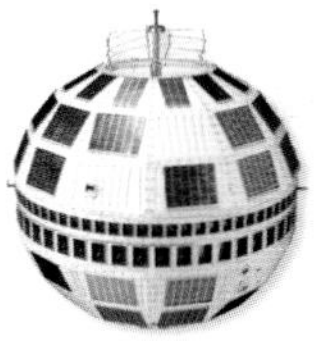

Telstar 1

Live footage of the moon landing in 1969 is broadcast to over 100 million TV viewers around the world.

A message is delivered between computers in 1969 by the Advanced Research Projects Agency Network. This paves the way for the invention of email.

First handheld mobile phone, weighing 4.4 lb (2 kg), is released in 1973, signaling the start of reliable mobile telephony.

Mobile phone

Cable News Network (CNN) begins 24-hour satellite news in 1980, ushering in a new era of around-the-clock news coverage.

Global Positioning Satellite (GPS) is launched in 1987, making satellite-led navigation possible.

The first online transatlantic chat, 1988, occurs between users in Finland and the US, giving birth to the modern chat service.

World Wide Web, launched by Tim Berners-Lee in 1989, today allows billions of people to communicate with each other.

Google, the internet search engine, is launched in 1998, promising a wider data search capability and faster speeds.

Wi-Fi symbol

Wi-Fi, or Wireless internet connectivity, becomes available in 1999, paving the way to remote online access.

2000

Wikipedia, a free, multilingual web-based encyclopedia founded in 2001, allows online users the freedom to edit its content.

Skype—an application that enables users to make free phone and video calls over the internet—is launched, 2003.

Facebook, an online social networking service that lets users communicate and share information with friends globally, opens to the public in 2006.

Apple iPhone, launched in 2007, has a large multi-touch screen as its main means of interaction, not a keypad or stylus.

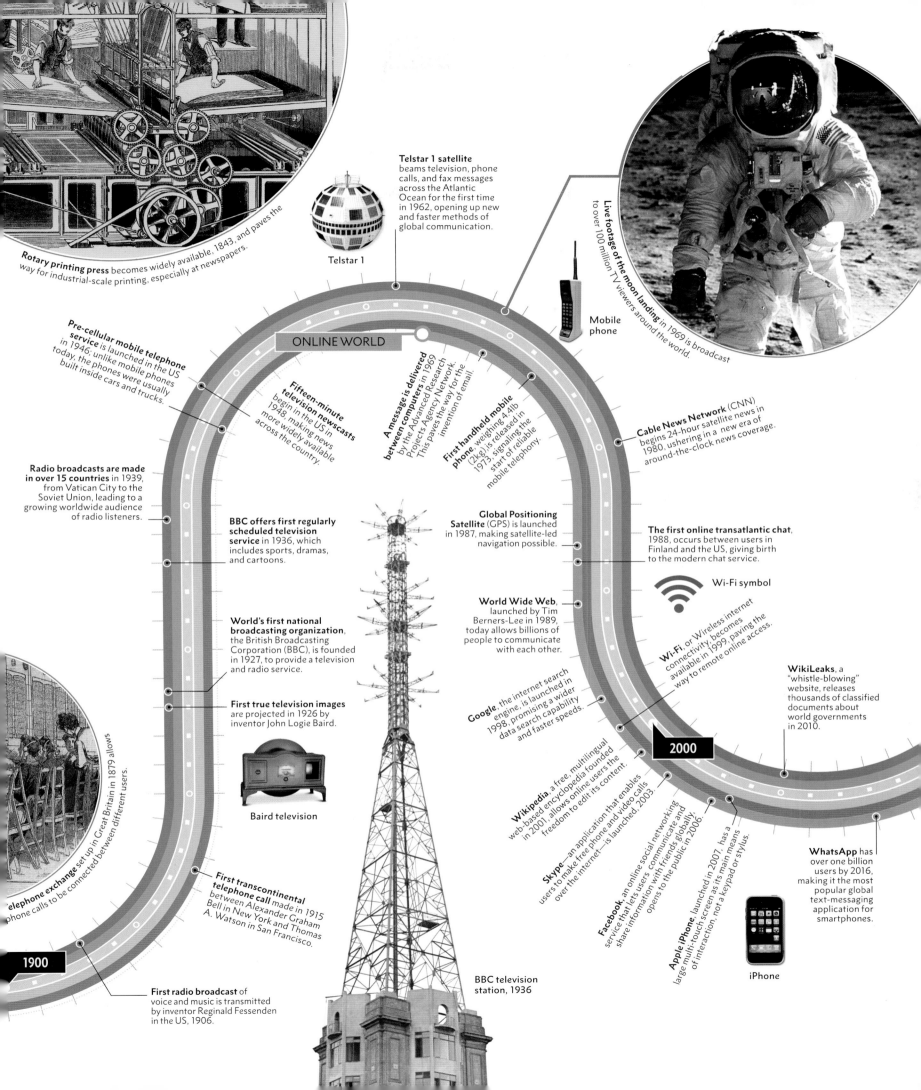

iPhone

WikiLeaks, a "whistle-blowing" website, releases thousands of classified documents about world governments in 2010.

WhatsApp has over one billion users by 2016, making it the most popular global text-messaging application for smartphones.

SOCIAL **NETWORKS EXPAND**

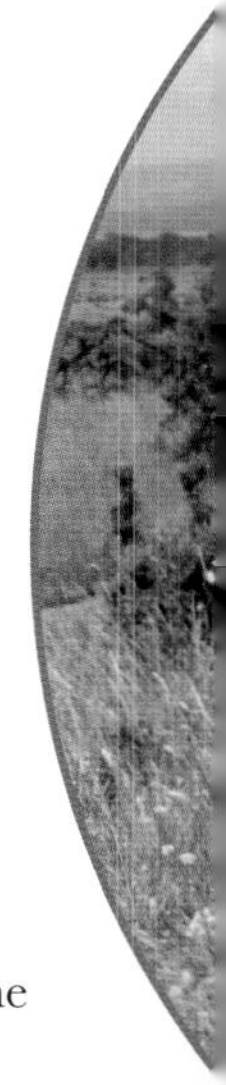

The first telephone, invented in 1876, connected two callers across seas and continents. Today, innovation has led to the creation of the smartphone, which can connect to wireless internet. This technology has resulted in the largest and most complex exchange network ever created.

The late 20th and early 21st centuries saw the arrival of breakthrough digital and communications technologies, each of which plays an important function in connecting people and spreading information and ideas. The internet disseminates news and information; social media connects individuals and enables them to organize; mobile phones make it possible to photograph and record what is happening and share it with a global audience. Social networking has become a global phenomenon: in 2015, of the 3.2 billion online users, 2.1 billion had social media accounts. At the most basic level, people use social media to keep in touch or to share their views with the world, but it has also grown to support diverse networks and even protest movements.

Unlike conventional news channels, the spread of ideas and images via social media can be beyond any authority's control. Social networks can be used to motivate individuals to support a collective. This was seen during the Arab Spring uprisings against the leaders of Tunisia, Egypt, Bahrain, and Libya. In Tunisia, in 2011, spontaneous protests broke out when street vendor Mohamed Bouazizi set fire to himself after being harassed by government officials and mobile phone footage of the protest was posted on Facebook. Sharing this footage encouraged others to join in, and the subsequent riots were coordinated via Twitter.

Countries with less developed transportation and communication infrastructures are also able to access social networks. They not only benefit from this technology but are able to innovate using it. In Kenya, an application called M-Pesa was invented to allow users to transfer, deposit, and withdraw funds via their smartphone. This enables them to send money directly to their village or remote family in minutes rather than the days it would take to travel.

> "BY GIVING **PEOPLE THE POWER TO SHARE** WE ARE MAKING **THE WORLD MORE TRANSPARENT.**"
>
> **Mark Zuckerberg**, cofounder of Facebook, 1984–

▼ Expanding networks
Since the invention of the first mobile phone in 1973, the speed of technological innovation increased and resulted in the creation of an array of devices that connect people all over the world in different ways.

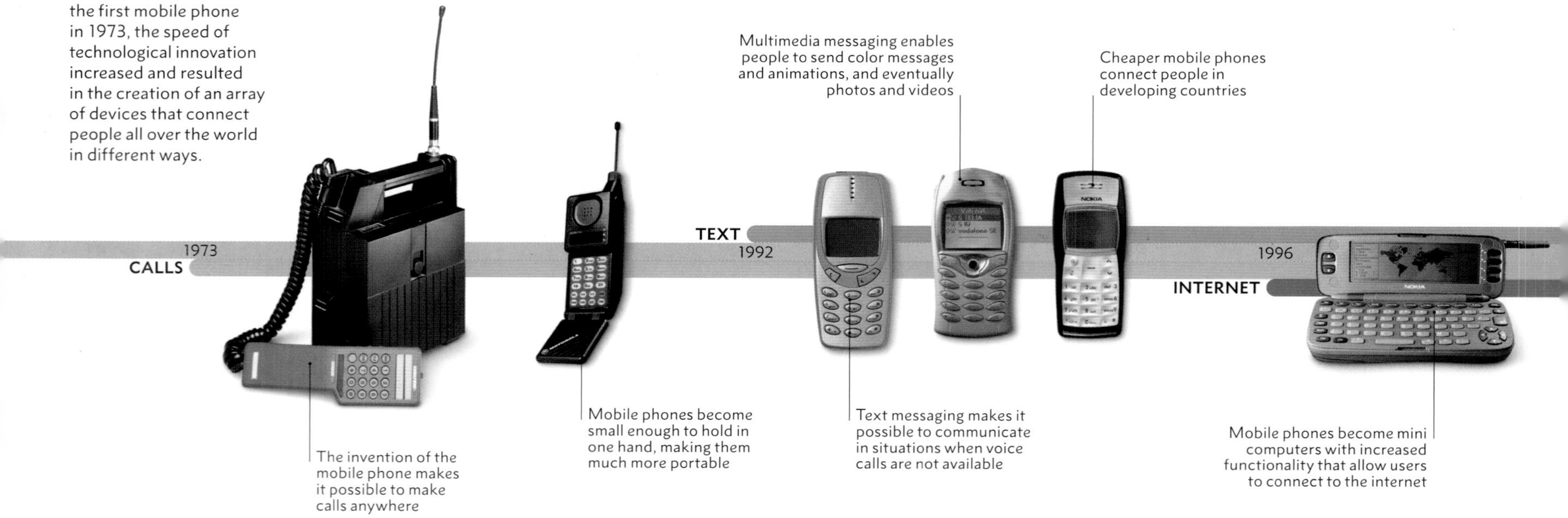
1973
CALLS
TEXT
1992
1996
INTERNET
Multimedia messaging enables people to send color messages and animations, and eventually photos and videos
Cheaper mobile phones connect people in developing countries
The invention of the mobile phone makes it possible to make calls anywhere
Mobile phones become small enough to hold in one hand, making them much more portable
Text messaging makes it possible to communicate in situations when voice calls are not available
Mobile phones become mini computers with increased functionality that allow users to connect to the internet

◄ Crowdfunding initiatives
Social media platforms have provided a new way for groups and individuals to secure investment. Crowdfunding has supported anything from art projects to innovative new products, such as a 3D printer.

▲ Political activism
Social media provides direct access to history-making events as they occur. It played a central role in the 2011 Occupy protests around the world from Wall Street, in New York, to Hong Kong as activists used social media both to organize themselves and to keep the world updated.

▲ New opportunities
Cheap mobile phones in countries with little or no landline infrastructure transformed communications. In Africa, 3G internet coverage has enabled new trade, online banking, and access to information about health and medicine, reducing the need for people to travel long distances.

◄ Saving lives
Medical appeals, such as campaigns for organ donation, often receive a generous response. In 2016, a plea on social media dramatically increased the stem cell donor list when internet users united to help a girl with leukemia, using the hashtag "Match4Lara."

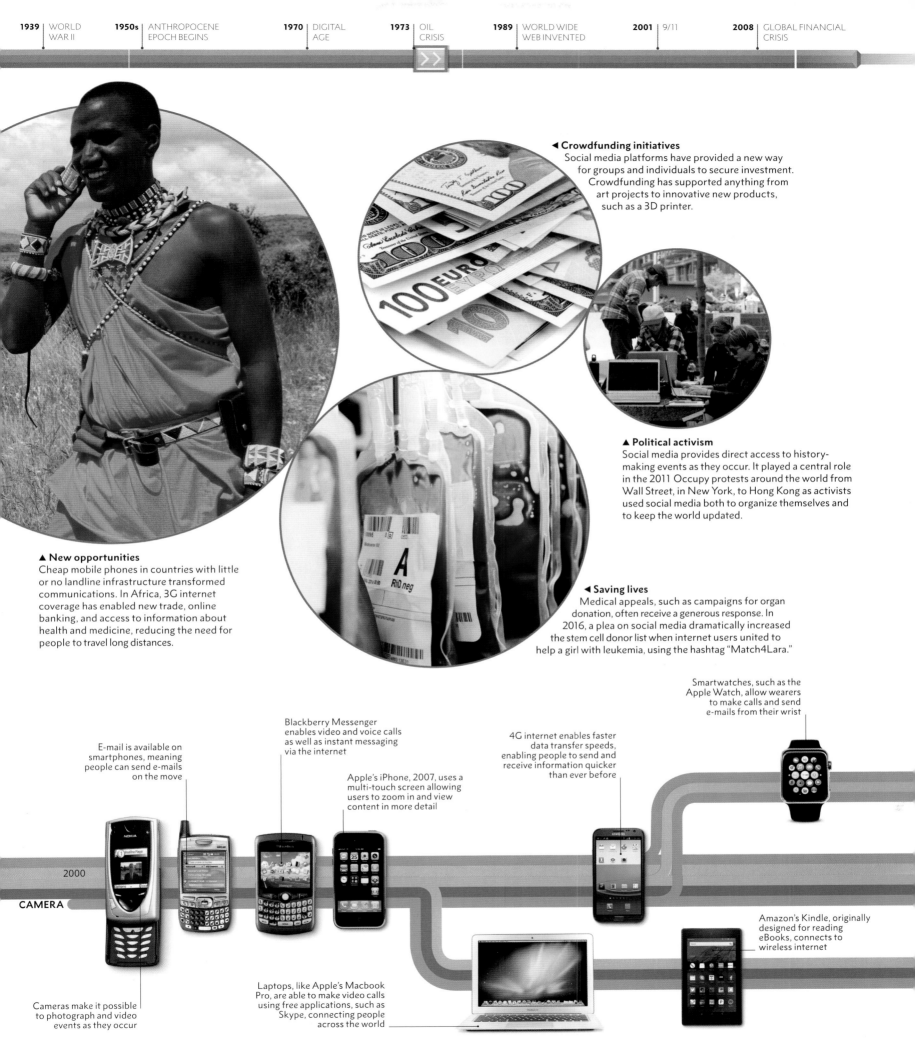

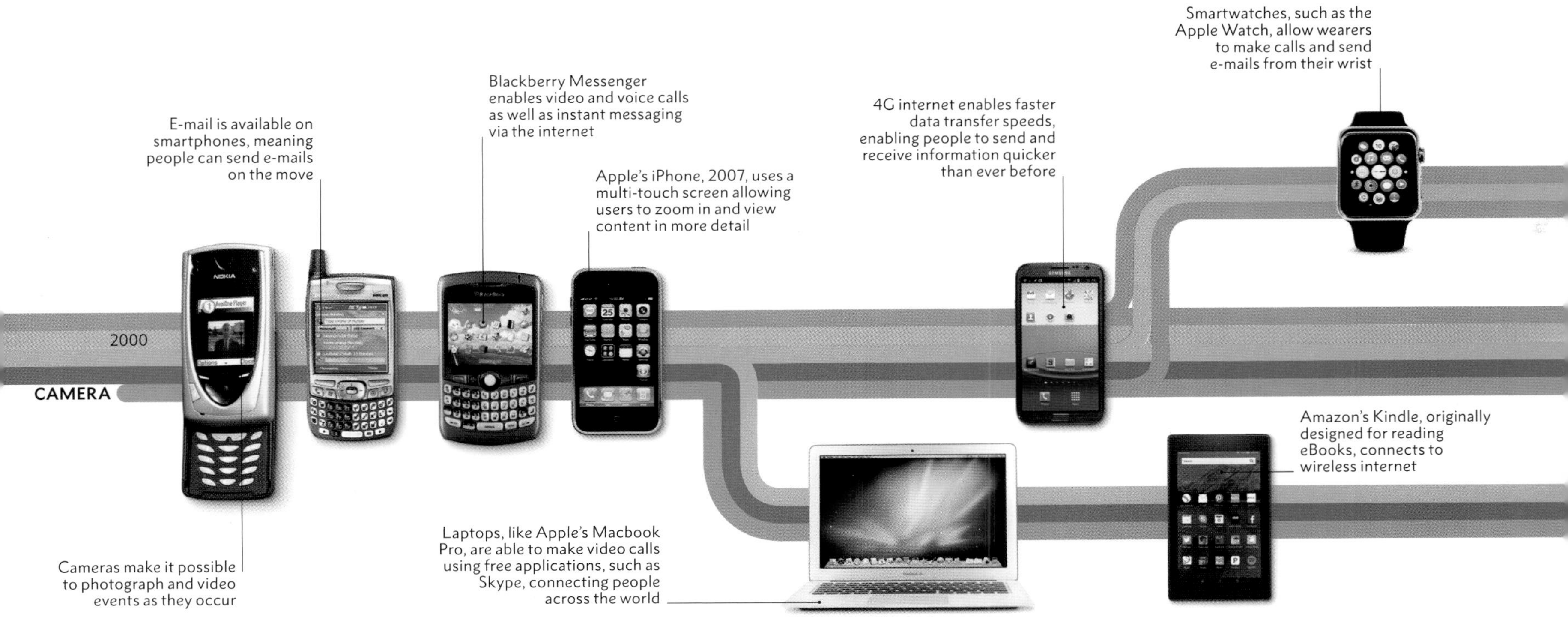
Smartwatches, such as the Apple Watch, allow wearers to make calls and send e-mails from their wrist
Blackberry Messenger enables video and voice calls as well as instant messaging via the internet
E-mail is available on smartphones, meaning people can send e-mails on the move
4G internet enables faster data transfer speeds, enabling people to send and receive information quicker than ever before
Apple's iPhone, 2007, uses a multi-touch screen allowing users to zoom in and view content in more detail
2000
CAMERA
Amazon's Kindle, originally designed for reading eBooks, connects to wireless internet
Cameras make it possible to photograph and video events as they occur
Laptops, like Apple's Macbook Pro, are able to make video calls using free applications, such as Skype, connecting people across the world

GROWTH AND **CONSUMPTION**

The 20th century was characterized by the sharp acceleration in the pace and scale of change. Industrialization and economic growth increased the ecological power of humans over the biosphere and led to extraordinary population growth and consumption of Earth's resources.

The extraordinary pace of change during the 20th century marked an entirely new period in human history and in the history of human relations with other species and with Earth itself. Population growth is a measure of a species' ecological power, as it is dependent on there being enough resources to support it. Over the last 250 years, the human population has been through a period of spectacular growth: in 1800, there were 900 million people in the world; by 1900 there were 1.6 billion; by 2000 there were 6.1 billion; and today the world population has reached over 7 billion. At the same time, people started living for longer and the average life expectancy doubled during the 20th century. This exceptional growth is partly because new innovation has increased our collective control over the resources of the biosphere. The acceleration of technological change is the primary cause of this transformation: innovation has made it possible to provide enough resources to sustain a growing population. One area where innovation and technological change has been crucial is food production.

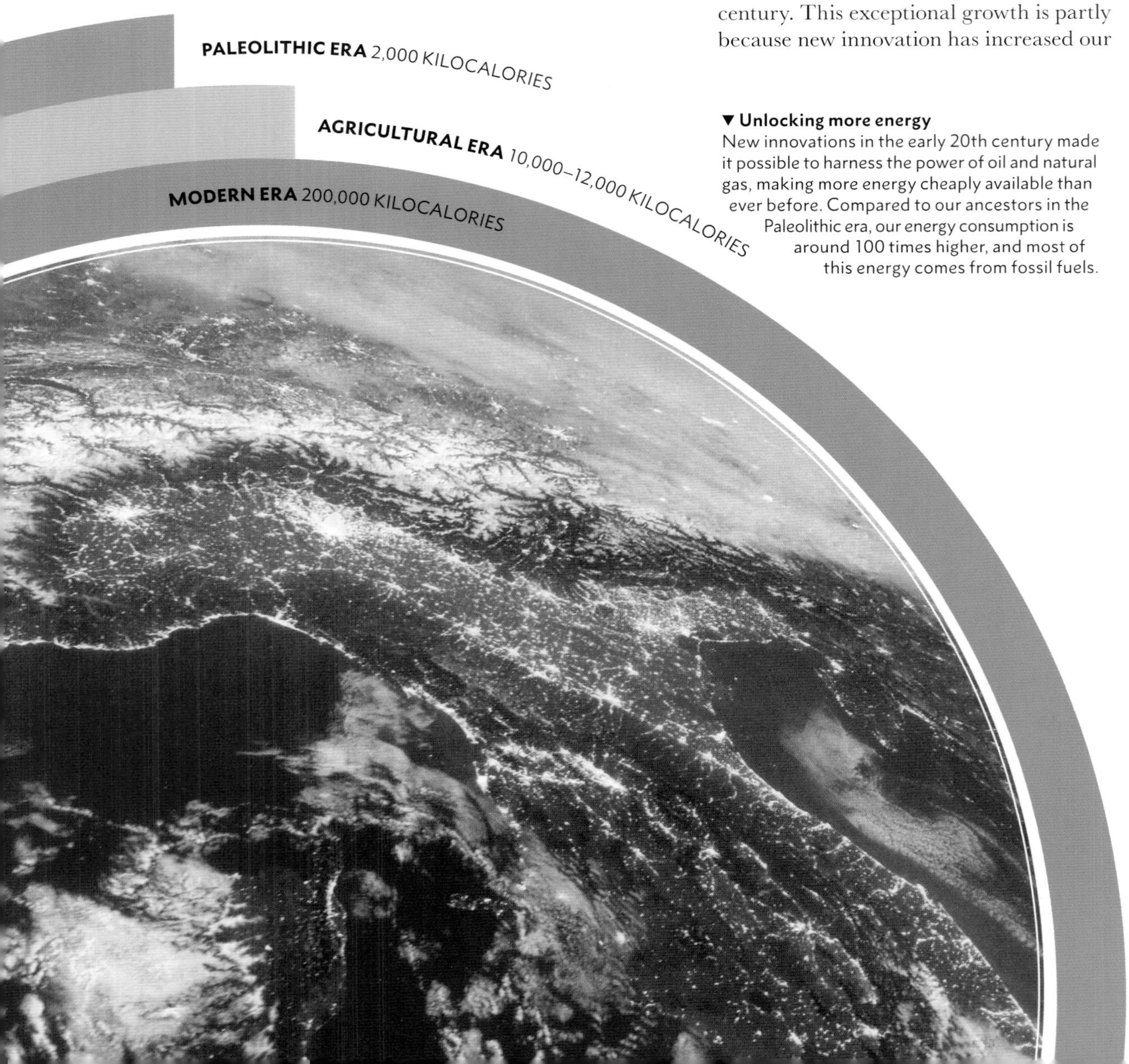

▼ Unlocking more energy
New innovations in the early 20th century made it possible to harness the power of oil and natural gas, making more energy cheaply available than ever before. Compared to our ancestors in the Paleolithic era, our energy consumption is around 100 times higher, and most of this energy comes from fossil fuels.

INNOVATIONS IN FOOD

Since 1900, food production has outpaced population growth and there has been a six-fold increase in grain production. This is because crops began to be farmed on an industrial scale: massive fossil-fuel-driven machines dug dams and irrigation canals. Chemical fertilizers increased the productivity of the land and enabled an area of arable land to produce around three times more food. Scientific innovation in the 1970s led to the creation of genetically modified grains that were engineered with useful genes from other species to produce crops that need less fertilizer or contain natural protection against pests.

In the agrarian era, most people were farmers and only a tiny elite—less than 5 percent of the population—consumed luxury goods. Today, around 35 percent of the global workforce works in agriculture and produces enough food to support the nonfarming communities in industrialized nations, where a new, much larger global middle class enjoys unprecedented wealth and consumer goods.

AS A **SPECIES** WE ARE USING **24 TIMES** AS MANY **RESOURCES** AS WE **USED 100 YEARS AGO**

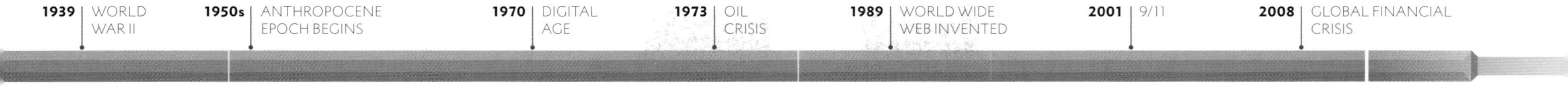

▶ **Fuel consumption**
Population growth has increased steadily in line with global energy use as humans unlocked the power of new forms of energy over time.

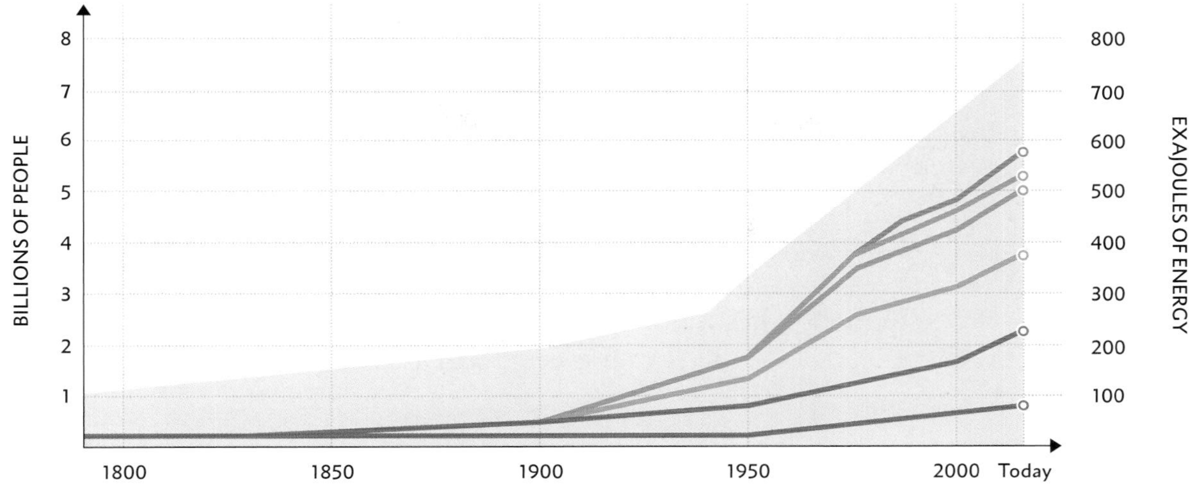

KEY
- Wood
- Coal
- Oil
- Natural gas
- Hydroelectric
- Nuclear
- Population growth

RISING CONSUMPTION

During the second half of the 20th century, rates of innovation accelerated so rapidly, and were so widespread, that the world was entirely transformed. One consequence of this change was consumer capitalism: populations of industrialized regions enjoyed high levels of wealth and material affluence. In 1900, oil lamps, steam-powered trains, and unrefrigerated goods were the norm. Within just 50 years, pipes and cables brought electricity into homes, providing light and heat and powering domestic technologies that have transformed modern life: washing machines, dishwashers, radios, televisions, stereos, telephones, and computers gradually became everyday items that were frequently marketed and sold to the workers who produced them. Advertising (see pp.316–17) and marketing convinced consumers to buy these products and bank loans made them available to those who could not otherwise afford such goods.

The fossil fuel revolution also brought electricity into factories, where further technological innovation meant that methods of production became cheaper. This made goods more affordable and expanded markets, which, in turn, encouraged investment in production and research. For example, the synthesis of plastic, a cheap new material, cut the costs of production. As more people were able to purchase once-expensive consumer goods, the cost of production fell, and even more people were able to buy them.

Today, not only are there more people than at any other point in human history, but they are also consuming more than ever before: the average consumption of each individual person is rising dramatically, all made possible by the energy from fossil fuels. Meanwhile, consumer products are cheaper, easier to purchase, and more disposable, all of which leads to huge amounts of waste. This waste includes materials such as plastics and electronic waste from computers, mobile phones, and televisions. The mass-manufacture of these items produces greenhouse gas emissions, and more emissions are created during the process of disposing of them.

> **INFINITE GROWTH** OF **MATERIAL CONSUMPTION** IN A **FINITE WORLD** IS AN **IMPOSSIBILITY**.
>
> **Ernst Friedrich Schumacher**, German economist, 1911–77

UNEQUAL GROWTH

One widely accepted measure of growth is Gross Domestic Product (GDP), which measures the total production of all countries. World GDP increased almost 12 times from 1913 to 1998. However, this growth has not always been equal: by 1900, the world had been divided into those countries that had industrial economies and those that did not (see pp.336–37). Industrialization raised the wealth of Europe and North America but led to a rapid decline in the wealth of East Asia.

Meanwhile, resources such as food are not distributed equally: 800 million people in the world, mostly people living in poor undeveloped countries in Asia and sub-Saharan Africa, do not have enough to eat. At the same time, around one-third of all food produced each year is wasted.

▼ **Waste products**
In 1900, the world produced about 0.55 million tons (0.5 million metric tons) of solid waste per day. In 2000, this had increased six-fold to around 3.3 million tons (3 million metric tons) per day.

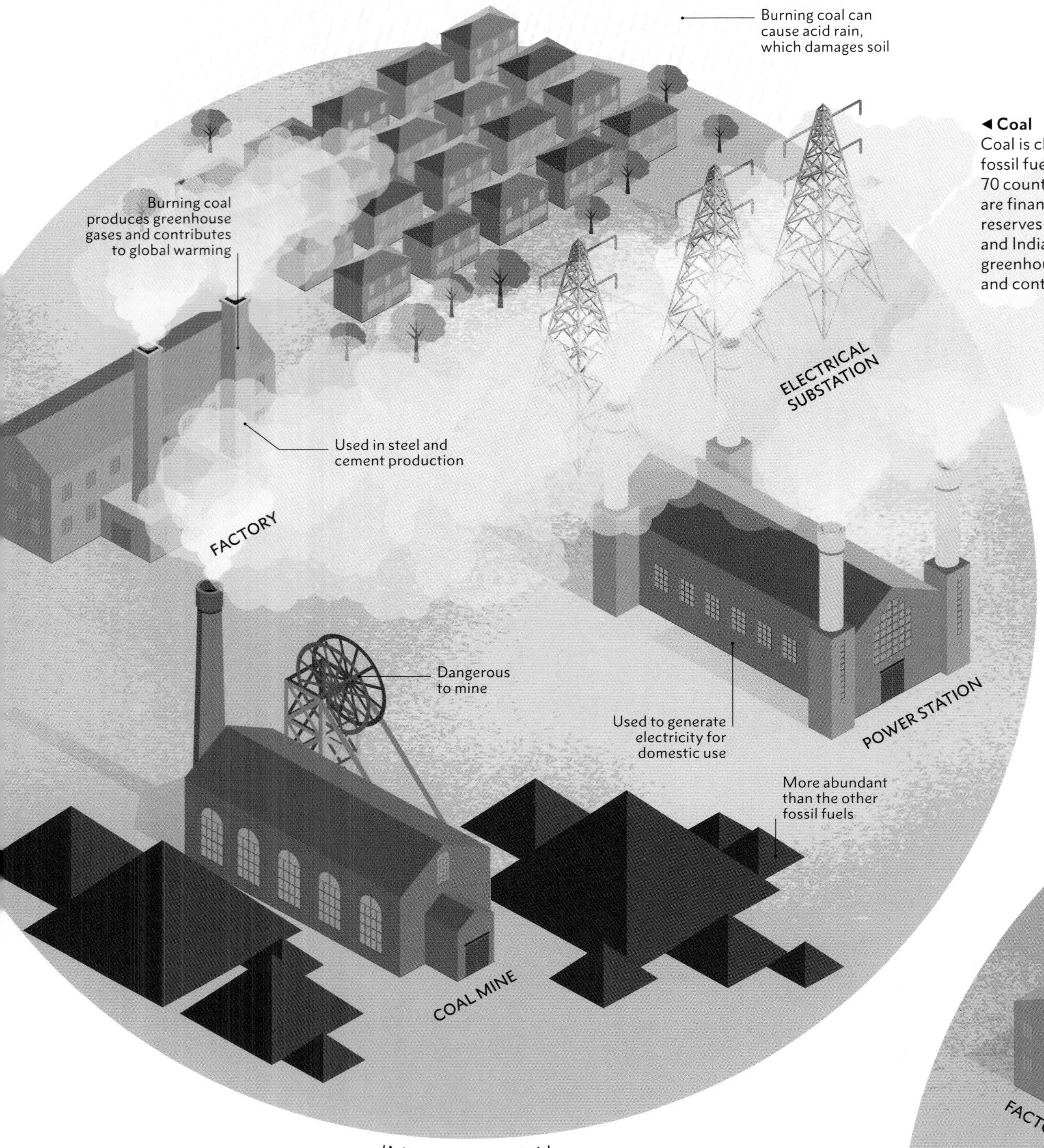
Burning coal can cause acid rain, which damages soil
Burning coal produces greenhouse gases and contributes to global warming
ELECTRICAL SUBSTATION
Used in steel and cement production
FACTORY
Dangerous to mine
Used to generate electricity for domestic use
POWER STATION
More abundant than the other fossil fuels
COAL MINE

IMPACT OF COAL

◄ Coal
Coal is cheaper to extract than the other fossil fuels and still relatively abundant: around 70 countries worldwide have coal reserves that are financially worth recovering. The biggest reserves are in the United States, Russia, China, and India. However, burning coal releases greenhouse gases that damage the environment and contribute to global warming.

▼ Oil
The most versatile fossil fuel, oil is also the one scientists predict we are running out of the fastest: some estimates say resources will run dry in just 55 years if we continue to use it at the current rate. Top producing oil countries include Saudi Arabia, Russia, the United States, Iran, and China.

IMPACT OF OIL

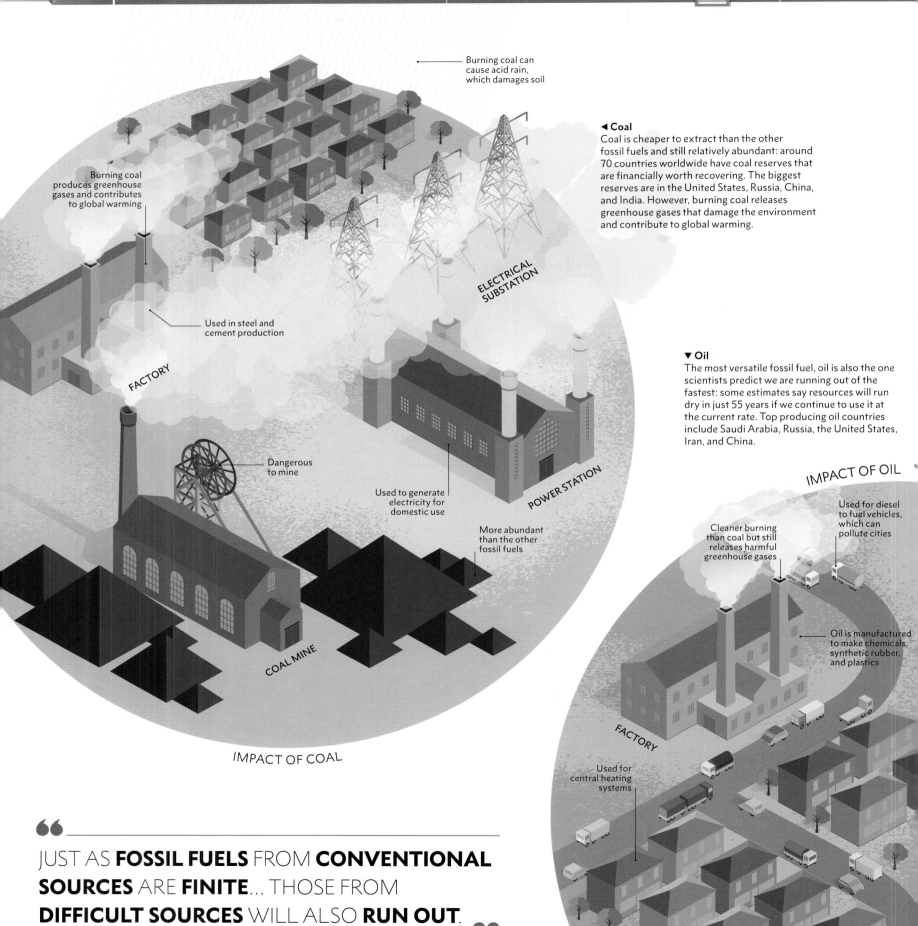

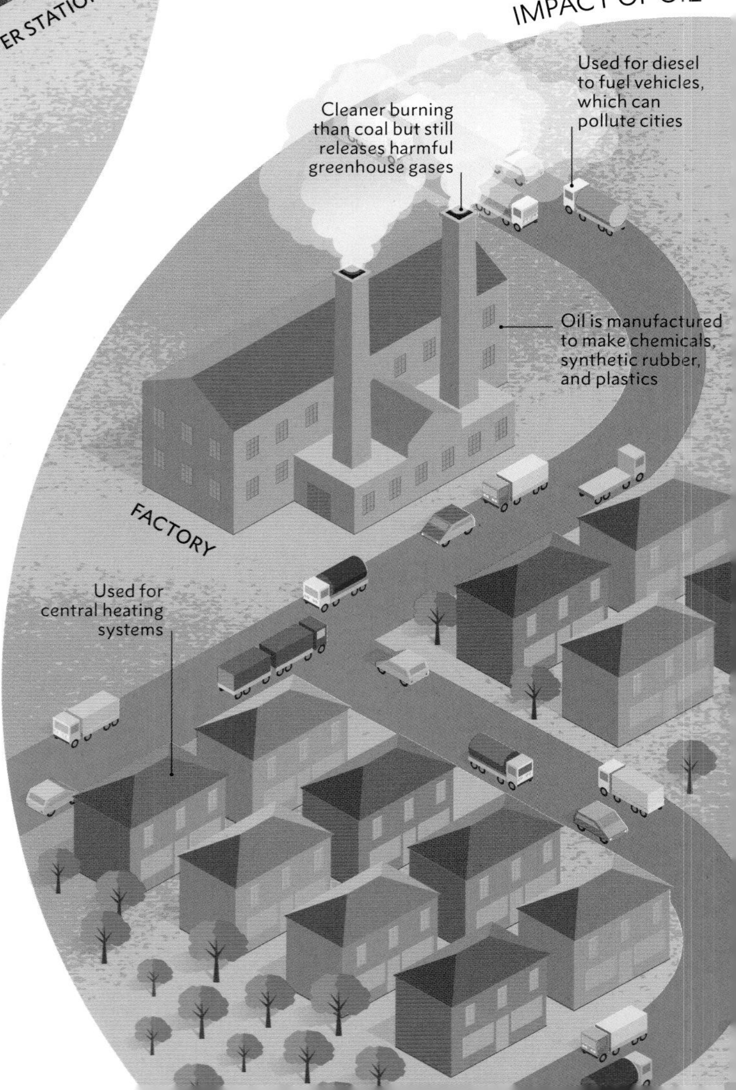
Used for diesel to fuel vehicles, which can pollute cities
Cleaner burning than coal but still releases harmful greenhouse gases
Oil is manufactured to make chemicals, synthetic rubber, and plastics
FACTORY
Used for central heating systems

> JUST AS **FOSSIL FUELS** FROM **CONVENTIONAL SOURCES** ARE **FINITE**… THOSE FROM **DIFFICULT SOURCES** WILL ALSO **RUN OUT**.
>
> **David Suzuki,** Canadian scientist and environmental activist, 1936–

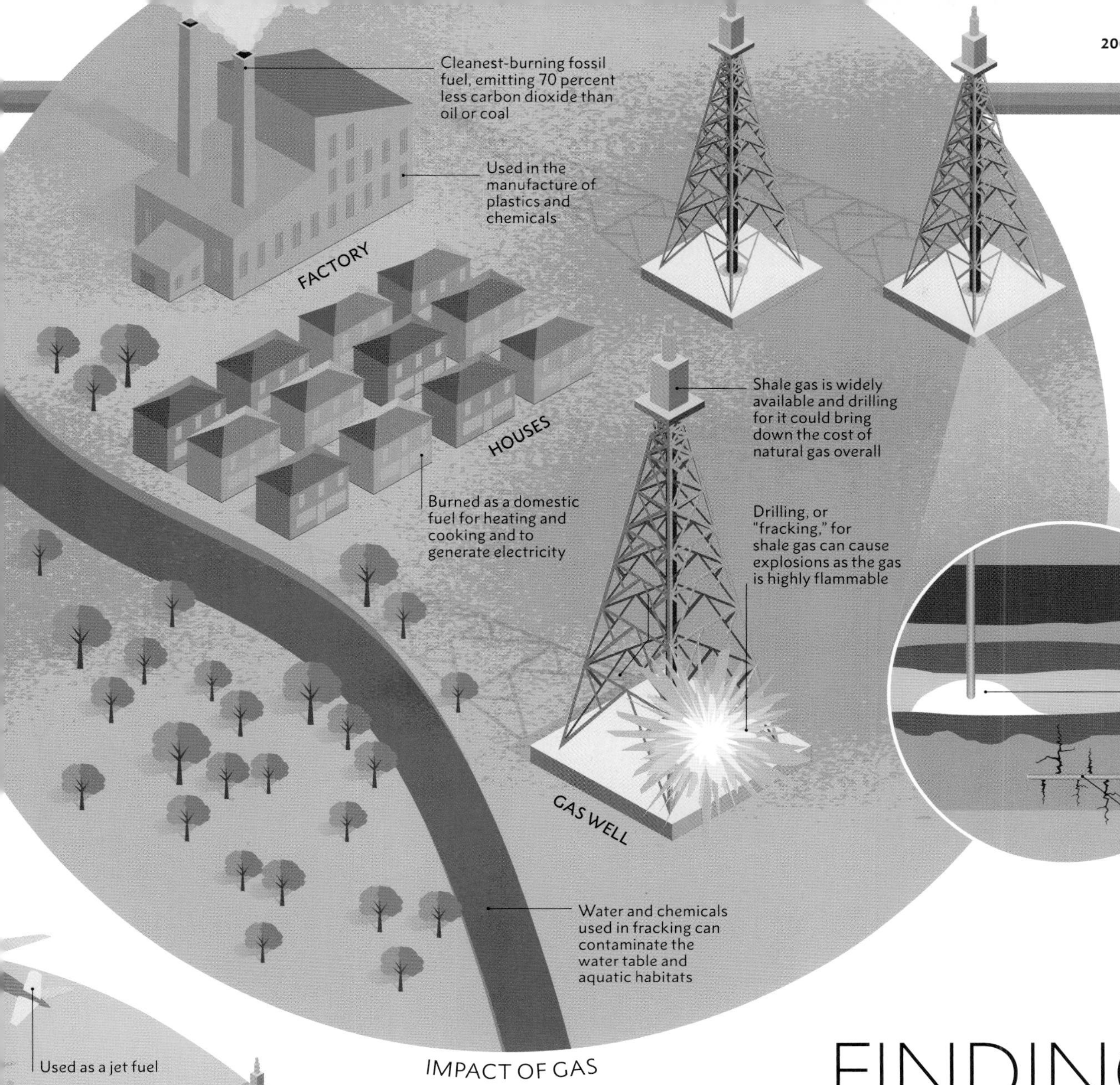
Cleanest-burning fossil fuel, emitting 70 percent less carbon dioxide than oil or coal
Used in the manufacture of plastics and chemicals
FACTORY
HOUSES
Burned as a domestic fuel for heating and cooking and to generate electricity
Shale gas is widely available and drilling for it could bring down the cost of natural gas overall
Drilling, or "fracking," for shale gas can cause explosions as the gas is highly flammable
GAS WELL
Water and chemicals used in fracking can contaminate the water table and aquatic habitats
IMPACT OF GAS

◀ Natural gas and shale gas
These two types of natural gas are found in many countries worldwide. The highest reserves of natural gas are located in Qatar, Iran, Russia, the United States, and Saudi Arabia. Leading producers of shale gas are China, the United States, Mexico, Australia, Argentina, Canada, and Algeria. Gas could alleviate dependence on foreign countries to supply energy but there are environmental and safety concerns about its extraction, and shale gas is not a renewable resource.

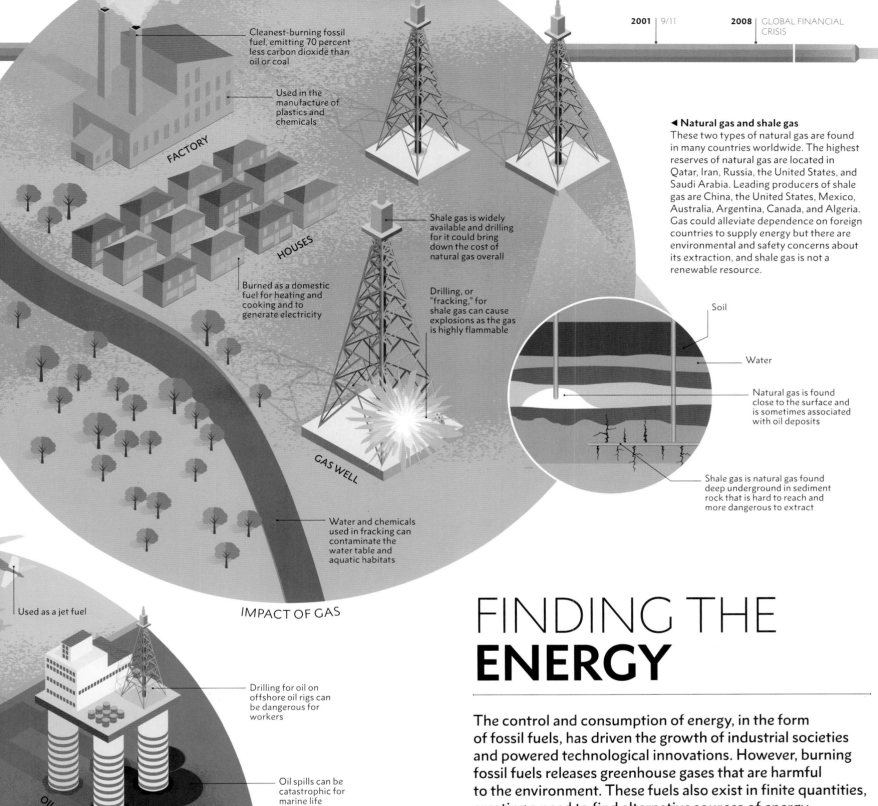

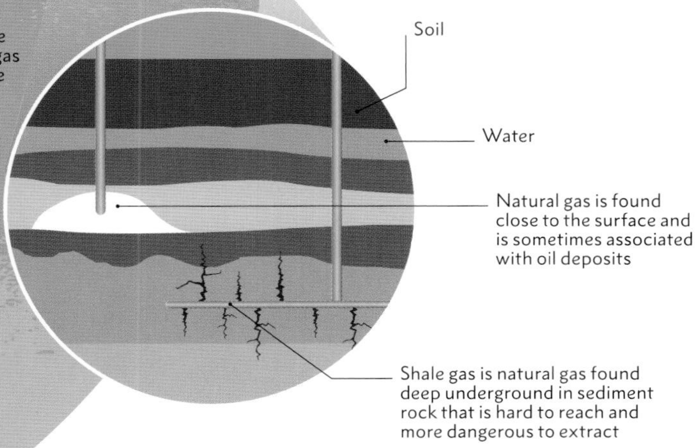
Soil
Water
Natural gas is found close to the surface and is sometimes associated with oil deposits
Shale gas is natural gas found deep underground in sediment rock that is hard to reach and more dangerous to extract

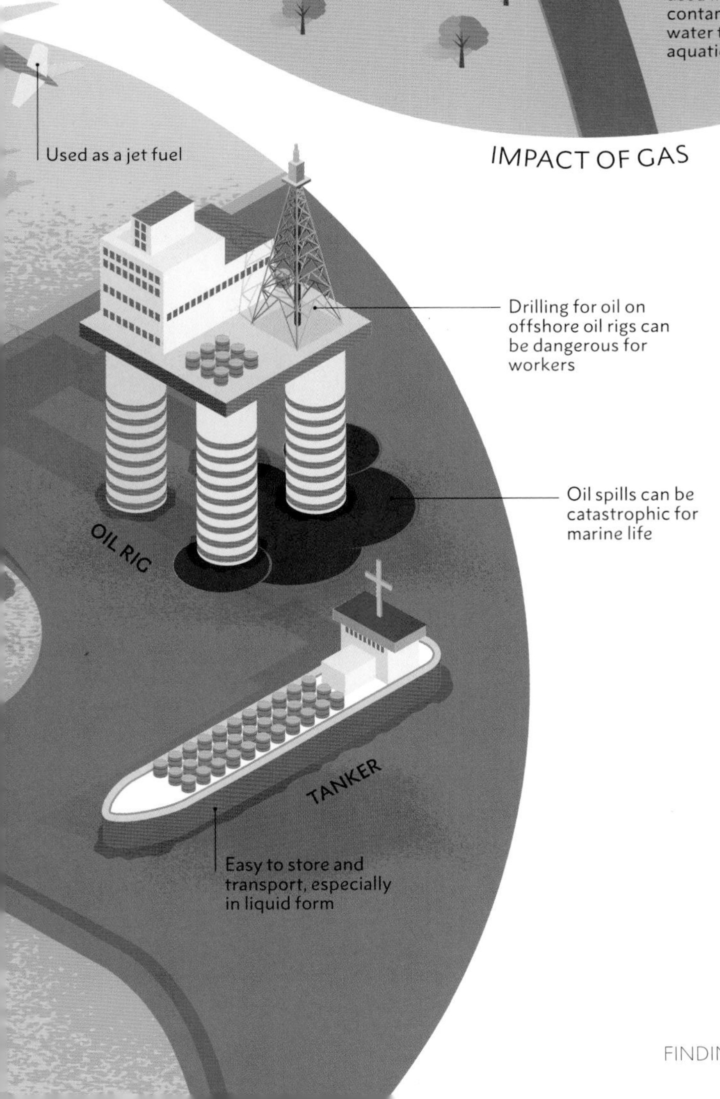
Used as a jet fuel
Drilling for oil on offshore oil rigs can be dangerous for workers
Oil spills can be catastrophic for marine life
OIL RIG
TANKER
Easy to store and transport, especially in liquid form

FINDING THE **ENERGY**

The control and consumption of energy, in the form of fossil fuels, has driven the growth of industrial societies and powered technological innovations. However, burning fossil fuels releases greenhouse gases that are harmful to the environment. These fuels also exist in finite quantities, creating a need to find alternative sources of energy.

Coal, oil, and natural gas are the three major fossil fuels—they are derived from plant and animal fossils that are millions of years old and take millions of years to form (see pp.148–49). Starting with coal, these fuels powered modern industrialization, but they are being depleted at an ever increasing rate. In the 20th century, when oil replaced coal as the world's leading fossil fuel, governments and industrialists joined forces to find and control new oil fields. The interdependence between governments, energy companies, and the supply and control of oil shapes world politics today. Meanwhile, shale gas, a form of natural gas, is predicted to become an important new source of energy. It is found domestically in many countries, and may reduce or even eliminate their potential dependence on foreign nations for energy.

NUCLEAR **OPTIONS**

During the 20th century, a global network of scientists discovered ways to harness nuclear energy, and in World War II deployed it with devastating immediate and long-term effects. In 2016, nuclear power provided almost 15 percent of the world's electricity.

Warfare often drives innovation. The atomic bombs dropped on the Japanese cities of Hiroshima and Nagasaki in 1945 demonstrated the terrifying power of the world's ultimate weapon. It remains the most devastating technology unleashed by one industrialized power on another, and the fear it inspired—that any nation with a bomb could destroy another at the touch of a button—helped to create the Cold War that dominated the late 20th century.

◄ **Radioactive energy source** Uraninite is a highly radioactive ore of uranium that is mined to provide an energy source that powers nuclear plants.

ALTERNATIVE ENERGY SOURCE

During the 1950s, concerns about an overreliance on fossil fuels brought peacetime usage of nuclear energy to the fore. The first electricity-producing nuclear power plant opened in the Soviet Union in 1954 and the industry spread rapidly in the 1960s. Nuclear power became even more politically important when skyrocketing prices brought about by the Middle East oil crises of the 1970s caused countries, such as France and Japan, to reduce their reliance on fossil fuels. By the year 2000, nuclear power accounted for 80 percent of France's electricity and 40 percent of Japan's.

Nuclear power has other important civil and commercial uses. By 2016, 240 smaller nuclear reactors were in operation in 56 countries worldwide, where they were used for research and training, materials testing, medicine, and industry.

ENVIRONMENTAL THREATS

The pros and cons of nuclear power remain a topic of heated debate. There are concerns that any country building a nuclear reactor has the power to create a nuclear weapon. Arguments that nuclear power stations have lower emissions than those run by fossil fuels are met by concerns about the disposal of radioactive waste and the toxic pollution created by mining uranium. Safety is also a concern, after serious accidents occurred in Fukushima, Japan, in 2011, and Chernobyl, Ukraine, in 1986. Chernobyl affected over 40,320,000 acres (16,316,900 hectares) of land and there are 148,274 invalids on the Chernobyl registry, while Fukushima displaced over 160,000 people. Nuclear accidents also devastate rural areas, as contaminated land can no longer be used for agriculture. Engineers are working on developing safer and more efficient power stations for the future.

“A WORLD **FREE OF NUCLEAR WEAPONS** WILL BE **SAFER AND MORE PROSPEROUS**.”

Ban Ki-moon, Secretary-General of the United Nations, 1944–

Atmospheric weapons test
Nuclear weapons have been tested above ground, underground, and underwater. Over 2,000 nuclear explosions were detonated worldwide between 1945 and 1996.

ENTERING THE **ANTHROPOCENE**

Human activity has become the most influential factor shaping life on Earth. The impact of industrialization and the pressures exerted by humankind have led to changes to the atmosphere, ecosystems, and biodiversity, while depleting many of Earth's resources. This has led scientists to propose we have entered a new geological epoch: the Anthropocene.

In 2000, Dutch scientist Paul Crutzen coined the term "Anthropocene" to describe a new geological epoch. He argued that the biosphere had been transformed by humans rather than by natural geological and climatic processes that defined previous epochs. Earth bears permanent signs of this human activity: airborne black carbon—the main component of soot produced by burning fossil fuels and biomass—is trapped in glacial ice; fertilizer chemicals linger in the soil; and plastics pollute both earth and water. All of these will likely leave a fossil record for future generations to discover. Population growth, more intensive agriculture, the destruction of biodiversity, and industrialization are among the main causes of environmental damage: they have completely reshaped Earth's ecology and biology.

The history of Earth is divided into geological time scales: epochs are periods spanning thousands of years. If the Anthropocene is officially accepted it will follow the Holocene epoch, which began after the last Ice Age around 11,700 years ago, when humans colonized new territories and populations first began to grow. As the species at the top of the food chain, humans began to make their mark on the world's fauna 50,000 years ago when they hunted many large mammals to extinction. Following the Ice Age, people started to settle in communities and began to develop agriculture. Scientists believe that deforestation to clear land for crops around 8,000 years ago released greenhouse gases into the atmosphere and created a spike in carbon dioxide (CO_2) levels. The effects of farming also changed the land; geologists can find agriculture's signature in European rock dating back to 900 CE.

During industrialization in the 19th century, Europe once again left an environmental mark and Crutzen believed that the Anthropocene started at this time. Other scientists suggest the Anthropocene began in the atomic era of the 1950s and the "Great Acceleration" that followed, which saw the rapid growth of economies, populations, and energy consumption. The Great Acceleration came after the detonation of the atomic bomb, the first nuclear weapon, which left a radioactive marker in sediments across the world, and marks the rise of truly global impacts caused by humans on the planet.

INDUSTRIAL IMPACT

While there is still some debate about the Anthropocene, few dispute the impact of industrialization upon the environment. Even in the early stages of Britain's industrial revolution, thick smog from the coal-burning factories spread into the atmosphere and created widespread health problems. These issues continued into the 20th century: a 1952 coal-fog left 4,000 dead from respiratory diseases in London in four days. In the United States, smog caused by car exhausts in California led to the discussion of a new environmental term: greenhouse gases.

SINCE THE **INDUSTRIAL REVOLUTION,** THE LEVEL OF **CO2 ON EARTH** HAS **INCREASED** BY **34 PERCENT**

Greenhouse gases, like carbon dioxide and water vapor, occur naturally in small quantities in Earth's atmosphere and prevent heat from escaping into space. Without them, Earth would be a frozen, arid planet. But in the last 250 years, intensified human activity—primarily burning fossil fuels for use in industry or electricity and transportation—has led to the highest concentrations of CO_2 in the atmosphere for

▼ **Burning fossil fuels** Industrialization was powered by the burning of coal, which released billions of tons of carbon dioxide into the atmosphere. After the 1880s, oil and gas drove further economic growth and released more carbon dioxide.

around 800,000 years. Carbon dioxide levels remained below 280 parts per million for thousands of years, but since the Industrial Revolution, they have risen at an increasing rate. Accelerating after the 1950s, they reached around 400 parts per million in the early 21st century. This is the main cause of global warming: a gradual increase in Earth's average temperature. More greenhouse gases trap more heat in the atmosphere and prevent it from escaping into space.

Scientists suggest a 50 percent reduction in global CO_2 emissions is needed by 2050 to prevent a global warming catastrophe. Global warming has already had serious effects, including glacial melting, a rise in sea levels, ocean acidification, warming surface temperatures, extreme weather, and the destruction of ecosystems.

Ecosystem destruction is also caused by widespread deforestation that began during the 19th century to provide wood and raw materials for industrialization. Trees were replaced with crops, such as coffee and tea, which could be grown on one plot of ground over consecutive years. Today, deforestation accounts for around one fifth of greenhouse gas emissions, as plants and trees absorb CO_2 during photosynthesis. Halting deforestation and replanting forests would help to reduce CO_2 levels.

DECLINING BIODIVERSITY

Cutting down forests has destroyed various ecosystems. As humans increasingly exploit the land, we leave less to sustain all other species, leading to a decline in wildlife diversity and abundance. Large numbers of plants and animals were destroyed in Africa, India, and the Pacific Islands in the 19th century during deforestation for industrialization.

Meanwhile, increasingly high levels of pollutants in the world's oceans have devastated marine life. Agricultural fertilizers and sewage can leak into waterways and contaminate freshwater, which eventually flows into the sea where it can create a dead zone. This is where algae form: when they sink to the seafloor and decompose, oxygen is removed from the water. The low levels of oxygen cause marine animals to leave or die. At the same time, around 88 million tons (80 million metric tons) of plastic litter have been dumped in the world's oceans, and around eight million more are added daily. Millions of animals and birds die annually when they mistake this plastic for food.

Every day, species' extinctions are continuing at up to 1,000 times or more the natural rate due to population growth, habitat conversion, urbanization, and over-exploitation of natural resources. In 2015, a study by the International Union for Conservation of Nature (IUCN) assessed 80,000 animal species and found nearly 25,000 of them to be under threat of extinction. If current trends continue, the Earth is on course for a sixth mass extinction on a scale not seen for 65 million years since the extinction of the dinosaurs.

The threat to biodiversity is the result of land-use changes, pollution, climate change, and rising CO_2 concentrations, and is now a matter for serious concern. Each creature has a supporting role in Earth's biosphere, which is an interdependent global ecosystem. This ecosystem provides essential services such as clean water, fertile soils, pollution absorption, storm protection, and recovery from natural disasters. Even the extinction of a creature as small as a bee has repercussions. Bees are the major pollinator of around one third of the world's food crops, but their numbers are in decline and severe food shortages are predicted as result of their dwindling populations.

REVERSING THE DAMAGE

Attempts are being made to help undo the centuries of human environmental damage. Since the 1970s, hundreds of environmental protocols and treatises have been adopted internationally; the countries signing up to them have agreed to implement targets linked to environmental concerns, but with varying degrees of success.

More recently, a set of 17 Sustainable Development Goals were adopted by the United Nations in 2015, which are expected to frame the policies of 193 nations until 2030. They aim to "end poverty, protect the planet, and ensure prosperity for all" by promoting "sustainable industrialization." This may become a defining theme for future generations to ensure environmental sustainability and protecting the world's ecosystems remain top priorities.

SINCE 1992, THE NUMBER OF **PROTECTED SITES WORLDWIDE** HAS **INCREASED TWENTY-FOLD**

" OUR **PLANET** IS BEING **TRANSFORMED** - NOT BY NATURAL EVENTS, BUT BY THE ACTIONS OF **ONE SPECIES: MANKIND**. "

Sir David Attenborough, 1926–

" THE **CLIMATE CRISIS** IS THE **GREATEST CHALLENGE HUMANITY** HAS EVER **FACED**. "

Al Gore, American politician and environmentalist, 1948–

HARD EVIDENCE

CLIMATE **CHANGE**

Earth's climate has fluctuated dramatically throughout its 4.5 billion year history, but scientists can now prove that human activities—such as burning fossil fuels and clearing land for agriculture—are also contributing to climate change.

Climate change is a long-term shift in weather conditions identified by changes in temperature, precipitation, winds, and other indicators. Climate science began over 100 years ago, when scientists first suggested that burning fossil fuels may cause global warming, which, in turn, contributes to climate change. In 2016, humans emitted carbon dioxide (CO_2) into the atmosphere 10 times faster than at any point in the last 66 million years, causing Earth to be at its warmest for 1400 years.

The effects of global warming have been monitored for decades: they include global temperature rises, the shrinking of glaciers and ice sheets, the thinning of the ozone layer, acidification and warming of the oceans, and rising sea levels. By comparing data on these events with past records, scientists try to predict the future impact of global warming. Climate change data is gathered by chemists, biologists, physicists, oceanographers, and geologists. They compare statistics on Earth's temperatures, weather, and greenhouse gases by feeding data into computerized climate change models. Air samples are analyzed to gauge the level of CO_2 in the atmosphere caused by natural sources compared to that of fossil fuels. Similar readings made from air bubbles trapped in Antarctic ice cores that are hundreds of thousands of years old tell us about past changes in Earth's climate (see pp.174–75). Plant fossils from Earth's crust tell us about species distribution during different atmospheric periods, which may indicate how they could react to higher levels of CO_2 in the future.

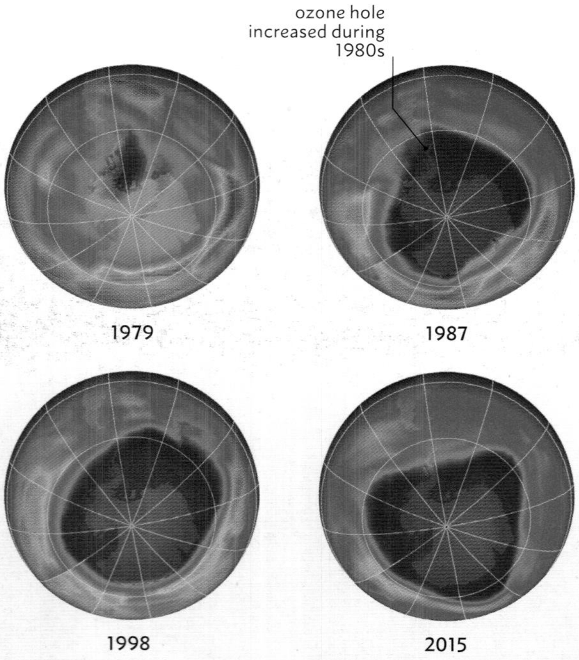

▲ Ozone depletion
The ozone layer stretches across the Earth's upper atmosphere and absorbs most of the sun's ultraviolet radiation. In the 1970s, robotic satellites showed a hole in the ozone layer. In 1987, the Montreal Protocol agreed to prohibit ozone-depleting chemicals, but the ozone hole is only predicted to return to 1980 levels by 2070.

Global temperature rise

To record global temperatures, scientists take air measurements from satellites, ships, and meteorological stations and then analyze the data. These measurements reveal that the average global temperature is 1.4°F (0.8°C) warmer than it was in 1880. In 2015, this caused heatwaves in Asia and Europe, flooding in Africa, droughts in South America, and an increase in extreme weather events: global storms, cyclones, and typhoons.

Rising sea levels

The Maldives are at risk

Tidal gauge readings, ice core samples, and satellite measurements have shown the global mean sea level has risen by 2¾in (7cm) in the last century. This rise is due to melting glaciers and polar ice caps, and the expansion of sea water as it warms. Rising sea levels have devastated low-lying coastal habitats, including a number of Pacific Islands.

Shrinking sea ice

Satellite images of the ice caps in Greenland and Antarctica reveal that as temperatures rise they are shrinking at a rate of 13.4 percent per decade. Sea ice reflects sunlight back into space. Without sea ice, the ocean absorbs 90 percent of the sunlight, which warms the water, adds to the Arctic temperature rise, and causes more melting in a process known as a positive feedback loop.

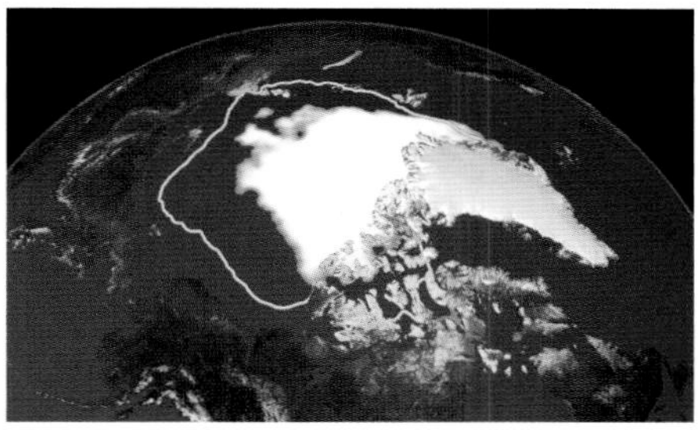

Sea ice at the smallest extent recorded, 2012

Ocean acidification

Scientists study ice core samples and the chemical composition of fossilized sea creatures assess the ocean's acidity over time. The acidity of the ocean's surface has risen by 30 percent over 200 years, as increased CO_2 in the atmosphere has been absorbed by the sea. Acidification prevents creatures such as corals, mussels, and oysters from absorbing the calcium carbonate they need to maintain their skeletons.

Warming oceans

Using robotic floats, scientists are able to show that the world's oceans have warmed by around 0.2°F (0.11°C) between 1971 and 2010. This has led to the destruction of ecosystems such as coral reefs. In 2016, warming ocean temperatures caused a global coral bleaching event. This is where coral lose the colorful algae that give them pigmentation and provide them with oxygen and nutrients. If the stress continues, the bleached coral will die.

Bleached coral

▼ Endangered elements
The periodic table of endangered elements shows the 44 elements facing supply restrictions in the future as well as the 17 Rare Earth Elements, three of which are also endangered.

KEY
- Limited availability—future risk to supply
- Rising threat from increased use
- Serious threat in the next 100 years
- Rare Earth Element

Lithium is used for the lithium-ion batteries that power personal electronics and electric cars today because they store more energy (in the same amount of space) than other technologies

Hafnium has a very high melting point, which is why it is used to make control rods for nuclear processors and nuclear submarines. It is also used as an insulator in microchips and found in computer circuitry.

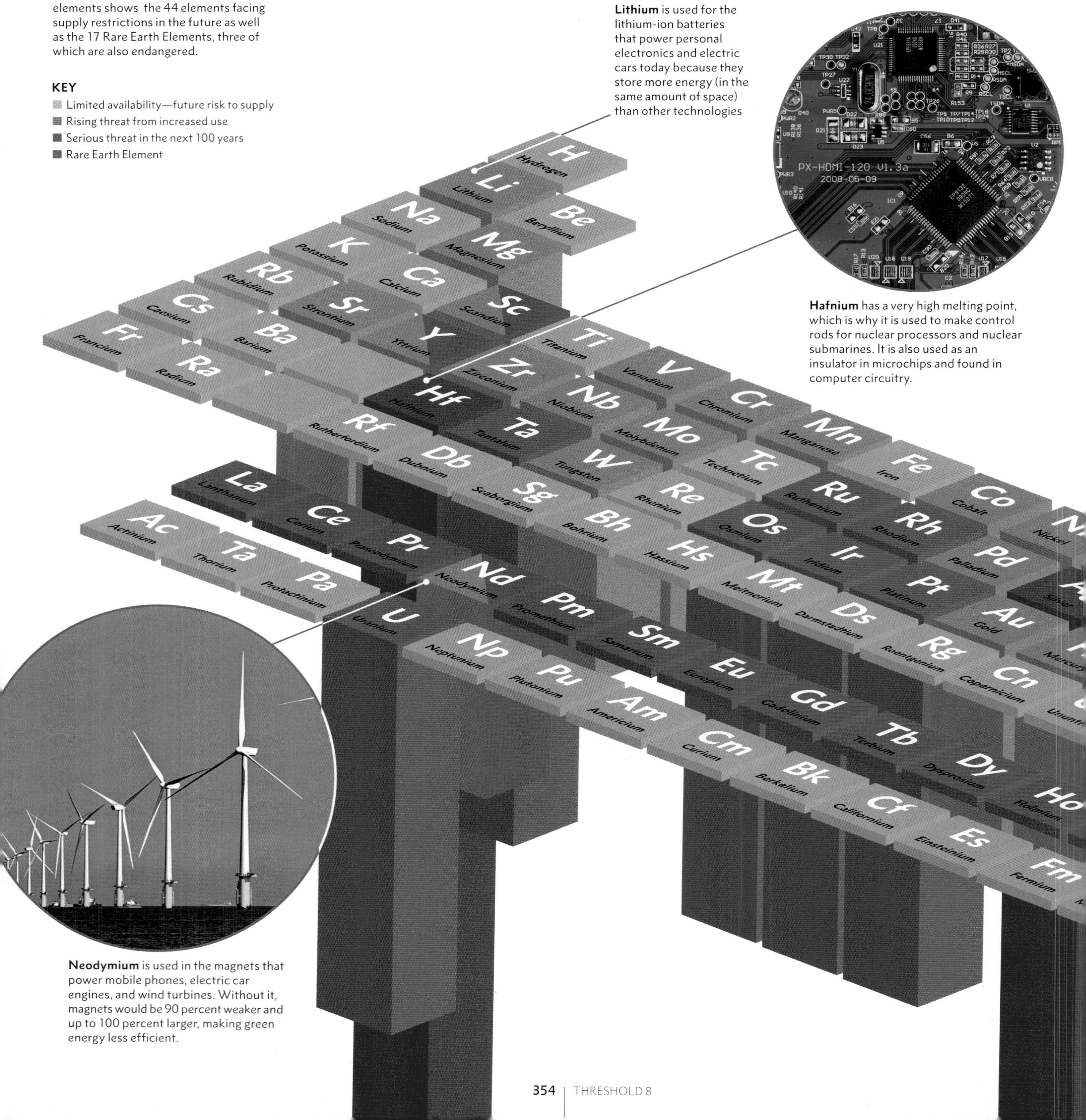

Neodymium is used in the magnets that power mobile phones, electric car engines, and wind turbines. Without it, magnets would be 90 percent weaker and up to 100 percent larger, making green energy less efficient.

ELEMENTS **UNDER THREAT**

The chemical elements that make up Earth occur on our planet in finite quantities. Of the 118 elements that have so far been identified, around 44 are considered endangered because demand for use in technology is predicted to outstrip supply.

IN 2010, **CHINA CONTROLLED 95 PERCENT** OF THE WORLD'S **RARE EARTH PRODUCTION**

Coal and oil are not the only natural resources at risk from current demand. Supplies of elements—including Rare Earth Elements (REEs) with magnetic, luminescent, and electrochemical properties vital for the latest technology—are also under threat. The reasons vary: some, like helium, occur in finite nonrenewable quantities. Others are hard to access: REEs are often widely dispersed and mixed with other minerals, which makes mining an expensive proposition, and refining them can create quantities of toxic waste. In addition, countries with economically viable mines may prefer to secure these resources for domestic use in medical and military equipment rather than export them to competing nations. As with oil, these countries are in a strong position to manipulate prices and protect their market share by controlling availability; it is possible to recycle REEs from old or obsolete electronics, such as computers and phones, but cheaper to extract them afresh. Technology may not work as well without them, but high prices and low supplies give manufacturers an incentive to innovate and create alternative products that use fewer—or no—REEs and endangered elements, and promote the sustainable use of what we have left.

Indium is used to make the touch-screen glass found in smartphones. It comes from zinc mines as it occurs in such small amounts that mining it is impractical. If the demand for zinc declines, it will have an impact on the availability of indium.

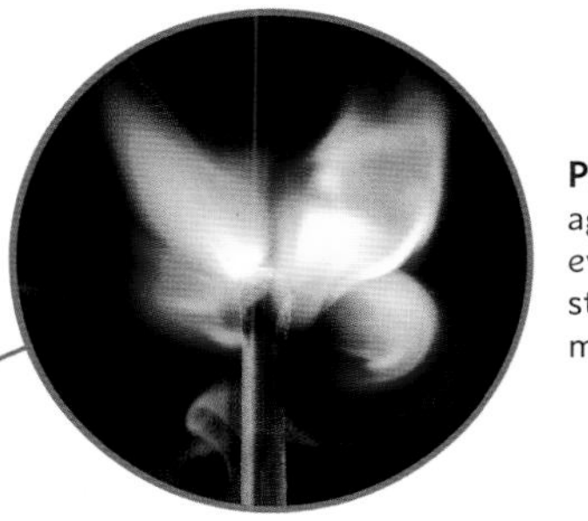

Phosphorus is an important ingredient in agricultural fertilizers. It is also used in everyday items, such as matches. Europe has started recycling phosphorus as a step toward a more sustainable supply.

Helium is the second most abundant element in the universe, but on Earth our extractable supplies are declining. It has many uses, including in MRI scanners

Solar-powered supertrees
Singapore's Gardens by the Bay, an innovative and energy-efficient space, contain supertrees inspired by their natural counterparts. They contain photovoltaic cells that convert sunlight into energy and use it for lighting.

THE QUEST FOR **SUSTAINABILITY**

Coal, oil, and gas have powered over 250 years of industrial progress, but these fossil fuels are in limited supply. Switching from nonrenewable to renewable sources of energy could lead to greater energy security and help to protect the environment.

In 2013, more than 80 percent of the world's energy came from coal, gas, and oil, while only 19 percent came from renewable energy sources. Researchers are seeking new forms of renewable energy as a matter of urgency.

GREEN TECHNOLOGY

The most common sources of renewable power are water, solar, wind, geothermal—which harnesses heat from Earth, such as hot springs—and biomass fuels created by burning decaying plant or animal material. Each has limitations. The construction of wind farms, solar panels, hydroelectric dams, and tidal barriers is expensive, and geothermal power is only available in volcanic areas. Burning biomass emits carbon dioxide, but it is carbon neutral when it is part of a sustainably managed program—for example, if new trees are planted to absorb the carbon dioxide released. Furthermore, new renewable technologies are developing fast and costs are coming down. With knowledge and experiences shared through global networks, we may be able to innovate to overcome the current limitations.

Many countries already use renewable energy. In Brazil, sugarcane is made into the biofuel ethanol; the country's gasoline includes a blend of 18–27 percent ethanol. Nearly 40 percent of Denmark's energy comes from wind power, over 26 percent of Germany's power comes from renewables, and some Chinese and Indian villages heat biomass material to generate electricity. In 2016, over 60 percent of global energy investment went into renewables, and green energy is predicted to overtake electricity generated by fossil fuels by 2030.

Renewable energy could create hundreds of thousands of jobs. It could also enable many countries to develop the long-term domestic energy security essential in the industrial world, and insulate them from the fluctuating prices of imported fuels. However, industrializing countries, such as China and India, continue to rely on coal. Fossil fuel subsidies are often high, which makes them cost effective. Despite these barriers to investment, renewables are catching up and, in some cases, are already cheaper than fossil fuels.

◀ Electric car
Instead of running on oil, electric vehicles are powered by a rechargeable battery, which means they emit less carbon dioxide.

" WE NEED TO ULTIMATELY **MAKE CLEAN, RENEWABLE ENERGY** THE **PROFITABLE** KIND OF **ENERGY**. "

Barack Obama, President of the United States (2009–2017), 1961–

WHERE **NEXT?**

Big History provides a unique perspective on the trends and themes that connect the story of humans. Can we use them to predict the future? Nothing is certain, but the themes of population growth, innovation, energy, and sustainability look set to recur for the next hundred years.

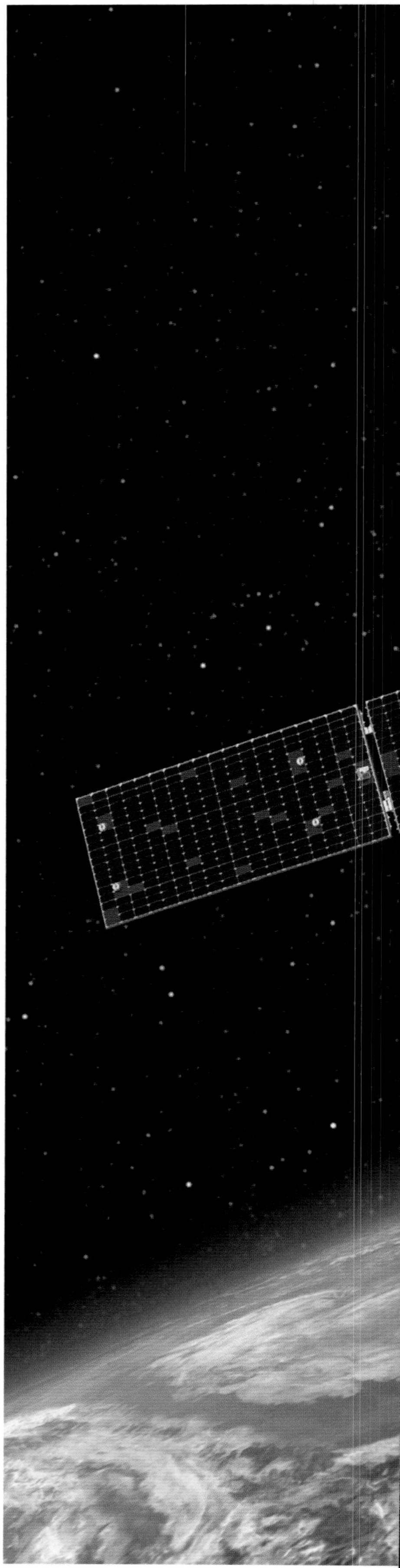

Population growth and innovation are the signatures of our success as a species. Our 18th-century ancestors combined thousands of years of collective learning with new agricultural technologies to put an end to the Malthusian crises that periodically reduced agrarian populations. Astonishing industrial innovation gave many people access to goods, services, and a quality of life previously unimaginable. Technological growth in the last century has outstripped that of all human history. Many of today's innovations, such as smartphones and the internet, would have seemed impossible as recently as the early 1980s. These technologies have connected the world in the most complex collective network ever known.

However, progress has come at a cost. It has led to increasing consumption of the dwindling resources of water and fossil fuels; it has brought about the mass-extinction of many plant and animal species; and it has led to an exponential rise in greenhouse gas emissions. It is now up to the global collective to reverse this damage and develop a less environmentally harmful existence for future generations.

> **BIG HISTORY** STUDIES **THE HISTORY OF EVERYTHING,** OFFERING A WAY OF **MAKING SENSE OF OUR WORLD** AND **OUR ROLE WITHIN IT**.
>
> **David Christian**, Big History historian, 1946–

A SUSTAINABLE FUTURE

The industrial revolution is sometimes described as the first in a series of waves of innovation: the first era of mechanization was succeeded by innovations from the steam age, the electrical age, the aviation and space age, and most recently the digital age. Today we are on the cusp of the sixth wave: sustainability, the great theme of our time. It aims to provide a high standard of living for the rising global population—predicted to reach 10 billion by 2050—while reducing our reliance on fossil fuels and using remaining resources efficiently. The ability to harness new forms of energy has defined past thresholds of human history; now our relationship with energy may determine the fate of our species.

There are signs that trends are shifting. Population growth rates have slowed in industrializing countries, such as India and China. This may be because economically developed countries tend to produce fewer children. But these children tend to be more highly educated, and as they join the billions of potential innovators already connected through the modern global communications network, this may be the key to saving our planet. Never before has collective learning been so accessible, integrated, and important.

The collective hub has already created important green innovations: electric cars, biofuels, solar-powered desalination of water, and zero-emission buildings, whose total energy consumption does not result in greenhouse gas emissions. In this sense, the near future is a place of limitless potential. The 21st century may be remembered as the dawn of global sustainability, attained through green innovation and powered by renewable energies. In a future still unwritten, all things are possible.

Innovation to predict climate change
The Orbiting Carbon Observatory detects where carbon dioxide is absorbed and how much is in the atmosphere, with the aim of improving predictions of climate change.

TIMELINES OF **WORLD HISTORY**

Human history is rich in detail. Civilizations have emerged, developed, and transformed, leaving behind a varied legacy and a huge body of knowledge for us to build on. Big History gives this detail context, and as we look back, connections, patterns, and forces start to appear, enhancing the story.

The spread of modern humans

- possible colonization route
- major site 100,000–10,000 BCE
- extent of ice sheet 18,000 BCE
- extent of ice sheet 10,000 BCE
- coastline 18,000 BCE
- ancient river
- ancient lake

▶ Mapping Prehistoric Migrations
Early humans migrated eastward from Africa around 100,000 years ago. They reached Australia 50,000 years ago, Japan 35,000 years ago, and crossed a land bridge over the Bering Strait to enter the Americas around 15,000 years ago. The first civilizations appeared between 5000 and 3000 BCE in areas where agriculture permitted a sufficient surplus to support large towns—in Mesopotamia, Egypt, northwest India, and around the Yangzi River in China.

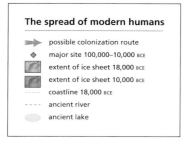

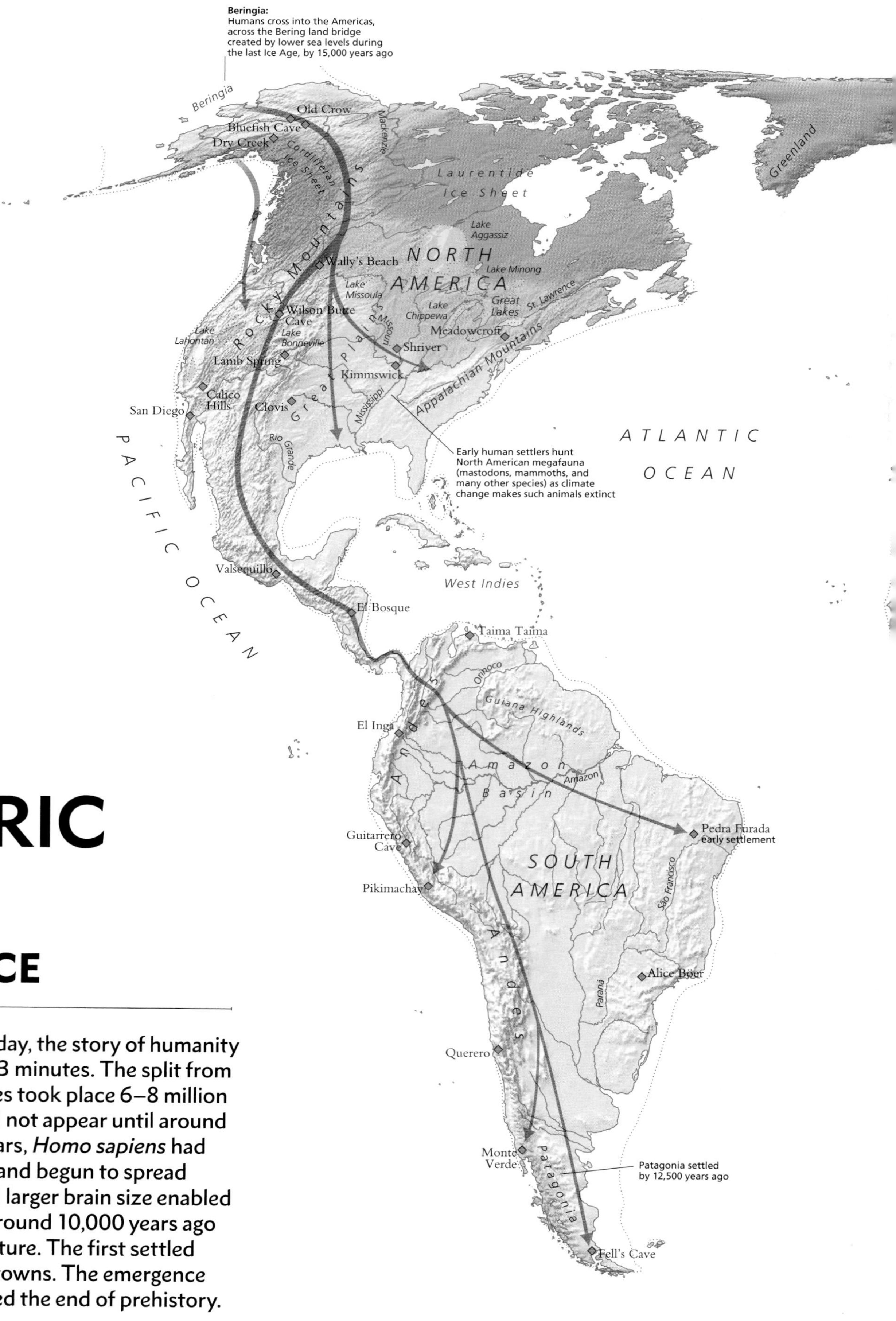

PREHISTORIC WORLD

8 MYA–3000 BCE

If Earth's 4.5 billion years were a single day, the story of humanity and its ancestors would occupy just 2–3 minutes. The split from the lineage that gave rise to modern apes took place 6–8 million years ago, but fully modern humans did not appear until around 200,000 years ago. Within 100,000 years, *Homo sapiens* had migrated from their African birthplace and begun to spread across the globe. Their adaptability and larger brain size enabled them to survive the Ice Age, and then around 10,000 years ago they developed the beginings of agriculture. The first settled villages appeared, which soon became towns. The emergence of increasingly complex societies marked the end of prehistory.

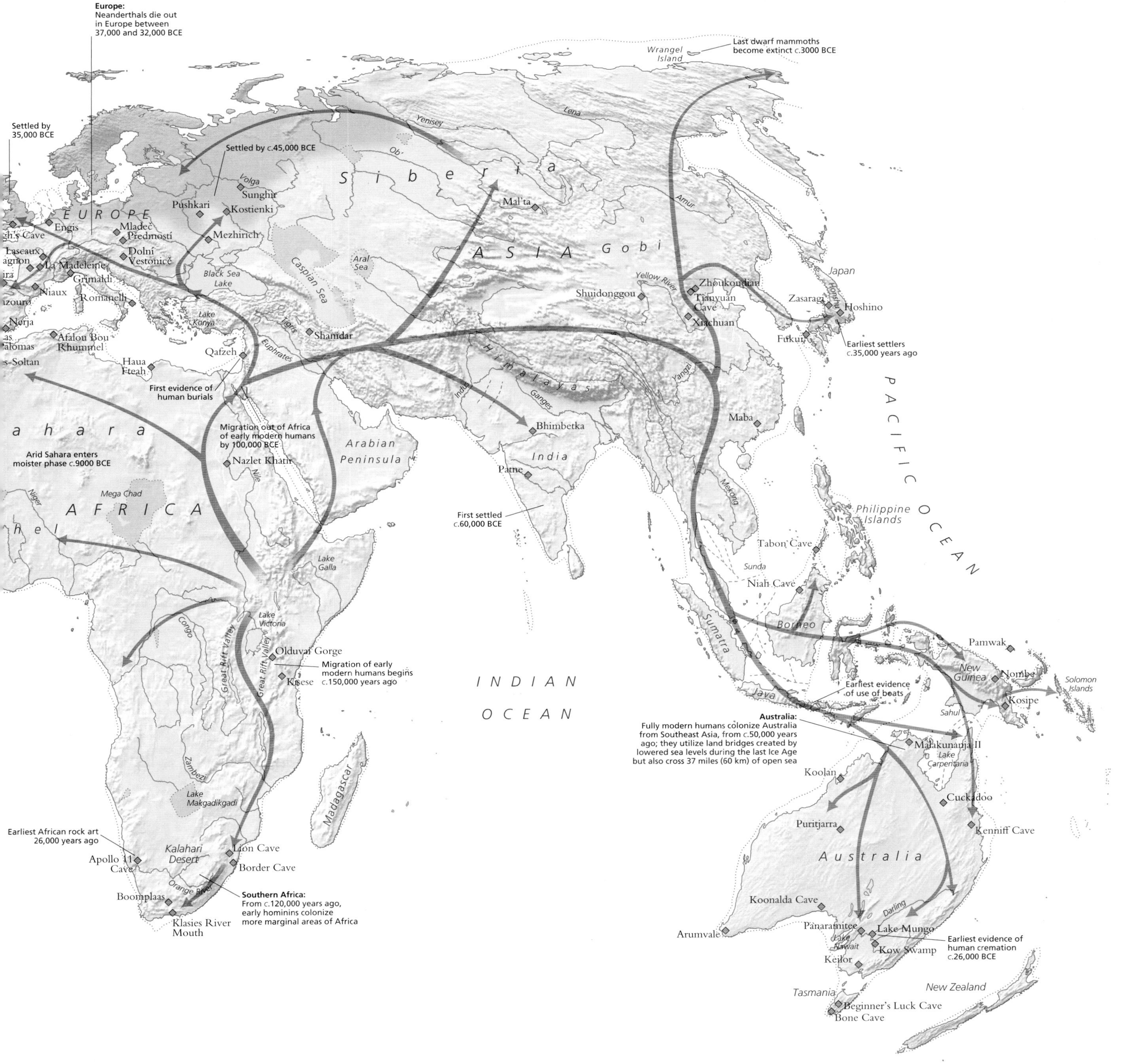
Europe:
Neanderthals die out in Europe between 37,000 and 32,000 BCE
Settled by 35,000 BCE
Settled by c.45,000 BCE
Wrangel Island
Last dwarf mammoths become extinct c.3000 BCE
Siberia
EUROPE
ASIA
Gobi
Pushkari
Sunghir
Kostienki
Mezhirich
Mladeč
Předmostí
Dolní Věstonice
Engis
Lascaux
La Madeleine
Grimaldi
Niaux
Romanelli
Nerja
Afalou Bou Rhummel
Haua Fteah
Qafzeh
Shanidar
Mal'ta
Zhoukoudian
Tianyuan Cave
Xiachuan
Shuidonggou
Zasaragi
Hoshino
Fukui
Japan
Earliest settlers c.35,000 years ago
First evidence of human burials
Migration out of Africa of early modern humans by 100,000 BCE
Nazlet Khatir
Arid Sahara enters moister phase c.9000 BCE
Mega Chad
AFRICA
Arabian Peninsula
India
Bhimbetka
Patne
First settled c.60,000 BCE
Maba
PACIFIC OCEAN
Philippine Islands
Tabon Cave
Niah Cave
Sunda
Borneo
Sumatra
Java
Pamwak
New Guinea
Nombe
Kosipe
Solomon Islands
Sahul
Earliest evidence of use of boats
Australia:
Fully modern humans colonize Australia from Southeast Asia, from c.50,000 years ago; they utilize land bridges created by lowered sea levels during the last Ice Age but also cross 37 miles (60 km) of open sea
Malakunanja II
Lake Carpentaria
Koolan
Cuckadoo
Puritjarra
Kenniff Cave
Australia
Koonalda Cave
Arumvale
Panaramitee
Lake Mungo
Lake Nawait
Kow Swamp
Keilor
Earliest evidence of human cremation c.26,000 BCE
Tasmania
Beginner's Luck Cave
Bone Cave
New Zealand
INDIAN OCEAN
Lake Galla
Lake Victoria
Olduvai Gorge
Kisese
Migration of early modern humans begins c.150,000 years ago
Congo
Great Rift Valley
Zambezi
Lake Makgadikgadi
Madagascar
Earliest African rock art 26,000 years ago
Apollo 11 Cave
Kalahari Desert
Lion Cave
Border Cave
Boomplaas
Orange River
Klasies River Mouth
Southern Africa:
From c.120,000 years ago, early hominins colonize more marginal areas of Africa
Volga
Ob'
Yenisey
Lena
Amur
Caspian Sea
Aral Sea
Black Sea Lake
Lake Konya
Tigris
Euphrates
Nile
Niger
Indus
Ganges
Himalayas
Yellow River
Yangzi
Mekong
Darling

THE PREHISTORIC WORLD

c.8–6 MYA (MILLION YEARS AGO)
The earliest human ancestors appear when hominins diverge from chimpanzee lineages. The fossil record from this period is fragmentary; as more evidence is uncovered, the relationship between species is reassessed, filling gaps in our knowledge of human evolution.

c.7–6 MYA
Sahelanthropus tchadensis in Central Africa is thought to walk upright but may predate the hominin-chimpanzee split, and may not be a direct human ancestor.

c.5.8–4.4 MYA
About the size of chimpanzees, two species of *Ardipithecus* in East Africa (*A. kadabba* c.5.8–5.2 MYA and *A. ramidus* c.4.4 MYA) may be able to stand upright and walk on two feet.

c.3.7–3 MYA
Australopithecus afarensis ("Southern ape-human") in East Africa walks upright but still climbs trees. Its brain is not much larger than that of a modern chimpanzee—around one-third the size of a modern human's brain.

c.3.6 MYA
Hominin footprints are preserved in volcanic ash at Laetoli in Tanzania. They indicate the upright, bipedal gait that allows Australopithecines to spread from forests to open savanna, giving them a wider food-gathering range than their competitors.

c.3.4 MYA
The earliest-known cut marks made by stone tools are on animal bones at Dikika, Ethiopia; the evidence suggests the animals were butchered and is linked to nearby finds of *Australopithecus afarensis.*

▲ **Olduvai Gorge** in Tanzania is one of the most important prehistoric sites in the world. The many fossils found here provide evidence of a remarkably diverse array of early hominins that flourished in East Africa's "cradle of mankind."

c.3.3 MYA
The oldest-known stone tools are discovered at Lomekwi, Kenya. Stone tools found with animal bones from c.2.6 MYA along the Gona River in Ethiopia suggest that meat is now a central part of the hominins' energy-rich diet.

c.3.18 MYA
"Lucy" is a young female *Australopithecus africanus* living in Kenya. She is one of 13 males and females in the first discovered family group of *A. africanus* fossils.

c.3.1–2 MYA
Australopithecus africanus—discovered in Taung, South Africa, in 1924—is the first African fossil to be identified as an early hominin. It walks on two feet but it is still adapted for climbing trees.

c.2.3–1.4 MYA
Paranthropus boisei ("Nutcracker Man") has powerful jaws and huge grinding teeth for chewing tough vegetable foods.

c.2.3 MYA
An early species of human, *Homo habilis* ("Handyman"), inhabits the Olduvai Gorge, Tanzania. It has a larger brain than that of the Australopithecines, and it is associated with simple stone tools and cut-marked animal bones.

c.1.95 MYA
The use of stone tools is now widespread. The prehistoric period known as the Stone Age lasts until the advent of copper c.5000–4500 BCE. It is split into three main periods: the Paleolithic (Old Stone Age), Mesolithic (Middle Stone Age), and Neolithic (New Stone Age). In the Paleolithic period, tool-makers first chip large stones and later make more sophisticated hand axes, blades, and scrapers. Paleolithic peoples are generally hunter-gatherers and live in small bands. The Paleolithic period ends between c.20,000 and c.9000 BCE for the Middle East, Europe, and East Asia (and later for South Asia, the Americas, and Africa), as the Mesolithic period begins.

c.1.8 MYA
Homo ergaster emerges in East Africa. Taller and more slender than its ancestors, it has a larger brain, is fully adapted to walking and running, and has lost its adaptations for climbing trees. It is skilled at tool-making and hunting, and inhabits a wide range of environments. A related species, *Homo erectus* or "Peking Man," is living in China by 1.6 MYA.

c.1.8–0.5 MYA
The earliest evidence for the deliberate use of fire is found in caves in South Africa and Israel.

c.1.7 MYA
Homo georgicus in Dmanisi, Georgia, is the earliest-known hominin found outside Africa; hominins have expanded their range into Eurasia.

c.1.65 MYA
Acheulean hand axes appear, made by *Homo ergaster*; these skillfully shaped tools mark a significant advance in human intelligence.

c.1.2 MYA
Paranthropines become extinct, *Homo habilis* and *Homo rudolfensis* continue, and *Homo antecessor*—the first Europeans—appear.

c.600,000 YA
Homo ergaster is well established in North Africa and the Middle East, while *Homo heidelbergensis* emerges in East Africa.

c.500,000 YA
Homo heidelbergensis flourishes in Central Europe and uses stone tools carefully flaked on both surfaces.

c.350,000 YA
Homo erectus continues to dominate in East Asia. In Europe, *Homo heidelbergensis* evolves into stockier and stronger *Homo neanderthalensis.*

c.350,000–200,000 YA
Neanderthals—the last major humanlike species before the evolution of fully modern humans, *Homo sapiens*—spread across Europe and into West Asia.

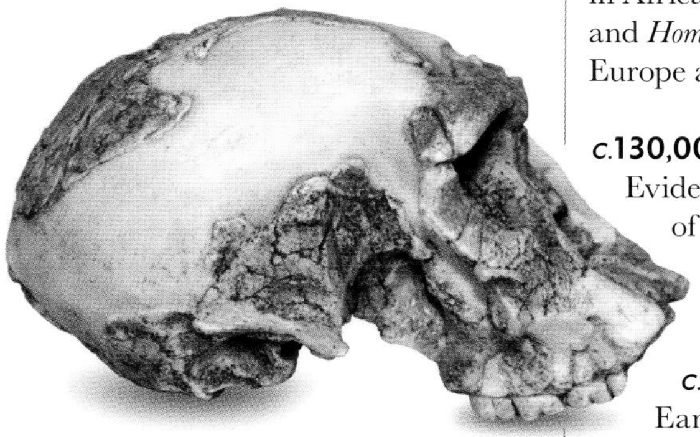

▲ **The most complete fossil skull** found at Olduvai Gorge, Tanzania, was unearthed in 1968. Crushed flat, the fragments were carefully reconstructed and identified as belonging to a young adult *Homo habilis*—nicknamed "Twiggy"—who lived *c.*1.8 MYA.

c.285,000 YA
Evidence for the use of red ochre, a natural pigment, is found at Kapthurin, Kenya.

c.280,000 YA
An incised pebble from Berekhat Ram in Israel may be an early example of art.

c.250,000–200,000 YA
Homo sapiens ("Wise Man") first appears at sites such as Omo in Ethiopia, Laetoli in Tanzania, and Jebel Irhous in Morocco. The first modern human has a smaller brain than the Neanderthals, but a lower larynx allows it to produce a bigger range of sounds.

c.200,000 YA
"Mitochondrial Eve" in Africa is the last common ancestor of all living humans, according to genetic research involving DNA samples from around the world.

c.186,000–40,000 YA
Highly skilled at making stone tools and heavy spears, Neanderthals engage in communal hunting and mass kills of large prey such as wild horses and bison; piles of animal bones are found in caves in Jersey, in the Channel Islands.

c.160,000 YA
Homo sapiens idaltu, a subspecies of modern humans, emerges in Africa; it is contemporary with *Homo ergaster* in Africa, *Homo erectus* in East Asia, and *Homo neanderthalensis* across Europe and the Middle East.

c.130,000 YA
Evidence of an increasing use of fish, shellfish, and marine mammals for food is found in South Africa.

c.110,000–90,000 YA
Early evidence of *Homo sapiens* in Southwest Asia indicates that they are migrating out of Africa.

c.90,000 YA
The oldest known harpoons, carved from bone, are found in Katanda, Democratic Republic of Congo.

c.77,000–50,000 YA
Homo sapiens spreads east from the Arabian peninsula to southern Asia, southern China, and across the ocean to the Philippines, New Guinea, and Australia.

c.75,000 YA
Advanced blade technologies, shell beads, and incised ochre are found at Blombos Cave in South Africa.

c.60,000–50,000 YA
An early burial site at La Chapelle-aux-Saints, France, contains the bones of two Neanderthal children and one adult.

c.45,000 YA
Homo sapiens has colonized much of Australia and also reaches Eastern Europe.

c.40,000 YA
Homo sapiens begins another major migration out of Africa, heading north into Western Europe and settling the whole of mainland Eurasia over the next 15,000 years. "Cro-Magnons" produce cave art and decorated artifacts in Europe. In China *Homo sapiens* may meet with late-surviving *Homo erectus.*

c.39,000 YA
The last known sites of *Homo neanderthalensis*, who is then driven to extinction by environmental changes and increased competition.

c.37,000 YA
A volcanic eruption in Campania, Italy, worsens the rapidly changing environmental conditions as ash falls across much of Europe.

c.35,000 YA
More advanced stoneworking techniques, known as Aurignacian, appear across Europe. The earliest-known figurative cave art is found in Sulawesi, Indonesia.

c.35,000 YA
Homo sapiens first spreads across northern Eurasia. Forced to retreat during the peak of the last Ice Age (21,000–18,000 YA), *Homo sapiens* returns around 15,000 YA.

c.33,000 YA
Two skulls of doglike canines are found in Siberia and Belgium; DNA studies suggest that the first domesticated dogs are descended from gray wolves in China.

c.32,000 YA
Homo sapiens has reached Japan. The Chauvet cave paintings, discovered in 1994 in a limestone cliff in the Ardèche region of southern France, depict hundreds of wild animals, both predators and prey.

▲ **This flint hand axe** is an Acheulean tool that has been skillfully shaped for a variety of tasks, including butchering meat.

c.29,000–21,000 YA
The Gravettian culture of Europe and Russia creates complex burial sites, shell jewelry, bone and antler sculptures, and clay figurines.

c.27,000 YA
Impressions left on dried silt suggest that early baskets are made at Pavlov, Czech Republic.

c.25,000–22,000 YA
Finds at Bluefish Cave in Canada suggest an early settlement of North America, by peoples originating from northern Eurasia.

c.22,000 YA
The Mesolithic period (*c.*20,000–10,000 BCE in Southwest Asia, 9000–3000 BCE in Europe) brings more sophisticated small chipped tools and greater numbers of settled communities exploiting the land more intensively for food.

c.22,000–19,000 YA
During the Last Glacial Maximum, ice caps are at their greatest extent, covering most of Northern Europe; on the grasslands further south, populations live by hunting and gathering, building shelters from scarce resources.

c.18,000 YA
The earliest-known pottery comes from Yuchanyan, China, and heralds a gradual revolution in the transportation and storage of food.

c.18,000–12,000 YA
In Europe, the Magdalenian culture hunts a wide range of species, especially reindeer, and creates beautiful art objects, engravings, and cave paintings.

c.17,000 YA
The Lascaux cave paintings in France include a wealth of animal images such as horses, stags, cattle, and bison (see p.366), as well as human and symbolic forms.

c.16,000–15,000 YA
Rising temperatures and retreating ice sheets begin to allow the gradual recolonization of areas of Northern

Europe that were abandoned during the worst climatic conditions.

c.15,000 YA
A major migration of *Homo sapiens* into North America uses a land bridge between Siberia and Alaska that is exposed by low sea levels at the height of the last Ice Age. Human populations move steadily southward, reaching Patagonia by 12,500 YA.

c.13,000–11,000 YA
North America is rapidly colonized by the "Clovis" people, who make distinctive stone tools that are flaked on both sides. They hunt large mammals to extinction and vanish within 1,000 years.

c.11,000 YA
Siberia becomes separated from North America by the Bering Straits due to higher temperatures, retreating ice sheets, and rising sea levels at the end of the last Ice Age.

c.10,800–9600 BCE
The Younger Dryas cold period, probably caused by melting ice sheets, leads to a steep decline in the availability of wild cereals in Northern Europe. A rapid rise in temperature occurs after 9600 BCE.

c.10,500 BCE
The earliest domesticated cereal is rye in Syria.

c.10,000 BCE
The Neolithic period begins in Southwest Asia, in the "Fertile Crescent" where humans learn to cultivate wild wheat rather than just gather it in; the domestication of plants and animals brings the first farming. Settlements grow larger and more permanent, and support specialty craft occupations. This final phase of the Stone Age ends with copper's discovery *c.*5000–4500 BCE.

c.9000–3000 BCE
Increased rainfall produces the "Green Sahara" of lakes, rivers, marshes, and grasslands across North Africa.

c.8500–7300 BCE
A stone wall is built around the large village of Jericho in the Jordan Valley, Palestine. Jericho is the world's oldest fortified town and one of the oldest continuously inhabited urban settlements.

c.8500–6000 BCE
Settled agriculture first develops in the Fertile Crescent of Anatolia (Turkey), the Middle East, and Mesopotamia (Iraq). The sheep and goat are domesticated in Mesopotamia.

c.8000 BCE
In China's Yellow River valley, communities are harvesting wild millet, which is fully domesticated by 6500 BCE. Wheat and barley are domesticated in West Asia. Squash is domesticated in Mesoamerica, squash and beans in Ecuador.

c.7400–6200 BCE
The early farming town of Çatalhöyük, Turkey, houses as many as 8,000 people.

c.7000 BCE
Farming begins to spread from Turkey into Southeast and Central Europe; Mediterranean hunter-gatherers gradually turn to agriculture, using imported West Asian crops and animals. Cattle are domesticated by hunter-fisher communities of the Green Sahara in North Africa and by farmers in the Indus Valley in Pakistan and northwest India. Banana, taro, and yam are cultivated in the highlands of New Guinea.

▲ **Jericho's high stone wall,** moat, and round tower were built to protect a small town of beehive-shaped houses that had stone foundations, plastered floors, and often their own courtyards and ovens.

c.7000–6000 BCE
Native (naturally occurring) copper and gold has been hammered into small objects since before 8000 BCE in West Asia, but now ores are smelted (heated with charcoal) to extract copper. Within 1,000 years, copper and lead are used to produce much more effective tools and weapons, such as copper ax heads.

c.6000 BCE
The invention of the plow in Egypt makes agriculture more productive. Villages along China's Yangzi River catch fish, grow rice, and raise pigs and chickens. Domesticated corn (maize) is developed from wild teosinte in Mexico, and becomes the principal cereal in the Americas.

c.5800 BCE
The "black mummies" of the Chinchorro people in coastal northern Chile and southern Peru are the world's first example of artificial mummification.

▼ **The Lascaux Caves**, discovered by four teenagers in the Dordogne, southwestern France, in 1940, contain some of the most impressive works in all of Paleolithic art. Cro-Magnon artists painted almost 2,000 figures on the walls, using reds, yellows, and blacks made from mineral pigments *c.*17000 YA. In the famous "Hall of the Bulls," a vast fresco covers the walls with wild horses, stags, and aurochs, or bulls.

c.5500 BCE
The world's first irrigation agriculture begins at Choga Mami in southern Mesopotamia, on the alluvial plains between the Tigris and Euphrates rivers. Irrigation is established along the Nile River in Egypt by 3100 BCE.

c.5500–4500 BCE
The Linearbandkeramik farming culture, known for its distinctive pottery, flourishes in Europe.

c.5100 BCE
Europe's earliest copper mines are at Ai Bunar, Bulgaria.

c.5000 BCE
Domestic animals are kept for milk and meat, and oxen for pulling plows, in West Asia, North Africa, and Europe. The llama, alpaca, and guinea pig are domesticated in the Andes, Andean coast, and tropical lowlands of South America, while potatoes are cultivated at high altitudes in the Andes.

c.4500 BCE
Irrigation techniques are introduced in the Indus Valley. The horse is domesticated in the Eurasian Steppe.

c.4200–3750 BCE
Emergence of the world's first city-states, in Mesopotamia; Uruk in Sumer, southern Mesopotamia, is possibly the first city and grows to house 50,000 people by 2800 BCE.

c.4000 BCE
In Egypt, farmers live in small communities in the Nile Valley, where the river's annual flooding deposits fertile silt on a broad strip along its banks. The vine and olive are domesticated in the eastern Mediterranean. In China, wet-rice cultivation in plowed irrigated fields begins.

c.4000–1000 BCE
The Old Copper Culture, a major network of industry and trade based on locally mined, cold-hammered native copper, flourishes in the Great Lakes region of North America.

c.3500 BCE
Solid wheels made of wood are first seen on wagons in Poland and the Balkans; disk wheels emerge in Sumer c.3200 BCE. Sumer is at the heart of a vast trading network that stretches from Egypt through West Asia to the Indus Valley, South Asia. Stamp seals are used in West Asia as a form of signature for economic and administrative purposes. The island of Crete gives birth to the Minoan civilization, Europe's first.

c.3300 BCE
The Bronze Age begins in Southwest Asia (reaching the Aegean by c.3200 BCE, China c.3000 BCE) when copper is alloyed with tin to create a much harder metal, bronze. Bronze Age societies are most often characterized by the development of writing, urban societies, large-scale architecture, and early forms of state. The period between the end of the Neolithic and the start of the Bronze Age is often known as the Chalcolithic (or Copper Age).

c.3300 BCE
Pictographic writing is invented at Uruk, Sumer, for recording property and commerce.

c.3200 BCE
Egypt's first hieroglyphic script is developed. The Indus Valley civilization dawns, with centers at Mehrgarh, Harappa, and Mohenjo-Daro. The first stone circles and lines of standing stones are built in Northern and Western Europe.

c.3100 BCE
Stonehenge in southern England begins as an earthwork enclosure.

c.3100 BCE
King Narmer completes the unification of Upper and Lower Egypt, becoming the first pharaoh.

c.3100–2900 BCE
Proto-Elamite script, an early Bronze-Age writing system, is in use across the Iranian plateau. Sumer develops a cuneiform script, making wedge-shaped marks with a sharpened reed on a soft clay tablet.

CALENDAR SYSTEMS

Most of the world uses the Gregorian calendar for noting day-to-day activities and historical landmarks, but some cultures use more ancient calendars, often for marking events of ritual and cultural significance. Historians also take account of older dating systems when analyzing source material.

CALENDAR	BASIS	SYSTEM
Hebrew	Lunisolar	12 months of 29/30 days and an intercalary (leap) month added seven times in a 19-year cycle. Era begins at 3761 BCE.
Maya	---	260-day and 365-day cycles combined to form "Calendar Round" of approximately 52 years.
Chinese	Lunisolar	12 months of 29/30 days, with a leap month every 2–3 years. Beginning of era is disputed: either 2697 or 2852 BCE.
Ancient Egyptian	Solar	365 days, with 12 months of 30 days, and 5 intercalary days at the end of the year.
Attic (ancient Athens)	Lunisolar	354 days, with 12 months of 28/29 days and a leap month of 30 days every third year.
Roman	Solar	355 days, 12 months of 28, 29, or 31 days, with a leap month added periodically. Era is dated from 753 BCE (date of foundation of Rome).
Japanese	Lunisolar	A system similar to the Chinese calendar was in official use until 1873. Era is dated from 660 BCE.
Gregorian	Solar	365 days, with a leap day every four years. Era begins at 1 CE (also known as AD 1).
Ethiopian/Ge'ez	Solar	12 months of 30 days, with a leap month of 5 days (6 days every fourth year). Era begins at 9 CE.
Coptic (Egypt)	Solar	Similar to Ethiopian, but era begins at 284 CE.
Islamic	Lunar	354 days (12 months of 29/30 days). Era begins at 622 CE (date of the Hegira—Muhammad's flight from Mecca).

◀ **This astrological table** from Mesopotamia was written in cuneiform on clay during the Uruk Era (c.3500–2900 BCE). Early calendars were devised by astronomer-priests to track seasonal turning points and rituals through the agricultural year.

The World in 750 BCE

- Greek cities and territories
- Phoenician cities and territories
- small Chinese states under the Eastern Zhou dynasty
- NOTE: Settlements in italics were not in existence in 750 BCE but were significant during this era.

▶ The Ancient World in 750 BCE
In 750 BCE, the Assyrian Empire was dominant in the Middle East, after a period of chaos c.1000 BCE. Egypt was weakened, while China's Zhou dynasty was in a state of near collapse following barbarian invasions. Greece was beginning to establish colonies in the Mediterranean. India's center of gravity shifted eastward after the collapse of the Indus Valley civilization. In the Americas, the Olmecs and Chavín had established the continent's first cities.

ANCIENT WORLD 3000–700 BCE

Writing first developed in Mesopotamia and Egypt around 5,000 years ago and reveals much about the city-states, kingdoms, and empires that emerged there and in China, India, Peru, and Mexico. Hierarchical societies had specialized classes of warriors, priests, merchants, artisans, and an agricultural underclass whose surplus supported the rest. Trade between cities increased prosperity, but the hunger for resources led to large-scale warfare for the first time. Egyptians, Hittites, Assyrians, and Babylonians battled over the Middle East, while in China and India local dynasties vied for supremacy.

Palaeosiberians
Lena
Yenisey
Samoyeds
Ob'
Siberia
British Isles
Germanic peoples
Baltic peoples
Finno-Ugrians
Volga
Slavs
Celts
Danube
Illyrians
Thracians
Ligurians
Etruscans
Caucasian peoples
Black Sea
Caspian Sea
Cimmerians
Gordium
PHRYGIA
URARTU
Tushpa
Nineveh
Nimrud
Tigris
Euphrates
ASSYRIAN EMPIRE
Rome
Cumae
Motya
Carthage
Mycenae
Aegean Sea
Knossos
Crete
Mediterranean Sea
Kadesh
Byblos
Tyre
Damascus
Babylon
Susa
ELAM
Ur
BABYLONIA
Bubastis
ISRAEL
AMMON
Jerusalem
Memphis
MOAB
JUDAEA
EGYPT
Thebes
KUSH
Napata
Red Sea
Semites
Arabian Peninsula
Persian Gulf
Iranians
Nilo-Saharan peoples
Chadians
Niger
Nile
Niger-Congo peoples
Kushites
Congo
Khoisan peoples
Zambezi
Kalahari Desert
Orange River
Madagascar
Amur
Altaic peoples
Gobi
Manchuria
Yellow River
Ainu
Japan
Japanese
Koreans
YAN
WEI
QI
LU
JIN
ZHOU
QIN
Erlitou
Zhengzhou
SONG
ZHENG
CHU
Yangzi
Wu
WU
YUE
Tibetans
Himalayas
Harappa
Indus
Mohenjo-Daro
Ganges
Pataliputra
INDIAN STATES
Dravidians
Sinitic peoples
Mon-Khmer peoples
Mekong
Philippine Islands
Malay
Borneo
Sumatra
Java
PACIFIC OCEAN
INDIAN OCEAN
Papuans
New Guinea
Australian Aborigines
Darling
New Zealand

THE ANCIENT WORLD

c.3100–2890 BCE
Egypt's First Dynasty, founded by King Narmer: his successors establish the capital at Memphis c.3000 BCE.

c.3000–2334 BCE
Mesopotamia's Early Dynastic Period: in Sumer, southern Mesopotamia, agricultural success leads to the world's first urban culture and city-states such as Uruk, Ur, and Eridu flourish. Mesopotamia and Egypt establish an extensive trade network.

c.3000 BCE
China's Longshan Culture emerges, with larger settlements defended by stamped earth walls, greater social complexity, and wealth marked by black pottery, jade carvings, and bronze artifacts.

c.2900 BCE
Early marble figurines are made by the Cycladic culture of Greek islands in the Aegean Sea.

c.2750 BCE
Europe's Bronze Age begins on the Greek islands of Crete and the Cyclades.

c.2700 BCE
Gilgamesh, celebrated in a later epic, may be the ruler of Uruk in Mesopotamia. In China, jade is being mined and carved into ritual vessels.

2686–2181 BCE
In Egypt, the Old Kingdom Period is an era of powerful kings, strong centralized government, and pyramid building—beginning with the Step Pyramid built as a royal tomb for King Djoser at Saqqara c.2650 BCE.

2613–2494 BCE
Egypt's 4th dynasty is the age of the first true pyramids—including those built at Giza for the pharaohs Khufu, Khafre, and Menkaure.

c.2600 BCE
The rich array of grave goods buried in the Royal Graves of Ur, Mesopotamia, indicates trade links extending as far as the Indus. The Indus Valley civilization flourishes; farmers are plowing their fields and dozens of towns and cities have emerged—Mohenjo-daro and Harappa have populations of 100,000 and 60,000 respectively.

c.2600 BCE
South America's first cities develop; several settlements with temple complexes, such as Norte Chico (or Caral), emerge in coastal Peru.

c.2550–2300 BCE
Megalithic stone circles are erected at the Stonehenge ceremonial complex in Britain.

c.2528 BCE
Egypt's Great Pyramid, a vast royal tomb, is completed at Giza, near Memphis.

c.2500 BCE
The Bronze Age reaches Central Europe, with the earliest bronze artifacts discovered in Poland. Metalworking, in the form of copper, spreads across Europe to the British Isles. The Bell Beaker culture spreads from Western to Central Europe, named after the shape of pottery vessels buried in graves.

◀ **This Cycladic figurine**, just 12.5in (32cm) tall, was carved in marble on the Aegean island of Amorgos c.3000–2000 BCE; many such figurines are found in graves, some with traces of paint.

▲ **Khufu's Great Pyramid** towers over others built at Giza in Egypt. The world's tallest building for more than 3,800 years—topping 480ft (145m)—is the oldest of the Seven Wonders of the Ancient World, and the only one that survives largely intact.

c.2500 BCE
The world's earliest town map is made on a small clay tablet at Ga Sur, Mesopotamia; the Babylonians later map the city of Nippur c.1500.

c.2500–2350 BCE
A border conflict between Umma and Lagash in Mesopotamia is the earliest international dispute to be recorded.

c.2500–800 BCE
The Arctic Small Tool people, ancestors of the Inuit, enter Alaska from Siberia, bringing the bow and arrow that replaces the spear and transforms hunting in North America; they spread across Canada and settle in Greenland until c.800 BCE.

c.2334 BCE
Sargon of Akkad conquers Sumer; his subsequent campaigns in Southwest Asia and Anatolia create the world's first empire—the Akkadian Empire—which stretches from the eastern Mediterranean to the Gulf.

c.2300 BCE
The city of Ebla, Syria, is destroyed: Ebla began as a small settlement in c.3500 BCE and gave rise to one of the earliest kingdoms in Syria, at the heart of a vast trading network.

c.2300 BCE
The Bronze Age reaches Southern Europe, beginning with the Balkans and Italy.

c.2205 BCE
The Xia dynasty is founded in eastern China by Yu the Great, according to Chinese tradition, and rules until 1766 BCE. However, the first ruling dynasty in China for which there is real historical evidence is the Shang dynasty (2070–1600 BCE).

2181 BCE
Egypt's 6th dynasty ends with the collapse of the Old Kingdom after natural disasters weaken its authority; the First Intermediate Period begins (to 2040 BCE).

c.2150 BCE
The collapse of the Akkadian Empire founded by Sargon sees the rise of powerful regional rulers of city-states—notably Gudea, ruler of Lagash in southern Mesopotamia.

c.2100 BCE
The 3rd dynasty of Ur witnesses a revival of Sumerian power; King Ur-Nammu builds the first ziggurat (stepped tower) at Ur.

c.2094–2047 BCE
The reign of Shulgi, son of Ur-Nammu: he completes the Great Ziggurat at Ur and builds roads and inns where travelers can rest; he also introduces the world's first national calendar—the Umma calendar—and standardizes timekeeping and weights and measurements in order to help organize the state's complex bureaucracy.

c.2050 BCE
The Minoans begin building great palaces as centers of power on the island of Crete, including Phaistos, Knossos, Malia, and Zakros.

2040 BCE
Mentuhotep II, ruler of Thebes, reunites Upper and Lower Egypt and initiates Egypt's Middle Kingdom (2040–1640 BCE). In reality Egypt is run by powerful officials, such as viziers, until strong centralized rule returns under Amenemhet I from 1985 BCE.

c.2000 BCE
The trading city of Ashur (or Assur) on the Tigris River is dominant in northern Mesopotamia.

c.2000 BCE
Maize, beans, and squashes are cultivated in southwestern North America, and long-distance trading routes are in place.

c.2000–1600 BCE
The Minoan civilization is at its height; Crete is home to several small kingdoms, exports pottery, gold, bronze, and other commodities, and establishes colonies around the Aegean.

1960 BCE
Senwosret (Sesostris) I of Egypt conquers Nubia and extends the southern frontier of Egypt to the second cataract of the Nile.

c.1900–1800 BCE
The city of Erlitou is built on the Yellow River in China's Henan province, where the Shang civilization later develops. The oldest-surviving Chinese writing appears on oracle bones, used in divination to consult ancestors.

c.1900–1700 BCE
The Indus Valley civilization is in decline as river flows alter and trade weakens; several cities are struck by disease and gradually abandoned.

c.1894 BCE
Babylon's First Dynasty is founded in southeastern Mesopotamia.

c.1813–1781 BCE
The reign of Shamshi-Adad, who unites northern Mesopotamia to create a short-lived precursor to the Assyrian Empire.

c.1800–1100 BCE
The Shang civilization centered along the Yellow River gives rise to many features that later come to characterize Chinese society, such as a strong bureaucracy and the worship of ancestors.

c.1800 BCE
Proto-Sinaitic script—the world's first partially alphabetic script—is developed among quarry workers in Sinai, Egypt.

c.1800 BCE
Long-distance trade networks are established in South America, allowing for the spread of pottery. There is large-scale cultivation on the Pacific coast, and substantial settlements such as El Paraiso and Sechin Alto in Peru are dominated by massive temple complexes.

c.1792–1750 BCE
The reign of Hammurabi, who establishes the Babylonian Empire in Mesopotamia through a series of conquests. His law code is displayed on monumental stelae (memorial stones) in temples throughout his empire.

c.1750 BCE
Linear A script comes into use in Crete (and has yet to be deciphered). At Sechin Alto, Peru, construction begins on the largest monument complex in the Americas of the second millennium BCE.

c.1725 BCE
Middle Kingdom Egypt begins to disintegrate; regional governors gain in power and civil war erupts.

c.1700 BCE
After several Minoan palaces are destroyed by fire, only that of Knossos is rebuilt, on a grand scale that suggests it now dominates Crete. Hittites establish the Old Kingdom in central Anatolia, with Hattusa as its capital.

c.1650–1550 BCE
Egypt's Second Intermediate Period: Lower Egypt is ruled by the Hyksos, a warrior elite who invade from the Levant; Upper Egypt remains ruled from Thebes by native kings.

c.1650 BCE
The Ebers Papyrus, the first major Egyptian medical treatise that survives, shows understanding of basic diagnosis and description of major diseases.

c.1628 BCE
The beginning of several years of global cooling, documented by tree rings. The climate change may be triggered by a major volcanic eruption—perhaps that of Vesuvius in Italy, or on the island of Thera (now Santorini) in the Aegean, where a catastrophic eruption buries the town of Akrotiri and other Minoan settlements.

c.1600 BCE
Mycenae in Greece emerges as a center of civilization in the Aegean; it uses Linear B script, the earliest form of Greek script. Egypt and Mesopotamia produce the earliest known hollow glass vessels.

c.1600–1046 BCE
China's Shang dynasty: tradition names King Tang as its founder.

1595 BCE
The Hittite king Mursili I sacks Babylon, ending Hammurabi's dynasty and the Old Kingdom.

c.1570s BCE
Kassites, the warrior elite of the fallen Old Kingdom, gain control over southern Mesopotamia.

c.1570–1070 BCE
Egypt's rulers are buried in rock-cut tombs in the Valley of the Kings near Thebes.

c.1550 BCE
Ahmose I drives the Hyksos from Lower Egypt and reunites Egypt under a new capital at Thebes; the New Kingdom begins (to 1069 BCE).

c.1500–1200 BCE
The Egyptians expand the Karnak temple complex at Thebes.

c.1500–900 BCE
Nomadic pastoralists migrate from Central Asia to northern India and begin to settle and cultivate crops by 1100 BCE; they speak Sanskrit, which is the language of early Indian sacred writings and the ancestor of modern languages such as Hindi and Urdu.

▶ **Carvings at Sechin Alto** in Peru (*c.*1750–900 BCE) depict human sacrifice and decapitation. These granite blocks, some weighing more than two tons, line a huge mound at the heart of the settlement.

▲ **The massive Lion Gate** (*c.*1250 BCE) guards the main entrance to the hilltop citadel of Mycenae in the Peloponnese, southern Greece. It is topped by a limestone slab carved to show two rearing lions, symbols of royal power, and was originally closed by a double door.

c.1500 BCE
The Hittite Old Kingdom of Anatolia declines; the Hurrian dynasty of Mittani emerges nearby in northern Mesopotamia. Mitanni, New Kingdom Hittites, and Egypt compete for control over the Levant in the eastern Mediterranean.

c.1500 BCE
Bronze-working is evident in Thailand and Vietnam. Copper is worked in the Sahara. Early metalworking exists in Peru and pottery in Central America.

c.1450 BCE
The Minoan palaces on Crete are destroyed and the island falls under Mycenaean control; the Mycenaeans are now at their peak, with trading links stretching from Sicily to the Levant.

c.1400–750 BCE
The Lapita people, named for the fine pottery found at Lapita, New Caledonia, migrate from Indonesia to Melanesia in the Pacific; from here they migrate 3,700 miles (6,000km) east, all the way to Samoa and Tonga; these skilled seafarers are the ancestors of the Polynesians.

c.1400–1300 BCE
Scribes at Ugarit in the Levant create an early alphabet; written in cuneiform (wedge-shaped marks on clay tablets), its 30 letters represent sounds (consonants and vowels) rather than whole words or concepts as hieroglyphs.

1390–1352 BCE
Egypt's New Kingdom reaches the zenith of its international and artistic power during the reign of Amenhotep III.

c.1352–1336 BCE
The reign of Amenhotep IV: he breaks with Egypt's old religion, worships the sun god Aten, and takes the name Akhenaten.

c.1350 BCE
The city of Ashur in northern Mesopotamia breaks free from Mittani; its rulers proclaim themselves kings of Assyria, founding the Middle Assyrian Empire (1350–1000 BCE).

c.1336–1327 BCE
The brief reign of Egypt's boy-king Tutankhamun: the priests of Amun restore the old religion and the city of Akhenaten is abandoned.

c.1300 BCE
The Urnfield Culture emerges in Danube region of Europe; named after its practice of cremating the dead and burying the remains in funerary urns, it spreads to Italy and to Central and Eastern Europe.

c.1300 BCE
China's Shang dynasty, fending off threats from nomadic tribes to the north, moves its capital from Zhengzhou to Xi'ang.

c.1279–1213 BCE
Ramesses II's long reign is a time of stability and prosperity for Egypt; he builds huge monuments such as the temple of Abu-Simbel and seeks to extend Egyptian influence.

c.1274 BCE
Ramesses II fights the Hittites at the battle of Qadesh; he negotiates a pioneering peace treaty with the new Hittite king, Hattusilis III, in 1259 BCE and later takes two Hittite princesses in marriage.

c.1200 BCE
The eastern Mediterranean enters a century of turmoil, as states in the Levant are raided by waves of migrants known as the Sea Peoples; the Hittite capital of Hattusas is destroyed and the Hittite Empire collapses by *c.*1180 BCE.

c.1200 BCE
The Iron Age begins when smiths in Anatolia devise forges that generate the high temperature needed to separate iron from iron ore. Iron tools are easier to make than bronze, and are lighter, stronger, and cheaper. Ironworking also develops in India, spreads into Europe by 1000 BCE, appears in West Africa *c.*800 BCE, and is found in China by 600 BCE. Iron-Age societies are more complex than those of the Bronze Age, often with developed bureaucracies and unified states controlling large areas. The end of the Iron Age is not marked by advances in technology but by political developments such as the Qin unification of China and the Roman conquest of Italy's peninsula, both in the 3rd century BCE.

c.1200 BCE
The Olmec culture, Mesoamerica's first civilization, develops in Mexico. The Chavín culture emerges in the Peruvian Andes and spreads to the coast.

1184–1153 BCE
The reign of Ramesses III, the last great pharaoh: when Egypt is invaded by the Sea Peoples, he drives them from Lower Egypt in 1178 BCE but cannot prevent them from colonizing the Levant. Egypt enters a period of economic and political decline.

1154 BCE
The Kassite dynasty of Babylon ends when the city is sacked by neighboring Elam.

c.1150 BCE
Mycenaean palaces' stronger defenses indicate fear of attack; by 1100 BCE, most are destroyed, the Mycenaean period ends, and Greece enters a Dark Age.

c.1100–1000 BCE
In the Levant, Canaanite port cities such as Tyre and Sidon establish trading posts and colonies across the eastern Mediterranean. Their most valuable export, a purple dye, earns the Canaanites their Greek name of Phoenicians (after the phoenix, a reddish-purple bird in Greek myth).

1070 BCE
The New Kingdom ends: Egypt enters a time of unrest called the Third Intermediate Period (1069–664 BCE); by 1000 BCE, Egypt loses all the territories it had won during the New Kingdom.

c.1050 BCE
Assyria loses territories to Aramaeans migrating into the Middle East, but survives as a state during the region's Dark Age.

1027 BCE
The Zhou dynasty supplants the Shang dynasty and rules China over three periods: Western Zhou (1027–771 BCE), Eastern Zhou or "Spring and Autumn" (771–476 BCE), and Warring States (476–221 BCE).

c.1000–300 BCE
Japan's Final Jomon Period: the prehistoric hunter-gatherer culture is named after the rope patterns (*jomon*) that decorate its pottery.

c.1000 BCE
The Western Zhou record the geography of China. Wet rice farming and bronze technology is exported to Korea. Polynesian culture evolves in Pacific.

◀ **This ritual wine vessel** was made in bronze by highly skilled metalworkers in China in the 10th century BCE, during the Western Zhou dynasty.

c.1000 BCE
The Adena culture starts to develop along the Ohio River in North America; it is characterized by ritual earthworks and burial mounds containing fine artifacts.

c.1000 BCE
Ironworking reaches central Europe, and hilltop forts are being built by Celts and others in central and western Europe.

c.1000 BCE
The Phoenicians continue to be a major maritime power, trading across the Mediterranean; their alphabet spreads to the Greeks, who are beginning to establish colonies in Aegean and West Asia.

c.1000 BCE
Egypt, Babylon, and Assyria are in decline, allowing the rise of the Kingdom of Israel; King David unites the Israelite tribes and makes Jerusalem his capital. South of Egypt, the Nubian kingdom of Kush is founded.

c.1000 BCE
In the Ganges valley in India, scribes record the ancient texts of the Vedas in Sanskrit; these sacred writings and hymns become the oldest texts of Hinduism.

c.965–928 BCE
The reign of King David's son, Solomon: he expands Israelite territory and builds a magnificent palace and temple in Jerusalem; Megiddo becomes an important fortress and administrative center. Upon Solomon's death, the kingdom splits in two; eventually Israel and then Judaea become part of the Assyrian Empire.

c.912 BCE
The Neo-Assyrian Empire is founded when Adad Nirari II comes to power; Assyria has regained lost territories and begins to re-emerge as a major power in Mesopotamia.

c.900 BCE
In Mesoamerica, the Olmec site of San Lorenzo is destroyed; La Venta becomes the leading Olmec center, dominated by a tall pyramid that is a forerunner of Mayan temples. The Olmec script of glyphs is the first in the region. Far to the south, the Chavín are now politically and culturally dominant in Peru; skilled engineers and architects build canals and level slopes into terraces for farming and construction.

c.900–700 BCE
Scythians adopt pastoral nomadism, spreading across the central Eurasian Steppe, and build kurgans (burial mounds).

883 BCE
Ashurnasirpal II inherits the Assyrian throne and moves the capital from Ashur to Nimrud in 880 BCE.

814 BCE
The traditional date for the founding of Carthage, a Phoenician colony on the North African coast (in Tunisia).

c.800 BCE
In the Peruvian Andes, Chavín de Huantar flourishes as the capital of the Chavín civilization and a pilgrimage center for a cult of supernatural beings that are half human, half animal; stone carvings line terraces, plazas, and galleries inside a massive, flat-topped temple.

c.800 BCE
In Western Europe, iron is replacing bronze for tools and weapons; in the early Iron Age culture centered around Hallstatt, Austria, chieftains live in hilltop forts and are buried with lavish grave goods. The first Greek colonies in the Mediterranean are founded at Ischia in Sicily and Al Mina in Syria.

800–480 BCE
The Archaic Period in Greece sees the rise of city-states (*poleis*), a rapidly growing population, the founding of colonies stretching from Asia Minor to North Africa and Spain, and the flowering of early philosophy, theater, and art; the Phoenician alphabet is adopted, and the works of Homer mark the birth of European literature.

776 BCE
The traditional date for the first Panhellenic games in Olympia, Greece; rivalry between Greek city-states is intense, but a distinct Greek culture is emerging and all Greeks are identified as "Hellenes."

771 BCE
China's Western Zhou dynasty collapses; the capital moves east to Luoyang, marking the start of the Eastern Zhou Period.

753 BCE
The traditional date for the founding of Rome by Romulus and Remus; Rome is one of several city-states in central Italy that are ruled by Latins, Sabines, and the Etruscans—Italy's first indigenous civilization.

c.750 BCE
The *Iliad* and *Odyssey*, epic poems traditionally attributed to Homer, are first written down, as are the works of the Greek poet Hesiod.

c.747 BCE
Piye, the Kushite ruler of Nubia, conquers Upper and Lower Egypt and unites them under Kushite rule.

745–727 BCE
The reign of Tiglath-Pileser III of Assyria; with a disciplined, technically advanced army and an efficient bureaucracy, Assyria recoups earlier losses.

721–705 BCE
Sargon II rules Syria; he conquers Babylon, the Armenian state of Uratu, and Israel; large numbers of Israelites are deported to northern Mesopotamia, known in the Bible as the "Lost Tribes of Israel."

701 BCE
Assyrians invade Judaea and lay siege to Jerusalem in 700 BCE.

▲ **Romulus and Remus** are suckled by a wolf in this bronze statue in Rome. Legend tells how the twin sons of the god Mars and a royal princess had been left out to die by a wicked uncle; fostered by a shepherd, they grew up to found the city of Rome.

CULTURE AND CREATIVITY

From prehistoric rock art to 3D film, writers, musicians, and artists of all kinds have produced works that capture our imagination and provide invaluable insight into the thoughts and aspirations of past civilizations and the world today.

- **c.540,000 YA** Geometric carving on a shell found in Java, Indonesia, may be the oldest piece of art made by our human ancestors.
- **c.75,000 YA** Pierced shells found in South Africa's Blombos Cave may be beads for jewelry.
- **c.40,000 YA** Flutes made from bone and ivory in the Geisenklösterle cave near Ulm in Germany are among the oldest known musical instruments.
- **c.32,000 YA** Figurative rock art found at the Chauvet Cave in France is part of a widespread artistic phenomenon that includes cave paintings at Lascaux and Altamira.
- **c.29,000–27,000 YA** The Czech Republic's *Venus of Dolní Věstonice* is the oldest known clay figurine.
- **c.18,000–12,000 YA** In France, artists of the Magdalenian culture carve bones, antlers, and spear throwers into animal shapes.
- **c.18,250–17,500 YA** The oldest known pottery comes from Yuchanyan in Hunan, China.
- **c.4700–4200 BCE** More than 3,000 pieces of gold jewelry are buried with the elite at Varna in Bulgaria.
- **c.2100–1900 BCE** The *Epic of Gilgamesh*, an epic poem from Mesopotamia, is the world's oldest surviving work of literature.
- **c.1500–1200 BCE** The *Rigveda*, an ancient Indian collection of Vedic Sanskrit hymns, is composed.
- **c.750 BCE** Two ancient Greek epic poems, Homer's *Iliad* and *Odyssey*, tell of the fall of Troy and Odysseus's journey home.
- **c.550–525 BCE** Greek potter and vase painter Exekias signs his masterpiece, a black-figure amphora showing Achilles and Ajax playing a board game.
- **458 BCE** The playwright Aeschylus, "father of tragedy," stages the *Oresteia* in Athens.
- **c.447–432 BCE** The Parthenon and its marble frieze are designed by Phidias, Athens' leading architect and sculptor.
- **c.400–300 BCE** The world's longest poem, India's *Mahabharata*, is compiled.
- **27 BCE–14 CE** The reign of Augustus sees the golden age of Latin literature, including Virgil's epic poem the *Aeneid*, Ovid's *Metamorphosis*, and Horace's *Odes*.
- **713–803 CE** The world's largest statue of the Buddha is carved out of a cliff face near Leshan in China.
- **c.1001–1010** The *Tale of the Genji* by Japanese noblewoman Murasaki Shikibu is one of the world's first novels.
- **1072** Chinese landscape artist Guo Xi paints *Early Spring* on a silk scroll.
- **c.1297–1300** Giotto di Bondone paints his fresco cycle of the life of the Virgin Mary and Jesus Christ in Padova, Italy.
- **c.1321** Dante Alighieri's *Divine Comedy* is Italy's greatest contribution to literature.
- **c.1360s** French musician Guillaume de Machaut composes his polyphonic *Mass of Our Lady* for Rheims Cathedral.
- **1387–1400** Geoffrey Chaucer's *The Canterbury Tales* presents a memorable group of pilgrims in medieval England.
- **1400–1450** Chinese ceramicists of the Ming dynasty perfect the art of blue-and-white porcelain.
- **c.1482** In Florence, birthplace of the Italian Renaissance, Sandro Botticelli paints *La Primavera* (Spring).
- **1501–1504** Michelangelo Buonarotti sculpts the statue of *David* in Florence.
- **c.1503–1506** Leonardo da Vinci paints the *Mona Lisa* in Florence.
- **1526** Albrecht Dürer, Germany's greatest artist of the Renaissance, makes the engraving *Knight, Death, and the Devil*.
- **c.1594–1595** England's William Shakespeare stages *Romeo and Juliet*.
- **1598** One of China's best-loved classical operas, *The Peony Pavilion*, is first staged.
- **1605** In Miguel de Cervantes' novel *Don Quixote*, a Spanish knight sets out with his squire to revive chivalry.
- **1607** Claudio Monteverdi's *L'Orfeo* is premiered in Mantua, Italy; it is the oldest opera still performed today.
- **1642** Dutch artist Rembrandt van Rijn paints *The Night Watch*.
- **1666** French playwright Molière stages his comedy of manners, *The Misanthrope*.
- **1678** John Bunyan's *The Pilgrim's Progress* is the first English novel.
- **1722** German composer Johann Sebastian Bach publishes *The Well-Tempered Clavier*.
- **1725** Italian composer Antonio Vivaldi publishes *The Four Seasons*.
- **1742** German composer George Frideric Handel's *Messiah* is first performed.
- **1791** Austrian composer Wolfgang Amadeus Mozart's opera *The Magic Flute* premiers in Vienna.
- **1797–1800** Spain's Francisco de Goya paints *The Naked Maja*.
- **1799** England's William Wordsworth begins his greatest poem, *The Prelude*.
- **1808** Ludwig van Beethoven finishes his *Symphony No. 5* in Vienna. Johann Wolfgang von Goethe's play *Faust* is published in Germany.
- **1813** English novelist Jane Austen's *Pride and Prejudice* is published.
- **1819** Austrian composer Franz Schubert writes his piano quintet *Die Forelle* (The Trout).
- **1825** Alexander Pushkin, founder of Russian literature, launches *Eugene Onegin*, a novel in verse.
- **c.1830–1832** Japanese artist Katsushika Hokusai's woodblock print *The Great Wave off Kanagawa* is one of his *Thirty-six Views of Mount Fuji*.

► ***Winged Victory of Samothrace*** is a masterpiece of Greek sculpture, carved in marble *c.*220–185 BCE.

1839 Britain's J.M.W. Turner paints *The Fighting Téméraire*. Polish composer Frédéric Chopin completes his piano sonata *The Funeral March*.

1851 American novelist Herman Melville's *Moby-Dick* is published.

1852 Britain's Pre-Raphaelite artist John Everett Millais exhibits *Ophelia*.

1853 Giuseppe Verdi's opera *La Traviata* is first performed in Venice, Italy. Hungarian composer and virtuoso pianist Franz Liszt completes his *Sonata in B Minor*.

1855 American poet Henry Longfellow's *Hiawatha* is published.

1860–1861 British novelist Charles Dickens publishes *Great Expectations*.

1865–1869 Russian novelist Leo Tolstoy's *War and Peace* is published.

1866 Britain's Julia Margaret Cameron treats photography as art in her portrait *Beatrice*. French novelist Victor Hugo writes *Les Miserables*.

1872 Denmark's Hans Christian Anderson publishes his *Fairy Tales*.

1874 French impressionists hold their first exhibition in Paris.

1875–1876 Russian composer Pyotr Ilyich Tchaikovsky writes *Swan Lake*.

1876 Richard Wagner's four-opera *Ring Cycle* premiers in Germany.

1879 Norwegian playwright Henrik Ibsen presents *A Doll's House*.

1884 In the US, Mark Twain writes *The Adventures of Huckleberry Finn*.

1886 American novelist Henry James's *The Bostonians* is published.

1888–1898 French sculptor Auguste Rodin creates *The Kiss*.

1889 Dutch artist Vincent van Gogh paints *The Starry Night*.

1899 French impressionist Claude Monet paints his first *Water-Lily Pond*.

1901 Russia's Sergei Rachmaninov premiers his *Piano Concerto No.2*.

1903 Irish playwright George Bernard Shaw writes *Man and Superman*.

1903–1905 French composer Claude Debussy writes *La Mer*.

1904 Russian playwright Anton Chekhov's *The Cherry Orchard* premiers.

1908–1909 Austrian composer Gustav Mahler writes his symphony *Das Lied von der Erde* (*The Song of the Earth*).

1909 Sergei Diaghlev's Ballets Russes make their debut in Paris.

1913 The premiere of Russian composer Igor Stravinsky's *The Rite of Spring* in Paris provokes a near riot in the audience.

1913–1927 French novelist Marcel Proust's *Remembrance of Things Past* is published in seven volumes.

1915 Czech novelist Franz Kafka's *The Metamorphosis* is published.

1917 Irish poet William Butler Yeats publishes *The Wild Swans at Coole*.

1922 Irish novelist James Joyce's *Ulysses* and Anglo-American poet T. S. Eliot's *The Waste Land* are published.

1924 British sculptor Henry Moore carves his first *Reclining Figure*.

1925 *The Great Gatsby* by novelist F. Scott Fitzgerald epitomizes America's Jazz Age.

1928 German composer Kurt Weill writes his *Threepenny Opera*, with lyrics by Bertolt Brecht. American composer George Gershwin writes *An American in Paris*.

1931 Spanish surrealist Salvador Dalí paints *The Persistence of Memory*.

1934 Choreographer George Balanchine's *Serenade* is first performed by the American Ballet in New York City.

This is the great picture upon which the famous comedian has worked a whole year.
6 reels of Joy.
Charles Chaplin
IN
"THE KID"

▶ Charlie Chaplin's famous Tramp made his debut in *The Kid* in 1921, when the British comic was already a global superstar of early film.

1936 Spanish playwright Federico García Lorca completes *The House of Bernarda Alba*.

1937 Spanish artist Pablo Picasso shows the horrors of war in *Guernica*. In the US, Walt Disney releases the first full-length animation film, *Snow White and the Seven Dwarfs*.

1939 German dramatist Bertolt Brecht writes *Mother Courage and her Children*. In the US, novelist John Steinbeck's *The Grapes of Wrath* is published and jazz singer Billie Holiday records *Strange Fruit*.

1941 Soviet composer Dmitri Shostakovich dedicates his Symphony No.7 to the city of Leningrad. In the US, filmmaker Orson Welles releases his epic *Citizen Kane*.

1942–1943 Dutch pioneer of abstract art Piet Mondrian paints his *Broadway Boogie-Woogie*.

1944 American composer Aaron Copland writes the ballet *Appalachian Spring*.

1947 French novelist Albert Camus' *The Plague* is published.

1949 British novelist George Orwell's *Nineteen Eighty-Four* is published. American playwright Arthur Miller's *Death of a Salesman* is premiered.

1952 Ernest Hemingway's novel *The Old Man and the Sea* is published in the US. French photographer Henri Cartier-Bresson publishes *The Decisive Moment*.

1953 Irish writer Samuel Beckett's play *Waiting for Godot* is premiered.

1956 Elvis Presley launches his career as the king of rock and roll with *Heartbreak Hotel*. Ravi Shankar and his sitar introduce western audiences to classical Indian music.

1957 Leonard Bernstein's *West Side Story* reinvents the American musical.

1962 Russian dissident writer Alexander Solzhenitsyn's *One Day in the Life of Ivan Denisovich* is set in a Soviet gulag. American pop artist Andy Warhol paints his *Marilyn Diptych*.

1965 American singer-songwriter Bob Dylan releases *Like a Rolling Stone*.

1966 Irish poet Seamus Heaney's *Death of a Naturalist* is published.

1967 British pop sensation the Beatles release their landmark album, *Sgt. Pepper's Lonely Hearts Club Band*.

1971–1974 American composer Philip Glass writes his minimalist cycle *Music in Twelve Parts*.

1976 In the US, William Eggleston's ground-breaking color photography is shown in New York and punk band The Ramones release their first album.

1982 Britain's first WOMAD (World of Music, Arts, and Dance) festival heralds the globalization of ethnic music.

1982 American pop icon Michael Jackson releases *Thriller*.

1985 Canadian Margaret Atwood's novel *A Handmaid's Tale* is published.

1987 American novelist Toni Morrison's *Beloved* is published.

2009 Canadian film director James Cameron's *Avatar* is a landmark in film's use of 3D technology.

The World in 1 CE

- Han Empire
- Roman Empire and client states
- Empire of Pontus under Mithridates Eupator, c.100 BCE
- Numidia under Masinissa from 201 BCE
- Burebista's Dacian Kingdom, 45 BCE

○ NOTE: Settlements in italics were not in existence in 1 CE but were significant during this era.

▶ The Classical World in 1 CE
By 1 CE, the Mediterranean World was dominated by the Romans. Their only real rival was the Parthian Empire to the east. China, united as a single state in 221 BCE, was still under the rule of the Han dynasty. India was fragmented into a number of smaller states, after the collapse of the Mauryan Empire in 185 BCE. In Central America, Maya city-states were beginning to emerge.

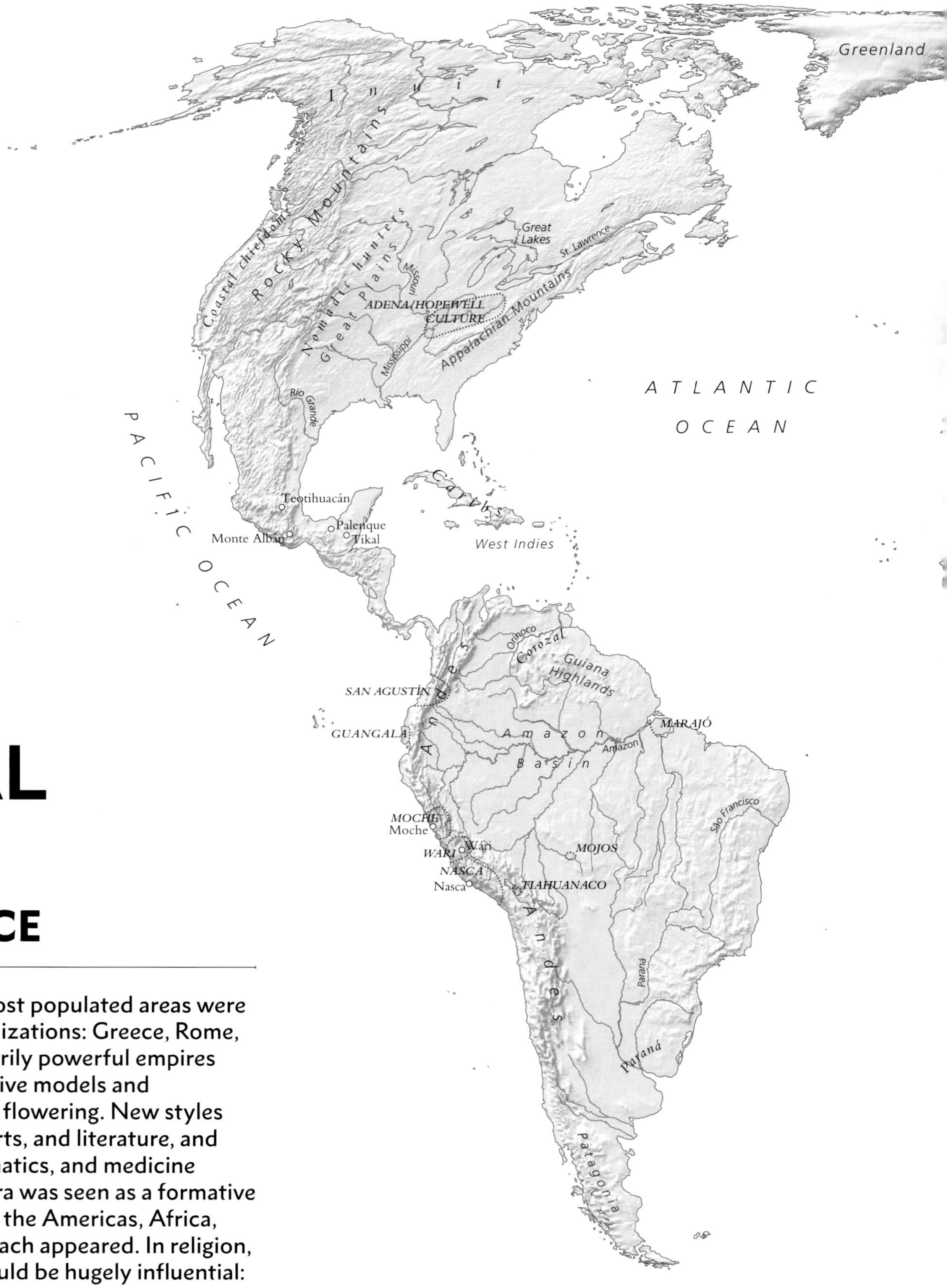

CLASSICAL WORLD 700 BCE–600 CE

From around 750 BCE, the world's most populated areas were dominated by a small number of civilizations: Greece, Rome, Persia, India, and China. These militarily powerful empires developed sophisticated administrative models and experienced an unparalleled cultural flowering. New styles emerged in architecture, the visual arts, and literature, and sciences such as astronomy, mathematics, and medicine became established. This Classical era was seen as a formative golden age in many later societies. In the Americas, Africa, and Japan, societies with a smaller reach appeared. In religion, a number of faiths appeared that would be hugely influential: Buddhism, Judaism, and Christianity.

Palaeosiberians
Lena
Yenisey
Samoyeds
Ob'
Siberia
Tungus
Amur
Finno-Ugrians
Volga
Germanic peoples
Baltic peoples
Slavs
Turks
Mongols
Pazyryk
Northern Xiongnu
Southern Xiongnu
Sarmatians
ROMAN EMPIRE
Lugdunum
Massilia
Numantia
Rome
Dacians
Danube
BOSPORAN KINGDOM
Caspian Sea
Caucasians
Gobi
Ainu
KOREA
Yellow River
JAPAN
THRACE
Thessalonica
Constantinople
Actium
Corinth
Athens
Carthage
Syracuse
ARMENIA
PONTUS
CAPPADOCIA
Antioch
Iranians
SOGDIANA
Kashgar
TOCHARIAN PRINCIPALITIES
Nisa
Bactra
BACTRIA
Ecbatana
PARTHIAN EMPIRE
Seleucia
PAHLAVAS
Taxila
Tibetans
Luoyang
Chang'an
Yangzi
RHODES
LYCIA
Palmyra
Euphrates
DECAPOLIS
Alexandria
Jerusalem
EGYPT
NABATAEA
Himalayas
SMALL STATES
Indus
Ganges
HAN EMPIRE
Berbers
GARAMANTES
Arabs
Sahara
SHAKAS
Ujjain
Pataliputra
Panyu
Arabian Peninsula
Nile
Saharan peoples
RUSH
Pratisthana
SATAVAHANAS
MAHA-MEGHAVAHANAS
Mon-Khmer peoples
ANNAM
Mekong
Chams
Niger
Meroe
AKSUM
HIMYARITES
Semites
SMALL STATES
Philippine Islands
Chadians
Kwa
Kushites
VIJAYANS
PACIFIC OCEAN
Nilotic Peoples
Congo
Malays
Borneo
Sumatra
INDIAN OCEAN
Bantus
Papuans
New Guinea
Java
Zambezi
Madagascar
Kalahari Desert
Khoisan peoples
Australian Aborigines
Darling
New Zealand

THE CLASSICAL WORLD

c.700 BCE
The screw pump (or Archimedes pump) is invented in Assyria for pumping low-lying water into irrigation channels. Nomadic Scythians from Central Asia begin to settle in Eastern Europe. In Greece, the Archaic Period continues with the rise of city-states. Agricultural villages appear in southeastern North America.

689 BCE
The Nubian king Taharqa becomes pharaoh of Egypt, founding the 25th dynasty. Babylon is destroyed by the Assyrian king Sennacherib.

664 BCE
Corinth loses to Corcyra in the earliest recorded naval battle between Greek city-states.

663 BCE
Assyrians sack Thebes, Egypt; their empire reaches its greatest extent.

660 BCE
Birth of Jimmu, the legendary first emperor of Japan.

c.650 BCE
The first coins are minted, in Lydia, Anatolia (Turkey). The age of tyrants begins in many Greek cities.

▼ **The Parthenon** built in 447–432 BCE was a temple to Athens' patron goddess; its huge ivory and golden statue of Athena was one of the Seven Wonders of the Ancient World.

630 BCE
Sparta wages war against the Messenians; it conquers most of the southern Peloponnese by 600 BCE. Thera founds the colony of Cyrene (in Libya), the first of five Greek cities in the region.

626 BCE
Nabolopassar secures Babylon's independence from Assyria and founds the Neo-Babylonian Empire.

621 BCE
Draco drafts Athens' first law code; Draconian law is later known for the severity of its punishments.

616 BCE
Tarquinius Priscus (Tarquin the Elder) is the first Etruscan king of Rome; construction begins on the Cloaca Maxima, one of the world's earliest sewage systems, and on the Circus Maximus, Rome's first stadium for chariot racing.

612 BCE
The Assyrian Empire crumbles with the sacking of Nineveh and Nimrud by the Medes and Babylonians.

604 BCE
Traditional date for the birth of Lao Tzu, founder of the Chinese religion Taoism.

c.600 BCE
Much of the Middle East falls to the Medes. Mesopotamia is dominated by the Neo-Babylonian Empire. The first known map of the world is made in Babylon. Ironworking technology reaches Zhou China.

▲ **Monte Albán** in Oaxaca, Mexico, is one of Mesoamerica's earliest cities. Founded by the Zapotecs c.500 BCE, and adorned with temples, plazas, ball courts, and canals, it was inhabited for more than 1,500 years.

594 BCE
Solon becomes archon (ruler) of Athens; in reforming its laws, he protects the property rights of the poor and bans debt-slavery.

587 BCE
Nebuchadnezzar II of Babylon destroys Jerusalem's temple and sends the Israelites into exile.

585 BCE
Miletus, a Greek city in western Anatolia, is a cradle of early philosophical thought; Thales of Miletus predicts a solar eclipse.

c.563 BCE
Traditional date of birth for the Buddha, Siddhartha Gautama.

c.560 BCE
Croesus succeeds to the throne of Lydia and begins its expansion.

c.551 BCE
Birth of Confucius, whose *Analects* provide the central philosophy of the Chinese way of life. In Persia, Zoroastrianism is the main religion.

c.550 BCE
Cyrus the Great defeats the Medes and founds the Persian Empire.

539 BCE
Cyrus quashes a rebellion in Babylon; the Babylonian Empire is absorbed by Persia, and exiled Jews are allowed to return home.

534 BCE
Tarquinius Superbus (Tarquin the Proud) becomes Rome's last king; the Etruscans are at their height.

525 BCE
Persia's Cambyses II annexes Egypt.

521–486 BCE
Reign of Darius I ("the Great") of Persia; under his rule, the Persian Empire reaches its greatest extent.

509 BCE
The Romans expel Tarquinius Superbus and set up a Republic; supreme authority now rests with two annually elected consuls.

507 BCE
Cleisthenes establishes democratic government in Athens.

c.500 BCE
Bronze coins are used in China. Ironworking spreads to Southeast Asia and East Africa. India's caste system is in place and the *Puranas* and parts of the epic *Mahabharata* are composed. The Nok culture flourishes in West Africa. The Zapotecs develop hieroglyphic writing in Mesoamerica.

499–491 BCE
Greek cities in Ionia, western Anatolia, revolt against Persian rule; their uprising is put down.

496 BCE
Rome defeats the Etruscan-led Latin League at Lake Regillus and signs its first treaty with Carthage.

490 BCE
Athenian Greeks defeat the Persians at the Battle of Marathon, ending the first Persian invasion of Greece.

481 BCE
China enters the Warring States Period (to 221 BCE), in which seven leading states jostle for supremacy.

480–479 BCE
Persian forces sent by Xerxes to invade Greece are defeated at Salamis, Plataea, and Mycale.

480 BCE
Xerxes' invasion marks the end of the Archaic Period in Greece. In the Classical Period (480–323 BCE), Greece is dominated in turn by Athens, Sparta, and Macedonia, and Greek culture is at its peak.

***c.*477 BCE**
Athens founds the Delian League of city-states to counter Sparta's Peloponnesian League.

***c.*450 BCE**
The Celtic La Tène culture emerges in Central Europe; supplanting the Halstatt culture, Celts expand east and south and into the British Isles. Steppe nomads are buried with spectacular grave goods at Pazyryk and Noin-Ula in Siberia. In Mexico, construction of the Zapotec city of Monte Albán begins.

447–432 BCE
Athens' ruler Pericles builds a new Parthenon to replace the temple destroyed by the Persians.

431–404 BCE
The Peloponnesian War sees the destruction of the Athenian Empire by Sparta and its allies.

***c.*401–399 BCE**
Xenophon leads an army of 10,000 Greek mercenaries supporting a Persian rebellion from Babylon to the Black Sea.

***c.*400 BCE**
Gallic Celts cross the Alps and settle in northern Italy. Carthage dominates the western Mediterranean. In Mesoamerica, the Olmec civilization enters its final phase, while the Zapotecs flourish in Monte Albán. The Moche culture emerges in Peru. Ironworking develops in Korea.

***c.*390 BCE**
Gallic Celts sack Rome; they soon leave, but Rome harbors a lasting fear of the Gauls.

***c.*380 BCE**
The Chu are dominant among China's Warring States.

371 BCE
The Theban general Epaminondas wins the Battle of Leuctra against Sparta; Thebes is now the dominant power in Greece, until Epaminondas dies in battle in 362 BCE.

370 BCE
Mahapadma Nanda founds the Nanda dynasty in Magadha, north India; he builds up a huge army and administrative system.

359–336 BCE
Philip II rules Macedonia and rapidly extends his power to win control of most of Greece.

356 BCE
Shang Yang, chancellor of the western Chinese state of Qin, makes wide-ranging reforms to create a powerful centralized kingdom.

343–342 BCE
A Persian invasion led by Artaxerxes III puts an end to Egypt's independence.

341–338 BCE
Rome defeats and dissolves the Latin League, moving closer to complete dominance of central Italy.

336 BCE
Philip of Macedon is murdered; he is succeeded by his 20-year-old son, Alexander the Great, who forces other Greek states into submission in 335 BCE, then crosses into Anatolia in 334 BCE to face the Persians.

332 BCE
Alexander conquers Egypt and founds Alexandria, one of many new cities across his empire.

331 BCE
At Gaugamela, Alexander defeats Darius III; the Persian Empire falls to Alexander, and his army pillages the capital at Persepolis.

326 BCE
Alexander pushes east and across the Indus into India, but has to retreat when his troops mutiny.

323 BCE
Alexander dies of a fever; his vast empire begins to disintegrate as his generals Ptolemy, Seleucus, and others fight for dominance.

321–297 BCE
Reign of Chandragupta Maurya, founder of the Mauryan Empire: he overthrows the last of the Nandas in *c.*320 BCE and unifies most of the Indian subcontinent, creating the largest empire in Indian history.

312 BCE
Rome's first aqueduct is built by Appius Claudius; he also begins the Via Appia, the first of Rome's network of roads across Italy.

***c.*300 BCE**
Europe's first Celtic states emerge. Alexander's empire is partitioned between Seleucid, Antigonid, and Ptolemaic dynasties. Rice farming reaches Japan from China.

290 BCE
With the defeat of the Samnites, Roman territory stretches across Italy to the Adriatic.

***c.*287 BCE**
China's northern states begin to build a "Great Wall" to keep out nomads from the Eurasian steppe.

***c.*273–232 BCE**
Reign of Ashoka, the Mauryan emperor of India; he embarks on imperial conquests and promotes the Buddhist concept of dharma (mercy) across his empire.

272 BCE
Rome defeats the invasion of Pyrrhus, the ruler of Epirus in Greece, establishing itself as a Mediterranean power.

264–241 BCE
First Punic War: Rome defeats Carthage and now controls the entire Italian peninsula and Sicily.

▼ **Alexander the Great** faces Darius III of Persia at the Battle of Issus (333 BCE), vividly portrayed in a Roman floor mosaic from Pompeii, southern Italy.

◀ **Lion-topped pillars** carrying the edicts of Ashoka (*c.*273–232 BCE) remain a national emblem of India.

c.247 BCE
King Devanampiya Tissa of Sri Lanka converts to Buddhism.

237 BCE
Hamilcar Barca revives Carthaginian rule in the Iberian peninsula.

221–210 BCE
Reign of the first Qin emperor, Shi Huangdi, who unites China; after his death, he is buried in a vast mausoleum with an army of 8,000 terracotta soldiers.

218–201 BCE
Second Punic War between Rome and Carthage: Hannibal Barca leads his army across the Alps, routs the Romans at Cannae, and captures much of southern Italy; he is defeated by the Roman general Scipio at Zama in North Africa.

206 BCE
The Qin dynasty is succeeded by the Han under Liu Bang; Han rule (to 220 CE) is seen as China's golden age.

c.200 BCE
Japan's Middle Yayoi Period (200–100 BCE) sees a large increase in the population and the consolidation of power around Naro in central Japan. In Ptolemaic Egypt, Alexandria is a major center of Greek trade, culture, and learning.

c.200 BCE
In eastern North America, Ohio's Adena culture is developing into the Hopewell culture. The Maya emerge in Mesoamerica as small communities on Mexico's Pacific coast merge and migrate northward to form larger states. The Nasca appear in Peru and create mysterious geoglyphs—long lines in the desert making abstract and animal shapes.

c.185 BCE
Pushyamitra Shunga, Hindu founder of the Shunga dynasty, takes power in India, assassinating the last Mauryan ruler and persecuting Buddhists.

171–138 BCE
Mithridates I conquers Greek-ruled kingdoms in Persia and founds the Parthian Empire.

167–160 BCE
Judah Maccabee and his brothers rebel against the Hellenization of Judaea under Seleucid ruler Antiochus IV, and re-establish Judaism.

149–146 BCE
Third Punic War: Rome destroys Carthage and creates the Roman province of Africa.

148–146 BCE
Rome defeats the Macedonians after a series of wars that began in 215 BCE. Roman forces sack Corinth, various leagues of Greek cities are dissolved, and Greece becomes the Roman province of Achaea.

142 BCE
The Maccabees free Jerusalem from Seleucid rule and make it the capital of the Hasmonaean kingdom; the dynasty rules Judaea until Jerusalem is seized by the Romans in 63 BCE.

129 BCE
Rome establishes the province of Asia, bringing the city-state of Pergamon under its control.

123–88 BCE
Under Mithridates II, the Parthian Empire reaches its greatest size.

107–104 BCE
The Roman general Marius organizes legions into cohorts and introduces professional service.

c.101 BCE
China's Han Empire reaches its largest extent under Emperor Wu; the Silk Road carries trade across Central Asia to the Mediterranean world, stretching from the Han capital at Chang'an to Antioch.

c.100 BCE
Celtic hill-forts in Europe are expanded into fortified towns. Maritime trade spreads Indian influence to Southeast Asia. Aksum (Axum, in Ethiopia) is on the rise.

91–89 BCE
Discontent among Italians without Roman citizenship erupts into the Social War; citizenship is granted to all Italians in 88 BCE.

73–71 BCE
A slave revolt led by the gladiator Spartacus in southern Italy is put down by Crassus, a Roman general.

64–63 BCE
Pompey, a Roman general, deposes the last Seleucid king, makes Syria a Roman province, then captures Jerusalem and annexes Judaea.

59–53 BCE
Julius Caesar becomes consul and joins with Pompey and Crassus in Rome's First Triumvirate.

58–50 BCE
Gallic Wars: Caesar conquers Gaul, creating a vast new province for Rome, and twice invades Britain.

49–44 BCE
Caesar crosses the Rubicon River into Italy, marches on Rome, and is proclaimed dictator. His Julian calendar introduces a 365-day year with a leap year every four years.

44–43 BCE
Caesar is assassinated. Cleopatra, the last of the Ptolemies, becomes ruler of Egypt. Octavian, Caesar's nephew and heir, forms the Second Triumvirate with Mark Antony and Lepidus to defeat Caesar's assassins, Brutus and Cassius.

30 BCE
Mark Anthony and Cleopatra, his ally and lover, commit suicide after their defeat by Octavian; Egypt is made a Roman province.

27–14 BCE
Octavian takes the title Augustus and rules as Rome's first emperor.

4 BCE
Probable birth date of Jesus Christ, in the Roman province of Judaea.

c.1 CE
Buddhism spreads in Southeast Asia. Nabataeans allied with Rome control Red Sea trade. In a global population of 300 million, 1 person in 7 lives in Roman territory.

9 CE
Roman forces are defeated by Germanic tribes in the Teutoburg forest and withdraw to the Rhine River, Rome's frontier for the next

▼ **Ohio's Great Serpent Mound** was likely built by the Adena or Hopewell cultures and renovated by later American Indian groups, but its origins and purpose are a mystery.

400 years. Wang Mang seizes the Chinese throne and establishes the Xin dynasty.

23 CE
Collapse of China's short-lived Xin dynasty; the restored Han dynasty establishes its capital at Luoyang.

c.30 CE
Crucifixion of Jesus Christ; his followers continue his teachings.

c.40 CE
South America's Arawak peoples migrate down the Orinoco River and settle in the Caribbean.

43 CE
Roman invasion of Britain, which becomes a Roman province.

46–57 CE
St. Paul, a Roman follower of Christ, visits fledgling Christian groups in Anatolia and Greece; by the late 4th century Christianity becomes the majority religion within the empire.

c.50 CE
Aksum is a major trading center.

c.60 CE
Kushans from Bactria (Afghanistan) invade northern India, founding the Kushan Empire. In Britain, Roman forces defeat the Iceni revolt and capture their queen, Boudicca.

c.64 CE
Fire destroys much of the city of Rome; Emperor Nero blames the Christians and many are martyred.

c.65 CE
Buddhism reaches China.

66–70 CE
First Jewish revolt against Roman rule: Roman forces lay siege to Jerusalem, destroy the Temple in 70 CE, and enslave thousands of Jews. Jewish resistance continues at Masada until 74 CE.

73 CE
China's Han dynasty is at its most powerful, controlling new stretches of Central Asia from Mongolia to Afghanistan.

79 CE
The volcanic eruption of Vesuvius (near Naples, southern Italy) buries Pompeii and Herculaneum.

c.100 CE
The Maya city of Teotihuacán in Mexico expands. The Moche culture flourishes in Peru.

c.105 CE
Paper is invented in China, for use by the highly complex bureaucracy.

117 CE
When Emperor Trajan dies, the Roman Empire is at its greatest extent; his successor Hadrian abandons Mesopotamia and Assyria and concentrates on better defense of the imperial frontiers.

122–126 CE
Hadrian's Wall is built along the Roman frontier in north Britain.

127–140 CE
The Kushan Empire expands enormously under Kanishka the Great, who conquers Magadha and campaigns against the Chinese in Central Asia; he promotes Buddhism across his empire and builds a huge stupa at Purusapura (Peshawar, in modern Pakistan).

132–135 CE
A second Jewish revolt against Roman rule is crushed and Jews are expelled from Jerusalem.

140 CE
Ptolemy of Alexandria uses astronomy to measure terrestrial locations with longitude and latitude for his world map.

c.150 CE
Han China regains its dominance of Central Asia. The Kushans become vassals of the Parthians. Christianity spreads to Roman North Africa. Nok Iron Age culture is at its peak in Nigeria.

167–180 CE
As Goths migrate southward, two Germanic tribes cross the Danube into Roman territory and invade Italy during the Marcomani War.

c.200 CE
Trade flourishes between India, China, and Rome. Chinese occupation of Korea ends: native Korean states emerge in Koguryo, Paekche, and Silla. Teotihuacán is the largest city in the Americas; the Maya city of Tikal also becomes prominent. The Hopewell mound-building culture continues to flourish in North America.

c.200–279 CE
Beginning with the *Mishnah*, scholars publish Jewish laws and doctrines in the Talmud.

c.220 CE
China's Han dynasty collapses and is replaced by the Three Kingdoms Period (220–280 CE) of the Shu, Wu, and Wei kingdoms.

226 CE
The Parthian Empire falls to Ardashir I, who founds the Sasanian (or Sassanid) dynasty and is made "king of kings" in Persia.

235 CE
Military anarchy erupts in Rome, with more than 20 emperors over the next 50 years; Germanic tribes attack Rome's frontiers along the Danube and Rhine, and invade Italy in 259 CE.

◄ **Julius Caesar** (100–44 BCE) personifies Rome's power in a classical statue created for Versailles in 1696.

c.250 CE
The Maya civilization enters its Classic Period (to 900 CE), with multiple city-states and the construction of pyramids and other monuments. The lodestone compass is invented in China.

260 CE
Postumus declares himself head of a breakaway Gallic empire made up of Gaul, Germany, Spain, and Britain; Roman rule is restored in 274 CE.

269–272 CE
Queen Zenobia of Palmyra (in Syria) takes Egypt and Syria from Rome, but is defeated and taken prisoner by Emperor Aurelian.

280 CE
China is reunited under the Western Jin dynasty.

293 CE
Emperor Diocletian sets up a tetrarchy of four co-emperors to secure Rome's borders in a radical reorganization of the empire.

c.300 CE
Armenia is the first to adopt Christianity as a state religion. In Africa, Aksum issues coins, while Bantu peoples begin to herd cattle. In the Pacific, the Rapa Nui people may already inhabit Easter Island (although ongoing research points to a much later date of *c.*1200 CE).

303 CE
Christians are persecuted in Rome by Emperor Diocletian.

304 CE
Xiongnu steppe nomads invade China, leading to the breakdown of order in the north and the start of the Sixteen Kingdoms Period (to 439 CE).

312–313 CE
Constantine the Great wins the Battle of Milvian Bridge to take control of the Western Roman

Empire; Licinius becomes sole ruler of the Eastern Roman Empire. Constantine confirms religious freedom for Christians with the Edict of Milan in 313 CE.

320 CE
Chandragupta I founds the Gupta Empire; his descendants rule northern India for 150 years.

330 CE
Constantine makes Byzantium (renamed Constantinople) capital of the Eastern Roman Empire.

***c.*350 CE**
The Yamato state emerges in Japan, building large burial mounds for its ruling class.

***c.*370 CE**
Nomadic Huns from Central Asia begin to invade eastern Europe; they defeat the Ostrogoths in Ukraine.

376–415 CE
Under Chandragupta II, the Gupta Empire is at its peak, dominating northern and central India.

378 CE
Goths defeat Rome's Eastern army and kill the emperor Valens in battle; a truce in 382 CE lets them settle in the empire in return for providing troops. In Mesoamerica, Teotihuacán deposes the Maya ruler of Tikal (in Guatemala).

386 CE
The Toba Wei reunify northern China and found the Northern Wei dynasty (to 534 CE).

395 CE
Honorius becomes emperor of the West and Arcadius of the East, marking the definitive division of the Roman Empire.

405 CE
The first translation of the Bible into Latin from Hebrew and Greek is completed by St. Jerome.

409 CE
Britain expels Roman officials and gains independence from Rome by 411 CE. Vandals invade Roman North Africa from Iberia, and eventually complete their conquest by taking Carthage in 439 CE.

410 CE
The sack of Rome by Visigoths under Alaric shakes the empire.

420 CE
The Moche in Peru build the Temple of the Sun. In China, the Eastern Jin dynasty is overthrown and replaced by the Liu Song dynasty (to 479 CE).

***c.*450 CE**
The population of Teotihuacán, Mexico, peaks at 250,000.

452 CE
Huns invade northern Italy under Attila, but turn back short of Rome.

475 CE
Visigoths under Euric rule Spain and southwest Gaul and gain independence from Rome.

476 CE
Romulus Augustulus, the last Roman emperor in the West, is deposed; the Western Empire is replaced by the barbarian states of the Vandals in North Africa, the Visigoths in Spain, and the Ostrogoths in Italy; the Eastern Empire consolidates its power.

477 CE
Buddhism becomes the state religion in China.

478 CE
The first shrine of the Shinto religion is built in Japan.

479 CE
The Liu Song dynasty falls and the short-lived Qi dynasty assumes power in southern China.

480 CE
Huns overthrow the Gupta Empire in India.

481–511 CE
Clovis I is the first king of the Franks in northwestern Gaul; he converts to Christianity in 496 CE and drives the Visigoths out of southwestern Gaul in 507 CE.

493 CE
Theodoric's Ostrogoths conquer Italy. China's Northern Wei dynasty moves its capital to Luoyang.

***c.*500 CE**
Angles, Saxons, and Jutes migrate to Britain; Celts survive in Wales and Ireland. Camel trains cross the Sahara from Ghana to North Africa; the Bantu reach southern Africa. Classic Maya civilization is at its height in Mesoamerica. The Tiahuanaco culture emerges in Bolivia; the Huari culture expands in the central Andes; the Paracas culture flourishes in southern Peru.

527–565 CE
The Byzantine (Eastern Roman) emperor Justinian the Great codifies Roman law, commissions the great church of Hagia Sophia, and reconquers Italy and the former Roman provinces in North Africa.

538 CE
Buddhism arrives in Japan.

568–572 CE
Byzantium concedes much of Italy to the Lombards.

***c.*570 CE**
Birth of Muhammad, prophet of Islam, in Mecca.

***c.*581–589 CE**
Yiang Jian, a Zhou general, seizes power to become the first emperor of the Sui dynasty (to 618 CE) and reunites China.

590–604 CE
Pope Gregory I ("the Great") asserts papal supremacy over the western and eastern churches.

597 CE
St. Augustine of Canterbury leads a papal mission to revive Christianity in Anglo-Saxon England.

▼ **On Easter Island,** Polynesian settlers constructed stone figures with enormous heads, called *moai*, which are believed to embody revered ancestors.

GREAT BUILDINGS

History more often preserves the names of those who commissioned the great buildings of antiquity—pharaohs, emperors, and kings—than their architects. Through the centuries, cultures worldwide have continued to create the human-made landscapes that surround and inspire us.

- **c.10,000 BCE** The temple at Göbekli Tepe, Turkey, is the world's oldest surviving monument.
- **c.2650 BCE** The Step Pyramid at Saqqara in Egypt is built by Imhotep, the world's earliest known architect.
- **c.2600 BCE** Construction begins on Khufu's Great Pyramid at Giza.
- **c.2300 BCE** The Stonehenge ceremonial complex in Britain is completed.
- **c.2100 BCE** The Great Ziggurat of Ur is built by King Ur-Nammu.
- **c.1700–1400 BCE** Minoans build a series of labyrinthine palaces at Knossos on the island of Crete.
- **c.1500–1200 BCE** The temple of Amun Re at Karnak near Luxor in Egypt is the largest in the world.
- **c.575 BCE** Babylon's Processional Way leads to the Ishtar Gate, lined with tiers of dragons and bulls.
- **c.447–432 BCE** The new Parthenon temple built on the acropolis in Athens, Greece, is adorned with marble sculptures by Phidias.
- **c.340–300 BCE** The open-air theater at Epidaurus, Greece, is built to seat 12,000 people.
- **c.250 BCE** Ashoka's Great Stupa at Sanchi in India is Buddhism's oldest surviving sanctuary.
- **c.200 BCE** The Great Wall of China, later rebuilt during the Ming dynasty, stretches 4,000 miles (6,400 km).
- **c.9 BCE – 40 CE** The Nabataeans build Al Khazneh (The Treasury) at Petra, Jordan.
- **80 CE** The Colosseum, Rome's first permament amphitheater, seats 50,000.
- **200 CE** At Teotihuacán, Mexico, the city's largest temple is the Pyramid of the Sun.
- **118–126 CE** Rome's Pantheon features the largest unreinforced concrete dome ever.
- **537 CE** Hagia Sophia in Constantinople is the world's largest church for 1,000 years.
- **c.675 CE** The Maya build the pyramid Temple of Inscriptions at Palenque.
- **692 CE** Jerusalem's Dome of the Rock on Temple Mount is completed; the site is sacred for Muslims, Christians, and Jews.
- **785 CE** Work begins on the Great Mosque of Cordoba in Moorish Spain.
- **c.1000** The Anasazi build Pueblo Bonito, their largest Great House in Chaco Canyon, New Mexico.
- **c.1150** Angkor Wat's temple complex in Cambodia covers 500 acres (200 ha).
- **1163–1345** The French Gothic cathedral of Notre Dame de Paris is built.
- **c.1300s** Built from almost a million granite blocks, the Great Enclosure encircles the Shona city of Great Zimbabwe.
- **1362–1391** The Court of Lions at the heart of the Alhambra palace in Granada, Spain, is an Islamic vision of Paradise.
- **1406–1420** The Forbidden City, the imperial palace and center of Chinese government until 1912, contains nearly 10,000 rooms in 980 buildings on 180 acres (72 ha) in Beijing.
- **1419–1446** Renaissance architect Filippo Brunelleschi crowns the Duomo in Florence with an octagonal dome.
- **c.1440s** The Incas build Machu Picchu, a sacred citadel in the Peruvian Andes.
- **1555–1561** Ivan the Terrible, "Czar of all the Russias," builds St. Basil's Cathedral in Moscow's Red Square.
- **1569–1575** Ottoman architect Mimar Sinan builds his finest mosque for Selim II in Edirne, Anatolia.
- **1626** St. Peter's Basilica in Rome is consecrated; it is the world's biggest church until 1989.
- **1601–1609** The "White Egret Castle" in Himeji is Japan's largest castle.
- **1634–1653** The Taj Mahal is built by Mughal emperor Shah Jahan in memory of his wife Mumtaz Mahal.
- **1675–1711** Christopher Wren rebuilds St. Paul's Cathedral after the Great Fire of London in 1666.
- **1682** Louis XIV of France makes the vast palace at Versailles the center of his court and government.
- **1869–1886** Ludwig II builds the fairytale castle of Neuschwanstein in the Bavarian Alps.
- **1889** The Eiffel Tower in Paris opens to the public on the centenary of the French Revolution.
- **1907** The reconstructed Great Mosque at Djenné in Mali is the world's largest clay building.
- **1930** Crowned in steel, New York's Chrysler Building is an Art Deco icon.
- **1931** The Empire State Building is New York's tallest skyscraper until the World Trade Center of 1973.
- **1937** San Francisco's Golden Gate suspension bridge measures 1.7 miles (2,735 m) end to end.
- **1949** Frank Lloyd Wright's Guggenheim Museum in New York is a masterpiece in concrete.
- **1973** Australia's Sydney Opera House spreads its spherical roofs like sails at the water's edge.
- **1977** The Pompidou Center in Paris puts its framework and elevators on the outside.
- **1997** Frank Gehry's Guggenheim Museum in Bilbao, Spain, is clad in titanium, glass, and limestone.
- **2010** Dubai's Burj Khalifa is the world's tallest building, at 2,722 ft (829.8 m).

◄ **Hagia Sophia in Istanbul** began as a church in 537 CE, and was later converted to a mosque in 1453 and a museum in 1935.

The World in 1300

- Byzantine Empire
- England and possessions
- Aragon and possessions
- Venetian Republic and possessions
- Mongol Empire on the death of Genghis Khan 1227
- controlled by Khwarizm Shah 1219
- Holy Roman Empire

○ NOTE: Settlements in italics were not in existence in 1300 but were significant during this era.

▶ **The Medieval World in 1300**
In 1300, the united tribes of the nomadic Mongols controlled much of Eurasia. European monarchies were beginning to extend their power and reach. North Africa and the Middle East were fragmented after the collapse of the unified Islamic Empire, while in the Americas the Aztecs of Mexico and the Incas of Peru were just beginning their phase of expansion.

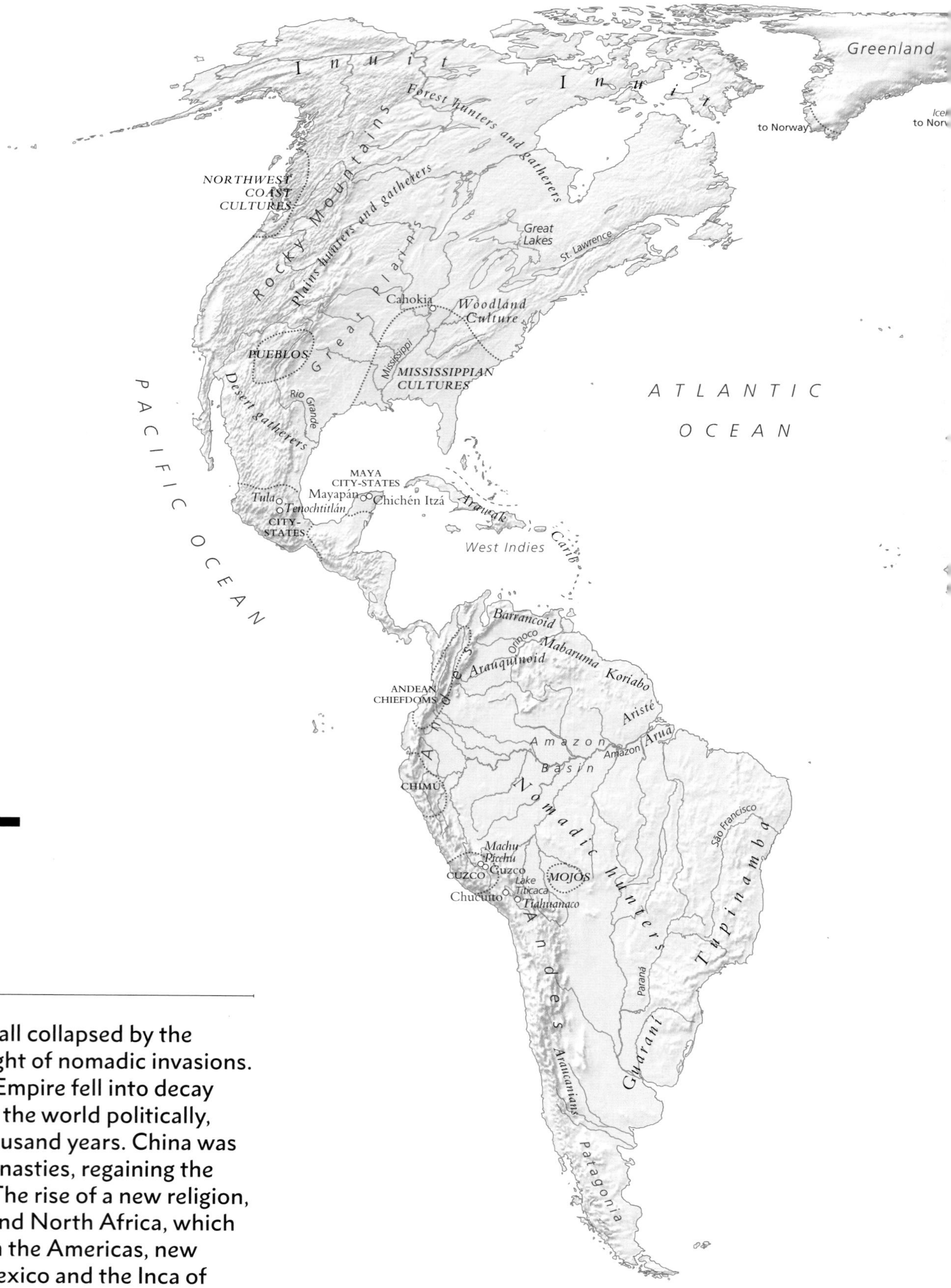

MEDIEVAL WORLD 600–1450

The Classical empires of Eurasia had all collapsed by the 7th century CE, largely under the weight of nomadic invasions. The urban civilization of the Roman Empire fell into decay and Europe lagged behind the rest of the world politically, economically, and culturally for a thousand years. China was reunited under the Tang and Song dynasties, regaining the military reach it had under the Han. The rise of a new religion, Islam, transformed the Middle East and North Africa, which were united under an Arab Empire. In the Americas, new empires emerged as the Aztecs of Mexico and the Inca of Peru united large areas under a single ruler for the first time.

Palaeosiberians
Lapps
Samoyeds
Ugrians
Tungus
Siberia
Ainu
NORWAY
SWEDEN
RUSSIAN PRINCIPALITIES
Copenhagen
DENMARK
TEUTONIC ORDER
LITHUANIA
ENGLAND
ANGEVIN EMPIRE
POLISH STATES
Kiev
KHANATE OF THE GOLDEN HORDE
Paris
BOHEMIA-MORAVIA
Venice
Genoa
HUNGARY
to Genoa
Karakorum
Gobi
PAPAL STATES
SERBIA
BULGARIA
GEORGIA
Caspian Sea
CHAGATAI KHANATE
Yellow River
Beijing
KORYO
JAPAN
Kyoto
Nara
ARAGON
Constantinople
BYZANTINE EMPIRE
ANATOLIA
TREBIZOND
Samarkand
Toledo
KINGDOM OF NAPLES
Sicily
MAJORCA
RUM
TURKISH BEYLIKS
LITTLE ARMENIA
Tabriz
EMPIRE OF THE GREAT KHAN
Luoyang
Kaifeng
Córdoba
GRANADA (NASRIDS)
VENETIAN REPUBLIC
Antioch
ZAYYANIDS
HAFSIDS
ACHAEA
ATHENS
CYPRUS
Damascus
Baghdad
Jerusalem
Euphrates
IL-KHANATE
Indus
Himalayas
TIBET
Yangzi
Hangzhou
PACIFIC OCEAN
MAMLUKS
Cairo
Bedouins
Delhi
SULTANATE OF DELHI
Ganges
SMALL DYNASTIES
Sahara
Nile
Medina
OMAN
PARAMARAS
Mekong
PAGAN
Tuaregs
GUJARAT
Mecca
Arabian Peninsula
YADAVAS
EASTERN GANGAS
ARAKAN
Chieng Mai
ANNAM
Laos
SMALL STATES
PEGU
Sukhothai
Timbuktu
KAKATIYAS
CHIENGMAI PHAYAO
SUKHOTHAI
Ayutthaya
CHAMPA
RASULIDS
ALWA
KANEM
DAJU
ETHIOPIA
Angkor
KHMER
Philippine Islands
HOUSA STATES
HOYSALAS
CERAS
PANDYAS
LAVO
IFAT
HADIYA
DAWARO
BALE
SMALL STATES vassals to Pandyas
Kwa
Congo
INTERLACUSTRINE STATES
MALAY STATES
Borneo
Bantus
SWAHILI CITY-STATES
East Indies
Papuans
Java
New Guinea
MAJAPAHIT
INDIAN OCEAN
Zambezi
Malays
Madagascar
GREAT ZIMBABWE
Great Zimbabwe
Kalahari Desert
Australian Aborigines
Khoisan peoples
Darling
Maoris
New Zealand
Lena
Yenisey
Ob'
Amur
Volga
Niger

◀ **Jerusalem's Dome of the Rock**, Islam's oldest monument, enshrines the rock from which the Prophet is said to have ascended to heaven.

THE MEDIEVAL WORLD

613
The prophet Muhammad starts preaching in Mecca; his teachings incur the authorities' hostility.

617
China's Sui emperor Yangdi is murdered; Li Yuan founds the Tang dynasty (618–906) and Xi'an becomes the capital.

619
Sasanian conquests of Syria, Mesopotamia, Palestine, and Egypt are complete, restoring the Persian Empire at the expense of Byzantium; Persian advances are reversed by the Byzantine emperor Heraclius by 627.

622
The Hegira (Muhammad's flight to Medina to escape persecution) marks the start of the Islamic era.

624
Muhammad's army defeats the Meccans at the Battle of Badr; they surrender when he takes possession of the Ka'aba, the holiest shrine in the Arabian peninsula, in 630.

628
Indian mathematician Brahmagupta refines a decimal place value system (which spreads to China, Egypt, and the Arab Empire by the 8th century) and is the first to describe the use of zero and negative numbers.

632
Death of Muhammad and the start of the caliphate. The first four caliphs ("successors") are the Rashidun ("rightly guided"), beginning with Abu Bakr; he suppresses a rebellion and cements Islamic dominion in Arabia, then invades Syria.

638
Islamic armies capture Jerusalem.

641
Islamic conquest of Egypt; advancing westward, Islamic armies take Tripolitana (in Libya) in 643.

642
Islamic conquest of Persia and the end of the Sasanian Empire.

646
Taika reforms in Japan centralize power and strengthen the position of the emperor.

651
The standardized version of the Qu'ran is issued by the third caliph, Uthman.

661
The fourth caliph, Ali (Muhammad's cousin and son-in-law) is murdered, ending the Rashidun caliphate; the Umayyad caliphate is established, but Islam is split between Sunni (pro Ummayad) and Shi'ite (pro Ali).

668
The Silla kingdom unifies Korea with the help of Tang China, ending the long Three Kingdoms Period.

670–677
First siege of Constantinople by Arab forces: it ends when the Byzantine fleet destroys the Arab fleet with "Greek fire" (an incendiary device) at the Battle of Syllaeum, securing a 30-year peace.

672
Resurgence of the Maya city-state of Tikal: construction begins on new causeways, pyramids, ball courts, observatories, and palaces.

683
Empress Wu becomes the only woman in Chinese history to rule in her own right (not as regent); she establishes the Zhou dynasty (to 705).

692
Commissioned by the caliph Abd al-Malik, the Dome of the Rock (or Qubbat as-Sakhrah) is completed on Jerusalem's Temple Mount, sacred to Islam, Judaism, and Christianity.

698
Islamic conquest of Carthage, North Africa's last Byzantine stronghold.

c.700
Rise of the Kingdom of Ghana and the Ife kingdom in West Africa. Teotihuacán, Mexico, is abandoned. Northern Peru is dominated by the Chimú. North American Indians replace spears with bows and arrows.

710–715
Islamic conquest of Sind in Pakistan.

711
Islamic invasion of Spain: Tariq ibn Ziyad, a Berber or "Moor" from North Africa, defeats the Visigoths and most of the Iberian Peninsula falls under Islamic rule.

715
Syria's Ummayad Mosque (the Great Mosque of Damascus) is completed.

715–720
The Lindisfarne Gospels are produced by the monk Eadfrith on Holy Island in northern England.

720
Arabs occupy Provence (in France) and Central Asia; the caliphate now extends from al-Andalus (Spain) to the borders of China.

c.725
Casa Grande, the Hohokam settlement in Arizona, is flourishing; irrigation allows a range of crops despite the desert environment, and Casa Grande is at the center of a trade network stretching from the Pacific coast to the Gulf of Mexico.

725
AD (*Anno Domini*, or "Year of Our Lord") Christian dating system is introduced in *On the measurement of Time*, a treatise by Bede, an Anglo-Saxon monk and scholar.

726–729
Byzantine emperor Leo III bans the worship of religious icons; known as iconoclasm, the ban is designed to curb monastic power.

732
Charles Martel's Frankish army defeats Arabic-Moorish forces in France, halting Islamic expansion into Western Europe.

740
Leo III's Byzantine army defeats the caliphate at the Battle of Akroinon, Anatolia, and expels the Umayyads from Asia Minor.

747–750
A revolt against Umayyad caliphs in Persian Khorasan leads to their defeat at the Battle of Zab and the foundation of Abbasid caliphate.

c.750
Maya city-states are at the peak of their power, controlling a trade network that stretches from California to South America; in Mexico, the Maya city of Tikal has a population of 90–100,000.

▲ **The Book of Kells** (*c.*800) ranks with the Lindisfarne Gospels (715–720) as one of the treasures of Celtic Christianity. Its illuminated pages were probably created by monks on the Scottish island of Iona.

c.750
The pre-Inca Andean city Tiwanaku in Bolivia is at its zenith, with extensive terracing and irrigation; as a ceremonial and trading center, its cultural and economic influence spreads through South America.

751
Pépin III, son of Charles Martel and father of Charlemagne, deposes the last Merovingian king of the Franks and founds the Carolingian dynasty.

754
Italy is invaded by Franks under Pépin, in support of Pope Stephen II against the Lombards who have conquered Ravenna—the last Byzantine territory in Italy.

756
An Umayyad emirate is established in Cordoba, Spain, the first to break away from a united Islamic caliphate.

760
The Indian system of numerals is adopted by the Abbasid dynasty, establishing the Arabic numerals (1–9) in widespread use today.

762
The Abbasid caliphate moves its capital from Kufa to Baghdad, Islam's first imperial city.

774
Lombards in northern Italy are defeated by Frankish king Charlemagne's forces.

782
Charlemagne conquers West Saxony, determined to convert pagan tribes to Christianity; he also launches a Carolingian cultural revival, attracting scholars to his court.

786
Haroun al-Rashid, immortalized in *The One Thousand and One Nights*, becomes the fifth Abbasid caliph; his rule (to 809) sees the cultural flowering of the Islamic world, as scholars begin translating ancient Greek and Roman texts into Arabic.

787
Emperess Irene brings iconoclasm in the Byzantine Empire to an end.

c.790
Vikings from Scandinavia begin raids against Western Europe, looting the rich monasteries of Lindisfarne (973) and Iona (795) in Britain and then Ireland in 795.

794
Emperor Kammu moves the Japanese capital from Nara to Kyoto.

800
Charlemagne is crowned emperor of the Romans by Pope Leo III. The Tibetan Empire expands to the Bay of Bengal. In Tang China, Zen Buddhism is pre-eminent. In Mesoamerica, the Maya civilization is in decline, its overpopulated cities gradually abandoned.

802
The Khmer Empire is founded by King Jayavarman II in Cambodia, Southeast Asia.

811
The first paper currency is issued in Tang China, known as "flying cash."

827
Islamic forces invade Sicily; Palermo falls in 831 and the rest of the island is conquered by 902.

832
Caliph Al-Ma'mun establishes Baghdad's House of Wisdom; scholars translate manuscripts from other cultures and older traditions, preserving ancient scholarship that would otherwise be lost.

843
The treaty of Verdun divides Charlemagne's Frankish Empire between his three sons: the west and east portions roughly correspond to modern France and Germany, and a middle kingdom is later known as Lotharingia (Lorraine).

c.850
The earliest reference to gunpowder is in China. In Burma the kingdom of Bagan is founded. The Cholas under King Vijayalaya gain power in India. Arab navigators perfect the astrolabe. Coffee is discovered by, legend has it, a goatherd in Ethiopia.

858–1159
The Fujiwara clan dominates Japanese politics of the Heian Period.

c.863
A Glagolitic alphabet—an early form of Cyrillic—is created in Moravia, eastern Europe, by the Byzantine missionary known as St. Cyril.

866
Vikings capture the city of York and establish a kingdom in northern England.

868
Diamond Sutra, the world's oldest surviving printed book, is a sacred Buddhist text hand-printed in China.

874
Vikings settle Iceland. The Islamic Samanid dynasty (to 999) is founded in Turkestan, Central Asia; its capital, Bukhara, becomes a center of Persian commerce and culture.

878
Alfred the Great, king of Wessex, defeats the Danes at Edington to halt their advance in England.

889
In Mesoamerica, Tikal is abandoned; as Maya city-states collapse in the south, the north (in Mexico's Yucatan Peninsula) sees the rise of Chichén Itzá, whose *cenotes* (water holes) are vital in this drought-prone region.

c.900
India's golden age of Hindu temple-building begins. The Toltecs found their capital at Tula in the Valley of Mexico; refugees from the collapsed

▲ **Chichén Itzá** in Mexico's Yucatán Peninsula became the leading Maya city-state in the 9th century. Dominating the city center is the massive pyramid temple of Kukulcán, a plumed serpent god that was known as Quetzalcoatl to the Aztecs and Toltecs.

Teotichuacán culture, they forge a militaristic empire that later inspires their Aztec descendants.

c.900–1600
The Thule culture emerges in coastal Alaska; these ancestors of the Inuit hunt caribou, whales, seals, and fish, and expand across the Canadian Arctic to reach Greenland by *c.*1200.

907
Collapse of China's Tang dynasty; several short-lived rival dynasties follow, until the Song rise in 960.

906
Magyars destroy Moravia (eastern Czech Republic) and begin to raid Western Europe.

910
The Benedictine abbey of Cluny is founded in Burgundy, France; as the center of a monastic empire in Europe, it governs *c.*10,000 monks.

909
Start of the Shi'ite Fatimid caliphate in Tunis, North Africa (to 1171); it takes Alexandria in 914, Sicily in 917.

911
The Norse chieftain Rollo is granted much of Normandy in France, and becomes a Christian.

916
Khitan Mongols establish the Liao dynasty, one of the Five Dynasties that control northern China.

930
The Althing assembly, the world's oldest national parliament, is established in Iceland.

932
The reunification of Muslim Spain is completed when Ummayad caliph Abd al-Rahman III captures Toledo.

935
Riven by civil war, Korea is reunified under the Koryo dynasty (to 1392).

937
Anglo-Saxon king Athelstan's victory over a coalition of Vikings, Welsh, and Scots at Brunanburh halts Viking expansion, and helps create England as a unified nation.

938
The kingdom of Dai Viet in Vietnam throws off Chinese rule.

946
Persian Shi'ite Buwayhids take Baghdad; Abbasid caliphs remain until 1258, but real power lies with the Buwayhid sultans in Shiraz.

955
Otto I, king of East Francia (Germany), defeats the Magyars at the Battle of Lechfeld, halting their westward expansion from Hungary.

960
China's Song dynasty is established, ending the anarchy of the Five Dynasties and Ten Kingdoms era.

▲ **Pueblo Bonito** is the biggest of 13 Great Houses built by the Anasazi in Chaco Canyon, New Mexico. Six stories high, with over 600 rooms, it was occupied in 828–1126.

962
Otto I ("the Great") is crowned emperor by the Pope, reviving the Carolingian Roman Empire in the west.

966
The Polish state is born when its ruler, Mieszko I, adopts Christianity.

969
The Fatimids of Tunisia conquer Egypt and make Cairo their new capital; claiming descent from the Prophet's daughter Fatima, the caliphate rules North Africa from the Atlantic Coast to the Red Sea.

▲ **The surrender of Jerusalem** to Saladin (Salah al-Din, Ayyubid sultan of Egypt and Syria) in 1187, during the Third Crusade, is depicted in a miniature from the 15th-century "Abbreviated chronicle" created by David Aubert for Philip Duke of Burgundy.

972
A unified Hungarian state forms under Duke Geza; his son Stephen is crowned first king of Hungary.

982
North Vietnamese Dai-Viet invade the Champa kingdom in the south and sack its capital, Indrapura.

986
Erik the Red begins the Viking settlement of Greenland, leading a group of colonists from Iceland.

987
Hugh Capet succeeds the last Carolingian king of the Franks; the Capetian dynasty rules France until 1328. In Mesoamerica,Toltecs conquer the Yucatan Maya and make the Maya city of Chichén Itzá their capital.

988
Vladimir the Great of Kiev converts to Christianity.

1000
Stephen, Grand Prince of Hungary, becomes its first king. Boleslav is crowned first king of Poland.

1001
The first Muslim raids into northern India are led by Mahmud of Ghazni (in Afghanistan).

1002
Leif Erikson (son of Erik the Red) is probably the first European to set foot in North America, landing at Vinland, northern Newfoundland.

1013
Danish invasion of England; Cnut the Great (King Canute) expands his empire to include England, Denmark, and Norway by 1030.

1014
Brian Boru, High King of Ireland, wins the Battle of Clontarf and breaks Viking dominion in Ireland.

1031
The Umayyad caliphate of Cordoba falls during the Christian reconquest of Spain.

1031
Seljuk Turks invade Khurasan in Persia; in 1040 they crush the Ghaznavids, laying the foundations for a new Islamic empire.

1040
Bantu expansion reaches its peak in central and southern Africa.

c.1041–1048
In China, printing with movable type is invented by Bi Sheng.

c.1050
The Ancestral Pueblo (Anasazi) civilization, centerd on Chaco Canyon in southwest North America, is at its peak, dominating trade routes; its pueblos (towns) feature multistory apartment housing built of stone or mud brick around a central plaza.

1054
Final schism between the Eastern and Western Churches, led by the patriarch in Constantinople and the pope in Rome. The Almovarid dynasty is founded in Morocco; it begins the Islamic conquest of West Africa in 1056.

1055
Seljuk Turks capture Baghdad, ending Persia's Buwayhid dynasty.

1066
William of Normandy (William the Conqueror) defeats Harold, the last Anglo-Saxon king of England, at the Battle of Hastings.

1071
Seljuk Turks defeat Byzantines at the Battle of Manzikert. The Norman conquest of southern Italy ends Byzantine presence in Italy; Sicily falls to the Normans in 1092.

1073
Persian poet and astronomer Omar Khayyam is invited to Isfahan to set up an observatory under Seljuk patronage; his major breakthroughs include measuring the length of the year accurate to six decimal places.

1075–1077
Investiture Controversy: Pope Gregory excommunicates the western (Holy Roman) emperor Henry IV for defying his ban on nobles appointing ("investing") church leaders.

1076
The empire of Ghana in West Africa falls to the Almoravids.

1081–1118
Alexius Commenus rules the Byzantine Empire, and partially restores its power after its defeats by Normans and Seljuk Turks.

1092
Death of the Seljuk sultan Malik Shah, followed by civil war and the disintegration of the Seljuk Empire.

1095
The First Crusade (1096–1099) is proclaimed by Pope Urban II; French crusaders seize Jerusalem from the Fatimid caliphate in 1099.

1119
The crusading order of the Knights Templar is founded in Jerusalem. Bologna University is also founded, the first in the Western world.

1122
The Concordat of Worms ends the Investiture Controversy, but papal-imperial rivalry continues. The Toltec city of Tula is burned down, marking the end of Toltec influence in the Yucatan.

1126
Jin from Manchuria overrun northern China and move the capital from Kaifeng to Beijing; this marks the end of the Northern Song dynasty, but Gaozong flees south and establishes the Southern Song dynasty in Hangzhou in 1127.

1144
The Second Crusade (1146–1160) is triggered when the crusader state of Edessa falls to Zengi, the atabeg (governor) of Mosul and Aleppo.

1147
Almohads seize the city of Marrakesh from the Almoravids, taking control of North Africa, and then invade Moorish Spain.

c.1150
Cahokia flourishes at the height of the Mississippian (or Cahokian) culture of the American Bottom. Home to c.20,000 people, the city features around 120 mounds and a huge central plaza; it is the largest pre-Columbian site in North America, and its influence spreads across the Midwest, before its rapid decline after 1200 CE.

1170
Robert "Strongbow" FitzStephen commands the English invasion of Ireland for Henry II.

1171
Saladin overthrows the Fatimid caliphate in Egypt and later seizes Damascus in Syria in 1174; now recognized as sultan of Egypt and Syria by the Abbasid caliphate in Baghdad, Saladin founds the Ayyubid dynasty.

1180–1185
The Gempei Wars between the Minamoto and Taira clans end Japan's Heian Period (794–1185) and usher in the Kamakura Period, named after the shogunate that is established at Kamakura in 1192; the emperor's power is curtailed, a samurai warrior class supplants the aristocracy, and Japan becomes a feudal society.

c.1181
Jayavarman VII reclaims the Khmer crown in Cambodia; he expels Cham invaders, restores the city of Angkor Wat, and builds a new capital at Angkor Thom.

1187
Saladin crushes Crusader forces at Hattin and seizes Jerusalem.

1189–1192
The Third Crusade ends when Richard I of England ("the Lionheart") is unable to liberate Jerusalem and concludes a peace treaty with Saladin.

1192
Minamoto Yoritomo becomes Japan's first shogun; his power base at Kamakura sidelines the imperial court, and shoguns rule as military dictators until 1867.

1192–1193
Ghurids of Persia defeat a Hindu rebellion in India; Muhammad of Ghur founds the Delhi Sultanate, India's first Muslim empire.

▲ **Taira no Tomomori**, son of the Taira clan head Taira no Kiyomori, was one of the clan's chief commanders in Japan's Gempei Wars of 1180–1185, which ended in Taira defeat at Dannoura.

▶ **A bronze head** commemorates an Oba (king) of Benin, which Ewuare the Great ruled from 1440 to 1473.

c.1200
Incas settle in the region around Cuzco in the Peruvian Andes.

1202–1204
The Fourth Crusade fails to win back Jerusalem; the crusaders fall out with the Byzantines and sack the city of Constantinople in 1204, all but ending Byzantine power.

1206
Temujin, having united all the tribes of Mongolia, takes the title Chinggis Khan ("ruler of the world"), or Genghis Khan.

1209
The Albigensian Crusade is launched against Cathar heretics in Languedoc, France; it leads to the Papal Inquisition set up in 1231.

1211
Genghis Khan invades Jin China; Mongol forces capture Zhongdu (Beijing), the Jin capital, in 1215, and the Jin dynasty falls in 1234.

1212
Alfonso VIII of Castile crushes the Almohads at Las Navas de Tolosa, a decisive battle in the Reconquista, the Christian reconquest of Iberia.

1215
England's Magna Carta ("Great Charter") is signed by King John after his heavily taxed barons revolt; stating that the king is not above the law, it is a milestone in human rights.

1217–1221
The Fifth Crusade targets Egypt but ends in retreat.

1219
The Mongol conquest of Islamic states in Eurasia begins when Genghis Khan invades the Khwarazmid Empire.

c.1225
In Nigeria, Great Zimbabwe's population reaches around 15,000; one of the great urban trading centers of sub-Saharan Africa, it later gives rise to, and is eclipsed by, the Mutapa Empire.

1227
Genghis Khan dies while quashing a revolt in China; his vast empire is divided between four sons.

1228–1229
In the Sixth Crusade, emperor Frederick II regains Jerusalem by treaty, without military action.

1231
Pope Gregory IX launches the Papal Inquisition, a campaign by the Church against heresy.

1235
Foundation of the kingdom of Mali, West Africa.

1237–1242
Mongol armies of the Golden Horde Khanate invade Russia and blaze across Central Europe; they reach Vienna by 1242 but return to Karakorum to elect a new leader.

1241
Foundation of the Hanseatic League: to protect Baltic trade routes, the German towns of Lubeck and Hamburg form an alliance that spreads to become a powerful trading confederation of up to 200 cities in northern Europe over the next 400 years.

1248–1254
The Seventh Crusade: Louis IX of France invades Egypt and is defeated, taken captive, and ransomed by his mother.

1250
Mamluks, slave soldiers in the Ayyubid army, murder the last Ayyubid sultan of Egypt; their commander marries his widow and founds the Mamluk dynasty.

c.1250
In Mexico, Aztecs settle at Chapultepec, but are soon expelled by Tepanecs in the wake of the Toltec collapse. In Peru, Incas under Manco Capac develop the city of Cuzco and begin their expansion, becoming the largest pre-Columbian empire in South America.

1256–1381
The Venetian-Genoese Wars, the struggle between the republics of Venice and Genoa for control of the Mediterranean trade routes, end in victory for Venice.

1258
The Abbasid caliphate falls as Mongols sack the city of Baghdad and execute the caliph.

1259
Möngke the Great Khan dies and the Mongol Empire starts to split into four khanates: the Golden Horde in the northwest, the Chagatai khanate in the center, the Il-Khanate in the southwest, and the Yuan dynasty in the east.

1260
Egypt's Mamluks defeat the Mongols at Ain Jalut, Palestine, and secure control over Syria and the Levant.

1261
Byzantines retake Constantinople, ending the Latin Empire formed by leaders of the Fourth Crusade.

1271
Kublai Khan, Great Khan of the Mongols since 1260, establishes the Yuan dynasty in China; he moves his capital from Shangdu (Xanadu) to Zhongdu (Beijing) in 1272.

1274
Led by Kublai Khan, Mongols try to invade Japan; a second unsuccessful attempt is made in 1281.

1275
Marco Polo, Venetian merchant and explorer, visits Kublai Khan's court in China; his memoir, written after his return to Italy in 1295, describes the Mongolian Empire at its peak.

1279
Yuan forces defeat the Southern Song dynasty and Kublai Khan becomes the first non-native emperor of all China.

c.1280
Maoris, Polynesians from Tahiti, arrive in New Zealand, the last landmass (apart from Antarctica) to be colonized by humans.

1291
Acre, the last major Crusader stronghold in Palestine, falls to the Mamluks; a few months later they take Beirut, ending the Christian presence in the Holy Land.

1297
Scots led by William Wallace rebel against English control; Robert the Bruce is crowned king of Scotland in 1306 and expels the English at the Battle of Bannockburn in 1314.

c.1300
Osman I, ruler of a Turkmen principality in Anatolia, founds the Ottoman state by attacking the frontiers of the fading Byzantine

▲ **Joan of Arc**, daughter of a French farmer, was 16 when she followed voices in her head to aid the French dauphin by leading his forces to relieve the siege of Orléans by the English in 1429.

Empire. The Chimú Empire emerges in Peru, with its capital at Chan Chan in the Moche Valley.

c.1320s–1330s
Forerunners of the Italian Renaissance, artists Giotto di Bondone and Andrea Pisano win major commissions in Florence.

1324
Europe's first known use of cannons made of iron is at the siege of Metz. Mansa Musa, emperor of Mali at the height of its prosperity, sets out on his Pilgrimage of Gold to Mecca.

c.1325
Aztecs found Tenochtitlán, on an island in Lake Texcoco, Mexico; the Aztec capital becomes the largest city in the pre-Columbian Americas.

1333
Japan's Kamakura shogunate ends with civil war; the Ashikaga shogunate takes power in 1336.

1336
A Hindu rebellion against Muslim rule establishes the Vijayanagara Empire as the dominant power in southern India, lasting until 1646.

1337
Start of the Hundred Years War between England and France.

1340
In the Reconquista of Muslim-held Spain and Portugal, the Battle of Rio Salado forces the Marinid sultan of Morocco to end the last significant Islamic incursion into Christian Iberia from Africa.

1347–1350
The Black Death pandemic reaches Europe from western Asia; it kills 75-200 million people across Eurasia.

1349
Chinese colonization of Singapore marks the start of Chinese settlement in Southeast Asia.

c.1350
Conflict between Inca and Chimú states in South America.

1354
Ottoman Turks seize Gallipoli from the Byzantines, gaining their first foothold in Europe.

1360
The Treaty of Calais ends the first phase of the Hundred Years War; with victories at Crécy and Poitiers, England now controls more territory in France than at any other period.

1368
The Ming dynasty is founded by Zhu Yuanzhang; the Ming take control of China in 1382 and expel the last emperor of the Yuan dynasty.

1378
The Great Papal Schism between rival popes in Rome and Avignon.

1386
Jagiello of Lithuania marries Jadwiga of Poland, creating one of Europe's largest states.

1388
John Wycliff's English translation of the Bible is published.

1389
The Battle of Kosovo sees Ottomans smash the Serbian Empire and gain control of the Balkans.

1392
The union of Japan's northern and southern imperial courts ends the Yoshino Period (1336–1392). Start of the Yi (Choson) dynasty in Korea (to 1910); neo-Confucianism replaces Buddhism as the state religion; the new capital is at Hanseong (Seoul).

1393
Turkic-Mongol warrior Timur Ling (Tamerlane) completes his conquest of the Il-Khanate; in 1395 he routs the Golden Horde in Central Asia; in 1398 he invades India and sacks Delhi, slaughtering 100,000 fugitives.

c.1400
The Aztec city of Tenochtitlán is at its height; intensive agriculture and extensive networks of trade and tribute support a population of up to 200,000. West Africa's Songhay Empire, centered on the trading hub of Gao, expands and eclipses the empires of Ghana and Mali.

▲ **The Inca citadel of Machu Picchu** was built in the 1440s, high above the Urubamba Valley in the Peruvian Andes. Home to no more than 1,000 people , is thought to have served as a ceremonial center and an impregnable stronghold for the Inca elite.

1400–1415
A Welsh rebellion led by Owain Glendower against English rule is suppressed.

1401
Timur massacres the population of Baghdad and launches an invasion of Syria; in 1402 he moves against the Ottomans, invading Anatolia and capturing the sultan.

1405
Timur plans to invade China, but dies en route and is buried at Samarkand, his capital; the vast Timurid Empire soon fragments.

1415
England defeats the French army at Agincourt in the Hundred Years War. The Portuguese capture the port of Ceuta, the first permanent European base in North Africa.

1428
The Mexica-Aztec Empire takes root with a triple alliance of city-states—Mexico-Tenochtitlán, Texcoco, and Tlacopan—that rules central Mexico until the Spanish conquest in 1521.

1428
Le Loi, leader of the Vietnamese resistance against Chinese rule, founds the Le dynasty and restores the kingdom of Dai Vet.

1429
Joan of Arc leads French troops to relieve the siege of Orléans during the Hundred Years War; she is captured, found guilty of heresy, and burned at the stake in 1431.

c.1430
Bruges, Flanders, is the commercial hub of northwest Europe; Philip the Good, Duke of Burgundy, holds court here, and his lavish patronage attracts merchants, bankers, and artists such as Jan van Eyck.

1434
Cosimo de Medici begins Medici dominance in Florence, Italy, using his huge banking wealth to control politics and become a major patron of Renaissance art and culture.

1436
Portuguese charting West Africa's coast become the first Europeans to cross the Tropic of Cancer.

1438
Inca conquests in South America begin under Pachacutec, the ninth *Sapa Inca* (Inca king); by the end of the century, the Inca Empire expands from Cuzco and Machu Picchu to almost all of western South America, from Quito to Chile.

1440
Moctezuma I becomes Aztec ruler at Tenochtitlán and starts to expand his empire. In the wealthy kingdom of Benin, Nigeria, Ewuare the Great becomes Oba (ruler) and begins to build a powerful empire.

1448
Johannes Gutenberg introduces the movable-type printing press to Europe at Mainz, in Germany.

PHILOSOPHY AND FAITH

The earliest inquiries into the nature and meaning of life come from the founders of the great Eastern religions. Since then, Western philosophers have journeyed to the outer limits of thought and understanding, posing questions that challenge our most fundamental beliefs.

Zoroaster, *c.*628–551 BCE, Persia Founder of the world's earliest monotheist religion, Zoroastrianism, which emphasizes individual responsibility and the struggle between *asa* (truth) and *druj* (lie).

Siddhartha Gautama (Buddha), *c.*563–483 BCE, India Founder of Buddhism as a path to nirvana (spiritual enlightenment) and release from the cycle of reincarnation.

Lao Tzu, *c.*6th century BCE, China Founder of Daoism and author of the *Dao De Jing* (Book of the Way), which posits the universal force of Dao and the individual's simple focus on the present as a route to virtue.

Confucius, *c.*551–479 BCE, China Founder of Confucianism, whose teachings on social harmony via social conventions and respect for others are compiled in the *Analects*.

Socrates, *c.*469–399 BCE, Greece One of the founders of Western philosophy, quoted as saying: "A life unexamined is not worth living." The Socratic method uses questions to test hypotheses in the quest for truth.

Plato, *c.*427–347 BCE, Greece Pupil of Socrates and founder of the Academy in Athens; Plato's *The Republic* describes all we perceive as a shadow of its abstract, ideal Form.

Aristotle, *c.*384–322 BCE, Greece Pupil of Plato, tutor to Alexander the Great, and author of *Metaphysics*. His wide-ranging interest in the nature of existence and logical classification influences Western science and philosophy for the next 2,000 years.

Plotinus, 205–270 CE, Roman Empire Founder of Neoplatonism, a development of Plato's original ideas; expounding on three fundamental principles of existence—the One, the Intellect, and the Soul—his treatises are collected in the *Enneads*.

St. Augustine of Hippo, 354–430 CE, North Africa/Roman Empire An early Church father who transmits Platonism and Neoplatonism through Christian theology and influences the medieval world view with *The City of God* and *Confessions*.

St. Thomas Aquinas, 1225–1274, Italy Pre-eminent medieval religious philosopher and author of *Summa Theologica*, Aquinas strives to reconcile philosophy and reason with theology and faith, arguing that they work in collaboration.

Niccolò Machiavelli, 1469–1527, Italy Renaissance diplomat, philosopher, and a founder of modern political science. In *The Prince*, he argues that the state should promote the common good, irrespective of any moral evaluation of its acts.

Francis Bacon, 1561–1626, England Philosopher, statesman, jurist, and father of scientific method and empiricism, insisting on the role of sensory experience and evidence: *Novum Organum* is a key work.

Thomas Hobbes, 1588–1679, England A founder of modern political philosophy. *Leviathan* sets out the theory of the social contract between rulers and their subjects.

René Descartes, 1596–1650, France Mathematician, scientist, and founder of rationalism who rejects empiricism; in *Meditations* he dismisses beliefs based on received wisdom, the senses, or logic, and seeks what can be known for certain—arriving at the conclusion *"Cogito, ergo sum"* ("I think, therefore I am").

Baruch Spinoza, 1632–1677, Dutch Republic One of the major rationalists, who argues that knowledge of the world is gained through reason; *Ethics* is a key work.

John Locke, 1632–1704, England Locke's philosophical empiricism in *An Essay Concerning Human Understanding* and his political liberalism in *Two Treatises of Government* influence the 18th-century Enlightenment and the Constitution of the United States.

Gottfried Wilhelm Leibniz, 1646–1716, Germany Mathematician and rationalist philosopher. In *Theodicy*, tackling the problem of evil in a world created by a good God, Leibniz concludes that we live in the "best of all possible worlds."

George Berkeley, 1685–1753, Britain Anglo-Irish bishop and empiricist in whose idealist metaphysical system reality is ultimately nonmaterial, and objects exist only as ideas in the minds of the perceivers: *Principles of Human Knowledge* states that "to be is to be perceived."

David Hume, 1711–1776, Britain Scottish economist, empiricist, and leading skeptic of metaphysics. Hume examines how the mind acquires knowledge in his *Treatise on Human Nature* and argues there can be no knowledge beyond experience.

Jean-Jacques Rousseau, 1712–1778, Switzerland Attests to the sovereignty of the citizen body; *The Social Contract* opens with "Man is born free, and everywhere he is in chains," influencing the Enlightenment, French Revolution, and the Romantic movement.

Immanuel Kant, 1724–1804, Germany Seeking to synthesize rationalism and empiricism, *Critique of Pure Reason* asserts the authority of reason but limits knowledge to the world we experience.

Thomas Paine, 1737–1809, Britain Anglo-American political activist, philosopher, and one of the founding fathers of the United States; *The Rights of Man* insists governments must safeguard their citizens' natural rights.

G. W. F. Hegel, 1770–1831, Germany The most systematic and influential of the German philosophical idealists: *The Phenomenology of Spirit* describes

▶ **Plato and Aristotle** take center stage among the great thinkers of ancient Greece in Raphael's *The School of Athens*, painted *c.*1509–1511 during the Renaissance.

how the human mind evolves from a state of mere consciousness to absolute knowledge.

- **Arthur Schopenhauer, 1788–1860, Germany** Espousing Kant's transcendental idealism, *The World as Will and Representation* sees the world we experience, our desires, and our actions as the product of a blind, aimless, metaphysical Will.
- **Søren Kierkegaard, 1813–1855, Denmark** A forerunner of existentialism who stresses the individual's unique position as self-determining agent. A major work is *Concluding Unscientific Postscript to Philosophical Fragments*.
- **Karl Marx, 1818–1883, Germany** Social theorist, economist, historian, revolutionary, and philosopher of Communism. *Das Kapital* aims to expose capitalism's production process and exploitation of labor.
- **Friedrich Nietzsche, 1844–1900, Germany** Rejects religious and metaphysical interpretations of the human condition for the principles of eternal recurrence, the Superman, and self-mastery; *Thus Spake Zarathustra* states "God is dead."
- **Bertrand Russell, 1872–1970, Britain** Philosopher, mathematician, and social reformer who rebels against idealism and founds analytic philosophy; logic is key to the search for truth, and central to Russell's *Principia Mathematica*.
- **Ludwig Wittgenstein, 1889–1951, Austria** Student of Betrand Russell and the pre-eminent analytical philosopher; *Tractatus Logico-Philosophicus* argues that the limits of language are the limits of philosophy and sets out the logic of language.
- **Jean-Paul Sartre, 1905–1980, France** Philosopher, writer, activist, and leader of the existentialist movement, which focuses on the totality of human freedom; *Being and Nothingness* is hugely influential.

MAJOR WORLD FAITHS

Originating from almost every corner of the globe, the world's great faiths are as diverse as its cultures. Some have their origins in prehistoric times, yet the 20th century saw the emergence of several new religions that have attracted followers in their millions.

◀ A Torah scroll contains Judaism's holiest text, made up of the five books of Moses.

NAME	PLACE/DATE	ADHERENTS	FOUNDER	TEXT
Chinese traditional religion	Unknown, prehistoric	400 million	Indigenous	n/a
Hinduism	India, prehistoric	900 million	Indigenous	The Vedas, Upanishads, and Sanskrit epics
Shinto	Japan, prehistoric	3–4 million	Indigenous	*Kojiki, Nihon-gi*
Voodoo	West Africa, unknown	8 million	Indigenous	n/a
Judaism	Israel, *c.*1300 BCE	15 million	Abraham; Moses	Hebrew Bible; Talmud
Zoroastrianism	Persia, 6th century BCE	200,000	Zoroaster	The Avesta
Daoism	China, *c.*550 BCE	20 million	Lao Tzu	*Dao De Jing*
Jainism	India, *c.*550 BCE	4 million	Mahavira	Mahavira's teachings
Buddhism	Northeast India, *c.*520 BCE	375 million	Siddhartha Gautama (Buddha)	Pali canon, Mahayana sutras
Confucianism	China, 6th/5th centuries BCE	5–6 million	Confucius	The Four Books and Five Classics
Christianity	Israel/Palestine, *c.*30 CE	2,000 million	Jesus Christ	The Bible (Old and New Testaments)
Islam	Saudi Arabia, revealed in the 7th century CE	1,500 million	n/a; Muhammad is the Prophet	The Qu'ran (scripture); Hadith (tradition)
Sikhism	Punjab, India, *c.*1500	23 million	Guru Nanak	Adi Granth (Guru Granth Sahib)
Church of Jesus Christ of Latter-day Saints	New York, United States, 1830	13 million	Joseph Smith	The Bible; *Book of Mormon*
Baha'i Faith	Tehran, Iran, 1863	5–7 million	Baha'u'llah	Writings of Baha'u'llah
Cao Dai	Vietnam, 1926	8 million	Ngo Van Chieu	Cao Dai Canon
Family Federation for World Peace and Unification	South Korea, 1954	3 million	Sun Myung Moon	*Sun Myung Moon, the Divine Principle*
Falun Gong	China, 1992	10 million	Li Hongzhi	Writings of master Li, including *Zhuan Falun*

The World in 1700

- Mughal Empire
- Ottoman Empire
- Qing Empire
- Russian Empire
- Safavid Empire
- England and possessions
- France and possessions
- Denmark and possessions
- Spain and possessions
- Portugal and possessions
- Netherlands and possessions
- Hohenzollern possessions
- Sweden and possessions
- Japan
- Venetian Republic and possessions
- Austrian Habsburg territories
- held temporarily by Netherlands during 17th century
- Holy Roman Empire

▶ The Early Modern World in 1700
By 1700, Spain and Portugal had carved out extensive colonies in South and Central America, while France and Britain vied for control of North America. The Ottoman, Safavid, and Mughal Muslim Empires controlled much of North Africa, the Middle East, and southern Asia. In East Asia, China's Qing dynasty controlled a territory smaller than that of the Tang or Han, while Japan had been unified a century earlier after 150 years of civil war.

EARLY MODERN WORLD 1450–1750

During the 15th and 16th centuries Europe—relatively stagnant for a thousand years—underwent radical change. The Renaissance saw the recovery of much Classical Greek and Roman learning and a renewed interest in science and technology. Europe's maritime states sent out voyages of exploration, followed by colonists who overwhelmed native states in the Americas and Africa. Advanced non-European cultures such as Ming (and then Qing) China, Mughal India, Safavid Persia, and Ottoman Turkey retained their political integrity, but were unable to stop Europeans from gaining political and economic footholds in their territories.

RUSSIAN EMPIRE
Siberia
SWEDEN
Stockholm
St Petersburg
COURLAND
Moscow
PRUSSIA
Warsaw
POLAND-LITHUANIA
ZAPOROGIAN COSSACKS
KHANATE OF CRIMEA
MOLDAVIA
WALLACHIA
Constantinople
DENMARK
Copenhagen
BRANDENBURG
NETHERLANDS
ENGLAND
GERMAN STATES
BOHEMIA
Paris
FRANCE
SWISS FEDERATION
AUSTRIA
Vienna
HUNGARY
PAPAL STATES
VENETIAN REPUBLIC
NAPLES
SARDINIA
SICILY
SPAIN
Madrid
ALGIERS
Oran
Ceuta
TUNIS
MALTA
Knights of St. John
OTTOMAN EMPIRE
TRIPOLI
FEZZAN
vassal of Tripoli
Sahara
Tuaregs
Nilo-Saharan peoples
Kazakhs
Turkmens
Caspian Sea
KHIVA
BUKHARA
KHANATE OF THE DZUNGARS
Chaldiran
Isfahan
SAFAVID EMPIRE
Beduins
Arabian Peninsula
OMAN
YEMEN
MUGHAL EMPIRE
Delhi
Agra
Himalayas
TIBET
BHUTAN
ASSAM
MANIPUR
NEPALESE PRINCIPALITIES
Cambay
Diu
Bassein
Bombay
Goa
Madras
Pondicherry
Calicut
Cochin
CEYLON
KINGDOM OF KANDY
POLYGAR KINGDOMS
Gobi
Manchuria
Beijing
Nerchinsk
KOREA
JAPAN
Kyoto
Nagasaki
QING EMPIRE
Formosa
Macao
BURMA
ARAKAN
CHIENGMAI
Chieng Mai
TRAN NINH
LAOS
Sukothai
SIAM
Lop Buri
ANNAM
CAMBODIA to Annam
Manila
PHILIPPINE ISLANDS
Spanish possessions
MALAY STATES
MAMPAVA
Malacca
ATJEH
Sumatra
Benkulen
BRUNEI
SULU
Malays
KUTEI
Moluccas
BANDJARMASIN
East Indies
Dutch possessions
Batavia
Java
Yogyakarta
MATARAM
PORTUGUESE TIMOR
Papuans
New Guinea
INDIAN OCEAN
Australian Aborigines
Maoris
New Zealand
Chukchi
Koryaks
Yenisey
Lena
Ob'
Volga
Amur
Yellow River
Yangzi
Indus
Ganges
Tigris
Euphrates
Nile
Danube
Niger
Congo
MOSSI KINGDOMS
MAMPRUSI
DAGOMBA
BORNU
DARFUR
WADAI
HAUSA STATES
NUPE
KWARARAFA
BORGU KINGDOMS
IGALA
BENIN
SMALL STATES
OYO
Fernando Po
DAHOMEY
ALLADA
SÃO TOMÉ
FUNJ
ETHIOPIA
HARAR
SMALL OROMO STATES
Kushites
TORO
BUNYORO
BUGANDA
BUSOGA
ANKOLE
KARAGWE
RWANDA
BURUNDI
Mombasa to Oman
Zanzibar to Oman
Kilwa to Oman
Bantus
TEKE
LUANGO
KAKONGO
NGOYO
KONGO
KIKONJA
LUBA
KUBA
KALUNDE
KANIOK
LUNDA
XINJE
SONGO
UPPER BEMBE
MATAMBA
ANGOLA
NDONGO
KASANJE
LOWER BEMBE
WILA
MUZUMBO A KALUNGA
St. Helena
UNDI
MOZAMBIQUE
KALONGA
LOZI
LUNDA
ROZWI
Malays
Madagascar
Mauritius
Bourbon
Kalahari Desert
THOVELA
Khoisan peoples
Delagoa Bay
Cape of Good Hope
DUTCH SOUTH AFRICA

THE EARLY MODERN WORLD

c.1450–1629
The Mwene Mutapa Empire in central southern Africa is at its peak; rich in gold, copper, and ivory, this Shona state controls lucrative trade routes from the interior to Arab kingdoms on the east coast until the Portuguese colonize the region.

1453
The city of Constantinople is captured by the Ottomans; the Byzantine Empire falls and Ottoman expansion continues in the Balkans and Greece. The Hundred Years War (1337–1453) ends when France recaptures Bordeaux; Calais is England's sole possession in France until the French retake it in 1558.

1454–1455
The Gutenberg Bible is printed in Mainz, Germany; it is the world's first mass-produced book.

1455–1485
The Wars of the Roses: the dynastic struggle between England's rival Plantagenet houses of Lancaster and York ends with Henry Tudor seizing the throne as Henry VII and founding the Tudor dynasty.

1467–1477
The Onin War opens the century-long Warring States Period; Japan is devastated as regional magnates (*daimyo*) strive to destroy their rivals.

1468
The Songhay Empire recaptures the wealthy city of Timbuktu (in Mali) from the Tuaregs and becomes the leading power in West Africa.

1469
The marriage of Isabella of Castile and Ferdinand of Aragon leads to a unified Christian Spain, which dominates 16th-century Europe.

c.1470
The Peruvian kingdom of Chimor is conquered by the Incas, ending Chimú culture; the Inca Empire extends 2,500 miles (c.4,000km).

▲ **The fall of Constantinople** in 1453 is illustrated in this miniature from *Voyage d'Outremer* by Bertrandon de la Broquière, a 15th-century pilgrim who described his travels to the Holy Land and Constantinople for Philip the Good, Duke of Burgundy.

1472
Ivan III (Ivan the Great), Grand Prince of Muscovy, marries Zoë, niece of the last Byzantine emperor; he gains independence from the Golden Horde, unites Russian principalities into a centralized state, and takes the title of Czar.

1473
The Aztecs conquer the state of Tlatelolco, expanding their empire with formidable military force.

1477
Charles the Bold, duke of Burgundy, dies; the Austrian Habsburgs acquire his territories in the Low Countries (the Netherlands), and Louis XI seizes his French territories.

1482
Portugal builds Sao Jorge da Mina (Elmina Castle) on the Gold Coast; Europe's first settlement in sub-Saharan Africa, it gives Portugal a monopoly of West Africa's gold trade.

1487
In Tenochtitlán, the Aztec Great Temple is rebuilt and opens with the ritual sacrifice of up to 5,000 people.

1488
Portuguese navigator Bartolomeu Dias sails into the southern Atlantic and rounds the Cape of Good Hope off the tip of southern Africa, discovering wind systems that link the Atlantic and Indian Oceans.

1492
The fall of Granada, the last Moorish territory in Spain, completes the Reconquista; Muslims and Jews are expelled from Spain. Backed by the Spanish crown, Christopher Columbus crosses the Atlantic in search of the riches of the East; he lands in the West Indies and "discovers" the New World.

1494
The Treaty of Tordesillas divides present and future discoveries in the New World between Spain and Portugal. In the Italian Wars (1494–1559), France and Spain vie for control of Italy, beginning with the invasion of Italy by Charles VIII of France to lay claim to Naples.

1497
Italian navigator John Cabot reaches Newfoundland; sponsored by the English king, Henry VII, he paves the way for the English exploration and settlement of North America.

1497–1498
Portugal's Vasco da Gama sails around the Cape of Good Hope and is the first European to reach India by sea; this fast new route between Europe and Asia transforms trade between the two contintents.

1498
Columbus is the first European to reach South America, during his third voyage to the New World.

1500
Pedro Alvares Cabral sights Brazil, claiming it for Portugal; Spain's Vicente Yanez Pinzon finds the mouth of the Amazon river.

1500
New Zealand's Maori culture enters its classic phase: tribes develop finely made bone tools and weapons, elaborate wood carvings, textiles, and tattoos, large hilltop forts and earthwork settlements (*pa*), and some

of the biggest war canoes ever built. This Stone Age society remains intact until Europeans arrive in 1642 and introduce metal technologies.

1501
Ismail I founds the Safavid dynasty and becomes Shah of Persia; he seizes Baghdad, makes Isfahan his capital, and goes east to Afghanistan and the edge of the Mughal Empire.

1501–1502
Amerigo Vespucci reaches the coast of Brazil and continues southward; he realizes this is not the eastern edge of Asia (as Columbus thought) but a separate continent. The Americas are later named after him.

1502
Start of the Atlantic slave trade between Europe, West Africa, and the Americas: the first shipment of African slaves is sent to Cuba to work in Spanish settlements.

1503–1506
In Florence, Italy, Leonardo da Vinci paints the *Mona Lisa*, which becomes the most famous work of the Renaissance, and Michelangelo unveils his statue of *David* in 1504.

1510–1512
Afonso de Albuquerque secures Goa as Portugal's principal base in India; he oversees the capture of Malacca, the first Portuguese settlement in Southeast Asia; he also sponsors the first Portuguese voyage to the Spice Islands (the Moluccas), which Francisco Serrao reaches in 1512.

1513
Juan Ponce de León reaches Florida, marking Spain's first contact with mainland North America. He also discovers the Atlantic's Gulf Stream; understanding this system of currents and winds in the Atlantic is a crucial aid to navigators in the age of sail.

1514
Ottomans crush the Safavid Persians at the Battle of Chaldiran, northwest Iran; Sultan Selim I then sweeps into Syria and Mamluk Egypt, greatly increasing Ottoman territories and securing almost all the Muslim holy places in Southwest Asia.

1517
Martin Luther publishes his *Ninety-Five Theses* in Wittenberg, Saxony, triggering the Reformation; the religious revolt against the Western Church leads to a lasting schism between Catholics and Protestants.

1519
Hernán Cortés lands in Mexico and marches on Tenochtitlán, the Aztec capital; the city falls in 1521, the Aztec emperor is executed in 1525, and the Spanish now dominate Central America. Charles I, the Habsburg king of Spain, is elected Holy Roman Emperor as Charles V.

1520
Financed by Spain, Portuguese explorer Ferdinand Magellan discovers a navigable route around the tip of South America, providing a western passage to the Spice Islands and cementing Spain's global role; his fleet completes the first circumnavigation of the globe in 1522. Smallpox strikes the Aztecs; over the next 100 years, diseases brought to the Americas from Europe kill up to 20 million people, or 95 percent of the population.

1526
Portuguese ships complete the first transatlantic slave voyage from Africa to the Americas; within 100 years, they transport around 10,000 slaves a year from Angola to Brazil.

1526
The Mughal Empire is founded in northern India: Babur, descended from Genghis Khan, defeats the Afghan sultan of Delhi, Ibrahim Lodi, and declares himself emperor.

1527
The sack of Rome by Charles V's imperial troops is the most shocking event of the Italian Wars; the papal Holy League is crushed.

1529
Under the Peace of Cambrai, a temporary truce in the Italian Wars, France relinquishes its rights in Italy, Flanders, and Artois; Charles V renounces his claims to Burgundy.

1529–1566
Suleiman the Magnificent's reign is the longest of the Ottoman Empire; his failed siege of Vienna in 1529 halts Ottoman expansion in central Europe, but he gains large territories in the Middle East and North Africa.

1531
Spanish conquistador Francisco Pizarro lands in Peru; in 1533, he executes the Inca emperor and conquers the Inca capital, Cuzco.

1534
Ignatius Loyola founds the Jesuits, a Catholic order of priests that leads the Counter Reformation. Henry VIII of England, denied a divorce by the pope, breaks with Rome and founds the Church of England, with the king as its supreme head.

1534
Jacques Cartier explores parts of Newfoundland and the Gulf of St. Lawrence, opening the way for French settlement of Canada.

1536
Wales and England are formally united by the Act of Union. Henry VIII begins the dissolution of monasteries in England, executes his second wife Anne Boleyn, and crushes a Catholic revolt.

1541
A Jesuit mission to Southeast Asia sets out from Portugal; led by Francis Xavier, it reaches Goa, the Spice Islands, China, and Japan.

▶ **Babur, India's first Mughal emperor,** was outnumbered by Ibrahim Lodi's men at the Battle of Panipat in 1526, but his firearms and field artillery scattered the enemy's terrified war elephants and gave him a decisive victory.

1543
A new scientific age begins: Polish astronomer Nicolaus Copernicus shows that Earth orbits the sun, not the other way around, in *On the Revolution of the Heavenly Bodies*; and Andreas Vesalius's *On the Fabric of the Human Body* presents the empirical evidence of human dissection.

1545–1547
The Council of Trent meets to counter the threat of Protestantism by reforming and remodeling the Catholic Church, and instigates the Counter Reformation. The Spanish discover huge silver deposits in Potosí, Bolivia; mines here and in Mexico finance the Spanish Empire.

1552
Henri II of France allies with Maurice of Saxony to drive Charles V from Germany.

1555
Charles V concedes the Peace of Augsburg, allowing German princes to be Protestant or Catholic.

1556
Akbar succeeds his father Humayun as Mughal emperor in India and presides over an enormous expansion

◀ **Queen Elizabeth I** displays England's power and wealth in this portrait at Hardwick Hall, Derbyshire.

of Mughal power. Under Ivan IV ("the Terrible"), Russia expands southward to destroy the Khanate of Astrakhan and control trade routes to Central Asia.

1558
Elizabeth I becomes Queen of England (to 1603). Ivan IV continues Russian expansion with settlements in the Khanate of Sibir (Siberia).

1565
The first Spanish colony in the Philippines is a settlement on Cebu.

1565–1572
Ivan IV's reign of terror destroys the Russian boyars (hereditary nobility) and seizes their estates.

1568–1648
In the Dutch Revolt, seven northern, largely Protestant provinces of the Low Countries, rebel against Catholic rule by Philip II of Spain; led by William the Silent of Orange, the northern provinces declare independence in 1581; the southern provinces remain loyal to Spain.

1569
Flemish cartographer Gerardus Mercator's world map shows the true compass bearing of every landmass for the first time; his projection remains the most widely used. In 1570, Abraham Ortelius produces the first modern atlas, *Theatre of the World*.

1571
The naval Battle of Lepanto off Greece is the last major clash between galley ships: a combined Christian fleet under the command of Don Juan of Austria, illegitimate son of Charles V, halts Ottoman expansion in the Mediterranean.

1572
The massacre of the Huguenots in Paris is the worst atrocity in the French Wars of Religion (1562–1598); more than 3,000 Huguenots are killed in Paris on August 24, and up to 20,000 across France during the following weeks.

1576
Mughal forces overrun Bengal in north India.

1580
Philip II of Spain unites the Spanish and Portuguese crowns following the death of the king of Portugal in battle in 1578.

1581
In the Netherlands, the Act of Abjuration renounces oaths of loyalty to Philip II of Spain, deposing him as ruler of the United Provinces and asserting their independence.

1582–1598
Toytomi Hideyoshi is shogun; he unifies Japan, forcing rival warlords to capitulate, and issues an edict to Jesuits, banning Christianity.

1585
Chocolate is introduced to Europe; the first commercial shipment of cacao beans arrives in Spain from the New World. Francis Drake establishes the first English colony in North America, in Roanoake, Virginia; it is abandoned in 1590.

1586
Drake raids Spanish Caribbean settlements; in 1587 he is sent by Elizabeth I to raid Portuguese and Spanish ships and ports, including Cadiz, southern Spain.

1588
Philip II's Spanish Armada of 130 ships sets out to conquer England, but is beaten by English seamanship and terrible storms.

1588–1629
Shah Abbas I rules Safavid Persia; he quashes political discord, reforms the army, regains territory that was lost to the Ottomans, Portuguese, and Mughals, and moves the capital to Isfahan, turning it into one of the most beautiful cities in the world.

1590
Under a treaty negotiated between Safavid Persia and the Ottomans, Ottoman frontiers extend to the Caucasus and the Caspian Sea.

1591
The sultan of Morocco invades the Songhay Empire, lured by the trans-Saharan gold trade; this and the dynastic disputes after the death of Emperor Askia Daud are key factors in the empire's decline. Toyotomi Hideyoshi completes his unification of Japan and moves his powerbase to Edo (Tokyo).

1592–1593
Japan's invasion of Korea is repulsed by Korea's heavily armed ships and by Chinese intervention.

1593–1606
The Long Turkish War between Habsburgs and Ottomans: clashes in Hungary and the Balkans shake the Habsburg–Ottoman frontier.

1595
Henri IV seeks to unite religious divisions in France by declaring war on Spain; his Edict of Nantes in 1598 allows Protestants to practice their religion, and marks the end of the French Wars of Religion.

1600
The Battle of Sekigahara gives Tokugawa Ieyasu control of Japan; he founds the Tokugawa shogunate that rules Japan in the Edo Period (1603–1867). The East India Company is granted a royal charter in London, to compete with Spain and Portugal for trade in the East Indies; it ends up trading mostly in India and China, becoming a diplomatic and military force in Britain's rapidly expanding empire.

1602
The Dutch East India Company is founded; it ousts the Portuguese from the Spice Islands in 1605, laying the foundations for the Dutch East Asian trading empire, and becomes the most successful commercial venture in the world for the next 100 years.

1606
Portuguese navigator Luis Vaez de Torres is the first European to sight the southern continent, Australia; Dutch navigator Willem Janszoon lands on the north Australian coast.

1607
Jamestown, Virginia, is the first permanent English settlement in North America; it exports its first tobacco crop in 1612.

1608
The invention of the telescope—credited to Dutch lens maker Hans Lippershey and others—is a major advance in scientific observation; in 1610, Italian scientist Galileo Galilei observes Jupiter's moons orbiting the planet, showing that not everything revolves around Earth.

1608
French explorer Samuel de Champlain founds Quebec, New France (Canada); expeditions west of the Great Lakes seek out rich hunting grounds for the fur trade; across North America, French, Dutch, and British traders vie to meet European demand for beaver and other pelts.

1612
Russia expels Polish invaders and elects the first Romanov czar, Mikhail, in 1613.

1616
Manchu leader Nurhaci declares himself Great Jin (khan) of China, founding the Qin dynasty; he declares war on Ming rule in 1618.

1618
A Bohemian Protestant revolt against Habsburg rule breaks out; Protestant and Catholic interventions ignite the Thirty Years War, which spreads across Europe. England's first shipment of African slaves to its American colonies arrives at Jamestown, Virginia.

1620
The *Mayflower* sails from Plymouth, southern England, carrying the Pilgrim Fathers—102 Puritans and Protestant separatists seeking a new life in the New World; they arrive near modern Boston in Cape Cod in November, endure the harsh winter with American Indian aid, and found the Plymouth Colony.

1623
Cardinal Richelieu becomes chief minister of France; he is determined to destroy the military power of the Habsburgs, Huguenot opposition, and the power of the French nobility.

1624
Shah Abbas takes Baghdad, Mosul, and Mesopotamia from the Ottomans; the Ottoman-Safavid conflict continues until 1639.

1625
The Dutch found a colony in North America named New Amsterdam (modern New York); they purchase Manhattan Island from the Lenape tribe for 60 guilders (*c.*$1,000).

1630
Sweden enters the Thirty Years War: Sweden's Lutheran king Gustavus Adolphus crushes the army of the Holy Roman Emperor at Breitenfeld in 1631, but dies in battle in 1632.

1633
Shogun Iemitsu restricts foreign travel and trade, beginning a long period of Japanese isolation (to 1853).

1635
The Treaty of Prague restores the 1627 boundaries of the Holy Roman Empire and gives Lutheranism privileged status as a religion; France enters the Thirty Years War, waging war on Habsburg Spain in 1635 and on the Holy Roman Empire in 1636.

▼ **A jewel of Mughal architecture,** the Taj Mahal in Agra, India, was built in 1634–53 by Shah Jahan as a mausoleum in memory of his beloved third wife, Mumtaz Mahal.

1638
Ottoman sultan Murad IV retakes Baghdad from the Safavids; the ensuing Treaty of Qasr-i-Shirin ends the Ottoman-Safavid conflict of 1628–1639, granting Mesopotamia to the Ottomans.

1642–1651
English Civil War: Charles I's insistence on the "Divine Right" of kings to rule with absolute power and to raise taxes without consulting parliament pits Parliamentarians (Roundheads) against Royalists (Cavaliers); Charles is tried for treason and executed in 1649; England becomes a Commonwealth under Oliver Cromwell, and the defeat of Charles's son ends the war.

1642
Dutch sailor Abel Tasman is the first European to discover Tasmania and New Zealand; he claims Tasmania for the Dutch, but is sent back to sea by New Zealand's Maori warriors.

1644
Manchus overthrow China's Ming dynasty and install the first Manchu Qing emperor in Beijing; they win control of all China by 1681, and the Qing dynasty rules China until 1912.

1648
The Peace of Westphalia that ends the Thirty Years War in Germany fatally undermines the authority of the Holy Roman Emperor.

1648–1652
France—exhausted by war and by peasant uprisings brought about by failed harvests and punitive taxes—collapses into civil war; the Fronde rebellion begins with parliament's protest against the king, then the aristocracy's, but the crown survives.

1652
Cape Town, South Africa, is founded as a supply camp for the Dutch East Indies Company. The first Anglo-Dutch War breaks out; ending in 1654, it is followed by two more in 1665–1667 and 1672–1674; all three are naval wars, fought for control of the seas and shipborne trade.

1659
The Peace of the Pyrenees ends the Franco-Spanish war; as Spain's New World revenues and influence fade, France becomes Europe's major power, embodied by Louis XIV ("the Sun King") and his lavish court.

1660
Restoration of England's monarchy follows the fall of the Commonwealth and the return of Charles II from exile in France. The Royal Society is founded in London to advance the understanding of science; granted its royal charter by Charles II in 1662, it is the world's oldest scientific society still in existence. The Peace of Oliva ends the Northern War between Brandenburg, Poland, Austria, and Sweden; Poland gives up its claims to Estonia and Livonia.

1661
Louis XIV assumes personal rule of France on the death of Cardinal Mazarin, his chief minister.

1664
The Second Anglo-Dutch War begins; the English seize New Amsterdam from the Dutch.

1666
The Great Fire of London destroys the medieval heart of the city. The French Academy of Sciences is founded in Paris; it becomes the Royal Academy of Sciences in 1699 and moves to the Louvre Palace.

1668
Portuguese independence is conceded by Spain. The Treaty of Aix-la-Chapelle ends the War of Devolution (1667–1668): the triple alliance of England, Sweden, and the Dutch forces Louis XIV to abandon French claims to the Spanish Netherlands. The English East India Company gains control of Bombay.

1673
French-Canadian explorer Louis Joliet and French Jesuit Jacques Marquette travel down the Mississippi River; they confirm that it leads to the Gulf, not to the Pacific as expected. English explorer Gabriel

Arthur crosses the Appalachian mountain chain via the Cumberland Gap; this becomes the principal route to the west in the 18th century.

1672–1678
Franco-Dutch War: backed by England and Sweden, Louis XIV invades the Dutch Republic; the war ends with the Dutch granting New Amsterdam (New York) to England while France gains border territories in the Spanish Netherlands.

1675–1676
First Indian War: Metacomet ("King Philip") leads Algonquian tribes against English colonists in New England; the Treaty of Casco awards tribes an annual measure of corn for each family settled on Indian lands.

1680
The Pueblo people in the colony of New Mexico revolt against the Spanish occupiers, who are determined to crush local religious practices; the Spanish return with overwhelming force in 1692.

1682
Philadelphia, Pennsylvania, is founded by the English Quaker and philosopher William Penn. Robert de La Salle navigates the Mississippi River to its mouth and claims Louisiana for France. Louis XIV moves the French court from Paris to Versailles; the vast palace symbolizes his absolute power as monarch.

1682
Peter I ("the Great") becomes czar of Russia; determined to reshape Russia as a western European power, he builds a new capital at St. Petersburg, abolishes the titles of boyars (nobles), centralizes government, introduces sweeping reforms to modernize every aspect of Russian society, restructures the army, creates a navy, and pursues an aggressive foreign policy.

1683
The Ottoman siege of Vienna is lifted by the Polish-Imperial army led by the Polish king Jan III Sobieski at the Battle of Kahlenberg; Ottoman power in the Balkans collapses.

1685
The Edict of Fontainebleau declares Protestantism illegal in France, overturning the 1598 Edict of Nantes; thousands of Huguenots flee France, chiefly to England, the Dutch Republic, and Prussia.

1686
The League of Augsburg is formed as a Protestant Grand Alliance to counter Louis XIV in the Nine Years War (1688–1697).

1687
English physicist Isaac Newton's *Principia Mathematica* establishes the universal law of gravity; his work dominates scientific thinking on the physical universe for *c.*300 years.

1688
In the Glorious Revolution, England's Protestant parliament deposes James II in favor of William III and Mary of Orange; James flees to France, lands in Ireland in 1689, and in 1690 is defeated at the Battle of the Boyne.

1689
William and Mary are crowned joint Protestant monarchs of England. The Treaty of Nerchinsk settles Sino–Russian borders.

1693
The eruption of Mount Etna triggers an earthquake that devastates Sicily, southern Italy, and Malta; the vast cloud of ash exacerbates the bitter winters and wet summers of Europe's Little Ice Age, causing crop failures and famine across Western Europe.

1694
The Bank of England is established, transforming the country's ability to finance wars and imperial expansion.

1690
China's Qing dynasty launches its conquest of Outer Mongolia, beginning a period of expansion that sees China's empire almost double in size by the end of the 18th century.

1696
The English establish a trading post of Fort William in Calcutta, India.

▶ **Louis XIV, the Sun King,** ruled France as absolute monarch for 72 years and is shown in his coronation robes in this 1701 portrait by Hyacinthe Rigaud.

1698
The London Stock Exchange is formed. English inventor Thomas Savery patents an early version of the steam engine.

1699
Mughals suppress a Sikh revolt in the Punjab; the Mughal Empire is now at its zenith, controlling all but the tip of India's subcontinent.

1700–1721
Great Northern War: Charles XII, just 15 when he came to the Swedish throne in 1697, blocks attempts by Baltic rivals to end Swedish pre-eminence by defeating Danish, Russian, and Polish-Saxon forces.

1700
Spain's childless king Charles II dies, having nominated Philip of Anjou, grandson of Louis XIV, to be his heir; Europe's alarm at the increase in French power escalates.

1701
The War of the Spanish Succession erupts (1713–1714): the English and Dutch back Austria's claims to the Spanish throne in a Grand Alliance against the unification of the French and Spanish thrones; Louis XIV sends troops into the Spanish Netherlands to defend them from the English and Dutch and recognizes James III, son of the exiled James II, as king of England in place of the Protestant William and Mary.

1701
West Africa's Asante (Ashanti) confederation of city-states defeats Denkyira to gain independence and establish the Asante kingdom; exploiting its military prowess and the Atlantic coastal trade, it grows into a powerful empire. The Agricultural Revolution begins in England when Jethro Tull invents the horse-drawn seed drill.

1704
British inventor Thomas Newcomen develops an improved version of Thomas Savery's steam engine; the first Newcomen engine is installed in 1712 to pump water from a mine, and the new technology develops rapidly to power the Industrial Revolution. *The Boston News-Letter*, North America's oldest continuously published newspaper, is launched; Boston is emerging as the New World's principal port for the Atlantic slave trade.

1707
The death of the emperor Aurangzeb heralds the decline of Mughal India. The Acts of Union unite Scotland and England as Great Britain.

1712
Pennsylvania Quakers petition for slavery to be abolished, but are refused; the first measure to free slaves is only passed in 1780.

1713–1714
The War of the Spanish Succession ends: the Treaties of Utrecht in 1713 and Rastatt in 1714 seek to balance

power in Europe by separating the French and Spanish crowns; Austria receives the Spanish Netherlands, and Britain is ceded Newfoundland, Nova Scotia, and Gibraltar.

1715
First Jacobite Rebellion: supporters of the deposed Stuart king James VII of Scotland (James II of England) oppose the succession of George I of Hanover to the British throne; they are defeated at Preston, and James's heir, Prince James Francis Edward Stuart (the "Old Pretender"), flees to France in 1716.

1715–1717
Yamasee War: the Spanish establish several missions in Texas and enourage the Yamasee and other tribes of American Indians to attack British settlers in South Carolina.

1717
First performance of Handel's *Water Music*, a masterpiece of the Baroque period, in London. English pirate Edward Teach (better known as Blackbeard) begins plundering ships in the Caribbean.

1720
The Treaty of the Hague ends the War of the Quadruple Alliance (1718–1720): Philip V of Spain abandons claims to Sicily and Sardinia, the French return territory in Pensacola, Florida and in the north of Spain to Philip, and Texas is confirmed a Spanish possession. In Tibet, Qing warriors oust Zunghar Mongols, install a new Dalai Lama, and make Tibet a tribute-paying protectorate.

1721
The Treaty of Nystad ends the Great Northern War (1710–1721): Sweden cedes Baltic ports to Russia. Peter the Great is proclaimed emperor of all Russians. In the Pacific, Dutch explorer Jakob Roggeveen discovers Easter Island and sights Samoa.

1722
Afghans rout Persian forces, overthrow the last Safavid shah, and assume control of the Persian Empire; in 1729 the Afghans are forced back to Kandahar by Afsharid Persians under Nader Shah.

1724
The Kingdom of Dahomey (in what is now Benin, West Africa) becomes the principal supplier of slaves to European traders. The Code Noir is introduced in the French territory of Louisiana: it stipulates basic rights for slaves, such as food and clothing, but it also legitimizes cruel punishments.

1727
Coffee growing begins to emerge in the Caribbean and in South America; the first coffee plantation is started in Brazil by the Portuguese, using seeds from French Guinea.

1728
Hindu Marathas, striking out into India's Deccan region, defeat the Nizam of Hyderabad. Danish-born Vitus Bering is commissioned by Russia to explore the Siberian coast and discovers the strait that separates Siberia from Alaska (now named the Bering Strait); he first sees the coast of Alaska on a second voyage in 1741.

1729
China tackles a steep rise in opium addiction by banning the sale and smoking of opium, which Britain is supplying from India in exchange for Chinese goods; European opium smugglers remain a major problem for China into the 19th century.

1730
The Arabian state of Oman expands its dominions in East Africa by driving the Portuguese from Kenyan and Tanzanian coasts and gaining control of Zanzibar.

1733
The colony of Georgia is founded; it is the last of the Thirteen Colonies established by Britain on North America's Atlantic coast. In England, John Kay invents the flying shuttle, which revolutionizes weaving and the textile industry. Prussia's King Frederick William I (the "Soldier King") institutes compulsory military service; the Prussian army grows to become the fourth largest in Europe.

1735
John Harrison of Britain unveils his marine chronometer; this is the first portable clock that can keep time at sea, enabling navigators to work out their longitude over a long voyage.

1736
Nader Shah, founder of the Afsharid dynasty, is crowned shah of Persia, ending Safavid rule in Persia. French explorer and scientist Charles Marie de La Condamine is the first European to discover the flexible rubber that the Maya have long been making in Ecuador; he sends samples of processed rubber to Paris.

1737
The Maratha Empire extends its control over northern India, at the expense of the Mughal Empire.

▲ **An Asanti comb** carved in wood shows women suckling infants, a symbol of fertility. The Asanti culture established a kingdom in Ghana, West Africa, in 1701.

1739
Nader Shah's Persian force defeats the Mughals at Karnal, occupies Delhi, and leaves laden with booty, including the Mughal emperor's Koh-i-Noor diamond; Persia now controls all territory north and west of the Indus River. The Ottoman and Habsburg Empires sign the Treaty of Belgrade, stabilizing the Ottomans' position in the Balkans. French brothers Pierre and Paul Mallet are the first Europeans to cross North America's Great Plains and open up a route from the Mississippi River to Santa Fe.

1743
French brothers Louis Joseph and Francois de La Vérendrye are the first European explorers to see North America's Rockies, during their bid to reach the Pacific coast.

1745
Second Jacobite rebellion of the Stuarts against Hanoverian rule in Britain: the French-backed "Young Pretender" (Prince Charles Edward Stuart, or "Bonnie Prince Charlie," grandson of the deposed James II) gathers support in Scotland, but his highlanders are massacred by British troops at Culloden in 1746.

1747
Nader Shah is murdered, weakening the Persian Empire; Ahmad Shah Durrani founds the independent state of Afghanistan. In West Africa, the Yoruba of the Oyo Empire invade the Kingdom of Dahomey and force it to pay tribute.

1748
Punjab is invaded by Afghans. European powers sign the Peace of Aix-la-Chapelle, ending the War of the Austrian Succession (1740–1748): Maria Theresa is confirmed as heir to the Habsburg thrones in Austria and Hungary.

1749
The Kingdom of Mysore rises to prominence in south India. Halifax is founded by Lieutenant General Edward Cornwallis, consolidating the British presence in Nova Scotia.

INVENTION AND DISCOVERY

Over the course of human existence, basic human needs—from the need to eat, keep warm, and survive, to the urge to obtain and apply knowledge—have inspired tens of thousands of inventions and discoveries. These have transformed the way we live, think, and communicate with one another. See also the scientific advances in astronomy and space described on p.411 and in medicine on pp.422–23.

- **c.3.3 MYA** Early human ancestors make stone tools.
- **c.1.8–0.5 MYA** Making fires enables early humans to cook, keep warm, and ward off predators.
- **c. 0.8 MYA** Rafts or boats are built for migrations to Australasian islands.
- **c.43,000 YA** The oldest known mines are dug in Africa and in Europe.
- **c.18,000 YA** Pottery vessels are used for food storage in China.
- **c.10,500 BCE** Agriculture begins in Mesopotamia, where wild cereals are domesticated; the first farm animals are sheep, domesticated *c.*9000 BCE.
- **c.8700 BCE** Copper is used to make small items such as a pendant found in Mesopotamia. Gold is known from *c.*4700 BCE, silver from *c.*3000 BCE.
- **c.5500 BCE** Irrigation is developed by Sumerian farmers in Mesopotamia.
- **c.3500 BCE** Carts first use solid wheels. Silk weaving begins in China.
- **c.3300 BCE** Writing is invented in Sumer, developing from pictograms to cuneiform script by 2900 BCE; Egypt invents hieroglyphs *c.*3200 BCE. Bronze, the first alloy, is made from copper and tin in Asia Minor.
- **c.2500 BCE** Plumbing in the cities of the Indus Valley (Pakistan) brings public water supply and sanitation.
- **c.2100 BCE** A lunisolar calendar is developed by the Babylonians.
- **c.1400 BCE** The first alphabetic script is devised by quarry workers in Sinai, Egypt.
- **c.1000 BCE** The smelting of iron becomes common in the Middle East.
- **c.600 BCE** A world map is incised on a Babylonian clay tablet.
- **c.575 BCE** The discovery of magnetism is attributed to Thales of Miletus, a Greek city in Anatolia.
- **384 BCE** The birth of Aristotle, the Greek philosopher and scientist who teaches that knowledge is gained from evidence.
- **c.300 BCE** Euclid sums up ancient Greek geometry in *The Elements*.
- **c.200 BCE** Steel is produced in India by smelting iron and other materials. The invention of the compound pulley is attributed to Archimedes, the Greek mathematician also credited with the laws of buoyancy and levers.
- **c.105 CE** Papermaking is credited to Cai Lun during China's Han dynasty.
- **876 CE** A symbol for zero is used by Indian mathematicians.
- **c.1041–1048** The first movable type system is invented by Bi Sheng in China.
- **1044** The earliest recorded recipe for gunpowder comes from China, and the first record of a magnetic compass.
- **1088** The first mechanical clock, designed by Su Song in China, uses a water wheel, not clockwork.
- **1202** The Italian mathematician Fibonacci introduces the Hindu numerals 0–9 to Europe and invents algebra.
- **c.1250** The first magnifying glass made for scientific study is designed by Roger Bacon in England.
- **c.1454–1455** The first major book to be produced on a printing press with movable type is the Gutenberg Bible in Germany.
- **1492** The oldest surviving globe is Martin of Bohemia's *Erdapfel* ("Earth apple").
- **c.1590** The compound microscope is invented by rival spectacle-makers Hans Lippershey and Zacharias Janssen in the Dutch Republic.
- **c.1600** Italian scientist Galileo Galilei demonstrates that the frequency of sound waves determines pitch.
- **1632** The slide rule invented by English mathematician William Oughtred allows complicated calculations to be easily carried out.
- **1634** Probability theory is developed by French mathematicians Blaise Pascal and Pierre de Fermat.
- **1638** Galileo formulates the law of falling bodies (that bodies of the same material falling through the same medium go at the same speed, regardless of mass).
- **1643** The mercury barometer is invented by Italy's Evangelista Torricelli.
- **1647–1648** Blaise Pascal demonstrates atmospheric pressure and proves his prediction that it decreases at higher altitudes.
- **1672** English physicist Isaac Newton describes light's spectrum of colors.
- **1687** Newton publishes his laws of motion and universal law of gravitation.
- **1752** The lightning rod is invented by Benjamin Franklin, scientist, politician, and founding father of the US.
- **1753–1758** Sweden's Carl Linnaeus formalizes binomial nomenclature in taxonomy (biological classification).
- **1766** British chemist Henry Cavendish discovers the gas hydrogen.
- **1769** Scottish inventor James Watt patents his version of the steam engine, launching the Industrial Revolution.
- **1771** British industrialist and inventor Richard Arkwright opens the first factory, a water-powered textile mill.
- **1778** Oxygen is named by French chemist Antoine Lavoisier, who identifies its role in combustion.
- **1779** Dutch physician Jan Ingenhousz discovers plant photosynthesis.
- **1783** The first manned flight is in a hot air balloon built by Joseph and Etienne Montgolfier in France.
- **1800** Italian scientist Alessandro Volta invents the battery.

▲ **Warriors prepare for battle** in an Assyrian relief from Ashurbanipal's palace at Nineveh, Mesopotamia (ancient Iraq), *c.*645 BCE. Disk wheels were invented in Mesopotamia *c.*3200 BCE; spokes made the wheels lighter, ideal for the fast horse-drawn chariots that first appeared *c.*2000 BCE.

1801 French weaver Joseph Marie Jacquard controls his new mechanical loom with cards punched with holes (later adapted for early computers).

1801–1804 The first steam railway locomotive is developed by British engineer Richard Trevithick.

1809 British chemist Humphry Davy demonstrates the arc lamp, the first electric light.

1820 Danish physicist Hans Christian Ørsted finds that electric currents produce a magnetic field.

1821 The first electric motor, built by British scientist Michael Faraday, uses electromagnetic rotation to turn electrical into mechanical energy.

1822 The first programmable computer is a difference engine designed as a calculator by British mathematician Charles Babbage.

1823 Human fossils are discovered by British geologist William Buckland.

1824 In France, the greenhouse effect is discovered by mathematician Joseph Fourier, and blind teen Louis Braille devises the Braille alphabet.

1826 The earliest surviving photographic prints are made in France by Joseph Nicéphore Niépce.

1827 French engineer Benoît Fourneyron invents the water turbine.

1831 Electromagnetic induction is discovered by British scientist Michael Faraday, who builds the first electric generator.

1834 The first refrigerator is built by American inventor Jacob Perkins.

1837 The electric telegraph is patented in the US by inventor Samuel Morse.

1859 The theory of evolution through natural selection is published by British naturalist Charles Darwin.

1864 Pasteurization is invented by French chemist and microbiologist Louis Pasteur

1866 The laws of heredity are explained by Austrian scientist Gregor Mendel.

1869 Russian chemist Dmitri Mendeleev presents his Periodic Table of elements.

1877 Thomas Edison's phonograph both records and reproduces sound in the US.

1878 The electric bulb is patented by Joseph William Swan in Britain; Edison soon develops electric lighting for public use.

1885 German engineers Gottfried Daimler and Karl Benz independently build the first automobiles powered by a high-speed internal combustion engine.

1893 Sigmund Freud and Joseph Breuer's *On the Physical Mechanism of Hysterical Phenomena* launches psychoanalysis.

1895 The wireless telegraph built by Italy's Guglielmo Marconi paves the way for the development of radio technology.

1898 Radium, a highly radioactive metallic element, is discovered by Polish-French physicist Marie Curie and husband Pierre.

1900 Quantum physics is born when German theoretical physicist Max Planck proposes that radiation comes in discrete packets (quanta) of energy.

1903 In the US, inventors Wilbur and Orville Wright make the first sustained controlled powered flight in an airplane.

1905 German physicist Albert Einstein's Special Theory of Relativity revolutionizes our understanding of space and time; his General Theory follows in 1915.

1907 Bakelite, an early plastic, is invented by the chemist Leo Baekeland in the US.

1910–1915 American geneticist Thomas Hunt Morgan links the inheritance of a specific trait with a particular chromosome.

1911 Nuclear physics begins when New Zealand-born Ernest Rutherford discovers the atomic nucleus.

▲ **Satellites orbiting Earth** play a vital role in global communications. The first satellite was launched in 1957; today, more than 2,000 satellites relay vasts amount of data for the internet, GPS navigation, weather forecasts, and more.

1912 The theory of continental drift is proposed by German geophysicist Alfred Wegener, paving the way for the understanding of plate tectonics.

1913 The structure of the atom is described by Danish physicist Niels Bohr. The first moving assembly line is installed by Ford Motor Company in the US.

1917 French physicist Paul Langevin demonstrates a forerunner of sonar to detect underwater submarines.

1926 Television transmission is first demonstrated by Scottish inventor John Logie Baird.

1935 The first practical radar system is designed by British physicist Robert Watson-Watt to detect aircraft.

1937 The first viable jet engine is trialed by British engineer Frank Whittle.

1940–1944 Scientists in the US discover that DNA is the hereditary molecule in most living organisms and is the chemical basis of genetic information.

1945 The atomic bomb is built in the US by physicist J. Robert Oppenheimer for the Manhattan Project.

1947 Radiocarbon dating is invented by American chemist Willard Libby.

1958 American scientist Charles Keeling begins to monitor atmospheric carbon dioxide; his findings reveal the impact of CO_2 emissions on our planet.

1960 The first laser is built in the US by physicist Theodore H. Maiman.

1961 The first industrial robot is used by General Motors in the US.

1971 In the US, computer programmer Ray Tomlinson sends the first email and Intel make the first microprocessor.

1973 IBM and Xerox each develop prototypes of personal computers, and the first telephone call from a handheld cellphone is made in the US.

1974 A barcode containing a UPC (Uniform Product Code) is first used in the US on a package of chewing gum.

1977 The world's ocean floors are mapped by American geologists Marie Tharp and Bruce Heezen.

1990 The World Wide Web is created by British computer scientist Tim Berners-Lee at CERN in Switzerland.

1994 The first smartphone is released by BellSouth Corporation in the US.

1996 The creation of Dolly the Sheep in Scotland is a milestone in cloning.

2009 The first Android-based tablet computers are released, followed by Apple's first iPad in 2010.

2012 Physicists at CERN find evidence that supports the existence of the Higgs boson, an elementary nuclear particle that gives all other particles their mass.

▶ The World of Empires in 1850
By 1850, Russia and the United States had expanded to span their respective continents. Most of Latin America had gained independence. European powers were colonizing Africa and had settled Australia and New Zealand, while Mughal India had fallen under British control. The Ottoman Empire still controlled much of North Africa and the Middle East. China's Qing Empire remained large but could not block new, European-controlled ports.

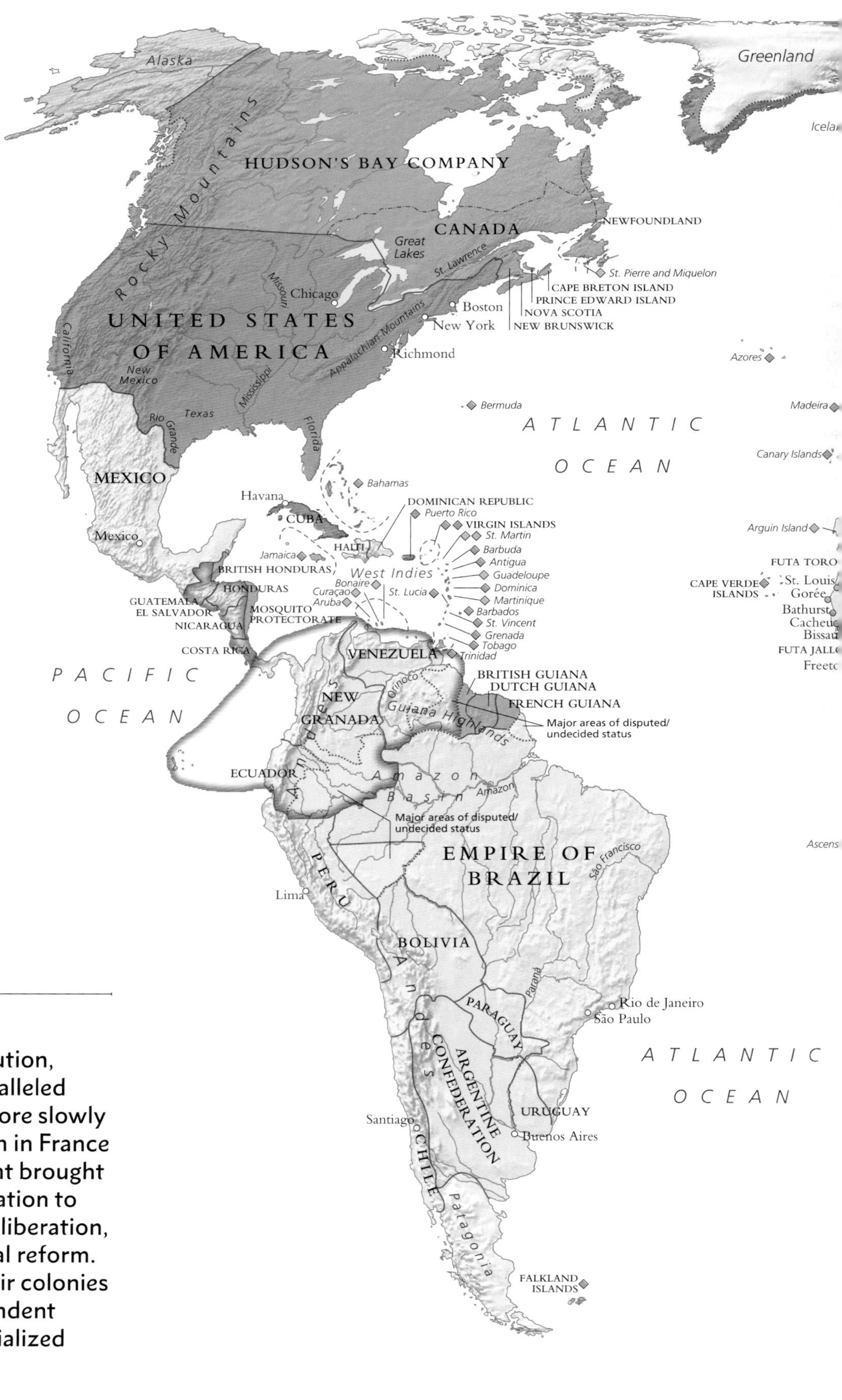

WORLD OF **EMPIRES** **1750–1914**

In an age of industrial, scientific, and political revolution, factories, steam power, and railways brought unparalleled prosperity—rapidly to Europe and the Americas, more slowly elsewhere. Traditional monarchies were overthrown in France and the United States. Swelling nationalist sentiment brought independence to much of Latin America and unification to Italy and Germany. Following demands for political liberation, governments conceded change and enacted political reform. European powers did not extend these rights to their colonies and made political and economic inroads in independent regions such as China and Japan, which had industrialized and reformed more slowly.

RUSSIAN EMPIRE
Siberia
Yenisey
Lena
Ob'
Volga
Amur
FINLAND
St. Petersburg
Moscow
NORWAY
SWEDEN
DENMARK
PRUSSIA
SAXONY
POLAND
WÜRTTEMBERG
BAVARIA
Berlin
Paris
Waterloo
Austerlitz
Vienna
FRANCE
AUSTRIAN EMPIRE
MOLDAVIA
SWITZERLAND
WALLACHIA
PAPAL STATES
SARDINIA
Rome
SERBIA
MONTENEGRO
Black Sea
Constantinople
SPAIN
Madrid
TUSCANY
KINGDOM OF THE TWO SICILIES
GREECE
GIBRALTAR
Ionian Islands
Malta
ALGERIA
Melilla
French conquest in progress
TUNIS
OTTOMAN EMPIRE
Tigris
Euphrates
Russian conquest in progress
Caspian Sea
Kazakhs
KHIVA
KOKAND
SMALL TURKMEN STATES
BUKHARA
KUNDUZ
BADAKHSHAN
BALKH
Tehran
PERSIA
AFGHANISTAN
BALUCHISTAN
Gobi
QING EMPIRE
Beijing
Yellow River
KOREA
JAPAN
Tokyo
Nanjing
Yangzi
Himalayas
NEPAL
BHUTAN
Indus
Delhi
Ganges
INDIA
Plassey
Chandernagore
Calcutta
Diu
Goa
Madras
Pondicherry
Ceylon
Nicobar Islands
Maldive Islands
Hong Kong
Macao
BURMA
LUANG PRABANG
ANNAM
Mekong
SIAM
Bangkok
TENASSERIM
CAMBODIA
Manila
PHILIPPINE ISLANDS
MAGUINDANAO
Mariana Islands
Marshall Islands
Caroline Islands
PACIFIC OCEAN
Labuan
SULU
SARAWAK
BRUNEI
MALAY STATES
ATJEH
MALAY STATES
Malacca
Singapore
Sumatra
Borneo
New Guinea
Papuans
DUTCH POSSESSIONS AND DEPENDENCIES
Batavia
Java
PORTUGUESE TIMOR
Chagos Islands
INDIAN OCEAN
ROCCO
Sahara
Berbers
Tuaregs
Cairo
BAHRAIN
Beduins
EGYPT viceroyalty
Nile
Arabian Peninsula
OMAN
YEMEN
Aden
Socotra
Nilo-Saharan peoples
Sudan
DAMARGAM
MARADI
GOBIR
Niger
MOSSI KINGDOMS
SOKOTO
WADAI
DARFUR
BORNU
ETHIOPIA
HARAR
Kushites
BORGU KINGDOMS
AMPRUSI
AGOMBA
ABUJA
IGALA
ASANTE
BENIN
DAHOMEY
Whydah
GOLD COAST
Fernando Po
SAO TOME AND PRINCIPE
RIO MUNI
Libreville
Bantus
Congo
BUNYORO
BUGANDA
BUSOGA
TORO
ANKOLE
KARAGWE
RWANDA
BURUNDI
TEKE
KUBA
KALUNDE
LUBA
KANIOK
KIKONJA
NGONI
LUNDA
KAZEMBE
ANGOLA
KASANJE
KIAKA
CHOKWE
NDULU
MBAILUNDU
BIHE
NGONI
LOZI
Zambezi
GALANGI
KAKONDA
WAMBU
Mombasa to Oman
Zanzibar to Oman
Kilwa to Oman
Seychelle Islands
Amirante Islands
Comoro Islands
Nossi-Bé Island
Ste Marie Island
PORTUGUESE EAST AFRICA
MERINA KINGDOM
Madagascar
Mauritius
Réunion
ST. HELENA
NDEBELE
Kalahari Desert
Khoisan peoples
Griqua
Swazi
Zulu
GAZA
REPUBLIC OF WINBURG-POTCHEFSTROOM
NATAL
BASUTO
Pondo
CAPE COLONY
Cape Town
AUSTRALIAN COLONIES
Darling
Perth
Adelaide
Sydney
Melbourne
Lord Howe Island
NEW ZEALAND

THE WORLD OF EMPIRES

1750
A new colonial boundary between Spain and Portugal in the New World recognizes the extent of Portuguese settlement in Brazil.

1751–1780
French philosopher Denis Diderot's *Encyclopédie* is a defining work of the Enlightenment, cataloging human knowledge in science, philosophy, politics, and religion.

1753
Swedish botanist Carl Linnaeus publishes *Species Plantarum*, laying the foundations for the system of biological classification used today.

1755
An earthquake in Lisbon, Portugal is one of the deadliest in history, killing 60,000–100,000 people.

1756–1763
In the Seven Years War, Hanover, Britain, and Prussia clash with France, Austria, Russia, Saxony, and Sweden.

1757
The Battle of Plassey wins Bengal for the British East India Company. Prussia defeats Austria at the Battle of Leuthen to control Silesia.

1758
Britain takes Fort Duquesne, Pennsylvania, from the French, as well as Senegal in West Africa. Russia's offensive into Prussia stalls after the Battle of Zorndorf.

1759
An Anglo–Prussian force defeats the French at Minden, north Germany; Britain takes the French West Indian island of Guadeloupe from France. General Wolfe leads British troops to capture Quebec and much of French Canada.

1760
A slave rebellion breaks out on the British colony of Jamaica, a large sugar producer in the West Indies; the revolt is suppressed in 1761. China's Qin dynasty establishes control over Mongolian tribes.

1762
Catherine the Great comes to the Russian throne; her reign brings wide-ranging reform and territorial expansion. French philosopher Jean-Jacques Rousseau publishes *The Social Contract*, questioning the relationship between governments and the governed.

1763
The Treaty of Paris ends the Seven Years War and confirms British supremacy in North America; the cost of the war has been exorbitant, and the participants' national debts are high, leading to increased taxation and discontent at home and in the colonies.

1764
The "Spinning Jenny" is invented by English weaver James Hargreaves, with multiple spindles for cotton; cloth production increases eightfold.

1768
The Russian–Ottoman War begins, over Russian interventions in Poland and the Crimea; it ends in 1774, giving Russia the right of free navigation in the Black Sea. English explorer James Cook begins his first voyage to the Pacific, on the *Endeavour*; he maps the complete coastline of New Zealand in 1769–1770 and the eastern coast of Australia in 1770.

1769
Egypt's de facto ruler Ali Bey al-Kabir deposes the Ottoman governor and attempts to declare independence, but is defeated by 1773. In North America, Spanish Franciscan friars start building a chain of missions along the Californian coast. Scottish inventor James Watt patents his improved version of the steam engine.

▲ **At the Boston Tea Party in 1773**, traders dresssed as American Indians dumped 342 chests of tea from an East India Company ship into the harbor, in protest at British regulations. The recent Tea Act allowed the export of tea from India to North America without it being taxed on arrival; this made it cheaper than the tea smuggled in by many colonial merchants, who now faced ruin.

1771
The first water-powered textile mill in England heralds the Industrial Revolution that transforms rural agricultural economies into ones based on manufactured goods, often made in cities and early factories.

1772
Poland is partitioned by Austria, Prussia, and Russia, who take over a third of its land.

1773
Captain Cook crosses the Antarctic Circle during his second voyage and circumnavigates the continent. At the Boston Tea Party in North America, merchants dump an East India cargo of tea into Boston Harbor in protest of British governance.

1774
Britain issues the "Intolerable Acts" in response to the Boston Tea Party, authorizing punitive measures against the 13 colonies; delegates from the 13 colonies form the First Continental Congress and decide to boycott British goods and trade.

1775–1783
The American War of Independence (or the American Revolutionary War) begins with battles at Lexington and Concord between British troops ("redcoats") and colonial forces led by General George Washington.

1776
Thomas Paine's *Common Sense* calls for American independence; the American Declaration of Independence is signed on July 4, asserting all men are equal and have the right to "life, liberty and the pursuit of happiness"—excluding enslaved Africans.

1777
The Treaty of San Idelfonso confirms Spanish possession of Uruguay and Portuguese possession of the Amazon basin.

1778
France enters the American War of Independence against the British. On Captain Cook's third Pacific voyage (begun in 1776), he makes the first European contact with the Hawaiian islands.

1779
Boer (Dutch-speaking) settlers in South Africa clash with Xhosa tribes over cattle raids and access to grazing land and water.

1780–1782
Tupac Amaru II leads *c.*75,000 Indians and Creoles against the Spanish colonial regime in Peru.

1781
At Yorktown, the last major battle in the American War of Independence, George Washington and his French allies defeat the British.

1783
The Russian Empire annexes Crimea. The first manned flights of a hot air balloon made by the Montgolfier brothers take place in Paris. Britain recognizes American independence in the Treaty of Paris.

1784
Under the India Act, the British government takes direct control of Indian territories, separates rule from trade, and places the East India Company under closer government scrutiny.

1787
Sierra Leone, West Africa, is established for freed slaves from Britain and its colonies, and the Committee for the Abolition of the Slave Trade is established in Britain. The American constitution is approved; it is the oldest written constitution still in use. The first steamboat sails on the Delaware River, designed by American inventor John Fitch.

1788
The First Fleet transports British convicts to Botany Bay, Australia; over the next 50 years, nearly 60,000 people settle in Australia.

1789
George Washington is elected first president of the United States. The French Revolution begins: the newly formed National Assembly takes the Tennis Court Oath, vowing to produce a constitution for France; Parisians storm the Bastille prison on July 14.

1791
Louis XVI and the royal family flee Paris but are caught at Varennes and sent back, losing the king his people's trust. A large-scale slave revolt begins in Haiti, in the French colony of Saint-Domingue, led by Toussaint Louverture.

1792
France abolishes the monarchy, pronounces itself a republic, and declares war on Austria, Prussia, and Piedmont. The monarchies of Holland, Spain, Austria, Prussia, and Russia form the First Coalition against revolutionary France, and are joined by Britain in 1793; the War of the First Coalition ends in victory and territorial gains for France in 1795.

1793
Louis XVI is guillotined in January, and then Queen Marie Antoinette in October; during Maximilien Robespierre's "Reign of Terror" (to 1794) 17,000 alleged enemies of the revolution are executed.

1793
Eli Whitney's cotton gin speeds up the process of combing cotton, greatly increasing the production of cotton in the American south.

1794
France abolishes slavery, ending the Haiti rebellion. A Polish uprising against further partition by Russia and Prussia leads to a third partition in 1795; Poland ceases to exist.

1795
The Directory governs France (to 1799), replacing the Committee of Public Safety. The British seize Dutch territory in the Cape of Good Hope, South Africa.

1798
The Society of United Irishmen stages the Wexford uprising against British rule in Ireland; French support is intercepted by British troops, and the revolt collapses. The War of the Second Coalition (to 1802) is led by Britain, Austria, and Russia against France. French commander Napoleon Bonaparte invades Egypt; his fleet is destroyed by the British navy under Horatio Nelson at the Battle of the Nile.

1799
End of the French Revolution: Napoleon stages a coup d'etat that overthrows the Directory and makes him First Consul. Britain assumes control of south India.

▲ **The storming of the Bastille** in 1789 was a pivotal moment in the French Revolution. Fearing that troops were on their way into Paris, a mob of around 600 broke into the prison, a hated symbol of a despotic monarchy, to seize its stores of ammunition.

c.1800
The Romantic cultural period in Europe emerges, in reaction to the rationalism of the 18th-century Enlightenment; Ludwig van Beethoven, the German early Romantic composer, finishes his popular *Moonlight Sonata* for solo piano in 1801.

1803–1805
The United States almost doubles in size when it buys French territory between the Mississippi River and the Rocky Mountains in the Louisiana Purchase; the territory is extended south to New Orleans in 1804. Meriwether Lewis and William Clark become the first American explorers to cross the new, western portion of the US, and reach the Pacific coast in 1805.

1803–1807
The War of the Third Coalition pits Britain—joined by Sweden, Russia, Austria, and Prussia—against France; it is one of a series of conflicts that make up the Napoleonic Wars (to 1815).

1804
Napoleon crowns himself Napoleon I and assumes the title Emperor of France; the Napoleonic Code is introduced throughout French territories and declares all men equal, ending hereditary nobility.

1805
To counter Napoleon's imperial ambitions in Europe, Austria joins Britain and Russia in the War of the Third Coalition (to 1807): Britain defeats the Franco–Spanish fleet at the Battle of Trafalgar; France defeats Russia and Austria at the Battle of Austerlitz.

▲ **Napoleon's retreat from Moscow** in the brutal Russian winter of 1812 marked a humiliating defeat. Decimated by extreme temperatures, food shortages, typhus, dysentery, and Russian attacks, only 30,000 of the 500,000-strong Grande Armée survived.

1806
Napoleon abolishes the Holy Roman Empire, goes on to defeat Prussia, and imposes the Continental System of embargoes against British trade.

1807
Russia is defeated and allies with France, ending the Third Coalition. Britain abolishes the slave trade.

1808
Napoleon, now in control of most of Western and Central Europe, declares his brother, Joseph Bonaparte, the new king of Spain; this triggers the Peninsular War (to 1814) between the Napoleonic empire and the allied powers of Britain, Spain, and Portugal.

1810
The outbreak of the Mexican War of Independence starts a series of revolts against Spanish rule; by 1826, all Spanish mainland colonies in South America gain independence. Russia reopens trade with Britain, withdrawing from Napoleon's Continental System.

1812
Napoleon invades Russia and occupies Moscow, but is forced to retreat. In the War of 1812 between Britain and the United States, British troops burn the White House in Washington, D.C. In Arabia, Egyptian forces recapture the holy city of Medina from the Wahhabi and reinstate Ottoman rule; they retake Jeddah and Mecca in 1813.

1813
At the Battle of the Nations at Leipzig, allies including Britain, Prussia, and Russia defeat France. In the South American fight for independence, Simón Bolívar invades Venezuela, captures Caracas, and is proclaimed *Libertador* (Liberator).

1814
Anti-French allies occupy Paris; Napoleon is exiled to Elba and Louis XVIII, brother of the beheaded Louis XVI, is placed on the French throne; the Congress of Vienna is convened to agree on the future of Europe.

1815
Napoleon escapes from Elba and returns to Paris at the start of the "Hundred Days"; defeated by British and Prussian forces at the Battle of Waterloo, he is exiled to St. Helena, where he dies in 1821; the French monarchy is restored.

1816
In South America the independent United Provinces of the Rio de la Plata are set up; Argentina declares its independence from Spain. In southeast Africa, Shaka becomes chief of the Zulu nation; he absorbs conquered tribes into a vast empire.

1818
Chile's independence from Spain is secured at the Battle of Maipu.

1819
The United States buys Florida from Spain. Colombia becomes a republic, with Simón Bolívar as president. Stamford Raffles founds the city of Singapore as a base for Britain's East India Company in the Malay peninsula, challenging Dutch dominance of the trading routes between China and India.

1821
The Greek War of Independence against the Ottomans begins, ending in 1832. Mexico achieves independence, Panama joins Gran Colombia, and José de San Martín declares Peru independent.

1823
The United Provinces of Central America (Guatemala, El Salvador, Nicaragua, Honduras, and Costa Rica) declare their independence as a federal republic. Simón Bolívar is named president of Peru; Spanish rule ends in 1824.

1825
Bolivia gains independence from Spain. In Britain, the Stockton to Darlington railway line is the first to use a steam locomotive, designed by George Stephenson, to pull a passenger train.

1826
French inventor Joseph-Nicéphore Niépce takes the first photograph.

1827
Britain, France, and Russia demand independence for Greece; when the Ottomans refuse, the allies sink three-quarters of the Ottoman fleet at the Battle of Navarino.

1828
Uruguay's independence is recognized by Brazil and Argentina. Russia acquires Armenia and declares war on the Ottomans.

1830
The French invade Algeria and take control. The July Revolution in Paris topples the French Bourbon monarch Charles X and crowns his cousin Louis-Philippe, Duke of Orléans. The Belgian Revolution begins, demanding independence from the Netherlands. A wave of rebellion and social unrest sweeps through much of Europe, calling for political reforms. In the United States, the Indian Removal Act strips American Indians of legal rights and forces them to leave desirable territory in the southeast and relocate along the "Trail of Tears" to west of the Mississippi.

1831
Belgium wins independence and elects Leopold I as king. Russia annexes the Duchy of Warsaw.

1832
The Treaty of London creates an independent kingdom of Greece and installs a monarchy.

1833
Britain's Slave Emancipation Act bans slavery in the British Empire and the Factory Act prohibits the employment of children under nine.

1836
Texan rebels are defeated at the Battle of the Alamo but recover and force Mexico to recognize the new republic of Texas; it becomes the 28th state of the United States in 1845. The first wagon train brings settlers west along the Oregon Trail; later wagon trains begin to head to California in 1841.

1838
Guatemala, Honduras, and Nicaragua become independent. At the Battle of Blood River, South Africa, Boers massacre Zulus.

1839–1842
In the First Opium War, China fails to block the British importing opium; the Treaty of Nanking opens five Chinese "treaty ports" to British trade, and cedes Hong Kong.

1839–1842
In the First Anglo-Afghan War, Britain seeks to curb Russia's growing influence over Afghanistan.

1840
The Treaty of Waitangi obliges Maoris to accept British rule in New Zealand, following the first shipload of British settlers in 1839.

1842
The Webster–Ashburton Treaty settles the US–Canadian border.

1844
In the United States, Samuel Morse send the first telegraph message; the first transatlantic telegraph cable is laid in 1858.

1845–1852
The Irish potato famine kills more than a million people.

1846
Japan refuses American demands to open its ports to foreign trade. The US–Mexican War over Texas begins; Mexico surrenders in 1847; in 1848 the Treaty of Guadalupe Hidalgo grants a vast swathe of Mexico's territory to the United States, including California.

1848
California finds gold, prompting the Gold Rush (to 1855). Karl Marx and Friedrich Engels publish *The Communist Manifesto*. The "Year of Revolution" sees political upheaval in France, the Austrian Empire, Germany, and Italian states: France overthrows the monarchy, forms the Second Republic, and elects Napoleon's nephew as president.

1850
The Taiping rebellion begins in China: fighting lasts till 1864 and claims 20 million lives. The Great Exhibition in London attracts six million people to view agricultural and manufacturing displays from around the world; Britain's empire is the largest in history, and covers one quarter of the globe by 1922.

1852
In South Africa, the independent Boer republic of Transvaal is recognized by Britain; the Orange Free State is recognized in 1854.

1853–1856
In the Crimean War, Russia disputes Ottoman territory; the Ottomans are joined by Britain and France, and win the war; field hospital reforms by British nurse Florence Nightingale in the Crimea lead to far fewer deaths from disease in wartime.

1856
Britain's Henry Bessemer invents a new process for producing steel, greatly reducing its price.

1856–1858
In the Indian Mutiny, native troops (sepoys) rebel against British officers; the revolt is crushed, the last Mughal emperor is exiled, and control of India passes from the East India Company to the British crown, ushering in the period known as the Raj (to 1947).

1856–1860
In the Second Opium War, Britain and France declare war on China; treaties call for China to open more ports and legalize opium imports.

1858–1871
The Wars of Italian Unification continue the Risorgimento led by Giuseppe Mazzini and Giuseppe Garibaldi in 1848: France and Piedmont-Sardinia form an alliance to end Austrian rule in Italy; Victor Emmanuel II of Piedmont-Sardinia is declared king of Italy in 1861; Rome becomes the capital in 1871.

1859
Charles Darwin's *On the Origin of Species by Means of Natural Selection* explains the process of evolution.

1861
Serfdom is abolished in Russia. Abraham Lincoln takes office as president of the United States; states in the rural south defy federal government and assert their right to own slaves; they secede from the Union and form a Confederacy under Thomas Jefferson.

1861–1865
THE AMERICAN CIVIL WAR
The war opens with a Confederate attack at Fort Sumter.
• 1862: Confederate defeat at Shiloh; Union defeat at the second Battle of Bull Run; Union victory at Antietam.
• 1863: President Lincoln signs the Emancipation Proclamation, abolishing slavery in the south; Union capture of Vicksburg and New Orleans; Confederate defeat at Gettysburg, the war's biggest battle.
• 1864: General Ulysses S. Grant is made commander-in-chief of the Union forces; Union general William T. Sherman's "march to the sea" destroys railway lines and towns from Atlanta to Savannah, Georgia.
• 1865: Union troops capture the Confederate capital of Richmond; Confederate general Robert E. Lee surrenders on April 9, ending America's deadliest war: 600,000 men are dead, 500,000 are injured. On April 14, a Confederate assassinates President Lincoln.

1862
Otto von Bismarck becomes prime minister of Prussia; he builds up the army and masterminds the unification of Germany in 1871. Japan expels foreigners.

1866
Prussia defeats Austria in the Seven Weeks War.

1867
Prussia unifies 22 states under the North German Federation, with Prussia at the helm. The dual monarchy of Austria-Hungary is established, under Franz Josef I of Austria. Karl Marx publishes the first part of *Das Kapital*. The US purchases Alaska from Russia.

1868
Japan overthrows the Tokugawa shogunate, ending the Edo Period, and restores imperial rule; during Meiji Tenno's reign (to 1912), Japan ends its isolationism, implements constitutional government, and starts to modernize its infrastructure.

1870
The Fifteenth Amendment to the US Constitution extends the right to vote to freed slaves. The Franco-Prussian

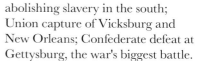

◄ **A Civil War eagle drum** displays the American bald eagle and the stars of the 13 federal Union states on a blue background for an infantry regiment.

War begins: Paris is beseiged, Napoleon III surrenders, and the Third Republic is proclaimed.

1871
The Franco–Prussian War ends. France cedes Alsace Lorraine to Germany; Wilhelm I of Prussia becomes emperor of Germany and names Bismarck as chancellor; the Paris Commune revolt is put down.

1876
At the Battle of Little Big Horn, Sioux and Cheyenne warriors kill Lt. Colonel George Custer and his troops, during the Indian Wars against forcible relocation.

1877–1878
Russia declares war on the Ottoman Empire; Serbia, Romania, and Montenegro gain independence; Bulgaria is granted some autonomy and put under Russian authority.

1881
Britain recognizes the Boers' South African Republic in Transvaal. The assassination of reformist Czar Alexander II triggers an anti-Jewish pogrom in Russia; many Jews to Western Europe, the United States, and Palestine.

1882
The Triple Alliance is formed by Germany, Austria-Hungary, and Italy against France. Unrest in Egypt prompts British occupation.

1884–1885
The Berlin Conference stimulates the "Scramble for Africa": Belgian king Leopold I acquires Congo, and the European powers in Berlin gradually partition almost all of Africa among themselves by 1914.

1885
The first automobiles are built by Gottlieb Daimler and Karl Benz.

1889
Brazil is declared a republic.

1893
New Zealand is first to grant voting rights to women.

1894
Ottomans massacre nationalist Armenians: by 1897, *c.*250,000 are killed.

1894–1895
After the First Sino–Japanese war over competing interests in Korea, China recognizes Korean independence and cedes Formosa (Taiwan) to Japan.

1895
The first motion picture, lasting 46 seconds, is shown by brothers Auguste and Nicolas Lumière in Paris, France.

▲ **The Wright Flyer takes to the air** at Kitty Hawk, North Carolina, in 1903. In the first successful sustained powered flight in an airplane, Orvilee Wright flew 40yd (37m) in 12 seconds; later that day his brother Wilbur flew 284yd (260m) in 59 seconds.

1896
Emperor Menelik II defeats the Italian army at the Battle of Adowa, securing Ethiopia's independence.

1898
The USS *Maine* is blown up in Cuba, triggering the Spanish–American War; Spain grants independence to Cuba, and cedes Puerto Rico, Guam, and the Philippines to the US. The US also annexes Hawaii.

1899–1902
In the South African (Anglo–Boer) War, up to 25,000 women and children die in British concentration camps; defeated by "scorched earth" attacks, the Boers recognize British sovereignty in South Africa.

1900
Immigration swells the population to *c.*75 million in the US. Nigeria becomes a British protectorate. In the Boxer Rebellion against foreigners in China, an international relief force occupies Beijing; Russia occupies southern Manchuria.

1901
Texas strikes "black gold" with an oil well at Spindletop; the US becomes the world's leading oil producer and soon enjoys the first modern consumer boom. The Commonwealth of Australia is proclaimed. Britain's Queen Victoria dies, after a 63-year reign.

1904
Britain and France sign the Entente Cordiale; in 1907 they sign the Triple Entente with Russia, to counter the Triple Alliance of 1882.

1904–1905
In the Russo–Japanese War, Japan attacks Russian warships and forces Russia to leave Manchuria.

1905
Revolution forces Czar Nicholas II of Russia to grant limited concessions, including the creation of an elected legislature, the Duma. Sweden recognizes Norway's independence.

1907
Australia, Canada, and New Zealand become self-governing dominions of the British Empire.

1908
Young Turks demand reform in the Ottoman Empire; as the empire falters, Austria-Hungary annexes Bosnia-Herzegovina and Bulgaria wins recognition of its independence.

1910
Portugal overthrows its monarchy and proclaims a republic. China invades Tibet. Japan annexes Korea. The Union of South Africa is founded as a dominion of the British Empire; it becomes a sovereign republic in 1961.

1910–1920
In the Mexican Revolution, calls for democracy and land reform spiral into civil war between revolutionary factions and the military, ending with a new constitution in 1917 and presidential elections in 1920.

1911
An uprising in China leads to the collapse of the Qin dynasty; Sun Yat-sen becomes the first president of the Chinese Republic.

1912
The luxury liner RMS *Titanic* sinks on its maiden voyage.

1912–1913
First Balkan War: backed by Russia, the Balkan League is formed by Serbia, Bulgaria, Greece, and Montenegro and defeats the Ottoman Empire, which loses all its remaining territory in Europe.

1913
Second Balkan War: the Balkan League implodes and Greece and Serbia defeat Bulgaria; Serbia's territorial ambitions and alliance with Russia alarm Austria-Hungary and Germany.

1914
The assassination of the Austrian archduke Franz Ferdinand in Sarajevo ignites World War I.

ASTRONOMY AND SPACE

Ever since early humans first gazed at the skies, we have speculated about Earth's origins and its relation to the sun, moon, and stars. We have built prehistoric monuments to mark the passage of time, developed theories to explain the motions of the planets, invented telescopes to peer into space, launched spacecraft, and investigated the forces that shape the universe.

- ***c.*****8000 BCE** The oldest known calendar is a megalithic site in Scotland that marks the 12 lunar months of the year.
- ***c.*****3000 BCE** Mesopotamia devises the earliest calendar with a 360-day year, consisting of twelve 30-day months.
- ***c.*****2000–1500 BCE** Egyptians and Babylonians produce star charts, mapping the heavens and dividing them into sections, or "zodiacs."
- ***c.*****360 BCE** Mathematical models of planetary motion around Earth are attributed to the Greek astronomer Eudoxus of Cnidus.
- ***c.*****150 CE** The ancient Greek model of a geocentric universe, placing Earth at the center of the solar system, is recorded by Roman astronomer and geographer Claudius Ptolemy in his *Almagest*, which is later translated into Arabic and Latin.
- **1420** Persian astronomer Ulugh Beg builds an observatory in Samarkand and measures the tilt of Earth's axis to within 1/100th of a degree.
- **1543** Polish astronomer Nicolaus Copernicus suggests that Earth orbits the sun and not vice versa, supplanting Ptolemy's geocentric universe with a heliocentric one.
- **1608** The first refracting telescopes are credited to Dutch spectacle-makers Hans Lippershey, Zacharias Janssen, and Jacob Metius.
- **1609–1619** German astronomer Johannes Kepler's laws of planetary motion describe the planets' elliptical orbits around the sun.
- **1610** In Italy, Galileo Galilei uses his superior telescope to observe Jupiter's moons and the phases of Venus as it orbits the sun, and supports heliocentrism.
- **1668** The first reflecting telescope is built by English physicist Isaac Newton.
- **1687** The universal law of gravitation is published by Isaac Newton in England.
- **1705** British astronomer Edmond Halley is first to predict a comet's return: Halley's comet makes a 76-year orbit of the sun.
- **1781** British astronomer William Herschel discovers the planet Uranus, using a homemade telescope; he also studies the evolution of stars and classifies them.
- **1846** Germany's Johann Gottfried Galle discovers the planet Neptune.
- **1905** German physicist Albert Einstein's special theory of relativity describes the relationship between space and time; he explains the effect of gravity on spacetime in his general theory of relativity in 1915.
- **1927** The Big Bang theory originates with Belgian astronomer Georges Lemaître.
- **1929** In the US, astronomer Edwin Hubble calculates that the Universe is expanding and has more than 100 billion galaxies.
- **1933** Radio astronomy begins in the US when telephone engineer Karl Jansky identifies radio waves from the Milky Way as a source of static; Grote Reber builds the first radio telescope dish in 1937.
- **1948–1949** The Big Bang theory is developed in the US by theoretical physicist George Gamow's description of cosmic nucleosynthesis and by his associates Ralph Alpher and Robert Herman's prediction of cosmic microwave background radiation.
- **1954** CERN, the European Organization for Nuclear Research, is established in Geneva, Switzerland, building high-energy particle accelerators to study the basic constituents of matter.
- **1957** Sputnik I, the world's first satellite, is launched into space by the Soviet Union, triggering a "space race" with the US.
- **1958** The first communications satellite is launched by the US, which also establishes the National Aeronautics and Space Administration (NASA).
- **1961** The first human in space, Soviet cosmonaut Yuri Gagarin orbits Earth.
- **1962** NASA's *Mariner 2* is the first spacecraft to study another planet, Venus.
- **1969** On NASA's Apollo 11 mission, American astronauts Neil Armstrong and Buzz Aldrin become the first humans to land on the moon.
- **1971** In Canada, Charles Thomas Bolton is first to discover a black hole.
- **1977** NASA launches the Voyager 1 probe; heading to the edge of the solar system, it travels farther into space than any other humanmade object.
- **1981** The first reusable spacecraft is NASA's space shuttle *Columbia*.
- **1990** The Hubble Space Telescope (HST) is launched into a near-Earth orbit to record clear images of deep space from 340 miles (547 km) above Earth's atmosphere.
- **1995** A satellite-based GPS (Global Positioning System) is completed by the US Department of Defense.
- **2000** The first crew arrives at the International Space Station (ISS) to conduct experiments in space.
- **2014** The European Space Agency's space probe Rosetta carries the lander Philae, the first to land on a comet.
- **2015** In the US, the Laser Interferometer Gravitational Wave Observatory (LIGO) detects a form of cosmic radiation predicted by Albert Einstein's general theory of relativity.

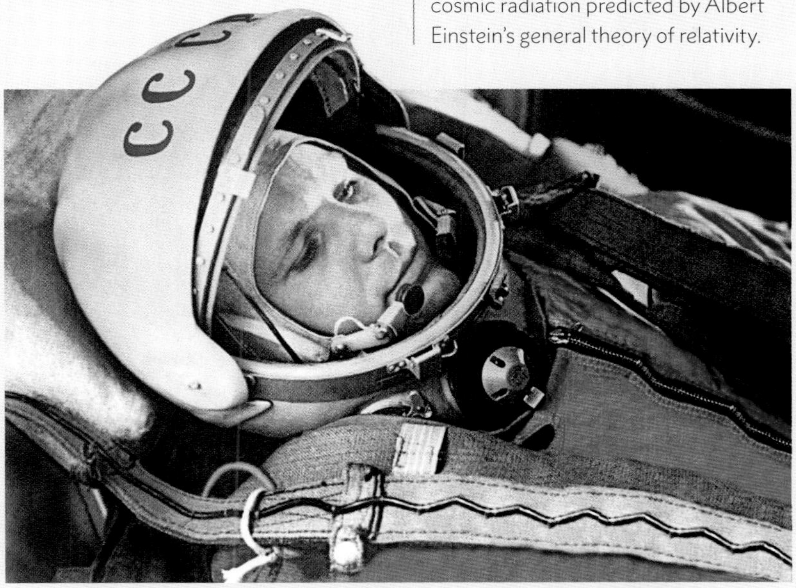

▲ **Cosmonaut Yuri Gagarin** made the first manned space flight in 1961, when Vostok 1 orbited Earth in 108 minutes. The first man in space, Gagarin was awarded his country's highest honor—the title "Hero of the Soviet Union."

The World in 1950

- United Kingdom and possessions
- France and possessions
- Denmark and possessions
- Spain and possessions
- Portugal and possessions
- Netherlands and possessions
- West Germany
- Japan and possessions
- Norway and possessions
- Belgium and possessions
- Italy and possessions
- New Zealand and possessions
- Australia and possessions
- US and possessions
- controlled by European Axis powers 15 November, 1942
- controlled by Japan 15 November, 1942

▶ The Modern World in 1950
European decolonization had begun, as India became independent in 1947 and Israel in 1948, but much of Africa remained under European control for a further 10 to 20 years. In Europe, the European Coal and Steel Community would be founded in 1952, beginning a process of integration that ultimately led to the European Union, composed of 28 countries by 2013.

MODERN WORLD 1914 ONWARD

European global dominance was shaken by two World Wars that devastated the continent. Unable to afford or defend them, European powers granted their colonies independence. Communism fueled the Russian Revolution and found fertile ground in countries such as China, which had suffered at the hands of European imperial ambitions. The Cold War between the communist USSR and its allies and capitalist-leaning countries such as the US threatened the world with nuclear annihilation for over 40 years. Following the collapse of the USSR in 1991, hopes for global peace were dashed as regional wars erupted, and new problems such as radical, violent movements within Islam challenged world leaders.

UNION OF SOVIET SOCIALIST REPUBLICS
Siberia
PEOPLES REPUBLIC OF CHINA
MONGOLIA
JAPAN
INDIA
AUSTRALIA
PACIFIC OCEAN
INDIAN OCEAN
BELGIAN CONGO
ANGLO-EGYPTIAN SUDAN
SAUDI ARABIA
ETHIOPIA
NIGERIA
FRENCH EQUATORIAL AFRICA
LIBYA
temporary UK-France administration
ERITREA
temporary UK administration
TIBET
Communist Chinese conquest in progress
Dairen
Chinese-USSR condominium
VIETNAM member of French Union
LAOS
member of French Union
CAMBODIA
member of French Union
Ryukyu Islands
US military administration
TRUST TERRITORY OF THE PACIFIC ISLANDS
SOMALIA
trusteeship
TANGANYIKA
trusteeship
RUANDA-URUNDI
trusteeship
CAMEROONS
trusteeship
BECHUANALAND
protectorate
TERRITORY OF NEW GUINEA
trusteeship
REPUBLIC OF CHINA (TAIWAN)
INDONESIA
PHILIPPINES
NEW ZEALAND
SOUTH AFRICA
MADAGASCAR
ANGOLA
MOZAMBIQUE
KENYA
IRAN
TURKEY
EGYPT
FRANCE
SPAIN
ITALY
NORWAY
SWEDEN
FINLAND
WEST GERMANY
EAST GERMANY
POLAND
CZECHOSLOVAKIA
HUNGARY
ROMANIA
YUGOSLAVIA
BULGARIA
ALBANIA
GREECE
AUSTRIA
SWITZ.
LUX.
BELG.
NETH.
DENMARK
UNITED KINGDOM
ALGERIA
FRENCH WEST AFRICA
SPANISH MOROCCO
GIBRALTAR
MALTA
CYPRUS
LEBANON
ISRAEL
SYRIA
IRAQ
JORDAN
KUWAIT
BAHRAIN
QATAR
TRUCIAL OMAN
OMAN
YEMEN
ADEN PROTECTORATE
AFGHANISTAN
PAKISTAN
NEPAL
SIKKIM
BHUTAN
BURMA
THAILAND
MALAYA
CEYLON
NORTH KOREA
SOUTH KOREA
BRITISH NORTH BORNEO
BRUNEI
SARAWAK
NETHERLANDS NEW GUINEA
PAPUA
PORTUGUESE TIMOR
FRENCH SOMALILAND
BRITISH SOMALILAND
UGANDA
ZANZIBAR
NYASALAND
NORTHERN RHODESIA
SOUTHERN RHODESIA
SWAZILAND
BASUTOLAND
SPANISH GUINEA
SAO TOME AND PRINCIPE
GOLD COAST
ST. HELENA
SEYCHELLES
COMORO ISLANDS
MAURITIUS
RÉUNION
MALDIVE ISLANDS
Chagos Islands
Cocos Islands
Christmas Island
Socotra
GUAM
Mariana Islands
Caroline Islands
Marshall Islands
Solomon Islands
NEW HEBRIDES
New Caledonia
Lord Howe Island
Hong Kong
Macao
Leningrad
Moscow
Warsaw
West Berlin
Vienna
Paris
Rome
Madrid
Tangier
Istanbul
Ankara
Jerusalem
El Alamein
Cairo
Tehran
Kabul
Lahore
Delhi
Calcutta
Bombay
Goa
Madras
Bangkok
Manila
Singapore
Jakarta
Shanghai
Beijing
Yan'an
Seoul
Tokyo
Hiroshima
Nagasaki
Sydney
Canberra
Auckland
Cape Town
Sahara
Sahel
Gobi
Arabian Peninsula
Caspian Sea
Black Sea
Himalayas
Sumatra
Borneo
Java
New Guinea
Yenisey
Ob'
Lena
Amur
Volga
Yellow River
Yangzi
Mekong
Ganges
Indus
Euphrates
Nile
Congo
Darling

THE MODERN WORLD

1914–1918 WORLD WAR I

Archduke Franz Ferdinand is shot by a Bosnian Serb in Sarajevo on June 28. World War I begins on July 28, when Austria–Hungary declares war on Serbia; Russia mobilizes in Serbia's defense. Germany declares war on Russia and France, and invades France via Belgium; Britain declares war on Germany and sends the British Expeditionary Force (BEF) to France.
• On the Western Front, in Flanders and France, British, French, and Belgian armies face those of Germany; on the Eastern Front in Poland, Galicia, and Serbia, Russian armies face those of Germany and Austria–Hungary.
• The conflict becomes global: the Ottoman Empire enters the war on the side of Germany and Austria–Hungary (the Central Powers); British troops invade Turkish-ruled Iraq and German East Africa; Japan joins the Allies (Britain, France, and Russia) by attacking German holdings in China and the Pacific.
• On the Eastern Front, German troops crush the Russian Second Army at Tannenberg.
• Unable to strike a decisive blow and advance on the Western Front, Allied and German troops start to dig trenches; by the end of the war, the trenches would stretch 25,000 miles (40,000km) if they were laid end to end.

1915

• German U-boats begin attacks on merchant ships in British waters; the liner RMS *Lusitania* is sunk off the west coast of Ireland and 1,200 crew and passengers are drowned, including 128 American citizens.
• Poison gas is first used by Germany during the Second Battle of Ypres.
• Allied troops land at Gallipoli in the Dardanelles, but face tough Turkish resistance, suffer almost 250,000 casualties, and begin to evacuate in December.
• Italy joins the Allies, declaring war on Austria–Hungary.
• The Armenian genocide begins: by 1922, the Turks massacre or deport around 1.5 million Armenians.

1916

• On the Western Front, huge offensives cause mass casualties (400,00 at Verdun, more than a million at the Somme), with little territory gained.
• The Sykes–Picot agreement between Britain and France agrees the postwar division of the Middle East in the event of the Ottoman Empire's defeat.

▲ **Known as the Harlem Hellfighers**, troops of the 369th Infantry Regiment from the US Army's 93rd Division arrived in France in 1917 and fought under French command in the trenches along the Western Front in 1918. More than one million American soldiers served in World War I, including nearly 380,000 African Americans.

► **Lenin points the way forward** above the red flag of communism in a Soviet poster idealizing the revolutionary fervor of 1917.

1917

• The US declares war on Germany after discovering the Zimmerman telegram from the German foreign minister urging Mexico to reclaim Texas, New Mexico, and Arizona from the US.
• The British begin a new push at Ypres; fought in a sea of mud, the campaign ends at Passchendaele.
• The Balfour Declaration commits Britain to a homeland for the Jews in Palestine, without consulting Britain's Arab allies.
• Russian revolutionaries proclaim a unilateral armistice to withdraw from World War I.

1918

• Under the Treaty of Brest-Litovsk between Russia and the Central Powers, Russia exits the war and loses Poland, Ukraine, Belarus, Finland, and the Baltic states.
• The Allies check Germany's Spring Offensive, which had reached the Marne, and launch a counter-offensive, which breaches the Hindenburg Line defenses.
• The Ottoman Empire signs an armistice on October 30, Austria on November 3, and Germany on November 11, ending World War I.
• The total number of military and civilian casualties is *c.*37 million, including 15 million dead.

1915

US marines occupy Haiti (to 1934), and the Dominican Republic in 1916 (to 1924), to restore order and secure American influence in the region.

1916

In the Mexican Revolution, Francisco Pancho Villa makes a cross-border raid into the United States, provoking a US military expedition into Mexico. The Irish nationalists' Easter Rising against British rule is crushed.

1917

The Russian Revolution erupts: food riots, strikes, and a military mutiny force the czar to abdicate; a liberal government is formed under Alexander Kerensky; civil war breaks out when the Bolsheviks under Vladimir Ilyich Lenin and Leon Trotsky seize power.

1918

Czar Nicholas II and his family are shot; Kaiser Wilhelm II abdicates and Germany becomes a republic; Charles I's abdication brings the Austrian Empire to an end.

1918–1919

The Spanish influenza global pandemic kills more than 50 million people.

1919

The Paris Peace Conference sets the terms for treaties with the Central Powers, dismantling the German, Austrian, and Ottoman Empires, redrawing national boundaries in Europe, and imposing substantial war reparations on Germany. A new German constitution establishes the Weimar Republic. The Turkish War of Liberation (to 1923) asserts Turkish independence.

1919

The Third International, or Communist International (known as Comintern), is set up to promote communism worldwide. Benito Mussolini founds the Fascist party in Italy. The first nonstop transatlantic flight is made by British aviators John Alcock and Arthur Whitten Brown. In Britain, Ernest Rutherford splits the atom in the world's first artificially created nuclear reaction.

1920
The US brings in prohibition (to 1933) and grants women the vote. In the Irish War of Independence (to 1921), Ireland declares independence from Britain, which deploys the "Black and Tans" against the Irish Republican Army (IRA). Russia invades Poland but the Red Army is driven back outside Warsaw. The League of Nations is established.

1921
Six million die in Russia's famine; Lenin's New Economic Policy (NEP) allows a partial return to capitalism. The Communist Party of China (CPC) is founded. Germany's war reparations are set at 132 billion Deutschemarks. The Anglo-Irish Treaty settles the division of Ireland into the Irish Free State and Northern Ireland, with separate parliaments in Dublin and Belfast.

1922
Russia's civil war between the "Reds" of the Bolshevik government and "White" counterrevolutionaries ends in victory for the Bolsheviks; the former Russian Empire becomes the Union of Soviet Socialist Republics (USSR). Turkish nationalists are at war with Greece and emerge victorious, with widespread deportations on both sides.

1922
Britain declares Egypt independent. The Anglo-Irish treaty is bitterly disputed during the Irish Civil War (to 1923) but the Irish Free State comes into effect, as a self-governing dominion of the Commonwealth; Northern Ireland remains part of the United Kingdom. James Joyce's modernist novel *Ulysses* is published in Paris. Mussolini marches on Rome with his Blackshirts and heads Italy's new Fascist government.

1923
The Turkish republic is founded; President Mustafa Kemal (Ataturk) embarks on a series of radical reforms to turn Turkey into a modern secular state. Germany is unable to pay war reparations; hyperinflation leads to currency collapse; Adolf Hitler, leader of the National Socialist (Nazi) Party, attempts to grab power in the Munich putsch and is briefly imprisoned in 1924. A military coup in Spain establishes a dictatorship under Miguel Primo de Rivero.

1924
Lenin dies; Josef Stalin is leader of the Soviet Union. Mohandas Gandhi, or Mahatma ("Great Soul"), becomes leader of India's National Congress in the drive for self-rule and an end to the British Raj.

1925
Italy becomes a one-party state under Mussolini, who takes dictatorial powers and the title *Il Duce* (The Leader). The Chinese nationalist leader Sun Yat-sen dies; he is succeeded by Chiang Kai-shek. In Syria and Lebanon, nationalists revolt against the French mandate but are suppressed by 1927. The Locarno Pact paves the way for the restoration of normal relations with Germany and for Germany to join the League of Nations in 1926.

1926
Scottish inventor John Logie Baird gives the first public demonstration of television images. Reza Khan Pahlavi crowns himself shah and sets out to modernize Iran; his dynasty rules until 1979. Military coups seize power in Poland and Portugal.

1927
Chiang Kai-shek purges Chinese communists. In the Soviet Union, Stalin has his rival Leon Trotsky expelled from the Communist Party; Trotsky is exiled in 1929 and assassinated in Mexico in 1940.

1927
American aviator Charles Lindbergh flies solo across the Atlantic. In the US, a consumer boom boasts one car for every six Americans. The first talking motion picture, *The Jazz Singer*, is released. Oil is discovered in Iraq.

1928
Stalin launches a Five Year Plan to transform the Soviet Union into a major industrial country. Kuomintang forces take Beijing and Chiang Kai-shek declares a national government of China; communist resistance led by Mao Zedong continues in remote rural areas.

1929
Riots in Palestine: Arab attacks on Jewish immigrants. The Wall Street Crash in the US ruins stock market investors and leads to a severe global economic downturn; the Great Depression lasts until the late 1930s.

1930
Peasants in the Soviet Union resist the mass collectivization of farms; hundreds of thousands of peasants are sent to gulags (forced labor camps). Ho Chi Min founds the Vietnamese Communist Party. Gandhi mounts a campaign of civil disobedience and leads the Salt March to challenge government monopolies and British rule in India.

1930
The collapse of the world economy triggers mass unemployment and political extremism: Hitler's Nazi party wins 107 seats in the Reichstag, becoming the second largest party in Germany; an army revolt in Brazil brings Getúlio Vargas to power.

1931
Revolution overthrows the monarchy in Spain. Japan occupies Manchuria; it installs a puppet government in 1932 under Pu Yi, China's last emperor, who was deposed in 1912.

1932
The kingdom of Iraq is declared independent of Britain and the kingdom of Saudi Arabia is founded. More than 12 million people are unemployed in the US: Franklin D. Roosevelt promises "a new deal" and is elected president. In Germany, the Nazis become the largest single party in fresh elections to the Reichstag but continue to be excluded from government.

1932–1933
Soviet collectivization results in the "Great Famine" in grain-growing regions; 6–8 million peasants die over the winter, including 4–5 million in Ukraine.

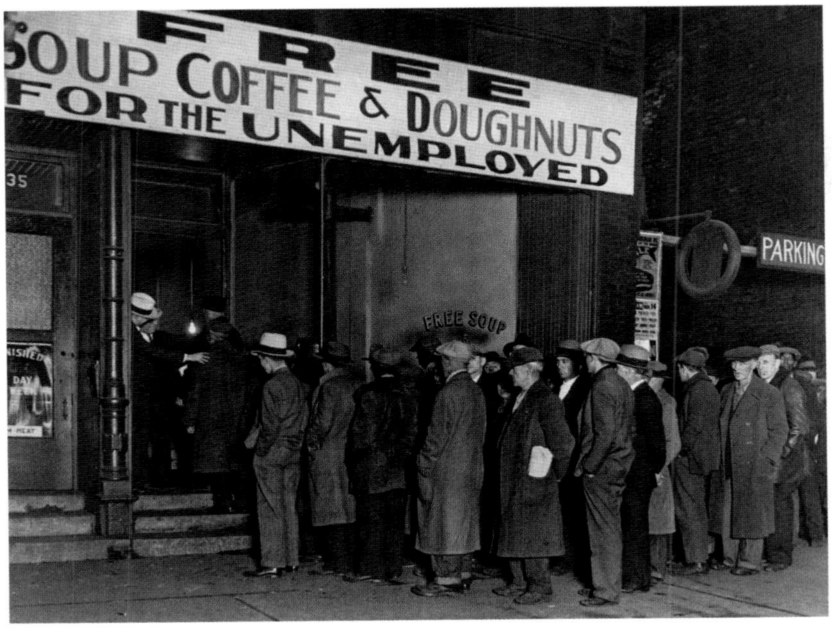

◄ **The first soup kitchen** for the hungry and homeless was opened by gangster Al Capone in Chicago during the Great Depression of the 1930s, when 12 million Americans (a quarter of the normal labor force) were unemployed.

1933
Hitler becomes chancellor and is soon given dictatorial powers. The Nazis organize a national boycott of Jewish shops and businesses; the first Nazi concentration camp opens at Dachau; the Gestapo secret police is formed; students burn "un-German" books; all other political parties are banned. Germany quits the League of Nations.

1933
In the US, the National Recovery Administration launches New Deal policies, such as work creation schemes, but tariffs and other nationalist policies block global recovery and the World Economic Conference fails to achieve a global response to the Depression.

1934
Hitler purges critics in the Night of the Long Knives; he becomes Führer (leader) of Germany after President Hindenburg dies. Austria's dictatorial chancellor Engelbert Dollfuss bans the Nazis in Austria and their attempted coup is blocked. The Soviet Union joins the League of Nations.

1934–1935
In the Long March in China, Mao Zedong and his communist forces evade Chiang Kai-shek's Nationalist army by retreating from Jiangxi to Shaanxi, a journey of around 6,000 miles (10,000 km).

1934–1937
On America's Great Plains, drought, intensive farming, and giant dust storms strip the land of topsoil in the Dust Bowl; thousands of destitute farmers migrate west to California.

1935
Hitler introduces conscription (banned by the Treaty of Versailles) and anti-Semitic Nuremberg Laws deprive Jews of German citizenship and ban marriage and sex between Jews and non-Jews. Italian troops invade Ethiopia; Emperor Haile Selassie leads a stout resistance and economic sanctions are imposed on Italy.

1936
Europe slides toward war: Germany sends troops into the demilitarized Rhineland; alarmed, Britain and France begin expanding their armed forces. Italy annexes Ethiopia; Haile Selassie warns the League of Nations "It is us today; it will be you tomorrow." Germany and Japan settle the Anti-Comintern Pact.

1936–1938
The Great Terror show trials of "Old Bolsheviks" launch Stalin's purge of the Communist party, government officials, army leaders, intellectuals, and peasants; between 680,000 and two million people are killed.

1936–1939
The Spanish Civil War: following the election of the left-wing Popular Front, a military revolt led by General Francisco Franco escalates into civil war between Franco's Nationalist rebels, backed by Germany and Italy, and Republicans loyal to the government, backed by the Soviet Union and International Brigades of volunteers from Europe and North America; Franco emerges as victor and dictator.

1936–1939
The Arab Revolt in Palestine, a British Mandate: in 1937, a British royal commission proposes a partition of Palestine between Arabs and Jews, but the plan is abandoned; instead, quotas are set in 1939 to restrict Jewish immigration.

1937
The Neutrality Act is one of several in the 1930s that seek to avoid American entanglement in foreign conflicts. Guernica is bombed by German aircraft supporting the Nationalists in Spain. The Second Sino-Japanese War breaks out: Japan's massacre of troops and civilians at Nanking, the Chinese nationalist capital, shocks the US. Brazil's President Vargas warns of a communist coup, brings in the *Estado Novo* (New State) constitution, and rules as dictator until 1945. Italy resigns from the League of Nations and joins the Anti-Comintern Pact.

▲ **Adolf Hitler** salutes his troops during the *Reichsparteitag* (National Party Convention) in Nuremberg, Bavaria, in 1938. The Nuremberg rallies staged annually from 1933 to 1938 were designed to show Germany united behind the Führer as he addressed more than half a million members of the Nazi party, the army, and the Hitler Youth.

1937
The *Hindenburg* airship explodes on arrival in the US, ending luxury airship travel; American aviator Amelia Earhart and copilot Fred Noonan disappear over the Pacific during their around-the-world flight; British engineer Frank Whittle successfully trials his turbo jet engine.

1938
Germany annexes Austria in the Anschluss (Unification), breeching the peace terms of 1919. Hoping to avoid war, Britain and France sign the Munich agreement allowing Germany to annex Sudetenland from Czechoslovakia. On Kristallnacht (the Night of Broken Glass), Jewish homes and businesses are attacked in Germany and Austria. Japan declares a New Order in East Asia; the US provides financial backing for Chiang Kai-shek's army against Japan.

1939–1945 WORLD WAR II
- Slovakia declares independence, Germany occupies Prague, and Czechoslovakia ceases to exist.
- Britain and France pledge to support Poland if it is attacked.
- Hitler and Mussolini sign the Pact of Steel military alliance.
- Japan turns to naval-led expansion in the Pacific and Southeast Asia.
- The Soviet Union and Germany sign the Molotov-Ribbentrop nonaggression pact.
- Germany invades Poland on September 1, sparking World War II.
- Britain and France declare war on Germany on September 3.
- Germany and the Soviet Union partition Poland.
- The Soviet Union attacks Finland.

1940
- German forces invade Denmark, Norway, the Netherlands, Belgium, Luxemburg, and France.
- Italy declares war on Britain and France.
- The first mass transport of prisoners arrives at the Auschwitz concentration camp in Poland.
- The French government signs an armistice with Germany and relocates to Vichy.
- In the Battle of Britain, the Royal Air Force (RAF) blocks the German Luftwaffe's attack; Hitler abandons plans for an invasion of Britain.
- The Blitz (*Blitzkrieg*, or lightning war) begins: the Luftwaffe's mass bombing raids on British cities continue until 1941; Allied bombers retaliate with raids on Germany.
- The US introduces the first peacetime conscription in its history.
- Japan allies with Germany and Italy in the Axis against the Allies.

1941
• Operation Barbarossa opens the German invasion of the Soviet Union; Leningrad is surrounded.
• On December 7, Japan attacks the US naval base at Pearl Harbor, Hawaii, and invades the Philippines and European colonies in Southeast Asia in order to dominate the Pacific; the US enters the war.

1942
• Berlin's Wannsee Conference briefs senior officials on the systematic deportation and extermination of Jews across Europe; six million Jews are killed by 1945, two-thirds of Europe's Jewish population.
• At the decisive Battle of Midway in the Pacific, the US navy repulses Japanese aircraft carriers.
• In the Battle of Stalingrad (to 1943), German troops enter the city, only to be encircled by Soviet troops; few survive their subsequent captivity. This is the biggest defeat in the history of the German army, and marks the turning point of the war.
• In North Africa, German troops under Erwin Rommel reach the borders of Egypt; they are defeated at El Alamein and pushed back toward Tunisia.

1943
• The Warsaw Ghetto Uprising in Poland is defeated.
• At Kursk, the largest tank battle in history ends in a decisive victory for the Soviets, leading to significant advances by the Red Army.
• US and British troops land in Sicily; Mussolini is removed from power and Italy surrenders.

1944
• Leningrad's 900-day siege is lifted.
• Polish troops take Monte Cassino, breaking through the German defensive line in Italy; Rome falls to the Allies.
• The Normandy landings on D-Day (June 6), launch the Allied invasion of occupied France.
• Hitler survives an assassination attempt by officers and high officials.
• Supported by French resistance fighters, General Charles de Gaulle's Free French forces liberate Paris.
• Facing superior US naval forces in the Pacific, Japanese naval pilots start to mount kamikaze suicide attacks.
• Hitler launches a final German offensive at the Battle of the Bulge in the Ardennes; German tanks break through the American front line but are defeated by a counter attack.

1945
• At the Yalta Conference, British prime minister Winston Churchill, US president Franklin D. Roosevelt, and Soviet premier Josef Stalin meet to discuss Europe's postwar reorganization.
• The Red Army reaches Berlin, Hitler commits suicide, and Germany surrenders on May 8, ending the war in Europe.
• The Potsdam Conference divides Germany, Austria, Berlin, and Vienna into four occupation zones and assigns parts of Poland, Finland, Romania, Germany, and the Balkans to Soviet control.
• The US drops the world's first atomic bombs on the Japanese cities of Hiroshima on August 6 and Nagasaki on August 9.
• Emperor Hirohito announces Japan's surrender on August 15, ending the war in the Pacific. World War II is the deadliest conflict of all time: more than 60 million civilians and military personnel are killed.
• The Nuremberg trials of leading Nazi war criminals begin.

1945
As World War II draws to a close, Ho Chi Minh, leader of the Viet Minh nationalist coalition, declares Vietnam an independent republic. The Indian Congress demands full independence from Britain. Josip Broz Tito, leader of Yugoslavia's resistance, becomes president of a socialist federal Yugoslavia. The United Nations (UN) forms; the UN General Assembly meets in 1946.

1946
Churchill describes the threat of communism as an "iron curtain" falling across Europe; communist governments are set up in Bulgaria and Albania (1946), Poland and Romania (1947), Czechoslovakia (1948), and Hungary (1949). Civil war resumes between China's communists and nationalists. Vietnam, Cambodia, and Laos rebel against French rule, triggering the First Indochina War (1946–1954).

1947
Pakistan and India become independent: their partition triggers widespread violence. The US sponsors the European Recovery Plan, or Marshall Plan, to boost postwar Europe's shattered economies. The UN agrees to the partition of Palestine into separate Jewish and Arab states. American test pilot Chuck Yeager is the first person to break the sound barrier.

1948
Gandhi is assassinated. South Africa introduces apartheid. The new state of Israel repulses invasions by five Arab states in the First Arab–Israeli War. The Soviet blockade of Berlin is the first crisis of the Cold War. Burma and Ceylon become independent. Korea is partitioned. The UN adopts the Universal Declaration of Human Rights.

1949
East and West Germany are set up, the former as part of the Communist Bloc. The North Atlantic Treaty Organization (NATO) is formed as an alliance for mutual defense. Eire becomes the Republic of Ireland. The Soviet Union tests its first atomic bomb. Indonesia gains independence after a four-year war against the Dutch. China's civil war ends in victory for the Communists; Mao declares the People's Republic of China; Chiang Kai-shek and his Nationalist troops flee to Taiwan.

► **A D-Day landing craft** approaches Omaha Beach, Normandy, on June 6, 1944 to begin the liberation of Occupied France. Around 130,00 Allied troops were landed at Utah, Omaha, Gold, Juno, and Sword as part of Operation Overlord.

1950
Senator Joseph McCarthy instigates a witch-hunt for alleged communists in the US (to 1954). North Korea invades South Korea: the Korean War (to 1953) is the first major armed confrontation of the Cold War, with US and UN troops supporting the South while the Soviet Union and China back the North. China occupies Tibet.

1952
East Germany closes its border with West Germany. A military coup takes power in Egypt. The Mau Mau Rebellion (to 1960) opposes British rule in Kenya. The US tests the first hydrogen bomb.

1953
Stalin dies. A coup driven by the US and UK ejects Iran's prime minister to strengthen the shah and privatize the oil industry. The double helix structure of DNA (deoxyribonucleic acid) is discovered by Francis Crick and James Watson. Edmund Hillary and Tenzing Norgay scale Mount Everest, the highest point on Earth.

1954
French rule in Indochina ends: Laos, Cambodia, and partitioned Vietnam become independent. USS *Nautilus* is the first nuclear-powered submarine.

◀ **Astronaut Buzz Aldrin** walks on the Moon during NASA's Apollo 11 mission of 1969, photographed by his fellow astronaut Neil Armstrong.

1955
The Warsaw Pact unites the Eastern Bloc in a military alliance to counter NATO in the West. A coup ousts President Juan Perón in Argentina. In the US, Rosa Parks electrifies the Civil Rights movement when she breaks Alabama's race laws by refusing to give up her seat on a bus for a white man. South Vietnam rejects reunification with communist North Vietnam and is declared a republic; the US declares its support against Viet Minh sympathizers, or "Viet Cong" (Vietnamese Communists), in the South; the Vietnam War lasts until 1974.

1956
Hungary's uprising is crushed by a Soviet invasion. Egypt's President Nasser nationalizes the Suez Canal, prompting the Suez Crisis; an Anglo–French invasion of the canal zone fails. Morocco and Tunisia gain independence from France, and Sudan from Britain.

1957
The Treaty of Rome establishes the European Economic Community (EEC), with six members. Malaya becomes independent. The Space Age begins with the Soviet launch of the Sputnik 1 satellite. President Suharto of Indonesia imposes martial law, nationalizes Dutch businesses, and expels all Dutch nationals. Ghana becomes the first British colony in Africa to gain its independence.

1958
Mao launches China's "Great Leap Forward"; forced industrialization plunges China into one of history's worst famines; at least 35 million are worked, starved, or beaten to death. General de Gaulle forms the Fifth Republic in France and is elected president. The North American Space Agency (NASA) is founded. Boeing 707, the first long-haul commercial jet airliner, begins flights across the Atlantic.

1959
In the Cuban Revolution, Fidel Castro becomes the first communist head of state in the Americas. A Tibetan uprising is crushed by China; the Dalai Lama and 80,000 Tibetans flee to India. Guerrillas from North Vietnam invade South Vietnam; two US soldiers are killed. Alaska and Hawaii become the 49th and 50th states of the US.

1960
The decolonization of Africa sees 12 French colonies gain independence, as well as Congo (from Belgium), and Nigeria and Somalia (from Britain). The Organization of the Petroleum Exporting Countries (OPEC) is founded. John F. Kennedy is elected US president.

1961
Soviet cosmonaut Yuri Gagarin is the first man sent into space. A military coup is staged in South Korea. South Africa leaves the British Commonwealth and becomes a republic. The Bay of Pigs invasion by anti-communist Cuban exiles backed by the US is repulsed. East German troops build the Berlin Wall, closing the border between East and West Berlin.

1962
France concedes independence to Algeria, as does Britain to Uganda, Jamaica, and Trinidad and Tobago. The first communication satellite relays live transatlantic television pictures. The Cuban Missile Crisis brings the US and the Soviet Union to the brink of nuclear war.

1963
The US–Soviet Test Ban Treaty ends nuclear testing in the atmosphere. In the US, Martin Luther King Jr. addresses 250,000 civil rights protestors at the March on Washington. President Kennedy is assassinated. The Federation of Malaya becomes Malaysia (including Singapore, Sarawak, and Sabah). Kenya gains full independence.

1964
The UN sends troops to Cyprus, where violence between Greek and Turkish Cypriots escalates. In the US, the Civil Rights Act creates equal rights for all, regardless of race, religion, or color. Nelson Mandela, a prominent figure in the anti-apartheid struggle, is sentenced to life imprisonment in South Africa.

1965
The US launches a massive bombing campaign against North Vietnam and lands troops for the first ground offensive in South Vietnam. The Indo-Pakistan War over Kashmir ends with a UN ceasefire.

1966
Mao launches the Cultural Revolution to rid China of "impure elements"; by 1976, 1.5 million people are killed and much of China's cultural heritage is destroyed.

1967
Army officers seize power in Greece and impose martial law. In the Six-Day (or Arab-Israeli) War, Israel takes Sinai, Gaza Strip, West Bank, Golan Heights, and Jerusalem. Nigeria plunges into civil war (to 1970) after the secession of Biafra.

1968
The Viet Cong's Tet Offensive convinces many Americans that the Vietnam War is unwinnable. The assassination of Martin Luther King Jr. sparks race riots across the US. Czechoslovakia's Prague Spring is crushed by Warsaw Pact troops. The Ba'ath party seizes power in Iraq. NASA's Apollo 8 mission is the first manned flight to orbit the moon.

1969
Yasser Arafat becomes leader of the Palestine Liberation Organization (PLO). Mu'ammar al-Gaddafi deposes King Idris to form the Libyan Arab Republic. Britain sends

◀ **In the Vietnam War** US Army Special Forces relied on helicopters as never before to carry troops deep into enemy territory.

troops to Northern Ireland when sectarian violence escalates; the Troubles continue until 1998. The first moon landing is made by Apollo 11's astronauts Neil Armstrong and Buzz Aldrin.

1970
The Nuclear Non-Proliferation Treaty comes into effect, ratified by the US, the Soviet Union, Britain, and 40 other countries. Salvador Allende's Marxist coalition wins the elections in Chile.

1971
Brutal dictator Idi Amin seizes power in Uganda (to 1979). Bangladesh becomes independent from Pakistan, and Qatar from Britain. The People's Republic of China joins the UN.

1972
On Bloody Sunday, British troops open fire on demonstrators in Londonderry, Northern Ireland. Black September terrorists kill Israeli hostages at the Munich Olympics.

1973
Under a ceasefire in Vietnam, US troops are withdrawn. The IRA extends its bombing campaign to mainland Britain. Chilean president Salvador Allende is killed during General Augusto Pinochet's US-backed military coup. Israel repulses Arab attacks led by Egypt and Syria in the Yom Kippur War; OPEC's oil embargo leads to oil shortages and a global recession. Denmark, Ireland, and the UK join the EEC.

1974
A bloodless coup restores democracy to Portugal. Turkey invades northern Cyprus; Greek Cypriots flee to the south. Ethiopia's emperor Haile Selassie is ousted in a coup. In the US, President Richard Nixon resigns over the Watergate bugging scandal.

1975
The Vietnam War ends with the fall of Saigon, renamed Ho Chi Minh City; North and South Vietnam are reunified in 1976. Civil war erupts in Lebanon (to 1990). Pol Pot's Khmer Rouge seize power in Cambodia and kill more than one million people by 1979. Indonesia invades the former Portuguese colony of East Timor. General Franco dies; democracy and monarchy are restored in Spain. Microsoft is founded by Bill Gates and Paul Allen.

1976
Mao dies and the Gang of Four stage a coup. Syrian peacekeeping troops enter Lebanon. During anti-apartheid protests in Soweto, South Africa, 176 people are killed in clashes with the police.

1977
Steven Biko, a prominent black rights leader in South Africa, is tortured to death in prison. A military coup overthrows Pakistan's government.

1978
Israeli troops enter Lebanon; Israel and Egypt sign the Camp David Accords in the US, leading to the Egypt–Israel Treaty in 1979. A communist coup takes power in Afghanistan. Vietnam invades Cambodia in response to border raids by the Khmer Rouge; peace terms are finally settled in 1991.

1979
Khmer Rouge is overthrown in Cambodia; Pol Pot flees across the Thai border and begins a guerrilla war. Idi Amin is driven out of Uganda. Left-wing Sandinistas overthrow the US-backed regime in Nicaragua. The Shah of Iran is ousted: Ayatollah Khomeini returns from exile and the Islamic republic of Iran is proclaimed; 63 hostages are seized from Tehran's US embassy. Soviet troops invade Afghanistan to suppress an Islamist revolt (to 1989).

1980
Black nationalists led by Robert Mugabe end white-minority rule in Rhodesia, now Zimbabwe. Saddam Hussein launches Iraq's invasion of Iran (to 1988); the war results in more than 1 million casualties. The anti-communist union Solidarity is formed in Poland—the first political mass movement in the Soviet bloc.

1981
Greece joins the EEC. President Anwar Sadat of Egypt is assassinated; Pope John Paul II and US president Ronald Reagan survive assassination attempts. Iran releases 52 US embassy hostages after 444 days. King Juan Carlos of Spain survives a military coup.

1982
In the Falklands War, Argentina's invasion of the British Falkland Islands is defeated. Israel invades Lebanon to expel the PLO but fails to prevent a Phalangist massacre of refugees. Poland's government bans Solidarity and imposes martial law.

1983
Civil war breaks out in Sri Lanka; government forces defeat the militant Tamil Tigers in 2009. Sudan sparks civil war by imposing Islamic Sharia (law) on the non-Muslim south; the war ends in 2005, and South Sudan becomes independent. In Lebanon, terrorists bomb the US embassy in Beirut, and later the French and US peacekeeping headquarters. The Soviets shoot down a Korean airliner. The US invades Grenada in the wake of a Marxist coup.

1983–1985
Ethiopia suffers famine: in one of the 20th century's deadliest disasters, more than 400,000 people die, millions are left destitute, and a long-running civil war hampers international relief efforts.

1984
India's prime minister Indira Gandhi is assassinated by Sikh extremists after troops storm the Golden Temple at Amritsar. A poison gas leak at a US-owned pesticide plant at Bhopal, India, is one of the worst industrial accidents in history. Scientists identify the HIV virus responsible for AIDS (Acquired Immune Deficiency Syndrome).

1985
Mikhail Gorbachev becomes Soviet leader and launches radical policies of *glasnost* (openness) and *perestroika* (restructuring). British scientists discover a hole in the ozone layer over Antarctica.

1986
Ferdinand Marcos, dictator of the Philippines, is toppled. The US bombs Tripoli after US soldiers die in a Berlin terrorist attack thought to have been ordered by Libya. An explosion at the Chernobyl nuclear power plant in Ukraine releases high levels of radioactive contamination that spreads from the Soviet Union to Western Europe. Portugal and Spain join the EEC, which legislates to create a single market for trade.

1987
The INF (Intermediate-range Nuclear Forces) Treaty reduces the superpowers' nuclear arsenals. In the First Intifada (to 1993), Palestinians fight against Israeli occupation of the West Bank and Gaza.

1988
PanAm flight 103 explodes over Lockerbie, Scotland, killing 270 people; in 2003, Libya accepts responsibility for the bombing.

1989
At Tiananmen Square in Beijing, China, troops and tanks fire on mostly student protestors calling for economic and political reform. The Iron Curtain collapses: Solidarity is legalized and is elected to power, ending communist rule in Poland; communist regimes fall in Hungary, Bulgaria, Czechoslovakia, Romania, and East Germany; the Berlin Wall is breached and the US and Soviet Union declare the Cold War is over.

1990
South African president F.W. de Klerk lifts the ban on the African National Congress (ANC) and frees Nelson Mandela. Iraq invades Kuwait: UN coalition forces are sent to the Persian Gulf. Germany is reunited. Solidarity leader Lech Walesa becomes Poland's first post-communist president.

1991
A UN coalition expels Iraqi forces from Kuwait, ending the First Gulf War. The Paris Peace Accords end the Cambodian–Vietnamese War. The Baltic republics of Estonia, Latvia, and Lithuania assert their independence from the Soviet Union. Boris Yeltsin becomes the first popularly elected president of Russia. After a failed military coup, the Soviet Union suspends all communist activities and dissolves into 15 countries; Russia, Ukraine, and Belarus form the Commonwealth of Independent States (CIS). Yugoslavia disintegrates: Slovenia and Croatia declare their independence; Slovenia repels the Serb-dominated Yugoslav People's Army in the Ten-Day War; fighting in Croatia ends in 1995.

1992
The Maastricht Treaty on the European Union (EU) is signed by 12 states, heralding common citizenship and common economic and defense policies. When Bosnia and Herzegovina seek independence from Yugoslavia, civil war erupts; Serbia attacks Bosnia's Muslim population, building prison camps as part of its "ethnic cleansing" policy.

1993
Czechoslovakia splits into the Czech Republic and Slovakia. Israel and the PLO sign the Oslo Accords, settling mutual recognition and limited Palestinian autonomy under the Palestinian Authority. Cambodia restores democracy: Prince Sihanouk is elected head of state.

1994
Civil war leads to genocide in Rwanda: Hutu extremists massacre 800,000 Tutsis; two million Hutus flee, fearing retribution. The ANC wins South Africa's first democratic elections; Nelson Mandela becomes the country's first black president. Russian troops enter the Muslim-dominated rebel region of Chechnya to keep it from breaking away. The US invades Haiti to restore democracy.

1995
A bomb planted in Oklahoma City by a Gulf War veteran kills 168 people. Israeli prime minister Yitzhak Rabin is assassinated. The Dayton Peace Accord ends the civil war in former Yugoslavia. Austria, Finland, and Sweden join the EU.

1996
Civil war erupts in Afghanistan: Taliban rebels capture Kabul and declare Afghanistan a fundamentalist Islamic state. Russia and Chechnya sign a ceasefire. Yeltsin wins Russian presidential elections, only narrowly defeating communist opposition to his privatization policy.

1997
Tutsi rebels attack Hutu refugee camps in Zaire; Zaire's government collapses, and thousands flee to Tanzania. Britain hands Hong Kong back to China. The Asian financial crisis leads to an economic slump in many developing countries. Under the Kyoto Protocol, industrialized nations agree to cut CO_2 and other greenhouse gas emissions to combat global warming.

◄ **Ecstatic crowds** from East and West Berlin celebrate the dismantling of the Berlin Wall in November 1989—a potent symbol of the collapse of communism.

1998
The Good Friday agreement ends the Troubles in Northern Ireland. India and Pakistan test nuclear weapons. Serbs and ethnic Albanians clash in Kosovo. Indonesia is hardest hit by the Asian financial crisis; President Suharto resigns. Oil drops to $11 per barrel, contributing to Russia's financial crisis. The US and Britain bomb Iraq after it ceases to cooperate with UN inspectors seeking to find and eliminate Iraqi weapons of mass destruction.

1999
Serbian ethnic cleansing of Kosovan Albanians is halted by NATO bombing. East Timor votes to secede from Indonesia; anti-independence rebels supported by the Indonesian military attack civilians; the government calls for international help. A military coup in Pakistan incurs international sanctions. Russia restarts the Chechen war.

2000
Israel ends its 22-year occupation of South Lebanon. In the Second Intifada, intense Israeli-Palestinian violence kills more than 3,000 people. The first draft of the human genome is completed.

2001
Four al-Qaeda terror attacks on September 11 kill 2,996 people in the US. The US and UK attack targets in Afghanistan thought to be hiding al-Qaeda leader Osama bin Laden; the war in Afghanistan ends in 2014.

2002
A single European currency, the Euro, is introduced. US-led forces begin a large-scale operation against the Taliban in Afghanistan. Civil war in Sierra Leone and Angola ends. East Timor wins independence from Portugal. Islamist terrorists

bomb a Bali nightclub, killing more than 200 people. Chechen militants besiege a Moscow theater; 118 people die. UN weapons inspectors return to Iraq; Iraq denies that it has weapons of mass destruction; US troops are ordered to the Gulf.

2003
Civil war erupts in Darfur, Sudan. Serbia and Montenegro emerge and Yugoslavia ceases to exist. In the Iraq War (to 2011) a US-led coalition invades Iraq and topples Saddam Hussein's government; occupying forces struggle to contain looting, insurgency, and sectarian violence between Sunni and Shia militias.

2004
Islamic terrorists bomb Madrid: 191 die. Ten countries join the EU, most of them former communist states. The most destructive tsunami in history kills more than 200,000 people in 11 countries across the Indian Ocean and Southeast Asia.

2005
Israel withdraws from Gaza; Syria withdraws from Lebanon. Islamist suicide bombings in London kill 52 people. Hurricane Katrina hits New Orleans in the US.

2006
Hamas wins parliamentary elections in Palestine. Iran announces it has produced enriched uranium to make nuclear fuel. In the Lebanon War, Israel launches strikes on Lebanon after Hezbollah captures two Israeli soldiers. North Korea holds its first nuclear test. Montenegro votes to split from Serbia.

2007
Bulgaria and Romania join the EU. Devolved government returns to Northern Ireland. Sectarian conflict between Sunni and Shia militias escalates in occupied Iraq. The US housing bubble bursts, triggering the financial crisis of 2007–2008.

2008
Kosovo declares independence from Serbia. Nepal abolishes its monarchy and becomes a republic. Russia and Georgia clash over South Ossetia's bid for independence. In the US, the Lehman Brothers investment bank collapses; banks worldwide face insolvency and stop lending, leading to a global recession. Barack Obama is the first African-American to be elected US president.

2009
Israeli troops invade Gaza to halt rocket attacks by Hamas.

2010
An earthquake devastates Haiti, where *c.*230,000 people die. The global recession continues: the International Monetary Fund (IMF) offers major loans to Greece and Ireland, imposing austerity measures to tackle government debt. Burma's military regime release pro-democracy leader Aung San Suu Kyi, ending 20 years spent under house arrest.

2011
In the Arab Spring—inspired by a Tunisian who burned himself to death in 2010 in protest at his treatment by police—pro-democracy rebellions erupt across North Africa and the Middle East. Mass protests in Egypt lead President Mubarak to hand power to the army; in Libya, rebels topple Gaddafi's regime but civil war continues as rival factions vie for control; Syria's civil war begins with a violent crackdown on civilian dissenters. South Sudan becomes independent of Sudan but faces attacks by rebel militias; civil war breaks out in 2013. US forces kill Osama bin Laden in Pakistan. The War in Iraq is formally ended.

2012
Egypt's presidential elections are won by Mohammed Morsi of the Muslim Brotherhood. The secular National Forces Alliance wins Libya's elections to form a new national congress. School girl Malala Yousafzai is shot by the Taliban in Pakistan but survives. Civil war breaks out in the Central African Republic. NASA's Curiosity Rover lands on Mars.

▲ **New York City's World Trade Center** is struck by the first of two airliners hijacked by al Qaeda terrorists on September 11, 2001. A third plane flew into the Pentagon in Washington D.C., and the fourth crashed into a field near Pittsburgh..

2013
France sends military forces to Mali to intervene against Islamist forces. Croatia joins the EU. Syria's government denies responsibility for a chemical attack on Ghouta, and pledges to hand over its chemical weapons for destruction. A military coup ousts Egypt's President Morsi, leading to widespread violence. Al-Shabaab Islamist militants attack a shopping mall in Nairobi, Kenya. Ethnic tensions fuel civil war in South Sudan.

2014
An outbreak of the Ebola virus in West Africa kills 11,000 people by 2016. Civil war erupts in Ukraine: the pro-Russian president is ousted; Russia annexes Crimea and invades eastern Ukraine; Malaysia Airlines Flight 17 is shot down over Ukraine, killing all 298 people on board. Israel launches airstrikes on Gaza, followed by a ground offensive, after three Israeli teenagers and one Palestinian are kidnapped and killed. The Islamic extremist group Boko Haram kidnaps 276 schoolgirls in Nigeria. Civil war resumes in Libya between the democratically elected government in Tobruk and rival Islamist factions in Tripoli and Benghazi. The terrorist group Islamic State of Iraq and the Levant (ISIL, or ISIS) seizes territory in northern Iraq and Syria; US and Arab air attacks target ISIS.

2015
Egypt begins airstrikes on ISIS in Libya. A Saudi-led Arab coalition attacks Iranian-backed Houthi rebels in Yemen. Al Shabaab carries out a mass shooting at Garissa University in Kenya. Iran agrees to limit its nuclear development program in exchange for sanctions relief. Russia begins airstrikes in support of the Syrian government; one of its fighter jets is shot down by Turkey. ISIS destroys ancient sites in Syria and carries out multiple attacks in Beirut and Paris. The US and Cuba resume diplomatic relations severed in 1961. A global pact commits all countries to reduce carbon emissions.

MILESTONES IN MEDICINE

Prehistoric fossils and mummies from the ancient world show evidence of early dentistry and surgery, and throughout history, physicians and scientists have continued to work to understand the human body, heal injury, and combat illness. Today, while many of the world's most dangerous diseases have been conquered, newly evolving ones continue to challenge us.

c.2700–2650 BCE The earliest known physicians are in Egypt: Merit Ptah—the first named woman in science—and Imhotep, who is made a deity of medicine after his death.

c.1600 BCE The *Edwin Smith Papyrus* is the first medical treatise to present a rational, scientific approach to medicine without magic. Some scholars attribute the text to Imhotep.

c.500 BCE In ancient Greece, physician Alcmaeon of Croton asserts that the brain, not the heart, is the seat of thought and emotions.

c.420 BCE Greek physician Hippocrates of Cos stresses the importance of observation and diagnosis; his theory of the four humors survives until the 19th century. Physicians still take the Hippocratic Oath to do their best to treat patients, protect them from harm, and share their medical knowledge with others.

c.280 BCE Herophilus of Alexandria, "father of anatomy," performs public dissections on human cadavers and describes the brain and the nervous system. Diocles of Carystus writes the first known work on anatomy.

c.40 CE Roman philosopher Cornelius Celsus publishes *On Medicine*, which covers diet, diseases (including heart disease), surgery, and pharmacology.

c.200 CE Galen of Pergamon, physician to the Roman emperor, describes the plague that killed as many as five million people in 165–180 CE. Galen's texts on anatomy and medicine dominate Western medicine for the next 1,300 years.

c.1012 Persian polymath Ibn Sina (Avicenna) publishes his *Canon of Medicine*, which is used as a medical textbook in the Islamic world and Europe until the 18th century.

c.1077 The first medical school in modern times is flourishing in Salerno, Italy, and attracts practitioners from Greek, Jewish, and Arabic medicine across Europe, North Africa, and Asia Minor.

1123 Europe's oldest hospital, St. Bartholomew's, is founded in London, UK.

1280 Syrian doctor Ibn al-Nafis links the pulse to the heartbeat and shows that blood circulates around the body.

c.1285 Eyeglasses (early spectacles) are in use in Italy.

1543 Flemish doctor Andreas Vesalius publishes *De Humani Corporis Fabrica* (On the Structure of the Human Body); his extensive study of dissections establishes the modern science of anatomy.

1552 The first anatomical theater is established in Padua, Italy.

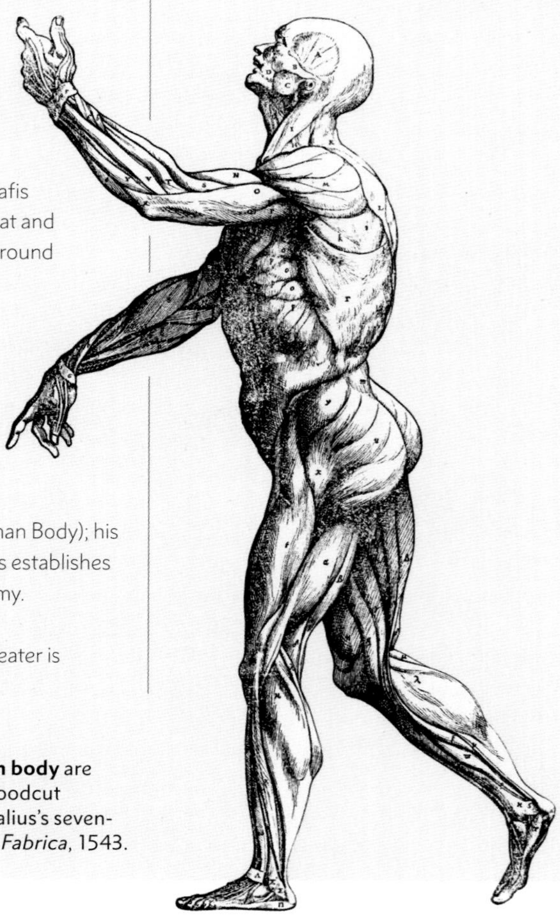

▶ **The muscles of the human body** are illustrated in profile in this woodcut engraving from Andreas Vesalius's seven-volume *De Humani Corporis Fabrica*, 1543.

c.1590 Dutch spectacle-maker Zacharias Janssen combines two lenses to create the first compound optical microscope.

1628 English physician William Harvey describes how blood is pumped around the body in a circuit, in *On the Movement of the Heart and Blood in Animals*.

c.1630 Obstetrical forceps are invented in England by a family of surgeons to the crown, for use in difficult childbirths.

1661 Italian physician Marcello Malpighi publishes his work on the lungs, describing capillaries and how these tiny blood vessels link arteries and veins.

1665 English physicist Robert Hooke coins the term "cell" to describe the smallest units of life, which he observes through his compound microscope.

1672 Dutch physician Regnier de Graaf describes the human reproductive organs.

1677 Cinchona (quinine) bark from the Andean rain forests is listed as a fever treatment in the London *Pharmacopoeia*.

1691 The microscopic structure of bones is described by the English physician Clopton Havers.

1695 Dutch microscopist Antoni van Leeuwenhoek describes human blood cells and sperm cells.

1714 The mercury thermometer is invented by German physicist Gabriel Fahrenheit; in 1724 he develops the temperature scale named after him.

1735 Claudius Aymand performs a successful appendectomy in London.

1796 "Father of immunology" British physician Edward Jenner carries out the first inoculation against smallpox.

1800 Nitrous oxide (an anaesthetic known as "laughing gas") is first used by British chemist Humphry Davy. French anatomist Marie-Francois-Xavier Bichat reveals that organs are made of groups of cells called tissues, and identifies 21 different types of tissue.

1816 The first stethoscope, invented by French doctor René Laënnec, is a long wooden tube with a single earpiece for listening to the heart.

1818 British obstetrician James Blundell performs the first successful transfusion of human blood to a patient.

1831 Chloroform is discovered in the US, France, and Germany; Scottish obstetrician James Young Simpson first uses it as a medical anaesthetic in 1847.

1840 British philanthropist and social reformer Elizabeth Fry founds the Institute of Nursing in London.

1841 American surgeon Dr. Crawford W. Long performs the first operation using diethyl ether as an anaesthetic.

1847 In Vienna, Ignaz Semmelweis introduces the regime of handwashing and sterilization of surgical instruments.

1851 The ophthalmoscope is invented by German physicist Hermann von Helmholtz for looking inside the eye.

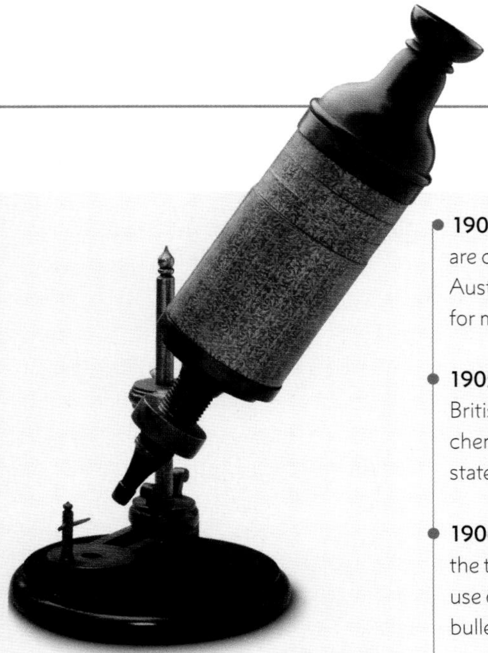

▲ **This microscope** belonged to English scientist Robert Hooke, author of the Royal Society's *Micrographia* (1665).

- **1854** British doctor and one of the founders of epidemiology John Snow traces the link between cholera and contaminated water from a pump on Broad Street, London.
- **1861** French chemist and "father of microbiology" Louis Pasteur publishes his germ theory, proving that airborne microbes cause decay and disease.
- **1865** British surgeon Joseph Lister pioneers modern antiseptics by using carbolic acid during surgery.
- **1866** A portable glass thermometer is developed by British physician Thomas Allbutt.
- **1869** Swiss surgeon Jacques Louis Reverdin performs a skin graft.
- **1880** The malaria parasite is identified by French doctor Charles Louis Alphonse Laveran.
- **1881** The sphygmomanometer is invented by Austrian physician Samuel Ritter von Basch to measure blood pressure.
- **1882** German physician Robert Koch discovers the bacterium that causes tuberculosis.
- **1895** German physicist Wilhelm Röntgen realizes that electromagnetic radiation can create X-rays.
- **1901** The first three major blood groups are described by Karl Landsteiner, an Austrian-American doctor, paving the way for more successful blood transfusions.
- **1905** The term "hormone" is coined by British scientist Ernest Starling to describe chemical mechanisms that maintain a stable state inside the body.
- **1906** German doctor Paul Ehrlich coins the term "chemotherapy" in predicting the use of chemical compounds as "magic bullets" to kill disease-causing organisms.
- **1919** The term "homeostasis" is first used by American psychologist Walter Cannon to describe the steady state achieved by the body through a number of cooperating mechanisms, such as hormones.
- **1922** Injections of insulin, the hormone that regulates blood sugar levels, are given to treat a diabetic patient.
- **1928** Scottish biologist and pharmacologist Alexander Fleming discovers penicillin, the first antibiotic.
- **1931** German physicist Ernst Ruska and electrical engineer Max Knoll invent the scanning electron microscope, which achieves far greater magnification than optical microscopes.
- **1932** American cardiologist Albert Hyman demonstrates an artificial pacemaker.
- **1938** The vaccine for Yellow Fever, developed by South African virologist Max Theiler in the US, is first used in Brazil.
- **1940s** British surgeon Archibald McIndoe pioneers plastic surgery to rebuild the faces of pilots burned during World War II.
- **1940** US army surgeon Austin Moore performs the first metallic hip replacement.
- **1942** German neurologist Karl Dussik describes the medical use of ultrasound as a diagnostic tool; Ian Donald develops the first ultrasound scanners in the 1960s.
- **1943** The first kidney dialysis machine is developed by Dutch physician Willem Johan Kolff.
- **1948** The World Health Organization (WHO) is set up in Geneva, Switzerland. The amniocentesis test is developed for fetal abnormalities and infections.
- **1952** The first open-heart surgery repairs a a hole in the heart of a five-year-old girl, using a technique to cool the heart developed by Canadian William Bigelow.
- **1953** The double-helix structure of DNA is discovered by British physicist Francis Crick and American biologist James D. Watson in Britain, incorporating work by Maurice Wilkins and Rosalind Franklin.
- **1954** The first successful kidney transplant between living patients is performed in the US by surgeon Joseph E. Murray on identical twins. The electronic clinical thermometer is invented.
- **1955** The first effective polio vaccine, developed by Joseph Salk, is used in a mass vaccination program in the US.
- **1956** The first plastic disposable syringe is patented by New Zealand pharmacist Colin Murdoch.
- **1958** American doctor Edward Hon and British doctor Ian Donald use ultrasound scanning to check the health of a fetus.
- **1961** American scientist Marshall Nirenberg cracks the genetic code of DNA. US surgeons successfully reattach the severed arm of a construction worker.
- **1967** Magnetic resonance imaging (MRI) is first used to see soft tissues inside the body. South African heart surgeon Christiaan Barnard performs the first heart transplant.
- **1972** Computed tomography (CT) scanning is used on a patient in London to produce images of human body organs.
- **1978** Louise Brown, the first "test-tube baby," is born, following conception by in vitro fertilization (IVF).
- **1980** Smallpox becomes the first infectious disease to be eradicated, following a major effort by the World Health Organization.
- **1981** AIDS (Acquired immune deficiency syndrome) is first identified in the US. Doctors perform minimally invasive "keyhole" surgery, making small incisions and using a fiber optic laparascope to see inside the body.
- **1982** The first artificial heart, invented by American scientist Robert Jarvik, is transplanted into a patient.
- **1984** French scientist Luc Montagnier discovers the human immunodeficiency virus (HIV) that destroys immune system cells, resulting in AIDS.
- **1986** The first heart, lungs, and liver transplant is performed in Britain.
- **2000** American philanthropists Bill and Melinda Gates launch the world's largest private foundation to enhance healthcare and reduce extreme poverty.
- **2001** Scientists perform the first germline gene transfer in animals, with the aim of preventing faulty genes being passed on to the next generation.
- **2002** Surgeons in the US view digital X-rays transmitted by satellite to guide a knee operation at the South Pole.
- **2003** Scientists publish the results of the Human Genome Project launched in 1990, identifying the DNA sequence of a full set of human chromosomes.
- **2006** A urinary bladder, grown in the laboratory from a patient's own cells, is successfully transplanted into the patient to replace a damaged organ.
- **2007** Previously thought to have no purpose, the appendix is shown to hold a back-up reservoir of bacteria essential to the workings of the large intestine.
- **2013** The US creates the first kidney grown in vitro, and Japan creates the first human liver grown from stem cells.
- **2015** A bionic hand reconstruction replaces a hand disabled by injury with a prosthetic that the patient controls with signals from the brain. Surgeons perform a full facial transplant.

INDEX

Page numbers in *italics* refer to illustrations and photographs, those in **bold** indicate the main information for the topic.

A

aardvarks *168*
Aborigines, Australian 212, 329
acacia trees *168*
accretion, of planets 71, *78*
acetate 106
Acheulian technology 211
acid rain 346
acidity, of ocean's surface 353
advertising **317**, 345
adzes 232, 284
aerobic respiration 116
Aetiocetus 171
"affluent foragers" *230*, 231
Africa
  colonization of 327, **328–329**
  continental drift *90*, 158, *159*
  development of writing 266
  early human species in 182, **194–195**, 199
  education in 332, *333*
  farming in *235*, 242–243
  habitats of *168*
  metallurgy in 280
  modern day 343, 345
  prehistoric map of 362–363
  "scramble for" 327
  see also slave labor
Afro-Eurasia (world zone) 235, 294, 336
  "Old World" 296, **297**, *298*
afterlife 275, **277**
Age of Discovery 336, 402–403
Age of Enlightenment 304, **319**, 332
Age of Exploration 275, 296–297, 396–397
Age of Fish 132
Age of Reptiles 154
aggression, in animals 240, 241
Agrarian Era 271, 294, 314, 344
'Ain Ghazal 256
air pollution 352
air travel 339
Akkadian Empire 288, 370
Alexander the Great 288, *289*, *379*
algae 100, **115**, *115*, 122, **137**
allantois *146*, 147
Allen, Horatio 313
Almagest, Ptolemy 23
alphabets 264–267
amber 150–151, *293*
Ambulocetus 170, *171*
American Indians 297, 299
American Indian groups *380*
American War of Independence 318, 320
Americas, the (world zone)
  colonization of 329
  development of writing 266–267
  early farming in 234, **242**, 248
  early human dispersal in *195*
  exploration of **297**, 298
  globalization in **299**, 336
  Mesoamerica 234, 244, 294
  "New World" **296–297**, 298, *298*
  prehistoric map of 362
  religion in 274
amino acids 59, 102
ammonia 102
amniotes 132–133, **147**, *147*
amoebas *114*, 115, *122*
amphibians 141, *141*, **153**
  evolution of eggs 147
  extinction of 162–163
anatomy, of prehistoric man 190, 199
Ancient Library of Alexandria 264, 266
ancient world map 368–369
Andromeda Galaxy 30
anaesthetic 334
angiosperms 160
Anglo-Saxons 277
Antarctica 158, *174*, *176*
Anthropocene Era 350
antibiotics 112, 335
antimatter 29, 39
antiparticles 34
apes 183, **186–187**
Apple Inc. 341
Aptian extinction event 163
Arab Spring uprisings (2011) 342, 421
Arabic records 264
Araucaria araucana 145
archaea **112**, *113*, 114
archaeological techniques 192, 197, 238
Archaeopteryx 156, *157*
Archean era 84, 85
Archimedes of Syracuse 23, *269*
  Archimedes' screw *269*
architecture 383
archosaurs 154
Ardipithecus ramidus 186, *187*
ard plows 248, *249*
arid habitats *147*, *152*, 153, 272
aristocracies 317
Aristotle **22–23**, 86, 172, 392
armadillos 167
armor *284*
art 374–375
  prehistoric 204–205, **212–213**
  see also cave art
arthritis 282, 283
arthropods *127*, *128*, 140, *142–143*
articulated bones 192
Asia
  colonization of 329
  development of writing 266
  early farming in *243*
  early human dispersal into 195
  prehistoric map of 363
  trade routes through 294–295
Askari soldiers *327*
aspirin 334
Assyrian Empire 368, 378
asteroids
  asteroid belt *74*, *75*
  meteorites 72–73, 86
  strikes *78–79*, 80, *103*, **154**
Astraspis 132
astronomical clocks 20–21
Atlantic Ocean 94
atlatls 204, *204*
atmosphere, planetary 71, *74*
  on Earth 80, *81*, 102, *102*
atomic bombs 348, *349*
atomic mass 63
atoms 22, **28–29**, 34, *102*
  inside stars 44, 58
  radiometric dating 88
Australasia (world zone) 235, 336
Australia 158, 195, 220, 299
  Aborigines 212, 329
  prehistoric map of 363
Australopithecus **184**, *184*, *186*, *189*, 206
automobiles 313, *338*, 339
aviation industries 339
Axial Age 274
axis, Earths 174
ayaté 278
Aztec empire **244**, *244–245*, 287, 298, 391
  map of *384*

B

babies 201, 259
Babylon, Mesopotamia 262, 373
back-boned animals *see* vertebrates
bacteria **112–113**, 114, 115
  evolution of 118
  reproduction of *120–121*

Bagha Qaghan *285*
Bahrām Chōbin *285*
bartering 291
basket weaving 278
bathymetry 94, *95*
bats *109*, 142, *142*, *143*
Bay of Fundy, Canada *82–83*
BBC (British Broadcasting Corporation) 341
Beaker people 254
beans, domesticated 236, *237*
Becquerel, Henri 86
bees 165, 240, 351
Beg, Ulugh 23, *23*, 245
Belgian Congo 329
Belgium 312–313, 320, 328
belief systems 274–275
  religions 86, 263, 295
Benz, Karl *338*
Bering Strait 195, *195*, 362
Bessel, Friedrich 29
Bethe, Hans 58
Bible 111, 264
Big Bang theory **34–35**, 37, 38–39
binary systems 57
biodiversity 220, 350, 351
biomass fuels 357
bipedal animals 142, 156, **186**, 201
birch bark tar 207, *216*, *217*
birdman of Lascaux 202–203
birds 133, 147
  evolution of wings 142–143, 156–157
birth control 335
Bismarck, Otto von *320*
bismuth 59
Black Death *292*, 293
black holes 47, **49**, 56
blood feuds 262
blood groups 335
blueshifts 29, *29*
blue whales *171*
Bohr, Niels *29*
Bolivar, Simón 319
Bolivia 299
bone (tools) 206, 207, 208, 214
bones 130, **192–193**, 218
  hyoid bones *192*, 202, *202*
  jaws 135
  limbs, hands, and feet *141*, 186–187
  wings *143*, *157*
bonobos 183
Book of Genesis 18
Boulton, Matthew 308–309
bows 209, 284, *285*
brachiopods 138
Brahe, Tycho 25
brain **126–127**, 202
  size in Hominins **188–189**, 201
Brazil 357
breastfeeding 201
breathing, during speech 202
Bridgwater Canal 312
Britain
  coal reserves in 307
  colonization by 328, 406–410
    map *404–405*
  education reform 332, *332*
  government 304, 314, 325
  industrialization in **304**, 310
  invasions of 382, 388
  manufacturing industry 308–309, **312–313**
  see also Reform Bill (1832)
British Broadcasting Corporation (BBC) 341
British Empire 328
Broca's area (brain) 202
bronze *217*, **280–281**, 291
Bronze Age 20–21, 280, 284, 370
bubonic plague *292*, 293, *293*
Buddhism 274, 275, 295
Burgess Shale, Canada 101, 129
burial practices 207, **218–219**, 221
  see also grave goods
Byzantine Empire 264, 295, 382

C

Caesar, Julius 380, *381*
Cable News Network (CNN) 341
"caching" (bodies) 218
calendars 18, 20–21, **244–245**, **367**
Cambrian-Ordovician extinction 162
Cambrian period
  beginning of life in 128–129, 128
  evolution of animals 130, **140**, 158
Cambridge University *105*
cameras 343
Campbell's monkeys *203*
canals (water) *268*, **269**, 309, 312
cannibalism 218
capitalism 322, **323**, 337
  consumerism 316, 345
carbolic acid 334
carbon *89*, 102, **148–149**
  within stars 56, 58–59
carbon dioxide (CO_2) 80
  from burning of coal 149
  levels in atmosphere 174, **350–351**, 352
  in photosynthesis 114
carbon emissions 348, 352, 357
Carboniferous period 140, **148**, *152*, 158, 176
  climate change in 176
  continental shift in 158
Carboniferous Rain forest Collapse 163
Carey, Samuel 91
carnivores 154, 156, 188, 211
  see also diets
Carolingian script 267
cars (automobiles) 313, *338*, 339
  electric *357*
cartilage 130, 132, 135
carts 246, *246*
carvings
  art 214, 218, *291*, *371*
  calendars *244–245*, *367*
  caves *204–205*, 212, *213*
CAT scanners 335
Catal Höyük, Turkey 256–257
Catholic Church 24, 25, 264
cattle 240
cave art 208, 209 212–213
  depicting hunting scenes *188*, *227*
  storytelling through *203*
cells
  complex cells, evolution of 100, **118-119**, 120
  multicellular organisms 100, **122–123**
  protocells 106, *106*, *107*
  single-cell organisms **112–113**, *119*
    reproduction of *120–121*
Celtic civilization 379, *387*
centipedes 140
Cepheid variables 29, **30**, *30*
cetaceans 170
Chaco Canyon 60
Chalicotherium 133
Chandragupta Maurya 288
Chan Muwan, King of Bonampak *260*
charcoal 149
chariots 284, *287*
Chauvet Cave, France 212, *212–213*
chemical elements **62–63,** 354–355
Chernobyl nuclear disaster 348
Chichén Itzá *387*
chieftains 259, 261, 277
childbirth 201
child labor *306–307*, *309*
chillis *296*
chimpanzees 170, 183
China *253*, 336
  astronomy in 18, 60
  coal reserves 307
  conflicts in **284**, 325
  development of law 263

China *cont.*
development of writing 266, 267
dynasties 368
education reform 332
emperors of **261**, 278, *279*
farming in 235, 248, 250, 269
industrialization 304
medieval period 386–391
money used in *290*, 291
religion in 275
renewable energy in 357
social status in 277, *277*, 278, *278*
trade in **298**, 299, 325
Silk Road **294–295**, 275
chlorophyll 114
chloroplasts 100, 118, *118*
choanoflagellates *122*
chocolate *317*
cholera 293, **331**, *331*, 334
chorion 147
Christianity 274, 275, *297*, 381
views on afterlife 277
views on evolution 110
churches, building of 383
Cigar Galaxy 60
cinema 375
citizenship 287, 315
city-states 269, **270–271**, *252*
clades, of species 173
classical world map 376–377
classification of species 172–172
clay 216, *217*, 254
climate change
prehistoric era 153, 158, **174–175**, *187*
early human dispersal, due to 195, 199, 221
ice ages **176–177**, 220, 226
leading to extinction 162
modern era 350, 352–353, *359*
clocks 244
see also calendars
clothing **214–215**, 282, 283
as status symbol **278–279**, 317
CNN (Cable News Network) 341
CO_2 (carbon dioxide) 80
from burning of coal 149
levels in atmosphere 174, **350–351**, 352
in photosynthesis 114
coal
formation of 148–149
in industrialization 304, 310, 312, 350
mining **306–307**, 308
reserves 307, *346*, 347
coastal habitats *152*, 158
coastal settlements 220, 226
Cockerill, William 313
coevolution 165
coffee 328
coins 291
Cold War 348
collagen 130
collective learning **204–205**, 288, 332
colonization 311, **328–329**
exploitation of, for trade 311, **322**, 336
opposition to 319, **327**
pre-industrialization 296–297, 394–395
Columbian Exchange 297
Columbus, Christopher 297, 298
combustion engines *338*, 339
comets 72, 74, 80
commandments 262, 263
commercial air travel 339
communication *see* language
communication technology 336, 340–341, 342–343
Digital Revolution 332, 403
communism 322
compensation 262
complex cells, evolution of 100, **118-119**, 120
multicellular organisms **122–123**
composite particles 34
Computerized axial tomography (CAT) scanners 335
computers 341, 343
Condition of the Working Poor, The, Engels 331
Confucianism **263**, 274, 275, 392
Confuciusornis 156
constitutions 320
consumerism 311, **316–317**, 339
leading to waste 345
continents
formation of **84–85**, 92
shift of **90–91**, 150, 158, *159*
convergent evolution 142
convergent plate boundary 92, *93*, 95
cooking, discovery of 216, *217*
Copernican Revolution 23
Copernicus, Nicolaus 25, 397
copper *216*, 280, 291
Copper Age 283
coprolites 238, 239
corals 120, 138, 139
core samples 187
corn (maize) *234*, *242*, 253
domestication of 236, *237*
Corpus Juris Civilis (Body of Civil Law) 263
Cortés, Hernán 296, 397
Cosmic Dark Ages 44, *44*
cosmic microwave background (CMB) 38, *38–39*
cosmological principle *39*
cotton 278, 309, **329**
cowpeas *235*
Cran Nebula 60
cratons 84, 85
creativity 374–375
Cretaceous-Paleogene extinction 163
Cretaceous period 154, 156
Crick, Francis *105*
crickets (katydids) *109*
crinoids 139
crocodiles 163
Cromford Mill 312
crops
corn (maize) *234*, *242*, 253
domestication of 236–237
grain 238–239, 250, 293
production and harvest of *249*, 344
rice *243*, 253
Crowdfunding 343
Crusades *388*, 389, 390
crust, Earth's 80–81, 84–85, **92–93**
Crutzen, Paul 350
crystals 73, **88–89**
currencies *see* money
cuticle 140
cyanobacteria 112, 114–115, 115
cynodonts *166*, 167
Cynognathus 159
Cyrus the Great 287
cytoplasm *113*

D

Dalton, John 28, *28*
Darby, Abraham 309
dark energy 38
dark matter 38, *38*, 44, *48*
Darwin, Charles 86, **110–111**, 172, 173
Darwin, George 86, 90
dating techniques 72, 86–87, 192
days, within calendars 244, 367
De Revolutionibus Orbium Coelestium, Copernicus 25
death
burial practices 207, **218–219**, 221
grave goods 21, 254, **276–277**
of Ötzi, mummified man 282–283
see also diseases
Declaration of Independence (1776) 318
Declaration of the Rights of Man and of the Citizen, Lafayette 318–319
decomposers 112, 115
deep-sea vents 106, *106*
see also ocean habitats
deforestation 221, 272, 299, 351
Deinonychus 157
Demetrius I, King of Macedon 262
democracy 318

Denisovans *194*, 197, 214
Denkania *145*
Denmark 357, *394–395*, *404–405*, *412–413*
deoxyribonucleic acid (DNA) *see* DNA
department stores *316*, 317
deserts
  creation of *152*, **153**, 272
  irrigation of 268–269
deuteron 58
Devonian period 135, *137*, 141, 162
diapsids 153, 154
diets
  of early farmers 293
  of Mayans 255
  of prehistoric man 184, **188**, 189
    hunter-gatherer groups 211
    Neanderthals **190**, 193
differentiation **78**, *79*, 80, 85
Digest 263
digestive systems *112*, 115, 169
Digges, Thomas 25
digging sticks 248
Digital Revolution 332
Dimetrodon 147
dinosaurs 133, **154**, *155*
  therapods 156, 157
diseases
  among early human species 193, 282
  from food shortages 248, 253
  plague *292*, 293
  presence of bacteria 112, *112*
  prevention of 334, 350
  spread of 295, 299, 331
disposable income 316
divergent plate boundary 92, *93*
divine laws 263
divine rights 274
DNA 102, **104–105**, 120–121
  analysis of 173, 196–197
  of simple and complex cell organisms *112*, *118*
  see also genetics
Döbereiner, Johann 63
dogs 167
domestication
  of animals 234, **240–241**, 242
    secondary products 246
  of plants 234, **236–237**
donkeys 246
double helix 104, *104–105*
draft animals 246, 248
dragons *279*
drilling, for oil and gas 347
droughts 269
Dunkleosteus *134*, 135, *135*
Dutch colonies 329
Dutch East India Company 398, 399
dwarf galaxies 45
dwarf planets *75*
dwarf stars 56, *57*
dyes (textile) 214, *215*, 278
dying stars *59*
dykes *268*, 269

E

Earth
  formation of **71**, *74*, 75, 78–79
  origin theories **18–19**, 46
    movement of 23, **24–25**
  layers 80–81, 84–85, 92–93
  meteorites found on 72
  calculating the age of 86
earthquakes 80, 92
earths (elements) 63
Easter Island 272, *273*, *382*
Ebola 335
eclipses 20
economic strength
  global **322–323**, 336, 337
  from industrialization 310, 311, 345
  see also money
ecosystems 140, 145, 351
  see also habitats
Ediacaran period 128, *128*
education 331, **332–333**, 358
eggs
  evolution of 146–147
  in reproduction **120**, 124, *124*, 145
Egypt
  in ancient world 370–373
  astronomy in 18
  development of writing 266, 267
  education reform 332
  farming in *249*, **250–251**, 269, **293**
  pharaohs of *258*, 261, *261*
  social hierarchy 258, 278
  tombs of 277
  written records in 264
Ehrlich, Paul 165, 335
Einstein, Albert 28, **32**, 47
Elizabeth I, Queen of England 398, *398*
El Gordo galaxy *38*
electoral reform 315, 317
electric cars *357*
electricity 345
electrons *28–29*, 34, **44**
electron-spin resonance (ESR) 192
elements, chemical 58–59, 62–63
embryos *122–123*, *146*, 147
emissions, carbon 348, 352, 357
Empedocles 22
empires 328–329
  populations of *252*
  rise and fall of 287, **288–298**
  see also colonization
endangered elements 355
End-Silurian extinction event 162
Engels, Friedrich 319, 331
engines
  combustion engines *338*, 339
  steam engines 307, **308–309**
England 396–400 *see also* Britain
Enlightenment, Age of 304, **319**, 332
Entreves, Alessandro d' 263
enzymes 114, *114*, 116
epidermis 137
Epoch of Recombination 44
equality 318–319
  women, loss of 259, *259*
Eratosthenes 264
Erithacus *156*
erosion
  of fossils 150
  of rock 86, *87*, 88
  of soil 272
ethanol 357
ether 112
Ethiopia *198*, 199, 327, *327*
Euglena 118, *118–119*
eukaryotes 100, *113*, **118**, 120
Eurasia
  conflicts in 284
  continental shift 158
  farming in 235, **242–243**, 250
  metallurgy in 280
  plague 253
  trade networks in 291, **294–295**
  see also Afro-Eurasia (world zone)
Europe
  development of writing in 266
  early farming in 246, 253
  imperialism in 288, 327, **328–329**
  plague in 293
  political and social reform 319, 320, 331, 332
  metallurgy in 281
  prehistoric map of 363
  trade markets in 317, 325
  world exploration by **296–297**, 298
European Industrial Revolution 312
European Organization for Nuclear Research (CERN) 37
European Space Agency (ESA) *76–77*

evolution, of life **108–109**, 128, 141
of eggs 147
history and theories of **110–111**, 173
of internal skeletons 130, 135
of mammals 169, 170–171
of humans **184**, *189*, 201
of plants 140, **145**, **160**, 165
of winged animals 142, 156
see also natural selection
exoplanets *76, 77*
"experimental" animals *100*
export trades 323
extinction 150, **162–163**, 351
due to continental shifts 158
due to ice ages 176
due to volcanic activity 154
of Hominin species 190, 221
of languages 297

F

fabrics *see* textiles
Facebook 341
factory production **308–309**, 310, 312, 345
assembly lines *339*
worker conditions *330*, 331
famine 248, 253, 293
farmers, hierarchy of *258*, 259
fats 114
feet, evolution of *186–187*
female organisms 124, 240
Fenton Vase *254–255*
fermentation 246
Fertile Crescent 234–235, 236, 280
fertilization (farming) 240, 344, 350
fertilization (reproduction)
of plants 145, 160, 165
of protocells *107*
sexual 111, **120–121**, 124
fibers (textiles) 278
financial institutions 311, 323, 337
fire
creation of 216–217
use in farming 232, *233*
fire-stick farming *220*, 221
slash-and-burn farming 232, *233*, 272
First Keck Telescope 27
fish **130–131**, 132, 141
extinction of 162–163
jawed 134–135
"fishapod" 141, *141*
fishing **189**, 206, 208, 231
exploitation of reserves 211, 220
flagellum *113*
flatworms *126–127*
fleas *293*
flightless birds *158*
flippers 142
flooding 269, 272, 293
flowering plants 101, 160, *161*, 165
Flying Shuttle 312
food chains 115, *135*
food shortages (famine) 248, 253, 293
foragers 211, 220, **230–231**
Ford, Henry 339
Ford Model T *339*
forest habitats 141, 150, 186
deforestation 221, 272, 299, 351
fossil fuels 345, **346–347**, 348, 350
leading to climate change 352, 357, 358
coal 149, 307
fracking 347
France 313, **328**, 348
colonies of 396–400, 406–410
maps *394–395*, *404–405*, *412–413*
medieval period 388, 390, *390*
revolution 314, **318**, 320
world wars 414–417
fraternity 318
free trade 317, 323, 325
freedoms 318–319
French Revolution (1789) 314, **318**, 320, 407, *407*
fruit, for plant reproduction 145, 160
fuel consumption *344*, *345*
Fukushima nuclear disaster 348
fundamental particles 34
funeral stones 264
fungi 115, 120, 124
fur, for clothing 214, *215*
fusion, nuclear 45, **56**, 58

G

Gaia satellite *76–77*
Galápagos Islands *110*, 111
galaxies
creation of 38, *45*, **48–49**
discovery of 30, *33*, 47, **50–51**
Galileo Galilei **25**, 26, 46
Gama, Vasco da 298
gametes 145
Ganesh *275*
Ganow, George 32
Gardens by the Bay, Singapore *356*
gas (elements) 63
gas (fossil fuel) 347
gaseous planets 71, 75
Gatling gun *326*, 327
gazelles *168*
GDP (Gross Domestic Product) 333, 345
General Theory of Relativity 32, 38, **47**, *47*
genetics **104–105**, 111
analysis of 196–197
and reproduction **120–121**, 124
see also DNA
genus, of species 173
geocentrism 24–25
geothermal energy 357
germ theories 334
Germany 320, *320*, 331
imperial power of 328
World Wars 414–417
germination 145, 236
gestures (communication) 202
gills 130, 135
glacial periods 176–177, 365
ice ages 220, 226
glaciers *152*, 176, 352
glass 217
gliding birds 156
global economies **322–323**, 336, 337
global exchange networks 297
globalization 333, 336–337
communication 342–343
of industry 312–313
populations during *252*
through religion 275
and trading 298–299
Global Positioning Satellite (GPS) 341
global warming 239, 351, 352
Glossopteris 159
gluons 34, 37
gods and goddesses 18, *256*, **274–275**
gold 281, *328*
cloth 278
trading of 291
Gondwana 158, *159*
Google 341
Gorham's Cave *194*
gorillas 183, *201*
Gould, John 111
governments 311, **314–315**
authority of 291, 316, **318**, 323
in Britain 304, 325
within empires 287, *289*
see political hierarchies
GPS (Global Positioning Satellite) 341
grain crops
diet of 293
domestication of 236, *237*
measurement of 250
pollen analysis 238–239
production and harvest of **236**, *249*, 344

granaries 250, *250, 251*
Grande Coupure 163
grasses 169
grasslands habitats *168*, 169
grave goods 21, 254, **276–277**
see also burial practices
gravitational lensing 47
gravity **46–47**, 71, *71*, 76
Earth's gravitational pull *78*, 80
moon's gravitational pull 82, *83*
stars, gravitational collapse of 44, 56
sun's gravitational pull 68
grazing animals 169
Great Acceleration 350
Great Britain *see* Britain
Great Dying 101
Great Exhibition (1851) *321*
Great Library of Alexandria 264, 266
Great Ordovician Biodiversification Event 162
Great Oxygenation Event 100, 116
Great Pyramid of Giza *370*
Great Wall of China 250
great white sharks *130–131*
Greece
astronomy in 18
classical era 378–380
coinage in 291
development of alphabet 264, 266, 267
independence of 320
influence on Romans 288
green energy 357
greenhouse gases 345, **350–351**, 352
Greenland 174, 176
Gross Domestic Product (GDP) 333, 345
guano (fertilizer) 248
gunboat diplomacy *324*
gunpowder 284
Gutenberg, Johannes 264
Guth, Alan 34

H

habitable zones (planets) 77
habitats 112, 116, 140, **186–187**
arid *147, 152*, 153, 272
coastal *152*, 158
grasslands *168*, 169
ocean 128, 128–129, 154
marine 351
reef 138, 139
rain forest 158, *233*
swamp **148**, 153
Hadean Era **78**, *79*, 82, *102–103*
Hadrian's Wall 86, *286*
hafnium 354
Haikouichthys 130
haloes (dark matter) 48
Hammurabi, King of Babylon 262
hands, evolution of 142, *143, 186–187*
Han dynasty 263, **294–295**, *377*
harvests 20, *233*, **236**, *249*
hearths *216*
Heezen, Bruce 91, 94
hekat 250
heliocentrism 24–25
helium 38, 63, 355
formation in stars 44, 56, 58
Hennig, Emil Hans Willi 173
herbivores 115, *135*, 169
dinosaurs 154
herding animals 169
Herschel, William *26*
Herto skull *198*
Hess, Harry Hammond 91
hierarchy, of society *258*, 259
hieroglyphs 254, 267
Higgs boson particles 34, **37**, *37*
high mass stars *57*
Hinduism 18, *19, 275*
Hipparcos Satellite 27
hippopotamus 170–171, *171*
Hitler, Adolf 415–417, *416*
Hittite Empire 372
Hohle Fels *Venus* 212, *213*
Holmes, Arthur 91
Holocene period 220, 226, 350
Homer 284, 373
Hominins **184–185**, 211, 220–221, 364–366
breeding of 196, 197, 201
burial practices 218
dispersal from Africa 194–195
evolution of **186–187**, *189*, 202
within the primate family 183
Homo antecessor 194, 195
Homo erectus 184, 185, 187, 216
brain size of *189*
dispersal of *194*, 195
intelligence of 202, **206**, **207**, 211
Homo ergaster 186
Homo floresiensis 185, 199, 214
Homo habilis 184, 185
brain size of *189*
dispersal of *194*, 195
intelligence of **206**, 211
Homo heidelbergensis 188, 195
Homo neanderthalensis 184, *189*
see also Neanderthals
Homo sapiens 198, **199**, *199*
burial practices of **218**, 221
clothing of 214
culture and language of 202, 203, 204
dispersal of *194*, 195
evolution of 189, *189*, 201
intelligence of **206**, **207**, 220
interbreeding with other Hominins 190, **196–197**
within primate family 183, *183*
honeybee 165
hoofed mammals 167, *168*, 170, *171*
Hooker Telescope 30, *31*
horses *169*, 246, *297*
domestication of 284
use in trade *295*
horticulture 232
household possessions *316*, 317
Hoyle, Fred 32
huarango trees 272
Hubble, Edwin 29, **30**, 32, *33*
Hubble Space Telescope *27*, 50–51
Human Genome Project 335
human rights 318–319
human sacrifices 277, *371*
humans *see Homo sapiens*
hummingbird hawk-moth *164*
hunter-gatherer groups *210*, 211
belief systems of 274
competing with farmers **232**, 242
diet and health of 190, **293**
social networks within 204
settlements of 228, **230–231**
Hutton, James 86
Huxley, Thomas Henry 111
hydroelectric energy 357
hydrogen 38, 63
formation in stars 44, 56, 58–59
formation of life 102, 114
hyenas *168*
hyoid bones *192*, 202, *202*

I

Iapetus Ocean 138
ice ages **176–177**, 220, 226
Bering Strait 195
ice cores 174–175
Iliad, Homer 274, 284
"Imilac" meteorite 72
imperialism *see* colonization
import trades 323
Inca Empire 248, 250, 274, 390
Machu Picchu *391*
map of *384*

India
continental shift of 158
development of writing 266
and globalization 336
imperialism in 288
renewable energy in 357
worship in 274, 275, 319
see also Indus civilization
indium 355
Indohyus *170, 171*
Indus civilization *246*, 266, 269
ancient world map 368–369
industrialization 304, *305*, **308–309**, **310–311**
effects on environment 350
globalization of 312
leading to consumerism 316–317
social impact of 331
wealth of industrialists 314, 323
inflation (cosmology) 35
inflation (economics) 291, 299
Information Age 332
inner core, Earth's 80, *80*
insects **142**, *143*
pollination by 160, **164–165**
interglacial periods **176**, 177, 190
internal combustion engines *338*, 339
internet 341, 342
interstellar cloud *68*
invertebrates 135, **141**, 158
marine 162
iron, as raw material 217, 280, 281, 372–373
Iron Age 281, 372
irrigation systems 248, **268–269**, 271
Islam 274, 275, 295
Islamic Empire 384, 386–387
terrorism by Islamic groups 420–421
Islamic Golden Age 267
islands, formation of 84, 85, *93*
isotopes 38, 72
Israel 216
Italy 320, 329

J

Jack Hills, Australia 88, 89, *89*
James Webb Space Telescope (JWST) 27
Japan 331, **348**, 389
Buddhism in 275
development of writing 267
education reform in 332
imperial power of 329, *404–405*
industrialization in 309, **313**
Jomon civilization *230–231*, 231, 372
trade markets 299, *324*, 325
jawless fish 130
jaws, evolution of 132, *134*, **135**
Jefferson, Thomas 318
jellyfish 126
Jenner, Edward 334
Jerusalem *386*, *388*
jewelry *190*, 208, 214
by metallurgy 280, 281
Jomon civilization *230–231*, 231, 372
Judaism 274, 381, *393*
Jupiter *74*, 75
Jurassic period 154, 163
justice 262–263
Justinian, Roman Emperor 263

K

Kaapvaal Craton 85
Kalahari bushmen *189*, *210*
katydid *109*
Kenya 211, 342
Kepler, Johannes 25
Kepler-452 system *77*
keratin 153
kings **261**, 262–263, 271, 274
grave goods of 277
pharaohs *258*, 261, *261*
kinship 204, 262
Koran 263
Kuiper Belt 75
Kushan Empire 294

L

labor (birth) 201
labor laws 331
lactose tolerance 246
Lafayette, Gilbert du Motier, Marquis de 318
Lagerstätte 138
Lamarck, Jean-Baptiste 111
land ownership **262**, 274, 314
language **202–203**, 216, 320
capabilities of Neanderthals 192
development of 204
extinction of 297
Large Hadron Collider (LHC) *36*, 37
Large Magellanic Cloud 60
larynx 202
Last Universal Common Ancestor (LUCA) *113*
Late Devonian mass extinction 135
Late Heavy Bombardment **75**, 80, 100
Latin 173, **264**, 287
Lau event 162
Laurasia 158
lava *102*
Lavoisier, Antoine-Laurent de 63
law and order 262–263
Law of Octaves 63
layers, rock *86*, *87*
leather 207, 214
Leavitt, Henrietta Swan *28*, 29
Legalism 263
Lemaître, Georges *32*
liberty 318
Lidgettonia *145*
life expectancy 344
light 32, 47, 50
from stars **44–45**, 60
see also telescopes
light-years **29**, 50
lignin 137, 141, 148
lignite 148
limbs
evolved from fins 132, *141*
wings evolve from 142, *143*
linen 214, 278
Linnaeus, Carolus (Carl von Linné) 172
lions *168*
literacy 267, **332**, *333*
literature 374–375
lithic mulching 272, *272*
lithium 354
llamas *234*
lobe-finned fish 141
Locke, John 318, 320, 392
London 304
looms 278, *309*
Lord of Sipán *276*, 277
low mass stars *57*
Lyell, Charles 86
Lystrosaurus *159*

M

machine guns *326*, 327
Magellan, Ferdinand 298
magma *79*, *84*, **92**, 94
magnesium 59
magnetic field 80, *81*, 91
magnetic resonance imaging (MRI) scanners 335
magnolias 160
Maillet, Benoît de 86
malaria 327
male organisms *123*, 124, 240
Malthus, Thomas 253
Malthusian cycles 253

mammals 147, **166–167**
extinctions of 163
evolution of 133, 142, 147
for domestication 240
hoofed *168*
manioc *296*
mantle, Earth's 80, 84, 92
manufacturing industries **308–309**, 310, 312, 345
assembly lines *339*
worker conditions *330*, 331
manure 248
manuscripts 267
Māori people 18, 329, 396
mapping, world *90*
marble 370, *370*
Mariana Trench 94
marine habitats 351
marine life 140, 158, 351
Mars 24, *74*
marsupials 158, 167
Marx, Karl 319, 393
mass spectrometer 88
Maxim, Hiram 327
Max Planck Institute for Astrophysics, Germany *61*
Maya civilization 376, 381
astronomy in 18
rulers of *260*, 261
technologies of 254–255
writing system 266
Mayr, Ernst 111
measles 293
measuring time (calendars) 18, 20–21, **244–245**, **367**
measuring volumes 250
meat-eaters (carnivores) 154, 156, 188, 211
medical advances **334–335**, 343, 422–423
medieval period 384–385
records 264, *265*
megalithic structures 221
meiosis 120
membrane *102*, 106, *107*, **112**, *118*
amnion 147
Mendel, Gregor 111
Mendeleev, Dmitri 63
mercantilism **298–299**, 315, 323
Mercury 47, *74*
Mesoamerica 234, 294
Aztec empire **244**, *244–245*, 287, 298
Maya civilization 376, 381
astronomy in 18
rulers of *260*, 261
technologies of 254–255
writing system 266
Mesolithic period 221, 226, 227, 365
Mesopotamia **271**, 288
early farming in **248**, 269
pottery production in 255
tombs of 277
trade tokens in *291*
written records in 264
Mesosaurus 159
Mesozoic Era 154, 158
Messel Lake, Germany 101, *138*
Mexico *378 see also* Maya civilization
metallurgy 280–281
meteorites 72–73, 86
see also asteroids
methane 102
microbes 100, **112–113**, 114, 116–117
evolution to complex cells **118**, 120, 122
evolution onto land 140
see also bacteria
micrometer 26
middle ages *see* medieval period
middle classes 315
and consumerism **316–317**, 344
demanding social reform 319, 331, 332
Middle East oil crisis 348
Middle Stone Age 226
Mid-Ocean Range 91
migration 337
Milankovitch cycles 174, *174*
military power
in empires 284, 288
technology 284, 311, *326*, 327
milk
production, in mammals *166*, 167
as secondary product 246, *247*
Milky Way Galaxy 30, *50*, 59
Miller, Stanley 102
millipedes 140
minerals 114
mining industry **306–307**, 308, 346, 355
Minoan civilization 371
mitochondria 118, *118*, **196**
mobile phones 341, 342
"molecular clock" 170–171
molecules 102, *102*, 104, 106
see also DNA
molten rock (magma) *79*, *84*, **92**, 94
monarchies 304, 318, 320
money *290*, 291, *298*
Mongol Empire 384, 390
monkey puzzle tree *145*
monkeys 183
monotheistic religions 275
monotremes 167
Montsechia vidalii 160
moon 20, 51, 78, **82–83**
morganucodonts 166
Moschops 153
Moshe, Neanderthal skeleton *192–193*
moths 278
motion, laws of 46, 47
moldboard (plow) 248
mountain ranges, formation of 90, 92
MRI scanners 335
Mughal Empire *394–395*, *397*
multicellular organisms 100, **122–123**
complex cells, evolution of **118-119**, 120
multituberculata 163
mummified tissues 197
music 374–375
musical instruments 208
mutations, of genes 108, 120, 170
natural selection **111**, 135, 142, 165
mutualism 165

N

Nagaoka, Hantaro *28*
NASA 50
nationalism 320, *321*
natural gas 347
Natural Law: An Introduction to Legal Philosophy, d'Entreves 263
natural selection **111**, 135, 142, 165
Nazca people 272
Neanderthals **190–191**, 198–197, 218
brain size *189*
clothing 214
dispersal of 194–195
language capability 202, 203
skeletal remains of 192–193
use of fire 216
Neander Valley *190*
Nebra Sky Disc 20–21
nebulae **29**, *29*, 30, 56
neocortex 189
neodymium 354
Neolithic Era 366
neon 59
Neptune 75, *75*
nervous systems 126, 126–127
Netherlands 329
maps of *394–395*, *404–405*, *412–413*
neutrons *34*, 35, **58–59**
neutron stars 56
New Guinea 235
"New World" **296–297**, 298, *298*
see also Americas the (world zone)
Newcomen, Thomas 307, 308
Newlands, John 63

news broadcasting 340, 342
Newton, Isaac 25, 46, *46*
Nice Model 75
nickel *80*
Nishinnoshima *85*
nitrogen 112
noble gases 63
nomadic groups 228, **295**, *294–295*
warfare by 284
non-domesticable animals 241
nonmetals 63
nonrenewable elements 355
nonrenewable energy *see* fossil fuels
North America 158, 234, 406
see also United States (US)
North Pole 176
see also polar regions
North River Steamboat 312
notochord 130, *130*
nuclear power 348
nuclear weapons 348, *349*
nuclei, atomic 29, **34**, 45
fusion 45, **56**, 58
nuclei, cell 118
nuclear DNA 197
nucleic acids 104, *105*

O

oceanic crust **84–85**, 92
oceans
climate change effects on 176, 352, 353
and continental shift 158
floor 94–95
oceanic crust **84–85**, 92
formation of 78, **80**, *81*, 89
habitats 128, 128–129, 154
see also tides
oil (fossil fuel) 313, **346–347**, 348
Oldowan technology *206*, 211
Olduvai Gorge *194*, *364*, *365*
"Old World" 296, **297**, *298*
Afro-Eurasia (world zone) 235, 294, 336
Olmec civilization 372, 373
online communication 341
On the Origin of the Species, Darwin 86, 111
On the Revolutions of the Celestial Spheres, Copernicus 25
Oort Cloud *75*
Opabinia *129*
Opium Wars 325, *325*
orangutans *182*, 183
orbit, planetary 47, **71**, **75**, 76
of the Earth **24–25**, 174
of the moon *83*
orbital velocity 82
Orbiting Carbon Observatory *359*
orders, of species 173
Ordovician-Silurian extinction events 162
origin stories 86
Orion Nebula 59, *59*
Ortelius, Abraham 90
Orthoceras *139*
ostracoderms 130
ostrich *158*
Ottoman Empire 275, 298, 299, 406, 408–409
maps of *394–395*, *404–405*
Ötzi, mummified man 282–283
outer core, Earth's 80, *80*
"overkill" hypothesis 220, 221
ovules 145
oxen 246, *249*
oxygen
formation on Earth 102, **116**, *116–117*, 148
formation in stars 58–59
high levels leading to extinction 135, 162, 351
low levels in water 351
in photosynthesis 114
ozone layer 352, *352*

P

Pacific Islands **235**, 272, 299, 336
Pacific Ocean 94
paddle wheel *269*
paganism 277
Paine, Thomas 318
pain relief 283
Paleolithic Era 203
art *188*, 212–213
burial practices 218–219
clothing 214–215, *214*
Paleo-Tethys Ocean *153*
Palissy, Bernard 86, *86*
pallasite meteorite 72
palynology 238
Pangaea *152*, **153**, 154, 167
continental shift **158**, 163
Parthenon, the *378*
paper money 291
parasites 112, 214
parchment 266
Parisii *290*
Parthian empire 294, *376–377*, 381
particle accelerators *36*, 37
patriarchy 259
Pax Romana 287
peat 148
pelvis 201
penicillin 335
pentaquark 37
periodic table **62–63**, *354–355*
Permian period 153
Permian-Triassic mass extinction 163
Persian Empire 275, *285*, 287, 295, 378
fall of *289*
Peru *see* Inca Empire
Petrie, Flinders 254
pharaohs *258*, 261, *261*
Philip of Macedon *290*
philosophers 392–393
"philosophes" 319
Phoenician civilization 264, 266, 278
phonographs 340
phosphorus 355
photons 34, 38, 44
photosynthesis 114, *115*, **116**, 116–117
in bacteria 112, 118
Pilbara Craton 85
pilus *113*
Pinwheel Galaxy 60
Pizarro, Francisco 296
placentalia 167
placentas 147
placoderms 135
plague *252*, 253
planetary nebula 56
planetesimals 71, *71*, 72, 73
planets *70*, 71, 76
see also Earth, formation of
plankton 129
plasma (matter) 58
plasmid *113*
plastics 345, 350
plate tectonics 82, 84, **94–95**
continental shift of 91, **92–93**
early theories of 86
Pleiades star cluster 20, 21
Pleistocene 220
plow, invention of 248
Plutarch 262, 263
Pluto *27*
poetry 374–375
polar regions 158, 174, 176, 352–353
political hierarchies 271
political reforms 320, 331
political revolution 314
French revolution 314, **318**, 320, 407, *407*
pollen grains 145, 160, **238–239**

pollination 145, **160**, 165
pollution 348, 351
see also emissions
polo (sport) *295*
Polynesia 235, 261, 272
polytheistic religions 274, 275
population growth, humans **252–253**, 344–345, 351
early species 195, 196, 199
farming, effects on 228, 234, 248
first states **259**, 271
spread of disease 293
pores, plant *136*
"portable" art 212
Portugal 297, *299*, 329, 391
colonies of 396–398
maps *394–395*, *404–405*, *412–413*
postal services 340
pottery 209, 231, *231*, **254–255**
predators 115, 135, *135*
natural selection by *109*
predatory birds 163
pregnancy 201
prehistoric world map 362–363
prepared-core technology 206, *207*
primates *182*, **183**, 201, 204
apes 186–187
brain size 188, 189
see also Hominins
primordial crust 84, *84*, 85
Principia, Newton 25
Principles of Biology, Spencer 111
printing press 264, 267, 341
prison reforms 319
Proconsul 186, *186*
prokaryotes 112, *113*
protectionism, by governments 323
proteins 102, 104, 114
protocells 106, *106*, *107*
protocontinents 85
protons 34, 35, 37
proton-proton chain 58
Proto-Sinaitic *266*
protostars 56
protosuns 68, *68*
Prototaxites *101*
protowings *143*
Prussia 312, 312, 406–409
map of *404–405*
pterosaurs 142, *142*, *143*
Ptolomy, Claudius **22–23**, 24, 25
Puerto Rico Trench 94, *94–95*
pumpkins *242*
pure metals 280
pyramids, Egypt 250

Q

Qianlong coin *290*
Qin dynasty 263
Qing Empire *394–395*, *404–405*
Qin Shi Huang, Emperor 277
quantum mechanics 28, 47
quarks **34**, 35, 37
quasars 38

R

racism 320, 327
radiation 38, 45, 49
from the sun 68, *69*, 114
radioactive waste 348
radioactivity 86
radio broadcasts 341
radiometric dating 72, 86, 88, *88*
radio telescopes 26
railways **308–309**, 313, 339
rain forest habitats 158, *233*
Raleigh, Sir Walter 297
Ram Mohun Roy, Raja 319
Rare Earth Elements (REEs) 355
Raven, Peter 165
Ray, John 172, 173
records, written 264–265
calendars 244, 367
recycling 221, 355
Red Deer Cave people *194*
red giant (stars) *57*
redshifts 29, *29*, 30
reef habitats 138, 139
reflectors (telescopes) 26, *26*
Reform Act (1832) 317
Reform Bill (1832) 331
reforms, social 311, 320, 331
refractors (telescopes) 26
reionization 45
Relativity, General Theory of 32, 38, **47**, *47*
religions 86, 274–275, 295, **393**
attitudes to law 263
renewable energy 357
reproduction
of plants 145, 160, 165
of protocells *107*
sexual 111, **120–121**, 124
reptiles 153, **154–155**, 162–163
evolution into birds and mammals 156, 167
evolution of eggs 147
winged 142
Republic of Letters 319
reservoirs *268*, 269
respiratory systems 114, 137
respiration, aerobic 116
retail 316, 317
revolution, political 314, 318, 331
Rheinische Zeitung 319
rhizoids 137
ribosome *113*
rice *243*, 253
domestication of 236, *236*
Richmond Union railway 313
rickets 293
Rights of Man, Paine 318
ritualistic burials 218
RNA 104, 104–105, 106
road networks 287, *287*, 294, **339**
rock erosion 86, *87*, 88
rock (cave) paintings 208, 209, 212–213
depicting hunting scenes *188*, *227*
storytelling through *203*
rocky planets 71
"rogue" planets 75
Roman alphabet 264
Roman Catholic Church 24, 25, 264
Roman Empire *286*, **287**, 294, 378–382
coal mining in 307
development of laws and justice 263
fall of 288
food and famine in 250, 253
map of 376–377
religion in 274, 275
trade in 291, 294
views on afterlife 277
Romulus and Remus, founders of Rome *373*
Rosetta Stone 266
Ross, Ronald 335
royal authority **261**, 262–263
Royal Mail 340
r-process 59
rule of law 262
rulers **261**, 262, 291
chieftains 259, 277
monarchies 304, 318, 320, 384
Russia 328, *404–405*, 414
see also Soviet Union
Russian Chemical Society *62*
Rutherford, Ernest 28, *29*

S

sacrifices
human 277, *371*
religious 274, 275
Sahelanthropus tchadensis 183, *184*, 186

salinization 272
Samurai warriors *324*
San bushmen *189, 210*
sanitation 334
Sargon of Akkad 288, *288*
satellites (communication) 341, *403*
Saturn *74*
sauropods 154
savanna habitat *168*
scala naturae 172
scanning electron micrograph (SEM) *136*
Schrödinger, Erwin *29*
screw pump (Archimedes' screw) *269*
scribes *258*
sculpture 208, *212*
  see also carvings, art
scurvy 293
sea crossing, earliest 195, 206
seafloor spreading 91
sea levels 86, 176, 352
seasons 82
secondary products, of animals 246
seeds 101, 158, 160
  in early farming 236
  evolution of 144–145
seismic waves *80*
seismology 95
semaphore 340
semi-domesticated animals 240
Semmelweis, Ignaz 334
sensory organs 126
settlements
  city states 270–271
  towns 256–257
  villages 220, 228–229, 230–231
sexual reproduction 111, **120–121**, 124
shaduf lifting system *269*
shale gas 347
shamanism *209*, 275
Shang Yang, Lord 263
Sharia law 263
sharks 130–131, 135, 162
sheep *235*, 246
shelled eggs 146–147
shipping industries 309, *309*
shopping 316, 317
signaling (communication) 340
silica 169
silicon 59, 102
silk, production of 278–279, 287, 295
Silk Road **294–295**, 275
silos 250
silver 281
  for trade **291**, 299, *299*, 325
single-cell organisms 112–113
  evolution into complex cells *119*, 122, *122*
  reproduction of *120–121*
skeleton *see* vertebrates
"skull cups" 218
Skype 341
slash-and-burn farming 232, *233*, 272
  fire-stick farming *220*, 221
slave labor **299**, 309, **317**, *317*
  abolition of 319
  in ancient Egypt 249
Slipher, Vesto 29, 30
slum towns 331
smallpox *297*, 334
smartphones 342
smartwatches 343
smelting 281
Smith, Adam 312, 323
smog 350
s-neutron-capture process 59
Snider-Pellegrini, Antonio 90, *90*
Snow, John 331
social groups **189**, *189*, 204, 241
social mobility 316–317
social networks 342–343
  Facebook 341
social reforms 311, 320, 331
social status 258, 259, 277, **278–279**, 317
societies
  laws and justice 262–263
  organization of *258*, 259
  prehistoric rulers of 261
Soho Manufactory **308–309**, 312
soil
  analysis of 238
  soil erosion and contamination 272
solar energy *356*, 357
solar-powered organisms 115, *117*
solar system
  calculating distances within 76–77
  formation of 71, *71*, 72, **74–75**
  mapping of 23, *24*
solar wind *74*, 80, *81*
soldering 281
solstices 20
sonar
  in bats *109*
  for ocean mapping 91, 94
Song Dynasty 307
sorghum *243*
South America 158, 232, 296
  trade in 299
South Pole 176
  see also polar regions
Soviet Union 339, 348
spacetime 47
space travel 339, 411
Spain 299, *299*, 329
  colonies of 396–398, 408
    maps *394–395, 404–405, 412–413*
special relativity 46
Species Plantarum, Linnaeus 173
specimens *172*
spectroscopes 26
speech 202, *202*
  see also language
Spencer, Herbert 111
sperm 120, 124, *124*
spices, trade of 298, 329
spinal cords **130**, *131*, 186
spinning machines 312
sponges 122, *122*, 138, 162
spores 145, 158, 238
squashes *242*
Standard Model 34
stars
  calculating distances of **29**, 76
  clusters 48–49
  early theories on 24–25
  elements formed in 58–59
  formation of **44–45**, 46–47
  life cycles of **56–57**, 60, *61*
  mapping of 20, **22–23**
  see also Sun
Statue of Liberty *318*
status, social 258, 259, 277, **278–279**, 317
steam engines 307, **308–309**
steamships 309, *309*, 313
steel 281, 313
stellar parallax 29
stems 137, *137*
stethoscopes 334
stomata 137
stone, as raw material
  money *291*
  from stone **204**, 211, 218, 221
    of early human species 188, 190, 199, 364
Stonehenge 244
storytelling 203, *202–203*
stromatolites *100, 114–115*, 115
subatomic particles 34, 37
subducting plates 84, 94
Sub-Saharan Africa 235, 328
subsistence farming 232
Sudan 327
suffrage 315
sugar cane 329

sugars, natural 104, 114, 246
Sumer 267, 271, *271*
sun 56–57, 58, **68**, *69*
as energy source 114
and formation of planets 71, **74–75**
origin theories 18, 22, **24–25**
use in calendars 20, 244
see also stars
sunboat 21
sundials 244
supermassive black holes 49
supernatural beliefs 261, 274
supernovas 45, **56–57**, *61*, 68
early documented 25, **60**
new elements in 59
superpowers 336, 337
surgical reforms 334
sustainability 351, 357, 358
Sutton Hoo ship burial 277
swamp habitats 140, **148**, 153
swidden farming 232, *233*
swim bladders 130, *131*, 141
sword making 280, 281
symbolism **204–205**, 221, 261
of freedom *318*
of money 291
symbols
as language 202, 203
written 206, 208
see also writing systems
synapsids 153, 166
Syria 293
Systema Naturae, Linnaeus 173

T

Taj Mahal, India *399*
taming, of animals 240
see also domestication, of animals
tanning (leather) 214
taphonomy 139, 150
taro *243*
tattoos 283
taxation 288, 315
tectonics, plate 82, 84, **94–95**
continental shift of 91, **92–93**
early theories of 86
teeth **135**, 184, *192*, *283*
telecommunication 340
telegraph systems 340
telephones 340, 341
telescopes 26–27
televisions 341
temperatures
due to climate change 351, 352
Earth's core 80
ice cores *174*
of oceans 353
of stars 44–45, 56, 58
of the sun 68, *68*
of universe, at Big Bang *35*
temples 383
termites *168*
Terracotta Army 277
terrorism 420–421
"test tube" babies 335
tetrapods 101, **132–133**, 141, *141*
text messaging 342
textiles 256, **278–279**
manufacturing of *305*, 309, 312, 313
prehistoric 209, 214, 246
tools for *257*
Tharp, Marie 91, 94
therapsids 166
Thermoluminescence (TL) 192
theropods 154, 156
Thomas, J.J. *28*
Thompson, William 86
tidal energy 357
tidal gauge readings 352
tides 82, *83*
Tiktaalik rosae *141*
tin 280
Toarcian turnover 163
tobacco *296*
tokens (money) 291
tombs *276*, 277
tongues 202
toolmaking **206–208**, 214, 231
agricultural *232*, 246, 248
for hunting 283
from metal 280
from stone **204**, 211, 218, 221, 364
of early human species 188, 190, 199
for textiles *257*
toothless beaks 157
Torah 263
tracking (hunting) 211
trade networks 287, **294–295**, 310
agreements within 325, 336
improved by formation money 291
international **298–299**, 316, 320
post-industrialization **322–323**, 336
pre-industralization 20, 256, 275
transform plate boundary 92, *93*, 95
transplants, surgical 335
transportation 337, *338*
canals (water) *268*, **269**, 309, 312
early modes 246, 297
networks 310
railways **308–309**, 313, 339
roads 287, *287*, 294, **339**
Trans-Siberian railway 313
trees 137, **140**, 150
formation of coal 148–149
Trevithick, Richard 312
"triangular trade" 317
Triassic period 154
Triassic-Jurassic extinction 163
tribal societies 259
see also nomadic groups
trilobites *128*, 138, *139*
triple alpha process *58*
tuberculosis 334
Tudor, House of 278
Turkey 256–257
Tutankhamun, Pharaoh 261, *280*
Twitter 342
Tycho's supernova 60
typewriters 340
typhoid 293

U

Ugaritic script *266*
UK *see* United Kingdom (UK)
ultraviolet radiation *44*, 45, 352
unionization 314
Union of Soviet Socialist Republics (USSR) 339, 348, 415
see also Russia
United Kingdom (UK) *412–413*, 416–417
see also Britain
United Nations (UN) 319, 351
United States (US)
American War of Independence 318, 320, *404–405*, 407–409
education reform 332, *332*
industrialization in 312–313, 331
manufacturing industry in 309
modern era 418–421, *421*
space travel 339, 411
trade by *324*, **325**, 328
world wars 414–417
Universal Declaration of Human Rights (1948) 319
Universal Law of Gravitation 46

Universe
Big Bang theory **34–35**, 37, 38–39
expansion of 30, **32**, *33*, 38
formation theories 19, **22–23**, 23
galaxies within 30, 48–49
light within 44
Ur, Mesopotamia *270–271*
uranium *88*
Uranus 75, *75*
urbanization 311, **331**, 334
Urey, Harold 102
Uruk, Mesopotamia 271
US *see* United States of America (US)
Ussher, Bishop James 86
USS Princeton 313
USSR (Union of Soviet Socialist Republics) 339, 348
see also Russia

V

Vaalbara 85
Varanops 133
variations (mutations), of genes 108, 120, 170
vegetation 137, 153, 169
Venus *74*
vertebrates 130, *131*, **132–133**
evolution of jaws 135
evolution of wings 142
move onto land 141, 153
Very Large Array (VLA) *27*
Vesalius, Andreas 172
villages, development of 228–229
settlements *230–231*, *268*
visible light telescopes 26
visual signaling 340
vizier *258*
volcanic activity **92**, *92*, *93*, 174
effects on life *103*, 106, 153, **163**
formation of continents 85, *85*
formation of Earth 80
underwater 94, 106
Voyager 1 (space probe) 60

W

Wallace, Alfred Russell 111
warfare 284, *285*
for religion 275
modern 327
warriors 261, 284, *284*, *402*
waste materials 221, **345**
management 331
toxic 348, 355
water, creation of **80**, *102*, 106
see also oceans
water contamination 293, 331, 351
water lilies 160
watering holes *169*
Watson, James *105*
Watt, James 307, **308–309**, *308*
wealth
in early societies 250, 256, *258*
of middle classes **316–317**, 344
of nations and empires 298–299, **314**, 329
after industrialization **323**, 325, 345
see also economic strength
Wealth of Nations, The, Smith 312
weather 352
see also climate change
weaving 214, **278**, *309*
spinning machines 312
Wedgwood, Josiah *316*, 317
weeks, within calendars 244
Wegener, Alfred 90, *90*
weirs *268*, 269
Wenlock limestone 138–139
whales *170–171*, 171
WhatsApp 341
wheat, domesticated 236, *236*, *237*
Wi-Fi 341
Wikileaks 341
Wikipedia 267, 341
wildebeest *168*
wind energy 357
wings, animals 142–143
evolution in birds 156, *157*
wireless telegraphy 340
women
education of 332
loss of equality 259, *259*
in workforce *309*
wood
as fuel 304, 307
for toolmaking 190, *232*, *249*
wool 246, 278
woolly mammoths *177*
working classes *330*, 331, **332**
World War I 414–415
World War II 337, 339, 416–417
World Wide Web 340, 341
worship 274–275
see also religions
writing systems
development of 266–267, 371
written symbols 206, 208
written records 264–265
calendars 244
Wu of Han, Emperor 263

X

XDF (Hubble eXtreme Deep Field) 50–51
Xenophanes 274
X-rays 49, *335*

Y

Yantarogekko 150–151
Yerkes Refractor 27
yolk sacs 147, *147*
Yongle Encyclopedia 267
Young, James 313

Z

Zaraysk bisons *213*
zebras *169*, 240
Zhou dynasty, China 368, *373*
ziggurats *271*
zircon crystals 88–89
Zoroastrianism 274, 275

ACKNOWLEDGMENTS

The publisher would like to thank the Big History Institute for their enthusiastic support throughout the preparation of this book—especially Tracy Sullivan, Andrew McKenna, David Christian, and Elise Bohan. Special thanks to the writers: Jack Challoner, Peter Chrisp, Robert Dinwiddie, Derek Harvey, Ben Hubbard, Philip Parker, Colin Stuart, and Rebecca Wragg-Sykes.

DK would also like to thank the following:
Editorial assistance: Steve Setford; Ashwin Khurana; Steven Carton; Anna Limerick; Helen Ridge; Angela Wilkes; and Hugo Wilkinson.
Design assistance: Ina Stradins; Jon Durbin; Saffron Stocker; Gadi Farfour; and Raymond Bryant.
Additional illustrations: KJA artists; Andrew Kerr.
Image retoucher: Steve Crozier.
Picture Research: Sarah Smithies.
Proofreaders: Katie John; Rebecca Warren.
Indexers: Elizabeth Wise; Jane Parker.
Creative Technical Support: Tom Morse.
Senior DTP Designers: Shanker Prasad; Sachin Singh.
DTP Designer: Vijay Kandwal.
Production manager: Pankaj Sharma.

PICTURE CREDITS
The publisher would like to thank the following for their kind permission to reproduce their photographs:

(Key: a-above; b-below/bottom; c-center; f-far; l-left; r-right; t-top)

1 Corbis: EPA.
2 Dorling Kindersley: Oxford University Museum of Natural History.
8–9 Getty Images: Fuse.
19 Getty Images: De Agostini.
20–21 Corbis: EPA.
20 Corbis: EPA (bc).
21 123RF.com: Boris Stromar (tr). **© LDA Sachsen-Anhalt:** Juraj Lipták (cr).
22–23 Bridgeman Images: Private Collection.
23 Alamy Stock Photo: Patrizia Wyss (tr).
26 Dorling Kindersley: The Science Museum, London (cl). **Fraunhofer:** Bernd Müller (ca). **Science Photo Library:** David Parker (br). **Thinkstock:** Photos.com (tl).
27 Alamy Stock Photo: PF-(bygone1) (bc); RGB Ventures (crb). **Corbis:** Design Pics / Steve Nagy (tl).
28 Alamy Stock Photo: Granger, NYC (cra).
30–31 Science Photo Library: Royal Astronomical Society.
30 Carnegie Observatories – Giant Magellan Telescope: (clb).
32 Getty Images: Bettmann (bl).
36–37 © CERN: Maximilien Brice.
37 © CERN: T. McCauley, L. Taylor (cra).
38 NASA: ESA, J. Jee (University of California, Davis), J. Hughes (Rutgers University), F. Menanteau (Rutgers University and University of Illinois, Urbana-Champaign), C. Sifon (Leiden Observatory), R. Mandelbum (Carnegie Mellon University), L. Barrientos (Universidad Catolica de Chile), and K. Ng (University of California, Davis) (bl).
38–39 ESA: and the Planck Collaboration (t).
45 ESO: M. Kornmesser (https://www.eso.org/public/images/eso1524a/) (tr).
46 Corbis: The Gallery Collection (cla).
47 Mark A. Garlick / www.markgarlick.com.
48–49 Image courtesy of Rob Crain (Leiden Observatory, the Netherlands), Carlos. Frenk (Institute for Computational Cosmology, Durham University) and. Volker Springel (Heidelberg Institute of Technology and Science, Germany), partly based on simulations carried out by the Virgo Consortium for cosmological simulations: (c). **Rob Crain (Liverpool John Moores University) & Jim Geach (University of Hertfordshire):** (b).
49 NASA: H. Ford (JHU), G. Illingworth (UCSC / LO), M.Clampin (STScI), G. Hartig (STScI), the ACS Science Team, and ESA (tr).
50 NASA: ESA, and Z. Levay, F. Summers (STScI) (cb).
50–51 NASA: ESA / G. Illingworth, D. Magee, and P. Oesch, University of California, Santa Cruz; R. Bouwens, Leiden University; and the HUDF09 Team.
51 NASA: illustration: ESA; and Z. Levay, (STScI) image: T. Rector, I. Dell'Antonio / NOAO / AURA / NSF, Digitized Sky Survey (DSS), STScI / AURA, Palomar / Caltech, and UKSTU / AAO (cra); ESA and John Bahcall (Institute for Advanced Study, Princeton) Mike Disney (University of Wales) (br).
59 123RF.com: Yuriy Mazur (tc). **NASA:** ESA, M. Robberto (Space Telescope Science Institute / ESA) and the Hubble Space Telescope Orion Treasury Project Team (crb).
60 Getty Images: John Livzey (bl). **Science Photo Library:** Konstantinos Kifonidis (tc).
60–61 Science Photo Library: Konstantinos Kifonidis
61 Science Photo Library: Konstantinos Kifonidis (tl, tc, tr).
62–63 Dorling Kindersley: The Science Museum, London / Clive Streeter.
68–69 ESO: L. Calçada / M. Kornmesser.
70 Alamy Stock Photo: Stocktrek Images, Inc.
72 ESA: Rosetta / NAVCAM – CC BY-SA IGO 3.0 (bl).
72–73 Heritage Auctions.
73 Alamy Stock Photo: dpa picture alliance (br). **Science Photo Library:** Chris Butler (crb).
76–77 ESA: D. Ducros, 2013.
79 Mark A. Garlick / www.markgarlick.com: (br).
82 Kevin Snair, Creative Imagery: (b).
83 Kevin Snair, Creative Imagery.
85 Getty Images: The Asahi Shimbun (cra).
86 Science Photo Library: Royal Institution Of Great Britain (bl); Sheila Terry (tr).
87 Alamy Stock Photo: Stefano Ravera.
88 Library of Congress, Washington, D.C.: David G. De Vries Call No: HAER CAL,41-MENPA,5–31 (c).
88–89 J. W. Valley, University of Wisconsin-Madison.
89 Alamy Stock Photo: B Christopher (tr); World History Archive (br).
90 Alfred-Wegener-Institute for Polar and Marine Research: (bl). **Science Photo Library:** UC Regents, National Information Service for Earthquake Engineering (c, ca).
90–91 Copyright by Marie Tharp 1977/2003. Reproduced by permission of Marie Tharp Maps LLC, 8 Edward St, Sparkill, NY 10976: (t).
92 Alamy Stock Photo: Nature Picture Library (clb).
100 Dorling Kindersley: Hunterian Museum and Art Gallery, University of Glasgow (c). **FLPA:** Frans Lanting (bl).
101 Dorling Kindersley: Hunterian Museum and Art Gallery, University of Glasgow (cla).
103 Peter Bull: based on original artwork by Simone Marchi.
105 Science Photo Library: A. Barrington Brown, Gonville and Caius College (b).
106 Alamy Stock Photo: World History Archive (ca).
108 Getty Images: Tom Murphy / National Geographic (b).
108–109 Piotr Naskrecki.
110 Alamy Stock Photo: Natural History Museum, London (t).
112 Science Photo Library: Professors P. Motta & F. Carpino, Sapienza – Università di Roma (bl).
114 Science Photo Library: Wim van Egmond (cl).
114–115 Anne Race.
115 NASA: SeaWiFS Project / Goddard Space Flight Center and ORBIMAGE (br).
120 Ardea: Auscape (bl).
122 Getty Images: Carolina Biological (cl). **Shuhai Xiao of Virginia Tech, USA:** (cra, cr).
123 Shuhai Xiao of Virginia Tech, USA.
124 Jurgen Otto.
125 Science Photo Library: Eye of Science (br).
126 Xiaoya Ma: Xianguang Hou, Gregory D. Edgecombe & Nicholas J. Strausfeld / doi:10.1038 / nature11495 fig 2 (bl).
134 Corbis: Scientifica.
136–136 Science Photo Library: Martin Oeggerli.
137 Chris Jeffree: (bc).
138 Corbis: Jonathan Blair (bl, clb, bc).
138–139 The Sedgwick Museum of Earth Sciences, University of Cambridge.
139 Alamy Stock Photo: Stocktrek Images, Inc. (br).
142 Robert B. MacNaughton: (bl).
144–145 Dorling Kindersley: Natural History Museum, London.
148–149 Alamy Stock Photo: imagebroker.
149 Dorling Kindersley: Natural History Museum, London (bc).
150–151 Deutsches Bernsteinmuseum: (b).
150 Alamy Stock Photo: Agencja Fotograficzna Caro (c).

152 Colorado Plateau Geosystems Inc: © 2016 Ron Blakey.
156–157 Science Photo Library: Natural History Museum, London.
158–159 Colorado Plateau Geosystems Inc: © 2016 Ron Blakey.
162–163 Alamy Stock Photo: dpa picture alliance (c).
163 Dorling Kindersley: Jon Hughes (cra); Natural History Museum, London (tc). **Science Photo Library:** Mark Garlick (cb).
164–165 123RF.com: Timothy Stirling.
165 FLPA: Biosphoto / Michel Gunther (b).
166 Roger Smith: Iziko South African Museum (cl).
167 Dorling Kindersley: Jon Hughes (tl).
172 Alamy Stock Photo: Natural History Museum, London (b).
173 TopFoto.co.uk: Ullsteinbild (ca).
174 Getty Images: National Geographic / Carsten Peter (br).
176 Colorado Plateau Geosystems Inc: © 2016 Ron Blakey (b).
182–183 Getty Images: Andrew Rutherford.
184–185 Kennis & Kennis / Alfons and Adrie Kennis: (all).
186 123RF.com: Stillfx (cr). **Science Photo Library:** MSF / Kennis & Kennis (bl).
187 123RF.com: Oleg Znamenskiy (cl). **Alamy Stock Photo:** RooM the Agency (cr).
188 Alamy Stock Photo: Heritage Image Partnership Ltd. (t).
189 Alamy Stock Photo: Nature Picture Library (tr).
190 Alamy Stock Photo: dpa picture alliance (bl). **Luka Mjeda:** (cra).
191 Science Photo Library: S. Entressangle / E. Daynes.
192–193 Alamy Stock Photo: Natural History Museum, London.
192 Kenneth Garrett: (clb). **Djuna Ivereigh www. djunapix.com:** (c).
193 Getty Images: National Geographic / Ira Block (br).
194 Alamy Stock Photo: National Geographic Creative (bc). **Darren Curnoe:** (cra). **Science Photo Library:** MSF / Javier Trueba (clb).
197 Robert Clark: (b).
198 Alamy Stock Photo: age fotostock (t); Hemis (bl).
199 Kennis & Kennis / Alfons and Adrie Kennis.
200–201 Alamy Stock Photo: Pim Smit.
202–203 Getty Images: Print Collector (b).
203 Alamy Stock Photo: Premaphotos (t).
204 Alamy Stock Photo: Arterra Picture Library (clb).
204–205 RMN: Grand Palais / Gérard Blot.
205 RMN: Grand Palais / Gérard Blot (crb).
206 Dorling Kindersley: Natural History Museum, London (cr); Robert Nicholls (tr). **Getty Images:** EyeEm / Daniel Koszegi (cb). **Wim Lustenhouwer / Vrije Universiteit University Amsterdam:** (crb). **Museum of Anthropology, University of Missouri:** (cla). **Science Photo Library:** P. Plailly / E. Daynes (bc). **SuperStock:** Universal Images Group (c).
207 Alamy Stock Photo: Granger, NYC (cra). **Science Photo Library:** John Reader (c). **Muséum de Toulouse:** Didier Descouens/Acc No: MHNT PRE.2009.0.203.1/ https://commons.wikimedia.org/wiki/File:Pointe_levallois_Beuzeville_MHNT_PRE.2009.0.203.2.fond.jpg (cla). **João Zilhão (ICREA/University of Barcelona):** (tr).
208 Alamy Stock Photo: Heritage Image Partnership Ltd. (cl). **Getty Images:** Field Museum Library (crb). **Pierre-Jean Texier:** (tc).
209 akg-images: Pictures From History (bc). **Alamy Stock Photo:** Arterra Picture Library (tl, tr); Natural History Museum, London (cra). **Joel Bradshaw, Honolulu, Hawaii:** (ca). **Dorling Kindersley:** University of Pennsylvania Museum of Archaeology and Anthropology (br).
210–211 Alamy Stock Photo: John Warburton-Lee Photography.
211 Alamy Stock Photo: Cindy Hopkins (crb).
212–213 Alamy Stock Photo: EPA / Guillaume Horcajuelo (b).
213 Paleograph Press, Moscow: (tr). **© urmu:** © uni tübingen: photo: Hilde Jensen (cla).
214 Getty Images: De Agostini (bl). **Lyudmilla Lbova:** (tr).
215 Science Photo Library: S. Entressangle / E. Daynes
216 Alamy Stock Photo: Penny Tweedie (bl). **Institut Català de Paleoecologia Humana i Evolució Social:** (ca).
216–217 Bridgeman Images: (t).
217 Alamy Stock Photo: EPA (cla). **Dorling Kindersley:** Board of Trustees of the Armouries (cra). **Rex by Shutterstock:** Isifa Image Service sro (bl).
218 Science Photo Library: Natural History Museum, London (bc).
218–219 Kenneth Garrett.
220 Ardea: Jean-Paul Ferrero.
226–227 Alamy Stock Photo: Archivio World 4.
229 Corbis: John Warburton-Lee Photography Ltd. / Nigel Pavitt (br).
231 Alamy Stock Photo: Peter Horree (bc).
232 Dorling Kindersley: Museum of London (cl).
232–233 SuperStock: Universal Images Group.
234 Getty Images: Bloomberg (ca); Morales (crb).
235 Alamy Stock Photo: Nigel Cattlin (clb); Danita Delimont (tr). **Dreamstime.com:** Jaromír Chalabala (cla)
236 Getty Images: Markus Hanke (ca).
238 Corbis: Imagebroker / Frank Sommariva. **Science Photo Library:** Ami Images (clb); Steve Gschmeissner (bl); Scimat (cb, bc).
238–239 Science Photo Library: Ami Images.
239 Dreamstime.com: Fanwen (tr).
240 Alamy Stock Photo: Kip Evans (bl).
242 Alamy Stock Photo: Chuck Place (cra). **Dreamstime.com:** Branex | (tl).
243 123RF.com: Samart Boonyang (cr). **Alamy Stock Photo:** Tim Gainey (cra). **Dreamstime.com:** Sergio Schnitzler (bc); Elena Schweitzer (tc).
244 SuperStock: De Agostini (b).
244–245 Science Photo Library: David R. Frazier.
245 Alamy Stock Photo: Sputnik (tr).
246 Corbis.
246–247 Alamy Stock Photo: India View.
248–249 The Trustees of the British Museum.
249 Alamy Stock Photo: The Art Archive (tl).
250 Photo Scala, Florence: Metropolitan Museum of Art (cl).
250–251 Alamy Stock Photo: Bert de Ruiter.
252–253 Alamy Stock Photo: Yadid Levy.
253 Getty Images: AFP (cb).
254 The Trustees of the British Museum.
254–255 The Trustees of the British Museum.
255 akg-images: Erich Lessing (cra). **Bridgeman Images:** Tarker (cr). **Reuters:** Chinese Academy of Science (br).
256 Getty Images: De Agostini (tc, cb).
257 Corbis: Nathan Benn (tc); Ottochrome / Nathan Benn (tr).
259 Artemis Gallery: (tl).
260–261 Alamy Stock Photo: Dennie Cox.
261 Corbis: Sandro Vannini (br).
262 Alamy Stock Photo: Glasshouse Images (br).
264 Alamy Stock Photo: Prisma Archivo (bc).
264–265 Alamy Stock Photo: The Art Archive.
266 Alamy Stock Photo: Eye35 (tl). **Bridgeman Images:** Louvre, Paris, France (cb). **Corbis:** Gianni Dagli Orti (clb). **Getty Images:** De Agostini (bl).
267 Alamy Stock Photo: Holmes Garden Photos (tc). **Bridgeman Images:** British Library (tr). **Wikipedia:** (br).
270 Alamy Stock Photo: Interfoto (b).
272 Alamy Stock Photo: age fotostock (b).
272–273 Getty Images: Yann Arthus-Bertrand.
275 Alamy Stock Photo: imagebroker (b).
276–277 Alamy Stock Photo: Heritage Image Partnership Ltd.
277 The Trustees of the British Museum: With kind permission of the Shaanxi Cultural Heritage Promotion Centre, photo by John Williams and Saul Peckham (cb).
278 Corbis: Burstein Collection (b).
278–279 Bridgeman Images: Museum of Fine Arts, Boston, Massachusetts, USA / Julia Bradford Huntington James Fund.
280 akg-images: (tr); Erich Lessing (cr). **Alamy Stock Photo:** Edwin Baker (ca).
281 Alamy Stock Photo: Ancient Art & Architecture Collection Ltd. (c). **Bridgeman Images:** Museo del Oro, Bogota, Colombia / Photo © Boltin Picture Library (tr). **Dorling Kindersley:** Gary Ombler (c) Fort Nelson (bc); University of Pennsylvania Museum of Archaeology and Anthropology (clb). **Dreamstime.com:** Xing Wang (cla)

282–283 Bridgeman Images: South Tyrol Museum of Archaeology, Bolzano, Wolfgang Neeb (b).
282 South Tyrol Museum Of Archaeology – www.iceman.it: (ca); Reconstruction by Kennis & Kennis © South Tyrol Museum of Archaeology, Foto Ochsenreiter (3) (tr).
283 Bridgeman Images: South Tyrol Museum of Archaeology, Bolzano, Wolfgang Neeb (cra). **South Tyrol Museum Of Archaeology – www.iceman.it**.
284 Getty Images: De Agostini (clb).
284–285 Alamy Stock Photo: 505 Collection.
286–287 Alamy Stock Photo: Clearview.
287 The Trustees of the British Museum.
288 Photo Scala, Florence: (ca).
288–289 Alamy Stock Photo: Peter Horree.
290 Dorling Kindersley: University of Pennsylvania Museum of Archaeology and Anthropology (t). **Numismatica Ars Classica NAC AG:** Auction 92 Pt 1 Lot 152 (bl); Auction 59 Lot 482 (bc); Auction 72 Lot 281 (br).
291 Alamy Stock Photo: Anders Ryman (b). **RMN:** Grand Palais (musée du Louvre) / Franck Raux (tr).
292–293 Bridgeman Images: Galleria Regionale della Sicilia, Palermo, Sicily, Italy.
293 Corbis: Demotix / Demotix Live News (cb).
294–295 Getty Images: AFP (b).
295 Dorling Kindersley: Durham University Oriental Museum (tc).
296 Alamy Stock Photo: John Glover (bl). **Getty Images:** De Agostini (cla); Clay Perry (clb).
297 Alamy Stock Photo: Melvyn Longhurst (crb). **Getty Images:** EyeEm / Cristian Bortes (cra); Tim Flach (br). **Science Photo Library:** Eye of Science (cr).
298 Courtesy of The Washington Map Society www.washmapsociety.org: original research published Fall 2013 issue (#87) of The Portolan, journal of the Washington Map Society (b).
299 Bridgeman Images: Pictures from History (t). **The Trustees of the British Museum**.
304–305 Getty Images: Robert Welch
307 Alamy Stock Photo: North Winds Picture Archives (br).
308 Science & Society Picture Library: Science Museum (tr).
308–309 SuperStock: Science and Society Picture Library (b).
309 Alamy Stock Photo: Granger, NYC (tr). **Getty Images:** Transcendental Graphics (br).
310 123RF.com: Jamen Percy (cla). **Alamy Stock Photo:** Granger, NYC (c, bc); Historical Art Collection (HAC) (tr). **Dorling Kindersley:** The Science Museum, London (cra).
311 Alamy Stock Photo: Chronicle (cr); Niday Picture Library (cl); North Winds Picture Archives (cb); Photo Researchers (cla).
312 Alamy Stock Photo: Liszt Collection (tr). **Getty Images:** Bettmann (cb); SSPL (ca).
312–313 Cyfarthfa Castle Museum & Art Gallery, Merthyr Tydfil: © Mervyn M. Sullivan (c).
313 Alamy Stock Photo: (cb); North Winds Picture Archives (cla). **Getty Images:** Hulton Archive (ca). **Library of Congress, Washington, D.C.:** LC-USZ62-136561 (br). **Science & Society Picture Library:** NRM / Pictorial Collection (clb).
315 Alamy Stock Photo: The Print Collection (br).
316–317 TopFoto.co.uk: The Granger Collection (b).
316 akg-images: Les Arts Décoratifs, Paris / Jean Tholance (tr).
317 Alamy Stock Photo: Granger, NYC (br); Pictorial Press Ltd. (tc).
318 Corbis.
319 Corbis: (bc).
320 Alamy Stock Photo: Lebrecht Music and Photo Library.
320–321 Alamy Stock Photo: Chronicle.
324 The Art Archive: British Museum, London.
325 Science & Society Picture Library: Science Museum (br).
327 akg-images: Interfoto (tr). **Bridgeman Images:** Private Collection / Archives Charmet (b).
328 123RF.com: Saidin B. Jusoh (crb); Pornthep Thepsanguan (cl). **Photoshot:** John Cancalosi (ca)
329 Corbis: Richard Nowitz (clb). **Getty Images:** John Wang (tl).
330–331 SuperStock: ACME Imagery.
331 Science & Society Picture Library: Science Museum.
332 Alamy Stock Photo: Granger, NYC (tr); Thislife Pictures (b).
333 Getty Images: Aldo Pavan (t).
334 Alamy Stock Photo: North Winds Picture Archives (tr); Photo Researchers (bc). **Science & Society Picture Library:** Science Museum (ca, crb, clb). **Science Photo Library:** Science Source / CDC (bl).
336 Science & Society Picture Library: National Railway Museum (tr).
337 Alamy Stock Photo: The Art Archive (tc); Lebrecht Music and Photo Library (tl); David Wells (bl); Cultura Creative (RF) (br). **Getty Images:** The LIFE Images Collection / James Whitmore (cl).
338 Alamy Stock Photo: Motoring Picture Library.
339 Alamy Stock Photo: Motoring Picture Library (cr).
340 Alamy Stock Photo: Granger, NYC (bc). **Corbis:** Araldo de Luca (br). **Science & Society Picture Library:** Science Museum (cb). **TopFoto.co.uk:** The Granger Collection (ca).
341 Alamy Stock Photo: The Art Archive (clb); Granger, NYC (tl); Everett Collection Historical (bc); D. Hurst (br). **Getty Images:** Bloomberg (cra). **NASA:** (tr). **Science & Society Picture Library:** National Museum of Photography, Film & Television (cb); Science Museum (ca).
342 Alamy Stock Photo: Annie Eagle (5/b). **Flickr.com:** Portal GDA/https://www.flickr.com/photos/135518748@N08/23153369341/in/photolist-BgZ548 (6/b); roach/https://www.flickr.com/photos/mroach/4028537863/sizes/l (4/b). **Science & Society Picture Library:** Science Museum (1/b, 2/b).
343 123RF.com: Adrianhancu (8/b); svl861 (ca); Norman Kin Hang Chan (5/b). **Alamy Stock Photo:** D. Hurst (4/b); Visions of America, LLC (tl); Frances Roberts (cra); Oleksiy Maksymenko Photography (3/b, 6/b). **Amazon.com, Inc.:** (7/b). **Getty Images:** Business Wire (2/b). **Rex by Shutterstock:** Langbehn (1/b). **Science Photo Library:** Thomas Fredberg (c).
344 ESA: NASA.
345 Getty Images: Yann Arthus-Bertrand (br).
348 Science Photo Library: Dirk Wiersma (c).
348–349 Getty Images: Galerie Bilderwelt.
350 Getty Images: EyeEm / William McClymont (bl).
351 Rex by Shutterstock: London News Pictures.
352 Alamy Stock Photo: Global Warming Images (bc). **NASA**.
353 Alamy Stock Photo: Reinhard Dirscherl (br). **NASA:** Goddard Space Flight Center Scientific Visualisation Studio (cra).
354 Alamy Stock Photo: fStop Images GmbH (tr). **Getty Images:** Davee Hughes UK (bl).
355 123RF.com: Andrii Iurlov (cra). **Dreamstime.com:** Emilia Ungur (cb).
356–357 Alamy Stock Photo: Sean Pavone.
357 Alamy Stock Photo: Oleksiy Maksymenko Photography (ca).
358–359 Alamy Stock Photo: Stocktrek Images, Inc.
364 Corbis: Wolfgang Kaehler (bl).
365 Alamy Stock Photo: Sabena Jane Blackbird (tl). **Dorling Kindersley:** Natural History Museum, London (bc).
366 akg-images: Erich Lessing (tr). **Getty Images:** Jerome Chatin (b).
367 akg-images: Erich Lessing (bc).
370 Bridgeman Images: Ashmolean Museum, University of Oxford, UK (bc). **Getty Images:** Ricardo Liberato (tr)
371 Corbis: Richard T. Nowitz (br).
372 Dreamstime.com: Witr (tl).
373 Alamy Stock Photo: B Christopher (tl); Peter Horree (br).
374 Bridgeman Images: De Agostini Picture Library (br)
375 Corbis: Christie's Images (tr).
378 Dreamstime.com: Sergio Bertino (bl). **Getty Images:** Guillermo Flores - www.memoflores.com (tr).
379 Getty Images: G. Nimatallah / De Agostini (br).

380 Alamy Stock Photo: Dinodia Photos (tl); Mark Burnett (br).
381 Alamy Stock Photo: Peter Horree (bc).
382 Corbis: Charles & Josette Lenars (b).
383 Dreamstime.com: Serhii Liakhevych (bc).
386 Getty Images: Gonzalo Azumendi (tl).
387 123RF.com: Luca Mason / lkpro (br). **SuperStock:** Interfoto (tl).
388 Corbis: Leemage (tr); Scott S. Warren / National Geographic Creative (bl).
389 akg-images: Cameraphoto (br).
390 Bridgeman Images: Centre Historique des Archives Nationales, Paris, France (crb). **Dorling Kindersley:** The University of Aberdeen (tl).
392 Alamy Stock Photo: World History Archive (br).
393 Alamy Stock Photo: imageBROKER (tr).
396 Corbis: Leemage (tc).
397 Rex by Shutterstock: Universal History Archive / UIG (br).
398 Alamy Stock Photo: The National Trust Photolibrary (tl).
400 Alamy Stock Photo: North Wind Picture Archives (tr).
401 Alamy Stock Photo: The Art Archive (cb).
402 Alamy Stock Photo: Heritage Image Partnership Ltd. (br).
403 Getty Images: Bjorn Holland (tr).
406 Rex by Shutterstock: Glasshouse Images (bl).
407 SuperStock: Iberfoto (tr).
408 Corbis: Christie's Images (t).
410 Getty Images: Popperfoto (tr).
411 Science Photo Library: Sputnik (br).
414 Corbis: (bl). **Mary Evans Picture Library:** Robert Hunt Library (tr).
415 Getty Images: Bettmann (bc).
416 Rex by Shutterstock: CSU Archives / Everett Collection (tr).
417 Getty Images: Galerie Bilderwelt (br).
418 NASA: (bl).
419 Corbis: Tim Page (tl).
420 Alamy Stock Photo: Agencja Fotograficzna Caro (bl).
421 Getty Images: Spencer Platt (tr).
422 Corbis: Heritage Images.
423 Science Photo Library: Science Source (tl).

Cover images: *Front:* **123RF.com:** Andrey Armyagov fcrb, Igor Dolgov crb; **Alamy Stock Photo:** Granger, NYC. cb/ (Map); **Bridgeman Images:** Biblioteca Monasterio del Escorial, Madrid, Spain cb; **Dreamstime.com:** Constantin Opris crb/ (Industry), Imagin.gr Photography clb/ (Acropolis), Sergeypykhonin cb/ (Steam); **Neanderthal Museum:** clb; **Science Photo Library:** Pascal Goetgheluck; *Back:* **123RF.com:** Nikkytok fclb, Pablo Hidalgo crb, Sergey Nivens fcrb; **NASA:** cb/ (Solar system), JPL-Caltech / UCLA cb, JPL-Caltech / University of Wisconsin-Madison / Image enhancement: Jean-Baptiste Faure clb; **Science Photo Library:** Pascal Goetgheluck

All other images © Dorling Kindersley
For further information see: **www.dkimages.com**